Computer Recognition Systems

Advances in Soft Computing

Editor-in-chief

Prof. Janusz Kacprzyk
Systems Research Institute
Polish Academy of Sciences
ul. Newelska 6
01-447 Warsaw, Poland
E-mail: kacprzyk@ibspan.waw.pl

Further books of this series can be found on our homepage: springeronline.com

Robert John and Ralph Birkenhead (Eds.)
Developments in Soft Computing
2001. ISBN 3-7908-1361-3

Mieczysław Kłopotek, Maciej Michalewicz and Sławomir T. Wierzchoń (Eds.)
Intelligent Information Systems
2001. ISBN 3-7908-1407-5

Antonio Di Nola and Giangiacomo Gerla (Eds.)
Lectures on Soft Computing and Fuzzy Logic
2001. ISBN 3-7908-1396-6

Tadeusz Trzaskalik and Jerzy Michnik (Eds.)
Multiple Objective and Goal Programming
2002. ISBN 3-7908-1409-1

James J. Buckley and Esfandiar Eslami
An Introduction to Fuzzy Logic and Fuzzy Sets
2002. ISBN 3-7908-1447-4

Ajith Abraham and Mario Köppen (Eds.)
Hybrid Information Systems
2002. ISBN 3-7908-1480-6

Przemysław Grzegorzewski, Olgierd Hryniewicz, María ç . Gil (Eds.)
Soft Methods in Probability, Statistics and Data Analysis
2002. ISBN 3-7908-1526-8

Lech Polkowski
Rough Sets
2002. ISBN 3-7908-1510-1

Mieczysław Kłopotek, Maciej Michalewicz and Sławomir T. Wierzchoń (Eds.)
Intelligent Information Systems
2002. ISBN 3-7908-1509-8

Andrea Bonarini, Francesco Masulli and Gabriella Pasi (Eds.)
Soft Computing Applications
2002. ISBN 3-7908-1544-6

Leszek Rutkowski, Janusz Kacprzyk (Eds.)
Neural Networks and Soft Computing
2003. ISBN 3-7908-0005-8

Jürgen Franke, Gholamreza Nakhaeizadeh, Ingrid Renz (Eds.)
Text Mining
2003. ISBN 3-7908-0041-4

Tetsuzo Tanino,Tamaki Tanaka, Masahiro Inuiguchi
Multi-Objective Programming and Goal Programming
2003. ISBN 3-540-00653-2

Mieczysław Kłopotek, Sławomir T. Wierzchoń, Krzysztof Trojanowski (Eds.)
Intelligent Information Processing and Web Mining
2003. ISBN 3-540-00843-8

Ahmad Lotfi, Jonathan M. Garibaldi (Eds.)
Applications and Science in Soft-Computing
2004. ISBN 3-540-40856-8

Mieczysław Kłopotek, Sławomir T. Wierzchoń, Krzysztof Trojanowski (Eds.)
Intellligent Information Processing and Web Mining
2004. ISBN 3-540-21331-7

Miguel López-Díaz, María ç . Gil, Przemysław Grzegorzewski, Olgierd Hryniewicz, Jonathan Lawry
Soft Methodology and Random Information Systems
2004. ISBN 3-540-22264-2

Kwang H. Lee
First Course on Fuzzy Theory and Applications
2005. ISBN 3-540-22988-4

Barbara Dunin-Kęplicz, Andrzej Jankowski, Andrzej Skowron, Marcin Szczuka
Monitoring, Security, and Rescue Techniques in Multi-Agent Systems
2005. ISBN 3-7908-23245-1

Marek Kurzyński, Edward Puchała, Michał Woźniak, Andrzej Żołnierek
Computer Recognition Systems
2005. ISBN 3-540-25054-9

Marek Kurzyński
Edward Puchała
Michał Woźniak
Andrzej Żołnierek (Eds.)

Computer Recognition Systems

Proceedings
of the 4th International Conference
on Computer Recognition Systems CORES '05

Springer

Prof. Marek Kurzyński
Dr. Edward Puchała
Dr. Michał Woźniak
Dr. Andrzej Żołnierek

Wroclaw University of Technology
Faculty of Electronics
Wybrzeze Wyspianskiego 27
50-370 Wroclaw
Poland

Library of Congress Control Number: 2005923764

ISSN 1615-3871
ISBN-10 3-540-25054-9 Springer Berlin Heidelberg New York
ISBN-13 978-3-540-25054-8 Springer Berlin Heidelberg New York

Springer is a part of Springer Science+Business Media
springeronline.com

Printed in Germany

Typesetting: Data conversion by the editors
Final processing by PTP-Berlin Protago-TEX-Production GmbH, Germany
Cover-Design: Erich Kirchner, Heidelberg
Printed on acid-free paper 62/3141/Yu - 5 4 3 2 1 0

Preface

This book contains papers accepted for presentation at the 4th International Conference on Computer Recognition Systems CORES'05, May 22-25, 2005, Rydzyna Castle (Poland), This conference is a continuation of a series of conferences on similar topics (KOSYR) organized each second year, since 1999, by the Chair of Systems and Computer Networks, Wroclaw University of Technology. An increasing interest to those conferences paid not only by home but also by foreign participants inspired the organizers to transform them into conferences of international range. Our expectations that the community of specialists in computer recognizing systems will find CORES'05 a proper form of maintaining the tradition of the former conferences have been confirmed by a large number of submitted papers. Alas, organizational constraints caused a necessity to narrow the acceptance criteria so that only 100 papers have been finally included into the conference program. The area covered by accepted papers is still very large and it shows how vivacious is scientific activity in the domain of computer recognition methods and systems. It contains various theoretical approaches to the recognition problem based on mathematical statistics, fuzzy sets, morphological methods, wavelets, syntactic methods, genetic algorithms, artificial neural networks, ontological models, etc. Most attention is still paid to visual objects recognition; however, acoustic, textual and other objects are also considered. Among application areas medical problems are in majority; recognition of faces, speech signals and textual information processing methods being also investigated. Some papers are also devoted to theoretical recognition problems. On behalf of the Program Committee of CORES'05 I would like to thank all Authors who submitted their papers to the conference, and to encourage all of them, in particular the young scientists, to participate in the next CORES conferences which, I hope, will be organized. We owe also deep thanks to the Reviewers of the papers for their generous work.

Wroclaw, February 2005. Juliusz L. Kulikowski

Preface

Dear Friends:

It is our honour and pleasure to welcome You to take a part in the 4th International Conference on Computer Recognition Systems - CORES'05, to be held from 22nd to 25th May, 2005 in Rydzyna Castle (Poland).

The CORES'05 Conference - endorsed by the International Association for Pattern Recognition - is organized by Chair of Systems and Computer Networks of Wroclaw University of Technology in co-operation with Association for Image Processing (Polish IAPR Member Society) and State School of Higher Vocational Education of Jan Amos Komenski in Leszno, under the auspices of President of Leszno Town Mr Tomasz Malepszy, Rector of Wroclaw University of Technology Prof. Tadeusz Luty and Deputy Chairman of Division IV (Technical Sciences) of Polish Academy of Sciences Prof. Adam Borkowski.

The first three editions of CORES conferences (1999, 2001, 2003) had the national character and - in our opinion - after 6 years has matured enough to become an international event.

The conference has attracted over 160 paper submissions. Each paper was reviewed by two independent and impartial referees and judged on clarity, significance and originality. From all submitted papers only 100 papers were accepted for presentation.

The conference proceedings published by Springer Verlag in the famous book series "Advances in Soft Computing", contain 5 invited papers and accepted regular papers, which were grouped into the following sessions: Features, learning and classifiers, Image processing and computer vision, Medical applications, Speech and text recognition, Fingerprint and face recognition, Various applications.

We would like to thank the members of the Conference Committees, invited speakers, the paper reviewers, the special events organizers, session chairmen, the authors who have contributed original research papers and all conference animators.

We are specially grateful to Prof. Janusz Kacprzyk, Editor-in-chief "Advances in Soft Computing" book series, for making the ASC available for the proceedings of CORES'05 and the whole editorial team for the time and effort they put into the publishing process.

Your participation is the key to the success of CORES'05. We hope that you will exchange and share yours knowledge and experience with colleagues from all over the world. To each participant we offer best wishes for a very productive event. We are confident that the conference will continue to grow both in terms of size and quality.

Be welcome and enjoy your stay in Rydzyna Castle!

Marek Kurzynski

Local Chairman of Program Committee
and Volume Editor

Michal Wozniak

Chairman of Organizing Committee
and Volume Editor

Wroclaw, February 2005

Contents

Part V SPEECH AND WORD RECOGNITION

Part VI FINGERPRINT AND FACE RECOGNITION

Part I

INVITED PAPERS

Neural Network-Based *Chaotic* Pattern Recognition – Part 2: Stability and Algorithmic Issues

D. Calitoiu[1], John B. Oommen[2], and D. Nussbaum[1]

[1] School of Computer Science, Carleton University, K1S 5B6, Canada, {dcalitoi, nussbaum}@scs.carleton.ca
[2] *Fellow of the IEEE*, School of Computer Science, Carleton University, oommen@scs.carleton.ca

Summary. Traditional Pattern Recognition (PR) systems work with the model that the object to be recognized is characterized by a set of features, which are treated as the inputs. In this paper, we propose a new model for Pattern Recognition (PR), namely, one that involves Chaotic Neural Networks (CNNs). To achieve this, we enhance the basic model proposed by Adachi [1], referred to as *Adachi's Chaotic Neural Network* (ACNN). Although the ACNN has been shown to be chaotic, we prove that it also has the property that the degree of "chaos" can be controlled; decreasing the multiplicity of the eigenvalues of the underlying control system, we can effectively decrease the degree of chaos, and conversely increase the periodicity. We then show that such a Modified ACNN (M-ACNN) has the desirable property that it recognizes various input patterns. The way that this PR is achieved is by the system essentially *sympathetically* "resonating" with a finite periodicity whenever *these* samples are presented. In this paper, which follows its companion paper [2], we analyze the M-ACNN for its stability and algorithmic issues. This paper also includes more comprehensive experimental results.

1 Introduction

Traditional Pattern Recognition (PR) systems work with the model that the object to be recognized is characterized by a set of features, which are treated as the inputs. We propose a now model of PR, namely *Chaotic* PR. This paper follows a companion paper [2] in which we earlier presented some analytic stability properties of the *Chaotic* PR systems, using Lyapunov Exponents. In [2], we had also presented some elementary PR results involving a few patterns, essentially the patterns discussed by Adachi [1]. In this paper, we analyze the stability of the model using the Routh-Hurwitz Criterion, we present algorithmic issues, and also the experimental results for a more "real-life" data set involving numerals.

Pattern Recognition (PR) is the study of how a system can observe the environment, learn to distinguish patterns of interest from their background, and make decisions about their classification or categorization. In general, a pattern can be any entity described with features, where the dimensionality of the feature space can range from being few to thousands. The four best approaches for PR are: template matching, statistical classification, syntactic or structural recognition, and Artificial Neural Networks (ANNs) [4],[5],[6],[7],[8]. The latter approach attempts to use some organizational principles such as learning, generalization, adaptivity, fault tolerance and distributed representation, and computation in order to achieve the recognition. The main characteristics of ANNs are that they have the ability to learn complex nonlinear input-output relationships, use sequential training procedures and adapt themselves to data. Some popular models of ANNs have been shown to be capable of associative memory and learning [9],[10],[11]. The learning process involves updating the network architecture and modifying the weights between the neurons so that the network can efficiently perform a specific classification/clustering task.

An associative memory permits its user to specify part of a pattern or key, and to thus retrieve the values associated with that pattern. One of the limitations of most ANN models of associative memory is the dependency on an external input. Once an output pattern has been identified, the ANN remains in that state until the arrival of an external input. This is in contrast to real biological neural networks which exhibit sequential memory characteristics. To be more specific, once a pattern is recalled from a memory location, the brain is not "stuck" to it, it is also capable of recalling other associated memory patterns without being prompted by any additional external stimulus. This ability to "jump" from one memory state to another *in the absence of a stimulus* is one of the hallmarks of the brain, and this is one phenomenon that we want to emulate.

The evidence that indicates the possible relevance of chaos to brain functions was first obtained by Freeman [13],[14] through his clinical work on the large-scale collective behavior of neurons in the perception of olfactory stimuli. Freeman developed a model for an olfactory system having cells in a network connected by both excitatory and inhibitory synapses. He described how a chaotic system state in the neighborhood of a desired attractor can fall on a stable direction when a perturbation is applied to a system parameter. From this model, he conjectured that the quiescent state of the brain is chaos, while during perception, when attention is focused on any sensory stimulus, the brain activity becomes more periodic. The periodic orbits observed can be interpreted as specific memories. If the patterns stored in memory are identified with an infinite number of unstable periodic attractors which are embedded in an attractor, then the transition from the quiescent state onto an "attention" state can be interpreted as the controlling of chaos. The controlling of chaos gives rise to periodic behavior, culminating in the identification of the sensory stimulus that has been received. Thus, mimicing this identification on a neural network can lead to a new model of pattern recognition, *which is the goal of this research endeavor*[3].

During its evolution, a CNN with fixed weights can be in one of the infinite states within the precomputed state space volume. In the case when one inserts one of the memorized patterns as an input in the network, we want the network to *resonate* with that pattern, generating *that* pattern with a certain periodicity. Between two consecutive appearances of the memorized pattern, the network can also be in an infinite number of states, but in none of the memorized ones.

The *resonance* with the memorized pattern given as input, and the *transition* through several states from the infinite set of possible states (even when the memorized pattern is inserted as input) represent the difference between this kind of pattern recognition and the classical type which corresponds to the strategies associated with statistical, syntactical or structural PR. This is explained in greater detail in [2] and [3], where we also show that in order to achieve recognition, one must decrease the level of chaos until periodic behavior is obtained.

1.1 Contributions of this paper

The primary contribution of this paper is the introduction of a PR system which is founded on the theory of chaotic networks. However, rather than relying only on the chaos of the system, we have shown that chaos and periodicity are, informally, "negotiable" quantities. A more chaotic network leads to a weak PR system and vice versa. In particular, by modifying Adachi's model, we anlyze the dynamics of a new model of chaotic neural networks, the M-ACNN with a PR behavior superior to that of the ACNN. We especially focus on the stability of the network and the retrieval characteristics in the transient dynamics of the network. The latter is analyzed by considering the frequencies of retrieval and the transitions among the stored patterns. This allows us to clarify the ability of the memory searching process. Adachi [1] explained that when the duration of the transient phase is long, the attracting state after such a long transient phase may not be useful for information processing. We have shown that by increasing the multiplicity of the eigenvalues of the ACNN, the PR property of the network can be enhanced, leading to the system resonating "sympathetically" whenever a reasonable version of a stored pattern is presented. Thus, the ACNN has a higher level of chaos then the M-ACNN but the recognition is superior in the latter, because the M-ACNN is more stable. Earlier, in [2] we presented some analytic stability properties of the *Chaotic* PR systems, using Lyapunov Exponents. In this paper, we analyze the stability of the model using the Routh-Hurwitz criterion,

---

[3]Unfortunately, if the external excitation forces the brain out of chaos completely, it can lead to an epileptic seizure, and a future goal of this research is to see how these episodes can be anticipated, remedied and/or prevented.

we present algorithmic issues and also the experimental results for a more "real-life" data set involving numerals.

2 Adachi model of chaotic neural networks: ACNN

The ACNN is composed of N neurons (Adachi set $N = 100$), topologically arranged as a completely connected graph i.e, each neuron communicates with every other neuron, including itself. The ACNN is modelled as a dynamical associative memory, by means of the following equations relating the two internal states $\eta_i(t)$ and $\xi_i(y)$, $i = 1..N$, and the output $x_i(t)$ as:

$$x_i(t+1) = f(\eta_i(t+1) + \xi_i(t+1)), \tag{1}$$

$$\eta_i(t+1) = k_f \eta_i(t) + \sum_{j=1}^{N} w_{ij} x_j(t), \tag{2}$$

$$\xi_i(t+1) = k_r \xi_i(t) - \alpha x_i(t) + a_i. \tag{3}$$

In the above, $x_i(t)$ is the output of the neuron i which has an analog value in [0,1] at the discrete time "t". The internal states of the neuron i are $\eta_i(t)$ and $\xi_i(t)$, f is the logistic function with the steepness parameter ε satisfying $f(y) = 1/(1 + exp(-y/\varepsilon))$. Additionally,

1. k_f and k_r are the decay parameters for the feedback inputs and the refractoriness, respectively,
2. w_{ij} are the synaptic weights to the ith constituent neuron from the jth constituent neuron, and
3. a_i denotes the temporally constant external inputs to the ith neuron.

While the network dynamics are described by Equation (2) and Equation (3), the outputs of the neurons are obtained by Equation (1). The feedback interconnections are determined according to the following symmetric auto-associative matrix of the p stored patterns as in:

$$w_{ij} = \frac{1}{p} \sum_{s=1}^{p} (2x_i^s - 1)(2x_j^s - 1), \tag{4}$$

where x_i^s is the ith component of the sth stored pattern.

3 A new model of chaotic neural networks: M-ACCN

We propose a new model of chaotic neural networks which modify the ACNN as below. In each case we give a brief rationale for the modification.

1. The M-ACNN has two global states used for all neurons, which are $\eta(t)$ and $\xi(t)$ obeying:

$$x_i(t+1) = f(\eta_i(t+1) + \xi_i(t+1)), \tag{5}$$

$$\eta_i(t+1) = k_f \eta(t) + \sum_{j=1}^{N} w_{ij} x_j(t), \tag{6}$$

$$\xi_i(t+1) = k_r \xi(t) - \alpha x_i(t) + a_i. \tag{7}$$

   After each step $t+1$, the global states are updated with the values of $\eta_N(t+1)$ and $\xi_N(t+1)$:

$$\eta(t+1) = \eta_N(t+1) \tag{8}$$

$$\xi(t+1) = \xi_N(t+1). \tag{9}$$

   Rationale: Note that at every time instant, when we compute a new internal state, we only use the contents of the memory from the internal state *for neuron* N. This is in contrast to the ACNN in which the updating at time *t+1* uses the internal state values of *all* the neurons at time t. Observe that this, as can be anticipated, could cause the CNN to be "less chaotic", as we shall see presently.
2. The weight assignment rule for the M-ACCN is the classical variant:

$$w_{ij} = \frac{1}{p} \sum_{s=1}^{p} (x_i^s)(x_j^s) \tag{10}$$

   This again, is in contrast to the ACNN which uses $w_{ij} = \frac{1}{p} \sum_{s=1}^{p} (2x_i^s - 1)(2x_j^s - 1)$.
   Rationale: We believe that the duration of the transitory process will be short if the level of chaos is low. Shuai [15] explained that a simple way to construct hyperchaos with all Lyapunov positive exponents is to couple N chaotic neurons, and to set the couplings between the neurons to be small when compared with their self-feedbacks, i.e $w_{ii} \gg w_{ij} (i \neq j)$. In the ACNN, if for any i,j, (where $1 \leq i, j \leq N$) the value $x_i^s = x_j^s = 0$ for all s, then $w_{i,i}$ will be unity. However, for the M-ACNN, the value of $w_{i,i}$ will be zero in the identical setting. Clearly, the M-ACCN has a smaller self-feedback effect than the ACNN.
3. The external inputs are applied in the M-ACNN, only to the neurons representing the stored pattern, by increasing their biases, a_i, from 0 to unity whenever $x_i^s = 1$. The biases to the other neurons remain to be 0. Thus

$$a_i = 1, \;\; \texttt{if} \;\; x_i^s = 1 \tag{11}$$

$$a_i = 0, \texttt{ otherwise.} \tag{12}$$

In other words, in our case $a_i = x_i^s$, as opposed to the ACNN in which $a_i = 2 + 6x_i^s$.
Rationale: The M-ACCN is more sensitive to the external input than the ACNN. The range of input values is between 0 and unity in the M-ACCN, in contrast with the range of input values being between 2 and 8 in the A-CNN. Thus, the M-ACNN will be more "receptive" to external inputs, leading to, hopefully, a superior PR system.

4 The M-ACNN orbital instability

The stability of the *Chaotic* PR system which we proposed, has been analyzed by two methodologies listed below. The first, which uses the Lyapunov Exponents and their properties, is given in the companion paper [2] and in [3]. The second, which utilizes the Routh-Hurtwitz criterion, is explained in great details below and in [3].

4.1 Analysis using Lyapunov Exponents

For a dynamical system, sensitivity to initial conditions is quantified by the Lyapunov exponents. For example, consider two trajectories with nearby initial conditions on an attracting manifold. When the attractor is chaotic, the trajectories, on average, diverge at an exponential rate characterized by the largest Lyapunov exponent. This concept is also generalized for the spectrum of Lyapunov exponents. The presence of positive exponents is sufficient for diagnosing chaos and represents local instability in particular directions [16]. In this regard, the M-ACNN has the following property.

Theorem 1. *The M-ACNN described by Equations (6) and (7) is locally more stable than the ACNN, as demonstrated by their Lyapunov spectrums.*

In the interest of brevity, the proof is found in [2], the companion paper.

4.2 Analysis using the Routh-Hurwitz Criterion

We consider a physical system described by a set of simultaneous equations

$$\frac{dA_i}{dt} = f_i(A_1, A_2, \cdots, A_r) \texttt{ with } i = 1..r, \tag{13}$$

where f_i are general nonlinear functions of the dependent variables $A_1, \cdots, A_r$. A state of equilibrium may be represented by a singular point or a limit cycle of Equation (13). The Routh-Hurwitz (RH) criterion is applicable only to an equilibrium point where all the derivates of $A_1, \cdots, A_r$ with respect to t are simultaneously zero. Under this condition we obtain:

$$f_i(A_1, A_2, \cdots, A_r) = 0 \texttt{ for all } i = 1..r; \tag{14}$$

If the system is linear, we obtain a single set of values for variables $\{A_i\}$ satisfying Equation (14). Hence the state of equilibrium is uniquely fixed. But since our system is nonlinear, Equation (14) may be satisfied for more than a single set of values for the variables $\{A_i\}$ inasmuch as nonlinear systems may have a *number* of equilibrium states. In order to investigate the stability of a system near a chosen equilibrium point, we apply a sufficiently small disturbance to the system by changing the A_i's from their equilibrium values. Then, if t increases infinitely and all the A_i's return to their original equilibrium values, the system is asymptotically stable at this equilibrium point. On the other hand, if some/all of the A_i's depart from their original stable values with increasing t, the system is unstable.

We now state some chaos-related properties of the M-ACNN using the RH criterion. The detailed proof can be found in [3].

Theorem 2. *The M-ACNN described by Equations (6) and (7) is locally unstable.*

Sketch of Proof: Let us denote a set of equilibrium values for the M-ACNN for the A_i's by $A_{10}, A_{20} \cdots A_{r0}$. Consider now small variations ε defined by:

$$A_1 = A_{10} + \varepsilon_1; A_2 = A_{20} + \varepsilon_2; \cdots A_r = A_{r0} + \varepsilon_r; \tag{15}$$

Substituting Equation (15) in Equation (13) and discarding terms of smaller significance than of the first order in ε we get:

$$\frac{d\varepsilon_1}{dt} = c_{11}\varepsilon_1 + c_{12}\varepsilon_2 + ... + c_{1r}\epsilon_r \ . \tag{16}$$

$$\frac{d\varepsilon_2}{dt} = c_{21}\varepsilon_1 + c_{22}\varepsilon_2 + ... + c_{2r}\epsilon_r \ . \tag{17}$$

$$\cdots$$

$$\frac{d\varepsilon_r}{dt} = c_{r1}\varepsilon_1 + c_{r2}\varepsilon_2 + ... + c_{rr}\epsilon_r \ . \tag{18}$$

where c_{ij} stands for $\frac{\partial(f_i)}{\partial(A_j)}$ at the equilibrium state $A_1 = A_{10}, \cdots A_r = A_{r0}$.

We know [12] that, if the real parts of the roots of the characteristics equation of the system Equation (16)-(18) are negative, the corresponding equilibrium state is stable, and conversely, if at least one root has a positive real part, the equilibrium is unstable. Consider now the characteristic equation given by Equation(16)-(18).

When expanded, this r^{th}-order determinant leads to an equation of the form:

$$c_0\lambda^r + c_1\lambda^{r-1} + ... + c_{r-1}\lambda + c_r = 0. \tag{19}$$

The determination of signs of the real parts of the roots of λ may be carried out by making use of the RH criterion. To apply this criterion, we first construct a set of r determinants set up from the coefficients of the rth -degree characteristic equation as shown in Equation (19).

The RH criterion states that the real part of the roots λ are negative provided that all the coefficients c_0, c_1, ... c_r are positive, and that all the determinants Δ_1, Δ_2, ... Δ_r are positive. Since the bottom row of the determinant Δ_r is composed entirely of zeros, except for the last element c_r, it follows that $\Delta_r = c_r\Delta_{r-1}$. Thus, for stability it is required that both $c_r > 0$ and $\Delta_{r-1} > 0$, and Δ_r need not actually be evaluated.

In the case of the M-ACNN, the Jacobian matrix for the system generates a characteristic equation:

$$\lambda^{2N} - (k_f + k_r)\lambda^{2N-1} + k_f k_r \lambda^{2N-2} = 0 \tag{20}$$

and

$$\Delta_1 = det(c_1) = -(k_f + k_r) \tag{21}$$

Clearly the sign of the Δ_1 depends on the magnitude of the coefficients k_f and k_r. This theorem follows since $k_f > 0$ and $k_r > 0$. □

Remarks:

1. The computation of Δ_1 is non-trivial for the ACNN. The first two terms of the the characteristic equation are : λ^{2N} and $(k_f + k_r)N(-1)^{N-1}\lambda^{2N-1}$ respectively. In this case, Δ_1, which is equal to $(k_f + k_r)N(-1)^{N-1}$, depends on the magnitude of the coefficients k_f and k_r, and the value of N. It appears as if Adachi *et al.*[1] proved the instability of the ACNN *empirically* and not *analytically*.
2. Adachi *et al.*[1] have found that the best parameters for their data set are $k_f = 0.2$ and $k_r = 0.9$. Our experiments confirm this.

5 Designing Chaotic PR Systems

To attempt to design PR systems based on the brain model suggested by Freeman [14],[13] is no easy task. Typically, PR systems work with the following model: given a set of training patterns, the PR system learns the characteristics of the class of the patterns, and this information is retained either parametrically or non-parametrically. When a testing sample is presented to the system, a decision of the *identity* of the class of the sample is made using the corresponding "discriminant" function, and this class is "proclaimed" by the system as the identity of the pattern. The same philosophy is also true for syntactic/structural PR systems.

As opposed to this, we do not expect chaotic PR systems to report the identity of the testing pattern with such a "proclamation". Rather, what we are attempting to achieve is to have the chaotic PR system continuously demonstrate chaos as long as there is no pattern to be recognized, or whenever a pattern that is not to be recognized is presented. But, when a pattern which is to be recognized is presented to the system, we would like the proclamation of the identity to be made by requiring that the system simultaneously *resonates sympathetically*.

To be more specific, let us suppose that we want the chaotic PR system to recognize patterns P_i and P_j. To accomplish this, we shall train the system using these patterns. It is interesting to observe what this training accomplishes. By a mere straightforward computation (as opposed to an *iterative* computation) this training phase assigns the weights between the neurons of the CNN. These weights effectively memorize the training patterns so that the network, in turn, effectively behaves as an "Associative Memory" system. Subsequently, on testing, if any pattern other than P_i or P_j is presented, the CNN must continue to be chaotic, since it is *not trained* to recognize such a pattern. However, if P_i or P_j, (or a pattern resembling either of them) is presented, the CNN must switch from being chaotic to being periodic. Note that as opposed to traditional PR systems, the output is not a single value. It is a *sequence* of values, which is chaotic (i.e., displays no periodicity) unless one of the trained patterns is presented. In the latter case, the system switches to being periodic, and by examining the periodicity in the system, the user must be able to infer that one of the stored patterns has been encountered, and thus infer the identity of the pattern.

Adachi *et al.* [1] had suggested, rather informally, that such a chaotic PR system could be developed. However, the mechanics of the system were not fully explained. The problem with Adachi's ACCN is that it is "extremely" chaotic, and there seems to be no easy way by which the level of chaos can be controlled. This is exactly what we can also deduce from the above two theorems.

In order to develop a PR system from Adachi's model, we must be able to decrease the level of chaos in a controlled manner while we simultaneously increase the stability. This is the rationale for the M-ACNN. By decreasing the number of k_f and k_r terms along the principal diagonal of the dynamical matrix, we can effectively increase the multiplicity of the eigenvalue "0". This multiplicity (of the eigenvalue "0") can be increased from the value 0 to the value $2N-2$ depending on the number of terms we choose to include along the principal diagonal. In the limit, we could design the CNN so as to have *only one* entry of k_f and k_r along the diagonal, thus forcing all the other eigenvalues to be exactly zero. Observe that by virtue of the theorems proven, the corresponding stability also increases. This will thus, in turn, lead to a chaotic system which can switch to become periodic and stable if it is presented with a testing sample resembling one for which it has been appropriately trained. This is exactly what we have achieved.

The formal procedure for the PR system is as explained above, and is found algorithmically as follows below:

Algorithm PR_using_M-ACNN

Begin_Module_Training
Input: The set of training patterns $S = \{X^1 \cdots X^p\}$ with $X^i = [x_1^i \cdots x_N^i]$.
Output: The weights of the M-ACNN.
Method:
/* Compute the weights using the set of training patterns */
FOR $i = 1$ to N
    FOR $j = 1$ to N
    $w_{ij} = 0$;
        FOR $s = 1$ to p
            $w_{ij} = w_{ij} + x_i^s x_j^s$
        ENDFOR
    ENDFOR
ENDFOR
End_Module_Training

Begin_Module_Testing
Input: A pattern Y
Output: A periodic sequence of one (or more) of the memorized patterns X^f if $Y = [y_1 \cdots y_N]^T$ is close X^f. The sequence must not contain any memorized pattern if Y is "far away" from any $\{X^s\}$ with $s = 1..p$. The output of the M-ACNN is given by $U = [u_1 \cdots u_N]$ obeying (2)-(4).
Criterion: Y is considered "close" to any X^s if the noise level is less than a predefined value, *Threshold.*
Method:
/*Read input pattern $Y = [y_1 \cdots y_N]$ */
FOR i=1 to N
    $a_i = y_i$
ENDFOR
/*Compute the output using the dynamical equations (2)-(4) */
$\eta(0) = 0$; $\xi(0) = 0$; $c_f = 0$;
/* initialize the periodicity counter for the training set */
FOR $f = 1$ to p
    $count(f, c_f) = 0$
ENDFOR
FOR $t = 0$ to N_{max}
    FOR $i = 1$ to N
        $\eta_i(t+1) = k_f \eta(t) + \sum_{j=1}^{100} w_{ij} u_j(t)$;
        $\xi_i(t+1) = k_r \xi(t) - \alpha u_i(t) + a_i$;
        $u_i(t+1) = f(\eta_i(t+1) + \xi_i(t+1))$;
    ENDFOR

```
η(t + 1) = η_N(t + 1)
ξ(t + 1) = ξ_N(t + 1)
/* Compute the distance between the output U and each pattern X^s */
   FOR s = 1 to p
      d_s(t) = 0;
   ENDFOR
   FOR s = 1 to p
      FOR i=1 to N
         d_s(t) = d_s(t) + |(u_i(t) − x_i^s)|
      ENDFOR
/* we accept a level of noise for Y, equal to (Threshold/N)% */
      IF d_s(t) ≤ Threshold
         f = s     /*index of recognized pattern X^f, close to Y */
         count(f, c_f) = t; c_f = c_f + 1;
      ENDIF
   ENDFOR
ENDFOR
/* Test the periodicity for only 2 cycles */
periodicity[f] = count(f, 2) − count(f, 1)
Report index f and periodicity[f].
```
End_Module_Testing
End_Algorithm PR_using_M-ACNN

6 Experimental results

In the training phase, as mentioned earlier, we present the system with a set of patterns, and thus by a sequence of simple assignments (as opposed to a sequence of iterative computations), it "learns" the weights of the CNN. The testing involves detecting a periodicity in the system, and then inferring what the periodic pattern is. We shall now demonstrate how the latter task is achieved.

In a simulation setting, we are not dealing with a real-life chaotic system. Indeed, in this case, the output of the CNN is continuously monitored, and the only way by which a periodic behavior can be observed, is to keep track of all the outputs as they come. Notice that this is an infeasible task, as the number of distinct outputs could be countably infinite. This is a task which the brain, (or, in general, a chaotic system), seems to be able to do, quite easily, and in multiple ways. However, since we have to work with serial machines, to demonstrate the periodicity, we have no choice but to compare the output patterns with the various trained patterns. Whenever the distance between the output pattern and *any* trained pattern is less than a threshold, we mark that time instant with a distinct marker characterized by the class of that particular pattern. The question of determining the periodicity of a pattern is now merely one of determining the periodicity of *these markers*.

To present our results in the right perspective, we have tested the schemes for two sets of data. The first was precisely the set which Adachi and his co-authors used [1]. These results are presented in [2], the companion paper. The second set is more realistic, and is one which involves the recognition of numerals. We report here only the results obtained from the second data set.

6.1 PR with a Numeral Data Set

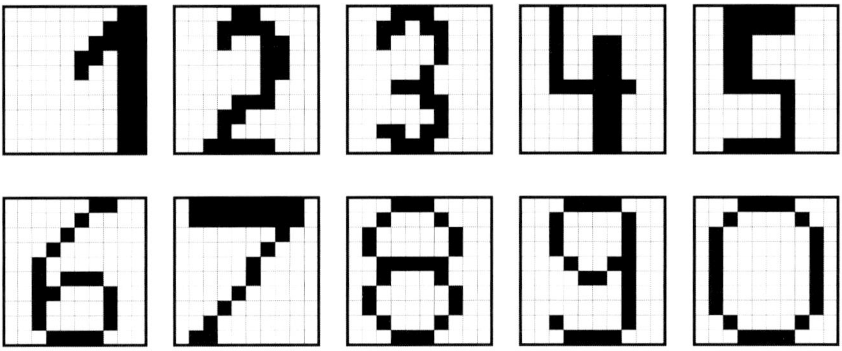

Fig. 1. The second set of patterns used in the PR experiments. These were the 10×10 bitmaps of the numerals $0 \cdots 9$. The initial state used was randomly chosen.

We conducted numerous experiments on a numeral dataset described below. The training set had 10 patterns, given in Fig. 1, and consisted of 10×10 bit-maps of the numerals $0 \cdots 9$. The parameters used were: $N = 100$ neurons, $\varepsilon = 0.00015$, $\alpha = 10$, $k_f = 0.2$ and $k_r = 0.9$ for Equations (5)-(7). The numeral data set was tested for cases when noise was included in the bitmaps. After the training, the system was presented with 10×10 binary-valued arrays which contained noisy versions of one of the numerals.The noise in each case was measured by the percentage of pixels which were modified from 0 to 1 and vice versa. Thus, if the noise was 15%, 15 (out of the 100) randomly chosen pixel values (say, x_i^p) of X^p were modified and were rendered different from those in the original pattern, X^p.

Numerous tests were done, but in the interest of simplicity, we merely mention the case when the noise was 15%, as presented in Fig. 2. After an initial (rather insignificant) non-periodic transient phase, with a mean length of 9.1 time units, the system resonated sympathetically. In this case, the PR accuracy was 100%. The actual values of the duration of the transitory phases and the respective periods are given in Table 1. In our opinion, the results are remarkable, especially when we observe the extremely poor quality of the testing samples.

Table 1. The transitory phase and the periodicity for M-ACNN, when the testing is done with patterns from the training set containing 15% noise. Note that some patterns have limit cycles with multiple periods.

Pattern	No of steps in transitory process	Periodicity
1	24	25
2	8	7,7,8
3	8	7,7,8
4	8	7,7,8
5	8	7,7,8
6	8	7,7,8
7	8	7,7,8
8	8	7,7,8
9	8	2,5,7,8
10	7	22

Fig. 2. The second set of patterns with 15% noise, used in recognition.

7 Conclusion

In this paper, we have proposed a new model for PR, namely one that involves Chaotic Neural Networks (CNNs). To achieve this, we enhanced the basic model proposed by Adachi [1], referred to as *Adachi's Chaotic Neural Network* (ACNN). Although the original ACNN has been shown to be chaotic, we have shown that it also has the fascinating property that it can be modified so that the degree of "chaos" can be controlled by decreasing the multiplicity of the eigenvalues of the underlying control system. By modifying the original ACNN, we have designed the Modified ACNN (M-ACNN) which "resonates" with a finite periodicity whenever the training samples (or their reasonable resemblances) are presented. Apart from analyzing the M-ACNN for its periodicity, stability and the length of the transient phase of the retrieval process, we have also demonstrated its PR capability by testing it on Adachi's dataset, and also for a real-life PR problem involving numerals. The accuracy in each case was a perfect 100%.

References

1. Adachi M, Aihara K (1997) Associative Dynamics in a Chaotic Neural Network, Neural Networks 10:83–98
2. Calitoiu D, Oommen BJ, Nussbaum D (2005) Neural Network-based Chaotic Pattern Recognition - Part 1: Lyapunov Stability and Periodicity Issues, Submitted for PRIP'2005 (Eight International Conference on Pattern Recognition and Information Processing), Minsk, Belarus
3. Calitoiu D, Oommen BJ, Nussbaum D, Periodicity and Stability Issues of a Novel Chaotic Pattern Recognition Neural Network, Submitted for Publication, Unabridged version of the Paper
4. Theodoridis S, Koutroumbas K (1999) Pattern recognition. Academic Press
5. Bishop C M, Bishop C(2000) Neural Networks for Pattern Recognition. Oxford University Press
6. Ripley B (1996) Pattern Recognition and Neural Networks. Cambridge University Press
7. Fukunaga K (1990) Introduction to Statistical Pattern Recognition. Academic Press
8. Friedman M, Kandel A (1999) Introduction to Pattern Recognition, statistical, structural, neural and fuzzy logic approaches. World Scientific
9. Schurmann J (1996) Pattern classification, a unified view of statistical and neural approaches. John Wiley and Sons, New York
10. Kohonen T (1997) Self-Organizing Maps. Springer, Berlin
11. Fausett L (1994) Fundamentals of Neural Networks. Prentice Hall
12. Minorsky N (1962) Nonlinear Oscillations. D.Van Nostrand Company
13. Skarda CA, Freeman WJ (1987) How brains make chaos to make sense of the world, Behavioral and Brain Science 10:161–165
14. Freeman WJ (1992) Tutorial in neurobiology: From single neuron to brain chaos, International Journal of Bifurcation and Chaos 2:451–482
15. Shuai JW, Chen ZX, Liu RT, Wu BX (1997) Maximum hyperchaos in chaotic nonmonotonic neuronal networks, Physical Review E 56:890–893
16. Rosenstein MT, Collins JJ, De Luca CJ (1993) A practical method for calculating largest Lyapunov exponents from small data sets ,Physica D 65:117–134
17. Geist K, Parlitz U, Lauterborn W (1990) Comparison of Different Methods for Computing Lyapunov Exponents, Progress of Theoretical Physics 83:875–893

A Brief Survey of Dynamic Texture Description and Recognition

Dmitry Chetverikov[1] and Renaud Péteri[2]

[1] Computer and Automation Institute, Budapest, Hungary
csetverikov@sztaki.hu
[2] Centre for Mathematics and Computer Science, Amsterdam, The Netherlands
Renaud.PETERI@mines-paris.org

1 Introduction

Dynamic, or temporal, texture is a spatially repetitive, time-varying visual pattern that forms an image sequence with certain temporal stationarity. In dynamic texture (DT), the notion of self-similarity central to conventional image texture is extended to the spatiotemporal domain. DTs are typically videos of processes, such as waves, smoke, fire, a flag blowing in the wind, a moving escalator, or a walking crowd.

In physics, there is a long-established tradition of measuring, quantifying and visualising fluid and other flows that can be viewed as special kinds of temporal textures. In particular, Particle Image Velocimetry [12] is a standard technique for making a flow visible and measurable by injecting many small particles that scatter light and show the fluid motion. A frame of a PIV sequence is a spatial texture; the whole sequence is a dynamic texture. Recently, there have been successful attempts to measure fluid flows using computer vision methods such as optic flow estimation [6] and feature tracking [5].

The mutual influence of physics and image processing is obvious. For these reasons, it would seem natural for the vision community to learn from physics how to mathematically describe processes and motion patterns presented as dynamic textures. However, in computer vision tasks and approaches are quite different from those in physics. In pattern recognition, the study of temporal textures dates back to early nineties when the pioneering paper by Nelson and Polana [18] was published. Nelson and Polana categorised visual motion into three classes [24]: activities, motion events and temporal textures. *Activities*, such as walking or digging, are defined as motion patterns that are periodic in time and localised in space. *Motion events*, like opening a door, do not show temporal or spatial periodicity. Finally, *temporal textures* exhibit statistical regularity but have indeterminate spatial and temporal extent. Computer vision aims at detection, segmentation and recognition of these three classes of visual motion, while in physics the emphasis is on the measurement and visualisation of physical processes.

This paper is a brief survey of approaches to description and recognition of dynamic textures. To our best knowledge, no such survey is currently available. Our survey is limited to temporal textures: we do not consider the other two classes of motion patterns. Even within DT area, our attention is further limited to characterisation and recognition only. In particular, we do not address DT modelling and synthesis, except for the case when model parameters are used for recognition. (For recent work on synthesis, see [16, 7, 8, 33].) Basically, we will deal with dynamic texture descriptors, or features, that have the potential of being used for DT detection, segmentation, recognition and indexing in video.

When video is not segmented, that is, when the exact spatiotemporal extent of a DT is unknown, the features should combine computational efficiency with robustness and descriptive power. In addition, when spatial orientation and scale are also unknown, the features have to be scale- and orientation-invariant, at least to a certain extent. For this reason, in our survey we will pay attention to spatiotemporal directionality (anisotropy), periodicity and scale as the basic structural features which are closely related to the desired properties.

2 Characterisation and recognition of temporal texture

Before discussing the existing approaches to dynamic texture, let us dwell on the tasks of DT analysis. These tasks are similar to those of the conventional spatial texture analysis: detection, segmentation, recognition, and indexing for retrieval. However, working with videos containing temporal textures of unknown spatiotemporal extent is different from working with static images. The difference is not just an additional dimension; it rather relates to the greater 'fuzziness' of dynamic textures.

Firstly, sequences showing physical processes like fire or smoke are difficult to segment: the visible spatial extent of such DT is permanently varying and less distinct. A dynamic texture (for example, smoke) can be partially transparent, so one may face the problem of motion-based *separation* of a DT from textured background. Problems like that are not addressed in traditional texture analysis. Secondly, in temporal textures we categorise both motion pattern and appearance. One may be interested in any flag flapping in the wind, or in flag of a certain country flapping in the wind. The categories are more general and more fuzzy, like 'gentle sea waves' or 'rough turbulent water'.

In static texture recognition, the classes are usually more strict and well-defined. However, the perceptual grouping of static textures may also occur at different levels. For example, one may speak of 'wood texture' in general, or of the texture of a certain type of wood. The well-known experiment by

Rao and Lohse [25] demonstrated that, in absence of any prior information and specific task to solve, the process of perceptual texture grouping in humans is driven by fundamental structural features such as directionality versus non-directionality, periodicity versus irregularity and, probably, structural complexity. (The latter is closely related to the level of detail, or scale.) When people are asked to repeatedly group texture patterns into more general categories by merging some of the previously obtained groups, the fundamental features play a dominant role. Their relevance to static texture analysis has been proved by numerous studies. It is reasonable to assume that they will be important in dynamic texture analysis as well, especially when a high degree of invariance is desired.

The existing approaches to temporal texture recognition can be classified into one of the following groups: methods based on optic flow, methods computing geometric properties in the spatiotemporal domain, methods based on local spatiotemporal filtering, methods using global spatiotemporal transforms and, finally, model-based methods that use estimated model parameters as features.

Methods based on **optic flow** [18, 24, 1, 9, 10, 20, 21, 17, 22, 23] are currently the most popular because optic flow estimation is a computationally efficient and natural way to characterise the local dynamics of a temporal texture. It helps reduce dynamic texture analysis to analysis of a sequence of instantaneous motion patterns viewed as static textures. When necessary, image texture features can be added to the motion features, to form a complete feature set for motion- and appearance-based recognition.

It is well-known [15, 30] that on a smooth moving contour one can locally determine only the velocity component normal to the contour; the tangential component cannot be obtained. (This is called the aperture problem.) In the case of optic flow, the *normal flow* can only be assigned to a pixel unless a larger region is considered and additional smoothness constraints are introduced. The normal flow is orthogonal to the contour and (anti-) parallel to the spatial image gradient. Its computation only needs the three partial derivatives of the spatiotemporal image function. Obtaining the complete flow vector requires more effort, and care should be taken not to enforce smoothness across motion discontinuities. Advantages and drawbacks of the two types of flow are discussed in section 3 in more detail.

In the early studies by Nelson and Polana [18, 24], the vector field of the normal flow was used to form features characterising the overall magnitude and directionality of motion, as well as local image deformations due to motion. Spatial co-occurrence matrices for normal flow directions within pixel neighbourhood were also considered to obtain directional difference statistics. The directionality was evaluated by accumulating a coarse histogram of flow directions and computing the absolute deviation from a uniform distribution. Local image deformations were described by the divergence and the curl of the normal flow field. The features were tested in a classification experiment [24] with seven motion sequences, including five DTs.

It is interesting that the early studies [18, 24] pay proper attention to the issue of invariance, while most of the later studies do not do that. To obtain a temporal and spatial scale-invariant measure, the average flow magnitude is scaled by its standard deviation, resulting in the 'peakiness' feature used later in [22, 23]. The directional difference features are also defined and normalised so as to provide invariance under translation, rotation and scaling in the image plane. The divergence and the curl are scale-dependent, but their ratio can be used to obtain an invariant feature.

The influence of the Nelson and Polana's work can be traced in more recent studies using optic flow to define DT features. In particular, this concerns the assumption that the normal flow is sufficient for adequate description of temporal texture dynamics, shared by most of the authors [1, 9, 10, 20, 21, 22, 23]. Fablet and Bouthemy published a series of studies [1, 9, 10] devoted to recognition of dynamic texture and other motion patterns. They introduced *temporal* co-occurrence that measures the probability of co-occurrence in the same image location of two normal velocities (normal flow magnitudes) separated by a certain temporal interval. In the early paper [1], three fixed intervals (1,4, or 8 frames) are considered and standard co-occurrence features [13] are used to discriminate between four different motion sequences, including one temporal texture.

Later on, the authors developed a more sophisticated approach [10] that accounts for both temporal and spatial aspects of image motion. The method captures the co-occurrence statistics using temporal multiscale Gibbs models. The temporal co-occurrence is defined for consecutive frames, the spatial co-occurrence for neighbouring scales. The maximum likelihood model is obtained for each class in the learning stage. The ML criterion is also used to classify a motion sequence in the classification stage. Eight classes are considered in the tests, including five DT categories: wind-blown trees and grass, gentle waves, turbulent flows, moving escalators. Each sequence is cropped to 'pure' dynamic texture.

The methods [1, 10] are limited in their capability to capture spatial and temporal periodicity of dynamic textures. The initial method [1] does not describe directionality at all; in the enhanced method [10], some information related to anisotropy may be hidden in the estimated model parameters. Although the issue of invariance is not addressed, both methods are probably rotation-invariant. Despite its sophistication, the enhanced method does not seem applicable to large sets of temporal textures or to non-segmented videos.

Peh and Cheong's work [20, 21] builds on that of Nelson and Polana [18] and Bouthemy and Fablet [1]. In [21], the magnitude and the direction of the normal flow are quantised into a small number of levels. Then two spatiotemporal maps are built that trace motion history through a number of previous frames. In each currently moving pixel, the magnitude map is set to the current flow magnitude. If the pixel has been stationary for τ previous frames, the map is set to zero. Otherwise, the map is set to the magnitude the flow had τ frames ago. A similar map is accumulated for the normal flow direction as well. In the experimental study, a fixed value $\tau = 5$ is used and 10 DT classes are considered. Each sequence is divided into subsequences of $\tau = 5$ frames each, used as samples. The magnitude and the direction maps of each sample are treated as image textures, and six conventional texture features are selected for classification tests. The classification results are compared to those achieved on the same data by the methods of Nelson and Polana [18] and Bouthemy and Fablet [1]. For most classes the reported success rates by the proposed method are higher than by the other two methods, with [1] being the least successful.

The approach [20, 21] is simple, fast and rotation-invariant up the to quantisation error in the flow direction. The features convey directional information; however, temporal periodicity analysis with fixed τ is impossible.

Recently, Péteri and Chetverikov [22, 23] have proposed a method that combines normal flow features with periodicity features, in an attempt to explicitely characterise both motion magnitude, directionality and periodicity. The normal flow features used are similar to [18] (peakiness, divergence, curl); however, a novel feature of orientation homogeneity of the normal flow field was also introduced. In addition, two spatiotemporal periodicity features were proposed based on the maximal regularity M_R, which is a measure of spatial periodicity of an image texture [4]. When applied to a dynamic texture, the method evaluates the temporal variation of spatial periodicity. For each frame t of a DT, M_R is computed in a sliding window. Then the largest value is selected, corresponding to the most periodic patch within the frame. This provides a largest periodicity value, $P(t)$, for each t. The mean and the variance of $P(t)$ are currently used as DT features. Some initial DT classification results are reported in [23].

The approach [22, 23] is rotation-invariant. Its periodicity-related part is affine-invariant. (See [4] for details.) Although promising, the temporal regularity method should be improved in at least two aspects: it should only be applied to the moving part of the image, for example, to thresholded optic flow; and the periodicity of $P(t)$ should be analysed.

The last optical flow-based approach we are going to mention is presented in the recent paper by Lu and co-authors [17]. This study is unique in that it uses *complete* not normal flow vectors. In addition, acceleration vectors are also computed. The 3D structure tensor technique (with spatiotemporal gradient) is used to obtain the complete flow vector by minimising an energy function in a neighbourhood of a pixel. To reduce the effect of the aperture problem, the eigenvectors of the tensor are calculated and combined into a measure of spatial 'cornerity' of the pixel. This measure is used as the weight in the histograms of velocity and acceleration: the higher the confidence of velocity estimation the larger the weight. These histograms are used to calculate the distance matrix for 7 DT sequences. To account for scale, a spatiotemporal Gaussian filter is applied to decompose a sequence into two spatial and two temporal resolution levels. The method [17] is rotation-invariant and it provides local directionality information; however, no higher-level structural analysis (e.g., periodicity evaluation) is possible.

The other four groups of methods are much less popular than the methods based on the optic flow. Methods computing **geometric properties in the spatiotemporal domain** are represented by two studies: the initial algorithm by Otsuka and co-authors [19] and its modification by Zhong and Sclaroff [34]. Otsuka and co-authors [19] assume that DTs can be represented by moving contours whose motion trajectories can be tracked. They consider trajectory surfaces within 3D spatiotemporal volume data and extract temporal and spatial features based on the tangent plane distribution. The latter is obtained using 3D Hough transform. Two groups of features, spatial and temporal, are then calculated. The spatial features include the directionality of contour arrangement and the scattering of contour placement. The temporal features characterise the uniformity of velocity components, the flash motion ratio and the occlusion ratio. The features were used to classify four DTs.

It is well-known that motion velocity is closely related to geometry in the spatiotemporal domain; it is also known that considering trajectories in this domain may help resolve ambiguities due to temporary occlusion [15]. However, for dynamic textures the assumption of good trajectory surfaces being available is not realistic. Zhong and Sclaroff [34] tried to avoid the problem by using 3D edges in the spatiotemporal domain. Their DT features are computed for voxels taking into account the spatiotemporal gradient. Unfortunately, the method and the results are not convincing enough, and research in this direction currently seems to have no continuation.

Methods based on **local spatiotemporal filtering** are represented by a single study by Wildes and Bergen [31] mentioned here for completeness. The analysis of local spatiotemporal pattern, its orientation and energy, is useful for *qualitative classification of local motion structure* into such categories as stationary, coherent, incoherent, flickering and scintillating. Results in [31] show correlation between the qualitative features and the character of motion, assuming that small and short DTs are considered. However, motion in different parts of a dynamic texture can be different. No method is given how to combine the local qualitative values into a global description, or how to characterise fundamental structural properties of entire DT.

Recently, there have been attempts [28] of video texture indexing using spatiotemporal wavelets. The emerging use of **global spatiotemporal transforms** indicates the necessity to characterise motion at different spatiotemporal scales. Spatiotemporal wavelets can decompose motion into local and global, according to the desired degree of detail. For example, a tree waving in the wind shows a coarse motion of trunk, a finer motion of branches and still finer motion of leaves. The periodicities of these motions are also different, resulting in energy maxima at different scales. These effects can hopefully be captured by spatiotemporal wavelets. Another argument in favour of wavelets is the fact that the MPEG-7 multimedia standard proposes the use of Gabor wavelet features for image texture browsing and retrieval [32]. An argument against global spatiotemporal transforms is the difficulty to provide rotation invariance.

Finally, let us briefly discuss studies devoted to **model-based DT recognition**. Impressive results have recently been achieved in DT synthesis using the framework based on a system identification theory which estimates the parameters of a stable dynamic model [16, 8, 33]. Saisan and co-authors [26] applied the dynamic texture model [7] to recognition of 50 different temporal textures. Despite this success, the applicability of the approach to real videos is doubtful for several reasons: it is time-consuming; it assumes stationary DTs well-segmented in space and time, and the accuracy drops drastically if they are not; it is difficult to define a metric in the space of dynamic models. Fujita and Nayar [11] modified the approach [26] by using impulse responses of state variables to identify model and texture. Their approach is applicable to multiple dynamic textures in different regions of the image, is faster than [26] and shows less sensitivity to non-stationarity. However, the problem of heavy computational load and the issues of scalability and invariance remain open.

3 Discussion and conclusion

Dynamic texture recognition is a new area whose history dates back to less than 15 years ago. It is natural that many of the proposed methods build on the experience gained in static texture analysis and try to combine optic flow with multiresolution histograms, co-occurrence and other known tools. And, probably, it is too early to make general conclusions about the development of the area. At the same time, we can already learn from the past and summarise some major issues that should be addressed in the future.

The first issue is the *normal* flow vector versus the *complete* flow vector. Optic flow is the basis for most of the current methods, and both versions have their advantages and drawbacks. As already mentioned, normal flow is purely local, fast to compute and does not tend to extend motion over discontinuities; it has been demonstrated to contain usable motion information. On the other hand, normal flow, even in its regularised form (smoothing, thresholding), is noise-sensitive. Its close relation to the spatial gradient, that is, to contours and shapes, implies that normal flow features correlate with appearance features. This was acknowledged in [24] as a negative aspect, but no real solution was proposed. Fablet and Bouthemy [10] even claim that the direction of normal flow contains no independent motion information; they only use magnitude. The applicability of normal flow to rotational motion (like the 'toilet' sequence of Szummer [29]) is questionable. In general, the examples of normal flow fields given in the literature do not reflect well the visual dynamics of the processes. The regularised complete flow field is much better in that respect. However, the iterative schemes for complete flow need more computation and tend to extend motion over discontinuities. (The latter is usually harmful, but sometimes it can be useful for overcoming short occlusions.) Both problems have been addressed in the recent research on optic flow estimation. Using modern multigrid numerical schemes, one can achieve near real-time performance on a general-purpose computer (27 fps for 200×200 size frames [3].) Motion borders can be preserved by using the total variation of the flow field (with the 3D gradient) as the smoothing term [2]. We plan to compare the two types of flow on a large database of dynamic textures.

The problem of combining *motion* features with *appearance* features is also open. As already mentioned, it is task-dependent. In some cases, we are only interested in the motion pattern; in other cases, we are interested in both motion and appearance. Imagine we search in videos for any flag waving in the wind. If normal flow strongly depends on appearance, learning on flag of a certain country (that is, with a specific picture) would not be wise.

A major open issue is that of capturing *temporal periodicity*. Many of DTs are quasi-periodic, and sometimes we humans recognise them due to this property. However, neither of the existing approaches treats the temporal periodicity properly. The reason is that recognition of periodicity requires correlating frames separated by an unknown and, possibly, large interval. This is computationally expensive, while for video processing one normally needs fast methods. Here, spatiotemporal multiscale (multiresolution) approaches may prove useful.

The question of *invariance*, both geometric and photometric, arises each time the viewing conditions are not constrained. When videos of outdoor temporal textures are taken, this is often the case. Ideally, we should be prepared to cope with perspective distortion, or, at least, with affine image distortion corresponding to the weak perspective model [14]. Currently, rotation and scaling in the image plane is the maximum we can handle. In that respect, it will be very useful to learn from the related recent efforts in 3D computer vision, such as [27].

Finally, much more attention should be paid to creating *test data* and designing experimental protocols for proper evaluation and comparison of the emerging techniques. What we typically have now is comparison on a few (maximum 10) randomly selected dynamic patterns, often from the obsolete and poor-quality Szummer dataset [29]. Classification experiments with such limited data are of limited significance: one could probably obtain similar accuracy by considering single frames instead of sequences, as the image textures involved are distinct enough. Also, we have to clarify if the DTs considered are pre-segmented or not. For example, Saisan and co-authors [26] use a large set of 50 pre-segmented DTs and report a good overall classification accuracy of almost 90%. When applied to a small set of only 5 DTs, but with 2 unsegmented patterns [11], the method yields the average of 58% because of complete failure in these two unsegmented cases.

To meet the need for a comprehensive database of dynamic textures, in the framework of the European FP6 Network of Excellence MUSCLE we are now creating a large dataset that will be available on the web site of the

NoE: http://www.muscle-noe.org/. This survey was also supported by the MUSCLE NoE.

We finish our survey and discussion by concluding that dynamic texture recognition is a novel, exciting and developing research area, where some progress has already been achieved, but a lot of work is still to be done.

References

1. P. Bouthemy and R. Fablet. Motion characterization from temporal cooccurrences of local motion-based measures for video indexing. In *Proc. Int. Conf. Pattern Recognition*, volume 1, pages 905–908, Brisbane, Australia, 1998.
2. T. Brox, A. Bruhn, N. Papenberg, and J. Weickert. High accuracy optical flow estimation based on a theory for warping. In *Proc. European Conference on Computer Vision*, volume 4, pages 25–36, Prague, Czech Republic, 2004.
3. A. Bruhn, J. Weickert, C. Feddern, T. Kohlberger, and C. Schnörr. Real-time optic flow computation with variational methods. In *CAIP 2003*, pages 222–229, Groningen, The Netherlands, 2003.
4. D. Chetverikov. Pattern regularity as a visual key. *Image and Vision Computing*, 18:975–986, 2000.
5. D. Chetverikov. Applying feature tracking to particle image velocimetry. *Int. Journal of Pattern Recognition and Artificial Intelligence*, 17:487–504, 2003.
6. T. Corpetti, É. Mémin, and P. Pérez. Estimating Fluid Optical Flow. In *Proc. Int. Conf. Pattern Recognition*, volume 3, pages 1045–1048, 2000.
7. G. Doretto, A. Chiuso, S. Soatto, and Y.N. Wu. Dynamic textures. *International Journal of Computer Vision*, 51:91–109, 2003.
8. G. Doretto, E. Jones, and S. Soatto. Spatially homogeneous dynamic textures. In *Proc. European Conference on Computer Vision*, volume 2, pages 591–602, Prague, Czech Republic, 2004.
9. R. Fablet and P. Bouthemy. Motion recognition using spatio-temporal random walks in sequence of 2D motion-related measurements. In *IEEE Int. Conf. on Image Processing, ICIP'2001*, pages 652–655, Thessalonique, Greece, 2001.
10. R. Fablet and P. Bouthemy. Motion recognition using nonparametric image motion models estimated from temporal and multiscale co-occurrence statistics. *IEEE Trans. PAMI*, 25:1619–1624, 2003.
11. K. Fujita and S.K. Nayar. Recognition of dynamic textures using impulse responses of state variables. In *Proc. Third International Workshop on Texture Analysis and Synthesis (Texture 2003)*, pages 31–36, Nice, France, 2003.
12. I. Grant. Particle image velocimetry: a review. *Proc. Institution of Mechanical Engineers*, 211 Part C:55–76, 1997.
13. R.M. Haralick, K. Shanmugam, and I. Dinstein. Textural features for image classification. *IEEE Trans. SMC*, 3:610–621, 1973.
14. R. Hartley and A. Zisserman. *Multiple View Geometry in Computer Vision.* Cambridge University Press, 2000.
15. B. Jähne. *Digital Image Processing.* Springer, 1997.
16. V. Kwatra, A. Schödl, I. Essa, G. Turk, and A. Bobick. Graphcut textures: Image and video synthesis using graph cuts. *ACM Transactions on Graphics, SIGGRAPH 2003*, 22:277–286, 2003.

17. Z. Lu, W. Xie, J. Pei, and J. Huang. Dynamic texture recognition by spatio-temporal multiresolution histogram. In *Proc. IEEE Workshop on Motion and Video Computing (WACV/MOTION'05)*, 2005.
18. R.C. Nelson and R. Polana. Qualitative recognition of motion using temporal texture. *CVGIP: Image Understanding*, 56:78–89, 1992.
19. K. Otsuka, T. Horikoshi, S. Suzuki, and M. Fujii. Feature extraction of temporal texture based on spatiotemporal motion trajectory. In *ICPR*, volume 2, pages 1047–1051, 1998.
20. C.H. Peh and L.-F. Cheong. Exploring video content in extended spatio-temporal textures. In *Proc. 1st European workshop on Content-Based Multimedia Indexing*, pages pp. 147–153, Toulouse, France, 1999.
21. C.H. Peh and L.-F. Cheong. Synergizing spatial and temporal texture. *IEEE Transactions on Image Processing*, 11:pp. 1179–1191, 2002.
22. R. Péteri and D. Chetverikov. Qualitative characterization of dynamic textures for video retrieval. In *Proc. International Conference on Computer Vision and Graphics (ICCVG 2004)*, Warsaw, Poland, 2004.
23. R. Péteri and D. Chetverikov. Dynamic texture recognition using normal flow and texture regularity. In *Proc. Iberian Conference on Pattern Recognition and Image Analysis (IbPRIA 2005)*, Estoril, Portugal, 2005.
24. R. Polana and R. Nelson. Temporal texture and activity recognition. In M. Shah and R. Jain, editors, *Motion-Based Recognition*, pages 87–115. Kluwer Academic, 1997.
25. A.R. Rao and G.L. Lohse. Identifying high level features of texture perception. *CVGIP: Image Processing*, 55:218–233, 1993.
26. P. Saisan, G. Doretto, Ying Nian Wu, and S. Soatto. Dynamic texture recognition. In *Proc. CVPR*, volume 2, pages 58–63, Kauai, Hawaii, 2001.
27. F. Schaffalitzky and A. Zisserman. Viewpoint invariant scene retrieval using textured regions. In R. C. Veltkamp, editor, *Proc. 2002 Dagstuhl Seminar on Content-based Image and Video Retrieval*, Lect. Not. in Comp. Sci., pages 11–24. Springer, 2004.
28. J.R. Smith, C.-Y. Lin, and M. Naphade. Video texture indexing using spatiotemporal wavelets. In *IEEE Int. Conf. on Image Processing, ICIP'2002*, volume 2, pages 437–440, 2002.
29. Martin Szummer. Temporal Texture Modeling. Technical Report 346, MIT, 1995.
30. E. Trucco and A.Verri. *Introductory Techniques for 3-D Computer Vision*. Prentice Hall, 1998.
31. R.P. Wildes and J.R. Bergen. Qualitative spatiotemporal analysis using an oriented energy representation. In *Proc. European Conference on Computer Vision*, pages 768–784, 2000.
32. P. Wu, Y.M. Ro, C.S. Won, and Y. Choi. Texture descriptors in MPEG-7. In W. Sharbek, editor, *CAIP 2001*, pages 21–28, Warsaw, Poland, 2001.
33. L. Yuan, F. Weng, C. Liu, and H.-Y. Shum. Synthersizing dynamic texture with closed-loop linear dynamic system. In *Proc. European Conference on Computer Vision*, volume 2, pages 603–616, Prague, Czech Republic, 2004.
34. J. Zhong and S. Scarlaroff. Temporal texture recongnition model using 3D features. Technical report, MIT Media Lab Perceptual Computing, 2002.

Open Issues in Pattern Recognition

Robert P.W. Duin[1] and Elżbieta Pekalska[1,2]

[1] ICT group, Faculty of Electr. Eng., Mathematics and Computer Science
Delft University of Technology, The Netherlands
{r.p.w.duin,e.pekalska}@ewi.tudelft.nl

[2] School of Computer Science, University of Manchester, United Kingdom
pekalska@cs.man.ac.uk

Summary. The area of pattern recognition has developed itself into a mature engineering field with many practical applications. This increased applicability, together with the development of sensors and computer resources, leads to new research areas and raises new questions. In this paper, old and new open issues are discussed that have to be faced in advancing real world applications. Some may only be overcome by brute force procedures, while others may be solved or circumvented either by novel and better procedures, or by a better understanding of their causes. Here, we will try to identify a number of open issues and define them as well as possible.

1 Introduction

Pattern recognition is the human ability to see regularities in observations. From the early development of computers, scientists and engineers tried to imitate this ability by mechanical means, either partially or in its entirety. Two main types of results have been obtained from these efforts so far.

First, a better understanding is reached of the human perception, reasoning and the ability to gain new knowledge and to apply it to a changing environment. This knowledge is partially formulated in physical and biological terms, giving more insight into the study of human senses and the neural system. To some extent, this knowledge is also partially expressed in mental, psychological and epistemological terms, describing how facts and observations are combined by reasoning, how uncertainty is handled and how conclusions are reached. Attempts to design sensors, computers and programs that simulate or mimic these processes bring an additional prospect to the investigation of possible biological models. An ever returning difficulty, however, is the relation of low level phenomena occurring in the senses and the nerves to a high level understanding and conceptual thinking.

Second, various pattern recognition systems have been developed that are of practical use, as for the assistance in medical diagnosis, industrial inspection, personal identification and man-machine interaction. Very often, they are not based on a detailed simulation of the human processes, but on independent approaches to the problem at hand.

In this paper, we will focus on the pattern recognition research aiming at the development of automatic systems as discussed above. We will especially deal with the possibilities of these systems to learn from sets of examples. We will also consider the process of going from a low level of single objects observed by sensors to a higher level of decision making, based on the global pattern of a class of objects. As already mentioned above, it is still little understood how this emergent process develops on the human level. Although some technical solutions and explanations exist, there is still the intuitive feeling that these are far from optimal. The basic question here asks how incidental observations, suffering from noise and considered in a particular context, can be integrated into general knowledge about classes of objects, independent of the noisy observations and the accidental circumstances.

A number of open issues will be discussed in the area of automatic pattern recognition. This is the field in which the development of recognition systems that learn from examples is studied. For some recent references, see the books of Webb [29] and Van der Heijden et al. [16] and the review by Jain et al. [18]. The more expert knowledge on the field of application is integrated, the better such a system is going to be. However, the ability to learn often conflicts with the implementation of detailed physical knowledge, as the first relies on flexibility, while the latter tries to reduce that. For this reason, we will focus on systems for statistical pattern recognition as they are concerned with learning from observations.

The issues to be described are just a selection of the many points which are not yet entirely understood. Some of them may be solved in the future by the development of novel procedures or by gaining an additional understanding. Others may remain an issue of concern to be dealt with in each application separately. In the subsequent sections, we will systematically describe them according to the following line of the advancement of a pattern recognition system:

- *Representation.* This is the way individual real world objects and phenomena are numerically described (or encoded) such that they can be related to each other in some meaningful mathematical framework. This framework has to allow the generalization to take place.
- *Design set.* This is the set of objects available or selected to develop the recognition system.
- *Adaptation.* This is usually a reduction of the representation such that it becomes more suitable for the generalization step.
- *Generalization.* This is the step in which objects of the design set are related such that classes of objects can be distinguished and new objects can be accurately classified.
- *Evaluation.* This is an estimate of the performance of a developed recognition system.

2 Representation

The problem of representation is a core issue for pattern recognition [5, 7]. It encodes the real world objects by some numerical description, handled by computers in such a way that the individual object representations can be inter-related. Based on that, later a generalization is achieved, establishing descriptions or discriminations between classes of objects. Originally, the issue of representation was almost neglected, as it was reduced to the demand of having good features provided by some expert. The learning is often believed to start at the given feature vector space. Indeed, many books on pattern recognition disregard the topic of representation, simply by assuming that objects are somehow already represented [2, 25].

A systematic study on representation [9, 24] is not easy, as it is application- or domain-dependent (where the word 'domain' refers to the nature or character of problems and the resulting type of data). For instance, the representations of a time signal, an image of an isolated 2D object, an image of a set of objects placed on some background, a 3D object reconstruction or the collected set of outcomes of a medical examination are entirely different observations that need separate approaches to find good representations. Anyway, if the starting point of a pattern recognition problem is not well defined, this cannot be improved later in the process of learning. It is, therefore, of crucial importance to study the representation issues seriously. Some of them are phrased in the subsequent sections.

2.1 The use of vector spaces

Traditionally, objects are represented by vectors in a feature vector space. This representation makes it very feasible to perform some generalization (with respect to this linear space), e.g. by estimating density functions for classes of objects. However, the object structure is lost in such a description. If objects contain an inherent, identifiable structure or organization, then relations between their elements, like relations between neighboring pixels in an image, are entirely neglected. This also holds for spatial properties such as Fourier coefficients or wavelets weights. These original structures might be partially re-discovered by deriving statistics over a set of vectors (representing objects), however, these are not included in the representation itself. One may wonder whether the representation of objects as vectors in a space in not too oversimplified to be able to reflect the nature of objects in a proper way. Perhaps objects might be better represented by convex bodies in a space or by some other structures. The generalization over sets of vectors, however, is heavily studied and mathematically well developed. How to generalize over a set of other structures is still an open question.

The essential problem of the use of vector spaces for object representation is originally pointed out by Goldfarb [13]. He prefers a structural representation in which the original object organization (connectedness of building structural elements) is preserved. However, as a generalization procedure for structural representations does not exist yet, Goldfarb starts from the evolving transformation systems [12] to develop a novel system [14].

Issue: How to overcome the fundamental inadequacy of vector space representations?

2.2 Compactness

An important, but seldom explicitly identified property of representations is compactness [1]. In order to consider classes, which are bounded in their domains, the representation should be constraint: objects that are similar in reality should be close in their representations. If this demand is not satisfied, objects may be described arbitrarily. Hence, there is no generalization.

This so-called *compactness hypothesis* puts some restriction on the possible probability density functions that classes may have in a vector space used for the representation, e.g. the feature space. This, thereby, also narrows the set of possible classification problems. A formal description of the probability distribution of this set may be of interest to estimate the expected performance of classification procedures for an arbitrary problem.

The *no-free-lunch theorem* claims that all expected performances for all classifiers are equal (in particular equal to the random assignment rule) [30]. This pessimistic result is based on an unbounded set of possible problems, not limited by the compactness hypothesis. If the latter is taken into account, the study on generalization abilities may be significantly improved.

Issue: What is the distribution of classification problems that is in agreement with the compactness hypothesis?

2.3 Representation types

The following representations are here distinguished:

- *Features.* Objects are described by a set of characteristic attributes. If these attributes are continuous, the representation is usually compact. Nominal and categorical attributes may cause problems. As a description by features is a reduction of objects to vectors, different objects may have the same representation. Consequently, classes may overlap.
- *Pixels* or other samples. A complete representation of an object may be approximated by its sampling. For images, these are pixels, for time signals, these are time samples and for spectra, these are wavelengths. A pixel representation is a specific, boundary case of a feature representation, as it describes the object properties in each point of observation.
- *Probability models.* Object characteristics may be related by some probabilistic model. Such models may be based on expert knowledge or trained from examples. Mixtures of knowledge and probability estimates are difficult, especially for larger models.

- *Structural models.* Instead of using probabilities, object models may also be based on a structural description. Automatic procedures aiming at the design of such descriptions from a set of examples are still in their childhood and mainly restricted to the estimation of model parameters. How to learn a structure is not yet clear. Some ideas can be found in [14].
- *Dissimilarities.* Instead of an absolute description by features, objects are relatively described by their dissimilarities to a collection of specified objects. These may be carefully selected prototypes, but also random subsets of the training set may work well [23]. The dissimilarities may be derived from raw data, such as images, spectra or time samples, from original feature representations or from structural representations such as strings or relational graphs. If the dissimilarity measure is nonnegative and zero only for two identical objects, always belonging to the same class, the class overlap may be avoided by dissimilarity representations.
- *Similarities.* In contrast to dissimilarities, similarities may be naturally additive with respect to the support (characteristics) for particular classes. Attributes that are in agreement with some class membership may increase the similarity to that class. For this reason, a similarity representation may be good to deal with missing values and partially characterized objects.
- *Conceptual representation.* Objects may be related to classes in various ways, e.g. by a set of classifiers, each based on a different representation, training set or model. The combined set of these initial classifications or clusterings constitute a new representation [24]. This is used in the area of combining clusterings [10] or combining classifiers [20].

Object descriptions in feature spaces and by dissimilarity representations constitute a good basis for generalization in some appropriately determined spaces. It is, however, difficult to integrate them with the detailed prior knowledge that one has on classes. On the other hand, probabilistic models and structural models, especially, are well suitable for this integration. They, however, constitute a weak basis for training general classification schemes. Usually, they are limited to assign objects to the class model that fits best based on the nearest neighbor rule.

Issue: Can representations be found that offer a good basis for modeling object structure and which can also be used for generalizing from examples?

2.4 Missing data problem

Recognition of partially characterized objects is important for many applications. Depending on the representation, many solutions are investigated, often trying to estimate some values for the missing data item. Here, we will just emphasize again the possibility of using a similarity representation for approaching this problem. Instead of estimating the missing values, it may be worth using the data example that are available.

Issue: Can similarities solve the missing data problem?

2.5 Optimized and trainable representations

The representation may be based on background knowledge of the recognition problem at hand. Some representations, like the conceptual one, are based on a generalization over other representations. Below we will discuss the adaptation of a representation to the optimal conditions following from a generalization procedure. In addition, it may be also possible to learn from the raw data what a good representation is, independently of the problem knowledge and independently of the generalization procedure.

Concerning the learning process, two types of representations are considered: optimized (or fixed) ones and trainable ones; see also [7, 24]. Optimized representations rely on some initial representations for which specific parameters are to be found. For instance, for dissimilarity representations, the measure itself is assumed to be given. It can be optimized with respect to a set of objects, but rather in a limited way such as the specification of a nonlinear transformation and the search for optimal parameters. This is also related to the adaptation step discussed below.

Trainable representations should be built on raw measurements and rely on some identified collection of sub-patterns that may be used to learn to describe the internal structure of objects. In the process of an active design, particular sub-patterns should be chosen together with some weights such that each class separately possesses a compact description and is well defined in the presence of other classes. This may be judged with respect to the chosen classification procedure. Another possibility to build a trainable representation is to consider a conceptual representation, based e.g. on a proximity of an object to a class. This proximity is related to the costs (weights of transformations) of generating an object from a set of primitives (basic descriptors) in the context of other objects within a class, as well as objects outside this class. Such an attempt is carried out in [14], where not only the essential transformations and the weights are learnt, but structural primitives (sub-patterns) as well.

Issue: Can good representations be learnt?

2.6 Spatial connectivity

In general, recognition problems may be context dependent. If this context is not incorporated to the representation (which often occurs in practice), the resulting conflicts have to be solved afterwards. An example is an image recognition system using pixels as features (hence a 16×16 image is represented as a point in a 256-dimensional space). The spatial connectivity between the neighboring pixels is not preserved in such a feature representation. It may be retrieved from the correlations between the features (pixels), but it is not included in the representation itself. Consequently, a statistical decision function built on this representation neglects the original structure of an image. This is inherent to the vector space inadequacy observed by

Goldfarb [13]. Feasible approaches incorporating the spatial connectivity to numerical representations have to be still developed. A possibility might be offered by proximity (similarity or dissimilarity) representations derived from intermediate structural description of objects [7, 24].

Issue: How to incorporate contextual relations into the representation?

3 Design Set

A pattern recognition problem is not only defined by a representation itself, but also by the set of examples given for training and evaluating a classifier in its various stages. The selection of this set and its usage strongly influence the overall performance of the final system. We will discuss some related issues.

3.1 Multiple use of the training set

The total design set or its parts are used in several stages during the development of a recognition system. Usually, one starts from the exploration of this set, which may lead to the removal of wrongly scanned or erroneously labeled objects. After gaining some insights into the problem, the analyst may select a classification procedure based on his/her observations. Next, the set of objects may go through some normalization. Additionally, the representation has to be optimized, e.g. by a feature selection or extraction algorithm. Then, a series of classifiers has to be trained and the best ones need to be selected or combined. An overall evaluation may result in a re-iteration of some steps leading to different choices.

In the entire process the same objects may be used a number of times for the estimation, training, validation, selection and evaluation. Usually, one estimates an average error by a cross-validation. It is well known that the multiple use of objects should be avoided as it biases the results and decisions. Re-using objects, however, is almost unavoidable in practice. A general theory about how much a training set is 'worn-out' by its use and which compensations or corrections may be possible does not exist, yet.

Issue: What is a general theory on the re-use of datasets for training?

3.2 Representativeness of the training set

Training sets should be representative for the objects to be classified by the final system. Usually, a randomly selected subset of the latter is used for training. Intuitively, it seems to be useless to collect many objects represented in the regions where classes do not overlap. On the contrary, in the proximity of the decision boundary, depending on its complexity (non-linearity) and the class overlap, a higher sampling rate seems to be advantageous. This is, of course, inherently related to the chosen classification procedure.

Such a representativeness can be only discussed for static problems, i.e. where raw measurements used to define representations do not significantly change over time. To rephrase it, we assume that the circumstances of the measurement collecting process are stable or if they change, the variable factors are identifiable and their influence on the construction of the final representation is negligible. In other situations, one should consider an active approach, where a classifier develops in time.

Issue: When is the training set sufficiently well sampled and representative for the recognition problem? Should a classifier develop over time?

3.3 Unknown or undetermined class distributions

For some problems, like in medical or machine diagnostics cases, the object distributions for one or more classes are badly defined or even undetermined. For instance, how the class of non-faces can be defined in the face detection problem? Or in machine diagnostics, what is the probability distribution of all casual events if the machine will be used for undetermined production purposes? Therefore, a training set that is representative for the class distributions cannot be found. An alternative may be to sample the domain of the classes such that all possible objects are approximately covered. This means that for any object that could be encountered in practice there exists a sufficiently similar object in the training set. 'Sufficiently similar' has to be defined in relation to the specified class differences. Moreover, as class density estimates can not be derived for such a training set, class posterior probabilities cannot be computed. For this reason such a type of domain based sampling is only appropriate for non-overlapping classes. In particular, this problem is of interest for non-overlapping (dis)similarity based representations [7].

Issue: Is domain sampling possible? Can from a given dissimilarity matrix be determined whether the sampling is sufficiently dense?

4 Adaptation

Once a recognition problem has been formulated by a set of example objects in some representation, the generalization over his set may be considered, finally leading to a recognition system. However, the selection of a proper generalization procedure may not be evident, or several mismatches may exist between the realized representation and the preferred generalization procedures. This occurs when e.g. the chosen representation needs a non-linear classifier and only linear decision functions are computationally feasible, or the space dimensionality is much too high with respect to the cardinality of the training set, or the representation cannot be perfectly embedded in a Euclidean space, while most classifiers demand that. For reasons like these, various adaptations of the representation may be considered. When class differences are explicitly preserved or emphasized, such an adaptation may be considered as a part of the generalization procedure. Some adaptation issues that are less connected to classification are discussed below.

4.1 Problem complexity

In order to determine which classification procedures might be beneficial for a given problem, Ho and Basu [17] proposed to investigate its *complexity*. This is yet an ill-defined concept. Some of its aspects include data organization, sampling, irreducibility (or redundancy) and the interplay between local and global character of the representation and/or of the classifier. Perhaps several other attributes are needed to define complexity such that it can be used to indicate a suitable pattern recognition solution to a given problem.

Issue: How can the complexity of a recognition problem be characterized?

4.2 Selection or combining

Representations may be complex, e.g. if objects are represented by a large amount of features or if they are related to a large set of prototype examples. A collection of classifiers can be designed to make use of this fact and later combined; see section 5. Additionally, also a number of representations may be considered simultaneously. In all these situations, the question arises whether a selection has to be made from the various sources of information, or whether some type of combination should be preferred. A selection may be made randomly, or be based on a systematic search procedure for which many strategies and criteria are possible. Combinations may sometimes be fixed, e.g. by taking an average, or a type of a parameterized combination like a weighted linear combination as a principal component analysis (PCA); see also [3, 21, 24].

The choice for some selection or combining procedure is sometimes dictated by economical arguments, minimizing the amount of necessary measurements or computations. If this is not an issue, the decision has to be made based on accuracy arguments. Selection neglects some information, while combination tries to use everything. The latter, however, may suffer from overtraining as weights or other parameters have to be estimated and may be adapted to the noise in the data. The recently popular sparse solutions offered by support vector machines [26] and sparse linear programming approaches [15, 11] constitute a way of compromise. How to optimize them is not yet clear.

Issue: What are the advantageous ways of optimizing a set of representations?

4.3 Nonlinear transformations

One way to build a simple classifier such as a linear function for a representation that demands a more complicated, nonlinear solution is to transform the representation in an appropriate way to emphasize the linear aspects. One special example is the transformation of a non-Euclidean dissimilarity representation such that it becomes embeddable in a Euclidean space. However, such a nonlinear transformation is not directly focused on finding the best classifier, as it just prepares the framework for future generalization. See [22] for a discussion. It is, thereby, doubtful whether this two-step procedure is better than a direct use of a nonlinear classifier on the original representation.

Issue: When are nonlinear transformations of the representation useful?

4.4 Class structure or class distribution

Assume we have some high-dimensional vector representation of a recognition problem. The training set consists of a set of vectors for each class. Do these vectors constitute a cloud of points, like we tend to draw on a piece of paper if we explain classification procedures in 2D spaces? This idea probably does not hold in high-dimensional spaces. Many representations simply do not posses as many degrees of freedom as the space dimensionality. Their intrinsic dimensionality tends to be much lower. Also a linear subspace, as e.g. the one found by the PCA, is often not a good model of a class description. The reason is that the linear interpolation between two vectors, representing objects of the same class, may produce representations for which no objects exist in reality. For instance, a linear interpolation between two images of different faces does not usually create a proper face image. Consequently, the class of face images, even of the same person with a slightly rotated head positions, yields neither a cloud of points, nor a linear subspace in some representation space. Most likely, classes constitute nonlinear manifolds in these high-dimensional spaces.

Issue: How to constitute classifiers making use of the fact that classes constitute non-linear structures when represented in high-dimensional spaces?

5 Generalization

The generalization over sets of points leading to class descriptions or discriminants was extensively studied in pattern recognition in the 60's and 70's of the previous century. Many classifiers were designed, based on the assumption of normal distributions, kernels or potential functions, nearest neighbor rules, multi-layer perceptrons, etcetera [4, 29, 18]. These types of studies were later extended by the fields of multivariate statistics, artificial neural networks and machine learning. However, in the pattern recognition community, there is still a high interest in the classification problem, especially in relation to practical questions concerning issues of combining classifiers, novelty detection or the handling of ill-sampled classes.

5.1 Classifier selection or classifier combination

The issue of selection or combining holds for representations as well as for classification. In the latter case, it is more clear and apparent as no further steps have to be considered. If a set of classifiers is computed in the process of developing a recognition system, do we take the best one, or do we combine them? Which analysis should produce the right answer? For selection some performance estimation is needed, e.g. based on an evaluation set, or on a systematic cross-validation scheme. Why is it not possible to use the same scheme for combining? See [20] for some ideas.

Issue: How to decide between a selection or combining a set of classifiers?
Issue: Which are good sets of classifiers to be combined?

5.2 Trained or fixed combining

A set of well trained classifiers may yield the classification outputs that are further compared and combined by a fixed procedure like averaging or product [19]. Imperfectly trained classifiers may be combined by a trained combiner. For this an additional training set is needed that may also be used to train the base classifiers better. If the same training set is used for training both the combiner and the base classifiers, then training of the combiner will suffer from bias and will not be representative for new objects, unless the base classifiers are undertrained. In this case, however, a fixed combiner may work well too. Consequently, it is not clear when and how a combiner should be trained [6].

Issue: When and how should a combining classifier be trained?

5.3 Sequential or parallel training

The architecture of a combined classifier with linear base classifiers is very similar to a neural network. The main difference refers to the training step, either classifier by classifier, or as an entire system at once. A similar difference exists between the feature selection followed by a linear classifier and training a sparse linear classifier that reduces the feature space by its design. Many more examples exist, like a PCA followed by a classifier or a regularized classifier. Is it possible to find a general rule that gives insights when either sequential or parallel training has to be preferred?

Issue: When should a recognition system be trained part by part and when in its entirety?

5.4 Classifier typology

Any classification procedure has its own explicit or built-in assumptions with respect to the class distributions or other characteristics. This implies that for a problem that exactly fulfils the assumption of a particular procedure, this procedure will generate a relatively well-performing classifier. Consequently, any classification approach has its problem for which it is the best. In some cases such a problem might be far from reality. The construction of such problems may reveal which the typical characteristics of a particular procedure are. Moreover, when new proposals are to be evaluated, it may be demanded that some examples of its corresponding typical classification problem are published, making clear what the area of application may be. See also [8].

Issue: A library of problems corresponding to the library of classifiers.

5.5 Generalization principles

What is the basic principle of generalization over a set of examples? How can we apply some rules to identify an unobserved property of an object that is different from all objects in the given set of examples? The two basic generalization principles are probabilistic inference, using the Bayes' rule and the minimum description length principle that determines the most simple model in agreement with the observations (Occam's razor). These two principles are essentially different. The first one is sensitive to multiple copies of an existing object in the training set, while the second one is not. Consequently, the latter is not based on densities, but just on object differences or distances. When should each principle be followed? Should this be decided from the start, in the selection of the design set and the way of building a representation, or is it possible to postpone it for later?

Issue: Bayes or Occam?

5.6 The use of unlabeled objects and active learning

The above mentioned principles are examples of statistical inductive learning, where a classifier is induced based on the design set and it is later applied to unknown objects. The disadvantage of such approach is that a decision function is in fact designed for all possible representations, whether valid or not. Transductive learning is an appealing alternative as it determines the class membership only for the objects in question, while relying on the collected design set or its suitable subset.

The use of unlabeled objects, not just the one to be classified, is a general principle that may be applied in many situations. It may improve a classifier based on just a labeled training set. If this is understood properly, the classification of an entire test set may yield better results than the classification of object by object.

Issue: How to make use of unlabeled data to construct classifiers?

5.7 Multi-class problems

Two-class problems constitute the traditional basic line in pattern recognition. It boils down to finding a discriminant or a binary decision. Multi-class problems can be formulated either as a series of two-class problems (this can be done in various ways, none of them is entirely satisfactory) or as a detection problem in which each of the possible classes is searched for. The one that fits best is used. This approach neglects all alternatives in the first step, but compares them later.

Issue: Should multi-class recognition be performed by detection or by classification?

5.8 One-class problems

In contrast, but also very similar to multi-class problems, are the one-class problems. In the latter case, a single class is well defined and all alternatives, outliers, classes (of any number) are just ill-sampled, not sampled at all or undefined. Discriminants seem to be inappropriate, while the class detectors do not use what might be available on the alternatives. Densities cannot be applied properly as they cannot be estimated for the outlier class; see [27, 28]. One-class classifiers can be constructed as examples of proximity-based conceptual representations [7, 24].

Issue: What is a proper one-class classifier?

5.9 Domain based classification

There are several reasons why densities cannot be used (properly) for constructing classifiers. Any classifier based on averages like the computation of a mean square error assumes a distribution of the training objects in agreement with a class distribution. If this does not exist, or if the training set does not agree with it, such classifiers cannot be formally used. As explained above, an appealing alternative is the use of distances and/or the minimum description length principle. Classifiers based on this cannot decide in class overlapping regions. So they need a representation that avoids this. The dissimilarity representation may be suitable, however, others may be used as well if the ground is found justifying why an overlap is avoided. So, there is a need for domain based classifiers [7]. These are classifiers that assume no class overlap and that do not require that the distribution of the training set follows the class distribution. The class domain, however, should still be sampled properly in such a way that it can be approximated or used to construct a decision function.

Issue: How to construct domain-based classifiers.

5.10 Object structure

We repeat here the issue already raised in the section on representation. Suppose that the object structure is incorporated to the representation. How do we generalize over a set of examples? Does this result describe a structure, a distribution, some domain in a space or a growth model? Again we refer to the proposal of Goldfarb [14] which seems to be a very general attempt to solve this problem.

Issue: How to learn the structure in objects?

6 Evaluation

Two questions are always apparent in the development of recognition systems. These are: how good is a particular system once it is trained and which are good recognition procedures in general? The first question has sometimes a definite answer, while the second one is open.

6.1 Recognition system performance

What do we mean if we wonder how good a system is? Is it its accuracy on average, computed over all objects we are going to classify or is it determined by the worst error it may be made? In the first case, we again assume that the set of objects to be recognized is well defined (in terms of distributions). Then, it can be sampled and the accuracy of the entire system can be estimated based on an evaluation set. We now neglect the issue that after using this evaluation set together with the training set, a better system can be found. A more interesting point is how to judge the performance of a system if the distribution of objects is ill-defined or if a domain based classification system is used as discussed above. Now, the largest mistake made becomes a crucial factor for this type of judgements. One needs to be careful, however, as this may refer to an unimportant outlier (resulting e.g. from invalid measurements).

Issue: How to evaluate the performance for ill-defined class distributions?

6.2 Prior probability of problems

How good is a recognition procedure in general? As argued above, any procedure has a problem for which it performs well. So, how large is the class of such problems? We cannot state that any classifier is better than any other classifier, unless the distribution of problems to which these classifiers will be applied is defined. Such distributions are hardly studied. What is done at most is that classifiers are compared over a collection of benchmark problems. Such sets are usually defined ad hoc and just serve as an illustration. The set of problems to which a classification procedure will be applied is not defined.

Issue: How to judge the expected performance of a recognition procedure?

7 Discussion and conclusions

Pattern recognition is a human activity that we try to imitate by mechanical means. There are no physical laws that assign observations to classes. It is the human consciousness that groups observations together. Although their connections and inter-relations are often hidden, by the attempt of imitating this process, some understanding might be gained. The human process of learning patterns from examples may follow along the lines of trial and error. It has, however, to be strongly doubted whether statistics play an important role in this process. Estimating probabilities, especially in multi-variate situations is not very intuitive for majority of people. Moreover, the large amount of examples needed to build a reliable classifier by statistical means is much larger than it is available for human learning.

In human recognition, proximities based on relations between objects seem to come before features are searched and may be, thereby, more fundamental. For this reason and the above observation we think that the study of dissimilarities, distances and domain based classifiers are of great interest. This is further encouraged by the fact that such representations offer a bridge between the possibilities of learning in vector spaces and the structural description of objects that preserve relations between object's inherent structure.

We think that the use of dissimilarities for representation, generalization and evaluation constitute the most intriguing issues in pattern recognition.

References

1. Arkedev AG and Braverman EM (1966) Computers and Pattern Recognition. Thompson. Washington, DC.
2. Bishop CM (1995), Neural Networks for Pattern Recognition. Clarendon Press.
3. de Diego IM, Moguerza JM, and Muñoz A (2004) Combining Kernel Information for Support Vector Classification. Multiple Classifier Systems. LNCS:3077. Springer-Verlag. 102-111.
4. Duda RO, Hart PE and Stork DG (2001). Pattern Classification 2nd. edition, John Wiley & Sons.
5. Duin RPW, Roli F, and De Ridder D (2002). A note on core research issues for statistical pattern recognition, Pattern Recognition Letters, 23:493-499.
6. Duin RPW (2002), The Combining Classifier: To Train Or Not To Train? ICPR2002 II:765-770.
7. Duin RPW, Pekalska E, Paclík P, and Tax DMJ (2004). The dissimilarity representation, a basis for domain based pattern recognition?, In: Pattern representation and the future of pattern recognition, Workshop ICPR2004. 43-56.
8. Duin RPW, Pekalska E, and Tax DMJ (2004) The characterization of classification problems by classifier disagreements. Proc. ICPR II:140-143.
9. Edelman S (1999), Representation and Recognition in Vision, MIT Press.
10. Fred A, and Jain AK (2002), Data clustering using evidence accumulation, ICPR2002. Quebec City, Canada. 276-280.

11. Fung, GM and Mangasarian OL (2004), A Feature Selection Newton Method for Support Vector Machine Classification. Computational Optimization and Aplications 28:185Ũ202.
12. Goldfarb L, (1990), On the foundations of intelligent processes – I. An evolving model for pattern recognition. Pattern Recognition. 23(6)595-616.
13. Goldfarb L, and Hook J (1998), Why classical models for pattern recognition are not pattern recognition models. In: International Conference on Advances in Pattern Recognition. Springer. 405-414.
14. Goldfarb L, Gay D, Golubitsky O, and Korkin D (2004), What is a structural representation? 2nd version. Faculty of Computer Science, UNB. Technical Report TR04-165.
15. Graepel T, Herbrich R, Schölkopf B, Smola A, Bartlett P, Müller KR, Obermayer K and Williamson R (1999) Classification on Proximity Data with LP-Machines. ICANN 1991, 304-309.
16. Van der Heijden F, Duin RPW, de Ridder D and Tax DMJ (2004) Classification, Parameter Estimation and State Estimation. An Engineering Approach using Matlab. John Wiley & Sons Ltd.
17. Ho TK, and Basu M (2002) Complexity measures of supervised classification problems. IEEE T-PAMI 24:289-300.
18. Jain AK, Duin RPW and Mao J (2000) Statistical Pattern Recognition: A Review. IEEE T-PAMI 22:4-37.
19. Kittler J, Hatef M, Duin RPW, and Matas J (1998) On Combining Classifiers, IEEE T-PAMI 20:226-239.
20. Kuncheva LI (2004) Combining Pattern Classifiers: Methods and Algorithms. Wiley. New York.
21. Pekalska E, Skurichina M, and Duin RPW (2004) Combining Dissimilarity Representations in One-class Classifier Problems. Multiple Classifier Systems. LNCS:3077. Springer-Verlag. 122-133.
22. Pekalska E, Duin RPW, Gunter S, and Bunke H (2004) On not making dissimilarities Euclidean. In: Structural, Syntactic, and Statistical Pattern Recognition. LNCS:3138. Springer Verlag, Berlin. 1145-1154.
23. Pekalska E, Duin RPW, and Paclík P (2004) Prototype Selection for Dissimilarity-based Classifiers. Pattern Recognition. accepted.
24. Pekalska E (2005) Dissimilarity representations in pattern recognition. Concepts, theory and applications. PhD thesis. Delft University of Technology.
25. Ripley BD (1996) Pattern Recognition and Neural Networks. Cambridge University Press. Cambridge.
26. Shawe-Taylor J and Cristianini N (2004) Kernel Methods for Pattern Analysis. Cambridge University Press.
27. Tax DMJ (2001) One-class classification. PhD thesis. Delft Univ. of Technology.
28. Tax DMJ, and Duin RPW (2004) Support vector data description. Machine Learning 54(1):45-56.
29. Webb A (2002) Statistical pattern recognition. Wiley. New York.
30. Wolpert DH (1995) The Mathematics of Generalization. Addison-Wesley.

The Role of Ontological Models in Pattern Recognition

Juliusz L. Kulikowski[1]

Institute of Biocybernetics and Biomedical Engineering PAS, 4 Ks. Trojdena Str., 02-109 Warsaw, Poland jlkulik@ibib.waw.pl

Summary. There are considered the role and applications of ontological models in advanced pattern recognition methods. Formal definition of ontological models, a general taxonomy, and specification of some typical ontological models are presented. Examples of a simple, a composite and an extended ontological model are given. The role of ontological models in composite patterns recognition is described and illustrated by examples.

1 Introduction

Let us start our considerations by reminding a historical fact: in the 40ths of the past century it arose a practical problem of recognition of radar echoes reflected by long-distanced flying objects observed on the background of apparent echoes caused by noise. A general solution of this problem was based on statistical decision methods [1], and this fact can be claimed a starting point of a statistical approach to machine-aided pattern recognition. This approach was historically prior to the one originated in the middle of 50ths by F. Rosenblatt by his works concerning the concept of *perceptron* - the earliest prototype of an artificial neural network recognizing simple graphical signs [2]. Moreover, since their very beginning both approaches to pattern recognition assigned a semantic interpretation to the recognized classes of signals. However, in statistical approach this interpretation was deduced from some assumed situations arising in the real world and as a consequence it also led to a pragmatic evaluation of decisions in the terms of their individual costs and a total risk. Otherwise speaking, the pattern recognition algorithm was considered in connection with a concept of a real world with existing in it mechanisms of available signals generation, as well as with the sources of disturbances and of noise affecting the signal analysis procedures. Such concepts, presented in a formalized form, are considered below as particular cases of *ontologies*. In classical philosophy the notion *ontology* meant a part of metaphysics concerning the nature and theory of existence. In a more narrow sense it is used in computer science for a common understanding of some domain [3] or an abstract view of the world we are modeling, describing the concepts and their relationships [4]. In such sense it is used as a tool for modern information, managing, educational, monitoring, control systems design, etc. An ontology characterizes, in general, many various aspects of the real world. It thus consists of several modules describing it from different points of view and in uniform manners. Such modules, components of ontologies, are called here *ontological models* (*OM*).

Example 1

A simple example of a conceptual model used in radar echoes detection and recognition is shown in Fig.1. It consists of: a) a module describing the rules of physical formation of radar echoes reflected from immobile and/or flying objects, b) a module describing the rules of physical disturbances and noise acting on radar signals and c) a module modeling the physical space penetrated by radar signals, including spatial distribution of immobile and flying objects, allocation of sources of noise, etc.

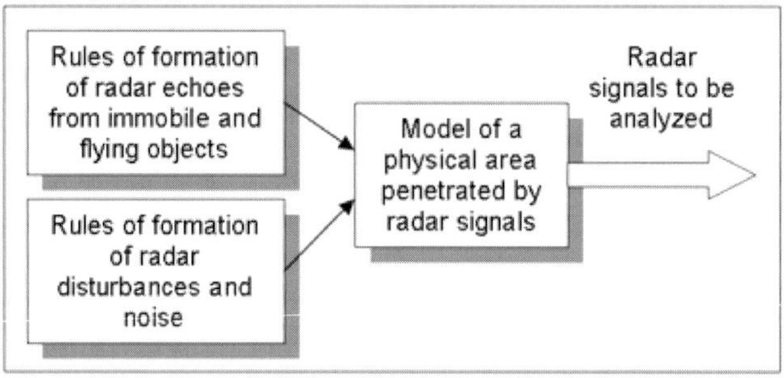

Fig. 1. Ontological model used in radar echoes recognition.

It should be remarked that the above-presented *model* is not a description of a given instance-situation arising in the real world. It is rather a formal description of situations considered as classes of possible instances that according to our primary knowledge may arise in the world and influence the form of signals used to a recognition of caused them situations. This conceptual model can be used as the first step to construction of an *OM*, as it will be shown below •

Till the pattern recognition methods were oriented to the recognition of relatively simple classes of objects (printed or hand-written characters, contours of irregular geometrical objects, edges, bifurcations and/or crosses of lines, well-defined physical signals, etc.), using advanced ontological models to the formulation of pattern recognition problems was not necessary. Attention was then paid mostly to releasing pattern recognition methods of strong primary assumptions concerning the properties of recognized classes of objects. This tendency led to elaboration of non-parametric statistical methods in pattern recognition, learning algorithms, etc. However, as more sophisticated pattern recognition problems were taken into consideration, the role of primary information about the external world played a more substantial role. For example, early detection of potential collisions arising in a large airways system would not be possible without taking into account primary information concerning the air network, types of airplanes, meteorological situation, etc. Such information introduced to an air-traffic control system constitutes a set of its *OM*s necessary for decision making. A necessity of taking into account some primary information about the external world arises also if advanced pattern recognition problems in road traffic surveillance, interpretation of aerial, satellite or geophysical observations, medical or technical diagnosis, etc. are considered. In such cases the following general problems arise:

- *What primary general information or what assumptions about the real world should be taken into account in order to formulate a pattern recognition problem appropriate to the needs of an observer (user);*
- *In what form this information should be presented in order to facilitate the solution of the pattern recognition problem.*

The aim of this paper is: 1st to indicate the role of *OM*s as general tools of presentation of the above-mentioned primary information, and 2nd to propose a taxonomy and specification of basic *OM*s used in composite pattern recognition.

2 Types of ontological models for pattern recognition

According to [3] an *ontology* can be formally represented by a quadruple:

$$O = [C,\ R,\ A,\ Top\ ] \tag{1}$$

where C is a non-empty set of *concepts* (including relation concepts and the *Top*), R is a set of all *assertions* in which two or more concepts are related to each other, A is a set of axioms and *Top* is *the highest-level concept* in the hierarchy.

According to some authors *concepts* can be defined as follows: ŞA concept can be anything about which something is said and can be abstract or concrete, elementary or composite, real or fictious, or the description of a task, function, action, strategy, reasoning process etc. [6].

In similar way, a wide interpretation of *assertions* is possible. It may denote any type of relations (including *functional* relations, *orderings, taxonomies*, etc.), *super-* and *hyper-relations, deterministic, probabilistic, fuzzy, rough*, etc.

Axioms specify the properties of concepts, in particular, as objects of assertions or as arguments of relations; they also may specify the type of assertions. Following (1), an *OM* within the given ontology O then can be defined as a quadruple:

$$OM_i = [C_i,\ R_i,\ A_i,\ Top_i], i = 1, 2, \ldots I, \tag{2}$$

such that: $C_i \subseteq C$ is a non-empty subset of concepts, $R_i \subseteq R$ is a subset of assertions concerning the elements of C_i and containing, in particular, a sub-taxonomy Ξ_i of the elements of C_i, $A_i \subseteq A$ is a subset of axioms concerning the relations in R_i and $Top_i \in C_i$ is the highest element in the sub-taxonomy Ξ_i.

An ontology thus can be represented in the form of a set:

$$O = OM_1,\ OM_2, \ldots,\ OM_I; R, A \tag{3}$$

consisting of several component OMs and of a subset R of super-relations between the OMs as well as of a subset A of axioms concerning the super-relations. In pattern recognition using a complete ontology of a domain is usually not necessary. In most cases it is enough to take into account selected OMs containing, in particular, the concepts of *objects* and of their *classes* of similarity.

Example 2

It will be defined an OM used to formulation of a simple problem of known binary signals recognition in the presence of noise.

Concepts: *probabilistic model, signal, admitted signals, admitted signal 0, admitted signal 1, noise, input signal, class of signals, class 0, class 1, random value, random value instance, probability, probability distribution.*

Assertions: *a priori probability distribution of admitted signals, conditional probability distribution of class 0 of signals, conditional probability distribution of class 1 of signals, probability distribution of input signals, taxonomy of signals, logical constraints imposed on signals.*

In the above-given case the following taxonomy of the concepts of signals has been assumed:

Signal:
- *admitted signal:*
  - *0*
  - *1*
- *input signal:*
  - *of class 0*
  - *of class 1*
- *noise*

In addition, the following logical assertion holds:

A simultaneous occurrence of signals 0 and 1 is impossible.

Axioms:

- *Admitted signals are random variables with given a priori probabilities;*
- *Noise is a random process with given a priori probability distribution;*
- *Input signal is a random process defined as a sum of admitted signal and noise;*
- *Two classes of input signals are random processes described by conditional probability distributions for fixed admitted signals.*

Top: probabilistic model.

The above-described *OM* is used as a basis of an approach to the solution of binary signal recognition problem widely known as a Bayesian approach [5]•

In similar way it can be constructed an *OM* used to formulation of a problem of any finite class of known signals recognition in the presence of noise. However, in practice only selected elements of such *OM*s are clearly specified, the rest ones remaining assumptive or undefined.

From the pattern recognition point of view the *OM*s based on specific concepts, concerning objects being to be recognized (*RO*s), are of particular interest. First of all, the concepts should specify the *nature* and/or *type* of *RO*s and their *similarity classes*. The *RO*s may have an *abstract* or a *real* nature. Examples of abstract objects are: functions, parameters, distributions, geometrical objects, graphs, symbolic or textual expressions, etc., which can be automatically recognized and/or evaluated by analysis of their numerical or graphical representations. Examples of real objects are: physical, chemical, biological or medical objects, phenomena or processes, human beings, economical, social and/or organizational states or processes, etc.

For pattern recognition methods and algorithms construction a classification of *OM*s based on formal properties of *RO*s also plays a substantial role. From this point of view there can be distinguished:

1. *Simple OM*s containing no assertions about the relationships between the *RO*s;
2. *Composite OM*s in which some relationships between the *RO*s are assumed;
3. *Extended OM*s in which some relationships not only between the *RO*s but also between the *RO*s and other objects having in the *OM*s a conceptual representation are assumed.

The above-given Example 2 concerns the case of a simple *OM*: two binary signals, *0* and *1*, each excluding the other one, have been assumed to be drawn at random. No statistical dependence between the results of drawing the signals in a series of experiments is also assumed.

A different situation would arise in the following case.

Example 3

Let us assume that an *OM* contains the following subset of medical concepts:

Inflammation:
  Process of inflammation:
    acute
    chronic
  Type of inflammation:
    focused
    fuzzy

Despite the fact that the classes of objects belonging to the same classification level are mutually disjoint, the classes belonging to different levels have non-empty intersections. As a consequence, partial decisions made on different pattern recognition levels would be, in general, dependent in the sense that decisions made on the higher level influence the ones made on the lover level However, this poly-hierarchical (two-dimensional) taxonomy can be scrolled into an uni-hierarchical one containing four mutually disjoint composite classes of objects [7]:

Inflammation:
acute focused,
acute fuzzy,
chronic focussed,
chronic fuzzy.

In practice it is easier to realize a pattern recognition system on the basis of the poly-hierarchical taxonomy leading to a two-level algorithm recognizing: 1st the *Process* and 2nd the *Type* of inflammation (see, for example [8])•

The above-illustrated situation is typical in the cases of poly-hierarchical taxonomies of *RO*s. For example, selected medical taxonomies of diseases consist of the following classification aspects:

1. International Classification of Diseases ICD:
   single classification aspect;
2. International Classification of Diseases in Oncology ICD-O:
   topography, morphology;
3. Systematic Medical Nomenclature SNOMED:
   topography, morphology, etiology, function,

where:

Topography relates the disease to the anatomy of human body;

Morphology characterizes the type of observed pathological changes or processes in cells, tissues or organs;

Etiology indicates the factors causing pathological changes or processes;

Function describes physical, functional, metabolic, psychical etc. abnormalities in the patient.

Each recognition of objects subjected to a poly-hierarchical taxonomy can be reached by a multi-level pattern recognition algorithm such that selected levels correspond to given classification aspects.

A substantial role in *OM*s play logical assertions concerning the *RO*s. In Example 2 a logical constraint prohibited a simultaneous occurrence of two different admitted signals, and, as a consequence, it made possible using a simple Bayesian model to the recognition of signals. A lack of such logical constraint in medical diagnostic problems leads, in general, to a class of composite *OM*s such that if they contain the concepts of diseases: $D_1, D_2, \ldots, D_k$ then their simultaneous occurrence in the combinations: $D_1 \cap D_2$, $D_1 \cap D_3$, $D_2 \cap D_k$, $D_1 \cap D_2 \cap D_k$ etc. is also admissible. This radically changes the algorithm of pattern recognition.

Before going to further considerations a simple example of an extended *OM* will be given.

Example 4

Let us assume that an *OM* contains a taxonomy of fingerprints collected in a computer-aided system of personal identification. On the first level of fingerprints classification the sub-classes *Identified* and *Unidentified* are distinguished. The *Identified* fingerprints are then subjected to a poly-hierarchical classification according to: 1st the *hand* (*left, right*), 2nd the *finger* (*thumb, forefinger, middle, ring, small*), 3rd *codes of characteristic features*, 4th *personal ascription*, 5th *ascription to registered cases*. In all sub-classes a category *unknown* is admitted. The *Unidentified* fingerprints are divided into the sub-classes: 1st *operational identifier*, 2nd *hand*, 3rd *finger*, 4th *codes of characteristic features*, and 5th *supposed personal ascription*. The recognizing system works as follows:

1. To an unidentified fingerprint f_i an *operational identifier* and, if possible, the values of *hand* and *finger* are assigned;
2. By a fingerprint analysis procedure a code of its characteristic features is assigned;
3. From a database of *Identified* fingerprints a subset S_i of fingerprints the most similar to the f_i is extracted;
4. The values of *personal ascriptions* of the S_i members are put into the *supposed personal ascriptions* of f_i;
5. If only one *supposed personal ascription* has been assigned to f_i on a sufficient credibility level, it is taken as a solution of the personal identification problem.

However, in the above-described identification procedure the category of *ascription to registered cases*, as not substantial for personal identification, has not been used. When the *supposed personal ascription* of f_i is established it indicates the corresponding *personal ascription* of an *Identified* object. As a consequence, through the category *ascription to registered cases* it becomes possible to associate, for the investigation purposes, the given person and his case with other cases and ascribed to them persons. Therefore, the *OM* due to its extension on the objects (*registered cases*) directly being not the *RO*s in the given pattern recognition problem makes possible a logical inference about the connections of the recognized object with external world •

Besides the above-given classification of *OM*s they also can be classified according to the formal nature of constituting them *Assertions*. From this point of view the following classification scheme can be proposed:

OMs:
- *Homogenous*:
  - *Deterministic*:
    - *Logical* (*classical logic*)
    - *Set-theoretical*
    - *Relational*
    - *Algebraic*
    - *Functional*

Linguistic (syntactical)
Geometrical
Topological
Other
Non-deterministic:
Logical (non-classical logic)
Probabilistic
Statistical
Fuzzy sets
Rough sets
Other
Heterogenous
(combinations of the above-given models)

From other point of view, if an interdependence between construction of an ontology and pattern recognition is taken into account, there can be also distinguished the:

*Closed OM*s whose type, form, contents etc. are definitely fixed, and

*Open OM*s, changing their type, form, contents etc. according to the current results of pattern recognition.

3 OMs for composite objects recognitio

- The problems of composite objects recognition arise in many application areas, like: robotics, air-traffic control, road-traffic surveillance, geophysical, meteorological, ecological and/or technological processes monitoring, medical diagnosis, physical and/or biological processes investigation, etc. In all the above-mentioned cases interest is paid to *RO*s satisfying the following general assumptions:
- Each *RO* consists of a certain number of components;
- The components satisfy a hierarchical relation of composition (*object - its component*), as well as (possibly) some other relations linking them within the *RO* as a whole.

Such *RO*s are called here composite *RO*s (*CRO*s). The *OM*s containing the concepts of *CRO*s should, in particular, characterize:

1. internal structure of a single compact *CRO*,
2. spatial configurations of collections of *RO*s constituting a *CRO*,
3. behavior of single compact *CRO*s or of collections of *RO*s in time.

For recognition of *CRO*s various structural methods have been developed. Generally, they can be divided into two groups:

a) *syntactic* (*formal-linguistic*) methods [9][10],
b) *relational* methods [11][12]

Relational approach leads to a particularly flexible set of tools, suitable to structural description of *CRO*s due to the following circumstances:

▷ it admits relations defined on any fixed number of arguments,
▷ extended algebra of relations can be used,
▷ a hierarchy of relations (relations of relations - super-relations) is admissible,
▷ parametric relations are admissible,
▷ the notion of hyper-relations can be defined.

However, it should be remarked that no direct one-to-one correspondence between a structural description of *CRO*s in an *OM* and the sets of relations describing their observed (visual) representation exists. For example, using pattern recognition methods it is easy to recognize and distinguish such geometrical forms as a square, a rectangle and a hexagon independently on their position and size. However, an *OM* containing geometrical assertions may justify a conclusion that some object are, in fact, visual 2D projections of 3D parallelepipeds. This example shows that additional geometrical assertions contained in the *OM* facilitate a correct interpretation of images, what in composite patterns recognition may be particularly important. Similar situations arise in medical images interpretation, analysis of scenes, etc. For example, in echosonographic cardiac imaging additional anatomic assertion helps in distinguishing between the correct left ventricles contour line and the better visible apparent one caused by papillary muscles. In road-scene analysis preliminary assertions are desirable in order to distinguish between the real objects of interest (pedestrians, cars, immobile objects etc.) and the apparent ones caused by shadows, light reflections etc. In fact, advanced scene analysis without well-defined *OM*s is not possible. This can be illustrated by the following example.

Example 5

Let us consider a road-traffic surveillance system whose goal consists in aiding navigation planning in autonomous vehicles [13]. The *OM*s on which the system is based specify the concepts of *road map, objects, obstacles, objects behavior, collisions, damages* etc. For this purpose using various mathematical tools is proposed. For example, predicting a possible collision can be reached by using probabilistic, statistical, fuzzy set, multi-valued logic and many other types of models. Detection of a collision caused by a given object X needs answering the questions like:

- What is the range of possible speeds that it can be going?
- What are its possible directions of travel?
- What is its possible rate of change of direction of travel?

etc. The questions are of a widely-interpreted pattern recognition area type. Replying to each of them needs using a special decision-making module in the system. However, in order to design a module for replying the first question, concerning possible speed of the object under observation it is necessary to :

1. distinguish an object on the road among other visible objects,
2. recognize it as a moving car,
3. identify its type,
4. take into account its technical characteristics.

For this the corresponding *OM*s are helpful. Answering other types of questions can be reached in similar way. Then, on a higher-level of decision making, identification of a possible collision between the identified objects becomes possible. The concept of a *collision* included into the *OM*is a starting point to the design of the corresponding decision module in the system •

4 Conclusion

The paper has shown some basic aspects of using ontological models in composite pattern recognition and interpretation. It seems evident that the problem needs much deeper investigation.

References

1. Peterson WW, Birdsall TG, Fox WC (1954) The theory of signal detectability, Trans. IRE, PGIT-4: 171-212
2. Rosenblatt F (1958) The perceptron, a probabilistic model for information storage and organization in the brain. Psychol Rev 68
3. Fernandez-Lopez M, Gomez-Perez A (2002) Overview and analysis of methodologies for building ontologies. The Knowledge Eng Rev **17**(2): 129-156
4. Gruber TR (1993) A translation approach to portable ontologies. Knowledge Acquisition **5**(2): 199-220
5. Kulikowski J.L. (1972) Cybernetic recognizing systems. PWN, Warsaw (in Polish).
6. Corcho O., Gomez-Perez A. (2000) Evaluating knowledge representation and reasoning capabilities of ontology specification languages. Proc. of the ECAI 2000 Workshop on Application of Ontologies and Problem Solving Methods.
7. Wingert F. (1997) SNOMED a systemized medical nomenclature. Using manual. IBIB PAN, Warsaw (in Polish, translated from German).
8. Kurzynski M. (1997) Pattern recognition. Statistical methods. OWPW, Wroclaw (in Polish).
9. Fu K.S. (1974) Syntactic methods in pattern recognition. Academic Press, New York.
10. Tadeusiewicz R., Flasinski M. (1991) Pattern recognition. PWN, Warsaw (in Polish).
11. Kulikowski J. (1971) Algebraic methods in pattern recognition. CISM, Courses and Lectures No 85. Springer Verlag, Wien.
12. Kulikowski J.L. (1992) Relational approach to structural analysis of images. Machine Graphics and Vision *1*(1/2): 299-309.
13. Schlenoff C., Balakirsky S., Uschold M., Provine R., Smith S. (2004) Using ontologies to aid navigation planning in autonomous vehicles. The Knowledge Eng. Rev. *18*(3): 243-255.

Current Feature Selection Techniques in Statistical Pattern Recognition

Pavel Pudil and Petr Somol

Dept. of Pattern Recognition, Inst. of Information Theory and Automation, Academy of Sciences of the Czech Republic, 182 08 Prague 8, Czech Republic
e-mail: {pudil, somol}@utia.cas.cz

Summary. The paper addresses the problem of feature selection (abbreviated FS in the sequel) in statistical pattern recognition with particular emphasis to recent knowledge. Besides over-viewing advances in methodology it attempts to put them into a taxonomical framework. The methods discussed include the latest variants of the Branch & Bound algorithm, enhanced sub-optimal techniques and the simultaneous semi-parametric probability density function modeling and feature space selection method.

1 Introduction

Pattern recognition can be with certain simplification characterized as a classification problem combined with dimensionality reduction of pattern feature vectors which serve as the input to the classifier. This reduction is achieved by extracting or selecting a feature subset which optimizes an adopted criterion.

2 Dimensionality Reduction

We shall use the term "pattern" to denote the D-dimensional data vector $\mathbf{x} = (x_1, \ldots, x_D)^T$ of measurements, the components of which are the measurements of the features of the entity or object. Following the statistical approach to pattern recognition, we assume that a pattern $\mathbf{x}$ is to be classified into one of a finite set of C different classes $\Omega = \{\omega_1, \omega_2, \cdots, \omega_C\}$. A pattern $\mathbf{x}$ belonging to class ω_i is viewed as an observation of a random vector $\mathbf{X}$ drawn randomly according to the known class-conditional probability density function $p(\mathbf{x}|\omega_i)$ and the respective *a priori* probability $P(\omega_i)$.

One of the fundamental problems in statistical pattern recognition is representing patterns in the reduced number of dimensions. In most of practical cases the pattern descriptor space dimensionality is rather high. It follows from the fact that in the design phase it is too difficult or impossible to evaluate directly the "usefulness" of particular input. Thus it is important to initially include all the "reasonable" descriptors the designer can think of and to reduce the set later on. Obviously, information missing in the original measurement set cannot be later substituted. The aim of dimensionality reduction is to find a set of new d features based on the input set of D features (if possible $d \ll D$), so as to maximize (or minimize) an adopted criterion.

- Dimensionality reduction divided according to the adopted strategy:
  1. *feature selection* (FS, in fact special case of the latter)
  2. *feature extraction* (FE, i.e., feature transformation).

The first strategy (FS) is to select the best possible subset of the input feature set. The second strategy (FE) is to find a transformation to a lower dimensional space. New features are linear or nonlinear combinations of the original features. The choice between FS and FE depends on the application domain and the specific available training data. FS leads to savings in measurements cost since some of the features are discarded and those selected retain their original physical meaning. The fact that FS preserves the interpretability of original data makes it preferable in, e.g., most problems of computer-assisted medical decision-making. On the other hand, features generated by FE may provide better discriminative ability than the best subset of given features, but these new features may not have a clear physical meaning.

- Alternative division according to the aim:
  1. *dimensionality reduction for optimal data representation*
  2. *dimensionality reduction for classification.*

The first aims to preserve the topological structure of data in a lower-dimensional space as much as possible, the second one aims to enhance the subset discriminatory power. In the sequel we shall concentrate on the FS problem only. For a broader overview of the subject see, e.g., [2], [9], [15], [22], [24].

3 Feature Selection

Given a set of D features, X_D, let us denote $\mathcal{X}_d$ the set of all possible subsets of size d, where d represents the desired number of features. Let J be some criterion function. Without any loss of generality, let us consider a higher value of J to indicate a better feature subset. Then the feature selection problem can be formulated as follows: find the subset $\tilde{X}_d$ for which

$$J(\tilde{X}_d) = \max_{X \in \mathcal{X}_d} J(X). \tag{1}$$

Assuming that a suitable criterion function has been chosen to evaluate the effectiveness of feature subsets, feature selection is reduced to a search problem that detects an optimal feature subset based on the selected measure.

One particular property of feature selection criterion, the *monotonicity property*, is required specifically in certain optimal FS methods. Given two subsets of the feature set X_D, A and B such that $A \subset B$, the following must hold:

$$A \subset B \Rightarrow J(A) < J(B). \tag{2}$$

That is, evaluating the feature selection criterion on a subset of features of a given set yields a smaller value of the feature selection criterion.

3.1 FS Categorisation With Respect to Optimality

Feature selection methods can be split into basic families:

1. *Optimal methods*: These include, e.g., *exhaustive search* methods which are feasible for only small size problems and accelerated methods, mostly built upon the Branch & Bound principle. All optimal methods can be expected considerably slow for problems of high dimensionality.
2. *Sub-optimal methods*: These methods essentially trade off the optimality of the selected subset for computational efficiency. These include, e.g., Best Individual Features, Sequential Forward and Backward Selection, Plus-l-Take Away-r, their generalized versions, genetic algorithms, and particularly the Floating and Oscillating Search.

Although the exhaustive search guarantees the optimality of a solution, in many realistic problems it is computationally prohibitive. The well known Branch and Bound (B&B) algorithm guarantees to select an optimal feature subset of size d without involving explicit evaluation of all the possible combinations of d measurements. However, the algorithm is applicable only under the assumption that the feature selection criterion used satisfies the monotonicity property (2). This assumption precludes the use of the error rate as the criterion. This is an important drawback as the error rate can be considered superior to other criteria [16], [7]. Moreover, the search algorithm becomes computationally prohibitive for problems of high dimensionality.

In practice, therefore, one has to rely on computationally feasible procedures which perform the search quickly but may yield sub-optimal results. A comprehensive list of sub-optimal procedures can be found, e.g., in books [1], [4], [24], [22]. A comparative taxonomy can be found, e.g., in [3], [6] or [8]. Our own research and experience with FS has led us to the conclusion that *there exists no unique generally applicable approach* to the problem. Some are more suitable under certain conditions, others are more appropriate under other conditions, depending on our *knowledge of the problem*. Hence continuing effort is invested in developing new methods to cover the majority of situations which can be encountered in practice.

3.2 FS Categorisation With Respect to Problem Knowledge

From another point of view there are perhaps two basic classes of situations with respect to *a priori* knowledge of the underlying probability structures:

- *Some a priori knowledge is available* – It is at least known that probability density functions (pdfs) are unimodal. In these cases, a probabilistic distance measure (like Mahalanobis, Bhattachaarya, etc.) may be appropriate as the evaluation criterion. For this type of situations we recommend either the recent prediction-based B&B algorithms for optimal search (see Section 4), or sub-optimal Floating and Oscillating methods (Section 5).
- *No a priori knowledge is available* – We cannot even assume that pdfs are unimodal. The only source of available information is the training data. For these situations we have developed two conceptually different alternative methods. They are based on approximating unknown conditional pdfs by finite mixtures of a special type and are discussed in Section 6.

4 Recent Optimal Search Methods

The problem of optimal feature selection (or more generally of subset selection) is difficult especially because of its time complexity. All known optimal search algorithms have an exponential nature. The only alternative to exhaustive search is the *Branch & Bound* (B&B) algorithm [11], [4] and ancestor algorithms based on a similar principle. All B&B algorithms rely on the monotonicity property of the FS criterion (2). By a straightforward application of this property many feature subset evaluations may be omitted.

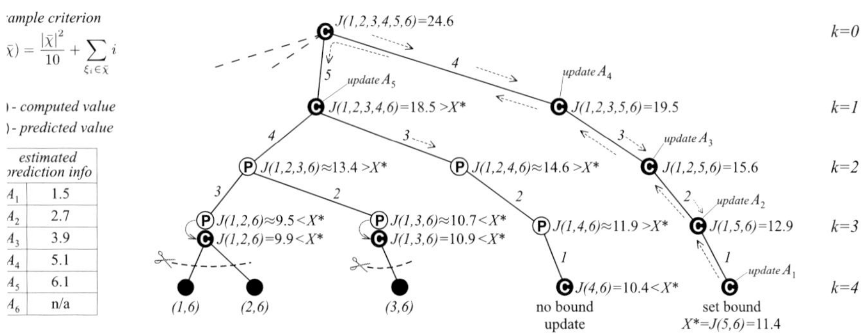

Fig. 1. Example of a "Fast Branch & Bound" problem solution, where $d = 2$ features are to be selected from a set of $D = 5$ features. Dashed arrows illustrate the way of tracking the search tree.

Before discussing more advanced algorithms, let us briefly summarize the essential B&B principle. The algorithm constructs a search tree where the root represents the set of all D features, X_D, and leaves represent target subsets of d features. While tracking the tree down to leaves the algorithm successively removes single features from the current set of "candidates" ($\bar{\mathcal{X}}_k$ in k-th level). The algorithm keeps the information about both the till-now best subset of cardinality d and the corresponding criterion value, denoted as the *bound*. Anytime the criterion value in some internal node is found to be lower than the current *bound*, due to condition (2) the whole sub-tree may be cut-off and many computations may be omitted. The course of the B&B algorithm can be seen in Fig. 1 (the meaning of `C`, `P` and A_i symbols is not important for now). This scheme in its simplest form is known as the "Basic B&B" algorithm. For details see [1], [4].

4.1 Branch & Bound Properties

When compared to the exhaustive search, every B&B algorithm requires additional computations. Not only the target subsets of d features $\bar{X}_{D-d}$, but also their supersets $\bar{X}_{D-d-j}$, $j = 1, \cdots, D-d$ have to be evaluated. The B&B principle does not guarantee enough sub-tree cut-offs to keep the total number of criterion computations lower than in exhaustive search.

To reduce the amount of criterion computations an additional node-ordering heuristic has been introduced in a more powerful "Improved B&B" (IBB) algorithm [1], [4]. IBB optimizes the order of features to be assigned to tree edges so that the *bound* value can increase as fast as possible and thus enables more effective branch cutting in later stages. Although IBB usually outperforms all simpler B&B algorithms, the computational cost of the additional heuristic can become a strong deteriorating factor. For detailed discussion of B&B drawbacks see [20]. In the following we present a recent effective framework for substantial B&B acceleration.

4.2 Prediction Mechanism Based Branch & Bound

The Fast Branch & Bound (FBB) [20] algorithm aims to reduce the number of criterion function computations in internal search tree nodes. A simplified algorithm description is as follows: FBB attempts to utilize the knowledge of past feature-dependent *criterion value decreases* (difference between criterion values before and after feature removal) for future prediction of criterion values without the need of real computation. Prediction is allowed under certain conditions only, e.g., not in leaves. Both the really computed and predicted criterion values are treated as equal while imitating the full IBB functionality, i.e., in ordering node descendants in the tree construction phase.

If the predicted criterion value remains significantly higher than the current *bound*, we may expect that even the actual value would not be lower and the corresponding sub-tree could not be cut-off. In this situation the algorithm continues to construct the consecutive tree level. However, if the predicted value comes close to the *bound* (and therefore there arises a chance that the real value is lower than the *bound*), the real criterion value must be computed. Only if real criterion values are lower than the current *bound*, sub-trees may be cut-off. Note that this prediction scheme does not affect the optimality of obtained results. The FBB algorithm course remains similar to

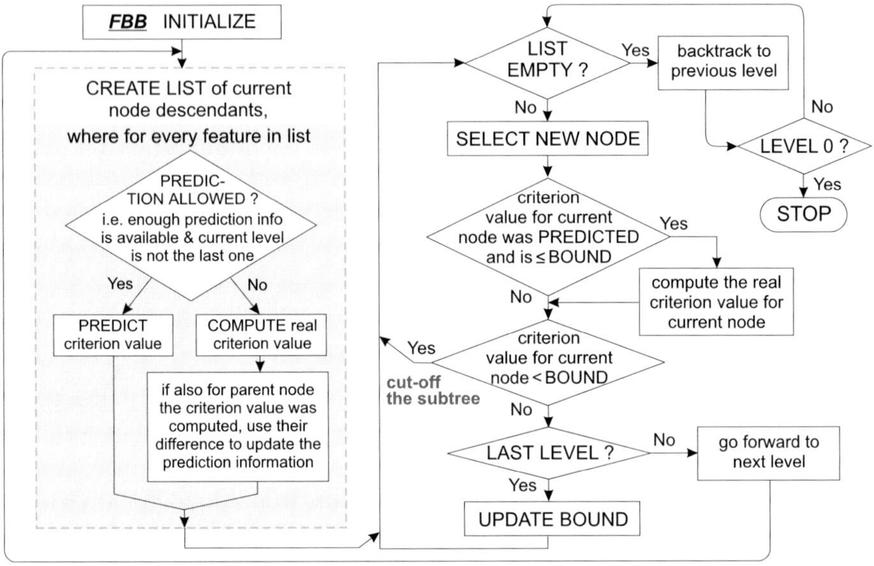

Fig. 2. A simplified diagram of the FBB algorithm

that of the IBB, possible sub-tree cut-offs are allowed according to real criterion values only. Possible inaccurate predictions may result in nothing worse than constructing sub-trees, which would have been pruned out by means of classical B&B algorithms. However, this situation is usually strongly outweighed by criterion computation savings in other internal nodes, especially near the root, where criterion computation tends to be slower.

The prediction mechanism processes the information about the averaged *criterion value decrease* separately for each feature. The idea is illustrated in Fig. 1. For a detailed and formal description of this rather complex procedure and other B&B related topics see [20].

4.3 Improving the "Improved" Algorithm

The FBB operates mostly the fastest among all B&B algorithms. However, it exhibits some drawbacks: it cannot be used with recursive criterion forms and there is no theoretical guarantee that extensive prediction failures won't hinder the overall speed, despite the fact that such faulty behaviour has not been observed with real data. The B& B with Partial Prediction (BBPP) [20] constitutes a less effective but robust alternative. While learning similarly to FBB, it does not use predictions to substitute true criterion values inside the tree. Predicted values are used only for ordering before features get assigned to tree edges. In this sense BBPP can be looked upon as a slightly modified IBB with the only difference in node ordering heuristics. The performance gain follows from the fact, that the original IBB ordering heuristics always evaluates more criterion values than is the number of features finally used. For detailed analysis of BBPP see [20]. Among other recent B&B related ideas the "trading space for speed" approach [5] deserves attention as an alternative that may operate exceptionally fast under certain circumstances. The BBPP and FBB algorithms are further investigated in [21], [23].

4.4 Predictive B&B Properties and Experimental Results

When compared to classical B&B algorithms the predictive algorithms always spend additional time for maintaining the prediction mechanism. However, this additional time showed not to be a factor, especially when compared to time savings arising from the pruned criterion computations. The algorithms have been thoroughly tested on a number of different data sets. Here we show representative results on 30-dimensional mammogram data (2 classes – 357 benign and 212 malignant samples, see [10]). We used both the recursive (where applicable) and non-recursive Bhattacharyya distance as the criterion function. Performance of different methods is illustrated in Fig. 3.

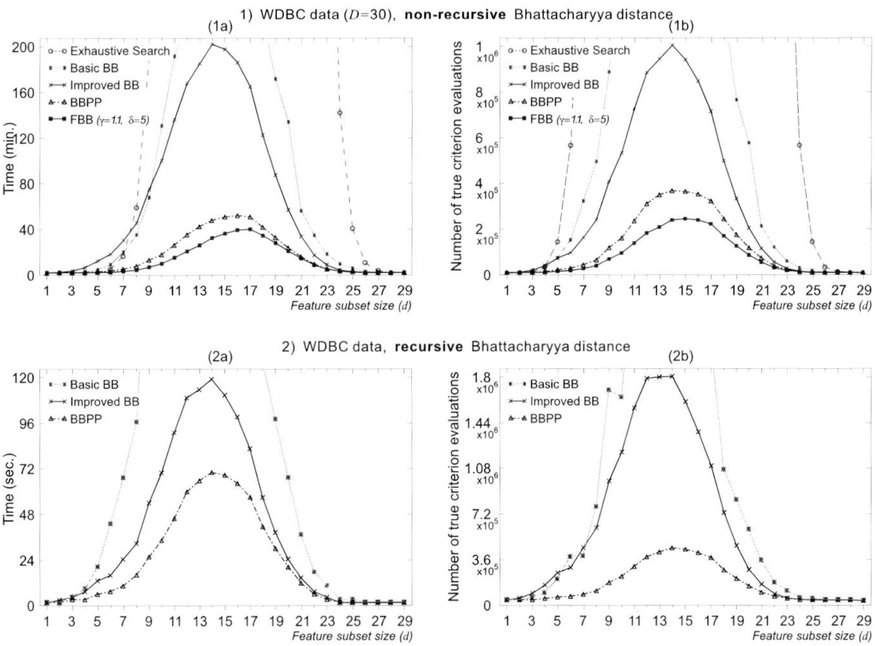

Fig. 3. Optimal subset search methods performance when maximizing the Bhattacharyya distance on 30-dimensional data (Wisconsin Diagnostic Breast Center).

We compare all results especially against the IBB algorithm [1], [4], as this algorithm has been long accepted to be the most effective optimal subset search method. Remark: Where applicable, we implement all algorithms to support "minimum solution tree" [25].

4.5 Summary of Recent Optimal Methods

The only optimal subset search method usable with non-monotonic criteria is the exhaustive (full) search. However, because of exponential nature of the search problem, alternative methods are often needed. Several recent improvements of the B&B idea especially in the form of prediction based FBB and BBPP resulted in a speed-up factor of 10 to 100 over the simplest B&B form, depending on particular data and criterion used.

It should be stressed that despite the shown advances all optimal methods remain exponential in nature. If there is no particular need to request absolute optimum of results, sub-optimal search methods offer greater flexibility and acceptable speed even for high-dimensional problems, while the solutions found are not necessarily much worse than optimal.

5 Recent Sub-optimal Search Methods

Despite the advances in optimal search, for larger than moderate-sized problems we have to resort still to sub-optimal methods. The basic feature selection approach is to build up a subset of required number of features incrementally starting with the empty set (*bottom-up* approach) or to start with the complete set of features and remove redundant features until d features retain (*top-down* approach). The simplest yet widely used *sequential forward (or backward) selection* methods [1], SFS (SBS), iteratively add (remove) one feature at a time so as to maximize the intermediate criterion value until the required dimensionality is achieved. Among the more interesting recent approaches the following two families of methods can be pointed out for general applicability and performance reasons:

1. *sequential floating search methods [13], [17]*
2. *oscillating search methods [18]*

Earlier sequential methods suffered from the so-called nesting of feature subsets which significantly deteriorated the performance. The first attempt to overcome this problem was to employ either the Plus-l-Take away-r [denoted (l,r)] or generalized (l,r) algorithms [1] which involve successive augmentation and depletion process. The same idea in a principally extended and refined form constitutes the basis of floating search.

5.1 Sequential Floating Search

The sequential forward floating selection procedure consists of applying after each forward step a number of backward steps as long as the resulting subsets

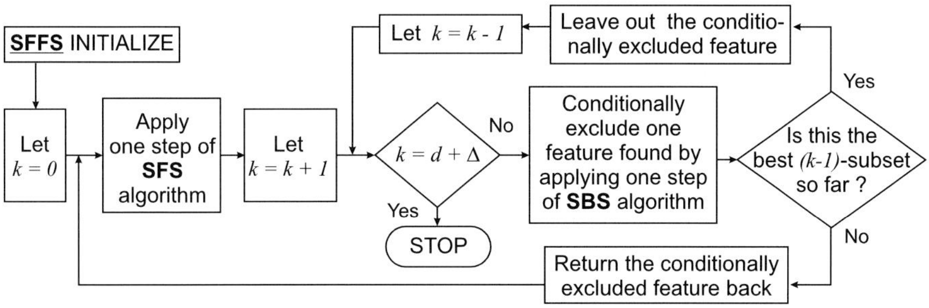

Fig. 4. Sequential Forward Floating Selection Algorithm

are better than previously evaluated ones at that level. Consequently, there are no backward steps at all if intermediate result at actual level (of corresponding dimensionality) cannot be improved. The same applies for the backward version of the procedure. Both algorithms allow a 'self-controlled backtracking' so they can eventually find good solutions by adjusting the trade-off between forward and backward steps dynamically. It is possible to say that, in a certain way, they compute only what they need without any parameter setting.

Formal description of this now classical procedure can be found in [13]. Neverheless, the idea behind is simple enough and can be illustrated sufficiently in Fig. 4. Condition $k = d + \delta$ terminates the algorithm after the target subset of d features has been found and possibly refined by means of backtracking (from dimensionalities $d + 1, \ldots, d + \delta$).

Floating search algorithms can be considered universal tools not only outperforming all predecessors, but also keeping advantages not met by more sophisticated algoritms. They find good solutions in all problem dimensions in one run. The overall search speed is high enough for most of solved problems.

5.2 Adaptive Floating Search

As the floating search algorithms have been found successful and generally accepted to be an efficient universal tool, their idea was further investigated. The so-called Adaptive Floating Search (AFS) has been proposed in [17]. The AFS algorithms are able to outperform the classical floating algorithms in certain cases, but at a cost of considerabe increase of search time and the necessity to deal with unclear parameters. Our experience shows that AFS is usually inferior to newer algorithms, which we focus on in the following.

5.3 Oscillating Search

The recent Oscillating Search (OS) can be considered a "higher level" procedure, that takes use of other feature selection methods as sub-procedures in its own search. The concept is highly flexible and enables modifications for different purposes. It has shown to be very powerful and capable of over-performing standard sequential procedures, including Floating Search. Unlike other methods, the OS is based on repeated modification of the current subset X_d of d features. In this sense OS is independent on predominant search direction. This is achieved by alternating so-called *down-* and *up-swings*. Both *swings* attempt to improve the current set X_d by replacing some of the features by better ones. The *down-swing* first removes, then adds back, while the *up-swing* first adds, then removes. Two successive opposite swings form an *oscillation cycle*. The OS can thus be looked upon as a controlled sequence of oscillation cycles. The value of o (denoted *oscillation cycle depth*) determines the number of features to be replaced in one swing. o is increased after unsuccessful oscillation cycles and reset to 1 after each X_d improvement. The algorithm terminates when o exceeds a user-specified *limit* Δ. The course of oscillating search is illustrated in Fig. 5.

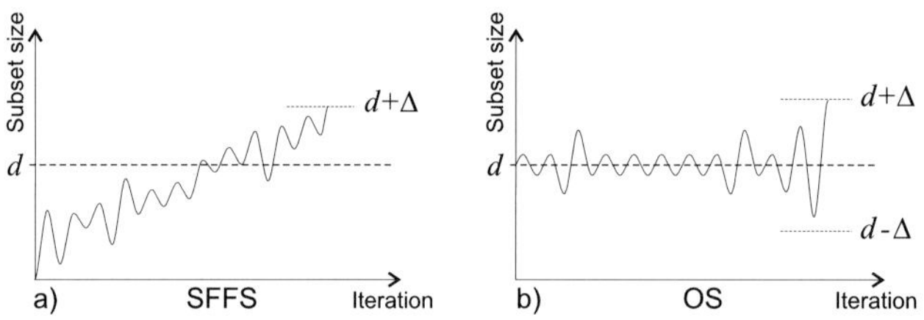

Fig. 5. Graphs demonstrate the course of search algorithms: a) Sequential Floating Forward Selection, b) Oscillating Search.

Every OS algorithm requires some initial set of d features. The initial set may be obtained randomly or in any other way, e.g., using some of the traditional sequential selection procedures. Furthermore, almost any feature selection procedure can be used in *up-* and *down-swings* to accomplish the replacements of feature o-tuples. Therefore, for the sake of generality in the following descriptions let us denote the adding / removing of a feature o-tuple by ADD(o) / REMOVE(o). For OS flow-chart see Fig. 6.

Oscillating Search – Formal Algorithm Description

Step 1: (*Initialization*) By means of any feature selection procedure (or randomly) determine the initial set X_d of d features. Let $c = 0$. Let $o = 1$.

Step 2: (*Down-swing*) By means of REMOVE(o) remove such o-tuple from X_d to get new set X_{d-o} so that $J(X_{d-o})$ is maximal. By means of ADD(o) add such o-tuple from $X_D \setminus X_{d-o}$ to X_{d-o} to get new set $X_d^{'}$ so that $J(X_d^{'})$ is maximal. If $J(X_d^{'}) > J(X_d)$, let $X_d = X_d^{'}$, $c = 0$, $o = 1$ and go to Step 4.

Step 3: (*Last swing has not improved the solution*) Let $c = c + 1$. If $c = 2$, then nor the last *up-* nor *down-swing* led to a better solution. Extend the

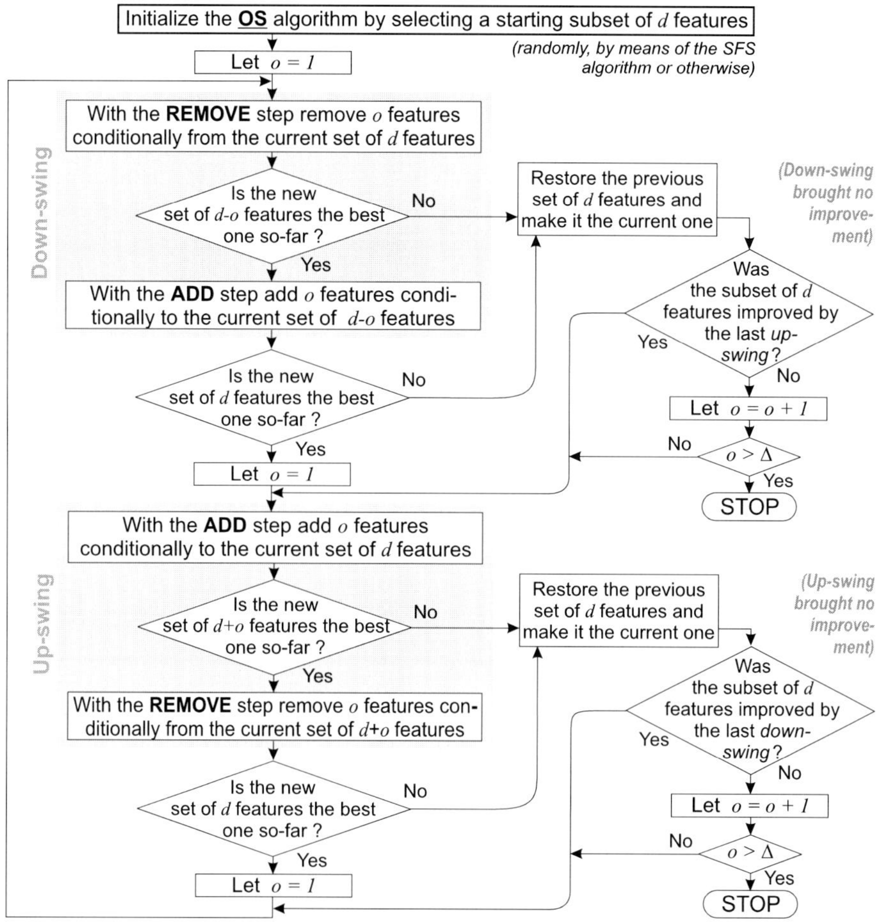

Fig. 6. Simplified Oscillating Search algorithm flowchart.

search by letting $o = o + 1$. If $o > \Delta$, stop the algorithm, otherwise let $c = 0$.

Step 4: (*Up-swing*) By means of ADD(o) add such o-tuple from $X_D \setminus X_d$ to X_d to get new set X_{d+o} so that $J(X_{d+o})$ is maximal. By means of REMOVE(o) remove such o-tuple from X_{d+o} to get new set $X_d^{'}$ so that $J(X_d^{'})$ is maximal. If $J(X_d^{'}) > J(X_d)$, let $X_d = X_d^{'}$, $c = 0$, $o = 1$ and go to Step 2.

Step 5: (*Last swing has not improved the solution*) Let $c = c + 1$. If $c = 2$, then nor the last *up-* nor *down-swing* led to a better solution. Extend the search by letting $o = o + 1$. If $o > \Delta$, stop the algorithm, otherwise let $c = 0$ and go to Step 2.

5.4 Oscillating Search Properties

The generality of OS search concept allows to adjust the search for better speed or better accuracy (lower Δ and simpler ADD / REMOVE vs. higher Δ and more complex ADD / REMOVE). In this sense let us denote *sequential OS* the simplest possible OS version which uses a sequence of SFS steps in place of ADD() and a sequence od SBS steps in place of REMOVE(). As opposed to all sequential search procedures, OS does not waste time evaluating subsets of cardinalities too different from the target one. The fastest improvement of the target subset may be expected in initial phases of the algorithm, because of the low initial cycle depth. Later, when the current feature subset evolves closer to optimum, low-depth cycles fail to improve and therefore the algorithm broadens the search ($o = o+1$). Though this improves the chance to get closer to optimum, the trade-off between finding a better solution and computational time becomes more apparent. Consequently, OS tends to improve the solution most considerably during the fastest initial search stages. This behavior is advantageous, because it gives the option of stopping the search after a while without serious result-degrading consequences. Let us summarize the key OS advantages:

- It may be looked upon as a universal tuning mechanism, being able to improve solutions obtained in any other way.
- The randomly initialized OS is very fast, in case of very high-dimensional problems may become the only applicable procedure. E.g., in document analysis for search of the best 1000 words out of a vocabulary of 50000 even the simple SFS may show to be too slow.
- Because the OS processes subsets of target cardinality from the very beginning, it may find solutions even in cases, where the sequential procedures fail due to numerical problems.
- Because the solution improves gradually after each oscillation cycle, with the most notable improvements at the beginning, it is possible to terminate the algorithm prematurely after a specified amount of time to obtain a usable solution. The OS is thus suitable for use in real-time systems.
- In some cases the sequential search methods tend to uniformly get caught in certain local extremes. Running the OS from several different random initial points gives better chances to avoid that local extreme.

5.5 Experimental Results of Sub-optimal Search Methods

All described sub-optimal sequential search methods have been tested on a large number of different problems. Here we demonstrate their performance on 2-class, 30-dimensional mammogram data (see [10]). The graphs in Fig. 7 show the OS ability to outperform other methods even in the simplest *sequential OS* form (here with $\Delta = d$ in one randomly initialized run). The ASFFS behavior is well illustrated here showing better performance than SFFS at a

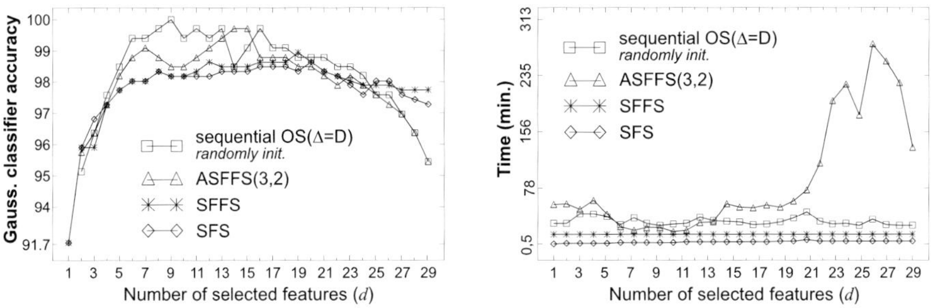

Fig. 7. Comparison of sub-optimal methods on classification problem.

cost of uncontrollably increased time. SFFS and SFS need one run only to get all solutions. SFFS performance is always better than that of SFS.

5.6 Summary of Recent Sub-optimal Methods

Concerning our current experience, we can give the following recommendations. Floating Search can be considered the first tool to try. It is reasonably fast and yields generally very good results in all dimensions at once, often succeeding in finding real optimum. The Oscillating Search becomes better choice whenever: 1) the highest quality of solution must be achieved but optimal methods are not applicable, or 2) a reasonable solution is to be found as quickly as possible, or 3) numerical problems hinder the use of sequential methods, or 4) extreme problem dimensionality prevents any use of sequential methods, or 5) the search is to be performed in real-time systems. Especially when repeated with different random initial sets the Oscillating Search shows outstanding potential to overcome local extremes in favor of global maximum.

It should be stressed that, as opposed to B&B, the Floating Search and Oscillating Search methods are tolerant to deviations from monotonic behaviour of feature selection criteria. It makes them particularly useful in conjunction with non-monotonic FS criteria like the error rate of a classifier (cf. Wrappers [7]), which according to a number of researchers seem to be the only legitimate criterion for feature subset evaluation.

6 Mixture Based Methods

For the cases when no simplifying assumptions can be made about the underlying class distributions we developed a new approach based on approximating the unknown class conditional distributions by finite mixtures of parametrized densities of a special type. In terms of the required computer storage this pdf estimation is considerably more efficient than nonparametric pdf estimation methods.

Denote the ωth class training set by $\mathbf{X}_\omega$ and let the cardinality of set $\mathbf{X}_\omega$ be N_ω. The modeling approach to feature selection taken here is to approximate the class densities by dividing each class $\omega \in \Omega$ into M_ω artificial subclasses. The model assumes that each subclass m has a multivariate distribution $p_m(\mathbf{x}|\omega)$ with its own parameters. Let α_m^ω be the mixing probability for the mth subclass, $\sum_{m=1}^{M_\omega} \alpha_m^\omega = 1$.The following model for ωth class pdf of $\mathbf{x}$ is adopted [14], [12]:

$$p(\mathbf{x}|\omega) = \sum_{m=1}^{M_\omega} \alpha_m^\omega p_m(\mathbf{x}|\omega) = \sum_{m=1}^{M_\omega} \alpha_m^\omega g_0(\mathbf{x}|\mathbf{b}_0) g(\mathbf{x}|\mathbf{b}_m^\omega, \mathbf{b}_0, \Phi) \tag{3}$$

Each component density $p_m(\mathbf{x}|\omega)$ includes a nonzero "background" pdf g_0, common to all classes:

$$g_0(\mathbf{x}|\mathbf{b}_0) = \prod_{i=1}^{D} f_i(x_i|b_{0i}), \quad \mathbf{b}_0 = (b_{01}, b_{02}, \cdots, b_{0D}), \tag{4}$$

and a function g specific for each class of the form:

$$g(\mathbf{x}|\mathbf{b}_m^\omega, \mathbf{b}_0, \Phi) = \prod_{i=1}^{D} \left[\frac{f_i(x_i|b_{mi}^\omega)}{f_i(x_i|b_{0i})}\right]^{\phi_i}, \quad \phi_i = \{0, 1\} \tag{5}$$

$$\mathbf{b}_m^\omega = (b_{m1}^\omega, b_{m2}^\omega, \cdots b_{mD}^\omega), \quad \Phi = (\phi_1, \phi_2, \cdots, \phi_D) \in \{0, 1\}^D.$$

The univariate function f_i is assumed to be from a family of normal densities. The model is based on the idea to identify a common "background" density for all the classes and to express each class density as a mixture of the product of this "background" density with a class-specific modulating function defined on a subspace of the feature vector space. This subspace is chosen by means of the nonzero binary parameters ϕ_i and the same subspace of $\mathcal{X}$ for each component density is used in all the classes. Any specific univariate function $f_i(x_i|b_{mi}^\omega)$ is substituted by the "background" density $f_i(x_i|b_{0i})$ whenever ϕ_i is zero. In this way the binary parameters ϕ_i can be looked upon as *control variables* as the complexity and the structure of the mixture (3) can be controlled by means of these parameters. For any choice of ϕ_i the finite mixture (3) can be rewritten by using (4) and (5) as

$$p(\mathbf{x}|\alpha_\omega, \mathbf{b}_\omega, \mathbf{b}_0, \Phi) = \sum_{m=1}^{M_\omega} \alpha_m^\omega \prod_{i=1}^{D} [f_i(x_i|b_{0i})^{1-\phi_i} f_i(x_i|b_{mi}^\omega)^{\phi_i}] \tag{6}$$

$$\alpha_\omega = (\alpha_1^\omega, \alpha_2^\omega, \cdots, \alpha_{M_\omega}^\omega), \quad \mathbf{b}_\omega = (\mathbf{b}_1^\omega, \mathbf{b}_2^\omega, \cdots, \mathbf{b}_{M_\omega}^\omega).$$

The EM ("Expectation-Maximization") algorithm can be extended to allow a mixture of the form (6) to be fitted to the data. It should be emphasized that although the model looks rather *unfriendly*, its form leads to a tremendous simplification [14] when we use normal densities for functions f. The use of this model (6) makes the process of feature selection a simple task.

So as to select those features that are most useful in describing differences between two classes, the Kullback's J-divergence defined in terms of the a posteriori probabilities has been adopted as a criterion of discriminatory content. The goal of the method is to maximize the divergence discrimination, hence the name "divergence" method (see [12]).

7 Summary

The current state of art in feature selection based dimensionality reduction for decision problems of classification type has been overviewed. A number of recent feature subset search strategies has been reviewed and compared. Recent developments of B&B based algorithms for optimal search led to considerable improvements of the speed of search. Nevertheless, the principal exponential nature of optimal search remains and will remain one of key factors motivating the development of sub-optimal strategies. Among the family of sequential search algorithms the Floating and Oscillating search methods deserve particular attention. Two alternative feature selection methods based on mixture modeling have been presented. They are suitable for cases, when no *a priori* information on underlying probability structures is known.

Acknowledgements

The work has been supported by grants of the Czech Ministry of Education CEZ:J18/98:311600001, ME567 and the Grant Agency of the Czech Republic No 402/03/1310.

References

1. Devijver PA, Kittler J (1982) Pattern Recognition: A Statistical Approach, Prentice-Hall
2. Duda RO, Hart PE, Stork DG (2000) Pattern Classification, 2nd Ed., Wiley-Interscience
3. Ferri FJ, Pudil P, Hatef M, Kittler J (1994) Comparative Study of Techniques for Large-Scale Feature Selection, Gelsema ES, Kanal LN (eds.) Pattern Recognition in Practice IV, Elsevier Science B.V., 403–413
4. Fukunaga K (1990) Introduction to Statistical Pattern Recognition, Academic Press
5. Chen X (2003) An Improved Branch and Bound Algorithm for Feature Selection, Pattern Recognition Letters 24(12):1925–1933
6. Jain AK, Zongker D (1997) Feature Selection: Evaluation, Application and Small Sample Performance, IEEE Transactions on Pattern Analysis and Machine Intelligence 19(2):153–158

7. Kohavi R, John GH (1997) Wrappers for Feature Subset Selection. Artificial Intelligence 97(1-2):273–324
8. Kudo M, Sklansky J (2000) Comparison of Algorithms that Select Features for Pattern Classifiers, Pattern Recognition 33(1):25–41
9. McLachlan GJ (1992) Discriminant Analysis and Statistical Pattern Recognition, John Wiley & Sons, New York
10. Murphy PM, Aha DW (1994) UCI Repository of Machine Learning Databases [ftp.ics.uci.edu]. University of California, Department of Information and Computer Science, Irvine, CA
11. Narendra PM, Fukunaga K (1977) A Branch and Bound Algorithm for Feature Subset Selection. IEEE Transactions on Computers 26:917–922
12. Novovičová J, Pudil P, Kittler J (1996) Divergence Based Feature Selection for Multimodal Class Densities, IEEE Transactions on Pattern Analysis and Machine Intelligence 18(2):218–223
13. Pudil P, Novovičová J, Kittler J (1994) Floating Search Methods in Feature Selection, Pattern Recognition Letters 15(11):1119–1125
14. Pudil P, Novovičová J, Kittler J (1994) Simultaneous Learning of Decision Rules and Important Attributes for Classification Problems in Image Analysis, Image and Vision Computing 12:193–198
15. Ripley B (1996) Pattern Recognition and Neural Networks, Cambridge University Press, Cambridge, Massachusetts
16. Siedlecki W, Sklansky J (1988) On Automatic Feature Selection, International Journal of Pattern Recognition and Artificial Intelligence 2(2):197–220
17. Somol P, Pudil P, Novovičová J, Paclík P (1999) Adaptive Floating Search Methods in Feature Selection, Pattern Recognition Letters 20(11,12,13):1157–1163
18. Somol P, Pudil P (2000) Oscillating Search Algorithms For Feature Selection, In: Proc 15th IAPR International Conference on Pattern Recognition, Barcelona, Spain, 406–409
19. Somol P, Pudil P (2002) Feature Selection Toolbox. Pattern Recognition 35(12):2749–2759
20. Somol P, Pudil P, Kittler J (2004) Fast Branch & Bound Algorithms for Optimal Feature Selection. IEEE Transactions on Pattern Analysis and Machine Intelligence 26(7):900–912
21. Somol P, Pudil P, Grim J (2004) On Prediction Mechanisms in Fast Branch & Bound Algorithms. Lecture Notes in Computer Science 3138, Springer, Berlin, 716–724
22. Theodoridis S, Koutroumbas K (2003) Pattern Recognition, 2nd Ed., Academic Press
23. Wang Z, Yang J, Li G (2003) An Improved Branch & Bound Algorithm in Feature Selection, Lecture Notes in Computer Science LNCS 2639, Springer, 549–556.
24. Webb A (2002) Statistical Pattern Recognition, 2nd Ed., John Wiley & Sons
25. Yu B, Yuan B (1993) A More Efficient Branch and Bound Algorithm for Feature Selection. Pattern Recognition 26:883–889

Part II

FEATURES, LEARNING AND CLASSIFIERS

Margin-based Diversity Measures for Ensemble Classifiers

Tomasz Arodź

Institute of Computer Science, AGH University of Science and Technology,
al. Mickiewicza 30, 30-059 Kraków, Poland
arodz@agh.edu.pl

Summary. The classifier ensembles have been used successfully in many applications. Their superiority over single classifiers depends on the diversity of the classifiers forming the ensemble. Till now, most of the ensemble diversity measures were derived basing on the binary classification information. In this paper we propose a new group of methods, which use the margins of individual classifiers from the ensemble. These methods process the margins with a bipolar sigmoid function, as the most important information is contained in margins of low magnitude. The proposed diversity measures are evaluated for three types of ensembles of linear classifiers. The tests show that these measures are better at predicting recognition accuracy than established diversity measures, such as Q or disagreement measures, or entropy.

1 Introduction

Classification methods are in focus of research in the field of pattern recognition. Among many supervised machine-learning methods, neural networks, support vector machines and boosting have been shown to achieve particularly good recognition accuracy in a wide range of tasks. The boosting classifier, apart from high accuracy, exhibits other properties that increase its popularity. Contrary to the artificial neural networks it is a non-parametric method. On the other hand, the SVMs rely on the dot product. For non-numerical data, e.g. strings, dedicated kernel functions have to be used. The boosting scheme can be used to enhance the performance of any existing classifier, independently of the nature of the features used by it.

The boosting classifiers [1] stem from the concept of training the ensemble of weak, or not very accurate, classifiers. The final classification decision is based on weighted majority voting of the trained ensemble. The training is done in rounds, and each of successive weak classifiers is given a different distribution of weights over the training set. In each round, the distribution is altered to focus the next weak classifier on examples that were misclassified by already trained weak classifiers. In another ensemble method, bagging [2], in each round, the weak classifier trains on the randomly selected subset of the training set. Another approach, known as the random subspace method [3] or the random feature subsets method [4], involves training in the randomly selected subset of feature set, using all the available examples.

Despite their conceptual simplicity, the exact factors leading to good learning capabilities for ensemble classifiers are still debated. For boosting, it has been argued that the ensemble is maximizing the margins of the combined classifier on the training set [5], i.e., the distances from the ensemble decision boundary to the examples. This accentuates the connection to other successful machine-learning method, the support vector machines [6]. The diversity of the ensemble is another factor linked to high classification accuracy [7], [8]. The trained group of classifiers has to yield different decision boundaries in order to increase the recognition accuracy above that of a single classifier.

Various methods for estimating the diversity of ensembles have been used. These techniques can be partitioned into three groups [9], depending on the information on the classification results for the examples they use. This can be support (e.g. probabilities) for class predictions, predicted classes and, finally, the oracle output, i.e., correctness of the class predictions. In ensemble classification framework, the latter two approaches have been studied more thoroughly. However, extensive evaluation of diversity measures has concluded [9] that the ensemble accuracy is not always linked to typical diversity measures, such as the Q statistics or the entropy.

In this paper, a new group of diversity measures is proposed. These methods are based on measuring the differences between the margins of the individual classifiers for the examples from the training set. They combine the knowledge of the correctness of the simple classifier's decision from the oracle-based group with the estimation of decision support from the probability-based group.

The rest of the paper is arranged in the following way. Section 2 introduces the proposed margin-based diversity measures. In Sect. 3, the framework for the tests of the methods is outlined. The tests involve three ensemble methods based on linear weak classifier and are conducted for 6 benchmark datasets. Next, in Sect. 4, the results of these experiments are presented and analysed. Finally, the conclusions of the paper are discussed in Sect. 5.

2 The Proposed Margin-based Diversity Measures

The margin is a concept defining the relation between the example, represented as a point in the feature space, and the decision boundary of the classifier in that space. Margins have been proven useful in analysing the effectiveness of boosting [5]. A margin of the ensemble h_{fin} on the example $\mathbf{x_i}$ belonging to class $c_i \in \{-1,+1\}$ is defined as $\rho_{\mathbf{x_i}}(h_{\mathrm{fin}}) = c_i h_{\mathrm{fin}}(\mathbf{x_i})$. The combined hypothesis $h_{\mathrm{fin}}(\mathbf{x_i}) \in [-1,1]$ is defined as a weighted (with weights $\{\alpha_j\}$) majority vote of the ensemble of T weak classifiers h_j, each yielding decision from $\{-1,+1\}$:

$$h_{\text{fin}}(\mathbf{x}) = \frac{\sum_{j=1}^{T} \alpha_j h_j(\mathbf{x})}{\sum_{j=1}^{T} \alpha_j}. \quad (1)$$

The sign of $h(\mathbf{x})$ represents the predicted class, while its magnitude the level of support for the decision. The margin of the combined hypothesis is also a number in range $[-1,1]$. The sign of the margin shows if the example has been classified correctly and the magnitude indicates the confidence in the decision. Good performance of typical boosting algorithm such as AdaBoost [1] is often attributed [5] to the maximisation of the margins and therefore reducing the number of errors and increasing the confidence in the classification.

We propose to use the concept of margin for studying the diversity of the boosting ensemble of classifiers. Diversity is usually studied using binary decisions of each weak classifier [9]. However, such an analysis discards the information on the degree of misclassification or the confidence in the correct decision. This information may influence the generalization of the classifier.

The proposed diversity measures require that the margin $\rho_{\mathbf{x_i}}(h_j)$ is defined for each of the weak classifiers h_j, representing the distance from the example $\mathbf{x_i}$ to the decision boundary of h_j. To measure this diversity, for each example $\mathbf{x_i}$, the difference in margins of each pair of trained weak classifiers could be used. However, contrary to the margin of the whole ensemble, the margin of the weak classifier is no longer bound to $[-1,1]$. This can downgrade the usefulness of the simple, margin difference-based measure if the training of the classifier results in large margins.

In measuring the diversity, the most important information is contained in margins of low magnitude. These correspond to examples lying close to the decision boundary, with low support for the decision. Differences between small-magnitude margins exhibited by two weak classifiers may influence the decision of the ensemble. This may happen in particular for examples not seen during training. Differences between two margins of the same sign and of relatively large magnitudes are less informative, even if they are large.

To take account of only the most informative fraction of the inter-margin differences, a function $a : \mathbb{R} \to [-1,1]$ is defined as:

$$a(x) = \frac{2}{1+\mathrm{e}^{-x}} - 1. \quad (2)$$

The function $a()$ is a simple bipolar sigmoid function, leading to suppressing the inter-margin distance if both margins are of the same sign and of large magnitude. Using the sigmoid function $a()$, the following pairwise margin-based method for measuring the diversity can be formulated:

$$PD_m = \frac{2}{T(T-1)} \sum_{i=1}^{m} \sum_{j=1}^{T} \sum_{k=j+1}^{T} \left| a\left(\rho_{\mathbf{x_i}}(h_j)\right) - a\left(\rho_{\mathbf{x_i}}(h_k)\right) \right|. \quad (3)$$

Evaluation of distances between all pairs of individual classifiers in the ensemble can be computationally expensive. To overcome this problem, a non-pairwise diversity measure could be used. One such diversity estimate, based on the margin sigmoid, is the variance, formulated as:

$$Var_m = \frac{1}{T-1} \sum_{i=1}^{m} \sum_{j=1}^{T} \left(a\left(\rho_{\mathbf{x_i}}\left(h_j\right)\right) - \overline{a_i} \right)^2 , \tag{4}$$

where $\overline{a_i} = \frac{1}{T} \sum_{j=1}^{T} a\left(\rho_{\mathbf{x_i}}\left(h_j\right)\right)$ is the mean value of $a\left(\right)$ for all h_j for a given example $\mathbf{x_i}$.

The Var_m non-pairwise measure is related to a pairwise measure similar to that of (3). For brevity, let $a_{ij} = a\left(\rho_{\mathbf{x_i}}\left(h_j\right)\right)$. Then,

$$\begin{aligned} Var_m &= \frac{1}{T-1} \sum_{i=1}^{m} \sum_{j=1}^{T} \left(a_{ij} - \overline{a_i}\right)^2 \\ &= \frac{1}{2T\left(T-1\right)} \sum_{i=1}^{m} \sum_{j=1}^{T} \sum_{k=1}^{T} \left(a_{ij} - \overline{a_i}\right)^2 + \left(a_{ik} - \overline{a_i}\right)^2 \\ &+ \frac{1}{2T\left(T-1\right)} \sum_{i=1}^{m} \sum_{j=1}^{T} \sum_{k=1}^{T} -2\left(a_{ij} - \overline{a_i}\right)\left(a_{ik} - \overline{a_i}\right) \\ &= \frac{1}{2} \frac{2}{T\left(T-1\right)} \sum_{i=1}^{m} \sum_{j=1}^{T} \sum_{k=1}^{T} \left(a\left(\rho_{\mathbf{x_i}}\left(h_j\right)\right) - a\left(\rho_{\mathbf{x_i}}\left(h_k\right)\right)\right)^2 . \end{aligned} \tag{5}$$

Thus, the variance of the sigmoid of the margin is equivalent to a pairwise diversity measure. However, instead of absolute values as in PD_m (3), squared values of differences $a_{ij} - a_{ik}$ are used in Var_m.

3 Experimental Setup

The behaviour of the proposed diversity measures has been studied for a group of ensemble methods based on linear weak classifiers. Thus, the estimation of the margin of individual weak classifier is straightforward. The ensemble types used are the AdaBoost-FLD, i.e., boosting with Fisher Linear Discriminant as a weak classifier, RSM-FLD, i.e., random subspace method with FLD as a simple classifier, and the AdaBoost-RFS-FLD, i.e., the fusion of the two above methods. In AdaBoost-RandomFeatureSubset-FLD, the FLD in each round of boosting is trained on a different, random feature subset. These classifier ensembles were discussed in detail in our study [10].

The tests for the AdaBoost-RSF-FLD and RSM-FLD were carried out for several feature subset sizes, as specified in Table 1 and for the number of rounds $T = 200$. The value of T is large enough for all ensembles to no longer increase the training accuracy significantly. As both the AB-RFS-FLD and RSM-FLD are based on random process of choosing the features, and the resampling scheme is used in FLD boosting, the results were averaged over 10 executions of each method.

3.1 Datasets Used in the Experiments

The above-mentioned classifiers and their diversities have been evaluated on a diverse set of real-world problems, originating from the UCI Machine Learning Repository[1], the ELENA repository[2] and the CMU Neural Networks Benchmark Collection[3]. The information on the number of training and testing samples and the number of features in the datasets is presented in Table 1. The datasets are two-class classification problems with exception of the UCI page-blocks set. To cast this five-class problem into two classes, a configuration involving recognition of class 0 from class 1 was used.

Table 1. Datasets used for the tests. Sizes of test and training sets are presented, with the total number of features and relative sizes of feature subsets used in training.

Dataset	Test Set	Training Set	Features	Feat. Subset Sizes [%]
ELENA-phoneme	1000	4404	5	40, 60, 80
UCI-page-blocks	1000	4473	10	20, 30, 40, 50, 75, 90
UCI-pima-diabetes	100	668	8	25, 37.5, 50, 75, 90
UCI-spambase	1000	3601	57	5, 10, 20, 30, 40, 50, 75, 90
UCI-ionosphere	100	251	34	10, 15, 20, 25, 30, 40, 50, 75, 90
NB-sonar	104	104	60	5, 10, 20, 35, 50, 75, 90

3.2 Comparison with other diversity measures

For each of the datasets and for each of the three ensemble classifiers, the two proposed diversity measures, PD_m and Var_m, were evaluated. For comparison, for the same datasets and ensembles, three reference diversity measures [9] were computed. These include two pairwise measures, the Q statistics and the disagreement measure Dis. The averaged Q statistics is defined for two-class problem as:

$$Q_{av} = \frac{2}{T(T-1)} \sum_{j=1}^{T} \sum_{k=j+1}^{T} \frac{N^{11}N^{00} - N^{01}N^{10}}{N^{11}N^{00} + N^{01}N^{10}}, \tag{6}$$

[1] http://www.ics.UCI.edu/~mlearn/MLRepository.html
[2] ftp://ftp.dice.ucl.ac.be/pub/neural-nets/ELENA/databases/
[3] ftp.cs.cmu.edu:/afs/cs.cmu.edu/project/connect/bench/

where the notation N^{01} indicates the number of examples that are incorrectly classified by weak classifier j and correctly by weak classifier k. The total disagreement measure Dis for a two-class classification takes a form similar to PD_m (3), and can be defined as:

$$Dis_{tot} = \frac{1}{T(T-1)} \sum_{i=1}^{m} \sum_{j=1}^{T} \sum_{k=j+1}^{T} \left| \operatorname{sgn}(\rho_{\mathbf{x_i}}(h_k)) - \operatorname{sgn}(\rho_{\mathbf{x_i}}(h_k)) \right| . \quad (7)$$

Moreover, as an example of non-pairwise diversity measure, the total entropy E was used. It is defined as:

$$E_{tot} = \frac{1}{T} \sum_{i=1}^{m} \sum_{j=1}^{T} \frac{1}{(T - \lceil T/2 \rceil)} \min \{T_i^c, T - T_i^c\} , \quad (8)$$

where T_i^c is the number of the weak classifiers from the whole ensemble that correctly recognise example $\mathbf{x_i}$.

4 Results of the experiments

The results of the tests for the chosen datasets are summarized in Table 2. For each dataset and for each of the three studied ensemble types, the average classification accuracy on the testing set is shown along with the two proposed diversity measures, pairwise PD_m and non-pairwise Var_m. Three reference diversity measures, pairwise Q_{av} and Dis_{tot} and non-pairwise E_{tot} are also presented. For AdaBoost-RFS-LDA and RSM-LDA, the results are shown for the feature subset sizes yielding the highest accuracy on the training set.

One can conclude from the results in Table 2, that the proposed measures behave in a different manner than the reference methods when comparing the two variants of boosting. The proposed methods are better at explaining the difference in accuracies between AdaBoost-FLD and AdaBoost-RFS-FLD in terms of differences in their diversities. For Var_m measure, in 5 out of 6 cases the increase in diversity is followed by increase in accuracy. For PD_m measure, in 4 out of 6 cases the one of these two classifiers which exhibited higher diversity achieved also higher accuracy. For two reference pairwise measures, Q_{av} and Dis_{tot}, and the non-pairwise E_{tot} measure, this relation holds for only 3 out of 6 cases, with additional one case inconclusive.

For comparison between RSM and boosting, the Q, Dis and E measures constantly show significantly lower diversity of the RSM-FLD over both AB-FLD and AB-RFS-FLD. This is true even for datasets where RSM-FLD exhibits higher accuracy than one of the classifiers based on boosting. The proposed methods also do not show correlation between the difference in RSM and AdaBoost accuracies and differences in their diversities. Thus, neither the reference nor the proposed measures allow for explaining the difference in accuracies between RSM and AdaBoost in terms of their diversities.

Table 2. Results of the tests. Values of diversity measures, proposed PD_m and Var_m and reference Q_{av}, Dis_{tot} and E_{tot}, as well as corresponding classification accuracies for three ensemble types: AdaBoost-RFS-FLD, AdaBoost-FLD and RSM-FLD. The ↓ signifies the diversity increases with the decrease of the diversity measure

Dataset	Classifier	Acc.[%]	$PD_m \uparrow$	$Var_m \uparrow$	$Q_{av} \downarrow$	$Dis_{tot} \uparrow$	$E_{tot} \uparrow$
ELENA-phoneme	AB-RFS-FLD	85.0	761	107.8	-0.007	2192.4	0.92
ELENA-phoneme	AB-FLD	82.5	855	131.1	-0.006	2198.5	0.93
ELENA-phoneme	RSM-FLD	76.1	253	15.1	0.863	551.1	0.17
UCI-page-block	AB-RFS-FLD	97.3	173	8.5	0.002	2207.2	0.88
UCI-page-block	AB-FLD	96.6	132	3.6	0.000	2207.2	0.88
UCI-page-block	RSM-FLD	94.0	167	5.5	0.958	201.1	0.06
UCI-diabetes	AB-RFS-FLD	73.1	73	6.6	-0.006	333.6	0.93
UCI-diabetes	AB-FLD	71.0	72	6.2	-0.006	333.9	0.94
UCI-diabetes	RSM-FLD	73.0	15	0.4	0.948	58.4	0.11
UCI-spambase	AB-RFS-FLD	94.0	75	3.0	0.007	1768.2	0.88
UCI-spambase	AB-FLD	92.3	76	2.9	0.026	1743.6	0.87
UCI-spambase	RSM-FLD	89.9	25	0.2	0.988	133.1	0.05
UCI-ionosphere	AB-RFS-FLD	94.0	77	20.1	-0.009	122.1	0.83
UCI-ionosphere	AB-FLD	88.0	31	4.3	0.078	105.2	0.64
UCI-ionosphere	RSM-FLD	87.4	16	0.8	0.969	13.5	0.08
NB-sonar	AB-RFS-FLD	82.2	14	1.5	-0.011	50.7	0.83
NB-sonar	AB-FLD	74.0	1	0.0	-0.585	9.6	0.10
NB-sonar	RSM-FLD	77.8	1	0.0	0.948	6.0	0.08

The results also show, that in 5 out of 6 cases the increase of PD_m is followed by the increase in Var_m. In the case which is exception to this trend, the Var_m measure predicted the accuracy better than the PD_m. Thus, the computationally more efficient non-pairwise measure can be successfully used instead of the more demanding pairwise measure.

5 Conclusions

In this paper, a new group of diversity measures has been proposed. The new measures combine the information from the existing support-based and oracle-based methods by using the differences in margins of the individual weak classifiers forming the ensemble. To extract the most informative fraction of the margin diversity, the sigmoid function of the margin is used before calculating either the sum of pairwise Manhattan distances or the variance. Both of these proposed measures, pairwise PD_m and non-pairwise Var_m, are better than existing methods at correlating the accuracies and diversities for two relatively similar ensemble classifiers: the AdaBoost-FLD and AdaBoost-RandomSubspace-FLD. However, like the exiting methods, they do not always show the correlation between differences in accuracy and diversity for less similar ensembles, i.e., the RSM-FLD and boosted FLD.

Acknowledgements

I would like to thank my PhD supervisor Prof. Dr. Witold Dzwinel for scientific guidance and inspiring discussions. The work is supported from the Polish State Committee for Scientific Research (KBN) Grant No. 3T11C05926.

References

1. Freund Y, Schapire R (1997) A decision-theoretic generalization of on-line learning and an application to boosting. Journal of Computer and System Sciences 55:119–139
2. Breiman L (1996) Bagging predictors. Machine Learning 24:123–140
3. Ho TK (1995) Random decision forests. In: Proc. of the 3rd Int'l Conference on Document Analysis and Recognition:278–282
4. Bryll R, Gutierrez-Osuna R, Quek F (2003) Attribute bagging: improving accuracy of classifier ensembles by using random feature subsets. Pattern Recognition 36:1291–1302
5. Schapire RE, Freund Y, Bartlett P, Lee WS (1997) Boosting the margin: a new explanation for the effectiveness of voting methods. In: Proc. 14th International Conference on Machine Learning:322–330, Morgan Kaufmann
6. Cortes C, Vapnik V (1995) Support-vector networks. Machine Learning 20:273–297
7. Brown G, Wyatt J, Harris R, Yao X (2005) Diversity creation methods: A survey and categorisation. Information Fusion Journal 6:5–20
8. Kuncheva L (2003) That elusive diversity in classifier ensembles. In: Proc. First Iberian Conference on Pattern Recognition and Image Analysis:1126–1138
9. Kuncheva LI, Whitaker CJ (2003) Measures of diversity in classifier ensembles and their relationship with the ensemble accuracy. Mach. Learn. 51:181–207
10. Arodz T (2005) Boosting the Fisher Linear Discriminant with random feature subsets. To appear in: IV International Conference on Computer Recognition Systems, CORES 2005, Advances in Soft Computing, Springer

Boosting the Fisher Linear Discriminant with Random Feature Subsets

Tomasz Arodź

Institute of Computer Science, AGH University of Science and Technology,
al. Mickiewicza 30, 30-059 Kraków, Poland
arodz@agh.edu.pl

Summary. Boosting increases the recognition accuracy of many types of classifiers. However, studies show that for the Fisher Linear Discriminant (FLD), a simple and widely used classifier, boosting does not lead to a significant increase in accuracy. In this paper, a new method for adapting the FLD into the boosting framework is proposed. This method, the AdaBoost-RandomFeatureSubset-FLD (AB-RFS-FLD), uses a different, randomly chosen subset of features for learning in each boosting round. The new method achieves significantly better accuracy than both single FLD and FLD with boosting, with improvements reaching 6% in some cases. We show that the good performance can be attributed to higher diversity of the individual FLDs, as well as to the better generalization abilities.

1 Introduction

Ensemble classification methods are one of the widely studied branches of pattern recognition. These machine-learning methods have proven to be successful in many real-world problems, from face recognition [1] to drug screening [2] and cancer diagnosis [3]. Apart from high learning capabilities, the popularity of the ensemble methods stems also from their conceptual simplicity. They involve training a number of classifiers instead of a single one. The final classification decision is based on majority voting of the simple classifiers forming the ensemble. However, to elevate the recognition accuracy above that of the single classifier, the classifiers forming the ensemble should yield different decision boundaries. To ensure this diversity, specific procedures for training each classifier in the ensemble are used.

There are several types of training, leading to different ensemble classifiers. In boosting [4], training is done in rounds, and each weak classifier is trying to minimize the error on the training set weighted with a distribution over the examples. The distribution is altered in each round to focus the next weak classifier on examples that were misclassified by already trained weak classifiers. Apart from boosting, other methods follow the ensemble of classifiers idea. In the bagging method [5], the weak classifier trains in each round on the randomly selected subset of the training set. Training on all the examples, but using randomly selected subsets from the whole feature set is also used, in random subspace [6] or the random feature subsets [7] methods.

A wide range of classifiers has been used in the boosting scheme. The first boosting algorithm, the AdaBoost [8], has originally been proposed for decision trees. A very simple classifier with decisions based on a single feature has been successfully boosted [1]. Neural networks are also used with boosting [9]. The nearest neighbour classification rule has also been adapted to the AdaBoost [8]. Recently, the Support Vector Classifier has been used as a simple classifier in boosting [10].

The widely-used FLD method has also been tested in boosting [11]. However, the results were discouraging. On the other hand, a modified versions of the FLD, the Null Space FLD and Principal Space FLD, have been collectively used with success in combined bagging and random subspace methods [12]. In this paper, a method to overcome the problems with boosting the FLD classifier is proposed. The method uses the classical FLD classifier within the AdaBoost scheme. However, in each turn, an independent, random subset of features is selected, and the FLD is trained using only these features.

The rest of the paper is arranged in the following way. Section 2 introduces the proposed AdaBoost-RandomFeatureSubset-FLD method. In Sect. 3, the method is tested on a number of benchmark datasets, and the results are compared with other FLD-based methods. Next, in Sect. 4, the results are analysed, and the reasons for the good training and generalization abilities are discussed. Finally, Sect. 5 summarises the conclusions of the paper.

2 Proposed Method for Boosting the FLD

When used within the AdaBoost scheme, the ensemble of FLD weak classifiers often gives results similar or worse than a single FLD classifier [13]. This is because classifiers trained in consecutive rounds of boosting are very similar [11]. Therefore, in order to successfully apply the FLD as a weak classifier, a method for forcing the weak classifiers to be more diverse has to be used. In this paper, diversity is achieved by introducing the concept of random feature selection to boosting. Specifically, in each turn, the weak classifier is trained using a different, randomly chosen subset of features.

The Algorithm 1 outlined below realizes the concept of training the group of FLD weak classifiers for the AdaBoost using a random subset of features. First, in line 1, the weights of the examples are initialised uniformly. Then, in each round, a random subset of features is selected (line 3). In the training of the FLD (line 6), the training set, with only the selected features, is used. To take account of the weights D_t of the training examples, the resampling is used (line 5). As soon as the new linear decision is obtained by FLD, the weighted classification error is evaluated (line 7) and the weights of the examples and of the trained weak FLD are calculated (line 10-13).

Algorithm 1 *The AdaBoost-RandomFeatureSubset-FLD algorithm*

INPUT:
Training set Tr of size m, $Tr = \{(\mathbf{x_1}, c_1), \ldots, (\mathbf{x_m}, c_m)\} \subset X \times \{-1, +1\}$
An integer number T - maximal number of training rounds
A number S - percentage of features to be used in each subspace
$\text{AdaBoost} - \text{RFS} - \text{FLD}\,(Tr, T, S)$
1 $\forall_{1 \le i \le m} D_1(i) = \frac{1}{m}$
2 **for** $t = 1$ **to** T
3 $FeatSset$ = **select randomly** S **distinct features from** X
4 $TrFtSset$ = **cast** Tr **into selected features** $FeatSset$
5 $TrFtSsetResmpld$ = **resample** $TrFtSset$ **according to weights** D_t
6 $h_t = \text{FLD}\,(TrFtSsetResmpld)$
7 $\varepsilon_t = \sum_{i: h_t(\mathbf{x_i}) \neq c_i} D_t(i)$
8 **if** $\varepsilon_t = 0$ **or** $\varepsilon_t \ge 0.5$
9 **exit loop**
10 $\beta_t = \frac{\varepsilon_t}{1 - \varepsilon_t}$
11 $\alpha_t = \log \frac{1}{\beta_t}$
12 $\forall_{i: h_t(\mathbf{x_i}) = c_i} D_{t+1}(i) = \beta_t D_t(i)$
13 $\forall_{i: h_t(\mathbf{x_i}) \neq c_i} D_{t+1}(i) = D_t(i)$
14 **return** $h_{\text{fin}}(\mathbf{x}) = \frac{\sum_{i=1}^{T} \alpha_i h_i(\mathbf{x})}{\sum_{i=1}^{T} \alpha_i}$

The effectiveness of boosting has been analysed with a number of methods, including the margin-based approach [14]. A margin of the combined classifier $h_{\text{fin}}(\mathbf{x_i})$ on example $\mathbf{x_i}$ is defined as $\rho_{\mathbf{x_i}}(h_{\text{fin}}) = c_i h_{\text{fin}}(\mathbf{x_i})$ and is a number in $[-1, 1]$. The sign shows whether the example has been classified correctly, while the magnitude of the margin represents the confidence in the decision. One of the reasons of good performance of AdaBoost is the maximisation of the margins. This leads to increase in the confidence of the classification and reduction of the number of errors.

Another factor which is argued [15] to influence the accuracy of the classifier ensembles is their diversity. Conventionally, the diversity is studied in terms of binary decisions of each weak classifier, e.g. using the Q statistics [16]. Thus, the information on the confidence in the correct decision or the degree of misclassification is lost in the analysis. We are using the concept of margin for studying the diversity of the boosting ensemble of classifiers. The margin-based pairwise distance PD_m is used, defined as the sum of Manhattan distances between bipolar sigmoid functions of margins of individual weak classifiers on all examples. Values of PD_m give insight into the generalization of the proposed classifier. To study the diversity the variance V_h of the normalized decision boundaries can also be used. Each linear decision boundary in the features space with f features is represented as a vector of

$f+1$ dimensions, including the decision threshold. In calculating V_h, each vector is normalized to unit length and variances of weights associated with each feature are aggregated.

3 Experimental Results

The behaviour of the new method has been studied for a set of real-world problems. Datasets with small to large number of features and examples, and with various features-to-examples ratios have been obtained from the UCI Machine Learning Repository[1], the ELENA repository[2] and the CMU Neural Networks Benchmark Collection[3]. The particular datasets used, along with information on the number of training and testing samples chosen and the number of features in the datasets, are presented in Table 1. All the datasets are two-class problems, except for the case of the UCI page-blocks set. For this five-class problem recognition of class 0 from class 1 is done.

Table 1. Datasets used in tests. Sizes of test and training sets are presented, with total number of features and relative sizes of feature subsets used in training.

Dataset	Test Set	Training Set	Features	Feat. Subset Sizes [%]
ELENA-phoneme	1000	4404	5	40, 60, 80
UCI-page-blocks	1000	4473	10	20, 30, 40, 50, 75, 90
UCI-pima-diabetes	100	668	8	25, 37.5, 50, 75, 90
UCI-spambase	1000	3601	57	5, 10, 20, 30, 40, 50, 75, 90
UCI-ionosphere	100	251	34	10, 15, 20, 25, 30, 40, 50, 75, 90
NB-sonar	104	104	60	5, 10, 20, 35, 50, 75, 90

For each of the datasets, the proposed AdaBoost-RandomFeatureSubset-FLD method has been tested. For comparison, the same tests have been carried out for the AdaBoost-FLD, RandomSubspaceMethod-FLD and single FLD classifiers. The tests for the AdaBoost-RSF-FLD and RSM-FLD were carried out for several values of the percentage of features used in the subset or subspace, as specified in Table 1. Each test has been repeated 10 times, and the results were averaged. The number of rounds T was chosen to be 200.

The results of the tests for the chosen datasets are summarized in Table 2. For each dataset, the average classification accuracy on the testing set is shown. For AdaBoost-RFS-FLD and RSM-FLD, the results are given for the feature subset sizes yielding the highest accuracy on the training set. The

[1] http://www.ics.UCI.edu/~mlearn/MLRepository.html
[2] ftp://ftp.dice.ucl.ac.be/pub/neural-nets/ELENA/databases/
[3] ftp.cs.cmu.edu:/afs/cs.cmu.edu/project/connect/bench/

optimal size of the feature subset for the proposed AdaBoost-RFS-FLD is also presented.

Table 2. Results of the tests. Classification accuracies for the proposed AdaBoost-RandomFeatureSubset-FLD compared to AdaBoost-FLD, RandomSubspaceMethod-FLD and classical FLD. Optimal size of the feature subsets for AB-RFS-FLD shown as percentage of total number of features

Dataset	AB-RFS-FLD[%]	AB-FLD[%]	RSM-FLD[%]	FLD[%]	Feats.[%]
ELENA-phoneme	85.0	82.5	76.1	75.9	60
UCI-page-blocks	97.3	96.6	94.0	94.9	50
UCI-pima-diabetes	73.1	71.0	73.0	73.0	75
UCI-spambase	94.0	92.3	89.9	90.0	20
UCI-ionosphere	94.0	88.0	87.4	87.0	15
NB-sonar	82.2	74.0	77.8	78.9	5

The results show that the usefulness of boosting for a simple FLD is limited, which is consistent with previous studies [11]. For some datasets, e.g. ELENA-phoneme and UCI-spambase, boosting results in an increase of accuracy, while for others it is not very useful or even counterproductive, as in e.g. UCI-pima-diabetes. Typically, it leads to a relatively small increase in accuracy. The behaviour of the RSM with FLD also varies with dataset. It ranges from no improvement to improvements higher than that of boosted FLD.

The proposed AdaBoost-RandomFeatureSubset-FLD method consistently outperforms the AdaBoost-FLD, the RSM-LDA and the classical FLD. In some cases, the increase in accuracy is significant, reaching 6% for UCI-ionosphere dataset.

4 Analysis of the AdaBoost-RandomFeatureSubset-FLD performance

The accuracy of the classifier on the test set depends on two features: the ability to minimize the training error and to generalize to previously unseen data. In order to inspect which of these two factors is responsible for the good performance of the AB-RFS-FLD in comparison to AB-FLD, the accuracy on the training set was evaluated. As the diversity is an important factor influencing the error of the ensemble, the Q statistics and the diversity of decision boundaries V_h were also calculated. The results in Table 3 show lower training error for the proposed method. The values of Q and, especially, of V_h show that increase in accuracy can be attributed to higher ensemble diversity. Moreover, the better training ability of the new method is reflected in achieving larger minimal margins of the ensemble on the training set, especially in cases where the margin is negative.

Table 3. Comparison of classification accuracies on the training set for the proposed AdaBoost-RFS-FLD and for AdaBoost-FLD. The Q and V_h diversity measures for AB-RFS-FLD in comparison to AB-FLD. Minimal margins $\min \rho$ of these ensembles

Dataset	AB-RFS-FLD				AB-FLD			
	Acc.[%]	Q	V_h	$\min \rho$	Acc.[%]	Q	V_h	$\min \rho$
ELENA-phoneme	88.9	-0.0074	0.419	-0.14	87.2	-0.0064	0.264	-0.15
UCI-page-blocks	98.6	0.0021	0.360	-0.06	97.7	0.0002	0.175	-0.07
UCI-pima-diabetes	93.3	-0.0059	0.364	-0.05	91.8	-0.0057	0.189	-0.07
UCI-spambase	96.0	0.0066	0.153	-0.09	95.2	0.0262	0.059	-0.08
UCI-ionosphere	100.0	-0.0089	0.361	0.09	100.0	0.0784	0.031	0.18
NB-sonar	100.0	-0.0111	0.363	0.12	100.0	-0.5845	0.005	0.84

To compare the generalization capabilities of the new AdaBoost-RFS-FLD with AdaBoost-FLD, the results on the test sets have been analysed in another way. For AdaBoost-RFS-FLD, the minimal round T_{exc} with training accuracy exceeding the training accuracy of the whole AdaBoost-FLD ensemble was found. Then, the ensemble classifier for AdaBoost-RFS-FLD was constructed by using only $T_{exc} - 1$ weak classifiers, from rounds 1 to $T_{exc} - 1$. Thus, the training accuracy of the constructed AdaBoost-RFS-FLD ensemble was approximately equal to the training accuracy of the full, 200 weak classifier AdaBoost-FLD ensemble, but not exceeded it. The accuracies of the AdaBoost-RFS-FLD ensembles of $T_{exc} - 1$ FLDs were compared to the accuracies of the full, 200 FLDs AdaBoost-FLD ensembles in Table 4.

As the training accuracies of both ensemble types were made similar, the differences in the testing accuracies capture the information on the relative generalization abilities of the two boosting methods. If, despite having similar training accuracies, one of the classifiers achieves higher test accuracy, we can assume it can generalize better. The tests show that the proposed method achieves higher testing accuracy for all datasets, indicating better ability to classify previously unseen data. This is, for most datasets, consistent with the margin-based diversity measure PD_m on the training set, which captures information related to the generalization abilities of the ensemble.

The results gathered in Tables 3 and 4 show that the AdaBoost-RFS-FLD method outperforms the AdaBoost-FLD in both training accuracy and generalization abilities. The higher accuracy on the training set may be attributed to the larger diversity of the individual weak classifiers, introduced by the ever-changing random subsets of features used in each round.

The better generalization accuracy can be justified theoretically. The bound on the margin-based generalization error [4] is dependent on the VC-dimension [17] ϑ of the weak classifier. For a linear classifier in the feature space defined by f features, this dimension is $\vartheta_{FLD} = f + 1$. In the AdaBoost-RandomFeatureSubset-FLD, the number of features used in FLD training is significantly reduced comparing to the AdaBoost-FLD. Thus, the

Table 4. Study of generalization ability. Classification accuracies for the tests of the proposed AdaBoost-RandomFeatureSubset-FLD with number of rounds $T_{exc}-1$ chosen to obtain training error approximately equal to that of 200 rounds AdaBoost-FLD. Pairwise margin-based diversity measure PD_m for whole AB-RFS-FLD and AB-FLD ensembles

Dataset	AB-RFS-FLD			AB-FLD		
	Acc. [%] for $T_{exc}-1$ FLDs		PD_m	Acc. [%] for 200 FLDs		PD_m
	Testing set	Training set		Testing set	Training set	
ELENA-phoneme	83.7	87.2	761	82.5	87.2	855
UCI-page-blocks	96.9	97.6	173	96.6	97.7	132
UCI-pima-diabetes	73.0	91.7	73	71.0	91.8	72
UCI-spambase	93.7	95.2	75	92.3	95.2	76
UCI-ionosphere	94.0	100.0	77	88.0	100.0	31
NB-sonar	82.2	100.0	14	74.0	100.0	1

VC-dimension of each weak classifier is reduced and the bound on the generalization error of the whole ensemble is lower.

5 Conclusions

In this paper, a method for overcoming problems with using the FLD as a weak classifier in the boosting scheme has been proposed. The method utilizes the concept of random feature selection. By training each FLD using a different feature subset, the diversity of the classifier is increased. The proposed method has been tested for a set of real-world data. In comparison with single FLD and FLD with boosting or random subspace ensembles, the AdaBoost-RandomFeatureSubset-FLD method shows significantly better accuracy. This stems from the better ability to minimize the training error and from higher generalization abilities.

Acknowledgements

I would like to thank my PhD supervisor Prof. Dr. Witold Dzwinel for scientific guidance and inspiring discussions. The work is supported from the Polish State Committee for Scientific Research (KBN) Grant No. 3T11C05926.

References

1. Viola P, Jones MJ (2004) Robust real-time face detection. Int. J. Comput. Vision 57:137–154

2. Svetnik V, Liaw A, Tong C, Wang T (2004) Application of breiman's random forest to modeling structure-activity relationships of pharmaceutical molecules. Lecture Notes in Computer Science 3077:334–343
3. Qu Y, Adam BL, Yasui Y, Ward MD, Cazares LH, Schellhammer PF, Feng Z, Semmes OJ, Wright GL (2002) Boosted decision tree analysis of surface-enhanced laser desorption/ionization mass spectral serum profiles discriminates prostate cancer from noncancer patients. Clinical Chemistry 48:1835–1843
4. Freund Y, Schapire R (1997) A decision-theoretic generalization of on-line learning and an application to boosting. Journal of Computer and System Sciences 55:119–139
5. Breiman L (1996) Bagging predictors. Machine Learning 24:123–140
6. Ho TK (1995) Random decision forests. In: Proc. of the 3rd Int'l Conference on Document Analysis and Recognition:278–282
7. Bryll R, Gutierrez-Osuna R, Quek F (2003) Attribute bagging: improving accuracy of classifier ensembles by using random feature subsets. Pattern Recognition 36:1291–1302
8. Freund Y, Schapire R (1996) Experiments with a new boosting algorithm. In: Proc. 13th International Conference on Machine Learning:148–156, Morgan Kaufmann
9. Schwenk H, Bengio Y (2000) Boosting neural networks. Neural Computation 12:1869–1887
10. Kim HC, Pang S, Je HM, Kim D, Bang SY (2003) Constructing support vector machine ensemble. Pattern Recognition 36:2757–2767
11. Skurichina M, Duin RPW (2000) Boosting in linear discriminant analysis. Lecture Notes in Computer Science 1857:190–199
12. Wang X, Tang X (2004) Multiple LDA classifier combination for high dimensional data classification. Lecture Notes in Computer Science 3077:344–353
13. Skurichina M, Duin RPW (2002) Bagging, boosting and the random subspace method for linear classifiers. Pattern Analysis and Applications 5:121–135
14. Schapire RE, Freund Y, Bartlett P, Lee WS (1997) Boosting the margin: a new explanation for the effectiveness of voting methods. In: Proc. 14th International Conference on Machine Learning:322–330, Morgan Kaufmann
15. Kuncheva L (2003) That elusive diversity in classifier ensembles. In: Proc. First Iberian Conference on Pattern Recognition and Image Analysis:1126–1138
16. Kuncheva LI, Whitaker CJ (2003) Measures of diversity in classifier ensembles and their relationship with the ensemble accuracy. Mach. Learn. 51:181–207
17. Vapnik V (1982) Estimation of Dependences Based on Empirical Data. Springer, NewYork

A Look-Ahead Branch and Bound Pruning Scheme for Trie-Based Approximate String Matching

Ghada Badr[1] and John B. Oommen[2]

[1] Carleton University
School of Computer Science, Carleton University, Ottawa, Canada : K1S 5B6.
gbadr@scs.carleton.ca

[2] Carleton University, *Fellow of the IEEE*,
School of Computer Science, Carleton University, Ottawa, Canada : K1S 5B6.
oommen@scs.carleton.ca

Summary. This paper deals with the problem of estimating a transmitted string X^* by processing the corresponding string Y, which is a noisy version of X^*. We assume that Y contains substitution, insertion and deletion errors, and that X^* is an element of a finite (but possibly, large) dictionary, H. The best estimate X^+ of X^*, is defined as that element of H which minimizes the Generalized Levenshtein Distance $D(X, Y)$ between X and Y such that the total number of errors is not more than K, for all $X \in H$. In this paper we present a new Branch and Bound pruning strategy that can be applied to dictionary-based approximate string matching when the dictionary is stored as a trie. The new strategy attempts to look ahead at each node, c, before moving further, by merely evaluating a certain local criterion at c. As opposed to the reported trie-based methods [10], [17], the pruning is done *a priori* before even embarking on the edit distance computations and thus it combines the advantages of partitioning the dictionary according to the string lengths, and the advantages gleaned by representing H using the trie data structure. The results demonstrate a marked improvement (even up to 33%) with respect to the number of operations needed on three benchmark dictionaries.

1 Introduction

We consider the traditional problem involved in the syntactic Pattern Recognition (PR) of strings, namely that of recognizing garbled words (sequences), and present a novel recognition strategy which involves tries, Branch and Bound pruning, and dictionary-based (as opposed to string-based) dynamic programming.

Let Y be a misspelled (noisy) string, of length M, obtained from an unknown word X^*, of length N, which is an element of a finite (but possibly, large) dictionary H, where Y is assumed to contain Substitution, Insertion and Deletion (SID) errors. Various algorithms have been proposed to obtain an appropriate estimate X^+ of X^*, by processing the information contained in Y, and the literature contains hundreds (if not thousands) of associated papers. We include a *brief* review here.

The trie is a data structure that offers search costs that are independent of the document size. Tries also combine prefixes together, and so by using tries in approximate string matching [10], [17], we can utilize the information obtained in the process of evaluating any one $D(X_i, Y)$, to compute any other $D(X_j, Y)$, where X_i and X_j share a common prefix. As opposed to this, in the field of AI Branch and Bound (BB) techniques [9] are well known, and have been used to prune paths for game trees etc. They are used when we want to prune paths that have costs above a certain threshold.

In this paper, we attempt to use the same data structure, the trie, for storing the strings in the dictionary so as to take advantage of the compact calculations for the distance matrix, by utilizing the common paths for the common prefixes. We then introduce a new BB pruning strategy that makes use of the fact that the length of the strings to be compared are known *a priori*. We thus propose to apply this new pruning strategy to the trie-based approximate search algorithm, which we call the Look-Ahead Branch and Bound (LHBB) scheme. By using these four features (the trie, BB, look-ahead, and dictionary-based dynamic programming), we can demonstrate a marked improvement, because this pruning can be done before we even start the edit distance calculations. LHBB helps us to search in portions of the dictionary where the word lengths are acceptable, without actually having to partition the dictionary, and at the same time make use of the effective properties of tries. The experimental results presented later shows improvements of up to 33% with small and large benchmark dictionaries. This high improvement is at the expense of just storing two extra memory locations for each node in the trie. Also, if the length of the noisy word is very far from all the acceptable words in the dictionary, i.e., those which can give an edit error smaller than K, the edit distance computations for this noisy word can be totally pruned with only a single comparative test. All of these concepts will be presented presently.

Wagner and Fisher [20] and others [16] proposed an efficient algorithm for computing the Levenshtein distance by utilizing the concepts of dynamic programming. This algorithm is optimal for the infinite alphabet case and it has $O(MN)$ worst case. Various amazingly similar versions of the algorithm are available in the literature, a review of which can be found in [4], [16], [18]. Ukkonen [19] designed solutions for cases involving other inter-*substring* edit operations which runs in $O(KN)$ worst case. String correction using GLD-related criteria has been done for noisy strings [4], [18], substrings [16], [18], and subsequences [13], and also for strings in which the dictionaries are treated as grammars [16], [18], [21]. The most recent survey on approximate string matching can be found in [11].

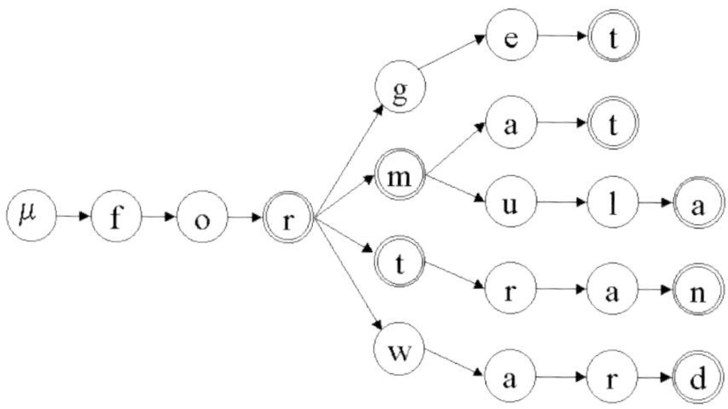

Fig. 1. An example of a dictionary stored as a trie with the words {for, form, fort, forget, format, formula, fortran, forward}.

All early algorithms proposed for estimating X^+, requires the separate evaluation of the edit distance between Y and every element of $X \in H$, and would thus unnecessarily repeat the same comparisons and minimizations for a substring and *all its prefixes*. Thus, most previous algorithms usually, have many redundant computations, which has been modified in [10] and [14].

1.1 Tries and Cutoffs

Tries offer text searches with costs which are independent of the size of the document being searched. The data is represented not in the nodes but in the path from the root to the leaf. Thus all strings sharing a prefix will be represented by paths branching from a common initial path. Figure 1 shows an example of a trie for a simple dictionary of words {for, form, fort, forget, format, formula, fortran, forward}. Shang *et al.* [17] used the trie data structure for exact and approximate string searching. They presented a trie-based method whose cost is independent of the document size. They proposed a k-approximate match algorithm on a text represented as a trie, which performs a Depth First Search (DFS) on the trie. The insight they provided was that the trie representation of the text drastically reduces the Dynamic Programming (DP) computations. The trie representation (see Figure 1) compresses the common prefixes into overlapping paths, and the corresponding column (in the DP matrix) needs to be evaluated only once.

In [17], the authors also applied a known pruning strategy called Ukkonen's cutoff [19] to abort unsuccessful searches. Chang and Lawler [5] showed that Ukkonen's algorithm evaluated $O(K)$ DP table entries. If the fanout of the trie is Σ, the trie method needs to only evaluate $O(K|\Sigma|^K)$ DP table entries, which is independent of the number of noisy words we are searching for. Their experiments showed that their method significantly out-performs the nearest competitor for $K = 0$ and $K = 1$.

2 Look-Ahead Branch and Bound Scheme

Given the fact that the dictionary is stored in a trie, any PR-related search for a word in H will have to search the entire trie. To minimize the computational burden, we shall now show how we can use concepts in AI to "reduce" the portion of the search space investigated. We do this by invoking the principles of Branch and Bound (BB) strategies. Due to space limitations, an overview of BB techniques as they are used in AI searching is not given here. They can be found in [9].

In the present case, we now investigate how we can use BB to prune and thus eliminate searching along parts of the *trie*. The differences between our pruning and the pruning applied by [12], [17] and [19] are given in [2].

2.1 The Three Components

The look-ahead component: The idea that we advocate is to prune, from the calculations, the sub-tries in which the strings stored are not within a pre-defined acceptable condition. When the edit cost for the symbols is of a 0/1 sort, the lengths of the string stored in $subtrie(c)$ can be directly related to the maximum edit distance allowed, and thus can simplify the equations and the condition that has to be tested per node even before we traverse the path. The maximum edit distance or error can give an indication about the maximum and minimum lengths of the strings allowed. We propose a strategy by which we will not traverse[3] the $subtrie(c)$ unless there is a "hope" of determining a suitable string in it, where the latter is defined as the string that could be garbled into Y with less than K errors, which as we shall see, can be determined *a priori*.

The dynamic component: Because we are considering costs of a 0/1 sort, the lengths of the *prefixes* to be processed can also be directly related to the maximum edit distance error K. The maximum and minimum allowed lengths for all strings stored in a $subtrie(c)$ are easily related to the length of Y, M, and to the error K, as:

$$max(length(X^+)) \leq M + K.$$

Further, if we are at node c and the length of the prefix calculated so far is N', and the length of any string in $subtrie(c)$ is N'', this constraint can be re-written as:

$$max(N' + N'') \leq M + K.$$

Since N' is constant per node c, this means:

$$max(N'') \leq M - N' + K. \tag{1}$$

[3] Observe that our method is distinct from the dictionary partitioning strategy which is also based on separately gathering strings based on their lengths [8].

Similarly, since K is the *absolute* number of errors,

$$min(N'') \geq M - N' - K. \tag{2}$$

Using these dynamic equations for the minimum and maximum lengths allowed for string eligible to be X^+, we can easily test at each node if the lengths of the suffixes stored are within these acceptable ranges, namely, $min(N'')$, and $max(N'')$. The corresponding inequalities which involves generalized edit distances is currently being derived.

The static component: To test if we are within acceptable ranges for the potential candidates for X^+, we need to store the information needed for these calculations within each node, so that the conditions can be tested locally (and quickly) within the corresponding node. Fortunately, this information is already known *a priori* and is easily calculated and stored. More specifically, we need to store two values at each node of the trie, which (for the 0/1 edit distances) are:

- *Maxlen*: A value stored at a node which indicates the length of the path between this node and the most distant node representing an element of the dictionary H.
- *Minlen*: A value stored at a node which indicates the length of the path between this node and the closest node representing an element in H.

When inserting a string in the trie, we already know the length of this string and only the values of *Maxlen* and *Minlen* need to be adjusted for the nodes along the path included in the insertion, by comparing their previous values with the length of the newly inserted string.

2.2 The overall heuristic

At each node of the trie, before we do any further computations, we test the following conditions, referred to as the LHBB conditions:

(a) $Minlen > M - N' + K$ *obtained by negating Eq.* (1), or

(b) $Maxlen < M - N' - K$ *obtained by negating Eq.* (2).

If (a) or (b) is true, it means there is no hope of finding a solution within the present subtrie, and so we prune the calculations for the subtrie. The LHBB, as its name implies, first looks forward at each node, and sees if it is expected to perform any further calculations. If at any time we reach a string X in the dictionary (which is thus an "accepting" node), we accept the string if the $D(X, Y) \leq K$. An example, the final pseudo-code for the technique, and the rationale for maintaining *Minlen* and *Maxlen* are included in [2].

3 Experimental Results

To investigate the power of our new method with respect to computation we conducted various experiments. The results obtained were remarkable with

Table 1. Statistics of the data sets used in the experiments.

	Eng	Dict	Webster
Size of dictionary	8KB	225KB	944KB
number of words in dictionary	964	24,539	90,141
min word length	4	4	4
max word length	15	22	21

respect to the gain in the number of computations needed to get the best estimate X^+. By computations we mean the addition and minimization operations needed, including the minimization operations required for calculating the LHBB criterion. The LHBB scheme was compared with the original trie-based work for approximate matching [17].

Three benchmark data sets were used in our experiments. Each data set was divided into two parts: a *dictionary* and the corresponding *noisy file*. The dictionary was the words or sequences that had to be stored in the Trie. The noisy files consisted of the strings which were searched for in the corresponding dictionary. The three dictionaries we used (whose statistics are shown in Table 1) were as follows:

- *Eng*[4]: This dictionary consisted of 964 words obtained as a subset of the most common English words [7] augmented with words used in computer literature.
- *Dict*[5]: This is a dictionary file used in the experiments done by Bentley and Sedgewick in [3].
- *Webster's Unabridged* Dictionary: This dictionary was used by Clement *et. al.* [1], [6] to study the performance of different trie implementations. The alphabet size is 54 characters.

Three sets of corresponding noisy files were created using the technique described in [15], and in each case, the files were created for a specific error value. The three error values tested were for $k = 1,\ 2,\ and\ 3$, as is typical in the literature [12], [17].

The two methods, Trie (the original method) [17] and our scheme, LHBB, were tested with the three sets of noisy words. We report a summary of the results obtained in terms of the number of computations (additions and minimizations) in Table 2. The results show the significant benefits of the LHBB scheme with up to 33% improvement. For example, for the *Websters* dictionary, when $K = 1$, the number of computations is 6849 million and 4776 million respectively, which represents an improvement of 30.26%. The improvement decreases as the number of errors increases. Additional experimen-

[4]This file is available at `www.scs.carleton.ca/~oommen/papers/WordWldn.txt`.

[5]The actual dictionary can be downloaded from `http://www.cs.princeton.edu/~rs/strings/dictwords`.

tal results and complexity conclusions are included in [2] and omitted here in the interest of brevity.

Table 2. The results obtained in terms of the number of operations needed for the original trie-based method (Trie) and the new proposed LHBB scheme (LH), when the number of errors permitted, K, are 1 *and* 2. The number of operations is given in millions. The results for $K = 3$ are shown in [2].

Error	**K=1**						**K=2**					
Operation	**Eng**		**Dict**		**Webster**		**Eng**		**Dict**		**Webster**	
	Trie	**LH**	**Trie**	**LH**	**Trie**	**LH**	**Trie**	**LH**	**Trie**	**LH**	**Trie**	**LH**
Additions	5.2	3.5	550	390	3360	2224	18.1	12.3	3654	2644	24901	16773
Improvement	32.69%		29.09%		33.80%		32.04%		27.64%		32.64%	
Minimizations	5.5	4.2	575	454	3489	2552	19.1	14.5	3809	3042	25830	19077
Improvement	23.63%		21.04%		26.85%		24.08%		20.13%		26.14%	
Total	10.7	7.7	1125	844	6849	4776	37.2	26.8	7463	5686	50805	35850
Improvement	28.03%		24.97%		30.26%		27.95%		23.81%		29.43%	

4 Conclusion

In this paper, we have presented a new Branch and Bound (BB) scheme that can be applied to approximate string matching using tries, which we have called a *Look-Ahead* Branch and Bound scheme or LHBB trie pruning strategy. The new scheme makes use of the information that the lengths of the strings stored in the dictionary are known *a priori*, and because we are using 0/1 costs for the inter-symbol edit distances, the lengths of the strings can be related to the edit distance costs. The heuristic that we propose works with a trie and has three characteristics, namely it has a static component, a dynamic component, and finally, it is of a look-ahead sort, as opposed to the cut-off methods already proposed [12], [19]. Several experiments were conducted using three benchmarks dictionaries for noisy sets involving different error values, $K = 1,\ 2,$ *and* 3. The results demonstrate a significant improvement, with respect to the number of operations needed, for approximate searching using tries which can be even as high as 33%. The new LHBB pruning can also be used together with Ukkonen's cutoff technique [19] to attain the highest performance that is still bounded by the $O(K|\Sigma|^K)$ worst case.

References

1. A. Acharya, H. Zhu, and K. Shen (1999) Adaptive algorithms for cache-efficient trie search. ACM and SIAM Workshop on Algorithm Engineering and Experimentation.

2. G. Badr and B. J. Oommen (2005) A look-ahead branch pruning scheme for trie-based approximate string matching. Unabridged version of the present paper.
3. J. Bentley and R. Sedgewick (1997) Fast algorithms for sorting and searching strings. Eighth Annual ACM-SIAM Symposium on Discrete Algorithms, New Orleans.
4. H. Bunke (1993) Structural and syntactic pattern recognition. In: Handbook of Pattern Recognition and Computer Vision. Edited by C.H.Chen, L.F.Pau and P.S.P. Wang, World Scientific, Singapore.
5. W. Chang and E. Lawler (1992) Approximate string matching in sublinear expected time. 13th Annual Symposium on Foundations of Computer Science, 116–124.
6. J. Clement, P. Flajolet, and B. Vallee (1998) The analysis of hybrid trie structures. Proc. Annual A CM-SIAM Symp. on Discrete Algorithms, San Francisco, California, 531–539.
7. G. Dewey (1923) Relative Frequency of English Speech Sounds. Harvard University Press.
8. M. Du and S. Chang (1994) An approach to designing very fast approximate string matching algorithms. IEEE Transactions on Knowledge and Data Engineering, 6(4):620–633.
9. M. Firebaugh (1988) Artificial Intelligence: A Knowledge-Based Approach. Boyd and Fraser.
10. R. L. Kashyap and B. J. Oommen (1981) An effective algorithm for string correction using generalized edit distances -i. description of the algorithm and its optimality. Inf. Sci., 23(2):123–142.
11. G. Navarro (2001) A guided tour to approximate string matching. ACM Computing Surveys, 33(1):31–88.
12. K. Oflazer (1996) Error-tolerant finite state recognition with applications to morphological analysis and spelling correction. Computational Linguistics, 22(1):73–89.
13. B. J. Oommen (1987) Recognition of noisy subsequences using constrained edit distances. IEEE Trans. on Pattern Anal. and Mach. Intel.,PAMI-9:676–685.
14. B. J. Oommen and G. Badr (2004) Dictionary-based syntactic pattern recognition using tries. Proceedings of the Joint IARR International Workshops SSPR 2004 and SPR 2004, 251–259.
15. B. J. Oommen and R. K. S. Loke (2003) Syntactic pattern recognition involving traditional and generalized transposition errors: Attaining the information theoretic bound. Submitted for Pubication.
16. D. Sankoff and J. B. Kruskal (1983) Time Warps, String Edits and Macromolecules: The Theory and practice of Sequence Comparison. Addison-Wesley.
17. H. Shang and T. Merrettal (1996) Tries for approximate string matching. IEEE Transactions on Knowledge and Data Engineering, 8(4):540–547.
18. G. A. Stephen (2000) String Searching Algorithms, volume 6. Lecture Notes Series on Computing, World Scientific, Sihgapore, NJ.
19. E. Ukkonen (1985) Algorithm for approximate string matching. Information and control, 64:100–118.
20. R. Wagner and A. Fischer (1974) The string-to-string correction problem. Journal of the Association for Computing Machinery (ACM), 21:168–173.
21. R. A. Wagner (1974) Order-n correction for regular languages. Comm. ACM, 17:265–268.

Feature Selection for High-Dimensional Data: A Kolmogorov-Smirnov Correlation-Based Filter

Jacek Biesiada,[1] Włodzisław Duch[2,3]

[1] Division of Computer Methods, Dept. of Electrotechnology, The Silesian University of Technology, Katowice, Poland
[2] Dept. of Informatics, Nicholaus Copernicus University, Toruń, Poland
[3] School of Computer Engineering, Nanyang Technological University, Singapore Google: Duch

Summary. An algorithm for filtering information based on the Kolmogorov-Smirnov correlation-based approach has been implemented and tested on feature selection. The only parameter of this algorithm is statistical confidence level that two distributions are identical. Empirical comparisons with 4 other state-of-the-art features selection algorithms (FCBF, CorrSF, ReliefF and ConnSF) are very encouraging.

1 Introduction

For large highly dimensional datasets feature ranking and feature selection algorithms are usually of the filter type. In the simplest case feature filter is a function returning a relevance index $J(\mathcal{S}|\mathcal{D}, C)$ that estimates, given the data $\mathcal{D}$, how relevant a given feature subset $\mathcal{S}$ is for the task C (usually classification or approximation of the data). The relevance index $J(\mathcal{S}|\mathcal{D}, C)$ is calculated directly from data, without any reference to the results of programs that are used on data with reduced dimensionality. Since the data $\mathcal{D}$ and the task C are usually fixed and only the subsets $\mathcal{S}$ varies an abbreviated form $J(\mathcal{S})$ is used. Instead of a simple function (such as correlation or information content) an algorithmic procedure, such as building a decision tree or finding nearest neighbors, may be used to estimate this index.

Relevance indices computed for individual features $X_i, i = 1 \dots N$ provide indices that establish a ranking order $J(X_{i_1}) \leq J(X_{i_2}) \cdots \leq J(X_{i_N})$. Those features which have the lowest ranks are filtered out. For independent features this may be sufficient, but if features are correlated many of them may be redundant. For some data distributions the best pair of features may not even include a single best feature [14, 2]! Thus ranking does not guarantee that the largest subset of important features will be found. Methods that search for the best subset of features may also use filters to evaluate the usefulness of subsets of features.

Although in the case of filter methods there is no direct dependence of the relevance index on the adaptive algorithms obviously the thresholds for feature rejection may be set either for relevance indices, or by evaluation of the feature contributions by the final system. Features are ranked by the filter, but how many are finally taken may be determined using adaptive system as a wrapper. Evaluation of the adaptive system performance (frequently cross-validation tests) are done only for a few pre-selected feature sets, but still this "filtrapper" approach may be rather costly. What is needed is a simple filter method that may be applied to large datasets ranking and removing redundant features, parameterized in statistically well-established way. Such an approaches is described in this paper.

In next section a new relevance index based on the Kolmogorov-Smirnov (K-S) test to estimate correlation between the distribution of feature values and the class labels is introduced. Correlation-based filters are very fast and may be competitive to filters based on information theory. Therefore in section 3 empirical comparisons between K-S filter, Pearson's correlation based filter and popular filters based on information gain is made on a number of datasets.

2 Theoretical framework

2.1 Correlation-Based Measures

For feature X with values x and classes C with values c, where X, C are treated as random variables, Pearson's linear correlation coefficient is defined as [11]:

$$\varrho(X,C) = \frac{E(XC) - E(X)E(C)}{\sqrt{\sigma^2(X)\sigma^2(C)}} = \frac{\sum_i (x_i - \bar{x}_i)(c_i - \bar{c}_i)}{\sqrt{\sum_i (x_i - \bar{x}_i)^2 \sum_j (c_i - \bar{c}_i)^2}}. \quad (1)$$

$\varrho(X,C)$ is equal to ± 1 if X and C are linearly dependent and zero if they are completely uncorrelated. The simplest test estimating significance of the differences in $\varrho(X,C)$ values is based on the probability that two variables are correlated [11]:

$$\mathcal{P}(X \sim C) = \mathrm{erf}\left(|\varrho(X,C)|\sqrt{N/2}\right), \quad (2)$$

where erf is the error function. The feature list ordered by decreasing values of the $\mathcal{P}(X \sim C)$ may serve as feature ranking. Non-parametric, or Spearman's rank correlation coefficients may be useful for ordinal data types.

An alternative approach is to use χ^2 statistics, but in both cases for large number of samples probability $\mathcal{P}(X \sim C)$ is so close to 1 that ranking becomes impossible due to the finite numerical accuracy of computations. For example, with $N = 1000$ samples small coefficients $\varrho(X,C) \approx 0.02$ lead to probabilities of correlation around 0.5. The $\varrho(X, C)$ or χ^2 thresholds for the significance of a given feature may therefore be taken from a large interval corresponding to almost the same probabilities of correlation.

Information theory is frequently used to define relevance indices. The Shannon information, or feature and class entropy, is:

$$H(X) = -\sum_i \mathcal{P}(x_i) \log \mathcal{P}(x_i); \qquad H(C) = -\sum_i \mathcal{P}(c_i) \log \mathcal{P}(c_i) \tag{3}$$

and the joint Shannon entropy is:

$$H(X, C) = -\sum_{i,j} \mathcal{P}(x_i, c_j) \log \mathcal{P}(x_i, c_j) \tag{4}$$

Mutual Information (MI) is the basic quantity used for information filtering:

$$MI(X, C) = H(X) + H(C) - H(X, C) \tag{5}$$

Symmetrical Uncertainty Coefficient (SU) has similar properties to mutual information:

$$SU(X, C) = 2 \left[\frac{MI(X, C)}{H(X) + H(C)} \right] \tag{6}$$

If a group of k features $\mathbf{X}_k$ has already been selected, correlation coefficient may be used to estimate correlation between this group and the class, including inter-correlations between the features. Denoting the average correlation coefficient between these features and classes as $r_{kc} = \bar{\varrho}(\mathbf{X}_k, C)$ and the average between different features as $r_{kk} = \bar{\varrho}(\mathbf{X}_k, \mathbf{X}_k)$ the relevance of the feature subset is defined as:

$$J(\mathbf{X}_k, C) = \frac{k r_{kc}}{\sqrt{k + (k-1) r_{kk}}}. \tag{7}$$

This formula has been used in the Correlation-based Feature Selection (CFS) algorithm [6] adding (forward selection) or deleting (backward selection) one feature at a time. A definition of predominant correlation proposed by Yu and Liu [16] for Fast Correlation-Based Filter (FCBF) includes correlations beetwen feature and classes and between pairs of features. The FCBF algorithm does a typical ranking using SU coefficient (eq. 6) to determine class-feature relevance, setting some threshold value $SU \geq \delta$ to decide how many features should be taken. In the second part redundant features are removed by defining the "predominant features".

A different type of selection method called ConnSF, based on inconsistency measure, has been proposed by Dash *et al.* [3] and will be used for comparison in Sec. 3. Two identical input vectors are inconsistent if they have identical class labels (a similar concept is used in rough set theory). Intuitively it is

```
set all weights W_xi = 0
for j=1 to m do begin
  randomly select instance X;
  find nearest hit H and nearest miss M;
    for i:=1 to k do begin
    W_xi <- W_xi - D(x_i, X, H)/m + D(x_i, X, M)/m
  end;
end;
```

Fig. 1. Sketch of the Relief algorithm.

clear that inconsistency grows when the number of features is reduced and that feature subsets that lead to high inconsistency are not useful. If there are n samples in the dataset with identical feature values x_i, and n_k among them belong to class k then the inconsistency count is defined as $n - \max_k c_k$. The total inconsistency count for a feature subset is the sum of all inconsistency counts for all data vectors.

A different way to find feature subsets is used in the Relief algorithm ([8] and [13]). This algorithm (see Fig. 1) estimates weights of features according to how well their values distinguish between data vectors that are near to each other. For a randomly selected vector X from a data set $\mathcal{S}$ with k features Relief searches the dataset for its two nearest neighbors: the nearest hit H from the same class and the nearest miss M from another class. For feature x and two input vectors X, X' the contribution to the weight W_x is proportional to the $D(x, X, X') = 1 - \delta(X(x), X'(x))$ for binary or nominal features, and $D(x, X, X') = |X(x) - X'(x)|$ for continuous features. The process is repeated m times, where m is a user defined parameter [8]. Normalization with m in calculation of W_x guarantees that all weights are in the $[-1, 1]$ interval. In our empirical studies (Sec. 3) we have used an extension of this agorithm for multiclass problems, called ReliefF [13].

2.2 Kolmogorov-Smirnov Correlation-Based Filter Approach

Equivalence of two random variables may be evaluated using the Kolmogorov-Smirnov (K-S) test [7]. The K-S test measures the maximum difference between cummulative distribution of two random variables. If a feature is redundant than the hypothesis that its distribution is equal to already selected feature should have high probability. n independent observations of two random variables X, X' are given in the training data, where for the K-S test to be valid n should be more than 40. The test for X, X' feature redundancy proceeds as follows:

- Discretization of feature values x into k bins $[x_i, x_{i+1}], i = 1 \ldots k$ is performed.

- Frequency f_i, f'_k of occurrences of feature values in each bin are recorded.
- Based on the frequency counts cumulative distribution functions F_i and F'_i are constructed.
- λ (K-S statistics) is the largest absoulte difference between F_i and F'_i, i.e,

$$\lambda = \sqrt{n/2} \max_i |F_i - F'_i| \qquad \text{for } i = 1, 2, \ldots k. \tag{8}$$

Probability that the maximum K-S distance λ_α is larger than observed may be calculated using K-S statistics for each parameter α [9] that has the meaning of statistical significance level. When $\lambda < \lambda_\alpha$ then the two distributions are equivalent with α significance level, and thus one of the features is redundant. Using typical significance values of 0.95 solves the problem of the threshold values for redundancy.

The Kolmogorov-Smirnov Correlation-Based Filter (K-S CBF) algorithm is presented below. First, the relevance is determined using the symmetrical uncertainty (other relevance criteria may also be used), and then K-S test applied to remove redundancy.

Algorithm K-S RBF:
Relevance analysis
1. Calculate the $SU(X, C)$ relevance indices and create an ordered list $\mathcal{S}$ of features according to the decreasing value of their relevance.
Redundancy analysis
2. Take as the feature X the first feature from the $\mathcal{S}$ list
3. Find and remove all features for which X is approximately equivalent according to the K-S test
4. Set the next remaining feature in the list as X and repeat step 3 for all features that follow it in the $\mathcal{S}$ list.

Fig. 2. A two-step Kolmogorov-Smirnov Correlation Based Fiter (K-S CBF) algorithm.

3 Empirical Studies

To evaluate the performance of the K-S CBF algorithm both artificial and real datasets have been used with a number of classification methods. Two artificial datasets, Gauss4, and Gauss8, have been used in our previous study [4]. Gauss4 is based on sampling from 4 Gaussian functions with unit dispersion in 4 dimensions, each cluster representing a separate class. The first function is centered at $(0, 0, 0, 0)$, the next at $(1, 1/2, 1/3, 1/4)$, $(2, 1, 2/3, 1/2)$, and $(3, 3/2, 3, 3/4)$, respectively. The dataset contains 4000 vectors, 1000 per each class. In this case the ideal ranking should give the following order: $X_1 > X_2 > X_3 > X_4$.

Gauss8 used here is an extension of Gauss4, adding 4 additional features that are approximately linearly dependent $X_{i+4} = 2X_i + \epsilon$, where ϵ is a uniform noise with a unit variance. In this case the ideal ranking should give the following order: $X_1 > X_5 > X_2 > X_6 > X_3 > X_7 > X_4 > X_8$ and the selection should reject all 4 linearly dependent features as redundant. K-S CBF algorithm and ConnSF [3] algorithm had no problem with this task, but FCBF [16] selected only 3 features, CorrSF [6] selected only first two and ReliefF [13] left only feature 1 and 5, giving them weight 0.154 (for features 2 and 6 the weight was 0.060, dropping to 0.024 for feature 3, 6 and to 0.017 for features 4, 8.

Title	Selected features					
	Full set	FCBF	CorrSF	ReliefF	ConnSF	K-S CBF
Features	1 to 8	1+2+3	1+2+5	1+5	1 to 4	1to 4
NBC	82.13	81.57	80.25	76.95	82.13	82.13
1NN	73.42	73.90	71.10	68.12	73.42	73.42
C4.5	78.30	79.12	78.95	76.15	78.70	78.70
SVM	81.88	81.70	80.90	76.95	81.73	81.73
Average	79.91	79.09	78.83	75.34	80.40	80.40

Table 1. Accuracy of 4 classifiers on selected subsets of features for the Gauss8 dataset.

In Table 3 results of Naive Bayes Classifier (NBC) (Weka implementation, [15]), the nearest neighbor algorithm (1NN) with Euclidean distance function, C4.5 tree [12] and the Support Vector Machine with a linear kernel are given (Weka and SVM, Ghostminer 3.0 implementation[4]).

Title	Features	Instances	Classes
Hypothyroid	21	3772	3
Lung-cancer	58	32	3
Promoters	59	106	2
Splice	62	3190	3

Table 2. Summary of the datasets used in empirical studies.

For the initial comparison on real data several datasets from the UCI Machine Learning Repository [10] and the UCI KDD Archive [1] were used. A summary of all datasets is presented in Table 3. For each data set all five feature selection algorithms are compared (FCBF [16], CorrSF [6], ReliefF [13], ConnSF [3], and K-S CBF) and the number of features selected by each algorithm is given. For data sets containing features with continuous values the MDLP discretization algorithm has been applied[5] [5]. 5 neighbors and 30 instances were used for ReliefF, as suggested by Robnik-Sikonia and Kononenko

[4]http://www.fqspl.com.pl/ghostminer/

[5]available from www.public.asu.edu/~huanliu/

[13]. For CorrSF and ConnSF forward search strategy has been used, and for FCBF, ReliefF, and the K-S CBF forward search strategy based on ranking.

Dataset	Selected features					
	Full set	FCBF	CorrSF	ReliefF	ConnSF	K-S CBF
Hypothyroid	21	5	**1**	*11*	6	6
Lung-cancer	58	6	*11*	8	4	**3**
Splice	62	*22*	**6**	19	10	14
Promoters	59	*6*	**4**	**4**	**4**	5
Average	50	*9.8*	**5.5**	10.5	6	7

Table 3. The number of selected features for each algorithm; bold face – lowest number, italics – highest number.

The overall balanced accuracy (accuracy for each class, averaged over all classes) obtained from 10-fold cross-validation calculations is reported. For datasets with significant differences between samples from different classes balanced accuracy is a more sensitive measure than the overall accuracy. Results of these calculations are collected in Table 3.

4 Conclusion

A new algorithm, K-S CBF, for finding non-redundant feature subsets based on the Kolmogorov-Smirnov test has been introduced. It has only one parameter, statistical significance or the probability that the hypothesis that distributions of two features is equivalent is true. Our initial tests are encouraging: on the artificial data perfect ranking has been recreated and redundant features rejected, while on the real data, with rather modest number of features selected results are frequently the best, or close to the best, comparing with four state-of-the-art feature selection algorithms. The new algorithm seems to work especially well with the linear SVM classifier. Computational demands of K-S CBF algorithm are similar to other correlation-based filters and much lower than ReliefF.

It is obvious that sometimes statistical significance at 0.05 level selected for our tests is not optimal and for the lung cancer data too few features have been selected, leading to a large decrease of accuracy. This parameter may be optimized in crossvalidation tests on the training set, but the method guarantees that each time only non-redundant subset of features will be selected. Various variants of the Kolmogorov-Smirnov test exist [11] and the algorithm may be used with other indices for relevance indication. These possibilities remain to be explored. Further tests on much larger bioinformatics data will be reported soon.

Acknowledgement. This work was financed by the Polish Committee for Scientific Research (KBN) grant (2005-2007).

Method	C 4.5 tree					
Dataset	Full set	FCBF	CorrSF	ReliefF	ConnSF	K-S CBF
Hypothyroid	**99.91**	*66.94*	85.92	98.94	93.45	98.68
Lung-cancer	*49.43*	63.87	**67.21**	63.87	63.22	64.76
Splice	94.35	94.05	*93.39*	**94.80**	93.63	94.05
Promoters	81.13	**85.84**	83.96	*65.09*	84.90	81.13
Average	81.21	*77.68*	82.62	80.68	83.80	**84.66**
Method	Naive Bayes					
Dataset	Full set	FCBF	CorrSF	ReliefF	ConnSF	K-S CBF
Hypothyroid	83.20	66.94	*58.48*	70.28	67.14	**85.83**
Lung-cancer	47.92	50.57	**73.48**	50.57	65.68	*41.74*
Splice	94.88	**96.03**	*93.31*	95.54	94.17	94.75
Promoters	90.56	**95.28**	93.39	*61.32*	93.39	87.35
Average	79.14	73.21	79.67	*69.43*	**80.10**	77.42
Method	1 Nearest Neighbor					
Dataset	Full set	FCBF	CorrSF	ReliefF	ConnSF	K-S CBF
Hypothyroid	*61.60*	83.89	71.72	78.92	**86.94**	83.93
Lung-cancer	39.94	*36.01*	**66.84**	53.13	65.41	45.70
Splice	*80.08*	84.45	**87.68**	84.04	86.77	83.82
Promoters	85.85	**92.45**	86.79	*59.43*	81.13	85.85
Average	*66.49*	74.28	78.26	69.01	**80.06**	74.83
Method	SVM					
Dataset	Full set	FCBF	CorrSF	ReliefF	ConnSF	K-S CBF
Hypothyroid	52.65	45.49	*44.07*	51.24	45.13	**84.31**
Lung-cancer	*41.37*	55.41	**66.07**	61.60	59.37	47.35
Splice	92.81	95.73	93.75	**95.75**	*90.08*	95.11
Promoters	**93.40**	91.50	77.36	*58.49*	87.33	93.11
Average	70.06	72.03	70.31	*66.77*	70.48	**80.04**

Table 4. Balanced accuracy for the 4 classification methods on features selected by each algorithm; bold face – best results, italics – worst.

References

1. S.D. Bay. *The UCI KDD archive.* Univ. of California, Irvine, 1999. http://kdd.ics.uci.edu.
2. T.M. Cover. The best two independent measurements are not the two best. IEEE Transactions on Systems, Man, and Cybernetics, 4:116–117, 1974.
3. M. Dash and H. Liu. Consistencyŋbased search in feature selection. Artificial Intelligence, 151:155–176, 2003.
4. W. Duch, T. Winiarski, J. Biesiada, and A. Kachel. Feature ranking, selection and disŋ cretization. In Proceedings of Int. Conf. on Artificial Neural Networks (ICANN), pages 251–254, Istanbul, 2003. Bogazici University Press.
5. U.M. Fayyad and K.B. Irani. Multiŋinterval discretization of continousŋvalued attributes for classification learning. In R. Bajcsy, editor, Proceedings of the

Thirteenth Internaŋ tional Joint Conference on Artificial Intelligence, Chambery, France, pages 1022–1027, San Francisco, CA, 1993. Morgan Kaufmann.
6. M.A. Hall. Correlationŋbased Feature Subset Selection for Machine Learning. PhD thesis, Department of Computer Science, University of Waikato, Waikato, N.Z., 1999.
7. R. Laha I. Chakravarti and J. Roy. Handbook of Methods of Applied Statistics. John Wiley and Sons, Chichester, 1967.
8. K. Kira and L.A. Rendell. A practical approach to feature selection. In Proceedings of the Ninth International Conference on Machine Learning (ICMLŋ92), pages 249–256, San Francisco, CA, 1992. Morgan Kaufmann.
9. N. Hastings M. Evans and B. Peacock. Statistical Distributions, 3rd. ed. John Wiley and Sons, Chichester, 2000.
10. C.J. Mertz and P.M. Murphy. The UCI repository of machine learning databases. Univ. of California, Irvine, 1998. http://www.ics.uci.edu.pl/mlearn/MLRespository.html.
11. W.H. Press, S.A. Teukolsky, W.T. Vetterling, and B.P. Flannery. Numerical recipes in C. The art of scientific computing. Cambridge University Press, Cambridge, UK, 1988.
12. J.R. Quinlan. C4.5: Programs for Machine Learning. Morgan Kaufman, San Mateo, CA, 1993.
13. M. RobnikŋSikonja and I. Kononenko. Theoretical and empirical analysis of relieff and rrelieff. Machine Learning, 53:23–69, 2003.
14. G.T. Toussaint. Note on optimal selection of independent binaryŋvalued features for pattern recognition. IEEE Transactions on Information Theory, 17:618–618, 1971.
15. I. Witten and E. Frank. Data minig – practical machine learning tools and techniques with JAVA implementations. Morgan Kaufmann, San Francisco, CA, 2000.
16. L. Yu and H. Liu. Feature selection for highŋdimensional data: A fast correlationŋbased filter solution. In Proceedings of the 12th International Conference on Machine Learning (ICMLŋ03), Washington, D.C., pages 856–863, San Francisco, CA, 2003. Morgan Kaufmann.

Linear Ranked Regression - Designing Principles

Leon Bobrowski[1,2]

[1] Faculty of Computer Science, Bialystok Technical University
[2] Institute of Biocybernetics and Biomedical Engineering, PAS, Warsaw, Poland
leon.bobrowski@ibib.waw.pl

Summary. A priori information about selected pattern recognition problem is often a necessary precondition to reach a sufficient quality of the problem solution. Such information could take the form of the ranked order in the referencing sets of objects or events. For example, we can encounter a case when one patient is suffering from a more advanced stage of a disease than the another one. In other cases, we can assume that some events took place earlier or later than the regarded one. A ranked regression task is aimed at designing such linear transformation of multivariate data sets on the line which preserves with the highest precision possible the ranked order. The convex and piecewise linear (CPL) criterion functions are used here for designing ranked linear models.

1 Introduction

Pattern recognition tools are based on particular form of data representation [1], [2]. It is assumed typically that data is represented as a set of the feature vectors $\mathbf{x}_j$ of the same dimensionality n. The vectors $\mathbf{x}_j$ can be treated as points in the n-dimensional feature space X. A priori information takes typically a form of a known division of the data set into learning subsets which contain elements $\mathbf{x}_j$ related to particular classes k.

A priori information can be given also in the form of the ranked order between selected feature vectors $\mathbf{x}_j$. In particular, the information about ordering particular pairs $\{\mathbf{x}_i, \mathbf{x}_j\}$ $(i < j)$ of the feature vectors $\mathbf{x}_j$ is taken into account. Such referencing pairs of the feature vectors are called dipoles [3]. The relative ranks of the vectors $\mathbf{x}_i$ and $\mathbf{x}_j$ constituting given dipole $\{\mathbf{x}_i, \mathbf{x}_j\}$ are linked to the geometric orientation of these dipoles in respect to the ranked line in the feature space X.

The designing procedure of the linear ranked models is described in the paper. The aim here is to design such linear transformation of the data set on the line, which preserve the ranked order between the feature vectors $\mathbf{x}_j$ in the best possible manner. The designing procedure can be based on the minimization of the convex and the piecewise linear (CPL) criterion functions, which are the sums of the positive and the negative CPL penalty functions [3]. These penalty functions are defined through differences between the feature vectors constituting referencing dipoles. In this manner, the task of the ranked model designing can be linked to the problem of the linear separability of two sets of vectors in a given feature space.

Designing the ranked models can be seen as an induction and the modelling of trends in data sets with taking into account a priori information about data ordering.

2 Feature vectors and dipoles' orientations

Let us take into consideration the data set C built from m feature vectors $\mathbf{x}_j = [x_{j1}, \ldots, x_{jn}]^T$ which have been numbered in a fixed manner

$$C = \{\mathbf{x}_j\} \ (j = 1, \ldots, m) \tag{1}$$

The component (*feature*) x_{ji} of the vector $\mathbf{x}_j$ is a numerical result of the i-th examination $(i = 1, ..., n)$ of a given object O_j $(j = 1, \ldots, m)$. The biomedical feature vectors $\mathbf{x}_j$ are often of a mixed type, because they represent both symptoms and signs $(x_i \in 0, 1)$ as well as the results of laboratory tests $(x_i \in R)$ performed on a given patient.

Let the symbol "$\prec$" mean the ranked relation "follows" which may be fulfilled between selected feature vectors $\mathbf{x}_j$

$$\mathbf{x}_j \prec \mathbf{x}_k \iff \mathbf{x}_k \ \textit{follows} \ \mathbf{x}_j \tag{2}$$

The relation "$\prec$" between the feature vectors $\mathbf{x}_j$ and $\mathbf{x}_k$ means that the pair $\{\mathbf{x}_j, \mathbf{x}_k\}$ is *ranked.* This relation should be determined on the basis of some additional a priori information about some (not necessary all) pairs of the vectors $\mathbf{x}_j$. For example, medical doctors can compare two patients with the same disease and conclude that one of them has developed a more serious stage of the disease than the other one. One of two enterprizes parameterized by the vectors $\mathbf{x}_j$ and $\mathbf{x}_k$ could be more profitable than the second one. One of two kinds of prehistoric animals characterized by the biometrical measurements $\mathbf{x}_j$ and $\mathbf{x}_k$ is known to be more ancient considering the evolutionary development (phylogenetic classification). Our aim is to design such transformation of the feature vectors $\mathbf{x}_j$ on the line $y = \mathbf{w}^T\mathbf{x}$ which preserves the relation "$\prec$" (2) as precisely as possible

$$y_j = y_j(\mathbf{w}) = \mathbf{w}^T\mathbf{x}_j \tag{3}$$

where $\mathbf{w} = [w_1, \ldots, w_n]^T$ is the vector of parameters.

The procedure of the ranked line designing can be based on the concept of the positively and negatively oriented dipoles $\{\mathbf{x}_j, \mathbf{x}_{j'}\}$ $(j < j')$.

Definition 1. *The ranked pair* $\{\mathbf{x}_j, \mathbf{x}_{j'}\}$ $(j < j')$ *of the feature vectors* $\mathbf{x}_j$ *and* $\mathbf{x}_{j'}$ *constitutes the* positively oriented dipoles $\{\mathbf{x}_j, \mathbf{x}_{j'}\}$ $((j, j') \in I^+)$, *if and only if* $\mathbf{x}_j \prec \mathbf{x}_{j'}$.

$$(\forall (j, j') \in I^+) \qquad \mathbf{x}_j \prec \mathbf{x}_{j'} \tag{4}$$

Definition 2. *The ranked pair* $\{\mathbf{x}_j, \mathbf{x}_{j'}\}$ $(j < j')$ *of the feature vectors* $\mathbf{x}_j$ *and* $\mathbf{x}_{j'}$ *constitutes the* negatively oriented dipole $\{\mathbf{x}_j, \mathbf{x}_{j'}\}$ $((j, j') \in I^-)$, *if and only if* $\mathbf{x}_{j'} \prec \mathbf{x}_j$.

$$(\forall (j, j') \in I^-) \qquad \mathbf{x}_{j'} \prec \mathbf{x}_j \tag{5}$$

Definition 3. *The line* $y(\mathbf{w}) = \mathbf{w}^T\mathbf{x}$ *(3) is fully consistent (ranked) with the dipoles* $\{\mathbf{x}_j, \mathbf{x}_{j'}\}$ *orientations if and only if*

$$\begin{aligned} &\forall (j, j') \in I^+) \quad y_j(\mathbf{w}) < y_{j'}(\mathbf{w}) \quad \text{and} \\ &\forall (j, j') \in I^-) \quad y_j(\mathbf{w}) > y_{j'}(\mathbf{w}) \end{aligned} \tag{6}$$

where I^+ *and* I^- *are the sets of the positively and negatively oriented dipoles* $\{\mathbf{x}_j, \mathbf{x}_{j'}\}$.

If the line (3) is fully consistent with the dipoles' $\{\mathbf{x}_j, \mathbf{x}_{j'}\}$ orientations, then the below implication holds (Fig. 1)

$$(\forall (j, k)) \quad (\mathbf{x}_j \prec \mathbf{x}_k) \Rightarrow y_j(\mathbf{w}) < y_k(\mathbf{w}) \tag{7}$$

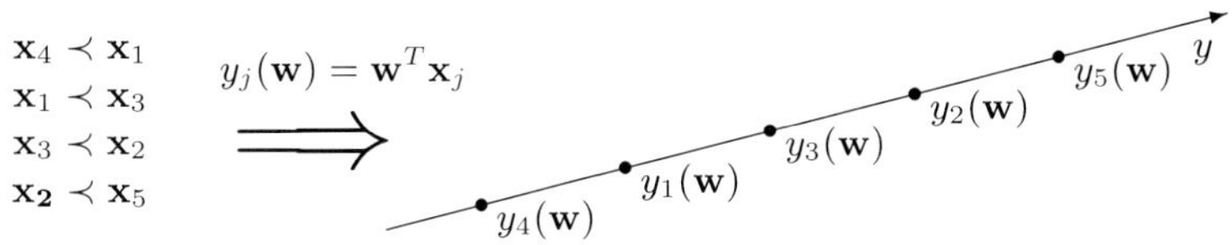

Fig. 1. An example of the order relations (2) and the ranked line (5), where $I^+ = \{(1, 3), (2, 5)\}$ and $I^- = \{(1, 4), (2, 3)\}$.

3 Designing the ranked lines

The relations (6) stand for the following desired inequalities on the line (3):

$$\begin{aligned} &\forall (j, j') \in I^+) \quad \mathbf{w}^T(\mathbf{x}_{j'} - \mathbf{x}_j) > 0 \\ &\forall (j, j') \in I^-) \quad \mathbf{w}^T(\mathbf{x}_{j'} - \mathbf{x}_j) < 0 \end{aligned} \tag{8}$$

The above conditions mean that two sets of the *differential vectors* $\mathbf{r}_{jj'} = (\mathbf{x}_{j'} - \mathbf{x}_j)$ are linearly separable by the hyperplane $H(\mathbf{w})$, which passes through the origin 0 of the feature space X.

$$H(\mathbf{w}) = \{\mathbf{x} : \mathbf{w}^T\mathbf{x} = 0\} \tag{9}$$

The separated sets C^+ and C^- of the vectors $\mathbf{r}_{jj'} = (\mathbf{x}_{j'} - \mathbf{x}_j)$ are given by

$$C^{+} = \{\mathbf{r}_{jj'} = (\mathbf{x}_{j'} - \mathbf{x}_{j}) : (j,j') \in I^{+}\} \qquad (10)$$
$$C^{-} = \{\mathbf{r}_{jj'} = (\mathbf{x}_{j'} - \mathbf{x}_{j}) : (j,j') \in I^{-}\}$$

Designing the separating hyperplane $H(\mathbf{w})$ can be carried out through the minimization of the convex and piecewise linear (CPL) criterion function $\Phi(\mathbf{w})$ similar to the perceptron criterion function [1]. Let us introduce for this purpose the positive $\varphi^{+}_{jj'}(\mathbf{w})$ and negative $\varphi^{-}_{jj'}(\mathbf{w})$ penalty functions (Fig.2)

$$(\forall (j,j') \in I^{+}) \quad \varphi^{+}_{jj'}(\mathbf{w}) = \begin{cases} 1 - \mathbf{w}^{T}\mathbf{r}_{jj'} & if\ \mathbf{w}^{T}\mathbf{r}_{jj'} < 1 \\ 0 & if\ \mathbf{w}^{T}\mathbf{r}_{jj'} \geq 1 \end{cases} \quad \text{and} \qquad (11)$$

$$(\forall (j,j') \in I^{-}) \quad \varphi^{-}_{jj'}(\mathbf{w}) = \begin{cases} 1 + \mathbf{w}^{T}\mathbf{r}_{jj'} & if\ \mathbf{w}^{T}\mathbf{r}_{jj'} > -1 \\ 0 & if\ \mathbf{w}^{T}\mathbf{r}_{jj'} \leq -1 \end{cases} \qquad (12)$$

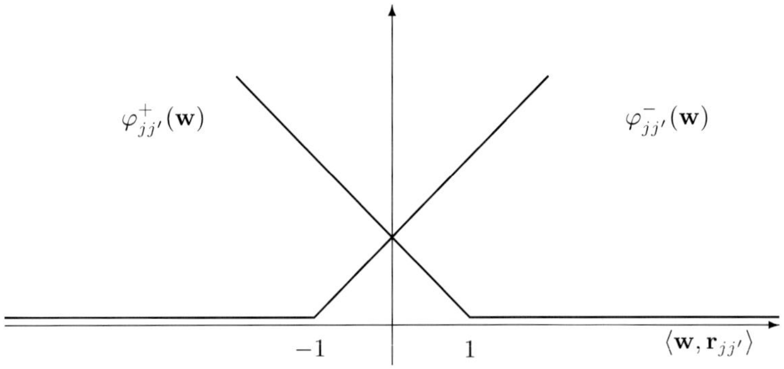

Fig. 2. The penalty functions $\varphi^{+}_{jj'}(\mathbf{w})$ (11) and $\varphi^{-}_{jj'}(\mathbf{w})$ (12).

The criterion function $\Phi(\mathbf{w})$ is the weighted sum of the CPL penalty functions $\varphi^{+}_{jj'}(\mathbf{w})$ and $\varphi^{-}_{jj'}(\mathbf{w})$

$$\Phi(\mathbf{w}) = \sum_{(j,j') \in I^{+}} \gamma_{jj'} \varphi^{+}_{jj'}(\mathbf{w}) + \sum_{(j,j') \in I^{-}} \gamma_{jj'} \varphi^{-}_{jj'}(\mathbf{w}) \qquad (13)$$

where $\gamma_{jj'}$ ($\gamma_{jj'} > 0$) is a positive parameter (*price*) related to the dipole $\{\mathbf{x}_j, \mathbf{x}_{j'}\}$ ($j < j'$).

The criterion function $\Phi(\mathbf{w})$ (13) is the convex and piecewise linear (CPL) as the sum of such type of the penalty functions $\varphi^{+}_{jj'}(\mathbf{w})$ and $\varphi^{-}_{jj'}(\mathbf{w})$. The basis exchange algorithms, similar to the linear programming, allow to find a minimum of such functions efficiently, even in the case of large, multidimensional data sets C^{+} and C^{-} [5]:

$$\Phi^{*} = \Phi(\mathbf{w}^{*}) = min_{\mathbf{w}} \Phi(\mathbf{w}) \geq 0 \qquad (14)$$

The optimal parameter vector $\mathbf{w}^{*}$ and the minimal value Φ^{*} of the criterion function $\Phi(\mathbf{w})$ (11) can be applied to a variety of data ranking problems. In particular, the vector $\mathbf{w}^*$ defining the best ranked line $y = (\mathbf{w}^*)^T\mathbf{x}$ (3) can be found this way.

Lemma 1. *The minimal value Φ^* (14) of the criterion function $\Phi(\mathbf{w})$ (13) is equal to zero if and only if there exists such a vector $\mathbf{w}$ that the ranking of the points $y_j(\mathbf{w})$ on the line (3) is fully consistent (Def. 3) with the relations "$\prec$" (4).*

Proof. If there exists such a vector $\mathbf{w}^*$ that the ranking of the points $y_j(\mathbf{w}^*)$ on the line (3) is fully consistent (Def. 3) with the relations "$\prec$" (4), then the sets C^+ and C^- (10) can be separated (8) by the hyperplane $H(\mathbf{w}^*)$ (9). In this case, the minimal value of the perceptron criterion function $\Phi(\mathbf{w})$ (13) is equal to zero as it results from the pattern recognition theory [1]. On the other hand, if the minimal value of the criterion function $\Phi(\mathbf{w})$ (13) is equal to zero in the point $\mathbf{w}^*$, then the values $\varphi^+_{jj'}(\mathbf{w}^*)$ and $\varphi^-_{jj'}(\mathbf{w}^*)$ of all the penalty functions (11) and (12) have to be equal to zero. It means, that the sets C^+ and C^- (10) can be separated (8) by the hyperplane $H(\mathbf{w}^*)$ (9). In the result, the ranking of the points $y_j(\mathbf{w}^*)$ on the line (3) is fully consistent (Def. 3) with the relations "$\prec$" (4). □

4 From linear independence to linear separability

The sets C^+ and C^- (10) are composed from the differential vectors $\mathbf{r}_{jj'} = (\mathbf{x}_{j'} - \mathbf{x}_j)$. The linear separability of these sets with the threshold equal to zero is important for the ranked models (7) designing.

Definition 4. *The sets C^+ and C^- (10) are linearly separable with the threshold equal to zero if and only if there exists such a parameter vector $\mathbf{w}^*$, that*

$$\begin{aligned}(\exists \mathbf{w}^*) \quad (\forall (j,j') \in I^+) \quad (\mathbf{w}^*)^T\mathbf{r}_{jj'} > 0 \\ (\forall (j,j') \in I^-) \quad (\mathbf{w}^*)^T\mathbf{r}_{jj'} < 0\end{aligned} \tag{15}$$

The above inequalities can be represented in the following manner

$$\begin{aligned}(\exists \mathbf{w}^*) \quad (\forall (j,j') \in I^+) \quad (\mathbf{w}^*)^T\mathbf{r}_{jj'} \geq 1 \\ (\forall (j,j') \in I^-) \quad (\mathbf{w}^*)^T\mathbf{r}_{jj'} \leq -1\end{aligned} \tag{16}$$

Let us introduce the hyperplanes $h^+_{jj'}$ and $h^-_{jj'}$ in the parameters space

$$\begin{aligned}(\forall (j,j') \in I^+) \quad h^+_{jj'} = \{\mathbf{w} : (\mathbf{w})^T\mathbf{r}_{jj'} = 1\} \\ (\forall (j,j') \in I^-) \quad h^-_{jj'} = \{\mathbf{w} : (\mathbf{w})^T\mathbf{r}_{jj'} = -1\}\end{aligned} \tag{17}$$

Definition 5. *The parameter vector $\mathbf{w}$ is situated on the* positive side *of the hyperplane $h^+_{jj'}$ if the inequality $(\mathbf{w})^T\mathbf{r}_{jj'} > 1$ is fulfilled. Similarly, the parameter vector $\mathbf{w}$ is situated on the positive side of the hyperplane $h^-_{jj'}$ if the inequality $(\mathbf{w})^T\mathbf{r}_{jj'} \leq -1$ holds.*

The penalty functions $\varphi_{jj'}^{+}(\mathbf{w})$ (11) and $\varphi_{jj'}^{+}(\mathbf{w})$ (12) are aimed to place the parameter vector $\mathbf{w}$ on the positive side of all the hyperplanes $h_{jj'}^{+}$ and $h_{jj'}^{+}$. Such solution $\mathbf{w}$ is possible if the sets C^{+} and C^{-} (10) are linearly separable. The linear independence of the difference vectors $\mathbf{r}_{jj'} = (\mathbf{x}_{j'} - \mathbf{x}_{j})$ constitutes sufficient condition for the linear separability of the sets C^{+} and C^{-} (10).

Let us take into consideration the matrix (the *basis*) $\mathbf{B}_k[m]$ of the dimension $m \times m$ with the rows constituted by m linearly independent vectors $\mathbf{r}_{jj'}[m] = (\mathbf{x}_{j'}[m] - \mathbf{x}_{j}[m])$ $((j,j') \in I_k)$. If $m < n$, then the m-dimensional vectors $\mathbf{r}_{jj'}[m]$ constituting the basis $\mathbf{B}_k[m]$ are built from the initial feature vectors $\mathbf{r}_{jj'}$ by neglecting the same features x_i. The i-th feature x_i can be neglected in all the vectors $\mathbf{x}_j$ (1) without changing the ordering (4) of the points $y_j(\mathbf{w}^*)$ on the line $y = (\mathbf{w}^*)^T\mathbf{x}$ (3) if the i-th component of the optimal vector $\mathbf{w}^* = [w_i^*, \ldots, w_n^*]$ (14) is equal to zero ($w_i^* = 0$).

$$\{w_i^* = 0\} \Rightarrow \{\text{the } i\text{–th feature } x_i \text{ can be neglected in all vectors } \mathbf{r}_{jj'}\} \quad (18)$$

The *reduced* feature vectors $\mathbf{r}_{jj'}[m]$ are generated by neglecting $(n-m)$ features x_i in accordance with the above rule.

Each basis $\mathbf{B}_k[m]$ built of the reduced vectors $\mathbf{r}_{jj'}[m]$ defines the *vertex* $\mathbf{w}_k[m]$ in the feature subspace $S_l[m]$

$$\begin{aligned} (\forall (j,j') \in I_k^{+}) \quad & (\mathbf{r}_{jj'}[m])^T \mathbf{w}_k[m] = 1\} \\ (\forall (j,j') \in I_k^{-}) \quad & (\mathbf{r}_{jj'}[m])^T \mathbf{w}_k[m] = -1\} \end{aligned} \quad (19)$$

where I_k^{+} and I_k^{-} are the sets of the indices (j,j') of the vectors $\mathbf{r}_{jj'}[m]$ constituting the basis $\mathbf{B}_k[m]$ $(I_k^{+} \cup I_k^{-} = I_k,\ I_k^{+} \cap I_k^{-} = \emptyset)$. The above set of the equalities can be given in the matrix form

$$\mathbf{B}_k[m]\,\mathbf{w}_k[m] = \mathbf{1}' \quad (20)$$

where $\mathbf{1}'$ is the vector with n components equal to 1 or -1 (19).

The vertex $\mathbf{w}_k[m]$ is the point of the intersection of m hyperplanes $h_{jj'}^{+}$ $((j,j') \in I_k^{+})$ and $h_{jj'}^{-}$ $((j,j') \in I_k^{-})$. It can be proved that the global minimum of the criterion function $\Phi(\mathbf{w})$ (13) is in one of the vertices $\mathbf{w}_k[m]$ [3]

$$(\exists \mathbf{w}_k^*[m])\ (\forall \mathbf{w}) \quad \Phi_n(\mathbf{w}) \geq \Phi_m(\mathbf{w}_k^*[m]) \quad (21)$$

where $\Phi_m(\mathbf{w}[m])$ is the criterion function (13) determined on the reduced vectors $\mathbf{r}_{jj'}[m]$ $(m < n)$.

The basis exchange algorithms allow to find the optimal vertex $\mathbf{w}_k[m]$ in an efficient manner even in the case of large data sets C (1) [5]. The basis exchange algorithms are iterative procedures, similar to the linear programming methods. One step of such procedure includes the exchange of the selected vector $\mathbf{r}_{jj'}[m]$ in the matrix $\mathbf{B}_k[m]$ $((j,j') \in I_k^{+} \cup I_k^{-})$ by another vector $\mathbf{r}_{ll'}[m]$ which has been so far outside the basis. As it results from the equalities (19), the basis vectors $\mathbf{r}_{jj'}[m]$ constituting the basis $\mathbf{B}_k[m]$ are linearly separable

$$(\forall (j,j') \in I_k^+) \quad (\mathbf{w}_k^*[m])^T \mathbf{r}_{jj'}[m] > 0 \qquad (22)$$
$$(\forall (j,j') \in I_k^-) \quad (\mathbf{w}_k^*[m])^T \mathbf{r}_{jj'}[m] < 0$$

Lemma 2. *If the sets C^+ and C^- (10) are formed by m linearly independent vectors $\mathbf{r}_{jj'}$ then these sets are linearly separable (15).*

Proof. If m vectors $\mathbf{r}_{jj'}$ are linearly independent then there exists at least one such m-dimensional feature subspace $S_l[m]$ with the basis $\mathbf{B}_k[m]$ built from m vectors $\mathbf{r}_{jj'}[m]$. The m-dimensional vectors $\mathbf{r}_{jj'}[m]$ are obtained from the primary vectors $\mathbf{r}_{jj'}$ by neglecting the same features x_i. The vectors $\mathbf{r}_{jj'}[m]$ used in the basis $\mathbf{B}_k[m]$ fulfil the equations (19). In the result, the inequalities (22) are also fulfilled and the linear separability (Def. 4) is assured. □

It can be proved that the linear separability is preserved during the feature space $S_l[n]$ increasing.

The ranked models $y = \mathbf{w}^T\mathbf{x}$ (3) can be designed by finding out such hyperplane $H(\mathbf{w}^*)$ (7) which separates the sets C^+ and C^- (10) in the best way.

Remark 1. If the sets C^+ and C^- (10) are linearly separable with the threshold equal to zero (15) in the reduced feature subspace $S_l[m]$, then the minimal value $\Phi_m(\mathbf{w}_k^*[m])$ (19) of the criterion function $\Phi_m(\mathbf{w}[m])$ (11) is equal to zero and the ranked model $y = (\mathbf{w}_k^*[m])^T\mathbf{x}[m]$ is fully consistent (Def. 3) with the dipoles $\{\mathbf{x}_j, \mathbf{x}_j\}$ orientations (4).

The minimal value $\Phi_m(\mathbf{w}_k^*[m])$ (21) of the criterion function $\Phi_m(\mathbf{w}[m])$ (13) is greater than zero if there does not exist such a line $y(w) = \mathbf{w}^T\mathbf{x}$ (3) which fulfils all the inequalities (6). The value $\Phi_m(\mathbf{w}_k^*[m])$ (21) can be used in the measure of the *linear consistency* in the ranked relations (4) and (5) between the feature vectors $\mathbf{x}_j[m]$ of the m-dimensional feature subspace $S_l[m]$

$$\Gamma_l = 1 - \Phi_m(\mathbf{w}_k^*[m]) \qquad (23)$$

The vector $\mathbf{w}_k^*[m]$ constituting the minimum (19) defines the ranked model (3)

$$y = (\mathbf{w}_k^*[m])^T\mathbf{x} \qquad (24)$$

The above model can be used in the *induction* of the ranked relation between the vectors $\mathbf{x}_j[m]$ and $\mathbf{x}_k[m]$ of the feature subspace $S_l[m]$

$$(\mathbf{w}_k^*[m])^T\mathbf{x}_j < (\mathbf{w}_k^*[m])^T\mathbf{x}_k \Rightarrow \mathbf{x}_j \prec \mathbf{x}_k \qquad (25)$$

5 Concluding remarks

The concept of ranked linear transformations (3) of the feature space X on the line is examined in the paper. Such lines reflect, to a possible extent, the relations "$\prec$ " (4) between the feature vectors $\mathbf{x}_j$ in the selected pairs $\{\mathbf{x}_j, \mathbf{x}_j\}$ ($(j, j') \in I^+$) or $(j, j') \in I^-$). It has been shown that the ranked linear transformations (3) are linked to the concept of the linear separability of some data sets with the threshold equal to zero (15).

Designing ranked linear transformations (6) can be based on the minimization of the convex and piecewise linear (CPL) criterion function $\Phi(\mathbf{w})$ (13). The basis exchange algorithms, similar to the linear programming, allow to find the minimum (21) of this function [5].

The proposed procedure of the ranked transformations designing allows for sequencing the feature vectors $\mathbf{x}_j$ in a variety of manners, depending on the choice of the ranked model $y = \mathbf{w}^T\mathbf{x}$ (3). The ranked models (24) could be defined on the basis of the selected dipoles sets I^+ (12) and I^- (13). The ranked model (24) can be used in the induction of the ranked order (25) between the feature vectors $\mathbf{x}_j$ and $\mathbf{x}_k$. Such model could be verified on the new basis of the dipoles from the testing sets.

The ranked linear transformations could have many applications. One of the most interesting applications could be sequencing of genomic data or phylogenetic classification. We apply a similar approach in designing tools for medical diagnosis support in the system *Hepar* [4].

Acknowledgements

This work was partially supported by the W/II/1/2005 and SPUB-M (COST 282) grants from the Bialystok University of Technology and by the 16/St/2005 grant from the Institute of Biocybernetics and Biomedical Engineering PAS.

References

1. Duda OR and Hart PE, Stork DG (2000) Pattern Classification, J. Wiley, New York
2. Fukunaga K (1990) Statistical Pattern Recognition, Academic Press, Inc., San Diego
3. Bobrowski L (1996) Piecewise-Linear Classifiers, Formal Neurons and separability of the Learning Sets. In: Proceedings of ICPR'96, pp. 224–228, (13th International Conference on Pattern Recognition, August 25-29, 1996, Vienna, Austria)
4. Bobrowski L, Wasyluk H (2001) Diagnosis supporting rules of the Hepar system, pp. 1309–1313 In: MEDINFO 2001, Petel VL, Rogers R, Haux R (eds), IOS Press, Amsterdam
5. Bobrowski L and Niemiro W (1984) A method of synthesis of linear discriminant function in the case of nonseparabilty. Pattern Recognition 17:205–210
6. Bobrowski L, Topczewska M (2003) Tuning of diagnosis support rules through visualizing data transformations, pp. 15–23. In: Medical Data Analysis, Perner P et al. (eds), Springer-Verlag, Berlin

Time Series Patterns Recognition with Genetic Algorithms

Marcin Borkowski

Warsaw University of Technology
Faculty of Mathematics and Information Science
Plac Politechniki 1, 00-661 Warsaw, Poland
marcinbo@mini.pw.edu.pl

Summary. The aim of this paper is to present applicable, working pattern recognition system, which can find and classify all useful dependencies between data entries in time series. The idea of predictor and its level of certainty are introduced in the work. Genetic algorithm has been deployed to prepare and govern a set of independent predictors. Practical part of solution consists of data fitting and prediction. Architecture of the system offers possibility to interleave learning phase with use. Analyzed data may be non continuous, and incomplete. In uncertain cases the system presents either more than one answer to processed data or no response at all. Early testing results, including prediction and fitting of simple time series with missing data amount ranging from 10 to 50 percent, are presented at the end of this paper.

1 Introduction

This work is oriented towards analysis of time series of unknown structure where classic and modern but highly specialized methods are inapplicable due to large percentage of missing data or non continuous nature of input. During the process only one input time series is presented to the system. The teaching data is obtained from the same input sequence that is being predicted or interpolated in later stage of the processing.

The practical uses of the idea include prediction and interpolation (data fitting) of time series based on patterns discovered in first stage. The first approach answers the question what is the next value in a sequence. The second supplies the values that are missing from sequence presented to the system.

Unlike most typical solution, in which single mechanism is taught to predict values from the whole time series, proposed method manages a set of homogeneous predicting tools. In the first case one predicting method (predictor) should know all data relations in order to predict requested values.

In the later case each predictor knows only limited number of relations (usually one) and can be used only in certain conditions where its knowledge is applicable.

In given prediction[1] case automatic selection among all predictors is possible due to numerical value of percentage relevancy, that is calculated along with predicted value. The idea behind this number is to reflect the relation between knowledge of predictor and currently tested part of input. The same predictor will present different relevancy when applied to different parts of data sequence. Relevancy factor is a natural outcome of applied method of learning and predictor representation.

In difficult situation a group of contradictory predictors can be activated at the same time. In such a case different final predictions may be proposed. It is expert's responsibility to analyze that condition. In test data such a situation represents inherent uncertainty of input. If expected value can not be predicted basing on given data, no eventual prediction will be made.

The process requires initial data. In case of prediction it is a time series up till the value being predicted, namely a sequence of k values from a_{-k} to a_{-1}, where a_0 is a value being predicted (2). Lower indexing represents consequentially numbered time periods for which values were observed. For data fitting it is a sequence of k numbers from a_{-k+1} to a_0 where some of entries are missing (1).

$$a_{-k+1}, a_{-k+2}, \dots, a_{-2}, a_{-1}, a_0 \ . \tag{1}$$

$$a_{-k}, a_{-k+1}, a_{-k+2} \dots, a_{-2}, a_{-1}, a_0, a_1, a_2, \dots \ . \tag{2}$$

The system may find reliable prediction for values a_i where $i > 0$ even without knowing any of a_j where $i > j \geq 0$. In practice it means that in some cases it is possible to predict more than one step ahead of input data.

Data fitting can be applied to prediction problem by supplying the series of $k+1$ entries $(a_{-k}, a_{-k+1}, \dots, a_{-1}, a_0)$ with missing a_0. Thus, only the second, more general method was considered in later work. Examples of traditional data fitting approach can be found in [1].

After one of missing values is supplied either by the prediction, data fitting or from outside environment, learning process can be continued without loosing previous knowledge. System can reenter teaching phase whenever current knowledge is insufficient. In test data it leads to better overall accuracy.

The concept of this work is to facilitate the power of Genetic Algorithm (GA) which can discover and manage a population of useful predictors. In order to achieve this functionality, niche techniques were adopted. Prediction and data fitting problem requires computational scaling. The more complex data the bigger number of independent predictors is required. Increasing number of predictions corresponds to the size of population of GA. Natural parallelism of GA is a base for good overall system scaling.

[1] For convenience terms predict and predictor also refer to data fitting.

2 Concept of Solution

The proposed solution consists of three cooperating modules. At the lowest level elementary predictors are applied to input. At the second level predictors are governed by GA algorithm. The top level module provides data fitting and is for testing purposes only. Input data for the problem is assumed to stay within $< -1, 1 >$ range.

2.1 Prediction with Patterns

Two general assumptions about elementary predictor refer to its computational simplicity and relevancy factor. Small computational complexity is important due to number of independent predictors maintained and taught by GA. Second one is critical for the final prediction maker, namely prediction program or expert.

In proposed approach a single predictor is a sequence of k pairs (value, offset). Each pair consists of the value within $< -1, 1 >$ range and a positive offset. The offset fixes positional relation between pattern values and input data. The shift of the first pair is always zero. The position of N-th pair is related to the first pair. Maximum offset distance between pairs and number k of pairs are parameters of the system. Formal notation is shown on Fig.1a. An example of pattern consisting of 3 pairs is shown on Fig.1b.

$$(a) : (v_1, 0)(v_2, o_2) \ldots (v_k, o_k) \quad \text{where}\, 0 < o_2 < \ldots < o_k \ .$$
$$(b) : (1.00, 0)(0.00, 2)(0.00, 4) \ .$$

Fig. 1. (a)Pattern notation, (b) Example

Application of the pattern Fig.1b to the sample input data is shown on Fig.2. Input values are marked on y-axis, time on x-axis. Obviously one pattern can be compared with input data on more than one position. Pattern is shifted in order to test each value (v_i, $1 \geq i \geq k$) as a prediction for unknown entry at $t = -1$. In the example, two other applications of pattern Fig.1b are possible, but in this example not useful.

Value positioned over missing entry represents prediction, the rest of values contributes to relevancy factor as (3) shows. The equation depends on position t of missing data in time line and position p of pair used as prediction in the pattern. In short, relevancy is in reverse proportion to the average percentage distance between all values in a pattern and input data entries corresponding to them.

$$\text{relevancy} = 1 - \sum_{i=1, i \neq p}^{k} \frac{\sqrt[4]{\frac{|v_i - a_{wi}|}{2}}}{k-1} \quad \text{where} \quad w_i = t - o_i + o_t \ . \tag{3}$$

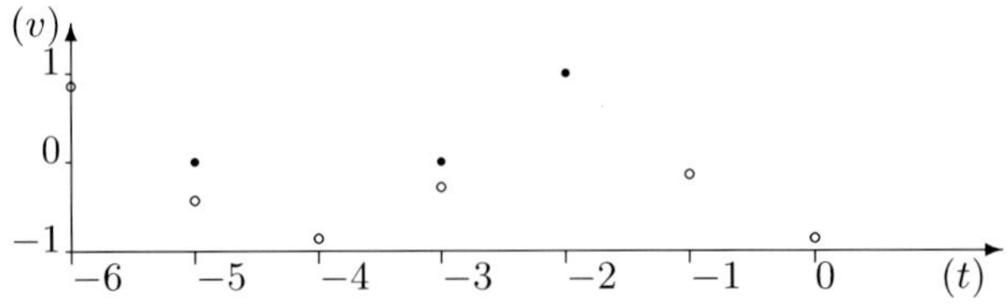

Fig. 2. Single Pattern applied to the time series

A single pattern may be applied to missing entry at all k positions. Higher level module is to decide, which of those k attempts should yield a prediction. At this point, a trivial algorithm is used to select position with the highest relevancy. During comparison, whenever non-predicting value from the pattern matches missing entry, relevancy factor is lowered proportionally to the number of such a conditions. Values outside of known input are all marked as missing. In practice, only predictions made basing on $k-1$ existing entries influence the final prediction.

Above example does not include pattern scaling. In scaled mode each pattern is changed according to linear transformation (4) before it can be applied to input data. Averages of pattern values and corresponding input entries are used to calculate shift $b = b_v - b_a$, estimated standard deviations of b_v and b_a yield the scaling factor a.

$$v' = av + b \quad \text{where } a \in < 0.005, 500 > \quad . \tag{4}$$

Scaling extends usability (pattern can be applied to the wider range of input data combinations) and generalization of single pattern, but at a cost of increased number of misleading predictions. The pattern that in any situation yields good relevancy but poor prediction is misleading for the system. Misleading examples are shown on Fig.3. Small and double circles represent the input data, black dots represent unscaled pattern, bold circles - the same pattern after scaling. It is a feature of GA learning process to eliminate misleading patterns.

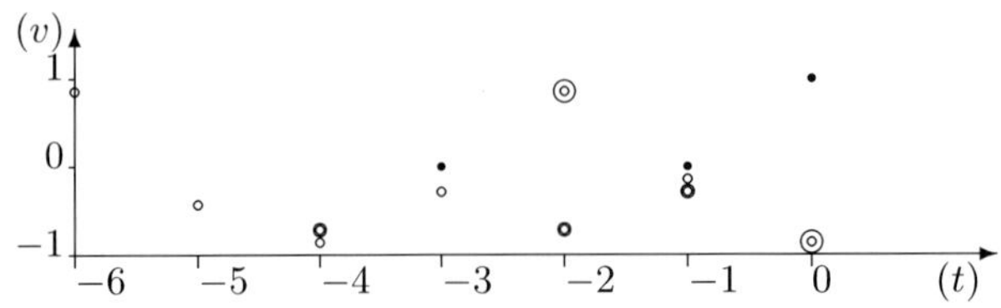

Fig. 3. Examples of misleading predictions

2.2 Predictions Managing

In order to make the system efficient and reliable a large number of independent predictors must be discovered and managed. In this work strength of GA is applied to this task. Typical GA techniques are described in many books including [2, 3]. To block domination of genetic population by one well fitted predictor, niche techniques were adopted. Other changes include modification of selection and scaling of mutation probability.

The GA population consists of fixed number of structures denoting prediction pattern in binary format. For each pair (see 2.1) both offset and value are encoded. Offset's maximum shift is limited by the number of bits devoted for it in structure. To extend the width of pattern and to avoid contradicting pairs (the same offset but different value), offset o_i is encoded relatively to the previous o_{i-1} and $o_i > 0$ (except the first one o_1, that is always zero). The value is limited to range $< -1, 1 >$ and its precision depends on number of bits it is encoded on.

Valuation of single predictor z is based on the input data and prediction algorithm (see 2.1). For each available entry (a_t where $t \in \gamma$) in the time series (data already known), prediction is made. If relevancy factor is high and prediction is correct, combined relevancy and prediction value are considered as local fitness value $F_l(z, a_t)$. If relevancy is high and prediction is false, misleading cases counter increases. Average of local fitness values, that are above automatically adjusted threshold, is considered as a base for fitness value. To obtain the final value, base is scaled down in linear proportion to misleading cases number and divided by crowd factor described below.

Applied niche technique was derived from work of Goldberg and Richardson [4] and Miller and Shaw [5]. The idea is based on the assumption that a group of similar predictors occupy the same area in the space of all predictions. If area is well fitted, most of GA population will eventually migrate to that location. To prevent this, crowd factor is introduced. The more predictors inhabit the same area, the higher crowd factor is. In practice it is equal to the number of predictors in the area. Fit values for subpopulation in given area are divided by crowd factor in proportion to their base fitness. Namely, the best predictor in the area is divided by 1, the second by 2 and the least is divided by full value of crowd factor.

Areas and crowd factor depend on similarity of predictors. Two predictors are similar if they yield the same predictions for the same input data. Average value of differences of local fitness (5) proved to be a good estimator of similarity. Predictors with similarity value above certain limit are considered to occupy the same area.

$$s(z_1, z_2) = 1 - \frac{\sum_{t \in \gamma} |F_l(z_1, a_t) - F_l(z_2, a_t)|}{2|\gamma|} \ . \tag{5}$$

Selection of a new population in GA is specific to the problem. Process of selection is divided into two phases. In the first one population is simply ordered according to the fitness values. The second part is combined with genetic operators. Algorithm uses modified mutation and one point crossover. A copy of current population is made to protect original structures from being overwritten by the operators. Crossover parents are taken from the copy, their offspring are placed in the population at the lowest available position. First two offspring replace the first and the second weakest structures in the original population. The second crossover replaces third and forth etc. Mutation acts in the same fashion, except it puts mutated structures after the results of crossover. Apart from classic bit mutation probability p_m, the operator requires structure mutation probability p_{sm}. First, the set of structures for mutation is chosen basing on p_{sm}, then bits for mutation are chosen for each structure in the set basing on p_m. Given the probabilities of structure mutation and crossover, expected number of changed structures can be calculated. The rest of population constitute a special space devoted to store and protect the best structures.

2.3 Pattern Application

Practical use of predictors found by the system requires a method to analyze different predictions obtained from various predictors. Although this stage of calculation may require a human expert knowledge, computer algorithm also produces satisfactory results.

Before any choice is made, prediction data is reorganized. Raw system output for prediction of unknown value a_i consists of series of triples (p_j, r_j, f_j) denoting prediction, relevancy and fitness obtained from application of j-th predictor at a_i. The number of triples is equal to GA population size. Preprocessing removes items with relevancy and fitness below given threshold and groups remaining basing on prediction similarity. Each group is described by quadruple $(p_j^{\mathrm{a}}, r_j^{\mathrm{a}}, f_j^{\mathrm{a}}, c_j)$ denoting average prediction, relevancy, fitness and number of predictions included in the group.

At this step either human expert may choose the best option or simple algorithmic choice can be made. In automatic choice a group with the highest value (6) is selected as final prediction. P is a constant used to reflect importance of c_j on final decision.

$$r_j^{\mathrm{a}} - (1 - f_j^{\mathrm{a}}) + c_j P \ . \tag{6}$$

3 Results

3.1 Testing Environment

Simple application prepared for testing purposes works in cycles. During each cycle one prediction (see 2.3) is made. Before that, GA must be initialized with fixed number of generations.

Initially there are many unknown entries in the time series, for some of them the system can produce prediction. Only the best prediction according to (6) is added to the time series. After each prediction GA reenters the learning mode. No previous knowledge is lost. GA continues with the same population, only the input data is more complete. Inter-prediction GA learning is limited to fixed number of generations usually much lower than during initialization. Predictions are made until any vacant entries remain, or process is manually interrupted. No human assistance is required for system to work. All test results were obtained automatically.

3.2 Sample Tests

A very restrained nature of current predictors limits the complexity of tests. More practical results are expected as soon as neural network predictors are applied. Only a very small part of test results is presented below. All test were conducted with the following parameters:

1. Number of bits denoting values - 12
2. Maximum distance between offsets - 4
3. Number of pairs in pattern - $< 3, 4 >$
4. Crossover probability - 35%
5. Structure mutation probability - 50%
6. Population size - 70
7. Initial GA generations - $< 300, 1000 >$
8. Inter-prediction GA generations - $< 100, 300 >$
9. Relevancy threshold - $< 70, 90 >$ %

In Table 1 results of data fitting with (a) unscaled and (b) scaled patterns are shown. Input data for the system was (a)10 and (b)5 times repeated sequence of $0; 1; 0; -1$ with (a)52 and (b)25 percent of missing data. The most useful prediction patterns found during the process are printed on Fig.4. In the scaled version system was forced to find the most universal pattern based on periodic nature of the input. Solution from unscaled test leads to misleading predictions in scaled environment.

Third test presented in Table 1c and on Fig.4c was performed on input of 50 consecutive values of sinus function taken proportionally from $< 0, 2\pi >$ range. Twelve percent of values were removed at random positions. Patterns were scaled during the process. All results in Table 1 are presented in the same sequence as they were discovered by the system. Brief analysis of Fig.4c shows that most significant patterns represent ascending and descending slope of sinus function, both in two variants of gradient. Results are significantly better than simple average of values surrounding the predicted value.

Table 1. Test results

Position	Expected	Predicted	Difference
(a)unscaled			
34	0.000	0.000	0.00%
27	-1.000	-1.000	0.00%
24	0.000	0.000	0.00%
25	1.000	0.999	0.06%
2	0.000	0.004	0.20%
6	0.000	0.001	0.05%
7	-1.000	-1.000	0.00%
21	1.000	0.998	0.08%
8	0.000	-0.002	0.10%
35	-1.000	-1.000	0.00%
37	1.000	1.000	0.00%
38	0.000	-0.002	0.08%
10	0.000	0.000	0.00%
13	1.000	0.981	0.93%
12	0.000	0.000	0.00%
20	0.000	0.000	0.00%
19	-1.000	-1.000	0.00%
16	0.000	-0.001	0.03%
18	0.000	0.000	0.00%
14	0.000	0.008	0.39%
0	0.000	0.000	0.00%
(b)scaled			
16	0.000	0.000	0.00%
9	1.000	1.000	0.00%
0	0.000	0.000	0.00%
2	0.000	0.000	0.00%
1	1.000	1.001	0.05%
(c)sinus			
1	-0.249	-0.256	0.36%
20	-0.482	-0.470	0.61%
44	0.588	0.575	0.63%
38	0.982	0.974	0.40%
46	0.368	0.373	0.25%
29	0.588	0.598	0.50%

(a): (1.000,0) (0.001,1) (-1.000,2) (1.000,4)
(-1.000,0) (0.001,1) (1.000,2) (-1.000,4) .
(b): (-0.531,0) (-0.531,4) (-0.531,8) .
(c): (-0.750,0) (-0.551,1) (-0.354,2)
(0.864,0) (0.725,1) (0.589,2)
(0.898,0) (-0.075,1) (-0.995,2)
(-1.000,0) (0.031,1) (0.996,2) .

Fig. 4. The best predictors in (a)unscaled, (b)scaled, (c)sinus tests

References

1. Press W H, Tenkolsky S A, Vetterling W T, Flannery B P (2002) Numerical Recipes in C. Cambridge University Press Chapter 15
2. Goldberg D E (1989) Genetic Algorithms in Search, Optimization, and Machine Learning. Addison-Wesley Publishing Company, Inc
3. Michalewicz Z (1996) Genetic Algorithms+Data Structures = Evolution Programs. Springer-Verlag Berlin Heidelberg
4. Goldberg D E, Richardson J (1987) Genetic Algorithms with Sharing for Multimodal Function Optimization. Proceedings of the Second International Conference on Genetic Algorithms :41–49
5. Miller B L, Shaw M J (1995) Genetic Algorithms with Dynamic Niche Sharing for Multimodal Function Optimization. IlliGAL Report No. 95010

Selection of Fuzzy-Valued Loss Function in Two Stage Binary Classifier

Robert Burduk[1]

Chair of Systems and Computer Networks, Wroclaw University of Technology, Wybrzeze Wyspianskiego 27, 50-370 Wroclaw, Poland,
e-mail: robert.burduk@pwr.wroc.pl

Summary. In this paper, a model to deal with Bayesian hierarchical classifier, in which consequences of decision are fuzzy-valued, is introduced. The model is based on the notion of fuzzy random variable and also on a subjective ranking method for fuzzy number defined by Campos and González. The Bayesian hierarchical classifier is based on a decision-tree scheme for given tree skeleton and features to be used in each inertial nodes. The influence of selection of fuzzy-valued loss function on classification result is given. Finally, an example illustrating this case of Bayesian analysis is considered.

1 Introduction

Some studies concerning the decision problem with fuzzy factors are presented in [1, 3, 4, 5, 6, 7, 8, 9, 12]. These papers describe only single-stage decision problems. This paper deals with a recognition problem, which - assuming a probabilistic model with a full information - values of a loss function are assumed to be fuzzy numbers. We will also consider the so-called Bayesian hierarchical classifier [10]. In this recognition problem the decision as to the membership of an object in a given class is not a single activity but is the result of a more or less complex decision process. This model has been stated so that, on one hand, the existence of fuzzy loss function representing the preference pattern of the decision maker can be established, on the other hand, a priori probabilities of classes and class-conditional probability density functions can be given.

In the further part, after the introduction of necessary symbols, we will calculate the set of optimal recognition algorithms for internal nodes, minimizing the global quality indicator. As a criterion of optimality we will assume the mean value of the fuzzy loss function (risk), which values depends on the stage of the decision tree, on which an error has occurred. The presented algorithm will be illustrated by a numerical example in which subjective method for ranking fuzzy numbers has been applied. This example present influence of selection of fuzzy-valued loss function on separation point of decision regions in parameter λ function.

2 Background

In the paper [11] the Bayesian hierarchical classifier is presented. The synthesis of multistage classifier is a complex problem. It involves specification of the following components:

- the decision logic, i.e. hierarchical ordering of classes,
- feature used at each stage of decision,
- the decision rules (strategy) for performing the classification.

The present paper is devoted only to the last problem. This means that we shall deal only with the presentation of decision algorithms, assuming that both the tree skeleton and feature used at each non-terminal node are specified.

The procedure in the Bayesian hierarchical classifier consist of the following sequences of activities. At the first stage, there are measured some specific features x_0. They are chosen from among all accessible features x, which describe the pattern that will be classified. These data constitute a basis for making a decision i_1. This decision, being the result of recognition at the first stage, defines a certain subset in the set of all classes and simultaneously indicates features x_{i_1} (from among x) which should be measured in order to make a decision at the next stage. Now at the second stage, features x_{i_1} are measured, which together with i_1 are a basis for making the next decision x_2. This decision – like i_1 – indicates features x_{i_2} necessary to make the next decision (at the third stage) and – again as at the previous stage – defines a certain subset of classes, not in the set of all classes, however, but in the subset indicated by the decision i_1, and so one. The whole procedure ends at the last N-th stage, where the decision made i_N indicates a single class, which is the final result of multistage recognition. Thus multistage recognition means a successive narrowing of the set of potential classes from stage to stage, down to a single class, simultaneously indicating at every stage features which should be measured to make the next decision in more precise manner.

3 Decision problem statement

Let us consider a pattern recognition problem, in which the number of classes is equal to M. Let us assume that classes were organized in a $(N+1)$ horizontal decision tree. Let us number all nodes of the constructed decision-tree with consecutive numbers of $0, 1, 2, \ldots$, reserving 0 for the root-node and let us assign numbers of classes from the $\mathcal{M} = \{1, 2, \ldots, M\}$ set to terminal nodes so that each one of them is labelled with the number of the class which is connected with that node. This allows the introduction of the following notation:

- $\mathcal{M}(n)$ – the set of numbers of nodes, which distance from the root is n, $n = 0, 1, 2, \dots, N$. In particular $\mathcal{M}(0) = \{0\}$, $\mathcal{M}(N) = \mathcal{M}$,
- $\overline{\mathcal{M}} = \bigcup_{n=0}^{N-1} \mathcal{M}(n)$ – the set of interior node numbers (non terminal),
- $\mathcal{M}_i \subseteq \mathcal{M}(N)$ – the set of class labels attainable from the i-th node ($i \in \overline{\mathcal{M}}$),
- $\mathcal{M}^i$ – the set of numbers of immediate descendant nodes ($i \in \overline{\mathcal{M}}$),
- m_i – number of direct predecessor of the i-th node ($i \neq 0$).

We will continue to adopt the probabilistic model of the recognition problem, i.e. we will assume that the class label of the pattern being recognized $j_N \in \mathcal{M}(N)$ and its observed features x are realizations of a couple of random variables $\boldsymbol{J}_N$ and $\boldsymbol{X}$. Complete probabilistic information denotes the knowledge of a priori probabilities of classes:

$$p(j_N) = P(J_N = j_N), \quad j_N \in \mathcal{M}(N) \tag{1}$$

and class-conditional probability density functions:

$$f_{j_N}(x) = f(x/j_N), \quad x \in X, \quad j_N \in \mathcal{M}(N) \ . \tag{2}$$

Let

$$x_i \in X_i \subseteq R^{d_i}, \quad d_i \leq d, \quad i \in \mathcal{M} \tag{3}$$

denote vector of features used at the i-th node, which have been selected from the vector x.

Our target now is to calculate the so-called multistage recognition strategy $\pi_N = \{\Psi_i\}_{i \in \overline{\mathcal{M}}}$, that is the set of recognition algorithms in the form:

$$\Psi_i : X_i \rightarrow \mathcal{M}^i, \quad i \in \overline{\mathcal{M}} \ . \tag{4}$$

Formula (4) is a decision rule (recognition algorithm) used at the i-th node, which maps observation subspace to the set of immediate descendant nodes of the i-th node. Equivalently, decision rule (4) partitions observation subspace X_i into disjoint decision regions $D^k_{x_i}$, $k \in \mathcal{M}^i$, such that observation x_i is allocated to the node k if $k_i \in D^k_{x_i}$, namely:

$$D^k_{x_i} = \{x_i \in X_i : \Psi_i(x_i) = k\}, \ k \in \mathcal{M}^i, \quad i \in \overline{\mathcal{M}}. \tag{5}$$

Let $\widetilde{L}(i_N, j_N)$ denote the fuzzy loss incurred when objects of the class j_N is assigned to the class i_N ($i_N, J_N \in \mathcal{M}(N)$). Our aim is to minimizes the mean risk, that is the mean value of the fuzzy loss function [6, 7]:

$$\widetilde{R}^*(\pi^*_N) = \min_{\Psi_{i_n}, \dots, \Psi_{i_{N-1}}} \widetilde{R}(\pi_N) = \min_{\Psi_{i_n}, \dots, \Psi_{i_{N-1}}} \widetilde{E}[L(I_N, J_N)]. \tag{6}$$

The π_N^* strategy we will call the globally optimal N-stage recognition strategy.

$\widetilde{R}^*(\pi_N^*)$ is a fuzzy-valued function on $\mathbb{R}$, taking on values on the set $\mathcal{F}_c(\mathbb{R})$ (set of normalizes convex fuzzy sets on $\mathbb{R}$ whose level sets and the closed convex hull of the support are compact).

For ranking fuzzy mean values we have selected the subjective method stated by Campos and González [2]. This method is based on the λ-average valued of a fuzzy number, which is defined for $\widetilde{A} \in \mathcal{F}_c(\mathbb{R})$ as the real number given by

$$V_S^\lambda(\widetilde{A}) = \int_0^1 [\lambda a_{\alpha 2} + (1-\lambda) a_{\alpha 1}] dS(\alpha) \tag{7}$$

where $\widetilde{A}_\alpha = [a_{\alpha 1}, a_{\alpha 2}], \lambda \in [0,1]$ and S being an additive measure on $Y \subset [0,1]$.

The parameter λ is a subjective degree of optimism-pessimism. In a loss context, $\lambda = 0$ reflect the highest optimism and $\lambda = 1$ reflect the highest pessimism. Then, the λ-ranking method to compare fuzzy numbers in $\mathcal{F}_c(\mathbb{R})$ is given by

$$\widetilde{A} \succeq \widetilde{B} \Leftrightarrow V_S^\lambda(\widetilde{A}) \geq V_S^\lambda(\widetilde{B}). \tag{8}$$

In the next chapter we will calculate globally optimal strategy for fuzzy loss function depends on the stage of the decision tree, on which an error has occurred.

4 The recognition algorithm

Let $\widetilde{L}(i_N, j_N)$ denote the fuzzy loss incurred when objects of the class j_N is assigned to the class i_N $(i_N, J_N \in \mathcal{M}(N))$. Let us assume now

$$\widetilde{L}(i_N, j_N) = \widetilde{L}^S_{d(w)} \tag{9}$$

where w is the first common predecessor of the nodes i_N and j_N. So defined fuzzy loss function means that the loss depends on the stage at which misclassification is made. Stage-dependent fuzzy loss function for the two-stage binary classifier are presented in Fig. 1.

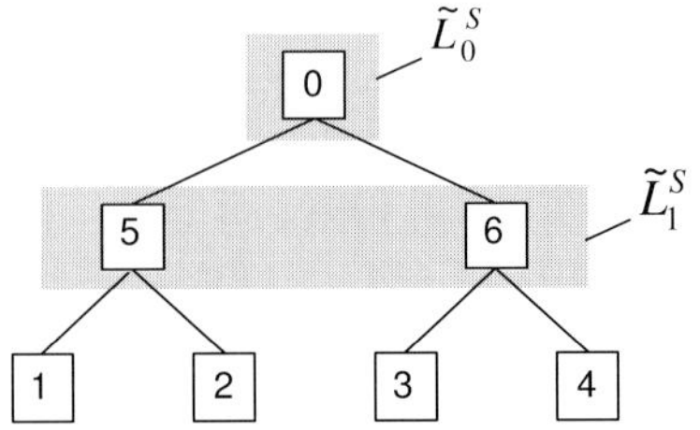

Fig. 1. Interpretation of dependent on the stage of the decision tree loss function

Putting (9) into (6) we obtain the optimal (Bayes) strategy, which decision rules are as follows:

$$\begin{array}{l} \Psi^*_{i_n}(x_{i_n}) = i_{n+1}, \\ (\widetilde{L}_{d(i_n)} - \widetilde{L}_{d(i_{n+1})})p(i_{n+1})f_{i_{n+1}}(x_{i_n}) + \\ + \sum\limits_{j_{n+2}\in\mathcal{M}^{i_{n+1}}} \left[(\widetilde{L}_{d(i_{n+1})} - \widetilde{L}_{d(j_{n+2})})q^*(j_{n+2}/i_{n+1}, j_{n+2})f_{j_{n+2}}(x_{i_n}) + \right. \\ \left. + \cdots + \widetilde{L}_{d(j_{N-1})} \sum\limits_{j_N\in\mathcal{M}^{j_{N-1}}} [q^*(j_N/i_{n+1}, j_N)p(j_N)f_{j_N}(x_{i_n})] \ldots\right] = \\ = \max\limits_{k\in\mathcal{M}^{i_n}} \left\{(\widetilde{L}_{d(i_n)} - \widetilde{L}_{d(k)}p(k)f_k(x_{i_n}) + \right. \\ + \sum\limits_{j_{n+2}\in\mathcal{M}^k} \left[(\widetilde{L}_{d(k)} - \widetilde{L}_{d(j_{n+2})})q^*(j_{n+2}/k, j_{n+2})p(j_{n+2})f_{j_{n+2}}(x_{i_n}) + \right. \\ \left.\left. + \cdots + \widetilde{L}_{d(j_{N-1})} \sum\limits_{j_N\in\mathcal{M}^{j_{N-1}}} [q^*(j_N/k, j_N)p(j_N)f_{j_N}(x_{i_n})] \cdots\right]\right\} \end{array}, \quad (10)$$

for $i_n \in \mathcal{M}(n)$ $n = 0, 1, 2, \ldots, N-1$, where $q^*(j_N/i_{n+1}, j_N)$ denotes the probability of accurate classification of the object of the class j_N in further stages using π^*_N strategy rules on condition that on the n-th stage the i_{n+1} decision has been made.

5 Illustrative example

Let us consider the two-stage binary classifier of Fig. Four classes have identical a priori probabilities which are equal 0.25. Class-conditional probability density functions of features $\mathbf{X}_0$, $\mathbf{X}_5$ and $\mathbf{X}_6$ are normally distributed in each class with the following class-conditional probability density functions:

$$f_1(x_0) = f_2(x_0) = N(1, 1), \quad f_3(x_0) = f_4(x_0) = N(3, 1),$$

$$f_1(x_5) = N(1, 1), \quad f_2(x_5) = N(3.5, 1),$$

$$f_3(x_6) = N(0, 1), \quad f_4(x_6) = N(1.8, 1).$$

The fuzzy loss function dependent on the stage of the decision tree are the following:

$$\begin{array}{lll} \text{case } a & \widetilde{L}^S_0 = (1, 2, 3)_T, & \widetilde{L}^S_1 = (0, 1, 2)_T, \\ \text{case } b & \widetilde{L}^S_0 = (1.5, 2, 2.5)_T, & \widetilde{L}^S_1 = (0.5, 1, 1.5)_T, \\ \text{case } c & \widetilde{L}^S_0 = (1.5, 2, 2.5)_T, & \widetilde{L}^S_1 = (0, 0.5, 1)_T, \\ \text{case } d & \widetilde{L}^S_0 = (1, 1.5, 2)_T, & \widetilde{L}^S_1 = (0, 0.5, 1)_T, \end{array}$$

and are described by a triangular fuzzy numbers.

Due to the peculiar distribution of $\mathbf{X}_5$ and $\mathbf{X}_6$, the decision rules Ψ^*_5 and Ψ^*_6, at the second stage of classification, are obvious. Their decision regions are following $D^{*(1)}_{x_5} \subset (-\infty, 2.25)$, $D^{*(2)}_{x_5} \subset (2.25, \infty)$, $D^{*(3)}_{x_6} \subset (-\infty, 0.9)$ and

$D^{*(4)}_{x_6} \subset (0.9, \infty)$. Let us now determine the rule at the first stage of classification. From (10) we obtain:

$$\Psi_0^*(x_0) = \begin{cases} 5 \text{ if} & \begin{aligned} &(\widetilde{L}_0^S - \widetilde{L}_1^S)p(5)f_5(x_0) + \\ &+ \widetilde{L}_1^S\big(g^*(1/5,1)p(1)f_1(x_0) + g^*(2/5,2)p(2)f_2(x_0)\big) > \\ &> (\widetilde{L}_0^S - \widetilde{L}_1^S)p(6)f_6(x_0) + \\ &+ \widetilde{L}_1^S\big(q^*(3/6,3)p(3)f_3(x_0) + g^*(4/6,4)p(4)f_6(x_0)\big), \end{aligned} \\ 6 \text{ otherwise} & \end{cases} .$$

Using the data from the example and subjective λ-method for comparison fuzzy risk, we finally obtain results presented in Fig. 2, where value of point x_0' (separation point for decision regions at the first stage) in function of parameter λ is presented.

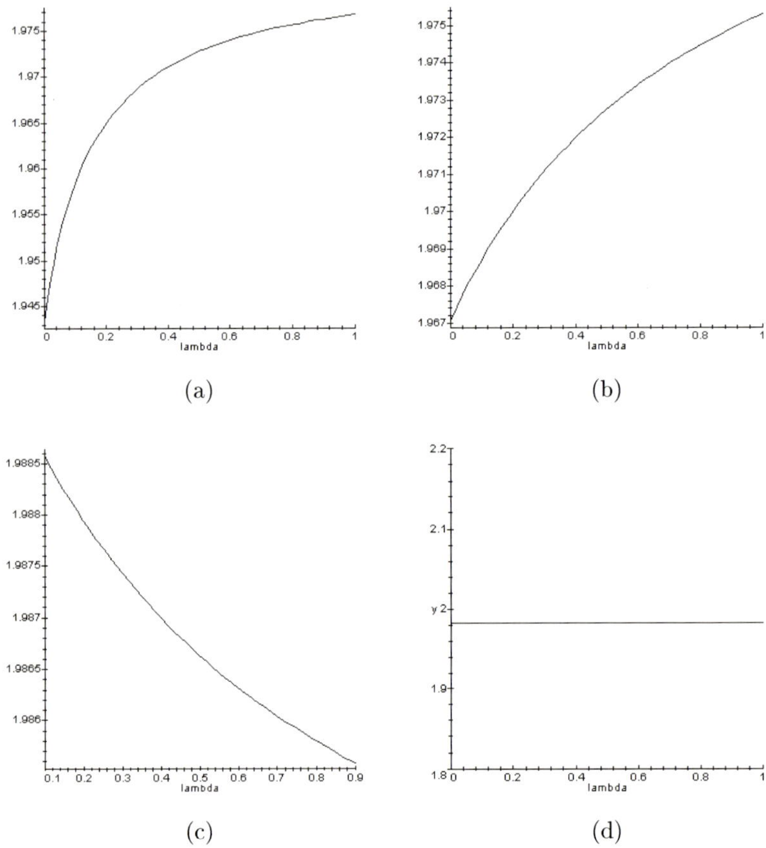

Fig. 2. Separation point x_0^* for decision regions - figure caption right to case

Interesting is the influence of selection of fuzzy-valued loss function on separation point x_0^* in parameter λ function. Let us denote $\widetilde{L}_0^S = (a_1, a_2, a_3)_T$ and $\widetilde{L}_1^S = (b_1, b_2, b_3)_T$. Then we have ascending function for $(a_1 + a_2)(b_2 + b_3) - (a_2 + a_3)(b_1 + b_2) < 0$ (case a and b), descending function for $(a_1 + a_2)(b_2 + b_3) - (a_2 + a_3)(b_1 + b_2) > 0$ (case c) and constant function for $(a_1 + a_2)(b_2 + b_3) - (a_2 + a_3)(b_1 + b_2) = 0$ (case d). In the latter case separation point x_0^* is independent from choice of parameter λ.

6 Conclusion

In the paper we have presented Bayes multistage classifier with a full probabilistic information. In this recognition model a priori probabilities of classes and class-conditional probability density functions are given. Additionally, consequences of wrong decision are fuzzy-valued and are represented by triangular fuzzy numbers. For ranking fuzzy numbers we use subjective method with parameter λ. In illustrative example we presented influence of selection of fuzzy-valued loss function on separation point of decision regions. In this example class-conditional probability density functions are normally distributed. In future work we can consider another distributions of random variable or fuzzy loss function forms.

References

1. Baas S, Kwakernaak H (1997) Rating and Ranking of Multi-Aspect Alternatives Using Fuzzy Sets. Automatica 13:47–58
2. Campos L, González A (1989) A Subjective Approach for Ranking Fuzzy Numbers. Fuzzy Sets and Systems 29:145–153
3. Casals M, Gil M, Gil P (1986) On the Use of Zadeh's Probabilistic Definition for Testing Statistical Hypotheses from Fuzzy Information. Fuzzy Sets and Systems 20:175–190
4. Corral N, Gil M (1984) The Minimum Inaccuracy Fuzzy Estimation: an Extension of the Maximum Likelihood Principle. Stochastica 8:63–81
5. Gertner G, Zhu H (1996) Bayesian Estimation in Forest Surveys when Samples or Priori Information Are Fuzzy. Fuzzy Sets and Systems 77:277–290
6. Gil M, López-Díaz M (1996) Fundamentals and Bayesian Analyses of Decision Problems with Fuzzy-Valued Utilities. International Journal of Approximate Reasoning 15:203–224
7. Gil M, López-Díaz M (1996) A Model for Bayesian Decision Problems Involving Fuzzy-Valued Consequences. Proc. 6th Int. Conf. on Information Processing and Management of Uncertainty in Knowledge Based Systems, Granada
8. Hung W (2001) Bootstrap Method for Some Estimators Based on Fuzzy Data. Fuzzy Sets and Systems 119:337–341
9. Jain R (1976) Decision-Making in the Presence of Fuzzy Variables. IEEE Trans. Systems Man and Cybernetics 6:698–703

10. Kurzyńsk M (1983) Decision Rules for a Hierarchical Classifier. Pattern Recognition Letters 1:305–310
11. Kurzyńsk M (1988) On the Multistage Bayes Classifier. Pattern Recognition 21:355–365
12. Viertl R (1996) Statistical Methods for Non-Precise Data. CRC Press, Boca Raton

A New Rejection Strategy for Convolutional Neural Network by Adaptive Topology

Hubert Cecotti and Abdel Belaïd

LORIA/CNRS, Campus Scientifique
BP 239, 54506 Vandoeuvre-les-Nancy cedex France
hubert.cecotti@loria.fr, abdel.belaid@loria.fr

1 Introduction

In pattern recognition, the recognition quality is directly connected to the rejection quality; the main goal of the system being to decrease the error rate. The rejection of input is usually based on confidence measure. If this measure exceeds a threshold, the inputs result is accepted otherwise it is rejected. The classifier properties are used to perform the rejection but also to try to correct rejected inputs [1,2,4,10]. In the classifier used, the features extraction takes place directly into the classification process and is supposed to give the best features set [7,8]. Although the choice of this kind of system looks more interesting, it needs a lot of patterns for learning the features and the features learnt are also limited to the topology of the system. The quality of a classifier depends of learning that depends of the number and of the quality of the different samples. The automatic learning of features inside a neural network is possible thanks to its topology. It is the topology that determines the geometric and morphological transformation allowing the extraction of features [9]. It is used as a guide in the classification process but it keeps fixed: the observation of the features extracted are fixed and well localized in the input space. After the learning of the classifier the network can be reused on transformed objects that were not directly learnt or improve the recognition rate of objects by performing a change of the topology. Without knowledge of the features extracted by the network, we can control the way these features are used by changing the topology. The new topology created between two layers works as a geometrical transformation: the input are not fixed in the time anymore, they can move in order to improve the recognition. The change of topology is driven by a self-organized map placed between two layers of the neural network that finds a better organization of the links between the layers. The goal of this system is to keep its own potential and to reinforce the system for rejected patterns. In the first part, we will describe the global system and its features. In a second part, the convolutional neural network used and the criteria of reject will be presented. In a third part, the change of the topology and its influence will be explained. Finally we will prove the interest of the method proposed for character recognition.

2 Global System

The system heart is composed of a convolutional neural network, a classifier based on a multi-layer perceptron with a specific topology that allows automatic features extraction [7,8]. The relations inter-layers represented in the system correspond to transformation of in the input. In the case of images, it corresponds to morphological and geometrical transformation; each of these transformations is translated by the links between two layers, where each layer can represent one or several representation of an image. The classification process is performed before the reject criteria that are a key in the system as it says if the pattern is well recognized or if the pattern needs to be treated again but with a change of the topology. In the case where the recognition rate is not enough, when the images are not well identified, the topology is changed for optimizing the way the image can be analyzed. Instead of normalizing the image directly before the classification process, the topology of the system is modified by supposed geometrical transformations. The advantage of such normalization is that it can be performed in each step in during the classification. The normalization inside the system can be done after several transformations common for every pattern for example. After having normalized the classifier, the classification process is performed again till the reject criteria are sure: the pattern is well recognized or the pattern is rejected, it's too much disturbed. Thus the classification process is not fixed in one step and doesn't need a new learning. If the classifier is learnt with straight clean images and then the system is tested on disturbed images, it cannot give a high confidence result except he learnt invariant features of the disturbed images proposed. With this classifier learnt with classical patterns, with reject criteria based on the probability of confidence rate, we can estimate if there was an error and what could be the different possible solution. The classification model proposed is based on a supervised part for the learning step and a non-supervised part during the testing step that transforms the topology of the neural network.

3 Convolutional Neural Network

3.1 Description of the topology

The goal of the topology based on convolutional neural network is to classify the image given in input by analyzing it through different "filter" learnt during the learning step. Each layer can be composed of several maps, each map corresponds to a transformation of the image: these transformation can put on features like edges, lines... The neural network used is composed of 5 layers:

- The first one corresponds to the input image. It is centered and reduced to the size to 29*29.
- The next two layers corresponds to the information extraction, performed by convolutions. They are described as follows:
  - The second layer is composed of 10 maps, each one corresponds to a specific image transformation by convolution and sub-sampling reducing its size. For each map, all the neurons have the same input link number and share their weights.
  - The third layer is composed of 50 maps; each map represents the convolution of a combination of 5 maps in the previous layer. Here also, in each map, the link weights are shared by all its neurons. A pivot neuron is considered in the map, which synthesizes the receptive field. In this case an input link is described by a special link to the value of the weight.
- The last two layers are fully connected.
- The last one corresponds to the output.

4 Rejection Method

The convolutional neural network outputs are not normalized and cannot be interpreted as probabilities. The Softmax function is used to normalize the output. The probabilities that the input is label as class i (in our case the ith neuron of the output layer $v(i)$), with N_{class} classes:

$$P(i) = \frac{e^{v(i)}}{\sum_{j=0}^{j=N_{class}} e^{v(j)}}$$

The reject criteria used use two types of reject: a relative and an absolute reject. The threshold for absolute reject are computed function to the confidence rate of each sample well recognized; the threshold for relative reject are computed function to the distance between the first and the second best choice in the case the first choice is the good one. The relative reject allows putting on a confusion problem whereas the absolute reject puts on the strength of the recognition. The test of a sample can have a result which emerges relatively compared to the others but which in the absolute does not represent a sufficient result: this phenomenon translates an error on all the classes. Let MS_{rel} the thresholds matrix which contains the mean distance between the two best first values where the first values corresponds to the output of the good class. In the case where no samples are present for computing a value of the matrix then its value is initialized to 1 meaning the maximum distance.

$$MS_{rel}(i;j) = \frac{1}{N_{image}} * \sum_{k=0}^{k=N_{image}} P_k(i) - P_k(j)$$

where $P_k(i) > P_k(j)$; $P_k(i)$ and $P_k(j)$ are respectively the first and the second best value; N_{image} is the number of images. Let $MS_{abs}(i,j)$ the thresholds vector which contains the mean confidence result for all well recognized samples.

5 Self-organized map adapted to neurons classification

5.1 Description of the map

The self-organized map used corresponds to the size of one layer of the neural network for classification. One of its layers corresponds to several transformation of the image given as input. Each neuron of this map corresponds to one neuron in the layer of the classifier. This map, of dimension 2 to match the image, conserves a fixed topology to keep the neighborhood relationship to keep the context between neurons: as the neurons represent pixels after several transformations. The idea here is to find a start with a standard configuration of the map and to find a new configuration, which correspond to a transformation. Let S the map of size $d * d$.

5.2 Features

A neuron of the map is defined by $n(i,j)$ where $0 \leq i < d$ and $0 \leq j < d$. It is defined by three components:

- x : X-coordinate real on the map, $0 \leq x < d$
- y: Y-coordinate real on the map, $0 \leq y < d$
- p : value of the neuron, it corresponds to an activation probability, $p \in 0;1$

As each neuron of the map corresponds to one neuron of the classifier, it becomes easy to establish for one neuron its new entry.

5.3 Initialization

The map is initialized relatively to a model, hypothesis of the class to identify. Each neuron of the map is defined by $n(i,j)$:

- $x = i$
- $y = j$
- p : Activation probability, gray density in (i,j)

With the difference of the traditional self-organizing maps like those of Kohonen [5], the semantics of the composition of the vectors of each neuron does not allow a traditional use of the algorithms of automatic classification. Our goal is to classify neurons compared to their position and a certain value corresponding to a density of gray or degrees of activation of a characteristic. A new distance would have to be chosen but this solution skews the results because it directly induces a relation between position and color. In our case, the neurons corresponding to a black color or the activation of a characteristic are the only neurons selected: they move and consequently involve their neighborhood with them at the time of their changes of position. Although the map physically contains as many neurons as on its corresponding layer in the network of classification, only the neurons exceeding a certain threshold are active.

5.4 Learning

The training database E for the map is composed of all the neurons that have a value near 1 from the layer to treat. A neuron $n(i,j)$ of value c belongs to the learning database if and only if $c = 1 \pm \epsilon$. The use of only this kind of neurons allows reducing the learning database relatively to the use of every neurons of the map. The first step consists to choose a sample: a neuron in the layer, and the nearest neuron in the map corresponding.

- A sample $s \in S$from the layer is selected randomly and we select the neuron $n(i_{win}, j_{win})$, the nearest of s. Let $(i,j) \in \{0..d-1\}^2$ the coordinate of one neuron in the map. Let D_{min} the distance between one neuron and a sample.

$$D_{min} = min(\sqrt{(n(i,j)_x - s_x)_2 + (n(i,j)_y - s_y)_2})$$

$$(i_{min}, j_{min}) = Argmin(\sqrt{(n(i,j)_x - s_x)_2 + (n(i,j)_y - s_y)_2})$$

- We select randomly a neuron (i_{win}, j_{win}) on the map such as $n(i_{win}, j_{win}) = 1 \pm \epsilon$ and we select the nearest sample $s \in S$ of $n(i_{win}, j_{win})$.

$$D_{min} = min(\sqrt{(n(i_{win}, j_{win})_x - s_x)_2 + (n(i_{win}, j_{win})_y - s_y)_2})$$

Let D_x and D_y such
$D_x = n(i_{win}, j_{win})_x - s_x$ and $D_y = n(i_{win}, j_{win}) - s_y$
Let define $D = D_{min} = \sqrt{x^2 + y^2}$ and $\theta = Arctan(y, x)$

The neighborhood is represented by a sector of a disk of center $(s_x, s_y$ with a radius of $D + \Delta r$. The weight of every neurons belonging to the sector must be changed. A neurone $n(i,j)$ of S belongs to the set of neurons V to change their weights if and only if

$$(D + \Delta r) \leq \sqrt{(n(i,j)_x - s_x)^2 + (n(i,j)_y - s_y)^2}$$

and

$$|\theta - Arctan(n(i,j)_y - s_y, n(i,j)_x - s_x)| \leq \Delta\theta$$

where $\Delta\theta$ corresponds to the maximum angle and Δr corresponds to the neighborhood around the neuron $n(i_{win}, j_{win}$.

The last step consists to update weights components corresponding to the position, for every neurons $n \in V$

$$n(i,j)_x \leftarrow n(i,j)_x + \alpha * (|\Delta x/D|) * (s_x - n(i,j)_x)$$

$$n(i,j)_y \leftarrow n(i,j)_y + \alpha * (|\Delta y/D|) * (s_y - n(i,j)_y)$$

where α corresponds to a factor depending the current iteration. The training step α is defined by a typical "mexican hat" function. The cone representing the set of neurons to update can be broken up into two parts. The first part corresponds to the inner core where the updated neurons are changed in a positive way, they are attracted by the sample. The second part is the periphery of the cone and represents the neurons to update negatively.

6 Adaptive Topology for rejected patterns

Each position of a neuron in the self-organized map corresponds to its real position in a layer of the main neural network for a standard links configuration: when there is no transformation. The components (x, y) relative to the position, for each neuron, represent their new localization. It means for the neurone in the upper layer, the components (x, y), for one link, represent the new observation. Consider the map of the same meaning of layer $c-1$ of the main neural network and each neuron n of the layer c taking as input neurons from the layer $c-1$. In the standard configuration, let the neuron n_c^k which has as input a neuron n_{c-1}^l, l representing the lth neurons of $c-1$ corresponding to (i,j) on the layer. Then after the use of self-organized map, the neuron n_c^k will have has input a neuron $n_{c-1}^{l'}$ where l' corresponds to the position $(n(i,j)_x, n(i,j)_y)$.

7 Experimental results

The system has been tested on the ModifiedNIST database: MNIST. This very well known database is composed of images of handwritten digits. These images have a size of $28 * 28$ and are in gray level on one byte [7,8]. The objective is to show the relevance of the system with respect to the rejection by the increase in the rate of confidence by the adaptive standardization of the classifier. As the database is composed of handwritten digits, many transformation and disturbance are present and it is not possible to learn them all. By only using the part of classification: the convolutional neural network, on the test database of MNIST containing 10000 images, we obtain 98.73% of recognition without rejection. The reject criterion that offers the best result in quality of reject is the one that combines relative and absolute threshold as shown in table 2.

If the classifier output checks the first or the second rule of rejection then the pattern is rejected. It is this solution that makes it possible to reject the maximum of errors while rejecting the minimum of good results. On the 1094 rejected samples, only 120 samples are rejected in an effective way. It is on this basis of 1094 samples, endemic part of the database, that our contribution will be shown. One will raise the ambiguity, which involves their rejection by the adaptation of the topology of the classifier while trying not to create new errors. As the majority of the rejected elements were initially well recognized without the rejection, the problem of the choice of the first assumption of model is fixed: it is the model corresponding to the class having the best rate of confidence. The first table shows that for a low error rate, hypotheses have to be chosen in the 3 best results.

Table 1. Recognition rate on MNIST

Top 1	Top 2	Top 3	Top 4	Top 5
98.73	99.72	99.91	99.97	99.99

Table 2. Recognition rate on MNIST with different reject criteria

Test abs	Test rel	Error	Reject	Reject criteria
87.47	99.97	0.03	12.50	abs
85.38	99.93	0.06	14.56	rel
83.86	99.97	0.02	16.12	abs and rel
88.99	99.92	0.07	10.94	abs or rel

Table 3. Recognition rate on MNIST function to the number of iteration

Test abs	Test rel	Error	Reject	Iteration
92.13	99.55	0.42	7.45	3
91.49	99.77	0.21	8.30	2
88.99	99.92	0.07	10.94	1

The results obtained show the interest of the method for reusing rejected patterns. The number of patterns recognized is bigger and many patterns initially rejected have been corrected and caught again. The number of images well recognized has increased of about 3% in relation to the default rejection case.

8 Conclusion

We presented a model combining supervised learning for the classification stage and a specific not-supervised model for the test stage. It allows changing the topology of the supervised part for improving the reject quality. Within the framework of the recognition of disturbed forms, our strategy was based on the adaptability of the classifier instead of, for example, making it learn more samples by artificially creating all the possible deformations, our approach consisted in defining the type of problematical transformation, by changing the localities observed by the principal classifier. The system makes it possible indeed to refine the results till it makes possible to raise ambiguities for certain confusions but there remains still depend on the effectiveness of the initial rejection and the assumptions concerning the choice of the model of class. On the level of the prospects, if the type of deformation is perfectly known: cut, rotation, problem of shift, it would be possible to introduce new properties inside the self-organizing map to speed up its convergence.

References

1. Akiyama K (1996) A new reject decision method for statistical pattern recognition. Proc. of IWFHR-5
2. Chow C.K (1970) On optimum recognition error and reject tradeoff. IEEE. Trans. Information Theory, vol 16, pp. 41-46
3. Dur Trier O, and Al (1996) Feature Extraction Methods for Character Recognition - A Survey. Pattern Recognition, vol. 4, nř 29, pp. 641-662
4. Gorksi N (1997) Optimizing error-reject trade off in recognition systems. 4th International Conference Document Analysis and Recogntion, pp. 556-559
5. Kohonen T (1982) Self-organized formation of topologically correct feature maps. Biological Cybernetics, 43:-69
6. Koerich L.A (2004) Rejection Strategies for Handwritten Word Recognition. to appear in 9th International Workshop on Frontiers in Handwriting Recognition (IWFHR-9)
7. LeCun Y, and Al (1998) Gradient-Based Learning Applied to Document Recognition. Proceedings of the IEEE, vol. 86, nř11, pp. 2278-2324
8. Simard P, and Al (2003) Best Practice for Convolutional Neural Networks Applied to Visual Document Analysis. International Conference on Document Analysis and Recognition, ICDAR, IEEE Computer Society, pp. 958-962
9. Teow L, and Al (2002) Robust vision-based features and classification schemes for off-line handwritten digit recognition. Pattern Recognition 35 (11): 2355-2364
10. Zimmermann M, and Al (2004) Rejection Strategies for Offline Handwritten Sentence Recognition. 17th International Conference on Pattern Recognition (ICPR) Volume II, p. 550-553

Fast PCA and LDA for JPEG Images

Weilong Chen[1], Meng Joo Er[1], and Shiqian Wu[2]

[1] School of Electrical and Electronic Engineering
Nanyang Technological University
50 Nanyang Avenue, Singapore 639798
wlchen@pmail.ntu.edu.sg, emjer@ntu.edu.sg
[2] Institute for Infocomm Research
21 Heng Mui Keng Terrace, Singapore 119613
shiqian@i2r.a-star.edu.sg

Summary. In this paper, we prove that the Principal Component Analysis (PCA) and the Linear Discriminant Analysis (LDA) can be directly implemented in the DCT (Discrete Cosine Transform) domain and the results are exactly the same as the one obtained from the spatial domain. In some applications, compressed images are desirable to reduce the storage requirement. For images compressed using the DCT, e.g., in JPEG or MPEG standard, the PCA and LDA can be directly implemented in the DCT domain such that the inverse DCT transform can be skipped and the dimensionality of the original data can be initially reduced to cut down computational cost.

1 Introduction

Principal Component Analysis (PCA) and Linear Discriminant Analysis (LDA) have been widely used for dimensionality reduction or feature extraction in pattern recognition [1]. They are also successfully employed in face recognition research area, i.e., Eigenfaces and Fisherfaces [3, 4]. Dimensionality reduction is essential for extracting effective features and reducing computational complexity in classification stage. However, dimensionality reduction using the PCA and LDA is also computationally expensive when the original dimensionality is high and the number of training samples is large. When the training database becomes larger, the training time and memory requirement will significantly increase. Moreover, systems based on the PCA or LDA should be retrained when new classes are added to obtain optimal projection results. Therefore, reduction in computational complexity is highly desirable. More recently, Discrete Cosine Transform (DCT) has been employed in face recognition for dimensionality reduction [5, 6]. The advantage of the DCT is that it is data independent (i.e., the basis images are only dependent on one image instead of on the whole set of training images) and it can be implemented using a fast algorithm. Nevertheless, only limited low-frequency coefficients are used as features if the DCT is employed for direct dimensionality reduction. Although the DCT is asymptotically equivalent to the Karhunen-Loeve Transform (KLT) (PCA) for Markov-1 signals with a correlation coefficient that is close to one [2], the PCA is more efficient for dimensionality reduction since it is the optimal dimensionality reduction method for Gaussian distributed data. Moreover, in practical applications, the training data are not necessarily Markov-1 signals with a correlation coefficient close to one. In this paper, we show that the PCA and LDA can be directly implemented in the DCT domain. In some applications, the DCT can be first employed to remove redundant information and the PCA or LDA can be subsequently implemented in the DCT domain such that the computational complexity can be significantly reduced. Furthermore, the PCA or LDA may be directly implemented in the block DCT domain which is widely used in the JPEG and MPEG standard. There are several advantages if the PCA or LDA can be implemented in the block DCT domain: 1) If the database is stored as compressed JPEG images to reduce storage requirement, DCT coefficients of JPEG images can be directly used such that the inverse DCT can be skipped to cut down computational cost; 2) In JPEG images, some redundant information has been removed by quantization such that the dimensionality of feature vectors can be initially reduced by removing coefficients with less information.

2 PCA in DCT Domain

2.1 PCA on Orthonormally Transformed Data

In the following section, we show that the PCA projection result is invariant for the orthonormally transformed data. Combining with the subsequent proof that the DCT is an orthogonal transformation, the PCA can be directly implemented in the DCT domain.

Theorem 1. *The PCA subspace projection result remains the same if the original data are transformed using an orthonormal transformation.*

Proof. Let $X = [x_1, x_2, \ldots, x_n]^T$ denote an n-dimensional random vector. The mean of the random vector X is denoted by $\bar{X} = E[X]$ and the covariance matrix of the random vector X is defined as $C_X = E[(X - \bar{X})(X - \bar{X})^T]$. The PCA projection matrix A can be obtained by eigen-analysis of the covariance matrix C_X, i.e.

$$C_X a_i = \lambda_i a_i, \; i = 1, 2, \ldots, m \tag{1}$$

where a_i is the ith largest eigenvector of C_X, $m < n$ and $A = [a_1, a_2, \cdots, a_m] \subset \Re^{n \times m}$. We have the following projection result:

$$Y = A^T (X - \bar{X}) \tag{2}$$

Let us assume that the original data are transformed by using an $n \times n$ orthogonal matrix Q, i.e., $Z = Q^T X$. The mean of the transformed random vector Z is given by

$$\bar{Z} = E[Q^T X] = Q^T E[X] = Q^T \bar{X} \tag{3}$$

We may obtain the covariance matrix C_Z of the transformed random vector Z as follows:

$$\begin{aligned} C_Z &= E[(Q^T X - Q^T \bar{X})(Q^T X - Q^T \bar{X})^T] \\ &= Q^T E[(X - \bar{X})(X - \bar{X})^T] Q = Q^T C_X Q \end{aligned} \tag{4}$$

Similarly, the PCA projection matrix $\tilde{A} = [\tilde{a}_1, \tilde{a}_2, \cdots, \tilde{a}_m] \subset \Re^{n \times m}$ for the orthonormally transformed data can be obtained by eigen-analysis of the covariance matrix C_Z and we have

$$C_Z \tilde{a}_i = \lambda_i \tilde{a}_i, \; i = 1, 2, \ldots, m \tag{5}$$

Substituting (4) into (5), we have

$$Q^T C_X Q \tilde{a}_i = \lambda_i \tilde{a}_i \tag{6}$$

Since Q is an orthogonal matrix, i.e., $Q^T = Q^{-1}$, equation (6) can be rewritten as

$$C_X Q \tilde{a}_i = \lambda_i Q \tilde{a}_i \tag{7}$$

Equation (7) shows that the covariance matrices C_X and C_Z have exactly same eigenvalues and the relation of their eigenvectors is $a_i = Q \tilde{a}_i$. Therefore, the PCA projection matrices A and $\tilde{A}$ also satisfy $\tilde{A} = Q^T A$. The projection result of Z is as follows:

$$\tilde{Y} = \tilde{A}^T (Z - \bar{Z}) = A^T Q (Z - \bar{Z}) = A^T (X - \bar{X}) = Y \tag{8}$$

From (8), we can conclude that for the PCA subspace projection, orthonormal transformation of the original data will not change the projection result.

2.2 Discrete Cosine Transform

The DCT can be written in a vector form as $y = C^T x$. The elements of the matrix C are

$$c_{n,k} = \frac{1}{\sqrt{N}}, \; k = 0, \; n = 0, 1, \ldots, N - 1$$

$$c_{n,k} = \sqrt{\frac{2}{N}} \cos\left[\frac{\pi(2n+1)k}{2N}\right], \; k = 1, 2, \ldots, N - 1, \; n = 0, 1, \ldots, N - 1$$

It has been proven that the transform matrix C is an orthogonal matrix [2], i.e., $C^T = C^{-1}$. The $M \times N$ 2D DCT and inverse DCT (IDCT) may be computed using the separable 1D row and column transformations, i.e.

$$Y = C_M^T X C_N \quad X = C_M Y C_N^T \tag{9}$$

For the PCA applied for images, a 2D image matrix is normally converted into a vector by concatenating the columns or rows of the matrix. Accordingly, we may convert the 2D DCT and IDCT shown in (9) into the following vector form:

$$\hat{Y} = G^T \hat{X} \quad \hat{X} = H^T \hat{Y} \tag{10}$$

where $\hat{Y}$ and $\hat{X}$ are MN-dimensional vectors, and G and H are $MN \times MN$ transformation matrices. If $\hat{X}$ is obtained by concatenating the columns of X, i.e., $\hat{X} = [x_{0,0} \ \cdots \ x_{M-1,N-1}]^T$ and $\hat{Y} = [y_{0,0} \ \cdots \ y_{M-1,N-1}]^T$, it can be derived from (9) that the transformation matrix is as follows:

$$G = \begin{bmatrix} c_{0,0}c_{0,0} & \cdots & c_{0,M-1}c_{0,0} & \cdots\cdots & c_{0,M-1}c_{0,N-1} \\ \vdots & & \vdots & & \vdots \\ c_{M-1,0}c_{0,0} & \cdots & c_{M-1,M-1}c_{0,0} & \cdots\cdots & c_{M-1,M-1}c_{0,N-1} \\ \vdots & & \vdots & & \vdots \\ \vdots & & \vdots & & \vdots \\ c_{M-1,0}c_{N-1,0} & \cdots & c_{M-1,M-1}c_{N-1,0} & \cdots\cdots & c_{M-1,M-1}c_{N-1,N-1} \end{bmatrix} \tag{11}$$

It can also be derived from (9) that the inverse transformation matrix H is exactly the transpose of G, i.e., $G = H^T$. Combining with (10), we have

$$G^{-1} = G^T \tag{12}$$

Equation (12) shows that the transformation matrix G is orthogonal. Furthermore, the sequence of elements in $\hat{Y}$ only corresponds to the sequence of column vectors in G. When the sequence of elements in $\hat{Y}$ is changed, matrix G remains orthogonal.

From Theorem 1 and (12), we can conclude that the projection result of PCA in the 2D DCT domain is exactly the same as the one in the 2D spatial domain.

2.3 Block DCT

In JPEG standard, images are first divided into rectangular blocks (8×8) [7]. The DCT is applied independently on each block. Given a $pn \times qn$ image which is divided into $n \times n$ subimages, the DCT is applied independently on each subimage. The transformation may be represented as follows:

$$[Y_{11} \cdots Y_{p1} \cdots\cdots Y_{pq}]^T = \mathrm{diag}[\mathrm{G}_{11} \cdots \mathrm{G}_{\mathrm{p1}} \cdots\cdots \mathrm{G}_{\mathrm{pq}}] \cdot [\mathrm{X}_{11} \cdots \mathrm{X}_{\mathrm{p1}} \cdots\cdots \mathrm{X}_{\mathrm{pq}}]^{\mathrm{T}} \tag{13}$$

Since the square matrices G_{ij} are all orthogonal, i.e., $G_{ij}^T = G_{ij}^{-1}$, we have

$$\mathrm{diag}[\mathrm{G}_{11} \cdots \mathrm{G}_{\mathrm{p1}} \cdots\cdots \mathrm{G}_{\mathrm{pq}}]^{\mathrm{T}} = \mathrm{diag}[\mathrm{G}_{11} \cdots \mathrm{G}_{\mathrm{p1}} \cdots\cdots \mathrm{G}_{\mathrm{pq}}]^{-1} \tag{14}$$

As shown in (14), the block diagonal transformation matrix is also an orthogonal matrix. Therefore, JPEG DCT coefficients can be directly used for the PCA approach. The inverse DCT can be skipped.

3 LDA in DCT Domain

Similarly, it can be shown that the LDA can be directly implemented in the DCT domain.

Theorem 2. *The LDA subspace projection result remains the same if the original data are transformed using an orthonormal transformation.*

Proof. For the LDA, the optimal subspace projection is determined as follows [1]:

$$E_{opt} = \arg\max_{E} \frac{|E^T S_b E|}{|E^T S_w E|} = [e_1, e_2, \ldots, e_m] \tag{15}$$

where S_b and S_w are between-class scatter matrix and within-class scatter matrix respectively, $[e_1, e_2, \ldots, e_m]$ is the set of generalized eigenvectors corresponding to the m $(m \leq c-1)$ largest generalized eigenvalues $\lambda_i,\ i = 1, 2, \ldots, m$, i.e.

$$S_b e_i = \lambda_i S_w e_i,\ i = 1, 2, \ldots, m \tag{16}$$

So, the projection result is

$$Y = E_{opt}^T X \tag{17}$$

Let us assume that the original data is transformed using an orthogonal matrix Q, i.e., $Z = Q^T X$. The between-class scatter matrix for Z is obtained as follows:

$$\begin{aligned} \tilde{S}_b &= \sum_{i=1}^{c} n_i (Q^T \bar{X}_i - Q^T \bar{X})(Q^T \bar{X}_i - Q^T \bar{X})^T \\ &= Q^T \sum_{i=1}^{c} n_i (\bar{X}_i - \bar{X})(\bar{X}_i - \bar{X})^T Q = Q^T S_b Q \end{aligned} \tag{18}$$

Similarly, the within-class scatter matrix for Z is given by

$$\tilde{S}_w = Q^T S_w Q \tag{19}$$

and

$$\tilde{E}_{opt} = \arg\max_{\tilde{E}} \frac{|\tilde{E}^T \tilde{S}_b \tilde{E}|}{|\tilde{E}^T \tilde{S}_w \tilde{E}|} = [\tilde{e}_1, \tilde{e}_2, \ldots, \tilde{e}_m] \tag{20}$$

We can obtain $\tilde{E}_{opt}$ by solving the generalized eigenvalues and eigenvectors of $\tilde{S}_b$ and $\tilde{S}_w$ from the following equation:

$$\tilde{S}_b \tilde{e}_i = \lambda_i \tilde{S}_w \tilde{e}_i,\ i = 1, 2, \ldots, m \tag{21}$$

Substituting (18) and (19) into (21), we have

$$Q^T S_b Q \tilde{e}_i = \lambda_i (Q^T S_w Q) \tilde{e}_i$$

$$S_b Q\tilde{e}_i = \lambda_i S_w Q\tilde{e}_i \tag{22}$$

Comparing (16) and (22), we can see that the generalized eigenvalues are the same and the generalized eigenvectors satisfy $e_i = Q\tilde{e}_i$, $i = 1, 2, \ldots, m$. Since Q is orthogonal, the optimal projection matrix satisfies $\tilde{E}_{opt} = Q^T E_{opt}$. Therefore, the projection result of Z is as follows:

$$\tilde{Y} = \tilde{E}^T Z = (Q^T E_{opt})^T Z = E_{opt}^T QZ = E_{opt}^T X = Y \tag{23}$$

From (23), we can conclude that for the LDA subspace projection, orthonormal transformation of original data will not change the projection result.

As shown in Section 2, the DCT or the block DCT transformation matrices are orthogonal. As a consequence, the result of LDA in the DCT domain is the same as the one in the spacial domain.

For LDA in face recognition, there will be singular problems in scatter matrices due to the high-dimensional data and small number of training samples. As a consequence, PCA + LDA (Fisherface) scheme is commonly used [4]. Since the DCT will not change the PCA projection result, the two-stage approach can also be implemented in the DCT domain and the projection result remains the same. If DCT coefficients can be initially discarded to avoid singular problems as well as keep important high-frequency features, the LDA can be directly implemented in the truncated DCT domain. This approach is more computationally efficient than the PCA + LDA. More recently, some variants of LDA have been proposed to solve the small sample size problem [8]. In fact, some variants of LDA can also be directly implemented in the DCT or block DCT domain.

4 Experimental Results

Although the proposed approach is not restricted in face recognition applications, in this section, we present some experimental results based on the FERET face database [9]. Cumulative match curves based on the Euclidean distance are employed to evaluate the performance. In the following experiments for the PCA, 50 principal components are used.

The coefficient feature vector can be formed by independently scanning each coefficient block in a zigzag manner and then concatenating them. Fortunately, in JPEG standard, DCT coefficients are also coded in a zigzag order pattern because DCT coefficients with great magnitude are mainly located at the upper-left corner of each block. It should be noted that in the experiments, the JPEG DCT coefficients are quantized using a typical quantization matrix. [7].

By using the PCA or LDA in the DCT domain, not only storage requirement is largely reduced, but also the computational complexity is simplified. Some redundant information may be first removed by truncating the DCT coefficients so that the dimensionality of the coefficient vectors can be reduced.

For the PCA and LDA, the computational complexity, which mainly depends on the dimensionality of the original data, can be largely reduced.

In each block, coefficients with less information can be initially discarded. The number of discarded coefficients can be determined according to the quantization procedure since a lot of coefficients become zero after quantization. However, we cannot simply discard the zero coefficients because their positions vary with different blocks. In our approach, coefficients are kept following the zigzag scan and the number of kept coefficients is the same for each block.

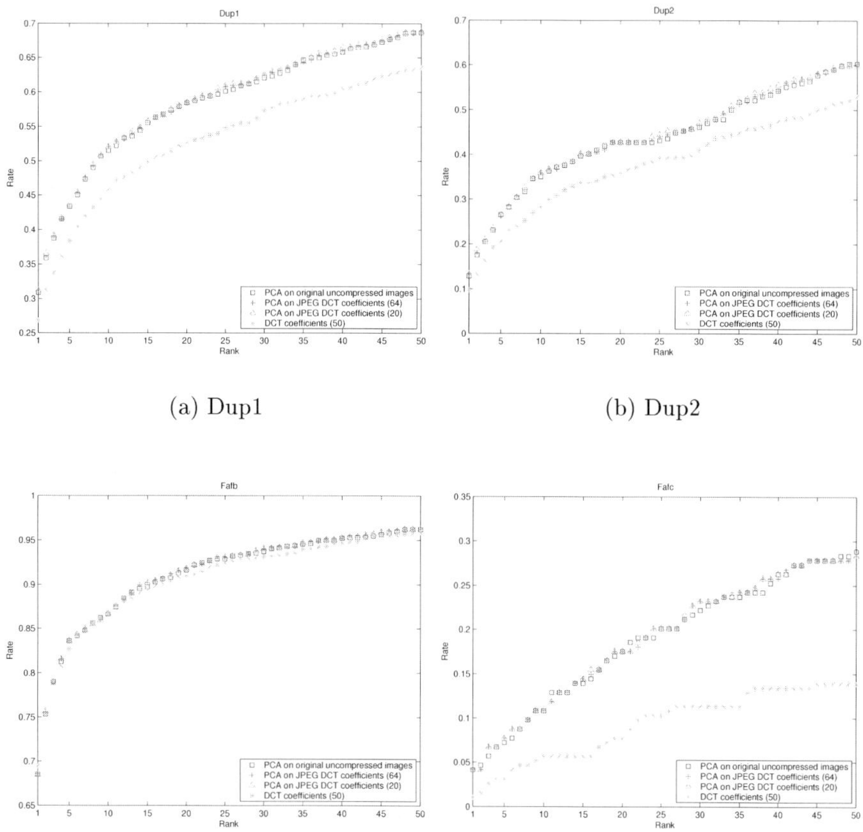

(a) Dup1 (b) Dup2

(c) Fafb (d) Fafc

Fig. 1. Performance of PCA in JPEG DCT domain with 20 coefficients and 64 coefficients of each block.

Fig. 1 shows the performance of the PCA on original uncompressed images and on quantized JPEG DCT coefficient vectors (64 complete coefficients and 20 coefficients of each block are respectively illustrated). Their performances are quite similar. The performance of the DCT for direct dimensionality reduction (i.e., feature vectors are constructed directly from low-frequency DCT coefficients) is also shown in Fig. 1. It is evident that the PCA implemented on the JPEG DCT coefficients outperforms the DCT for direct dimensionality reduction. For the experimental results shown in Fig. 1, the dimension of feature vectors are all 50. Therefore, dimensionality can be first reduced by appropriately discarding DCT coefficients with less information to save the computational cost for the PCA and LDA.

5 Conclusion

This paper presents an efficient way of reducing storage requirements and computational complexity of the PCA and LDA for pattern recognition applications. It has been proven that the PCA and LDA can be directly implemented in the DCT domain or block DCT domain. Therefore, JPEG images can be used to reduce storage requirements and by initially discarding an appropriate number of DCT coefficients with less information, the computational complexity of the PCA and LDA can be significantly reduced. Our approach can be adopted for real-time pattern recognition systems.

References

1. Fukunaga K (1990) Introduction to statistical pattern recognition, second edition. Academic Press
2. Rao K R, Yip P (1990) Discrete Cosine Transform: Algorithms, Advantages, Applications. Academic Press, Boston
3. Turk M A, Pentland A P, (1991) Eigen Faces for Recognition. Journal of cognitive Neuroscience. 3:71–86
4. Belhumeur P N, Hespanha J P, Kriegman D J, (1997) Eigenfaces Versus Fisherfaces: Recognition Using Class Specific Linear Projection. IEEE Trans. Pattern Analysis and Machine Intelligence. 19(7):711–720
5. Pan Z, Adams R, Bolouri H, (2000) Image Redundancy Reduction for Neural Network Classification Using Discrete Cosine Transforms. Proc. of The IEEEŋINNSŋENNS International Joint Conf. on Neural Networks, Como, Italy. 3:149–154
6. Hafed Z M, Levine M D, (2001) Face Recognition Using the Discrete Cosine Transform. International Journal of Computer Vision. 43(3):167–188
7. Gonzalez R C, Woods R E, (1992) Digital Image Processing. AddisonŋWesley
8. Yu H, Yang J, (2001) A Direct LDA Algorithm for Highŋdimensional Data ŋ with Application to Face Recognition. Pattern Recognition. 34, 2067–2070.
9. Phillips P J, Wechsler H, Huang J, Rauss P, (1998) The FERET Database and Evaluation Procedure for Face Recognition Algorithms. Image and Vision Computing J. 16(5):295ŋ-306

A Hybrid ε-Insensitive Learning of Fuzzy Systems

Tomasz Czogala[1] and Jacek M. Leski[12]

[1]Institute of Medical Technology and Equipment.
Roosevelt St. 118A, 41-800 Zabrze. Poland. tomaszcz@itam.zabrze.pl
[2]Institute of Electronics. Silesian University of Technology.
Akademicka 16, 44-100 Gliwice. Poland. jleski@polsl.pl

Summary. Initially, it is shown that ε-insensitive learning of a fuzzy system may be presented as a combination of both an ε-insensitive gradient method and solving a system of linear inequalities. Then, a hybrid learning algorithm is introduced. Example is given of using this algorithm for design a fuzzy model of real ECG data. Simulation results show an improvement in the generalization ability of a fuzzy system learned by the new method with respect to the traditional and other ε-insensitive learning methods.

1 Introduction

The support vector machine (SVM) is historically the first method based on the main result of the statistical learning theory, i.e. the generalization ability of a machine depends on both empirical risk on a training set and complexity of this machine. SVM is successfully applied to a wide variety of classification and regression problems.

In the last few years, there has been increasing interest in fuzzy systems which incorporate tools well-known from the statistical learning theory. An ε-insensitive approach to the learning of neuro-fuzzy systems has been introduced in [7] and extended in [6]. This approach is based on the premise that human learning, as well as thinking, is tolerant to imprecision. Instead of the usually used quadratic loss function, an ε-insensitive loss function is used which assumed a zero loss for the difference between a model and the reality less than some pre-set value, noted ε. If this difference is greater than ε, then the loss increases linearly. In the previous works the ε-insensitive learning was used for the consequences of if-then rules only. The premises of if-then rules were selected using preliminary fuzzy clustering in the input space. Such selected premises remain unchanged in the learning process. However, in the traditional approach to the fuzzy (or neuro-fuzzy) modeling both premises and consequences of if-then rules are adjusted during the process of learning

[1]. Thus, the main goal of this work is to introduce ε-insensitive learning of a fuzzy system, in which both premises and consequences are adjusted during learning. The next goal is to investigate the generalization ability of the fuzzy system obtained by means of new learning methods for real ECG data.

2 Fuzzy systems with parametric consequences in if-then rules

Let us assume that I fuzzy if-then rules with t-input and one-output are given. The ith rule in which the consequent is represented by a fuzzy singleton may be written in the following form

$$R^{(i)} : \text{IF } \mathbf{x} \text{ is } \mathbf{A}^{(i)}, \text{ THEN } y = \mathbf{p}^{(i)\top}\mathbf{x}', \tag{1}$$

where $\mathbf{x} \in \mathrm{I\!R}^t$ is the input variable, $y \in \mathrm{I\!R}$ is the output variable, $\mathbf{x}' \triangleq \left[\mathbf{x}^\top, 1\right]^\top$ is the augmented input vector, $\mathbf{p}^{(i)} = \left[\widetilde{\mathbf{p}}^{(i)\top}, p_0^{(i)}\right]^\top \in \mathrm{I\!R}^{t+1}$ is the vector of consequent parameters of the ith rule; $p_0^{(i)}$ denotes a bias of the ith model. The antecedent fuzzy set of the ith rule, $\mathbf{A}^{(i)}$ has a membership function $\mu_{\mathbf{A}^{(i)}}(\mathbf{x}) : \mathrm{I\!R}^t \to [0,1]$. In case of the Gaussian membership functions and the algebraic product used as the t-norm, the fuzzy antecedent is defined as

$$\mu_{\mathbf{A}^{(i)}}(\mathbf{x}_n) \triangleq \exp\left[-\frac{1}{2}\sum_{j=1}^{t}\frac{\left(x_{nj} - c_j^{(i)}\right)^2}{s_j^{(i)}}\right], \tag{2}$$

where x_{nj} denotes the jth component of the nth input data, and the parameters $c_j^{(i)}$, $s_j^{(i)}$, $i = 1, 2, \cdots, c$; $j = 1, 2, \cdots, t$ are centers and dispersions of the membership functions for the ith rule and the jth input variable, respectively.

For the input $\mathbf{x}_0$, the overall output of the fuzzy model is completed by a weighted averaging aggregation of individual rules conclusions, as [1]

$$y_0(\mathbf{x}_0) = \frac{\sum_{i=1}^{I} \mathcal{F}_i(\mathbf{x}_0)\, \mathbf{p}^{(i)\top}\mathbf{x}'_0}{\sum_{k=1}^{I} \mathcal{F}_k(\mathbf{x}_0)}, \tag{3}$$

where $\mathcal{F}_i(\mathbf{x}_0) = \mu_{\mathbf{A}^{(i)}}(\mathbf{x}_0)$ denotes the firing strength of the ith rule. The output value y_0 may be considered to be a linear combination of unknown parameters $\mathbf{p}^{(i)}$. If we introduce the following notation:

$$S^{(i)}(\mathbf{x}_0) = \frac{\mathcal{F}_i(\mathbf{x}_0)}{\sum_{k=1}^{I} \mathcal{F}_k(\mathbf{x}_0)}, \tag{4}$$

$$\mathbf{d}(\mathbf{x}_0) = \left[S^{(1)}(\mathbf{x}_0)\,\mathbf{x}_0'^{\top}, S^{(2)}(\mathbf{x}_0)\,\mathbf{x}_0'^{\top}, \cdots, S^{(I)}(\mathbf{x}_0)\,\mathbf{x}_0'^{\top}\right]^{\top}, \tag{5}$$

$$\mathbf{P} = \left[\mathbf{p}^{(1)\top}, \mathbf{p}^{(2)\top}, \cdots, \mathbf{p}^{(I)\top}\right]^{\top}, \tag{6}$$

then (3) may be written in the form

$$y_0(\mathbf{x}_0) = \mathbf{d}(\mathbf{x}_0)^{\top}\mathbf{P}. \tag{7}$$

Usually, the learning of fuzzy systems presented above is executed using the following scheme. The parameters from the premises and the consequents of if-then rules are adjusted separately. First, the premises parameters are adjusted using unsupervised learning — clustering of the input data using the fuzzy c-means method. Second, the consequents parameters are adjusted by means of the gradient descent method or the least squares method.

The LS learning methods use the quadratic loss function to match reality and a fuzzy model. In this case only perfect matching between reality and the model leads to a zero loss. The approach to fuzzy modeling presented in this paper is based on the premise that human learning as well as thinking is tolerant to imprecision. Hence, an ε-insensitive loss function is used and learning methods based on this loss function lead to so-called ε-insensitive fuzzy modeling [4], [5], [6], [7]. We seek consequence parameters vector $\mathbf{P}$ on the basis of a set of i.i.d. data pairs called a training set $\mathcal{T}^{(N)} = \{(\mathbf{x}_i, y_i)\}_{i=1}^{N}$, where N is the data cardinality and each independent datum $\mathbf{x}_i \in \mathbb{R}^t$ has a corresponding dependent datum $y_i \in \mathbb{R}$.

Using the ε-insensitive loss function [8], $\rceil\xi\lceil_{\varepsilon} = \max(|\xi| - \varepsilon, 0)$, where $\varepsilon \geq 0$ denotes the insensitivity parameter, the learning criterion function has the following form [6]

$$\min_{\mathbf{P}\in\mathbb{R}^{I(t+1)}} I(\mathbf{P}) \triangleq \sum_{n=1}^{N} \rceil y_n - \mathbf{P}^{\top}\mathbf{d}(\mathbf{x}_n)\lceil_{\varepsilon} + \frac{\tau}{2}\widetilde{\mathbf{P}}^{\top}\widetilde{\mathbf{P}}, \tag{8}$$

where $\widetilde{\mathbf{P}} = \left[\widetilde{\mathbf{p}}^{(1)\top}, \widetilde{\mathbf{p}}^{(2)\top}, \cdots, \widetilde{\mathbf{p}}^{(I)\top}\right]^{\top}$ is a narrowed vector $\mathbf{P}$, with excluded components corresponding to the biases. The second term in (8) is related to the minimization of the Vapnik-Chervonenkis dimension (complexity) of the regression model [9]. Parameter $\tau > 0$ controls the trade-off between the regression model complexity and the amount up to which errors are tolerated.

If we define $\mathbf{X}_d \triangleq [\mathbf{d}(\mathbf{x}_1), \mathbf{d}(\mathbf{x}_2), \cdots, \mathbf{d}(\mathbf{x}_N)]^{\top} \in \mathbb{R}^{N\times I(t+1)}$, $\mathbf{y} = [y_1, y_2, \cdots, y_N]^{\top} \in \mathbb{R}^N$ the minimization problem (8) can be rewritten in the matrix form

$$\min_{\mathbf{P}} \; I(\mathbf{P}) = \rceil\mathbf{y} - \mathbf{X}_d\mathbf{P}\lceil_{\varepsilon} + \frac{\tau}{2}\mathbf{P}^{\top}\widetilde{\mathbf{I}}\mathbf{P}, \tag{9}$$

where $\widetilde{\mathbf{I}} = \mathrm{diag}\left(\left[\mathbf{1}^{\top}, 0, \mathbf{1}^{\top}, \cdots, 0, \mathbf{1}^{\top}, 0\right]^{\top}\right)$ and $\mathbf{1}$ denotes the vector of respective dimension with all entries equal to 1.

The procedure of seeking optimal $\mathbf{P}$, called the ε-insensitive Learning by Solving a System of Linear Inequalities (εLSSLI) has been introduced in [5] and may be summarized in the following steps:

1. Fix $\tau \geq 0$, $0 < \rho < 1$ and $\mathbf{D}^{[1]} = \mathrm{diag}(\mathbf{1})$. Initialize $\mathbf{b}^{[1]} > \mathbf{0}$. Set iteration index $k = 1$,
2. $\mathbf{P}^{[k]} = \left(\mathbf{X}_d^\top \mathbf{D}^{[k]} \mathbf{X}_d + \frac{\tau}{2}\widetilde{\mathbf{I}}\right)^{-1} \mathbf{X}_d^\top \mathbf{D}^{[k]} \left(\left[\mathbf{y}^\top - \varepsilon \mathbf{1}^\top, -\mathbf{y}^\top - \varepsilon \mathbf{1}^\top\right]^\top + \mathbf{b}^{[k]}\right)$,
3. $\mathbf{e}^{[k]} = \mathbf{X}_d^{[k]} \mathbf{P}^{[k]} - \left[\mathbf{y}^\top - \varepsilon \mathbf{1}^\top, -\mathbf{y}^\top - \varepsilon \mathbf{1}^\top\right]^\top - \mathbf{b}^{[k]}$,
4. $\mathbf{D}^{[k+1]} = \mathrm{diag}\left(1 \Big/ \left|e_1^{[k]}\right|, \cdots, 1 \Big/ \left|e_N^{[k]}\right|, 1 \Big/ \left|e_{N+1}^{[k]}\right|, \cdots, 1 \Big/ \left|e_{2N}^{[k]}\right|\right)$,
5. $\mathbf{b}^{[k+1]} = \mathbf{b}^{[k]} + \rho \left(\mathbf{e}^{[k]} + \left|\mathbf{e}^{[k]}\right|\right)$,
6. if $\left\|\mathbf{b}^{[k+1]} - \mathbf{b}^{[k]}\right\|_2 > \xi$, then $k = k + 1$, go to 2., else stop.

Constant ξ is a pre-set value.

3 ε-insensitive learning of rules premises

Let us observe that for Gaussian membership functions of rules premises (2) the following unknown parameters should be determined: $c_j^{(i)}$, $s_j^{(i)}$ for $j = 1, 2, \cdots, t$ and $i = 1, 2, \cdots, I$. Usually, the above mentioned unknown parameters are estimated by means of the fuzzy c-means clustering [5]. Indeed, in our case we have I clusters. So, the name fuzzy I-means method will be better. As a result of preliminary clustering of the training set, the following assumption for initialization of the premises of parameters is made:

$$c_j^{(i)} = \frac{\sum\limits_{n=1}^{N} u_{in}\, x_{nj}}{\sum\limits_{n=1}^{N} u_{in}} \tag{10}$$

and

$$\left(s_j^{(i)}\right)^2 = \frac{\sum\limits_{n=1}^{N} u_{in} \left(x_{nj} - c_j^{(i)}\right)^2}{\sum\limits_{n=1}^{N} u_{in}}, \tag{11}$$

where u_{in} denotes a membership degree of a vector $\mathbf{x}_n = [x_{n1}, x_{n2}, \cdots, x_{nt}]^\top$ from the training set to the ith cluster (to the premise of ith if-then rule).

Frequently, the above described method of obtaining the premises of rules is used for initialization of a gradient descent method [3]. The measure of the error of system output value may be defined for a single pair from the training set as

$$E_n = L\left(y_n - y_0\left(\mathbf{x}_n\right)\right), \tag{12}$$

where y_n, $y_0(\mathbf{x}_n)$ denote the desired (target) and actual value of system output for $\mathbf{x}_n$, respectively. Function $L(\cdot)$ stands for a loss function. Most frequently a quadratic loss function is used, that is $L(\cdot) = \frac{1}{2}(\cdot)^2$. The ε-insensitive loss function will be used in this paper. For the entire training set, we define the error function as the average of E_n

$$E = \frac{1}{N}\sum_{n=1}^{N} E_n. \tag{13}$$

In the so-called batch mode of learning, parameters are updated after the presentation of all examples from the training set that is called an epoch. Thus, the minimization of error E is made iteratively (for parameter $\boldsymbol{\theta} \in \left\{c_j^{(i)}, s_j^{(i)}\right\}_{i,j=1}^{j=t,i=I}$):

$$\boldsymbol{\theta}_{\mathrm{new}} = \boldsymbol{\theta}_{\mathrm{old}} - \eta \left.\frac{\partial E}{\partial \boldsymbol{\theta}}\right|_{\boldsymbol{\theta}=\boldsymbol{\theta}_{\mathrm{old}}}, \tag{14}$$

where $\eta > 0$ is the learning rate parameter.

Taking into account (3) we may express the partial derivatives of error E_n with respect to the unknown parameters from premises of rules as:

$$\frac{\partial E_n}{\partial c_j^{(i)}} = \mathcal{A}_n \frac{\left[y^{(i)}(\mathbf{x}_n) - y_0(\mathbf{x}_n)\right]\mathcal{F}_i(\mathbf{x}_n)}{\sum_{k=1}^{I}\mathcal{F}_k(\mathbf{x}_n)} \frac{x_{nj} - c_j^{(i)}}{\left(s_j^{(i)}\right)^2}, \tag{15}$$

$$\frac{\partial E_n}{\partial s_j^{(i)}} = \mathcal{A}_n \frac{\left[y^{(i)}(\mathbf{x}_n) - y_0(\mathbf{x}_n)\right]\mathcal{F}_i(\mathbf{x}_n)}{\sum_{k=1}^{I}\mathcal{F}_k(\mathbf{x}_n)} \frac{(x_{nj} - c_j^{(i)})^2}{\left(s_j^{(i)}\right)^3}, \tag{16}$$

where

$$\mathcal{A}_n = \left.\frac{\partial E_n}{\partial y_0(\mathbf{x}_0)}\right|_{\mathbf{x}_0=\mathbf{x}_n}. \tag{17}$$

Indeed, for the quadratic loss function, we obtain $\mathcal{A}_n = -(y_n - y_0(\mathbf{x}_n))$. Using the ε-insensitive loss function the measure of the error for the nth example has the form [9]

$$E_n = \left\rceil y_n - y_0(\mathbf{x}_n)\right\lceil_\varepsilon \triangleq \begin{cases} 0, & |y_n - y_0(\mathbf{x}_n)| \le \varepsilon, \\ |y_n - y_0(\mathbf{x}_n)| - \varepsilon, & |y_n - y_0(\mathbf{x}_n)| > \varepsilon. \end{cases} \tag{18}$$

In the above case, quantity $\mathcal{A}_n$ takes the form

$$\mathcal{A}_n = \left.\frac{\partial E_n}{\partial y_0(\mathbf{x}_0)}\right|_{\mathbf{x}_0=\mathbf{x}_n} = \begin{cases} 0, & |y_n - y_0(\mathbf{x}_n)| \le \varepsilon, \\ \mathrm{sgn}(y_n - y_0(\mathbf{x}_n)), & |y_n - y_0(\mathbf{x}_n)| > \varepsilon, \end{cases} \tag{19}$$

where sgn$(\cdot)$ denotes a signum function.

A very simple operation speeding up the convergence is proposed by Jang et al. [2]. In (14), the learning rate parameter is selected in a special way:

$$\boldsymbol{\theta}_{\text{new}} = \boldsymbol{\theta}_{\text{old}} - \frac{\upsilon}{\sqrt{\sum_{i=1}^{r} \left(\frac{\partial E}{\partial \theta_i} \right)^2_{\theta_i = (\theta_i)_{\text{old}}}}} \left. \frac{\partial E}{\partial \boldsymbol{\theta}} \right|_{\boldsymbol{\theta} = \boldsymbol{\theta}_{\text{old}}}, \tag{20}$$

where r denotes the number of optimized parameters, in our case $r = 2tI$; $\upsilon > 0$ is the so-called step size. If in four successive steps of gradient descent learning error E increases and decreases commutatively, then the step size is decreased, that is, multiplied by $n_D < 1$. However, if in four successive steps of gradient descent learning error E decreases, then the step size is increased, that is, multiplied by $n_I > 1$.

The following hybrid learning algorithm is proposed. First, the parameters of premises of rules are obtained using the fuzzy I-means algorithm. Next, the parameters of consequents of rules are obtained using εLSSLI. Then, the above initial parameters of rules are adjusted iteratively. Each iteration consists of ε-insensitive gradient descent modification of the parameters of premises and determining the parameters of consequences by εLSSLI method. The iterations are stopped when parameters in successive iterations differ imperceptibly. The above algorithm may be called the ε-I-ε-gradient and may be summarized in the following steps:

1. Fix $\varepsilon \geq 0$, $\tau > 0$, $I \geq 2$.
2. Repeat 25 times the fuzzy I-means algorithm for a different random initialization of the partition matrix $\mathbf{U}$.
3. Initial values of parameters of the premises $c_j^{(i),[1]}$, $s_j^{(i),[1]}$ are determined using (10) and (11) from clustering corresponding to the minimal value of the Xie-Beni index[10].
4. Set the iteration index, $\ell = 1$ and the step size, $\upsilon = 0.01$.
5. Determine the parameters of consequences $w^{(i),[\ell]}$ using the εLSSLI algorithm with $c_j^{(i),[\ell]}$, $s_j^{(i),[\ell]}$.
6. Cumulate gradients with respect parameters $c_j^{(i),[\ell]}$, $s_j^{(i),[\ell]}$.
7. Update the premises parameters $c_j^{(i),[\ell+1]}$, $s_j^{(i),[\ell+1]}$ using (20).
8. Determine error measure (13) for ℓth iteration, $E^{[\ell]}$.
9. If $\ell > 4$ and $E^{[\ell]} < E^{[\ell-1]}$ and $E^{[\ell-1]} < E^{[\ell-2]}$ and $E^{[\ell-2]} < E^{[\ell-3]}$ and $E^{[\ell-3]} < E^{[\ell-4]}$ then $\upsilon \leftarrow 1.1\,\upsilon$.
10. If $\ell > 4$ and $E^{[\ell]} < E^{[\ell-1]}$ and $E^{[\ell-1]} > E^{[\ell-2]}$ and $E^{[\ell-2]} < E^{[\ell-3]}$ and $E^{[\ell-3]} > E^{[\ell-4]}$ then $\upsilon \leftarrow 0.9\,\upsilon$.
11. If $\ell > 1$ and $\left|E^{[\ell]} - E^{[\ell-1]}\right| < 1 \cdot 10^{-4}$ then stop else $\ell \leftarrow \ell + 1$, go to 5.

4 Numerical experiments

In all experiments, $\mathbf{b}^{[1]} = \mathbf{1}$ and $\rho = 0.9$ were used in the εLSSLI method. The iterations for the εLSSLI method were stopped as soon as the Euclidean norm in a successive pair of $\mathbf{b}$ vectors was less than $\xi = 10^{-4}$. For the ε-insensitive learning the standard fuzzy c-means clustering method was used with the weighted exponent $m = 2$. The iterations were stopped as soon as the Frobenius norm in a successive pair of partition matrices was less than 10^{-3}. The purpose of experiments was to compare the generalization ability of a fuzzy system learned using hybrid algorithm and the classical (zero-tolerance) learning as well as the state-of-the-art method based on the Support Vector Regression (SVR) machine [9]. The support vector regression (SVR) machine from Matlab Support Vector Machine Toolbox by S. Gunn was used. This toolbox has been obtained by Internet — *http://www.isis.ecs.soton.ac.uk/resources/svminfo*. The following benchmark database was used. ECG signal from the MIT-BIH database – the record numbered 100. The sampling frequency of that signal is equal to 360 Hz and the quantization step size is 5μV. The learning process was conducted for the first 450 samples. The testing set consists of 5000 samples. The order of the model was equal to 4 and a nonlinear one-step predictor was build. Thus, the following vectors were used as input: $\mathbf{x}_n \triangleq [x(n-1),\, x(n-2),\, x(n-3),\, x(n-4)]^{\top}$ and output $y_n \triangleq x(n)$. In all experiments parameter τ was taken from a set $\{0.001, 0.01, 0.05, 0.1, 0.5\}$ and ε was changed in the range from 0 to 0.5 (step 0.05). The number of if-then rules was changed from 2 to 6. After the training stage using the training part of data, the generalization ability of the designed model was determined as a root mean squared error (RMSE) on the test set. The training stage was repeated for each combination of the above values of parameters. Table 1 shows RMSE for the tested algorithms. It should be noted that despite the number of if-then rules, learning tolerant to imprecision leads to a better generalization comparing to zero-tolerant learning. The best generalization for each number of rules is obtained for parameters ε and τ value different from zero. It must also be noted that we observe an improvement in the generalization ability for algorithms with gradient modification of premises parameters. The following results were obtained using SVR machine – RMSE $= 0.0263$ (for $\sigma = 2.8$, $\varepsilon = 0.02$, $C = 327$).

5 Conclusions

This work presents a new approach to fuzzy modeling with learning tolerant to imprecision. A hybrid learning method for premises and consequences of if-then rules is proposed. This method consists of the following steps. First, the parameters of premises of rules are obtained using the fuzzy c-means algorithm. Next, the parameters of consequents of rules are obtained using the εLSSLI. Then, the above initial parameters of rules are adjusted iteratively.

Table 1. RMSE obtained on the testing part of ECG signal by the 0-insensitive learning (LS), the ε-I learning (without gradient modification of premises) and the ε-I-ε-gradient method.

I	LS learning	ε-I	ε-I-ε-gradient
2	0.03138	0.02953, $\varepsilon = 0.45$, $\tau = 0.01$	0.02578, $\varepsilon = 0.2$, $\tau = 0.05$
3	0.02793	0.02582, $\varepsilon = 0.55$, $\tau = 0.05$	0.02442, $\varepsilon = 0.2$, $\tau = 0.05$
4	0.02620	0.02615, $\varepsilon = 0.1$, $\tau = 0.05$	0.02331, $\varepsilon = 0.1$, $\tau = 0.05$
5	0.03390	0.02029, $\varepsilon = 0.55$, $\tau = 0.05$	0.02015, $\varepsilon = 0.4$, $\tau = 0.05$
6	0.02213	0.02382, $\varepsilon = 0.55$, $\tau = 0.01$	0.02068, $\varepsilon = 0.3$, $\tau = 0.001$

The method of adjusting premises is based on the gradient descent approach. An example is given of using the proposed hybrid learning of parameters of premises and consequences of if-then rules for designing fuzzy models of the ECG data. Simulation results show an improvement of the generalization ability of the fuzzy system with respect to the traditional as well as previously introduced ε-insensitive learning methods.

References

1. Czogala E, Leski JM (2000) Fuzzy and neuro-fuzzy intelligent systems. Physica-Verlag, Heidelberg.
2. Jang JSR, Sun CT, Mizutani E (1997) Neuro-fuzzy and Soft Computing. A Computational Approach to Learning and Machine Intelligence. Prentice-Hall, Upper Saddle River.
3. Leski JM, Czogala E (1999) A new artificial neural network based fuzzy inference system with moving consequents in if-then rules and its applications. Fuzzy Sets and Systems, Vol.108, No.3, pp.289–297.
4. Leski JM (2001) Neuro-fuzzy modeling with ε-insensitive learning, Methods of Artificial Intelligence in Mechanics and Mechanical Engineering, Gliwice, pp.133–138.
5. Leski JM (2002) ε-insensitive learning techniques for approximate reasoning systems (Invited Paper). International Journal of Computational Cognition, Vol.1, No.1, pp.21–77.
6. Leski JM (2003) Neuro-fuzzy system with learning tolerant to imprecision. Fuzzy Sets and Systems, Vol.138, No.2, pp.427–439.
7. Leski JM (2004) ε-insensitive fuzzy c-regression models: introduction to ε-insensitive fuzzy modeling. IEEE Trans. Systems, Man and Cybernetics – Part B: Cybernetics, Vol.34, No.1, pp.4–15.
8. Vapnik V (1995) The nature of statistical learning theory. Springer, New York.
9. Vapnik V (1998) Statistical learning theory. Wiley, New York.
10. Xie XL, Beni G (1991) A validity measure for fuzzy clustering. IEEE Trans. Pattern Analysis and Machine Intelligence, Vol.13, No.8, pp.841–847.

The Incremental Method for Discovery of Association Rules

Damian Dudek and Aleksander Zgrzywa

Institute of Applied Informatics, Wroclaw University of Technology
Wybrzeze Wyspianskiego 27, 50–370 Wroclaw, Poland
{damian.dudek, zgrzywa}@pwr.wroc.pl

Summary. We present a new method for incremental discovery of association rules, which is highly general and independent of a mining algorithm. The heart of the method is the rule maintenance algorithm, which keeps the base of discovered rules as if they were mined in a single run through the whole transaction database. For more general and flexible results we take into account thresholds of rules statistical significance and influence of time. The method can be used as a learning model in knowledge-based systems with bounded resources, e.g. software agents.

1 Introduction

Mining association rules is an important domain within knowledge discovery in databases (KDD), which has been developed for more than ten years. At the beginning these algorithms were used for analysis of large commercial databases, primarily concerning supermarkets and shopping carts. By the time they gradually started to be recognized as a more general tool for data processing, suitable for various application domains. Since the pioneer work by Agrawal et al. [1, 2], there have been a huge body of research in the field of association rules, including [3, 4, 5, 8, 9, 10, 12, 13, 15, 16, 17, 18]. The list is not exhaustive.

Consider a knowledge-based system, where historical facts are collected and analysed in order to find interesting association rules and use them to improve performance. Assume, that life-cycle of the system consists of performance phases interlaced with stand-by phases. During the former phases the system performs actions (e.g. retrieves and displays documents to the user), while during the stand-by phase it simply waits for user's input. We would like the system to accumulate interaction experience during the performance phases (not to disturb current activities) and analyse it from time to time (preferably in stand-by phases while system resources are idle or little utilized). Once a portion of stored facts is analysed, it is removed from the database. We proposed a method, which satisfies all these requirements and provides such an incremental discovery of association rules that results in similar rule set as if it was obtained from the whole training set.

The remainder of the paper is as follows. In Section 2 we introduce association rules more precisely, presenting a formal model and problem description. Section 3 presents the proposed method of incremental discovery of association rules, including assumptions and detailed description of the rule maintenance algorithm. Finally, we conclude pointing out major results, possible applications, related research and unresolved issues.

2 Association Rules

Association rules express knowledge that given values of some attributes in an examined database occur together with some values of other attributes [6, 14]. Every association rule consists of two parts – lists of attributes: the antecedent and the consequent. A given attribute can appear only in one part of a rule. An association rule represents information that if some values of antecedent attributes occur in a training example, then given values of consequent attributes also often appear. We present the extended formal model of association rules, based on works by Agrawal et al. [1, 2] and Goethals [9].

Let $U = \{I_1, I_2, ..., I_n\}$ be a set of attributes, each of which has domain $\{0, 1\}$. We define a *transaction schema over a set* U as a pair $T = (TID, TS, I)$, where TID is a set of transaction identifiers, TS is a set of time points, and I is a set of attribute, such that $I \subseteq U$. A *transaction of a schema* T is a vector $t = (tid, ts, i_1, i_2, ..., i_m)$, where $tid \in TID$, $ts \in TS$, and $i_k \in \{0, 1\}$ for $k \in [1; m]$, is value of an attribute I_k in a transaction t, which we write $t(I_k)$. We say, that a set $X \subseteq U$ is *satisfied* by a transaction t, which we denote $t \vdash X$, if $t(x_j) = 1$ for each attribute $x_j \in X$. A *transaction database* B is a set of transactions over a schema T. The *frequency* of a set X in a transaction base B, which we denote $freq(X, B)$, is a fraction: $freq(X, B) = |\{t : t \vdash X\}|/|B|$

An *association rule* is an expression of the form $X \Rightarrow Y$, where $X \subset U$, $Y \subset U$ and $X \cap Y = \emptyset$. We define the *base number of a rule* r, denoted by $b(r)$, as the number of facts, which were analysed resulting in discovery of this rule. We define *the mean time of a rule* r, denoted by $t_m(r)$, as average of time-stamp values ts of the analysed transactions for the rule r. We call two rules $r : X_r \Rightarrow Y_r$ and $p : X_p \Rightarrow Y_p$, *semantically equal*, denoted by $p \equiv r$, if $X_r \equiv X_p$ and $Y_r \equiv Y_p$. The *support* of a rule $r : X \Rightarrow Y$ in a transaction database B, denoted by $sup(r, B)$ is equal to $freq(X \cup Y)$. The support can be viewed as estimated probability that the sum of sets X and Y is satisfied by a random transaction from the database B [11]. A rule $r : X \Rightarrow Y$ is a *frequent rule* in the database B, if $sup(r, B) \geq \sigma$, for the minimal support threshold $\sigma \in [0; 1]$. The *confidence* of a rule $r : X \Rightarrow Y$ in B is a fraction:

$$con(r, B) = \frac{freq(X \cup Y, B)}{freq(X, B)} = \frac{sup(r, B)}{freq(X, B)} \tag{1}$$

The confidence represents estimated conditional probability $Pr(Y|X)$ that a set Y is satisfied by a random transaction from the database B, provided that X is satisfied by this transaction [11]. A rule $r : X \Rightarrow Y$ is called *confident* in a transaction database B, if its confidence is equal or greater than a given minimal confidence threshold $\gamma \in [0; 1]$. The problem of association rules mining can be defined as follows [9]: given an attribute set U, transaction database B, thresholds of minimal support σ and minimal confidence γ, find a set of all the frequent and confident rules $R(B, \sigma, \gamma) = \{X \Rightarrow Y : X, Y \subset U \wedge X \cap Y = \emptyset \wedge sup(X \Rightarrow Y, B) \geq \sigma \wedge con(X \Rightarrow Y, B) \geq \gamma\}$.

Many algorithms have been worked out to resolve the problem above. The most well-know one is Apriori, proposed by Agrawal et al. [1]. It has two phases: (1) finding frequent itemsets, (2) generating rules over the found itemsets. During the first stage, at the beginning single-attribute frequent itemsets are found, and then they are extended to larger itemsets, consisting of two and more attributes. This process is based on assumption, that for every nonempty sets $A, B \subseteq U$, if $A \subseteq B$, then $freq(A) \geq freq(B)$ [11]. In the second phase rules are generated from each found frequent itemset. The support of the rules is equal to the support of a given itemset and confidence is computed based on frequency of antecedents.

3 Incremental Discovery of Association Rules

Most algorithms for mining association rules run in the so called batch mode, i.e. they are intended to analyse the whole available transaction database and when new facts are added to the database, the discovery process needs to be restarted from scratch. The aim of our method is to enable an incremental discovery process, which is composed of possibly many subsequent runs. In each run mining covers a portion of transactions, that have been accumulated so far. When association rules are found in the current portion, they are added to the rule base, using a rule maintenance algorithm, which ensures, that the resulting set of rules is highly similar to the one, that would be obtained if the whole transaction database was analysed in a batch mode (see Fig. 1).

3.1 Overview

The incremental discovery runs in 3 main steps: (1) the current portion of transactions is processed by any algorithm for associations discovery (e.g. Apriori); (2) the discovered rules are added to the rule base using the maintenance algorithm; (3) the analysed facts are disposed. We assume that input data – transaction portions are in an acceptable format for a mining algorithm. KDD issues like data selection, sampling and preprocessing are beyond the scope of this paper.

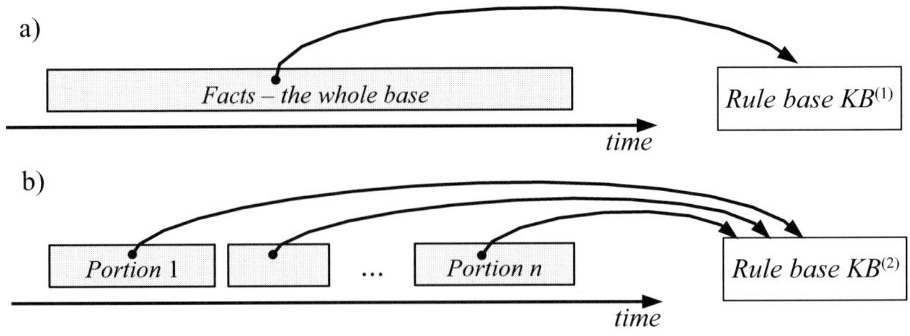

Fig. 1. The discovery of association rules in: (a) batch mode, (b) incremental mode. The resulting rule bases $KB^{(1)}$ and $KB^{(2)}$ are highly similar. The actual size of each fact portion depends on the situations when: (i) the learning system has accumulated a sufficient, representative number of new observations, due to an appropriate threshold defined by the designer; (ii) the system has an opportunity for analyzing the facts, i.e. it is able to use not utilized resources during a stand-by phase.

Basic maintenance of rules. The heart of our method is its second phase, when the rule maintenance algorithm is used, that updates the rule base after every run so that it remains approximately the same, as if all the facts analysed so far were processed in a single run. Our algorithm is based on observation that portions of transactions can be treated like sets and not sequences. This simplifies the problem of incremental maintenance of the rule base into appropriate computation of statistical measures of rules' significance (i.e. support and confidence), based on frequency proportion formulas. Assume that the mining process up to some moment was based on b_1 facts and resulted in discovery of a rule r with $sup_1(r)$ and $con_1(r)$. After another run over b_2 new facts the same rule r was found, but with new $sup_2(r)$ and $con_2(r)$. It can be proved, that the updated measures with respect to the whole transaction base analysed so far, should be:

$$sup'(r) := \frac{b_1 sup_1(r) + b_2 sup_2(r)}{b_1 + b_2} \tag{2}$$

$$con'(r) := \frac{con_1(r) con_2(r) \left(b_1 sup_1(r) + b_2 sup_2(r)\right)}{b_1 sup_1(r) con_2(r) + b_2 sup_2(r) con_1(r)} \tag{3}$$

Adding significance thresholds. The above formulas (2) and (3) ensure that the rule base is perfectly the same in incremental and non-incremental analysis, provided that no constraint of support and confidence are imposed on the discovered rules. However, in most KDD systems we need to filter the great number of mined rules, choosing only these ones, which satisfy given requirements, e.g. minimal support or confidence. Taking such constraints into account complicates the incremental maintenance of the rule base. Let us consider the following example. We set minimal support and confidence, σ and γ respectively, and start two runs of data mining – each on a separate portion of facts. In both runs a rule r is discovered, but in the first one it does not satisfy our thresholds and it is dropped out. Then the formula (3) fails to give any reasonable result, whereas the formula (2) produces incorrect outcome, as the support sup_1 is taken as 0 (zero), instead of a non-zero value from the range $(0;\sigma)$. To improve this situation, we proposed using two estimators $\hat{\sigma}$ and $\hat{\gamma}$, which represent the expected support and confidence of a random rule that is dropped. We leave decision to a designer, whether these estimators are to be set arbitrary or computed on real data during the mining phase.

Adding time influence. As stated before, the algorithm for rule maintenance is based on frequency proportion formulas. If the incremental mining process is run for a long period, we can observe an unfavourable effect: new rules from a single run have less and less chance to be added to the rule base, as they are based on a number of facts that is incomparably smaller than the number of all the facts analysed so far. Thus, even though the application environment changes, which should be reflected in the discovered rules, the rule base remains very resistant to any modifications. To amend this drawback we propose time influence function f_T, which placed in our maintenance formulas can balance significance of past and recent facts. As the actual shape of the function will depend on a given application, we only require that it satisfies the following: (i) the domain is the continuous interval $[0;+\infty)$; (ii) the values are real numbers within $[0;1]$; (iii) for every pair $x_1, x_2 \in [0;+\infty)$, if $x_1 < x_2$, then $f_T(x_1) \geq f_T(x_2)$. Example functions are shown in Fig. 2.

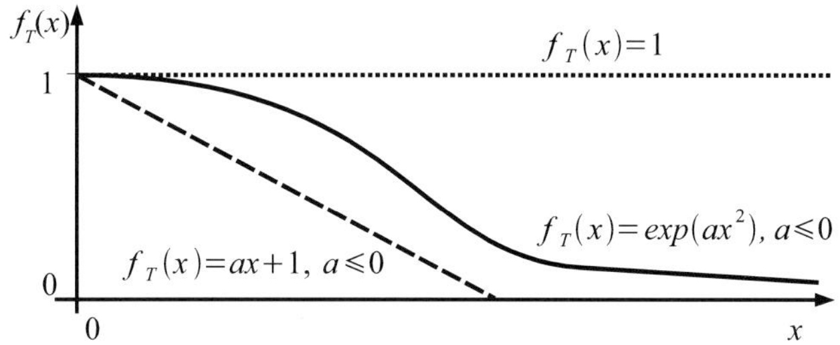

Fig. 2. Examples of the time influence function f_T. Argument x is time, that has passed from a given moment in the past until now.

The function f_T is not applied to the time-stamp of each single fact, which is analysed, but instead to the mean time of all the facts in a given portion. This approach makes the resulting shape of the function somewhat rough, but it is justified for the sake of performance.

3.2 The Algorithm For Rule Maintenance

Below we present the complete rule maintenance algorithm, which handles significance thresholds and time influence mentioned before.

Algorithm : association rule maintenance

Input : time function f_T, the current time t_{now}, parameters concerning the previous runs: rule base KB_R, minimum support threshold σ_g, minimum confidence threshold γ_g, expected support $\hat{\sigma_g}$, expected confidence $\hat{\gamma_g}$, mean time of facts t_{mg}, number of analysed facts b_g; parameters concerning the recent run: the set of new rules R, minimum support threshold σ_c, minimum confidence threshold γ_c, expected support $\hat{\sigma_c}$, expected confidence $\hat{\gamma_c}$, mean time of facts t_{mc}, number of analysed facts b_c.

Output : updated KB_R, σ_g, γ_g, $\hat{\sigma}_g$, $\hat{\gamma_g}$, t_{mg}, b_g.

1. Update threshold parameters: $\sigma_g := \frac{\sigma_g b_g+\sigma_c b_c}{b_g+b_c}$ $\gamma_g := \frac{\gamma_g b_g+\gamma_c b_c}{b_g+b_c}$.
2. For each rule $r_i \in R$ repeat the steps 3–6.
3. If there is a rule $p_j \in KB_R$, such that $p_j \equiv r_i$, then leave p_j in KB_R and update its parameters:
   $sup(p_j) := \frac{b(p_j)f_T(t_{now}-t_m(p_j))sup(p_j)+b(r_i)f_T(t_{now}-t_m(r_i))sup(r_i)}{b(p_j)f_T(t_{now}-t_m(p_j))+b(r_i)f_T(t_{now}-t_m(r_i))}$

   $con(p_j) := \frac{con(p_j)con(r_i)(b(p_j)sup(p_j)f_T(t_{now}-t_m(p_j))+b(r_i)sup(r_i)f_T(t_{now}-t_m(r_i)))}{b(p_j)sup(p_j)f_T(t_{now}-t_m(p_j))con(r_i)+b(r_i)sup(r_i)f_T(t_{now}-t_m(r_i))con(p_j)}$

   $b(p_j) := b(p_j) + b(r_i)$ and $t_m(p_j) := \frac{b(p_j)t_m(p_j)+b(r_i)t_m(r_i)}{b(p_j)+b(r_i)}$

4. If the updated $sup(p_j) < \sigma_g$ or $con(p_j) < \gamma_g$, then remove p_j from KB_R.
5. If there is no rule $p_j \in KB_R$, such that $p_j \equiv r_i$, update parameters for r_i:
   $sup(r_i) := \frac{b_g f_T(t_{now}-t_{mg})\hat{\sigma}_g+b(r_i)f_T(t_{now}-t_m(r_i))sup(r_i)}{b_g f_T(t_{now}-t_{mg})+b(r_i)f_T(t_{now}-t_m(r_i))}$
   $con(r_i) := con(r_i)$, if $\hat{\gamma}_g = 0$
   $con(r_i) := \frac{\hat{\gamma}_g con(r_i)(b_g\hat{\sigma}_g f_T(t_{now}-t_{mg})+b(r_i)sup(r_i)f_T(t_{now}-t_m(r_i)))}{b_g\hat{\sigma}_g f_T(t_{now}-t_{mg})con(r_i)+b(r_i)sup(r_i)f_T(t_{now}-t_m(r_i))\hat{\gamma}_g}$,
   if $0 < \hat{\gamma}_g \leq 1$.

6. If the updated $sup(r_i) \geq \sigma_g$ and $con(r_i) \geq \gamma_g$, then add r_i to KB_R with:
   $b(r_i) := b_g + b(r_i)$, and $t_m(r_i) := \frac{b_g t_{mg}+b(r_i)t_m(r_i)}{b_g+b(r_i)}$

7. For each rule $p_j \in KB_R$, such that $\neg\exists r_i \in R.p_j \equiv r_i$, repeat the steps 8–9.
8. Update parameters of the rule p_j:
   $sup(p_j) := \frac{b(p_j)f_T(t_{now}-t_m(p_j))sup(p_j)+b_c f_T(t_{now}-t_{mc})\hat{\sigma}_c}{b(p_j)f_T(t_{now}-t_m(p_j))+b_c f_T(t_{now}-t_{mc})}$
   $con(p_j) := con(p_j)$, if $\hat{\gamma}_c = 0$
   $con(p_j) := \frac{con(p_j)\hat{\gamma}_c(b(p_j)sup(p_j)f_T(t_{now}-t_m(p_j))+b_c\hat{\sigma}_c f_T(t_{now}-t_{mc}))}{b(p_j)sup(p_j)f_T(t_{now}-t_m(p_j))\hat{\gamma}_c+b(r_i)\hat{\sigma}_c f_T(t_{now}-t_{mc})con(p_j)}$,
   if $0 < \hat{\gamma}_c \leq 1$
   $b(p_j) := b(p_j) + b_c$, and $t_m(p_j) := \frac{b(p_j)t_m(p_j)+b_c t_{mc}}{b(p_j)+b_c}$

9. If the updated $sup(p_j) < \sigma_g$ or $con(p_j) < \gamma_g$, then remove p_j from KB_R.
10. Return updated: $KB_R, \sigma_g, \gamma_g, \hat{\sigma}_g, \hat{\gamma}_g, t_{mg}, b_g$.

4 Conclusions

Results and possible applications. We proposed an original method for incremental discovery of association rules, that is independent of a data mining algorithm, if only it supports appropriate input and output data. The method was intended to be applied within an agent architecture [7], but it is also suitable for other AI systems that collect numerous facts in some kind of history and use these records to improve performance. The epoch mode of the method makes it especially well suited for systems, where performance phases interlace with stand-by phases (e.g. personal assistants). Moreover, the method can be beneficial, if we want to keep system's history in a constrained size and avoid its uncontrolled growth, as after every run the analysed facts are removed.

Related work. The solutions to the problem of incremental association rule mining and maintenance were proposed, among others, in the works [3, 8, 12, 13, 16, 18]. Most of them deal with the issue of changing, large transaction databases and provide considerable computation reduction of association rules discovery when insertions, updates or deletions of facts occur. However, our work differs from the mentioned approaches and provides some value added: (i) it is independent on the mining algorithm; (ii) it takes into account the issue of time influence; (iii) our method maintains the final rule base and not large itemsets, which is especially important for knowledge-based and AI systems; (iv) the analysed facts are deleted, so that the transaction database can remain small (important for resource-bounded systems).

Future research. Experimental verification of the method is necessary to study efficiency issues. As the work presented here is still in progress, we have prepared the test environment based on the Apriori algorithm, the plan of experiments and measures for evaluating the method with respect to both the quality of maintained rules and efficiency of the incremental runs. The key concept is that we take a training set of facts and perform the analysis in the batch mode (the whole set in one run), and then in the incremental mode (dividing the set into portions). Then we compare the resulting rule sets and times of analysis in both cases. As the method presented is highly general, it also needs to be examined for concrete application domains in order to provide successful improvement of performance. For every system optimal settings should be experimentally determined, including among others: (i) the range of portion size for a single run; (ii) attribute selection for analysis; (iii) minimal support and confidence thresholds; (iv) the shape of the time function.

References

1. Agrawal R, Imielinski T, Swami A (1993) Mining association rules between sets of items in large databases. In: Proceedings of ACM SIGMOD Conference on Management of Data (SIGMOD'93) 207–216

2. Agrawal R, Srikant R (1994) Fast Algorithms for Mining Association Rules. Proceedings of the 20th International Conference on Very Large Databases, Santiago, Chile
3. Aumann Y, Feldman R, Lipshtat O (1999) Borders: An Efficient Algorithm for Association Generation in Dynamic Databases. Journal of Intelligent Information Systems, Vol. 21:61–73
4. Bayardo Jr. RJ, Agrawal R, Gunopulos D (2000) Constraint-Based Rule Mining in Large, Dense Databases. Data Mining and Knowledge Discovery Journal, Vol. 4, No. 2/3:217–240
5. Chen G, Wei Q, Liu D, Wets G (2002) Simple association rules (SAR) and the SAR-based rule discovery. Computers & Industrial Engineer., Vol. 43:721–733
6. Cichosz P (2000) Learning Systems. Science and Technology Publishers (WNT), Warsaw (in Polish)
7. Dudek D, Zgrzywa A (2003) The APS Learning Method of a Belief-Desire-Intention Agent. In: Bubnicki Z, Grzech A (eds.): Knowledge Engineering and Expert Systems (IWSE 2003), Vol. 2. Wroclaw University of Technology Press, Wroclaw 237–244 (in Polish)
8. Fong J, Wong HK, Huang SM (2003) Continuous and incremental data mining association rules using frame metadata model. Knowledge-Based Systems, Vol. 16:91–100
9. Goethals B (2002) Efficient Frequent Pattern Mining. PhD Thesis, Transnational University of Limburg, Diepenbeek, Belgium
10. Harms SK, Deogun JS (2004) Sequential Association Rule Mining with Time Lags. Journal of Intelligent Information Systems, Vol. 22, No. 1:7–22
11. Hastie T, Tibshirani R, Friendman J (2001) The Elements of Statistical Learning. Data Mining, Inference, and Prediction. Springer-Verlag, New York Berlin Heidelberg
12. Lee G, Lee KL, Chen ALP (2001) Efficient Graph-Based Algorithms for Discovering and Maintaining Association Rules in Large Databases. Knowledge and Information Systems, Vol. 3:338–355
13. Lee SD, Cheung DW, Kao B (1998) Is Sampling Useful in Data Mining? A Case in the Maintenance of Discovered Association Rules. Data Mining and Knowledge Discovery, Vol. 2:233–262
14. Mannila H (1997) Methods and problems in data mining. A tutorial. In: Afrati F, Kolaitis P (eds.) Proceedings of the International Conference on Database Theory (ICDT'97), Delphi, Greece 41–55
15. Shen L, Shen H, Cheng L (1999) New algorithms for efficient mining of association rules. Information Sciences, Vol. 118:251–268
16. Tsai PS, Lee C-C, Chen AL (1999) An Efficient Approach for Incremental Association Rule Mining. In: Zhong N, Zhou L (eds.): PAKDD'99, Lecture Notes in Artificial Intelligence (LNAI), Vol. 1574. Springer-Verlag, Berlin Heidelberg 74–83
17. Zaki MJ (2000) Scalable Algorithms for Association Mining. IEEE Transactions on Knowledge and Data Engineering, Vol. 12, No. 3:372–390
18. Zhou Z, Ezeife CI (2001) A Low-Scan Incremental Association Rule Maintenance Method Based on the Apriori Property. In: Stroulia E, Matwin S (eds.): AI 2001, Lecture Notes in Artificial Intelligence, Vol. 2056. Springer-Verlag, Berlin Heidelberg 26–35

Feature Extraction with Wavelet Transformation for Statistical Object Recognition

Marcin Grzegorzek *, Michael Reinhold, and Heinrich Niemann

Chair for Pattern Recognition,
University of Erlangen-Nuremberg,
Martensstr. 3, 91058 Erlangen, Germany,
{grzegorz,reinhold,niemann}@informatik.uni-erlangen.de

Summary. In this paper we present a statistical approach for localization and classification of 3-D objects in 2-D images with real heterogeneous background. Two-dimensional local feature vectors are computed directly from pixel intensities in square gray level images with the wavelet multiresolution analysis. We use three different resolution levels for the feature computation. For the first one local neighborhoods of size 8×8 pixels, for the second one 4×4 pixels, and for the third one 2×2 pixels are taken into account. Then we define an object area as a function of 3-D transformations and represent the feature vectors as density functions. Our localization and classification algorithm uses a combination of object models created for the three different resolutions in the training phase. Experiments made on a real data set with 42240 images show that the recognition rates are much better using the resolution combination of the wavelet transformation.

1 Introduction

The automatic localization and classification of objects in real environment images is becoming more important lately. Object recognition systems can be applied for example: to face classification, to localization of obstacles on the road with a camera mounted on a driving car, to service robotics [10], to handwriting recognition, and so on. There exist two main approaches for localization and classification of 3-D objects in 2-D gray level images: based on results of a segmentation process [5], or directly on the object appearance [3, 8]. The appearance-based methods compute feature vectors from pixel intensities in images without previous segmentation process. Some of them use only one global feature vector for the whole image (e.g. eigenspace approach [2]), other describe objects with more local features (e.g. neural networks [7]).

*This work was partly funded by the German Research Foundation (DFG) Graduate Research Center 3D Image Analysis and Synthesis

In the present work two-dimensional local feature vectors are computed directly from pixel intensities (appearance-based approach) using the wavelet multiresolution analysis [6] and modeled by density functions [4]. The main advantage of the local feature vectors is that a local disturbance only affects the feature vectors in a small region around it. In contrast to this a global feature vector can totally change, if only one pixel in the image varies. We introduce feature extraction on three different resolution levels in each image and create three statistical object models for each object class in the training phase. Our new algorithm for object localization and classification uses a combination of the object models obtained for these different resolution levels, which significantly improves the recognition rates.

In Sect. 2 the training of statistical object models with all its steps, especially the computation of feature vectors, is presented. Beginning with the computation of the object density value, through the recognition algorithm for one resolution, until the combination of object models for different resolutions Sect. 3 describes the whole recognition phase. The experimental evaluation of the new approach made on a large image data set can be found in Sect. 4. Sect. 5 closes our contribution with conclusions.

2 Training of Statistical Object Models

In order to learn object models we preprocess training images (Sect. 2.1), compute feature vectors in the preprocessed images (Sect. 2.2), define an object area (Sect. 2.3), and model the feature vectors by density functions (Sect. 2.4). At the end of the training process we get three statistical models for each object class, because the feature vector calculation is applied for three different resolutions of the wavelet transformation.

2.1 Image Sample Set for Training

First we define a set of object classes $\Omega = \{\Omega_1, \ldots, \Omega_\kappa, \ldots, \Omega_k\}$ and take training images of them on a dark background. The original training images are preprocessed by converting them to gray level images sized $2^n \times 2^n$ pixels, where $n \in \{6, 7, 8, 9\}$. Then we set one image $\mathbf{g}_{\kappa,i}$ for each object class Ω_κ as a reference image. With a pose of an object in the image $\mathbf{g}_{\kappa,j}$ we denote the 3-D transformation (translations and rotations) that maps the object in the reference image $\mathbf{g}_{\kappa,i}$ to the object in $\mathbf{g}_{\kappa,j}$. The 3-D transformation can be described with translations $\mathbf{t} = (t_x, t_y, t_z)^{\mathrm{T}}$ and rotations $\phi = (\phi_x, \phi_y, \phi_z)^{\mathrm{T}}$. The x and y axes lie in the image plane, and the z axis is orthographic to the image plane. With rotation around the x and y axes as well as with translation along the z axis (scaling) change the size and appearance of the object in the image. These are the so called external transformation parameters ($t_{ext} = t_z$ and $\phi_{ext} = (\phi_x, \phi_y)^{\mathrm{T}}$). The remaining transformation parameters are called internal and do not change the object size and appearance. Until the end of Sect. 2 the number of object class κ is omitted, because the training is identical for all object classes.

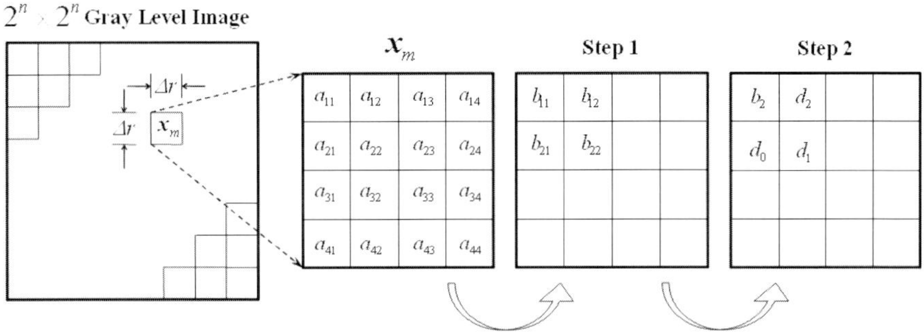

Fig. 1. Computation of a feature vector at a grind point $\mathbf{x}_m$ with a Haar wavelet (scale $s = 2$). In the first step low-pass coefficients $b_{ij} = 0.25 \cdot \sum_{k=0}^{1} \sum_{l=0}^{1} a_{k+2i-1,l+2j-1}$ are computed from the gray values a_{ij}. After the second step $b_2 = 0.25 \cdot \sum_{k=1}^{2} \sum_{l=1}^{2} b_{kl}$ is the low-pass coefficient. The other coefficients result from combinations of low-pass and high-pass filtering ($d_0 = 0.25 \cdot [-(b_{11}+b_{12})+(b_{21}+b_{22})]$, $d_1 = 0.25 \cdot [-(-b_{11}+b_{12})+(-b_{21}+b_{22})]$, $d_2 = 0.25 \cdot [(-b_{11}+b_{12})+(-b_{21}+b_{22})]$)

2.2 Computation of Feature Vectors

In all preprocessed training images feature vectors are computed using the wavelet transformation [1]. For the calculation of these vectors a grid with the size $\Delta r = 2^s$, where s is the index of the scale, is laid on an image (Fig. 1). On each grid point $\mathbf{x}_m$ a two-dimensional local feature vector $\mathbf{c}_m = \mathbf{c}(\mathbf{x}_m)$ is calculated. For this purpose we perform s-times the wavelet multiresolution analysis [6] using Haar wavelet. The components of the feature vector $\mathbf{c}_m$ are given by:

$$\mathbf{c}_m = \mathbf{c}(\mathbf{x}_m) = \begin{pmatrix} \ln(2^{-s} \, |b_{s,m}|) \\ \ln[2^{-s} \, (|d_{0,s,m}| + |d_{1,s,m}| + |d_{2,s,m}|)] \end{pmatrix} \tag{1}$$

$b_{s,m}$ is the low-pass coefficient and $d_{0\ldots2,s,m}$ result from combinations of low-pass and high-pass filtering. An illustration for the computation of a feature vector for $s = 2$ can be seen in Fig. 1 (indexes m and s are omitted). Our algorithm works with three resolution levels of the wavelet transformation: L_3 ($s = 3$), L_2 ($s = 2$), L_1 ($s = 1$). For each of these resolutions object models are created. The following training steps are identical for all resolution levels.

2.3 Modeling of Object Area

For the object model we want to consider only those feature vectors that belong to the object and not to the background. For each feature vector $\mathbf{c}_m$ in each external training pose $(\phi_{ext,t}, t_{ext,t})$ (for each training image) a discrete assignment function is defined [8]:

$$\widehat{\xi}_m(\phi_{ext,t}, t_{ext,t}) = \begin{cases} 1 & \text{if} \quad c_{m,1}(\phi_{ext,t}, t_{ext,t}) \geq S_t \\ 0 & \text{if} \quad c_{m,1}(\phi_{ext,t}, t_{ext,t}) < S_t \end{cases} \tag{2}$$

S_t is chosen manually. In the test images objects appear not only in the training poses, but also between them. In order to localize such objects we construct a continuous assignment function $\xi_m(\phi_{ext}, t_{ext})$ using values of $\widehat{\xi}_m(\phi_{ext,t}, t_{ext,t})$ by interpolation with trigonometric functions. The set of feature vectors belonging to the object for the given external pose (ϕ_{ext}, t_{ext}) (called object area $O(\phi_{ext}, t_{ext})$) can be now determined with the following rule:

$$\xi_m(\phi_{ext}, t_{ext}) \geq S_O \Longrightarrow \mathbf{c}_m(\phi_{ext}, t_{ext}) \in O(\phi_{ext}, t_{ext}) \tag{3}$$

The threshold value S_O is also chosen manually. In the case of internal transformations the object area does not change the size and can be translated and rotated with these transformations. So, we can write the object area as a function of all transformation parameters: $O(\phi, \mathbf{t})$.

2.4 Density Functions of the Feature Vectors

All feature vectors computed in the training phase (1) are interpreted as random variables. The object feature vectors are modeled with the normal distribution [4]. For each object feature vector $\mathbf{c}_m \in O$ we compute a mean value vector μ_m and standard deviation vector σ_m. The density of the object feature vector can be written as: $p(\mathbf{c}_m) = p(\mathbf{c}_m|\mu_m, \sigma_m, \phi, \mathbf{t})$. The feature vectors, which belong to the background are modeled with the equal distribution $p(\mathbf{c}_m) = p_b$.

3 Localization and Classification

After for each object class Ω_κ three corresponding object models $\mathcal{M}_{\kappa,s}$ ($s \in \{1, 2, 3\}$) were created, we can localize and classify objects in test images. At the beginning test images are preprocessed and feature vectors are computed (1) with the same method as in the training phase (Sect. 2.1 and 2.2). Then we start our recognition algorithm that uses only one of the trained object models for each object class (Sect. 3.2). After that the results are refined by using the combination of object models for different resolutions (Sect. 3.3). In both cases object density values (Sect. 3.1) for many pose and class hypotheses are needed.

3.1 Object Density Value

In order to compute the object density value for the class Ω_κ in the pose $(\phi, \mathbf{t})$ for a given test image we determine the set of feature vectors that belong to the object $C = \{\mathbf{c}_1, \mathbf{c}_2, \ldots, \mathbf{c}_M\}$ (object area $O_\kappa(\phi, \mathbf{t})$, Sect. 2.3) and compute their values. Then we compare the calculated object feature vectors with the corresponding density functions stored in the object model $\mathcal{M}_{\kappa,s}$ and read density values for these vectors $(p(\mathbf{c}_1), p(\mathbf{c}_2), \ldots, p(\mathbf{c}_M))$. The object density value for the object class Ω_κ in the pose $(\phi, \mathbf{t})$ can be computed as:

$$p(C|\mathbf{B}_{\kappa,s}, \phi, \mathbf{t}) = \prod_{i=0}^{M} \max\{p(\mathbf{c}_i), p_b\} \tag{4}$$

$\mathbf{B}_{\kappa,s}$ comprehends the trained mean value vectors and standard deviation vectors from $\mathcal{M}_{\kappa,s}$ and p_b is the background density value (Sect. 2.4).

3.2 Recognition Algorithm for One Resolution

The localization and classification algorithm for one resolution (one object model) is realized with the maximum likelihood estimation [9] and can be described with the following equation:

$$(\widehat{\kappa}, \widehat{\phi}, \widehat{\mathbf{t}}) = \operatorname*{argmax}_{\kappa}\{\operatorname*{argmax}_{(\phi,\mathbf{t})} G(p(C|\mathbf{B}_{\kappa,s}, \phi, \mathbf{t}))\} \tag{5}$$

$\widehat{\kappa}$ is the classification result and $(\widehat{\phi}, \widehat{\mathbf{t}})$ is the localization result. First the object density (normalized by G) is maximized according to the pose parameters $(\phi, \mathbf{t})$ and then to the class κ. The norm function G is defined by:

$$G(p(C|\mathbf{B}_{\kappa,s}, \phi, \mathbf{t})) = \sqrt[M]{p(C|\mathbf{B}_{\kappa,s}, \phi, \mathbf{t})} \tag{6}$$

M is the number of feature vectors belonging to the object area $O_\kappa(\phi, \mathbf{t})$. This norm function decreases the dependency between the maximization result and the object area size.

3.3 Combination of Object Models for Different Resolutions

Our recognition algorithm uses a combination of object models obtained for different resolutions of the wavelet transformation. We start for the resolution level L_3 ($s = 3$), where the feature vectors are computed from local neighborhoods of size 8×8 pixels. According to Sect. 3.2 we find a class and pose of the object in the scene $(\widehat{\kappa}_3, \widehat{\phi}_3, \widehat{\mathbf{t}}_3)$ (5). Then the maximum likelihood estimation is applied for all object classes for the resolution level L_2 ($s = 2$) in the small neighborhood of the localization result from L_3 $(\widehat{\phi}_3, \widehat{\mathbf{t}}_3)$ [1]. A refined recognition result $(\widehat{\kappa}_2, \widehat{\phi}_2, \widehat{\mathbf{t}}_2)$ is obtained. Analogical for the resolution level L_1 ($s = 1$) we maximize the object density (normalized by the function G (6)) only in the small neighborhood of $(\widehat{\phi}_2, \widehat{\mathbf{t}}_2)$ and get the finally recognition result $(\widehat{\kappa}_1, \widehat{\phi}_1, \widehat{\mathbf{t}}_1)$.

4 Experiments and Results

We verified our approach on a 3D-REAL-ENV image data base (Sect. 4.1). Using the combination of object models for different resolutions (Sect. 3.3) the execution time increases (Sect. 4.2), but we obtain better localization and classification rates (Sect. 4.3).

4.1 Image Data Base

3D-REAL-ENV (Image Data Base for 3-D Object Recognition in Real World Environment) consists of 10 objects depicted in Fig. 2. We made the experi-

[1] The small neighborhood of $(\widehat{\phi}_s, \widehat{\mathbf{t}}_s)$ is defined for rotations with $\pm 5(s-1)[^\circ]$, and for translations with $\pm 2^{s-1}$ pixels.

Fig. 2. 10 object classes used for experiments. In the first row examples of test images with "more heterogeneous" background can be seen. From left: bank cup, toy fire engine, green puncher, siemens cup, nizoral bottle. The second row contains examples of test images with "less heterogeneous" background. From left: toy passenger car, ricola container, stapler, toy truck, white puncher.

ments using gray level images of size 256×256 pixels. The pose of an object is defined with external rotations and internal translations $(\phi_x, \phi_y, t_x, t_y)^{\mathrm{T}}$. For the training we took 3360 images of each object with two different illuminations. The objects were put on a turntable ($0° \leq \phi_{table} \leq 360°$) and a robot arm with a camera was moved from horizontal to vertical ($0° \leq \phi_{arm} \leq 90°$). The angle between two adjacent training viewpoints amounts to 4.5°. For the tests 2880 images with homogeneous, 2880 images with "less heterogeneous", and 2880 with "more heterogeneous" background were taken. In the test images with "less heterogeneous" background the objects are easier to distinguish from the background than in the test images with "more heterogeneous" background. The object poses and the illumination in the test images were different from the training viewpoints and illuminations.

4.2 Execution Time

In Table 1 we compare the execution time in the recognition phase for different resolution levels and their combinations. The finest resolution level L_1 is very time consuming and can be used only in combination with L_2 and L_3.

Table 1. Execution time of the localization and classification algorithm for L_3, L_2, L_1, and combinations L_3–L_2, L_3–L_2–L_1.

Pentium 4 2.66 GHz 512 MB RAM	**L_3**	**L_2**	**L_1**	**L_3–L_2**	**L_3–L_2–L_1**
Recognition in 1 Test Image	3.6s	124.7s	73.7m	24.0s	96.5s

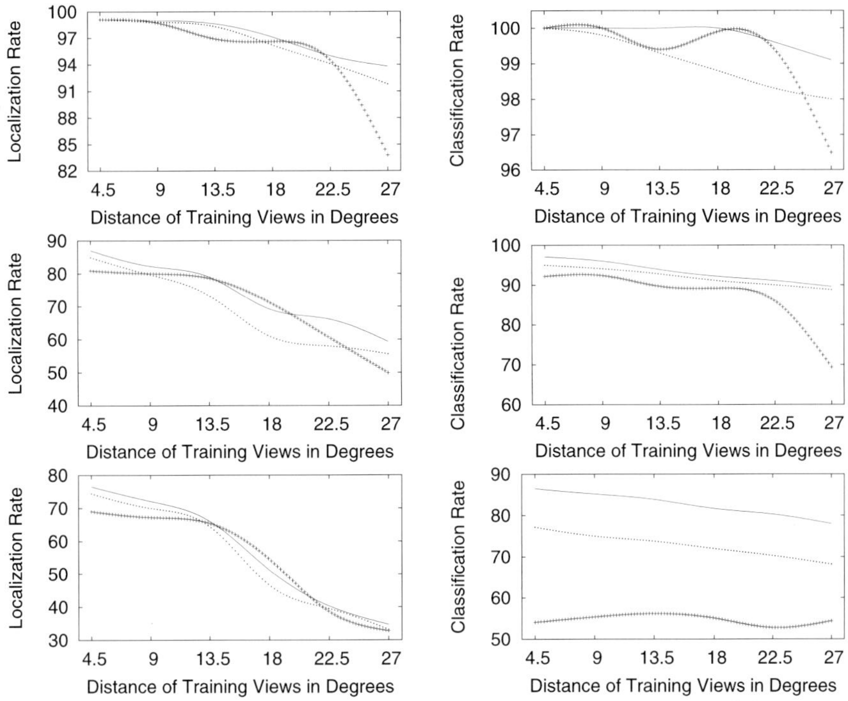

Fig. 3. Localization and classification rates depending on the distance of the training views for test images with homogeneous (first row), "less heterogeneous" (second row), and "more heterogeneous" background (third row). (— combination of L_3–L_2–L_1; $\cdots$ combination of L_3–L_2; +++ resolution level L_3).

4.3 Localization and Classification Rates

We count a localization result as correct, if the error for the external rotations (ϕ_x, ϕ_y) is not bigger than 15° and the error for the internal translations is not bigger than 10 pixels. Fig. 3 presents the recognition rates depending on the distance of the training views for test images with homogeneous, "less heterogeneous", and "more heterogeneous" background. The advantage of the combination of the resolution levels is visible especially for classification. Table 2 contains the recognition rates for 4.5° distance of training views.

Table 2. Recognition rates for 4.5° distance of training views.

Distance of Training Views 4.5°	Localization			Classification		
	Hom. Back.	Less Het. Back.	More Het. Back.	Hom. Back.	Less Het. Back.	More Het. Back.
L_3	99.1%	80.9%	69.0%	100%	92.2%	54.1%
L_3–L_2	99.1%	84.9%	74.4%	100%	95.0%	77.2%
L_3–L_2–L_1	99.1%	87.0%	76.5%	100%	97.1%	86.5%

5 Conclusions

In this article a powerful statistical, appearance-based approach for 3-D object localization and classification in images with heterogeneous background is presented. After computation of local feature vectors for three different resolutions of the wavelet transformation we define an assignment function, which assigns the features to the object or to the background, and statistically model them by density functions. Our new algorithm for localization and classification of objects uses a combination of the statistical models obtained for the three resolutions. In the experiments we showed that the new algorithm brings an improvement of the recognition rates, especially for the classification, in relatively short execution time. In the future we will introduce color modeling to our system, because the color information of objects is presently lost in the image preprocessing step.

References

1. C. Chui. *An Introduction to Wavelets.* Academic Press, San Diego, USA, 1992.
2. Ch. Gräßl, F. Deinzer, and H. Nieman. Continuous parametrization of normal distribution for improving the discrete statistical eigenspace approach for object recognition. In V. Krasnoproshin, S. Ablameyko, and J. Soldek, editors, *Pattern Recognition and Information Processing 03*, pages 73–77, Minsk, Belarus, Mai 2003.
3. R. Gross, I. Matthews, and S. Baker. Appearance-based face recognition and light-fields. *IEEE Transactions on Pattern Analysis and Machine Intelligence*, 26(4):449–465, April 2004.
4. M. Grzegorzek, F. Deinzer, M. Reinhold, J. Denzler, and H. Niemann. How fusion of multiple views can improve object recognition in real-world environments. In T. Ertl, B. Girod, G. Greiner, H. Niemann, H.-P. Seidel, E. Steinbach, and R. Westermann, editors, *Vision, Modeling, and Visualization 2003*, pages 553–560, Munich, Germany, November 2003. Aka/IOS Press, Berlin, Amsterdam.
5. J. Kerr and P. Compton. Toward generic model-based object recognition by knowledge acquisition and machine learning. In *Proceedings of the Eighteenth International Joint Conference on Artificial Intelligence*, pages 9–15, Acapulco, Mexico, August 2003.
6. S. Mallat. A theory for multiresolution signal decomposition: The wavelet representation. *IEEE Transactions on Pattern Analysis and Machine Intelligence*, 11(7):674–693, July 1989.
7. S. Park, J. Lee, and S. Kim. Content-based image classification using a neural network. *Pattern Recognition Letters*, 25(3):287–300, February 2004.
8. M. Reinhold. *Robust Probabilistic Appearance-Based Object Recognition.* Logos Verlag, Berlin, Germany, 2004.
9. A. R. Webb. *Statistical Pattern Recognition.* John Wiley & Sons Ltd, Chichester, England, 2002.
10. M. Zobel, J. Denzler, B. Heigl, E. Nöth, D. Paulus, J. Schmidt, and G. Stemmer. Mobsy: Integration of vision and dialogue in service robots. *Machine Vision and Applications*, 14(1):26–34, 2003.

Digital Filter Design with Constraints in Time and Frequency Domains

Norbert Henzel

Institute of Electronics, Silesian University of Technology, 44-101 Gliwice, Poland
nhenzel@polsl.pl

Summary. This paper describes a new method for design of linear phase finite impulse response (FIR) filters. This new approach, based on the ε-insensitive loss function, allows the design process to take into account not only constraints specified in the frequency domain, but also constraints on the output, time domain, signal. The performances of the proposed approach are shortly illustrated with a design of a highpass filter used for ECG baseline wander reduction.

1 Introduction

Digital filters are integral parts of many signal and information processing systems.The digital filter design process starts with some specifications coming from the considered application. Typical specifications, in the simplest case of linear phase lowpass FIR filter, includes the passband and stopband edges, desired stopband attenuation, maximum passband deviation, maximum filter length, etc. Due to common presence of digital processing systems and an increasing number of applications involving digital signal processing and digital filtering, there is a growing need for flexible digital filter design techniques that accept sophisticated specifications.

This paper describes a new digital filter design method that uses not only the constraints on the designed filter's frequency response but takes also into account the constraints on the output, time domain, signal. This approach exploits the ε-insensitive loss function, that plays recently an important role in a vast range of intelligent processing systems, e.g. [1, 2, 3, 4]. The proposed method is a very useful complement to existing method for filter design.

2 Digital Filter Basics

A digital filter is a linear time-invariant system, operating on an input sequence $x(n)$ to produce an output sequence $y(n)$. This system can be completely described by the impulse response sequence $h(n)$. The input-output relation for digital filter is given by

$$y(k) = \sum_{m=-\infty}^{\infty} x(m)h(k-m), \tag{1}$$

or in an equivalent form as

$$y(k) = \sum_{m=-\infty}^{\infty} x(k-m)h(m). \tag{2}$$

Typically, $h(m) = 0$ for $0 > m > N-1$, so we obtain

$$y(k) = \sum_{m=0}^{N-1} x(k-m)h(m). \tag{3}$$

The number of impulse response coefficients, N, is said to be the length of the filter, and the quantity $N-1$ is called the order of the filter.

The frequency response $H\left(e^{j\omega}\right)$ of an FIR filter is given by the discrete-time Fourier transform of its impulse response $h(n)$:

$$H\left(e^{j\omega}\right) = \sum_{n=0}^{N-1} h(n)\, e^{-j\omega n} \tag{4}$$

where the frequency variable ω is in radians.

If the impulse response $h(n)$ of the FIR filter has even symmetry, $h(n) = h(N-1-n)$, or odd symmetry, $h(n) = -h(N-1-n)$, the phase response of the designed filter is linear and the resulting design problem is real-valued. In this case, the frequency response function $H\left(e^{j\omega}\right)$ can be written as

$$H\left(e^{j\omega}\right) = e^{-j(N-1)/2\omega} e^{-j\beta} H_0(\omega) \tag{5}$$

where $H_0(\omega)$ is a real-valued function, called amplitude response and the constant β satisfies $\beta = 0$ or $\beta = \pi/2$. In the first case, $\beta = 0$, the filter amplitude response is given by

$$H_0(\omega) = \begin{cases} \sum_{n=0}^{(N-1)/2} b_n \cos(\omega n) & \text{for } N-1 \text{ even,} \\ \sum_{n=0}^{N/2} b_n \cos\left(\omega\left(n-\frac{1}{2}\right)\right) & \text{for } N-1 \text{ odd.} \end{cases} \tag{6}$$

where the coefficients b_n are related to $h(n)$ in as follows:

$$b_n = \begin{cases} h\left(\frac{N-1}{2}\right) & \text{for } N-1 \text{ even, } n=0, \\ 2h\left(\frac{N-1}{2}-n\right) & \text{for } N-1 \text{ even, } n \neq 0, \\ 2h\left(\frac{N}{2}-n\right) & \text{for } N-1 \text{ odd.} \end{cases} \tag{7}$$

Similar expressions can be developed for $\beta = \pi/2$.

The linear phase response of a FIR filter is a very demanded property in some applications, e.g. processing of an electrocardiographic (ECG) signals.

The relation between linear-phase FIR filter amplitude response $H_0(\omega)$ and coefficients b_n for a given set of frequency points ω_i, $1 \leq i \leq L$, distributed evenly over the frequency domain can be compactly represented in matrix form. For example, the first case in (6) can be written as

$$\mathbf{H}_0 = \mathbf{Tb} \tag{8}$$

where $M = \frac{N-1}{2}$, $\mathbf{b} = [b_0\ b_1\ \cdots b_M]^T$, $\mathbf{H}_0 = [H_0(\omega_1)\ H_0(\omega_2)\ \cdots\ H_0(\omega_L)]^T$ and $\mathbf{T} = \begin{bmatrix} \cos(0 \cdot \omega_1) & \cos(\omega_1) & \cdots & \cos(M \cdot \omega_1) \\ \cos(0 \cdot \omega_2) & \cos(\omega_2) & \cdots & \cos(M \cdot \omega_2) \\ \vdots & \vdots & \ddots & \vdots \\ \cos(0 \cdot \omega_L) & \cos(\omega_L) & \cdots & \cos(M \cdot \omega_L) \end{bmatrix}$.

Let us denote that equation (3) can be rewritten in the following forms:

$$y(k) = \sum_{m=0}^{M-1} h(m)\big(x(k-m) + x(k-(N-1)+m)\big) + h(M)x(k-M), \tag{9}$$

or equivalently as

$$\begin{aligned} y(k) &= \sum_{m=1}^{M} \frac{1}{2} b_m \big(x(k-M+m) + x(k-M-m)\big) + b_0 x(k-M) \\ &= \sum_{m=0}^{M} \frac{1}{2} b_m \big(x(k-M+m) + x(k-M-m)\big). \end{aligned} \tag{10}$$

These input–output relations can be presented in a matrix form:

$$\mathbf{Y} = \frac{1}{2}\mathbf{X} \cdot \mathbf{b}' + \mathbf{X_0} \cdot b_0, \tag{11}$$

where: $\mathbf{b}' = [b_1, \ldots, b_M]^T$, $\mathbf{b} = [b_0,\ \mathbf{b}'^T]^T$, $\mathbf{X_0} = [x(M+1), \ldots, x(K-M)]^T = [x_1, \ldots, x_{K-2M}]^T$, $\mathbf{Y} = [y(2M+1), \ldots, y(K)]^T = [y_1, \ldots, y_{K-2M}]^T$, $\mathbf{X} = [\mathbf{x}_1^T, \ldots, \mathbf{x}_{K-2M}^T] = \mathbf{S_1} + \mathbf{S_2}$, $\mathbf{S_1} = \begin{bmatrix} x(M+2) & \cdots & x(2M+1) \\ \vdots & \ddots & \vdots \\ x(K-M+1) & \cdots & x(K) \end{bmatrix}$, $\mathbf{S_2} = \begin{bmatrix} x(M) & \cdots & x(1) \\ \vdots & \ddots & \vdots \\ x(K-M-1) & \cdots & x(K-2M) \end{bmatrix}$.

3 New Linear Phase FIR Design Method with Constraints

Many recent digital filter design methods aim at minimizing a given error criterion. For formulating the design problems it is useful to define an error function

$$E_c(\omega) = H\left(e^{j\omega}\right) - D\left(e^{j\omega}\right) \tag{12}$$

where $H\left(e^{j\omega}\right)$ and $D\left(e^{j\omega}\right)$ are the actual and the desired frequency response of the filter, respectively. This constrained digital filter design may be described in terms of a so called ε-insensitive loss function. The ε-insensitive loss function introduced by Vapnik in [5, 6], for a vector argument, $\mathbf{g} = [g_1, g_2, \cdots, g_L]^T$, can be defined as $\rceil \mathbf{g} \lceil_\varepsilon \triangleq \sum_{i=1}^{L} \rceil g_i \lceil_\varepsilon$ where $\rceil g_i \lceil_\varepsilon \triangleq \begin{cases} 0, & |g_i| \leq \varepsilon, \\ |g_i| - \varepsilon, & |g_i| > \varepsilon. \end{cases}$

The constrained digital filter design problem can be view as a solution of a linear regression model described by (11) and

$$\mathbf{D_0} = \mathbf{Tb}, \quad \mathbf{b} \in \mathbb{R}^{M+1} \tag{13}$$

$$\mathbf{D_0} = \mathbf{Tb}, \quad \mathbf{b} \in \mathbb{R}^{M+1} \tag{14}$$

where $\mathbf{D_0} = [d_1, d_2, \ldots, d_L]^T$ specifies the desired amplitude response d_i at frequency point ω_i, L is a number of frequency points ω_i, $\mathbf{T} = [\mathbf{t}_1, \mathbf{t}_2, \ldots, \mathbf{t}_L]^T$, $\mathbf{t}_i = [\cos(0\omega_i), \cos(\omega_i), \ldots, \cos(M\omega_i)]^T$.

Using the ε-insensitive loss function, the parameters vector $\mathbf{b}$ is obtained as a solution of the following problem

$$\underset{\mathbf{b}\in\mathbb{R}^{M+1}}{\text{minimize}}\ E_{\varepsilon,\delta}(\mathbf{b}) \triangleq \frac{1}{2}\mathbf{b}^T\mathbf{b} + C\sum_{i=1}^{L} \rceil d_i - \mathbf{t}_i\mathbf{b} \lceil_\varepsilon V \sum_{k=1}^{K} \rceil y_k - \mathbf{x}_k\mathbf{b} \lceil_\delta, \tag{15}$$

where the parameters $C, V > 0$ controls the trade-off between the coefficient energy and the amount of frequency and time domains bound errors.

To cooperate with (possibly) infeasible constraints we introduce slack variables $\xi_i^+, \xi_i^-, \mu_k^+, \mu_k^- \geq 0$. For all impulse response points d_i, input x_k and output y_k, the criterion (15) can then be written in the following (primal) form

$$\underset{\mathbf{b},\xi_i^+,\xi_i^-,\mu_k^+,\mu_k^-}{\text{minimize}}\ \tfrac{1}{2}\mathbf{b}^T\mathbf{b} + C\sum_{i=1}^{L}\left(\xi_i^+ + \xi_i^-\right) + V\sum_{k=1}^{K}\left(\mu_k^+ + \mu_k^-\right),$$

$$\text{subject to} \begin{cases} d_i - \mathbf{t}_i\mathbf{b}' - b_0 \leq \varepsilon_i + \xi_i^+, \\ \mathbf{t}_i\mathbf{b}' + b_0 - d_i \leq \varepsilon_i + \xi_i^- \\ \xi_i^+ \geq 0,\ \xi_i^- \geq 0 \\ y_k - \mathbf{x}_k\mathbf{b}' - x_k b_0 \leq \delta_k + \mu_k^+, \\ \mathbf{x}_k\mathbf{b}' + x_k b_0 - y_k \leq \delta_k + \mu_k^- \\ \mu_k^+ \geq 0,\ \mu_k^- \geq 0 \end{cases} \tag{16}$$

where $i = 1, \ldots, L,\ k = 1, \ldots, K$. This kind of optimizations problems can be solved more easily in its dual formulation. The key idea to solve this optimization problem is to construct a Lagrange function from both the objective function and the constraints, by introducing a dual set of variables. It can be shown, that such function has a saddle point with respect to the primal (original) and dual variables at the optimal solution. The Lagrangian function of (16) is

$$\begin{aligned}\mathscr{L} = &\frac{1}{2}\mathbf{b}^T\mathbf{b} + C\sum_{i=1}^{L}\left(\xi_i^+ + \xi_i^-\right) + V\sum_{k=1}^{K}\left(\mu_k^+ + \mu_k^-\right) + \\ &-\sum_{i=1}^{L}\alpha_i^+\left(\varepsilon_i + \xi_i^+ - d_i + \mathbf{t}_i\mathbf{b}' + b_0\right) - \sum_{i=1}^{L}\alpha_i^-\left(\varepsilon_i + \xi_i^- + d_i - \mathbf{t}_i\mathbf{b}' - b_0\right) \\ &-\sum_{k=1}^{K}\beta_i^+\left(\delta_i + \mu_k^+ - y_k + \mathbf{x}_k\mathbf{b}' + x_k b_0\right) - \sum_{i=1}^{L}\left(\eta_i^+\xi_i^+ + \eta_i^-\xi_i^-\right) \\ &-\sum_{k=1}^{K}\beta_i^-\left(\delta_i + \mu_k^- + y_k - \mathbf{x}_k\mathbf{b}' - x_k b_0\right) - \sum_{k=1}^{K}\left(\lambda_k^+\mu_k^+ + \lambda_k^-\mu_k^-\right).\end{aligned} \tag{17}$$

where $\alpha_i^+, \alpha_i^-, \eta_i^+, \eta_i^-, \beta_k^+, \beta_k^-, \lambda_k^+, \lambda_k^- \geq 0$ are the Lagrange multipliers.

The dual optimization problem of (16) we get by setting the derivatives (with respect to the primal variables) of (17) equal to zero

$$\begin{cases} \frac{\partial\mathscr{L}}{\partial\mathbf{b}} = \mathbf{b} - \sum\limits_{i=1}^{L}\left(\alpha_i^+ - \alpha_i^-\right)\mathbf{t}_i - \sum\limits_{k=1}^{K}\left(\beta_k^+ - \beta_k^-\right)\mathbf{x}_k = \mathbf{0}, \\ \frac{\partial\mathscr{L}}{\partial b_0} = \sum\limits_{i=1}^{L}\left(\alpha_i^+ - \alpha_i^-\right) + \sum\limits_{k=1}^{K}\left(\beta_k^+ - \beta_k^-\right)x_k = 0, \\ \frac{\partial\mathscr{L}}{\partial\xi_i^+} = C - \alpha_i^+ - \eta_i^+ = 0, \\ \frac{\partial\mathscr{L}}{\partial\xi_i^-} = C - \alpha_i^- - \eta_i^- = 0, \\ \frac{\partial\mathscr{L}}{\partial\mu_k^+} = V - \beta_k^+ - \lambda_k^+ = 0, \\ \frac{\partial\mathscr{L}}{\partial\mu_k^-} = V - \beta_k^- - \lambda_k^- = 0. \end{cases} \tag{18}$$

Substituting (18) into the Lagrangian (17), results in the following optimization problem:

$$\begin{aligned} &\underset{\alpha_i^+,\alpha_i^-,\beta_k^+,\beta_k^-}{\text{maximize}} \quad \mathscr{L} \\ &\text{subject to} \quad \begin{cases} \sum\limits_{i=1}^{L}\left(\alpha_i^+ - \alpha_i^-\right) + \sum\limits_{k=1}^{K}\left(\beta_k^+ - \beta_k^-\right)x_k = 0, \\ 0 \leq \alpha_i^+, \alpha_i^- \leq C, \\ 0 \leq \beta_k^+, \beta_k^- \leq V. \end{cases} \end{aligned} \tag{19}$$

where the Lagrangian is given by the expression:

$$\mathscr{L} = -\frac{1}{2}\Big(\sum_{i=1}^{L}(\alpha_i^+ - \alpha_i^-)\,\mathbf{t}_i + \sum_{k=1}^{K}(\beta_k^+ - \beta_k^-)\,\mathbf{x}_k\Big)\cdot$$
$$\Big(\sum_{i=1}^{L}(\alpha_i^+ - \alpha_i^-)\,\mathbf{t}_i + \sum_{k=1}^{K}(\beta_k^+ - \beta_k^-)\,\mathbf{x}_k\Big)^T - \sum_{i=1}^{L}(\alpha_i^+ + \alpha_i^-)\,\varepsilon_i$$
$$+\sum_{i=1}^{L}(\alpha_i^+ - \alpha_i^-)\,d_i - \sum_{k=1}^{K}(\beta_k^+ + \beta_k^-)\,\delta_k + \sum_{k=1}^{K}(\beta_k^+ - \beta_k^-)\,y_k. \quad (20)$$

and $\alpha_i^+, \alpha_i^-, \eta_i^+, \eta_i^-, \beta_k^+, \beta_k^-, \lambda_k^+, \lambda_k^- \geq 0$.

At the saddle point, for each Lagrange multiplier, the Karush-Kühn-Tucker (KKT) conditions must be satisfied

$$\begin{cases} \alpha_i^+\big(\varepsilon_i + \xi_i^+ - d_i + \mathbf{t}_i\mathbf{b}' + b_0\big) = 0, \\ \alpha_i^-\big(\varepsilon_i + \xi_i^- + d_i - \mathbf{t}_i\mathbf{b}' - b_0\big) = 0, \\ \beta_k^+\big(\delta_k + \mu_k^+ - y_k + \mathbf{x}_k\mathbf{b}' + x_k b_0\big) = 0, \\ \beta_k^-\big(\delta_k + \mu_k^- + y_k - \mathbf{x}_k\mathbf{b}' - x_k b_0\big) = 0, \\ (C - \alpha_i^+)\xi_i^+ = 0, \\ (C - \alpha_i^-)\xi_i^- = 0, \\ (V - \beta_k^+)\mu_k^+ = 0, \\ (V - \beta_k^-)\mu_k^- = 0. \end{cases} \quad (21)$$

The parameters b' could be determined form the first condition of (18)

$$\mathbf{b}' = \mathbf{t}^T(\alpha^+ - \alpha^-) + \mathbf{x}^T(\beta^+ - \beta^-), \quad (22)$$

and the parameter b_0 can be determined from the following equations, derived from KKT conditions (21), by taking arbitrary condition for which the corresponding condition is met

$$b_0 = \begin{cases} d_i - \mathbf{t}_i\mathbf{b}' - \varepsilon_i, & \text{for } 0 < \alpha_i^+ < C, \\ d_i - \mathbf{t}_i\mathbf{b}' + \varepsilon_i, & \text{for } 0 < \alpha_i^- < C, \\ (y_k - \mathbf{x}_k\mathbf{b}' - \delta_k)/x_k, & \text{for } 0 < \beta_k^+ < V, \\ (y_k - \mathbf{x}_k\mathbf{b}' + \delta_k)/x_k, & \text{for } 0 < \beta_k^- < V. \end{cases} \quad (23)$$

4 Results

The usefulness of the proposed filter design method were investigated using some standards developed by the International Electrotechnical Commission

(IEC) within the European project "Common Standards for Quantitative Electrocardiography" in order to test the accuracy of ECG signal processing methods. These Common Standards are very well suited to analyze ECG system's software performance in term of baseline removal, line frequency suppression, waveform detection, localization of fiducial points, measurement of ECG parameters, etc. They establish also exact procedures for testing hardware aspects of ECG systems, for example, calibration, amplifier linearity, gain factors, etc.

As a possible approach to evaluate methods used for ECG baseline wander reduction, the IEC committee suggests artificial signal composed of triangular waves. The triangular wave is 1.5 mV high and has 80 ms base width. This signal shall not produce an output signal with an offset from the isoelectric line greater than 20 μV, and shall not produce a slope greater than 50 $\mu V/s$ in a 200 ms region following the impulse and a slope of 100 $\mu V/s$ anywhere outside the region of the impulse. On the other hand, the amplitude response of the required low-pass filter in the 0.67 – 40 Hz passband, should not have ripples greater than ±10%. In other frequency bands the constraints are less important. The described specifications represents a highpass filter with a very narrow transition band and constraints in both frequency and time domains. Using classical filter design methods [7, 8, 9] it was impossible to find filter that fulfills all the constraints. This is caused primary by the fact, that the time domain constraints could not be used during the design process.

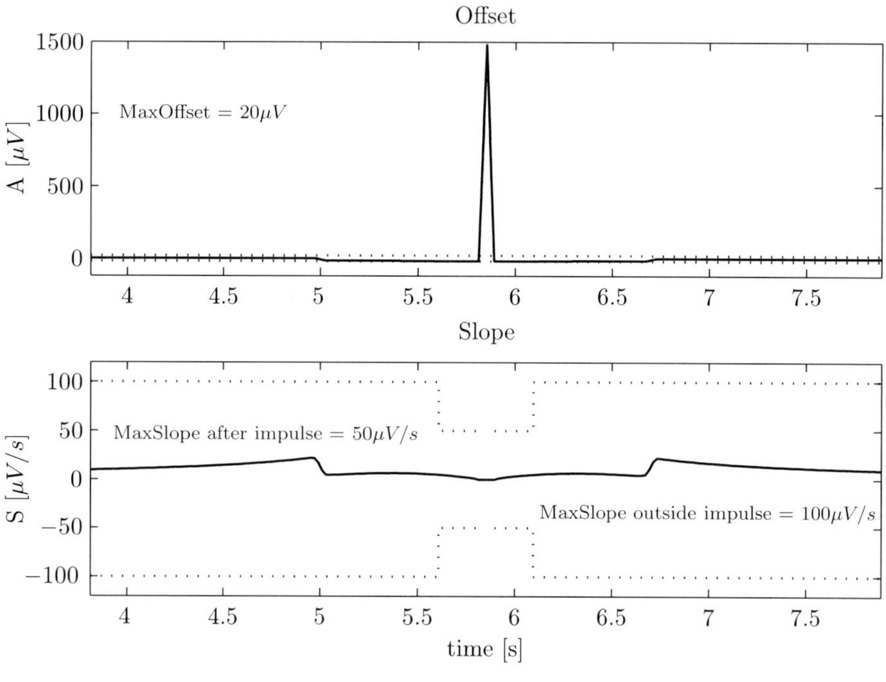

Fig. 1. Time domain response of the designed filter.

The proposed filter design method was used to calculate the required filter coefficients. The minimum filter order, sufficient to fulfill the specified constraints was found to be equal 890. The parameters ε_i $i = 1, \ldots, L$ and δ_k $k = 1, \ldots, K$ was chosen according to the frequency and time domain constraints, defined above.

Figure 1 presents the time domain response of the designed filter applied to the described above triangular-wave signal. All constraints concerning the maximum distortion and slope were fulfilled. Also, the frequency response of the filter, not shown here, does not exceed the predefined limits.

5 Conclusions

In this paper a new method for digital filter design was presented. This method allows to define the filter's specification not only in the frequency domain, but also with respect to the required output signal. The possibilities offered by this new method was illustrated with design of ECG highpass filter for baseline wander reduction.

References

1. Łęski J (2002), Computationally effective algorithm to the ε-insensitive fuzzy clustering, *System Science*, 28(3):31–50.
2. Łęski J (2003), Towards a robust fuzzy clustering. *Fuzzy Sets and Systems*, 137:215–233.
3. Schölkopf B, et al. (1997), Comparing support vector machines with gaussian kernels to radial basis function classifiers, *IEEE Trans. Sign. Processing*, 45:2758–2765.
4. Drucker H, et al. (1997), Support vector regression machines. In M. Mozer et al., (eds), *Advances in Neural Information Processing Systems 9*, MIT Press, Cambridge.
5. Vapnik V (1982), *Estimation of Dependences Based on Empirical Data.* Springer-Verlag, Berlin.
6. Vapnik V (1998), *Statistical Learning Theory.* Wiley, New York.
7. Parks T, McClellan J (1972), A program for the design of linear phase finite impulse response digital filters. *IEEE Transactions on Audio and Electroacoustics*, 20(3):195–199.
8. McClellan J, Parks T (1973), A united approach to the design of optimum FIR linear-phase digital filters. *IEEE Transactions on Circuits and Systems*, 20(6):697–701.
9. McClellan J, Parks T, Rabiner L (1973), A computer program for designing optimum FIR linear phase digital filters. *IEEE Transactions on Audio and Electroacoustics*, 21(6):506–526.

Efficient Implementation of Nearest Neighbor Classification*

José R. Herrero and Juan J. Navarro

Computer Architecture Department, Universitat Politècnica de Catalunya,
Jordi Girona 1-3, Mòdul D6, E-08034 Barcelona, (Spain)
{josepr,juanjo}@ac.upc.es

Summary. An efficient approach to Nearest Neighbor classification is presented, which improves performance by exploiting the ability of superscalar processors to issue multiple instructions per cycle and by using the memory hierarchy adequately. This is accomplished by the use of floating-point arithmetic which outperforms integer arithmetic, and block (tiled) algorithms which exploit the data locality of programs allowing an efficient use of the data stored in the cache memory.

1 Introduction

The Nearest Neighbor (NN) classification procedure is a popular technique in pattern recognition, speech recognition, multitarget tracking, medical diagnosis tools, etc. A major concern in its implementation is the immense computational load required in practical problem environments. Other important issues are the amount of storage required and the data access time. In this paper, we address these issues by using techniques widely used in linear algebra codes. We show that a simple code can be very efficient on commodity processors and can sometimes outperform complex codes which can be more difficult to implement efficiently.

1.1 Nearest Neighbor Classification

The classification problem consists in assigning a class from ℓ classes $C_1, C_2, \ldots, C_\ell$ to each of the D_{size} unclassified vectors $\mathbf{X^j} = [x_1^j, x_2^j, \ldots, x_{V_{size}}^j]$ with length V_{size}, for $j = 1, \ldots, D_{size}$. The NN classification uses a set of vectors $\mathbf{P^k} = [p_1^k, p_2^k, \ldots, p_{V_{size}}^k]$, for $k = 1, \ldots, P_{size}$, called a set of prototypes, whose class is known. Then, an unclassified vector $\mathbf{X^j}$ is classified in the same class as $\mathbf{P^s}$ if $\mathbf{P^s}$ is the prototype with minimum distance to $\mathbf{X^j}$, that is

$$d(\mathbf{X^j}, \mathbf{P^s}) = \min_{k=1,..,P_{size}} d(\mathbf{X^j}, \mathbf{P^k})$$

*This work was supported by the Comissionat per a Universitats i Recerca of the Generalitat de Catalunya (BE94/Annex 3-642 and BE94-A1/110) and the Ministerio de Educación y Ciencia of Spain (TIN2004-07739-C02-01) where, in our examples, the distance function is defined as the square of the Euclidean distance:

$$d(\mathbf{X^j}, \mathbf{P^k}) = \sum_{i=1}^{V_{size}} (x_i^j - p_i^k)^2$$

Hence, the distance between the vector $\mathbf{X^j}$, which is to be classified, and all the vectors $\mathbf{P^k}$, $k = 1, \ldots, P_{size}$, in the prototype set must be computed. The time needed to classify a set of D_{size} vectors is, consequently, proportional to $(V_{size} \times D_{size} \times P_{size})$.

In the algorithms we use, which are shown below, the set of unclassified vectors is kept in matrix $D(V_{size}, D_{size})$ where $x_i^j = D(i,j)$. The set of prototypes is kept in matrix $P(V_{size}, P_{size})$, where $p_i^k = P(i,k)$. Vector $ClassP(P_{size})$ indicates the class the prototypes belong to, i.e. $ClassP(k) = r$ if $\mathbf{P^k}$ belongs to class C_r. The result of the classification is stored in vector $ClassD(D_{size})$, where $ClassD(j) = r$ if $\mathbf{X^j}$ is classified as belonging to class C_r.

Figure 1a shows the common brute force algorithm, which we label as *jki* form due to the loop ordering. In this algorithm, all of the V_{size} components of an unclassified vector $\mathbf{X^j}$, stored in a column of matrix D, are compared to the correspondent components of each vector in the prototype set, stored as columns of matrix P. Often, the *jki* code has been modified to exit the loop that computes the distance when the current distance exceeds the running minimum found, reducing the number of computations required for classification, while keeping the same accuracy as the brute force algorithm. Figure 1b shows the modified *jki* loop, henceforward called *jki_exit* algorithm.

```
MAX = 2 ** 30
DO J = 1, Dsize
 mindis = MAX
 DO K = 1, Psize
  distance = 0
  DO I = 1, Vsize
   sub = D(I,J) - P(I,K)
   distance = distance + sub*sub
  ENDDO
  IF (distance.LT.mindis) THEN
   mindis = distance
   mincla = ClassP(K)
  ENDIF
 ENDDO
 ClassD(J) = mincla
ENDDO
```
a)

```
    MAX = 2 ** 30
    DO J = 1, Dsize
     mindis = MAX
     DO 2 K = 1, Psize
     distance = 0
      DO 1 I = 1, Vsize
       sub = D(I,J) - P(I,K)
       distance = distance + sub*sub
       IF (distance.GT.mindis) GO TO 2
1     CONTINUE
      mindis = distance
      mincla = ClassP(K)
2    CONTINUE
     ClassD(J) = mincla
    ENDDO
```
b)

Fig. 1. Codes for the **a)** *jki* and **b)** *jki_exit* forms

1.2 Related Work

A major concern in the implementation of the NN technique is the immense computational load associated with it and the large amount of computer memory required when large prototype and data sets exist. These problems have been addressed at length and many alternatives proposed: special-purpose hardware, such as systolic arrays [9] and several approaches have shown to be computationally advantageous over the brute force method:

- Modified metrics as alternative distance measures to the Euclidean distance used in classical NN classifiers [3, 5, 10, 12, 15, 18, 19, 21].
- Selection of a design subset of prototypes from a given set of prototype vectors [5, 8, 11, 18, 21] and generation of prototype reference vectors [7].
- Use of fuzzy logic and Self Organizing Maps [4, 14].

A paper [2] compared RISC-based systems to special purpose architectures for Image Processing and Pattern Recognition (IPPR). They concluded that although a lot of progress had been achieved in RISC technology, low advantages could be obtained for IPPR due to the difficulty of producing efficient code for such machines. In this paper, however, we study ways of improving the efficiency of Nearest Neighbor classification on general purpose RISC-based High Performance Workstations since their price can make them cost-effective. Our approach aims to maximize speed maintaining the accuracy of the brute force method by means of an efficient codification of the algorithm using floating-point arithmetic, which increases speed, and a block algorithm, which reduces the number of misses in the cache memory. Such techniques have often been used in numerical applications [1, 20] but never, to our knowledge, to NN classification.

1.3 Processor Overview

Our tests have been carried out on two high performance workstations, which incorporate superscalar processors: an HP PA-7150 [13] and a DEC Alpha AXP-21064 processor [6] respectively. Both implement Integer/Floating-Point two-way superscalar operation, i.e. one integer and one floating-point instruction can be issued each cycle. Loads and stores of floating-point registers are treated as integer operations. The CPU can read two consecutive data words (a total of 8 bytes) every cycle from the external data cache.

The PA-7150 cache has 256 Kbytes, with a line size of 32 bytes. The number of elements in a line (L) is therefore 32 for byte, 8 for simple and 4 for double precision floating-point data types. According to our experiments, a cache miss produces a penalty of 35 cycles. Consequently, the number of Cycles Per Miss (CPM) is 35. The AXP-21064 incorporates separate 8 Kbyte on-chip instruction and data caches, and a 1 Mbyte off-chip unified cache. All of them have line size equal to 32 bytes. A first level cache hit has a 3 cycle latency while a miss which hits in the second level cache is available in 11 cycles for the first word, and 18 for the following one.

1.4 Performance Metrics

In order to compare different codes that solve the same problem, CPU_{time} is a clear candidate to be used as a metric. However, when problem size is changed from execution to execution it is advisable to use a metric normalized to the size of the problem. In this paper we introduce Normalized Cycles (NC), which for our classification problem is computed by:

$$NC = \frac{CPU\_time\_in\_cycles}{V_{size} \cdot P_{size} \cdot D_{size}} \qquad (1)$$

We model the NC with the following expression:

$$NC = NC(cpu) + NC(mem) \qquad (2)$$

where $NC(cpu)$ is the component obtained considering no misses in the memory hierarchy (caches, TLBs, page faults) and $NC(mem)$ represents the penalty cycles due to the misses in the memory system. In the analytical models we develop in this paper, we do not consider the misses produced by instruction fetches since a separate instruction cache exists and the programs we evaluate are sufficiently small so that no instruction misses occur. We include only misses produced by load accesses to matrices since they constitute almost all the data accesses. Experimental results of the NC for different codes are reported in sections 2 and 3. All our programs are written in Fortran.

2 Algorithm Analysis

In this section we present the results obtained from the execution of several codes using distinct data representations. For certain applications floating-point arithmetic is required. In other cases, however, vector elements can be coded as bytes. We have implemented the NN codes using three different data types for the vector elements: byte, simple and double precision floating-point numbers, which require 1, 4 and 8 bytes of storage space respectively.

For any of the data types used, results depend on problem size. For small problems, the sizes of both the prototype and data to be classified have been defined to be small enough to fit into the cache simultaneously. Data are brought into the cache the first time they are referenced. Subsequent references will hit in cache, since all data are kept in it. Executing a code many times and dividing the execution time by the number of times it is performed hides the misses from the first execution. Therefore, $NC(mem) \approx 0$ for a small problem executed many times and NC is approximately $NC(cpu)$:

$$NC_{SmallProblem} \approx NC(cpu) \qquad (3)$$

When the problem size is big enough so that all the data do not fit in the cache at the same time, cache misses arise and data are flushed from the cache between uses. Locality is not well exploited resulting in a poor cache utilization. The $NC(mem)$ component in large problems can be easily estimated by:

$$NC(mem) \approx NC_{LargeProblem} - NC_{SmallProblem} \qquad (4)$$

since $NC(cpu)$ are the same for both large and small problems.

Table 1. NC obtained for different problem sizes

	PA-7150				AXP-21064			
	Small Problem		Large Problem		Small Problem		Large Problem	
Data type	*jki_exit*	*jki*	*jki_exit*	*jki*	*jki_exit*	*jki*	*jki_exit*	*jki*
byte	6.1	15.0	7.1	16.3	24.0	24.0	25.1	25.1
simple float	4.5	3.6	6.0	8.4	22.1	11.9	23.9	19.7
double float	4.5	3.6	7.0	13.3	23.0	20.6	29.2	21.0

To analyze the consequences data size has on performance, two different problem sizes have been tested. Considering the PA-7150's 256 Kbytes data cache, as a small problem we used a database of 200 vectors — 100 for the prototype set and 100 used as data for classification — where each vector has 80 elements. Experiments on a large problem were carried out on a database of 20852 vectors — 10426 for the prototype set and 10426 used as data for classification — each also having 80 components. Table 1 shows the NC measured for a small and a large problem for the different algorithms and data types. These results show that the use of floating-point data is always worthwhile.

Data initialization is important since it impacts on the performance of the *jki_exit* algorithm. A distribution obtained from a real application has been used. The figure on the right shows the probability distribution of the number of iterations of the inner loop computed before it is exited. The mean value of the number of iterations is $\overline{x} = 22$.

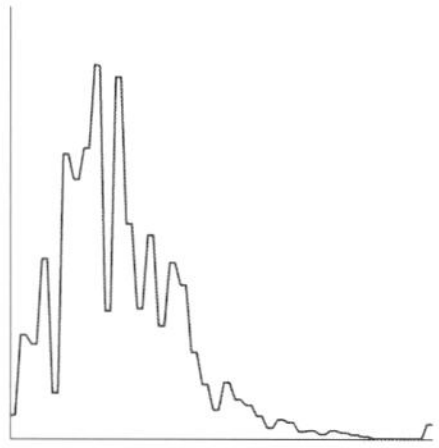

2.1 The $NC(cpu)$ Component

Despite considerably increasing the memory requirements, using simple (4 bytes) or double (8 bytes) floating-point data is better than a simple byte (assuming a datum can be represented in a single byte). This is so because the PA-RISC 7150 processor can issue one load of a floating-point value together with one floating-point multiplication and one floating-point addition (or subtraction) per cycle. On the other hand, when integer arithmetic (byte or integer data type) is used, just one instruction - a load, a multiplication, an addition or a subtraction - can be issued each cycle. Moreover, several data conversions are performed, since all the arithmetic is performed on 32 bit data. Furthermore, the multiplication is computed on the floating-point unit which requires data movements from a general purpose register to a floating-point register, and vice-versa, through memory. From these results we infer that the use of floating-point arithmetic is always beneficial.

In the *jki* code, the compiler applies software pipelining [16] producing an instruction scheduling which circumvents the problem introduced by data dependencies. When a conditional branch is present in the loop body, as in the *jki_exit* code, no software pipelining is applied. Consequently, due to the dependencies between instructions plus the extra instructions implementing the "if" statement, the $NC(cpu)$ of the *jki_exit* becomes larger than that of the *jki* even for the reduced number of iterations of the *jki_exit* inner loop; e.g. an average of 22 for *jki_exit* in our experiments as opposed to 80 for *jki*.

The same is basically true for the AXP-21064. The overhead of the *jki_exit* code is so large that this algorithm is outperformed by the simpler *jki* code. However, since the compiler we had available was not able to perform software pipelining, the instruction scheduling obtained was not as effective as that of the PA-7150. We will thus center our attention on the latter although results of a hand coded software pipelined version developed for the former processor will be shown at the end of the paper.

2.2 The $NC(mem)$ Component

For a small problem the $NC(mem)$ is negligible. As the problem size grows, the $NC(mem)$ component of the NC increases while the $NC(cpu)$ remains constant (equations (3) and (4)). In order to predict the number of cache misses a code produces, it is important to take several aspects into consideration. We analyze the jki and jki_exit codes and make comments upon the relevant points.

Spatial Locality

For each inner loop iteration one data element and one prototype element are referenced. It is important to note that for both data structures, accesses to consecutive addresses (column accesses for both D and P) are performed in consecutive iterations of the innermost loop I, exploiting the spatial locality. When the first element in a line which is not present in cache is referenced, a cache miss is produced with a penalty of CPM cycles. However, since the whole line, containing L elements, is brought into the cache, the subsequent $L-1$ accesses to elements in that line are cache hits introducing no extra penalty cycles.

Temporal Locality

The elements of P as well as those of D are reused through the algorithm. Each element of D is reused once for each iteration of the middle loop K, while each element of P is reused for each iteration of the outer loop J.

For each iteration of the middle loop K a new column of P is referenced. However, a fixed column of D is reused for each iteration of loop K and will only be evicted from the cache, due to conflicts with elements of P, every $\frac{C}{E_{size} \cdot V_{size}}$ iterations of loop K, where C is the cache size in bytes and E_{size} is the element size: 1, 4 or 8 for byte, simple or double precision floating-point data respectively. Therefore, we can be reasonably certain that the elements of D are rarely involved in cache misses.

For each iteration of the outermost loop J, all the elements in P are referenced. However, for large matrices, when a new line of P is referenced in iteration $J = j$ a cache miss occurs. Despite having been accessed in iteration $J = j - 1$, the line has already been evicted from cache because accesses to the whole matrix P have been performed and conflicts appeared among its elements due to its large size.

An Analytical Model for $NC(mem)$

Taking into consideration the spatial and temporal locality of the algorithms presented above, we conclude that, for the *jki* algorithm, a total of $D_{size} \cdot P_{size} \cdot \frac{V_{size}}{L}$ misses occur for the prototype set P. Thus, from equation (1) we obtain:

$$NC(mem) \approx \frac{CPM}{L} \qquad (5)$$

All the statements asserted above are valid for the *jki_exit* code with the only difference that matrix P is not completely referenced within an iteration of loop J. Given a mean number of iterations $\overline{x}$ before the inner loop is exited, approximately $P_{size} \cdot V_{size} \cdot \frac{\overline{x}}{V_{size}}$ elements of P will be used. Assuming these data are still too many to fit in cache, $D_{size} \cdot P_{size} \cdot \frac{V_{size}}{L} \cdot \frac{\overline{x}}{V_{size}}$ misses occur. Therefore, for the *jki_exit* algorithm

$$NC(mem) \approx \frac{\overline{x}}{V_{size}} \cdot \frac{CPM}{L} \qquad (6)$$

The two left columns in table 2 show the NCs obtained using our theoretical model. For these data, an estimation of the $NC(cpu)$ is obtained from the NC of the small problem (equation (3)) shown in table 1. Then, applying the results in equations (2), (3) (5) and (6) we obtain an estimation of the NCs for a large problem which are very close to the empirical results shown in table 1. It is important to note that for each data type used (byte, simple and double precision floating-point) the $NC(mem)$ differs since L changes (32, 8, 4) — see equations (5) and (6). For this reason the usage of simple floating-point data produces better NCs than the use of double floating-point data since both have the same $NC(cpu)$. Despite producing a lower $NC(mem)$, the use of byte data results in worse NCs since its $NC(cpu)$ component is too large to make it competitive.

3 Block Algorithm

In this section we present a new code which exploits the locality in the algorithm considerably better, also producing an $NC(mem) \approx 0$ for large problems. Figure 2a shows the code of a block algorithm we propose for substituting the *jki* algorithm. The same idea can be applied to the *jki_exit* algorithm. In this code the number of loops has increased, but the arithmetic operations are the same. Consequently, the accuracy is exactly the same as that of the non-blocked algorithms. Figure 2b shows the Data and Computation Diagram [20] for the *block* algorithm. The shaded area shows cached data that can be reused. In the *block* code, the same column of P is referenced for each iteration of loop J , while a new column of D is accessed. The probability of cache hits for the data in D is very high, and we will assume no misses appear. For each iteration of loop K all the elements in a block of $V_{size} \times B_{size}$ elements of D are referenced. The block size B_{size} will be dimensioned so that the block fits into the cache: $B_{size} \times V_{size} < \frac{C}{E_{size}}$. If a large B_{size} is chosen, the number of intrinsic misses of P will decrease but the number of conflicts will grow. The optimal block size is considered to be approximately half the cache size C [17]. Consequently, the data in the block remain in the data cache during all the iterations of loop K. Some conflicts will arise, but their number and influence is so low to be considered null. Consequently, the $NC(mem)$ of a block algorithm is very low and can be considered insignificant.

```
MAX = 2 ** 30
DO JJ = 1, Dsize, Bsize
  DO ind = 1, Bsize
    Vaux(ind)=MAX
  ENDDO
  DO K = 1, Psize
    ind = 0
    DO J = JJ, MIN(JJ+Bsize-1,Dsize)
      ind = ind+1
      distance = 0
      DO I = 1, Vsize
         sub = D(I,J) - P(I,K)
         distance = distance + sub*sub
      ENDDO
      IF (distance.LT.Vaux(ind)) THEN
         Vaux(ind) = distance
         ClassD(J) = ClassP(K)
      ENDIF ENDDO ENDDO  ENDDO
```
a)

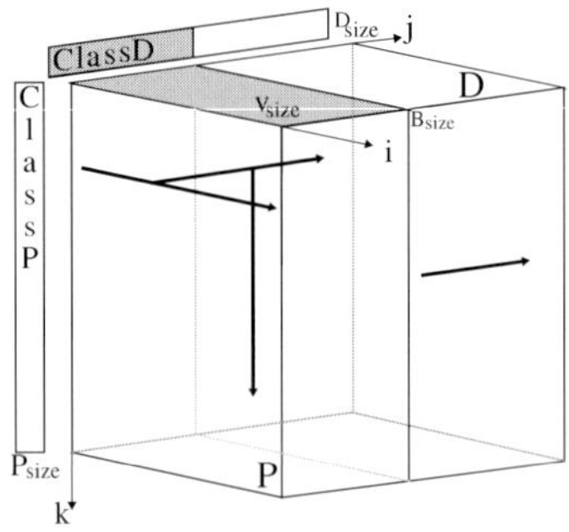

b)

Fig. 2. a) Code for the *block* form. **b)** Data and Computation Diagram for the *block* form.

The last two columns in table 2 show the experimental measures obtained for a large problem using the block algorithm proposed above for a block size $B_{size} = \frac{1}{2} \cdot \frac{256 \cdot K}{V_{size} \cdot E_{size}}$. It should be noted that the NCs obtained are almost identical to those shown in table 1 corresponding to a small problem.

Table 2. NC on the PA-7150 for large problems without and with block algorithms

Data type	*jki_exit*	*jki*	*block_exit*	*block*
byte	6.4	16.1	6.6	15.2
simple float	5.7	8.0	4.8	3.7
double float	7.2	12.3	4.9	3.9

When the byte data type is used, the improvement obtained by the use of blocks is minor, because there is a large $NC(cpu)$ which was already high for the code without blocks which cannot be lowered by blocking. Moreover, there exists high spatial locality due to the fact that the number of elements in a cache line is big ($L = 32$). Simple floating-point data produced slightly better results than double floating-point data (see figure 3) due to more efficient use of the cache line (higher spatial locality).

Finally in table 3 we show the results obtained when we hand-optimized the FORTRAN code on the Alpha AXP-21064 by the application of software pipelining techniques to improve the instruction scheduling for a better instruction level parallelism, and tiling to improve the use of data locality. We show only the results obtained for the single precision float data type since that was the one producing the best performance. SP means Software Pipelining, *num* Bl, means a *num* number of square blocks was applied, Pc implies

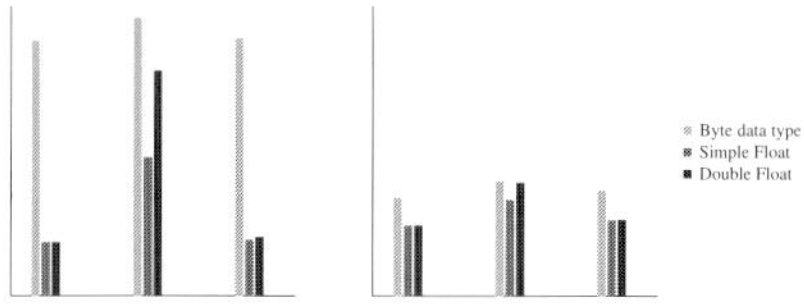

Fig. 3. Comparison of NC for different problem sizes (grouped by data types)

that data precopies were done; BRL stands for Blocking at the Register Level meaning that tiling was also applied in order to improve the reuse of registers in the inner loop. The data in the table shows that the combination of well-known techniques applied to the Nearest Neighbor Algorithm produces significant improvements in performance.

Table 3. NC obtained with hand optimized code on the AXP-21064

	No SP			With SP		
Problem Size	*1 Bl.*	*2 Bl.*	*2 Bl + Pc*	*1 Bl*	*1 Bl + Pc + BRL*	*2 Bl + Pc + BRL*
small	13.7	15.0	11.0	8.1	4.8	4.6
large	14.7	15.1	11.2	9.4	5.6	4.6

4 Conclusions

NN classification has the significant drawback of requiring a large number of computations and data accesses which make it slow if the advantages that current computer architectures offer are not used to full advantage. Frequently, the byte data type has been used in an attempt to reduce the memory usage. In order to decrease the number of computations, an IF statement has often been added to the inner loop to test whether a better solution has already been found. In our experiments, this improved performance by a factor larger than 2. However, this is not the best solution available. Due to processor characteristics, the usage of floating-point arithmetic outperforms the use of integer arithmetic. The resulting machine code can run faster because the instruction level parallelism is higher and no data conversions are needed.

The disadvantage introduced by the use of floating-point data is the larger amount of memory used. This issue can be overcome easily by means of block algorithms. When these kinds of algorithms are used, the temporal locality of programs is better exploited resulting in low number of cache misses, allowing the computations to proceed at full speed. The use of simple floating-point data produces fewer misses than the use of double precision floating-point data due to better usage of spatial locality. However, the difference from the latter is almost negligible because of the reduced number of cache misses incurred when a block algorithm is used (see figure 3). In our experiments, the results obtained when a block algorithm and simple precision floating-point data are used are between 2 and 4 times faster than the algorithms which use integer arithmetic although they require 4 or 8 times more data storage. These results can be generalized for other superscalar architectures.

References

1. E. Anderson, J. Dongarra, LAPACK User's Guide, SIAM, Philadelphia, 1992.
2. P. Baglietto, M. Maresca, M. Migliardi, Image Processing on High-Performance RISC Systems, Proceedings of the IEEE, 84(7):917–930, July 1996.
3. S. Bandyopadhyay and U. Maulik Efficient prototype reordering in nearest neighbor classification, Pattern Recognition 35(12):2791–2799, Dec. 2002.
4. Z. Chi, J. Wu and H. Yan, Handwritten numeral recognition using self-organizing maps and fuzzy rules, Pattern Recognition, 28(1):59–66, 1995.
5. B.V. Dasarathy, Nearest Neighbor (NN) Norms: NN Pattern Classification Techniques, IEEE Computer Society Press, 1991.
6. Digital Equip. Corp., DECchip 21064 and DECchip 21064A Alpha AXP Microprocessors - Hardware Ref. Manual, 1994.
7. C. Decaestecker, Finding Prototypes for Nearest Neighbor Classification by Means of Gradient Descent and Deterministic Annealing, Pattern Recognition, 30(2):281–288, 1997.
8. A. Djouadi, E. Bouktache, A Fast Algorithm for the Nearest-Neighbor Classifier, IEEE Trans. on Pattern Analysis and Machine Intelligence, 19(3):277–282, 1997.
9. J. Fu, T.S. Huang, VLSI for Pattern Recognition and Image Processing, Springer-Verlag, Berlin, 1984.
10. P.J. Grother, G.T. Candela and J.L. Blue, Fast Implementation of Nearest Neighbor Classifiers, Pattern Recognition, 30(3):459–465, 1997.
11. Y. Harnamoto, S. Uchimura and S. Tornita, A Bootstrap Technique for Nearest Neighbor Classifier Design, IEEE Trans. on Pattern Analysis and Machine Intelligence, 19(1):73–79, 1997.
12. R. Van Der Heiden and F.C.A. Groen, The Box-Cox Metric for Nearest Neighbor Classification Improvement, Pattern Recognition, 30(2):273–279, 1997.
13. Hewlett Packard, PA-RISC 1.1 Architecture and Instruction Set Reference Manual, 1994.
14. T. Kohonen, The self-organizing map, Proc. of the IEEE 78(9):1464–1480, 1990.
15. M. Kudo, N. Masuyamaa, J. Toyamaa and M. Shimbob, Simple termination conditions for k-nearest neighbor method, Pattern Recognition Letters 24(9-10):1203–1213, June 2003.
16. M. Lam, Software Pipelining: An Effective Technique for VLIW Machines, Proc. of the SIGPLAN'88, pp 318–328.
17. M.S. Lam, E.E. Rothberg and M.E. Wolf, The Cache Performance and Optimizations of Blocked Algorithms, ASPLOS 1991, pp. 67–74.
18. E.W. Lee and S.I. Chae, Fast Design of Reduced-Complexity Nearest-Neighbor Classifiers Using Triangular Inequality, IEEE Trans. on Pattern Analysis and Machine Intelligence, 20(5):562–566, 1998.
19. C.-L. Liu, H. Sako, H. Fujisawa, Performance evaluation of pattern classifiers for handwritten character recognition, Int. J. on Document Analysis and Recognition 4:191–204, 2002.
20. J.J. Navarro, A. Juan and T. Lang, MOB Forms: A Class of Multilevel Block Algorithms for Dense Linear Algebra Computations, ACM Int. Conf. Supercomputing, 1994, pp. 354–363.
21. F. Ricci and P. Avesani, Data Compression and Local Metrics for Nearest Neighbor Classification, IEEE Trans. on Pattern Analysis and Machine Intelligence, 21(4):380–384, 1999.

Inductive Development of Customer e-Loyalty Theory with Bayesian Networks

Waldemar Jaroński[1], Koen Vanhoof[1] and José Bloemer[2]

[1] University of Limburg, Universitaire Campus, Gebouw D, 3590 Diepenbeek, Belgium waldemar.jaronski@luc.ac.be; koen.vanhoof@luc.ac.be
[2] University of Nijmegen, P.O. Box 9108, 6500 HK Nijmegen, The Netherlands j.bloemer@nsm.kun.nl

Summary. The main objective of this paper is to show the use of Bayesian networks in inductive research applied in an e-loyalty study and to investigate whether e-loyalty theories can be discovered by means of Bayesian networks.

1 Introduction

Two well-known routes for scientific discovery exist: in the deductivist approach, we start by making speculations about a theory, forming assumptions and advancing hypotheses; next, we proceed by proposing a hypothetical model, and ultimately, we can deduce generalizations [5]; the inductive research in the strict sense starts typically with making observations about the world and recording them as data; next, the data are rearranged and analysed so as to "bring order out of chaos"; lastly, lawlike generalizations or patterns are induced [5]. We will not follow the strict inductivist route literally but we will adapt it so that it fits in the Bayesian network framework. In our implementation of the inductive research, we start with making observations and recording data on online visitor bases of four different web portals; next, for each data set we develop a specific theoretical model; and lastly, we arrange the findings from these four different models into one overall theoretical model of e-loyalty.

We will adopt the definition of theory according to which it is "a systematically related set of statements, including some lawlike generalizations, that is empirically testable" [9]. The purpose of theory is to increase scientific understanding through a systemized structure capable of both explaining and predicting phenomena. Furthermore, we accept the position that a model is a formal representation of a theory.

The remainder of this paper has the following structure. In Section 2 we present the details on Bayesian network approach for inductive research. Data issues, such as collection and pre-processing are addressed in Section 3. In

Section 4 we discuss the results. We conclude in Section 5, in which we also draw implications for researchers and practitioners, and provide limitations of the presented research.

2 Bayesian network approach for inductive research

2.1 Experimental design

As an underlying assumption, we will presume that the observed data have been generated by an a priori unknown process that can be represented with some Bayesian network model. This process concerns the way that describes the e-loyalty phenomenon. Consequently, the aim of the structural learning with the Bayesian network methodology is to try to recover this process by inferring the best structure, in the form of a Bayesian network, from the observed data. In the course of the model generation many models will be evaluated with a scoring metric. The scoring metric that we will use to find the best structure has the property that given sufficient data it scores the generative model, i.e., the model from which the data could be sampled, higher than any other model that is not equivalent to the generative model [4]. So we should end up with a model that is the best model that could possibly generate the observed data at hand.

Learning the Bayesian network model of any domain involves selection of a good network structure and estimation of the model's probabilistic parameters. We have taken the Bayesian learning approach, which is suited both for the selection of the best-fitting network structure (structural learning) and estimation of the model's parameters (parameter learning). First, data are used to choose the network structure with the largest posterior probability. The posterior probability of a candidate Bayesian network structure B_s can be according to the Bayes' rule expressed as

$$p(B_s|D) = \frac{p(B_s)p(D|B_s)}{p(D)} \tag{1}$$

where $p(B_s)$ is the prior probability of the model structure B_s, $p(D)$ is the probability of the observed data D, and $p(D|B_s)$ is the likelihood of the data given the Bayesian network structure B_s. We can obtain the probability of the data $p(D)$ by summing up the nominators in Formula 1 for each possible Bayesian network structure B_{s_i}:

$$p(D) = \sum_i p(B_{s_i})p(D|B_{s_i}) \tag{2}$$

It follows from Equation 1 that if the prior probabilities $p(B_s)$ of the candidate hypothetical models B_s are equal then the posterior probability of the model $p(B_s|D)$ is uniquely identified by the likelihood $p(D|B_s)$ of the data given the model B_s. We can ignore the probability $p(D)$ of the data since this value is constant and independent of any particular model B_s. The likelihood $p(D|B_s)$ can be considered the complete likelihood of the model structure B_s obtained by summing the likelihoods for all the possible instantiations of probabilistic parameters θ in the model. This can be expressed as an integral over all the possible instantiations of probabilistic parameter values θ contained in conditional probability tables:

$$p(D|B_s) = \int p(D|B_s, \theta) d\theta \tag{3}$$

The probability of data $p(D|B_s)$ given the model is referred to as the *marginal likelihood* of data D given Bayesian network structure B_s to denote that all the parameter values have been marginalized out of the model. For us, the most important consequence of the marginal likelihood in Formula 3 is that the higher this likelihood is, the higher the chance that the model B_s has generated the data D. Under certain conditions, the calculation of the marginal likelihood can be carried out quite efficiently [2].

3 Data issues

The data have been created by means of software that acts as a layer between the web and the web browser and is aimed to monitor behaviour of web users while surfing the web. The data can be categorised in three groups: sociodemographic data, behaviour data, and opinion data. The data concerning the socio-demographic profile is gathered during the installation. Each piece of information concerning the behaviour, such as URL address, date, exact time and duration of viewing is registered locally in a database on the user's computer. As regards the opinion data, occasionally, when the user is in the process of visiting a specific web site, the website address triggers a poll regarding user's opinions and judgments towards the site and the user is invited to complete it. Completed participation in the opinion polls is rewarded with bonus points.

As regards the representativeness of the sample, it might be more representative for heavy Internet users, who surf online a lot we note that the sample might suffer from the self-selection bias. We have also performed a quick comparison of the samples in the current study with the respondents from other Web-based studies that we have found in the literature in terms of the basic demographic attributes [10]. As regards gender, proportion of males varied between 66.3-71.3% across the selected sites. In conclusion, we can presume that our respondent base is representative of the entire population of web surfers.

The websites ranged over many types, including among others news, portal, financial, e-commerce, and adult sites. We have decided to consider portal sites for two reasons; firstly, there was more usable data on portals than on other website types; secondly, portals constitute a relatively homogeneous group. We have extracted data describing four of the most often visited portal sites in the Netherlands in 2000: WorldOnLine, MSN, Ilse, and Freeler. The surveying and behaviour data for this sample concern the period between September 2000 and April 2001.

The screening of the database resulted in a set of variables that we ultimately have taken into further investigation. Among the sociodemographic variables there are age, gender, education and position in the household. Position in the household indicates whether he/she is financially a head of a family, or rather a consumer of financial resources. Education is a selection of the highest education level reached by the respondent, and contains seven categories: high school, college, etc. The second group of variables that we have eventually taken into further consideration were three potential dimensions of the web site quality. In this context, we have examined look and feel, layout, and the ease of navigation. Another variable that we include is the overall rating about the website. This is a theoretically more complex, more equivocal, and more capacious concept than the other three mentioned above, since it tries to capture the user's overall perception of the website. As such, it can be regarded as the user attitude. The responses of each of these were all recorded with one item on 5-value rating scales.

In line with our multidimensional theoretical definition of the e-loyalty, we have included two variables that allow for the operationalization of the e-loyalty. The Likelihood to return refers to the user's subjective level of certainty that he/she will visit the website again. The Stickiness was computed as a quotient of the effective time spent on the website by a user (see section on data collection) and the number of sessions at the given website during the measurement period.

The data required additional cleaning and pre-processing to allow for efficient application of our approach and to reduce the data bias in the results. These operations resulted in the ultimate size of 409, 169, 140 and 215 records, for MSN, WOL, Ilse and Freeler respectively.

4 Results

Scoring every possible model structure by means of the marginal likelihood $p(D|B_s)$ would be in practice infeasible. A possible way around would be to use a search algorithm that takes only a subset of possibly highly scored Bayesian network structures into consideration. For this purpose, we applied the greedy search algorithm known as K2 [2]. This method requires an ordering of the variables. The selection of ordering was inspired by making the assumption that independent sociodemographic variables can act as moderators or determine the perception of website quality features. These perceptions can in turn act as determinants of the attitude towards the website. All these variables can at last be antecedents of customer e-loyalty.

4.1 Qualitative analysis

The results of the structural learning procedure for all four datasets are shown in Fig. 1. The models shown can be regarded as the most probable structures of dependencies between the variables involved in this e-loyalty study for each website given the assumptions stated above. For instance, let's take a look at the consequences of the found structure for the WOL data. The variables Gender and Pos_Household seem to be related only to each other, whereas Age seems neither to be relevant to loyalty nor to any other feature. We can see that Education is directly related with Layout. Look&Feel influences directly Navigation and Attitude. Navigation is also a determinant of Attitude. Furthermore, there is a link between Attitude and Return. In this model, Attitude can be thus regarded as a classical mediating variable, since it mediates the link between perceptions of website quality attributes and intentional measure of e-loyalty. We can see also that the variables that we have conceptualised as measures of customer e-loyalty, i.e. Return and Stickiness, are interdependent - a positive result (because as measures of one underlying concept they should ideally be somewhat correlated). The model suggests that Stickiness and other variables in the domain are independent given the value of Return, so that once the value of Return is known, our beliefs regarding Stickiness are not altered.

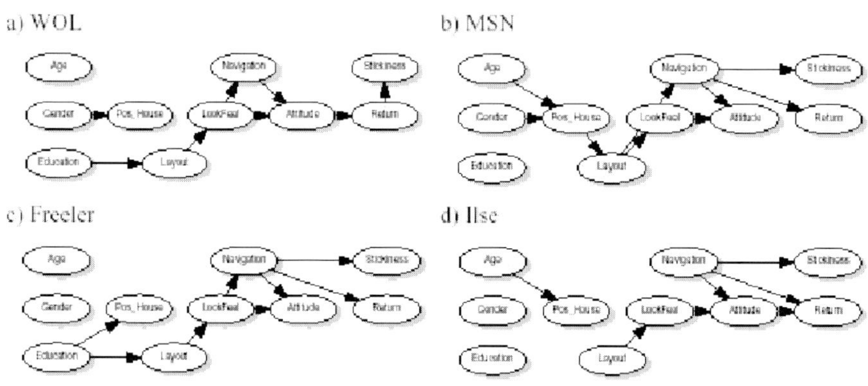

Fig. 1. The most likely models for each dataset.

Let's take a look at the consequences of the found structure. The variables Gender and Pos_Household seem to be related only to each other, whereas Age seems neither to be relevant to loyalty nor to any other feature. We can see that Education is directly related with Layout. Look&Feel influences directly Navigation and Attitude. Navigation is also a determinant of Attitude. Furthermore, there is a link between Attitude and Return. In this model, Attitude can be thus regarded as a classical mediating variable, since it mediates the link between perceptions of website quality attributes and intentional measure of e-loyalty. We can see also that the variables that we have conceptualised as measures of customer e-loyalty, i.e. Return and Stickiness, are interdependent - a positive result (because as measures of one underlying concept they should ideally be somewhat correlated). The model suggests that Stickiness and other variables in the domain are independent given the value of Return, so that once the value of Return is known, our beliefs regarding Stickiness are not altered. These above models can be regarded as causal networks under the assumption that every statistical association derives from causal interaction, and that there are no hidden common causes that could play a role in the domain [4].

4.2 Statistical validation

In the statistical sense, the validation of the structure of the above models is based on the measure of the posterior probability of the model. On the basis of this measure we can conclude that these models are valid conceptualisations of the domain in question. Furthermore, we conclude that the links are significant. Keeping in mind the constraints of the prior ordering and the greedy nature of the algorithm, we can conclude that the models outperform any other alternative model in its ability of explaining the data. To be more precise, our conclusion is that these are probably the best models in that they best explain the given data, and that other models might also provide good explanation but are less probable. We cannot however conclude categorically whether these models are significant in the absolute sense using the Bayesian approach.

4.3 Overall model of customer e-loyalty

The most likely model structures that we have found thus far were specific for each website separately. One way to build an overall model would be to pool all the data together, and repeat the same procedure as above. However, the results would be then biased by the big size of MSN data. Furthermore, the approach that we here apply enables drawing conclusions from different studies when no original data are available.

To construct a possible overall model of e-loyalty resulting from the four website specific models in question, we can sum up all the occurrences of direct dependencies between various nodes. This approach is not a good measure since it does not take the probability of each dependency into account.

The probabilistic nature of dependencies enables the construction of the overall model consistently with the probabilistic framework. In this case, we can resort to the Bayes factor between various dependencies. The Bayes' factor can be used to compare different dependencies given the prior probabilities of each dependency are equal. As a result, the most likely general model of dependencies in the e-loyalty domain is presented in Fig. 2.

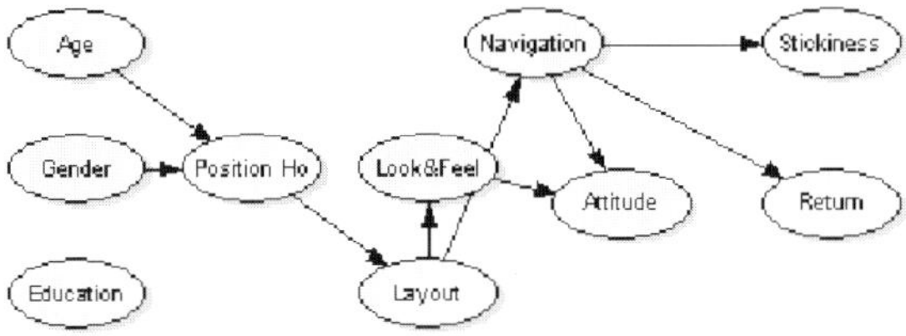

Fig. 2. The most probable overall model of e-loyalty theory found in the study.

Let us consider this overall model in light of the extant theory of e-loyalty. We note that these results can be easily accepted intuitively. The link between ease of navigation and intention to revisit the website is supported in the literature. As a matter of fact, recent research in drivers of e-loyalty shows that ease of navigation is one of the most important website characteristics that could contribute to e-loyalty [3, 7, 11]. In [1], support was found for the theoretical relation of shopping efficiency and loyalty intention, however they didn't find support for the hypothesis that website navigation is a dimension of shopping efficiency. Consumer characteristics such as gender, age, income, are often considered as potentially having influence on customer perceptions and evaluations of service delivery [11, 8], so the presence of the link between Position in the Household and the perception of Layout is theoretically sound. The indirect link between demographics such as Age and Gender and Layout through Position in the household is also very likely. In conclusion, we find that the discovered overall model of customer e-loyalty is to a large extent consistent with the extant e-satisfaction and loyalty literature.

5 Conclusions and future research

We have performed an investigation into potential of inductive development of e-loyalty theory by means of Bayesian networks. The result was positively surprising: the learned models are very similar to each other in terms of theoretical consequences. We can thus observe that the inductive search with the Bayesian network approach makes very reliable inference from data. Hence, we conclude that the results obtained are generic, in the sense that the differences that exist in all possible aspects of each portal site considered, and most importantly web users' perception thereof, do not have any influence on the underlying theoretical model of e-loyalty. This finding suggests also that there exists an overall model of e-loyalty that is valid generally.

Even more interestingly, we found this overall model likely to be theoretically sound given the existing e-loyalty literature. Unfortunately, since the e-loyalty literature is scarce, we were not able to find that they are fully confirmed by this literature. Our contribution to the e-loyalty phenomenon is that attitude is not so much important given the website quality.

From the Bayesian network modelling perspective, we must conclude that not only the greedy nature of the algorithm that searches for the most likely model, but also the marginal likelihood score itself, as a measure of goodness of fit, proved appropriate in developing theory of customer e-loyalty.

The approach applied here is based on a set of assumptions that should be taken into account when interpreting the results. One of the main limitations is a requirement of a prior ordering of variables as it can influence the results to a large extent; there are various approaches to circumvent this limitation that should be evaluated. Another topic for further work is to analyse the impact of different schemes of category aggregation on the results of structural learning, in terms of favouring the existence of links between constructs or the lack thereof. Similarly, studies of its impact on the strength and the character of these relationships should also be undertaken.

References

1. Chen Z, DeVaney SA, Liu S (2003) Consumers' Value Perception of an E-Store and Its Impact on E-Store Loyalty Intention. In: Proc. of the 7th Triennial AMS/ACRA Retailing Conference 2003, November 6-9, Columbus
2. Cooper G, Herskovits E (1992) A Bayesian Method for the Induction of Probabilistic Networks from Data. Machine Learning 9:309–347
3. Gommans M, Krishnan KS, Scheffold KB (2001) From Brand Loyalty to E-Loyalty: A Conceptual Framework. Journal of Economic and Social Research 3:43–58
4. Heckerman D, Meek C, Cooper G (1999) A Bayesian Approach to Causal Discovery. In: Glymour C, Cooper G (eds) Computation, Causation, and Discovery. MIT Press, Cambridge, MA
5. Hunt SD (1991) Modern Marketing Theory: Critical Issues in the Philosophy of Marketing Science. South-Western Publishing Company, Cincinnati
6. Larranaga P, Kuijpers C, Murga R, Yurramendi Y (1996) Learning Bayesian network structures by searching for the best ordering with genetic algorithms. IEEE Transactions on System, Man and Cybernetics 26:487–493
7. Loiacono ET, Watson RT, Goodhue DL (2000) WebQualTM: A measure of website quality. Available online: http://www.terry.uga.edu/cisl/includes/pdf/webqual.pdf [Accessed 6 December 2002]
8. Ranaweera, C., G. McDougall, and H. Bansal, 2005, "A Model of online Customer Behavior: Moderating Effects of Customer Characteristics." Marketing Theory (forthcoming)
9. Rudner R (1966) Philosophy of Social Science. Englewood Cliffs, NJ. Prentice-Hall, Inc.
10. Szymanski DM, Hise RT (2000) e-Satisfaction: an Initial Examination. Journal of Retailing 76:309–322
11. Zeithaml VA, Parasuraman A, Malhotra A (2002) Service Quality Delivery Through Web Sites: A Critical Review of Extant Knowledge. Journal of the Academy of Marketing Science 30:362–375

Reference Set Size Reduction for 1-NN Rule Based on Finding Mutually Nearest and Mutually Furthest Pairs of Points

Adam Jóźwik[1,2] and Paweł Kieś[3]

[1] Institute of Biocybernetics and Biomedical Engineering, Polish Academy of Science, 02-109 Warsaw, 4 Trojdena Str.
adamj@ibib.waw.pl

[2] Technical University of Łódź, Computer Engineering Department, Al. Politechniki 11, 90-924 Łódź

[3] Institute of Fundamental Technological Research, Polish Academy of Science, 00-049 Warsaw, 21 Świętokrzyska Str.
pkies@ippt.gov.pl

Summary. Two algorithms for reference set size reduction are presented and tested on an actual data set of a large size. The first one, as most of existing procedures, is based on the consistency idea, which means that all points from the primary reference set are correctly classified by 1-NN rule operating with the reduced set. The second algorithm requires division of the reference set into some subsets and replacing these subsets by their gravity centers. These gravity centers assume the same label as the majority of points of the corresponding subset. This algorithm enables the condensation of the reference set to the desired size, however, the resulting sets do not offer as good classification quality as other existing methods.

As opposed to the first algorithm, the second algorithm does not enable the control of the reduced set size. It is shown that combining both of these algorithms promises as good of a performance while allowing the control of the size of the obtained condensed sets.

1 Introduction

The k-NN classifier [1] offers a good performance and its training requires determination of k. For a fixed feature set, the value of k is the only parameter that ought to be established on the basis of the training set. The value of k is chosen experimentally in such a way as to minimize the probability of misclassification, which can be estimated with the *leave-one-out* method.

However, there are some applications with high requirements pertaining to the speed of classification. If the parameter k is greater than one then certain acceleration can be achieved by an approximation of a k-NN classifier by a 1-NN one. To make this approximation, it is sufficient to reclassify the primary reference set (the training set) by applying the k-NN rule and then to use the 1-NN rule with this reclassified reference set. Often, the difference between the performance offered by the optimum k, in the sense of minimum misclassifications, and that obtained by the 1-NN rule is not significant and the reclassification phase can be omitted.

More effective acceleration can be achieved by a reduction of the reference set [2, 3]. Most procedures for reference set reduction known in the literature are based on the consistency idea, i.e., all points from the primary reference set must be correctly classified by the 1-NN rule with the reduced reference set. However, none of the algorithms promises the reduction of the reference set to a minimum possible size.

2 Reference Set Size Reduction Algorithms

Two most popular algorithms for the reference set reduction are described below.

2.1 Hart's Algorithm

The first point from the reference set is qualified to the (initially empty) reduced reference set. Next, the remaining points of the primary reference set are classified by the 1-NN rule with the current reduced reference set. Each misclassified point is added to the reduced reference set. Such classification of all points from the primary reference set is repeated as long as m subsequent classifications do not increase the size of the reduced reference set, where m denotes the numerical force of the original reference set. Usually from 3 to 6 repetitions are sufficient. The first points selected to the reduced reference set can lie far away from the class boundary (in the Bayes classifier meaning). This disadvantage of the Hart's algorithm has been eliminated by the Gowda-Krishna modification.

2.2 Gowda-Krishna Algorithm

A mutual distance measure $mdm(\boldsymbol{x})$ is associated with each point $\boldsymbol{x}$ of the primary reference set. The $mdm(\boldsymbol{x})$ is calculated in the following way: for a point $\boldsymbol{x}$, the nearest point $\boldsymbol{y}$ from the opposite class is found. The number of points from the same class as $\boldsymbol{x}$ that lie closer to $\boldsymbol{y}$ than $\boldsymbol{x}$ is the value of $mdm(\boldsymbol{x})$. Next, all the points of the primary reference set are arranged according to the growing values of $mdm(\boldsymbol{x})$. Finally, the Hart's algorithm is applied to the reference set arranged this way.

2.3 Proposed Algorithm 1 for Reference Set Reduction

The Gowda-Krishna mutual distance measure is used to arrange the points of the reference set in a way that ensures the presentation of the points lying on the class boundaries as the first ones. This approach allows to avoid recruitment of points that lie far from the class borders. The class border, as it has already been mentioned, is understood as boundaries of decision regions defined by the Bayes classifier.

We propose another way to avoid the above-mentioned disadvantage of the Hart's procedure. For simplicity, we consider the two-class problem. Our approach is based on finding all such pairs of points $\boldsymbol{x}$ and $\boldsymbol{y}$ from the opposite class in the reference set, so that $\boldsymbol{y}$ is the nearest neighbor of $\boldsymbol{x}$ in the opposite class of $\boldsymbol{x}$ and vice versa, $\boldsymbol{x}$ is the nearest neighbor of $\boldsymbol{y}$ in the opposite class of $\boldsymbol{y}$. All points from such pairs are automatically selected to the reduced set. However, these points do not ensure the consistency. The consistency can be obtained by addition of new points from the reference set by the Hart's procedure.

Following is an explanation of how the pairs of mutually nearest points from the opposite classes can be determined. Let $\boldsymbol{X}$ and $\boldsymbol{Y}$ represent different classes in the reference set. To find the first pair of the mutually nearest points we start with the point $\boldsymbol{x}_1$ in $\boldsymbol{X}$ and find its nearest neighbor $\boldsymbol{y}_1$ in $\boldsymbol{Y}$, then we find $\boldsymbol{x}_2$ as the nearest neighbor of $\boldsymbol{y}_1$ in $\boldsymbol{X}$, then we find $\boldsymbol{y}_2$ as the nearest neighbor of $\boldsymbol{x}_2$ in $\boldsymbol{Y}$, then $\boldsymbol{x}_3$ as the nearest neighbor of $\boldsymbol{y}_2$ in $\boldsymbol{X}$, and so on, until $\boldsymbol{x}_i = \boldsymbol{x}_{i+1}$ (or $\boldsymbol{y}_i = \boldsymbol{y}_{i+1}$). It can be noticed that in order to determine another such pair we ought to start with the point from the set $\boldsymbol{X} := \boldsymbol{X}$-$\{\boldsymbol{x}_1, \boldsymbol{x}_2, ..., \boldsymbol{x}_i\}$ or from the set $\boldsymbol{Y} := \boldsymbol{Y}$-$\{\boldsymbol{y}_1, \boldsymbol{y}_2, ..., \boldsymbol{y}_i\}$. Starting with a point $\boldsymbol{x} \in \{\boldsymbol{x}_1, \boldsymbol{x}_2, ..., \boldsymbol{x}_i\} \cup \{\boldsymbol{y}_1, \boldsymbol{y}_2, ..., \boldsymbol{y}_i\}$we would obtain the same pair of mutually nearest points from the opposite classes, which has already been found previously. The above-mentioned property makes the procedure of finding such pairs rapid. A more formal description of the procedure for finding the mutually nearest points from the opposite classes and the illustrating example (Fig. 1) are given in the Appendix.

2.4 Proposed Algorithm 2 Based on Finding Mutually Furthest Points

The concept of the proposed algorithm is to divide the reference set into some subsets. These subsets are later replaced by their gravity centers, which assume the same membership labels as the majority of the points in the corresponding subsets. Ties can be broken by assigning the class that is most heavily represented in the training set and then by indicating the class with the lower index in the assumed sequence of classes.

The general idea of the proposed procedure is as follows. We start with the condensed set containing only one point, *i.e.* the gravity center of the whole reference set, with the label of the majority of points. Next, the reference set is divided into two subsets and the gravity centers of these subsets are found. They assume the same labels as the majority of points of the corresponding subsets. The single gravity center of the divided subset (*i.e.* the whole reference set) is replaced by the two new gravity centers. In this way the number of subsets as well as the size of the current condensed reference set increases from 1 to 2. The condensed reference set contains now two points. In the subsequent step one of the subsets is divided again. Its gravity center is replaced by the two gravity centers of the newly obtained subsets. The labels of the new gravity centers are assigned in the earlier-mentioned manner, *i.e.* by the majority rule. In this way the size of the condensed reference set increases until it reaches the desired value.

To complete the algorithm definition it is necessary to establish which of the currently existing subsets will be divided in the subsequent step and provide a method of this division.

The selection of the subset to be divided requires finding one pair of mutually furthest points in each of the currently existing subsets. Two points $\boldsymbol{x}$ and $\boldsymbol{y}$ from the considered subset form a pair of mutually furthest points if $\boldsymbol{y}$ is the maximally distanced point from $\boldsymbol{x}$ and $\boldsymbol{x}$ is the maximally distanced point from $\boldsymbol{y}$. The subset with the largest distance between mutually furthest points is recruited as the one to be divided in the next step.

Let us now explain how the pair of mutually furthest points is found. We start with finding the point $\boldsymbol{x}_1$ maximally distanced from the gravity center of the reference set. Then we find the point $\boldsymbol{x}_2$ maximally distanced from $\boldsymbol{x}_1$, next the point $\boldsymbol{x}_3$ maximally distanced from $\boldsymbol{x}_2$, and so on. We stop when $\boldsymbol{x}_{i-1}=\boldsymbol{x}_{i+1}$ for certain i. The points $\boldsymbol{x}_{i-1}$ and $\boldsymbol{x}_i$ form a pair of mutually furthest points. A more formal description of the method for finding the mutually furthest points in the subset of the reference set and the illustrating example (Fig. 2) are presented in the Appendix.

The hyperplane orthogonal to the straight line joining $\boldsymbol{x}_{i-1}$ and $\boldsymbol{x}_i$ and passing through the point $(\boldsymbol{x}_{i-1}+\boldsymbol{x}_i)/2$ divides the considered subset into two smaller subsets.

We call this approach a reference set condensation algorithm because the obtained set is not a subset of the primary reference set.

3 Experimental Results

The analyzed data set comes from ultrasound images that are sections of certain 3D objects found in a human body (liver). Pattern classification is used for segmentation of the images. The most usable information is contained in the gray level distribution of the investigated pixel neighborhood. By an application of the orthogonal discrete wavelet transforms, 13 features were extracted. Only two classes of pixels were taken into account, *i.e.* class 1 that represents the objects (metastasis) of interest and class 2 that denotes the background (liver areas without metastasis). The data set contained 80800 pixels, 10100 and 70700 from the first and the second class respectively. This data set was divided into the training part with 40000 pixels and the testing one with 40800 pixels. In each part the ratio of pixels in the classes was 1:7, *i.e.* exactly the same as in the whole analyzed data set.

Let us come to the explanation of the experiments with discussed algorithms and their combinations mentioned in Tab. 1.

Algorithm 2 (see experiment I in Tab. 1) allows to condense the reference set to the desired size. The minimum error rate of 0.0162 was obtained for the primary, not condensed, reference set containing 40000 pixels. We can observe a gradual increase of the error rate as the size of the condensed set decreases (experiment I). Application of one of the algorithms based on consistency idea to the condensed sets received by the algorithm 2, *i.e.* the Hart's algorithm (experiment II), the Gowda-Krishna procedure (experiment III) or the algorithm 1 (experiment IV) gives very significant reference set size reduction. As we can see from the results (experiment V), direct use of algorithm 2 offers few times greater error rates.

Table 1. The results for condensed and reduced reference sets of different size: 1st sub-row: error rate, 2nd sub-row: size of reduced/condensed set

I. Algorithm 2	0.0480	0.0294	0.0226	0.0204	0.0190	0.0176	0.0169	0.0162
condensation	5000	10000	15000	20000	25000	30000	35000	40000
II. Hart reduction	0.0528	0.0340	0.0272	0.0249	0.0234	0.0211	0.0199	0.0196
	539	942	1185	1411	1595	1683	1838	1947
III. Gowda-Krish-	0.0521	0.0347	0.0261	0.0233	0.0238	0.0211	0.0199	0.0189
na reduction	471	852	1071	1230	1382	1463	1593	1699
IV. Algorithm 1	0.0534	0.0361	0.0277	0.0243	0.0233	0.0213	0.0206	0.0242
reduction	496	885	1121	1287	1456	1540	1655	1890
V. Algorithm 2	0.1174	0.1058	0.0984	0.0964	0.0932	0.0924	0.0906	0.0869
condensation	500	900	1200	1300	1500	1550	1700	1900

The best results, from among the considered algorithms based on the consistency idea, as Tab. 1 implies, offers the Gowda-Krishna procedure, i.e. the error rates and the reduced set sizes are the smallest ones. Comparing the results of the experiments II and IV, it can be seen that algorithm 1 offers smaller reduced set sizes but slightly higher error rates. Both algorithms, as well as the Gowda-Krishna, require several presentations of the primary reference set.

We have already mentioned that Hart's algorithm selects at the beginning the points, which may lie far from the class boundaries. The Gowda-Krishna modification has eliminated this disadvantage. However, there exists another way to avoid selection of points lying far from the class boundaries. It consists of multiple use of the Hart's algorithm and each time (except the first one) first selecting those points that were selected to the reduced set in the previous use as the last ones. It means that the points ought to be presented in the reverse order than they were recruited to the reduced set by the previous pass of the Hart's procedure. This modification concerns also the proposed algorithm 1 that could use the Hart's procedure as the subsequent stages after the first stage of selecting the mutually nearest points. A comparison of these modifications with Gowda-Krishna algorithm deserves further study.

Tab. 2 shows how many points are recruited at the start stage and after each presentation of the whole primary reference set by the Hart's procedure in its original and modified (algorithm 1) version.

It was observed (see again Tab. 2) that algorithm 1 requires a smaller number of passes as compared to the original Hart's procedure. We expect that algorithm 1 is less dependent on the primary arrangement of the original reference set.

This algorithm selects majority of points to the reduced set in the initial stage, which does not depend on the primary arrangement of points in the reference set.

Table 2. Comparison of the reduced set sizes after each stage of the Hart's procedure and algorithm 1. The primary reference set contained 16000 pixels

Algorithm type	Start stage	Numbers of points in the reduced set after subsequent passes			
Hart's algorithm	1	908	1092	1108	1108
Algorithm 1	997	1065	1065	-	-

4 Conclusions

In the presented paper we have brought attention to a promising approach to the reference set reduction that consists in combining different types of the reference set size reduction algorithms. Subsequent use of the algorithm based on the division of the primary reference set and then use of the algorithm based on the consistency idea offers low error rate and simultaneously enables to control the compromise between the quality and the speed of classification.

The described approach signalizes a few problems that require further examination. We have applied algorithm 2 as the first followed by algorithm 1 (eventually original Hart's procedure or Gowda-Krishna algorithm). Probably it would be better to inverse this sequence. The approach based on the primary reference set division can be performed with the use of data clustering methods. The following question arises: which of the known clustering approach is most suitable? The authors intend to continue studies to develop more effective algorithms that would enable the flexible control between the speed and the quality of classification.

References

1. Fix E, Hodges JL (1952) Discriminatory Analysis: Nonparametric Discrimination Small Sample Performance, Project 21-49-004, Report Number 11, USAF School of Aviation Medicine, Randolph Field, Texas: 280–322, reprinted in the book: Dasarathy BV (1991) NN Pattern Classification Techniques, IEEE Computer Society Press: 40–56.
2. Hart PE (1968) The condensed nearest neighbor rule, IEEE Trans. Information Theory, Vol. 14, No. 3: 515–516.
3. Gowda KC, Krishna G (1979) The condensed nearest neighbor rule using the concept of mutual nearest neighborhood, IEEE Trans. Information Theory, Vol. 25, No. 4: 488–490.

Acknowledgements

This work was supported by the grant 4 T11E 017 22 from the Polish State Committee for Scientific Research in 2002-2005.

A Appendix

A.1 Procedure of finding pairs of mutually nearest points from the opposite classes

Notations:

```
Z := X∪Y={z_i}_{i=1}^m - the training set,
S_t - a set containing the points from the pairs
      of mutually nearest points,
O(u) - the opposite class of u.
```

Procedure:

```
R_v:=∅; S_nea:=∅;
for i:=1 to m do
begin
   if z_i not in R_v then
   begin
      u:=z_i; v_1:=u; v_2:=0; v_3:=0; R_v:=R_v∪{z_i};
      repeat
         find in O(u) the nearest point z_j to u, breaking
               the ties by selecting z_j with the lowest i;
         R_v:=R_v∪{z_j};
         v_1:=v_2; v_2:=v_3; v_3:=z_j
      until (v_1=v_3);
      S_nea:= S_nea∪{ v_1,v_2}
   end
end
```

End of the procedure.

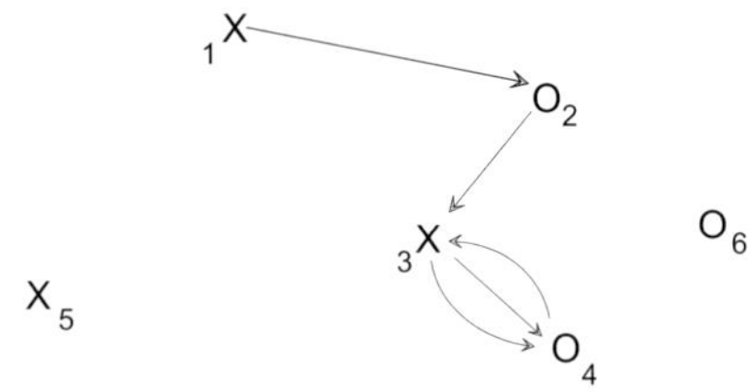

Fig. 1. An example illustrating the procedure of finding the pair of mutually nearest points from the opposite classes

A.2 Procedure of finding pairs of mutually furthest points

Notations:

$S_{ref}=\{z_i\}_{i=1}^{ms}$ - a subset of the reference set, where ms is a numerical force of S_{ref}.

Procedure:

```
u:=z_i; v_1:=u; v_2:=0; v_3:=0;
repeat
   find in S_ref the furthest point z_j to u,
         break the ties by selecting z_j with the lowest i;
   v_1:=v_2; v_2:=v_3; v_3:=z_j; u:=z_j;
until (v_1=v_3);
the points v_1 and v_3 are the mutually furthest points in
S_ref.
```

End of the procedure.

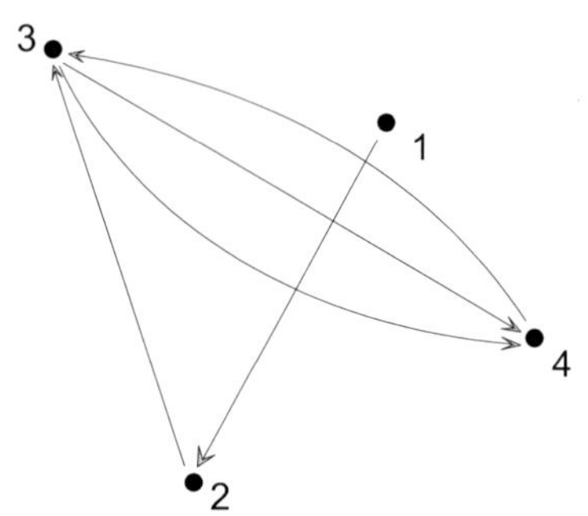

Fig. 2. An example illustrating the procedure of finding a pair of mutually furthest points

Learning from a Test Set

Piotr Juszczak and Robert P. W. Duin

Information and Communication Theory Group,
Faculty of Electrical Engineering, Mathematics and Computer Science,
Delft University of Technology, The Netherlands
p.juszczak@ewi.tudelft.nl, r.p.w.duin@ewi.tudelft.nl

Summary. Classification of partially labeled data requires linking the unlabeled input distribution $P(\mathbf{x})$ with the conditional distribution $P(y|\mathbf{x})$ obtained from the labeled data. The latter should, for example, vary little in high density regions. The key problem is to articulate a general principle behind this and other such reasonable assumptions. In this paper we provide a new approach to semi-supervised learning based on the stability of estimated labels for the unlabeled dataset, e.g a large test set, and the maximization of the mutual label relation. No clustering assumptions are required and the approach remains tractable even for continuous marginal class densities. We demonstrate the approach on synthetic examples and UCI repository datasets.

1 Introduction

In many classification problems there is an easy access to unlabeled objects and a specified cost, in time or money, to label them. Therefore, usually we label a small number of objects and hope, that they are sufficiently representative for the classification problem. However, to benefit from remaining unlabeled objects, one must exploit implicitly or explicitly the link between density $P(\mathbf{x})$ over objects $\mathbf{x}$ and the conditional $P(y|\mathbf{x})$ representing the posterior probability of the labels y.

Most classification methods do not attempt to explicitly model or incorporate information from the density $P(\mathbf{x})$. However, some classification algorithms such as density based algorithms as the Parzen classifier [9] or transductive SVM [13] have a possibility to relate $P(\mathbf{x})$ to $P(y|\mathbf{x})$; the decision boundary is biased to fall preferentially in low density regions of $P(\mathbf{x})$.

In such algorithms, the unlabeled objects, e.g a large test set to be classified, provide additional information about the structure of the domain while the few labeled objects identify the classification task expressed in this structure. A tacit assumption in this context is to associate high-density clusters in data with pure classes. When this assumption is appropriate, it is only required to label a single object per cluster to classify the whole dataset.

The presented problem is in broad terms related to a number of other problems like maximum entropy discrimination [7], data clustering by information bottleneck [12], and minimum-entropy data partitioning [10].
In this paper we investigate label propagation from a small labeled set over a large unlabeled set for density based classifiers in the semi-supervised learning framework, using as an example the Parzen classifier. The main difference between the various semi-supervised learning algorithms proposed in literature, such as spectral methods [4], random walks [11], graph mincuts [3] and transductive SVM [13], lies in the way of realizing the assumption of the labels consistency. However, the following three assumptions are often made about the representation space where the classification problem is present:

1. nearby objects are likely to have the same label,
2. objects on the same structure, e.g. a cluster or a manifold are likely to have the same label,
3. the decision boundary should lie in regions of low density [1].

The semi-supervised learning method proposed in this paper is based on the stability of estimated labels for unlabeled objects. In contradiction to the mentioned methods, in particular [4, 3, 11], there is not an implicit clustering step involved in the label propagation process. Therefore, there is no necessity to specify or optimize the number of clusters beforehand.
The layout of this paper is as follows. In section 2, the formal notation and the problem description are introduced, and the proposed algorithm is presented. Section 3 shows advantages and disadvantages of the proposed algorithm based on experiments on artificial and real-world data. Sections 4 presents the discussion and final conclusions.

2 Problem description

Given a partially labeled data set $\{(\mathbf{x}_1, y_1), \dots, (\mathbf{x}_l, y_l), \mathbf{x}_{l+1}, \dots, \mathbf{x}_N\} \subset R^m$, the first l objects are labeled X_l and the remaining objects $\mathbf{x}_i \in X_u$ $(l+1 \leq i \leq N)$ are unlabeled. The goal is to predict the label of the unlabeled objects. The example of such a problem is presented in figure 1.
Our classification model assumes that each data example has a label, for $\mathbf{x} \in X_l$ or a distribution $P(y|\mathbf{x})$ over the class labels for $\mathbf{x} \in X_u$ [2]. These distributions are unknown and represent the parameters to be estimated. Given an object $\mathbf{x}_k$, which may be labeled or unlabeled, we interpret its label as a weighted sum of crisp and soft labels of its neighbors N_G:

---

[1] The third assumption is related to the second. An example is handwritten digit recognition where one tries to classify e.g. 2 and 5. The probability of having a digit which is between 2 and 5 should be lower than the probability of a distinct 2 or 5.

[2] $P(y|\mathbf{x})$ are also called soft labels

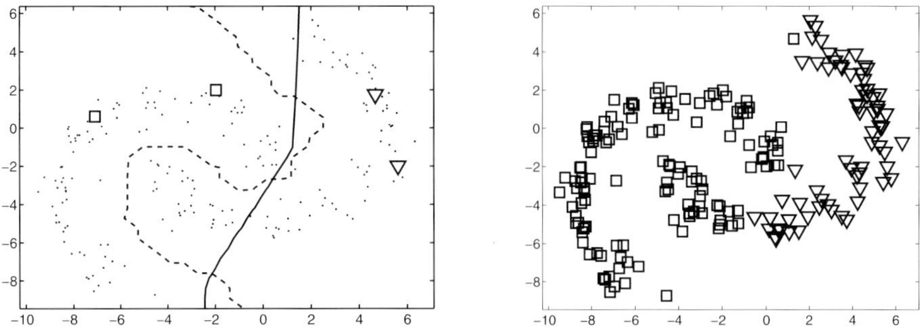

Fig. 1. On the left: A classification problem with four labeled objects denoted by (∇ , $\square$) and many unlabeled objects denoted by ($\cdot$). The continuous line denotes a classifier trained just on the labeled set and the dashed line a classifier trained on labeled and unlabeled objects. On the right: the corresponding classification labels for the classifier trained just on X_l.

$$P'(y_i|\mathbf{x}_k) = \sum_{\mathbf{x}_i \in N_G(\mathbf{x}_k)} P(y_i|\mathbf{x}_i) p_{MLR}(y_i|\mathbf{x}_i, \mathbf{x}_k) \tag{1}$$

where $p_{MLR}(y_i|\mathbf{x}_i, \mathbf{x}_k)$ is the measure of the mutual label relation between the set of examples $\mathbf{x}_i \in N_G(\mathbf{x}_k)$ and the object $\mathbf{x}_k$. In other words $p_{MLR}(y_i|\mathbf{x}_i, \mathbf{x}_k)$ is the measure of the contribution of $\mathbf{x}_i$ to the probability that $\mathbf{x}_k$ has the label y_i. $P(y_i|\mathbf{x}_k)$ is computed over ϵ - neighborhood of $\mathbf{x}_k$ defined as follows:

$$N_G(\mathbf{x}_k) = \{\forall \mathbf{x}_i \in \{X_l \cup X_u\} \mid p_{MLR}(y_i|\mathbf{x}_i, \mathbf{x}_k) \geq \epsilon\} \backslash \{\mathbf{x}_k\} \tag{2}$$

In general, $P(y|\mathbf{x}_i)$ are only available for labeled objects X_l and have to be estimated for unlabeled objects X_u. We will now discuss how to estimate $P(y|\mathbf{x}_i)$ for the set X_u and how to choose the measure of the mutual label relation $p_{MLR}(y_i|\mathbf{x}_i, \mathbf{x}_k)$.

2.1 Estimation of soft labels $P(y|\mathrm{x})$

We propose to estimate $P(y|\mathbf{x}_i)$ using the conditional maximum log-likelihood as the criterion. The $P(y|\mathbf{x}_i)$ is estimated for unlabeled objects for the fixed value of $p_{MLR}(y_i|\mathbf{x}_i, \mathbf{x}_k)$:

$$\max_{P(y|\mathbf{x}_i)} \sum_y^C \log \sum_{\mathbf{x}_i \in N_G(\mathbf{x}_k)} P(y|\mathbf{x}_i) p_{MLR}(y_i|\mathbf{x}_i, \mathbf{x}_k) \tag{3}$$

where for the two-class problem, $C = 2$, $P(y|\mathbf{x}_i) = \{0, 1\}$ for labeled objects and $0 \leq P(y|\mathbf{x}_i) \leq 1$ for unlabeled objects. Since $p_{MLR}(y_i|\mathbf{x}_i, \mathbf{x}_k)$ are fixed this objective function is jointly convex in the free parameters and has a unique maximum value. This convexity also guarantees that this optimization is easily performed via the EM algorithm.

2.2 Estimation of the mutual label relation $p_{MLR}(y_i|\mathbf{x}_i, \mathbf{x}_k)$

In the previous subsection we assumed that the mutual label relation $p_{MLR}(y_i|\mathbf{x}_i, \mathbf{x}_k)$ was known and fixed. In this section we compute and optimize $p_{MLR}(y_i|\mathbf{x}_i, \mathbf{x}_k)$ for the set of label and unlabeled objects in the maximum likelihood sense for the known and fixed sets of soft labels $P(y|\mathbf{x})$.
Consider a set of points $\{\mathbf{x}_1, \ldots, \mathbf{x}_N\}$ with a metric $d(\mathbf{x}_k, \mathbf{x}_i) = \|\mathbf{x}_k - \mathbf{x}_i\|$. Since close objects have high value of $p_{MLR}(y_i|\mathbf{x}_i, \mathbf{x}_k)$ about their labels and objects far away low value of $p_{MLR}(y_i|\mathbf{x}_i, \mathbf{x}_k)$ we can relate the mutual label relation to the distances between objects e.g. as follows $p_{MLR}(y_i|\mathbf{x}_i, \mathbf{x}_k) = \exp(-\frac{\|\mathbf{x}_k - \mathbf{x}_i\|}{2\sigma^2})$ [3]. The new estimate of the soft labels $P'(y|\mathbf{x}_k)$ of $\mathbf{x}_k$ can be defined now as:

$$P'(y_i|\mathbf{x}_k) = \sum_{\mathbf{x}_i \in N_G(\mathbf{x}_k)} P(y_i|\mathbf{x}_i) \exp\left(-\frac{\|\mathbf{x}_k - \mathbf{x}_i\|}{2\sigma^2}\right) \qquad (4)$$

which is related to the weighted Parzen density estimator. $P'(y_i|\mathbf{x}_k)$ is computed for all labels $y_i \in C$ and normalized to $\sum_{y_i \in C} P'(y_i|\mathbf{x}_i) = 1$.
Now we define how the information about the label of an object $\mathbf{x}_i$ influences the label of an object $\mathbf{x}_k$. In particular, the mutual label relation should decreases when the distance between objects $\mathbf{x}_k$ and $\mathbf{x}_i$ increases. This is related to the choice of σ. For large σ more distant objects have the influence on the soft labels of an object $\mathbf{x}_k$ and for small σ only nearby objects influence the soft labels of an object $\mathbf{x}_k$. We computed σ in equation (4) based on a leave-one out maximum likelihood estimation [6, 8]. The initial estimate σ_l of σ is optimized for just the labeled objects $\mathbf{x} \in X_l$. The final σ_{lu} is optimized for both labeled and unlabeled objects $\mathbf{x} \in \{X_l, X_u\}$. In a series of n EM algorithms σ takes the values:

$$\sigma_l > \sigma_2 > \ldots > \sigma_{n-1} > \sigma_{lu}$$

The change in σ from large, σ_l, to small, σ_{lu}, values, during learning of soft labels, changes the stress between the global labels consistency and the local labels consistency.

2.3 Proposed algorithm

The proposed algorithm of the classification with the partially labeled dataset is summarized in algorithm 1.
In the initial step of the algorithm the soft labels are computed using only labeled objects, $P(y|\mathbf{x}) = 0\ \forall \mathbf{x} \in X_u$. In the second step based on the current estimation of σ_t soft labels are optimized $P'(y|\mathbf{x})$ for $\mathbf{x} \in X_u$ using the maximum likelihood criterion. Next, the equation (4) is recomputed using both

[3] $p_{MLR}(y_i|\mathbf{x}_i, \mathbf{x}_k)$ can be defined in several ways e.g. as the L_1 norm, a wavelet function or a Gaussian process

crisp and soft labels. Step 4 is repeated until the difference between the current estimated labels $P'(y|\mathbf{x}_i)$ and the previous estimated labels $P(y|\mathbf{x}_i)$ is smaller than γ. The procedure is repeated for n different σ-s.

soft-PARZEN.(

1. set a number of EM algorithms to n; compute $\sigma_1 = \sigma_l$ and $\sigma_n = \sigma_{lu}$; set $t = 1$ and a stopping criterion γ;

2. compute: $\sigma_1 > \sigma_2 > \ldots > \sigma_{n-1} > \sigma_n$ for each EM algorithm;
   set $P(y|\mathbf{x}) = 0 \quad \forall \mathbf{x} \in X_u$

while $t \leqslant n$

3. optimize soft labels $P'(y|\mathbf{x})$ based on equation (3) with a fixed σ_t; using the soft labels $P(y|\mathbf{x})$ from the step $t-1$ as the initialization of the labels;

4. repeat 3 until stopping criterion is reached
   e.g. $\sum_i |P'(y|\mathbf{x}_i) - P(y|\mathbf{x}_i)| < \gamma$; $t = t + 1$;

end

)

Algorithm. 1

3 Experiments

Consider an example (figure 2) of classification with the proposed algorithm. We are given 2 labeled objects per class and 196 unlabeled objects in an intertwining two banana shape patterns. This pattern has a manifold structure where distances are locally but not globally Euclidean, due to the curved arms. Therefore, the pattern is difficult to classify for traditional algorithms using locally defined relations, such as 1-nearest neighbor; figure 1b. We used the proposed algorithm, described in algorithm 1, to incorporate unlabeled data into the Parzen density estimator and scale the Euclidean distance between objects using their soft labels. Figure 2 shows three different timescales. At $t = 5$ the σ is overestimated, therefore there are large, Gaussian clusters and $P(y|\mathbf{x})$ are estimated ruffly. At $t = 15$ because σ becomes smaller local mutual label relations in marginal regions start to change the soft labels. At $t = 20$ almost all objects, apart of one, have correct labels.

Next, we evaluated the performance of the presented algorithm on some of the UCI repository datasets [1]: *waveform*, *satellite*, *letter*, *ecoli*. Datasets were divided into two parts: labeled set X_l and the unlabeled set X_u constituted from remaining objects, the ratio $\frac{X_l}{X_u}$ is indicated by numbers on the abscissa. The label propagation was performed on X_u and the obtained classifier was

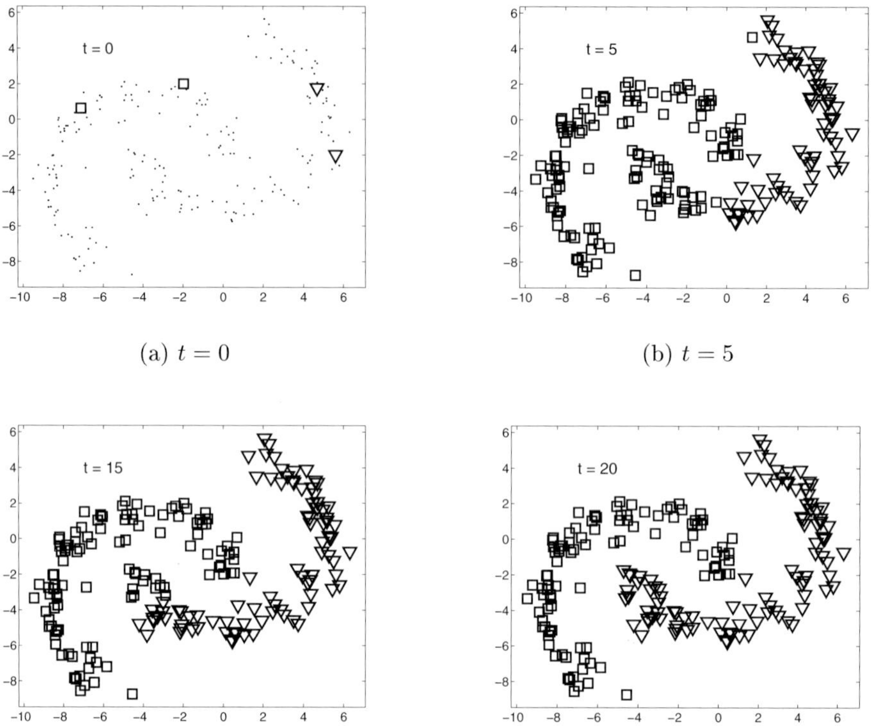

(a) $t = 0$ (b) $t = 5$

(c) $t = 15$ (d) $t = 20$

Fig. 2. Label estimates for the *soft*-PARZEN algorithm for the banana shape dataset. Labeled (soft and crisp labels) objects denoted by (∇ , $\square$) and unlabeled objects denoted by ($\cdot$).

tested on the same set of unlabeled data X_u. The random division was repeated 50 times for each ratio $\frac{X_l}{X_u}$. The performance of the proposed algorithm (*soft*-PARZEN) is compered with the 1-nearest neighbor label propagation (1-NNLP) [2, 3] and the Parzen classifier trained just on labeled objects (PARZEN). The mean error and the standard deviation are shown in figure 3. It can be seen, that the proposed *soft*-PARZEN algorithm outperforms both: 1-NNLP and the Parzen classifier trained on just labeled objects, on considered classification problems. In case of *waveform* and *ecoli* the performance of *soft*-PARZEN is close to 1-NNLP and for *satellite* and *letter* there is significant improvement.

The *soft*-PARZEN and 1-NNLP perform similar if distances between objects in pure clusters and between clusters differ significantly. However, if in the

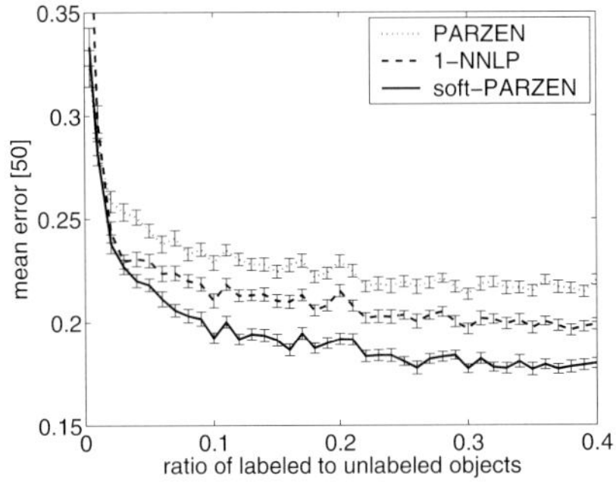

(a) waveform(500, 21, 3)

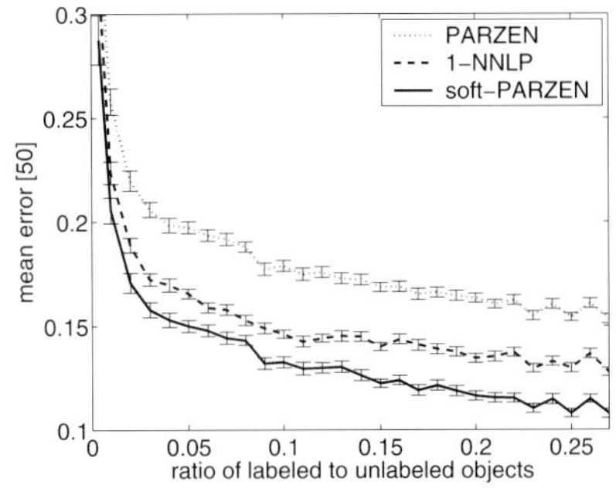

(b) satellite (500, 36, 6)

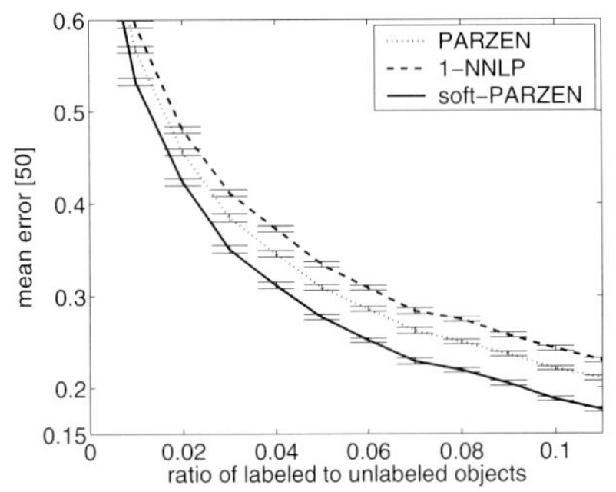

(c) letter(500, 16, 26)

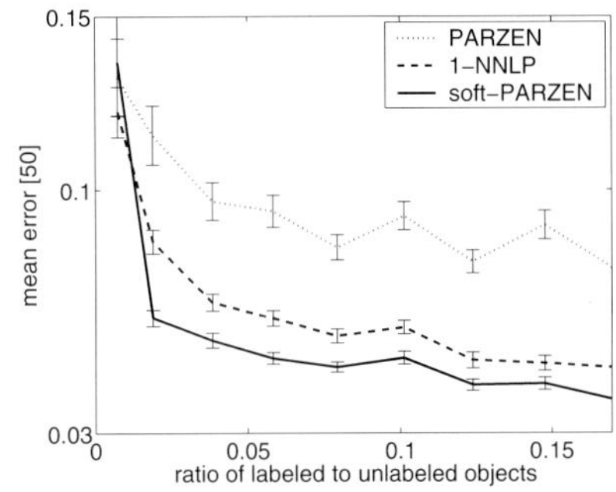

(d) ecoli(272, 7, 3)

Fig. 3. Mean square error and standard deviation for *soft*-PARZEN, 1-NNLP compared with a classifier trained just on a labeled dataset PARZEN for UCI repository datasets: *waveform*, *satellite*, *letter*, *ecoli*. Numbers in brackets indicate the size of X_u and the number of features and classes in a dataset.

data there is not a clear cluster structure the *soft*-PARZEN might outperform the 1-NNLP significantly.
The performance of the proposed method depends on the quality of the labeled data and their relation to the structure of the unlabeled dataset. If the clusters of the unlabeled data are not related to the class information it is hard to expect that the proposed method performs well. For a broader discussion about merits and disadvantages of the semi-supervised learning we point reader to the paper [5].

4 Conclusions

The proposed algorithm based on expectation maximization of soft labels and the mutual label relation *soft*-PARZEN provides a robust variable resolution approach to classifying data sets with significant cluster structure and very few labels. When the cluster structure in absent or unrelated to the classification task, the proposed method can be expected to derive particular but small improvement over a classifier trained just on the labeled dataset. In such cases the performance is strongly related to the quality of the already labeled set.

In future work we will test the proposed algorithm on large, high-dimensional datasets and explore theoretical connections to network information theory.

Acknowledgments

This work was partly supported by the Dutch Organization for Scientific Research (NWO).

References

1. C. L. Blake and C. J. Merz. UCI repository of machine learning databases, 1998.
2. A. Blum and S. Chawla. Learning from labeled and unlabeled data using graph mincuts. In *ICML*, pages 19–26. Morgan Kaufmann, San Francisco, CA, 2001.
3. A. Blum, J. Lafferty, M. R. Rwebangira, and R. Reddy. Semi-supervised learning using randomized mincuts. In *ICML*, 2004.
4. O. Chapelle, J. Weston, and B. Schöelkopf. Cluster kernels for semi-supervised learning. In *NIPS*, volume 15, pages 585–592, 2002.
5. I. Cohen, F.G. Cozman, N. Sebe, M.C. Cirelo, and T.S. Huang. Semi-supervised learning of classifiers: theory, algorithms, and their application to human-computer interaction. *PAMI*, 26(12):1553–1566, 2004.
6. R. P. W. Duin. On the choice of the smoothing parameters for parzen estimators of probability density functions. *IEEE Transactions on Computers*, 25(11):1175–1179, 1976.
7. T. Jaakkola, M. Meila, and T. Jebara. Maximum entropy discrimination. In *NIPS*, volume 12, pages 470–477, 1999.
8. Tsvi Lissack and King-Sun Fu. Error estimation in pattern recognition via $L^\alpha-$distance between posterior density functions. *IEEE Transitions on Information Theory*, 22(1):34–45, 1976.
9. E. Parzen. On the estimation of a probability density function and the mode. *Annals of Mathematical Statistics*, 33:1065–1076, 1962.
10. S. Roberts, C.C. Holmes, and D. Denison. Minimum entropy data partitioning using reversible jump markov chain monte carlo. *PAMI*, 23(8):909–914, 2001.
11. Martin Szummer and Tommi Jaakkola. Partially labeled classification with markov random walks. In *NIPS*, volume 14, pages 945–952, 2001.
12. N. Tishby and N. Slonim. Data clustering by markovian relaxation and the information bottleneck method. In *NIPS*, pages 640–646, 2000.
13. V. N. Vapnik. *Statistical learning theory*. Wiley, NY, 1998.

Image Clustering with Median and Myriad Spatial Constraint Enhanced FCM

Jacek Kawa[1] and Ewa Pietka[1]

Silesian University of Technology, Department of Biomedical Engineering, Gliwice, Poland jkawa@polsl.pl

Summary. In the current study two approaches to the clustering problem have been tested. First, a sequential analysis of filtering and fuzzy c-means (FCM) method is performed. Then, the standard FCM has been modified by adding to the objective function a second term that formulates a spatial constraint. In both approaches mean, median, and myriad are implemented. The analysis has been performed on a synthetic image and clinical images.

1 Introduction

Medical image segmentation employing fuzzy set theory [1] has been an important field in the research in the past decades. Especially the fuzzy c-means (FCM) algorithm [2] has been widely used in many clustering approaches. Its advantages include a conceptual and computational simplicity and the ability to model uncertainty within the data. FCM features also several weaknesses. It does not incorporate spatial context information which makes it sensitive to noise and image artifacts.

In the current study two approaches to the clustering problem have been tested. First, a cascade analysis has been designed in which an image is subjected to a median (or myriad) filter and then a standard FCM is implemented. In the second approach, the FCM objective function is modified by adding a second term that formulates a spatial constraint based on the median (or myriad) estimator, respectively. In Section 2 the definition and features of median and myriad are given. Section 3 discusses the modification of the FCM objective function. Evaluation of presented clustering methods and their implementation in medical image analysis can be found in Section 4.

2 Median and myriad filtering

Median of a set $\{x_1, \ldots, x_n\}$ is an M-estimator of location, with a cost function given as [3]:

$$J(\Psi) = \sum_{i=1}^{N} |x_i - \Psi| \tag{1}$$

i.e. $median(\{x_1, \ldots, x_N\}) = \widehat{\Psi} = \arg\min_{\Psi} J(\Psi)$.

Median of an ordered data set $A = \{x'_1, x'_2, \ldots, x'_N\}$ is defined as:

$$median(A) = \begin{cases} x'_{(k+1)/2}, & k = 1, 3, 5, \ldots \\ 0.5(x'_{k/2} + x'_{k/2+1}), & k = 2, 4, 6, \ldots \end{cases} \tag{2}$$

Myriad is defined as a parameter Θ in Cauchy distribution probability density function [4]:

$$f(x) = \frac{K}{\pi} \frac{1}{K^2 + (x - \Theta)^2} \tag{3}$$

Myriad can also be defined [3, 4] as an M-estimator of location:

$$\widehat{\Theta_K} = myriad\{x_1, x_2, \cdots, x_N; K\} = \arg\min_{\Theta} \sum_{k=1}^{N} \ln[K^2 + (x_k - \Theta)^2] \tag{4}$$

where K is called a *myriad linearization parameter.*

Parameter K defines the behavior of a myriad. Two special cases (Eq. 5 and 6) with respect to K can be denoted [5]:

$$lim_{K\to\infty} \widehat{\Theta}_K = \frac{1}{N} \sum_{k=1}^{N} x_k \tag{5}$$

$$\widehat{\Theta}_0 = lim_{K\to 0} \widehat{\Theta}_K = \arg\min_{x_j \in M} \prod_{k=1, x_k \neq x_j}^{N} |x_k - x_j| \tag{6}$$

where M is a set of the most common values.

In other words for $K \to \infty$, myriad is equal to the mean value of data, and for $K \to 0$, myriad is equal to one of the most common value of the data set.

In image processing methods an implementation of median (myriad) running-window filtering, replaces each data sample by its spatial neighborhood function, defined as:

$$MEDF(x; Z) = median(S) \tag{7}$$

$$MIRF(x; Z; K) = myriad\{S; K\} \tag{8}$$

respectively, where $S = neighborhood(x, Z)$, and Z is the size of the mask.

As shown in next section, both myriad and median filtering can be used as an additional penalty into cost function of FCM.

3 FCM modifications

Clustering methods partition a set of observed data vectors $\mathbf{x}_k$ into c-clusters. Each cluster is represented by its prototype $\mathbf{v}$.

Let $\mathbf{x}_k = (x_i, \ldots, x_n)$ be an observed data vector of $\{\mathbf{x}_k\}_{k=1}^N$ data set in feature space $\mathbf{F}^n$. Standard FCM is derived to minimize the objective function:

$$J(\mathbf{U}, \mathbf{V}) = \sum_{i=1}^{c} \sum_{k=1}^{N} u_{ik}^m ||\mathbf{x}_k - \mathbf{v}_i||^2 \qquad \mathbf{x}_k, \mathbf{v}_i \in \mathbf{F}^n \tag{9}$$

with respect to the membership function u_{ik} and the center $\mathbf{v}_i$ for a given fuzzyfication level m $(1 \leq m < \infty)$.

The membership function values u_{ik} are positive and fulfill: $\sum_{i=1}^{c} u_{ik} = 1 \forall k,\ u_{ik} \leq 1$.

The FCM clustering is performed iteratively, starting with a set of c initially given prototypes and fuzzyfication level m. In each step a new matrix of $\mathbf{U}$ is created. For samples that belongs also to the set of prototypes, values of u_{ik}, that minimize the corresponding part of cost function are:

$$u_{ik} = \begin{cases} 1, \mathbf{x}_k = \mathbf{v}_i \\ 0, \text{otherwise} \end{cases} \tag{10}$$

And for remaining samples:

$$u_{ik} = \frac{||\mathbf{x}_k - \mathbf{v}_i||^{\frac{-2}{m-1}}}{\sum\limits_{j=1}^{c} \left(||\mathbf{x}_k - \mathbf{v}_j||^{\frac{-2}{m-1}} \right)} \tag{11}$$

The membership matrix $\mathbf{U}$ is later employed to compute a new set of prototypes as a weighted mean of points and corresponding membership function values:

$$v_i = \frac{\sum\limits_{k=1}^{N} u_{ik}^m \, \mathbf{x}_k}{\sum\limits_{k=1}^{N} u_{ik}^m} \tag{12}$$

The procedure is repeated until the desired accuracy of $\mathbf{V}$ is obtained, i.e. $max(|\mathbf{V}_i' - \mathbf{V}_i|) < \epsilon$.

In image processing, a lack of spatial context of information in FCM reduces its robustness in a presence of noise. Outliers influence both membership function and prototypes calculations.

Chen and Zhang [6] have modified the objective function of FCM by introducing an element, that depends on the mean value of neighboring pixels, into Eq. 9:

$$J(\mathbf{U}, \mathbf{V}) = \sum_{i=1}^{c} \sum_{k=1}^{N} u_{ik}^m ||\mathbf{x}_k - \mathbf{v}_i||^2 + \alpha ||\tilde{\mathbf{x}}_k - \mathbf{v}_i||^2 \tag{13}$$

where $\tilde{\mathbf{x}}_k$ is a mean of pixels within a predefined neighborhood $\mathbf{x}_k$, and α controls the 'strength' of modification.

Since the modification propagates features of a mean filtered image into the clustering results, blurred edges is one of the most important disadvantages of the method.

In order to reduce the drawbacks, median and myriad estimators have been added into the objective function. Thus, the following objective functions have been obtained:

$$J(\mathbf{U}, \mathbf{V}) = \sum_{i=1}^{c} \sum_{k=1}^{N} u_{ik}^m ||\mathbf{x}_k - \mathbf{v}_i||^2 + \alpha ||MEDF(\mathbf{x}_k; Z) - \mathbf{v}_i||^2 \quad (14)$$

$$J(\mathbf{U}, \mathbf{V}) = \sum_{i=1}^{c} \sum_{k=1}^{N} u_{ik}^m ||\mathbf{x}_k - \mathbf{v}_i||^2 + \alpha ||MIRF(\mathbf{x}_k; Z; K) - \mathbf{v}_i||^2 \quad (15)$$

Necessary conditions for Eq. 14 and Eq. 15, to be a local minimum are:

$$u_{ik} = \frac{(||\mathbf{x}_k - \mathbf{v}_i||^2 + \alpha ||\mathbf{y}_k - \mathbf{v}_i||^2)^{\frac{-1}{m-1}}}{\sum_{j=1}^{c} \left((||\mathbf{x}_k - \mathbf{v}_j||^2 + \alpha ||\mathbf{y}_k - \mathbf{v}_j||^2)^{\frac{-1}{m-1}}\right)} \quad (16)$$

$$v_i' = \frac{\sum_{k=1}^{N} u_{ik}^m (\mathbf{x}_k + \alpha\, \mathbf{y}_k)}{(1 + \alpha) \sum_{k=1}^{N} u_{ik}^m} \quad (17)$$

where $\mathbf{y}_k$ denotes $MEDF(\mathbf{x}_k; Z)$ for median, and $MIRF(\mathbf{x}_k; Z; K)$ for myriad modification, respectively.

In all cases, $\mathbf{Y}$ matrix can easily be computed before the partitioning starts, thus the computational cost of introduced modification for each iteration is relatively small.

4 Experimental results and conclusions

In this section experimental results of fuzzy clustering employing standard as well as modified version of FCM are presented.

For testing purposes a 2-cluster image has been created (Fig. 1).[1] The test image has later been corrupted with Gaussian noise with the mean value set to 0 and $\sigma = 0.1$ (Fig. 1), and salt & pepper noise of 2% density (Fig. 2). Seven different methods have been examined: standard FCM (FCM), FCM applied to mean-filtered image (AVG), FCM with mean-based constraint (FAVG), FCM applied to median-filtered image (MED), FCM with median-based constraint (FMED), FCM applied to myriad-filtered image (MYR), and FCM with myriad-based constraint (FMYR).

[1]Please note, that some borders of objects are blurred.

Clustering of the test image has been performed. The image has been partitioned to 2 or 3 classes ($c \in \{2, 3\}$), the neighborhood mask has been set to 3×3 or 5×5 pixels, and constraint weight has been tested for $\alpha \in \{0.5, 1, 2, 3, 4\}$. Fuzzyfication level m has been chosen experimentally and set to 2. Initial prototypes have been uniformly distributed between the lowest and the highest gray-level value in the histogram. Myriad filtering has been performed with myriad linearization parameter equal to 0.

All obtained clusters for each method have been examined and an error rate (ER) has been computed as a percentage of misclassified pixels in the image (IM):

$$ER = \frac{\bar{A}}{\bar{IM}} 100\% \qquad (18)$$

$$A = \{y \in IM : min(|u_{iy}^{orig} - u_{jy}^{cor}|) \geq Threshold; i, j = 1, \ldots, c\} \qquad (19)$$

where u_{iy}^{orig} and u_{iy}^{cor} refer to the original and the corrupted image, respectively.

The best results for image corrupted with Gaussian noise (boldfaced in Table 1) are shown in Fig. 1 and for image corrupted with salt & pepper noise in Fig. 2.

Fig. 1. Clustering results of an image corrupted with a Gaussian noise. Shown left-to-right and top-to-bottom are: image, corrupted image, clustering results for FCM, FAVG, AVG, FMED, MED, FMYR and MYR

Results have shown a significant improvement when clustering the image with modified FCM, with respect to employing standard FCM to filtered image. Myriad-modified FCM is more sensitive to the variation of parameters chosen in the evaluation process, than median and mean FCM modifications, yet better results can be obtained.

Table 1. Best error rates for an image corrupted with Gaussian noise

c	Z	α	FCM	FAVG	AVG	FMED	MED	FMYR	MYR
2	5×5	0.5	25.5700	6.3450	1.9450	6.3200	1.8050	5.7200	17.3400
2	5×5	1.0	-	3.2700	-	2.9150	-	4.0000	-
2	5×5	2.0	-	1.4900	-	1.2350	-	16.5600	-
2	5×5	3.0	-	1.1900	-	**1.0900**	-	16.9100	-
2	5×5	4.0	-	1.1900	-	1.1150	-	17.0850	-
3	5×5	0.5	**8.5650**	2.8150	5.5750	2.6050	2.2650	9.4750	3.7350
3	5×5	1.0	-	1.7350	-	1.5800	-	4.0800	-
3	5×5	2.0	-	1.2750	-	1.2650	-	1.9800	-
3	5×5	3.0	-	1.2150	-	1.2750	-	1.8200	-
3	5×5	4.0	-	1.2950	-	1.2800	-	1.9000	-
2	3×3	0.5	-	6.0200	**1.5350**	6.4100	2.0600	6.0300	**1.5450**
2	3×3	1.0	-	2.9350	-	2.9450	-	2.9550	-
2	3×3	2.0	-	1.2500	-	1.4450	-	1.2600	-
2	3×3	3.0	-	0.9750	-	1.2100	-	**0.9600**	-
2	3×3	4.0	-	**0.9650**	-	1.2250	-	0.9850	-
3	3×3	0.5	-	2.6000	1.6700	2.5350	**1.6650**	2.6050	1.5800
3	3×3	1.0	-	1.5250	-	1.5000	-	1.5000	-
3	3×3	2.0	-	1.0000	-	1.1250	-	0.9950	-
3	3×3	3.0	-	1.0000	-	1.1450	-	0.9800	-
3	3×3	4.0	-	1.0250	-	1.1700	-	0.9600	-

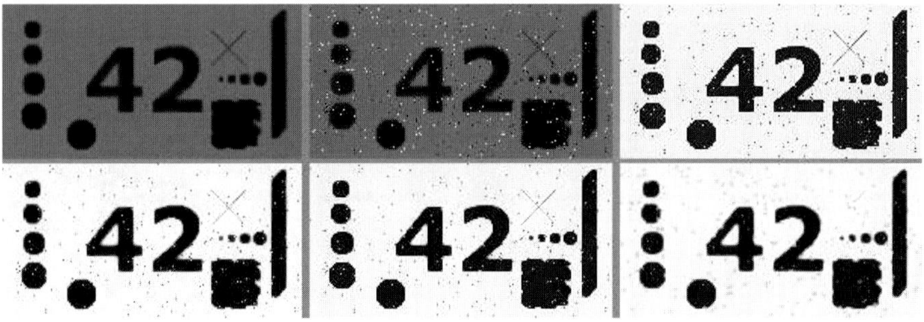

Fig. 2. Clustering results of an image corrupted with a salt & pepper noise. Shown left-to-right and top-to-bottom are: image, corrupted image, clustering results for FCM, FAVG, FMED and FMYR

The second experiment has been performed on a medical image of a hand radiograph (Fig. 3a). Using a 3-class partitioning approach, three structures are marked separately in the image. The first one refers to the image background and allows for an extraction of the entire hand image. The second class shows the soft tissue within the phalanges (back area in Fig. 3b, d, f, h). The third class reflects the bony structures within the phalangeal region (Fig. 3c, e, g, i). A comparison of four methods shows a better performance of the FMIR clustering method in the upper region (tip of fingers), and in

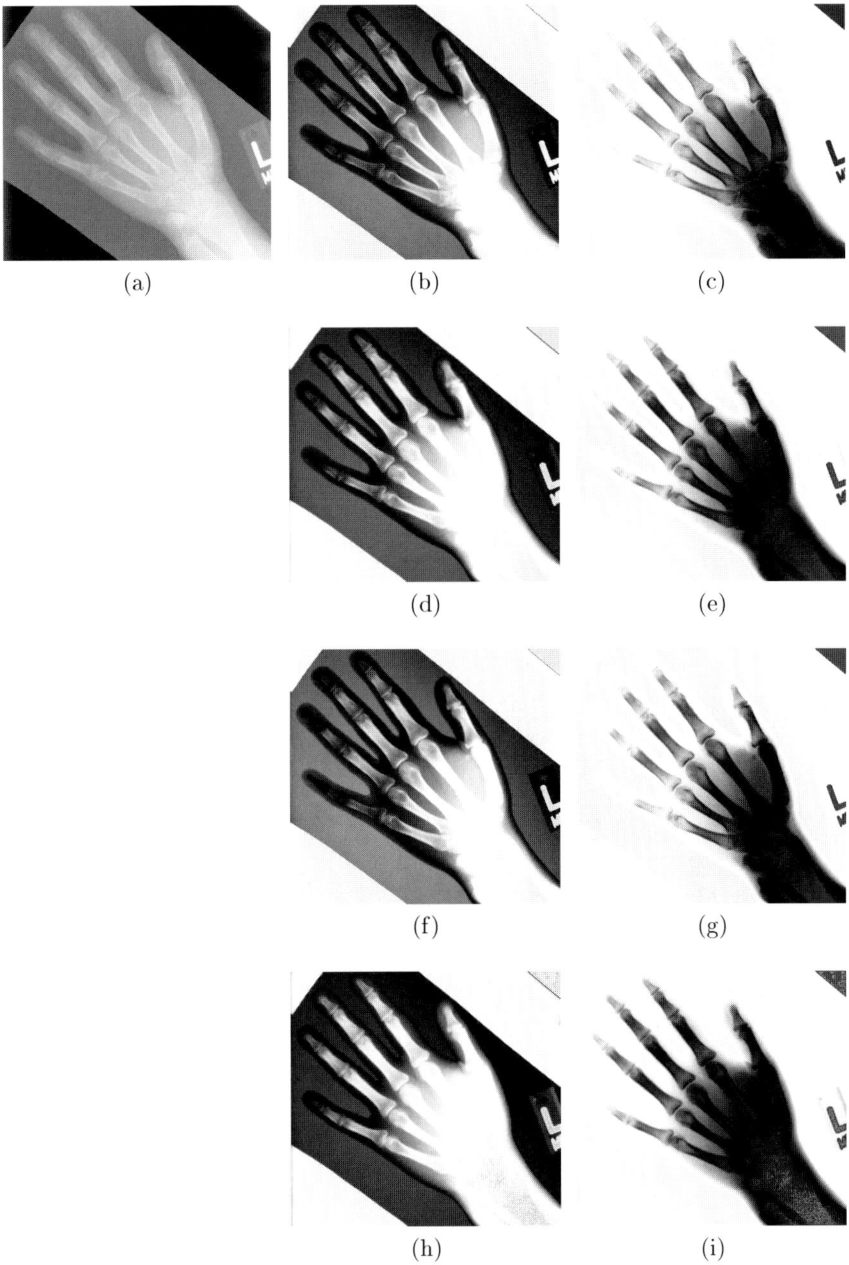

Fig. 3. Hand radiograph (a) and selected clustering results for FCM (b, c), FCMAVG (d, e), FCMMED (f, g), FCMMYR (h, i)

the radius and ulna. The FCM and FMED partitions with better accuracy in the metacarpal area. Since the phalangeal region is an important area in medical diagnosis of skeletal maturity and bone malformations, a segmentation of distals, middles and proximals becomes very important.

References

1. Zadeh LA (1965) Fuzzy sets. Information and Control 8:338–358
2. Bezdek JC (1981) Pattern Recognition with Fuzzy Objective Function Algorithms. Plenum Press, New York
3. Kalluri S, Arce GR (2000) Fast algorithms for weighted myriad computation by fixed-point search. IEEE Trans. Signal Processing 48:159–171
4. Przybyla T (2004) Robust fuzzy data-clustering methods. PhD Thesis, Silesian University of Technology, Gliwice (in Polish)
5. Gonzales JG, Arce GR (2001) Optimality of the myriad filter in practical impulsive-noise environment. IEEE Trans. Signal Processing 49:438–444
6. Chen S, Zhang D (2004) Robust image segmentation using FCM with spatial constraints based on new kernel-induced distance measure. IEEE Trans. System, Man, and Cybernetics–Part B. 34:1907–1916

Feature Selection in Unsupervised Context: Clustering Based Approach

Artur Klepaczko[1] and Andrzej Materka[2]

[1] Technical University of Lodz, ul. Wolczanska 223, 90-924 Lodz
aklepaczko@p.lodz.pl
[2] Technical University of Lodz, ul. Wolczanska 223, 90-924 Lodz
materka@p.lodz.pl

Summary. In this paper we present a novel feature selection method that is applicable in unsupervised learning tasks. The method is based on clustering quality measures, which reflect different aspects of clustering performance. Sequential Floating Forward Search algorithm is employed to search through the original feature space for the best possible subset. Main stress has been put on the objectivism of the new technique, so that it could be applied in various classification tasks. Results of experiments with texture images are presented in order to confirm effectiveness of the method.
The work was supported by the European Community grant COST B21.

1 Introduction

While solving various classification tasks, we often face situations where large number of data features leads to classification accuracy drop and intractable computational complexity. In such cases, reducing data dimensionality – either through feature extraction or selection – is indispensable. Main advantage of the latter is that parameters chosen from the original feature space provide some physical interpretation, thus giving us knowledge on the intrinsic nature of the specific problem. Moreover, after the selection phase, measuring data can be limited to those few selected ones, which saves memory resources.

This paper focuses on the problem of selecting features that are best to improve classifier performance in unsupervised manner. It is particularly crucial for clustering – unsupervised classification procedure. In this context, feature extraction has one considerable limitation. Commonly used Principal Component Analysis (PCA) preserves most of the total variance of data set, but partitioning of it does not become more explicit. Feature space obtained through transformation is composed of *most expressive features* (MEF) [2, 4], in contradiction to *most discriminative* ones (MDF), that result from Linear Discriminant Analysis (LDA), as an example. LDA however, cannot be run without knowledge about data samples classes.

On the other hand, constructing feature selection method which would not utilize such information seems troublesome. Machine–learning–standard algorithms – like Fisher coefficient or Probability of Error and Average Correlation Coefficients (POE + ACC)[2] – estimate feature's relevancy based directly on data category labels. Since they are not available in cluster analysis, such estimation should be grounded on some other, possibly indirect, assumptions. Let us recall here two very recent examples that attempt to solve the problem.

In [6] one can find description of a feature selection method based on spectral clustering. The method is referred to as the weighted-based approach due to the analogy to graph theory. The key matrix of the algorithm is the *affinity* matrix, in which each (i, j) element is an inner product between appropriate data samples. Features are multiplied by weights indicating their relevancy, modified during optimization. After convergence (to local optimum), non-zero weights determine most relevant features. Every optimization step, clustering is performed and its quality assessed by calculating clusters *coherences*, proportional to the eigenvalues of the affinity matrix. The algorithm proceeds in the direction of the biggest step toward maximal accumulated coherence. One indirectly presumes that the most relevant features assure the highest clustering quality. Note however, that coherence is the only measure used to estimate relevancy. Its credibility is restricted to spectral clustering. In other words, partitioning of some data set evaluated as the best according to cluster coherences may not be confirmed by some other quality measures. Moreover, as we will show further on, there are different aspects that need to be reflected while evaluating quality of clustering results. Consequently, to make feature selection method more objective, one should use more quality measures, with at least one of them being independent from the clustering algorithm.

In an entirely different approach presented in [1] feature selection is performed simultaneously with mixture model based clustering. It is assumed that features of data points belonging to the same cluster are derived from the same probability distributions. Parameters of these probabilistic models, as well as cluster labels treated as the missing variables, have to be found by the Expectation Maximization procedure. An additional set of parameters that have to be estimated, is composed of the coefficients indicating features saliency. Despite attractiveness of the method, two remarks can be raised. Firstly, one has to presume the form of the distribution function. Frequently, the Gaussian one proves effective, but it still remains only an assumption, and does not have to occur proper choice in general. Secondly, one presumes conditional independence of data features. However, very often features are correlated with one another and disregarding it violates final results.

Considering all above, we have decided to develop a new feature selection method for the use in unsupervised context. Our main concept is basically similar to the one incorporated in the weighted based approach: features of the best discriminative power give the highest quality of clustering results. What makes our method exceptional is not only different clustering algorithm, but – above all – that it reflects deeper understanding of what high quality of clustering results is. Moreover, our method is more objective, since one of the quality measures we use is universal and can be applied to evaluate effects of any clustering procedure.

In the subsequent sections of this paper we discuss three different aspects of clustering quality (Section 2), provide detailed description of our feature selection method (Section 3) and present some experimental results (Section 4). At last, we conclude with a couple of summarizing remarks (Section 5).

2 Clustering Quality Assessment

In this section we describe three major elements of clustering quality:

- quality of clusters themselves,
- quality of partitioning, and
- credibility of clustering results.

Before clarifying the above, two introductory comments should be made. First of all, one of the basic difficulties in cluster analysis is determining correct number of clusters. This of course affects feature selection process based on clustering. Some of the coefficients that we describe below serve primarily for this purpose. To adopt them as clustering quality measures, we simplify the problem and postulate that the number of clusters is known. Then, by the help of these coefficients, we may compare different clustering results, obtained for various feature subsets.

Second issue that is worth explanation is our choice of clustering method. We decided to engage agglomerative hierarchical algorithm with single linkage and standardized Euclidean norm, due to its superior performance over other methods in our experiments with textured image data.

As we have already stated, feature selection based on clustering becomes more objective if it uses at least one universal quality measure. Compactness coefficient defined in (1) satisfies this requirement. It constitutes therefore the main factor of our selection technique [7]:

$$\zeta = \sum_{i=1}^{c} d_i \times \frac{n_i}{n}, \tag{1}$$

where c denotes number of clusters, n – data set size, n_i – number of data points in i-th cluster, and d_i – cluster *diameter* (average distance between points in a cluster):

$$d_i = \sum_{j=1}^{n_i} \sum_{k=1}^{n_i} \frac{\|\mathbf{x}_j - \mathbf{x}_k\|^2}{n_i(n_i - 1)}. \tag{2}$$

The compactness coefficient measures the quality of clusters themselves and promotes those which encapsulate as many data points as possible with minimal average distance. The best partitioning should thus minimize (1).

Second factor of clustering quality estimates how well clusters are separated from one another. We call this *quality of partitioning* as it relates only to distances between identified clusters and says nothing about within–cluster scatter. In hierarchical clustering one can use for this purpose *inconsistency* coefficient [3], which can be computed for every link in the hierarchical tree:

$$\gamma_m = \frac{l_m - \mu_m}{\sigma_m}, \tag{3}$$

where l_m is the length of a link m, μ_m and σ_m denote mean and standard deviation of link m and links lying directly underneath. Actually, this coefficient is used to determine the most probable number of clusters: the mostly inconsistent link indicates where to cut the tree, what in consequence allows forming final groupings. Since we demand that correct number of clusters is known, interpretation of inconsistency changes. When the cutoff link occurs consistent, then we may assert that partitioning at this level leads to false groupings and this in turn suggests that current feature subset is irrelevant.

The third coefficient is called *cophenetic correlation.* It reveals credibility of clustering results by estimation of a degree of correspondence between point-to-point distances in the hierarchical tree and in the input space [3]:

$$\varphi = \frac{\sum_{j<k}(Y_{jk} - y)(Z_{jk} - z)}{\sqrt{\sum_{j<k}(Y_{jk} - y)^2 \sum_{j<k}(Z_{jk} - z)^2}}, \tag{4}$$

where Y_{jk} and Z_{jk} are the distances between j-th and k-th point in the original input space and in the tree, y and z are the means, and $\varphi \in (0, 1]$. Note that distance between any two points in the tree is equal to the length of a link which connects them. Relevant feature subsets are expected to maximize (4).

To illustrate importance of all three aspects of clustering quality, let us analyze a simple example. Consider two sample images depicted in Fig. 1. They represent one texture but with different brightness level. We generated 16 samples of each class and used MaZda suite (software for examining textured images [2]) to describe them by various texture parameters (270 features altogether). By the help of Fisher coefficient we selected 10 best, 10 second best and 10 worst features and constructed three 2–cluster sets. To verify features saliency we employed 1-NN classifier, which misclassified one sample for both 10 best and 10 second best features, while the worst subset produced 21 errors.

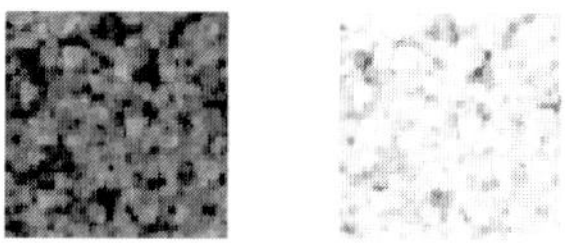

(a) (b)

Fig. 1. Beach sand texture samples with lower (**a**) and higher (**b**) brightness level

Next, we executed hierarchical algorithm and calculated quality coefficients described above. Table 1 shows the results. As it can be seen, compactness for the worst subset is significantly higher than for the best ones. This obviously implies rejection of these features. On the other hand, the minimal compactness is achieved for the second best subset. The other two coefficients however, are too low and this indicates that best feature subset determined by Fisher coefficient is also most relevant one in terms of clustering quality.

Table 1. Quality estimation for different feature subsets

Quality measure	10 best features	10 second best features	10 worst features
Compactness	22.8463	0.000194782	17458.4
Inconsistency	1.15466	0.707107	0.924314
Cophenetic correlation	0.991897	0.818368	0.664835

3 Clustering Quality Based Feature Selection Method

The main principle of the feature selection method which we introduce is already made clear: cluster data samples described by different feature subsets, calculate compactness, inconsistency and cophenetic correlation coefficients, then point at the most relevant subset which achieves the highest score on all three quality measures. The rule itself is quite simple. More problematic is answering how the search through the feature space should be conducted. In [5] there is a concise description of several sequential and genetic search algorithms for automation of feature selection methods. The one that we decided to use and that proved efficient in our experiments is called Sequential Forward Floating Search (SFFS).

Before launching the search we declare the desired number q of relevant features. During the operation we construct q t–element feature subsets Φ_t, each consisting of the most relevant combination of size t. At the initiation, $t = 0$, Φ_0 is empty and all features are gathered in the candidates set Υ. The algorithm terminates when $t = q$. Every step of the procedure is divided into two stages. During the first one, the most relevant feature from the set Υ is added to the current set Φ_t (and deleted in Υ), which will constitute set Φ_{t+1} (t is incremented). The most relevant feature is the one that maximizes the so–called *evaluation function* $J(\Phi_t)$, defined below. In the second stage we look for the least significant feature among elements of current set Φ_t. This will be the one which after being removed from Φ_t gives the highest value of $J(\Phi_{t-1})$. If it occurs that $J(\Phi_{t-1}) > J(\Phi_t)$, the least significant feature is removed from Φ_t and placed back in Υ (t is decremented). This second stage is repeated until removal of the least significant feature does not help achieving higher values of evaluation function.

Now we may define the evaluation function J. It should encapsulate all three quality measures and be maximized by the most relevant feature subset. As it occurred in our tests, simple multiplication of inconsistency and cophenetic correlation coefficients divided by the compactness value does not work well. The reason for this lies in the nature of compactness factor. It cannot be treated as a precise qualifier comparable with the other two factors or with Fisher coefficient. Clusters that have compactness value of the order 10^2 and 10^{-2} are actually equally good. We can assume poor quality of clusters only when compactness goes over 10^3. Therefore, this coefficient can be used only as an argument of a function that does not differentiate between values higher than specified threshold. Our choice is the *unipolar sigmoid* function:

$$\widehat{\zeta} = \frac{1}{1 + \exp(-\frac{1}{\eta\zeta})}, \tag{5}$$

where ζ is given by (1) and a real constant η determines the slope angle of the function curve in the neighborhood of 0 ($\eta = 100$ in our method). We put ζ in the denominator, so greater values of (5) indicate better feature subsets.

The last problem is the range of inconsistency coefficient. Its comparative capabilities are limited due to the fact, that it is not bounded. It may happen that two different feature subsets allow obtaining maximal inconsistency coefficients for the cutoff links, but they are not equal. Therefore, we relate the value of cutoff link inconsistency coefficient to the maximum obtained for a given tree. Then we will always gain a number between 0 and 1 and can objectively compare clustering results. The coefficient that we constructed this way, we called *relative cutoff link inconsistency* coefficient:

$$\rho_{cutoff} = \frac{\gamma_{cutoff}}{\max_m \gamma_m}. \tag{6}$$

Consequently, final form of the evaluation function is given by (7):

$$J = \varphi \times \rho_{cutoff} \times \widehat{\zeta}. \tag{7}$$

4 Experimental Results

To verify efficiency of our clustering-based feature selection method we carried out a series of experiments with textured images. Figure 2 depicts samples of 6 examined textures. Every texture was represented by 16 such samples. Again, by the use of MaZda, samples were described by 270-feature vectors. Next, we composed all possible 2– and 3–cluster data sets (35 combination) and determined best features for these sets according to Fisher and POE+ACC coefficients. Afterwards, we selected features using our unsupervised method, every time requiring 10-best subsets. The verification criteria were as follows. For every analyzed data set we determined:

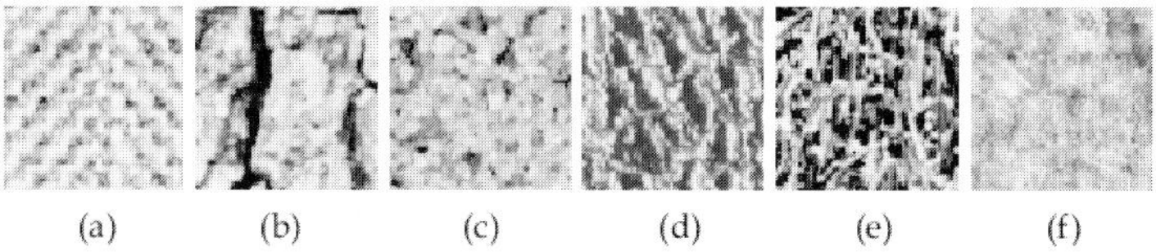

Fig. 2. Texture samples used in verification test: (**a**) herringbone weave, (**b**) tree bark, (**c**) beach sand, (**d**) calf leather, (**e**) grass, (**f**) woolen clothe

1. number of features simultaneously selected by Fisher coefficient and clustering-based method (crit_1),
2. number of misclassified samples described by 10 best features according to Fisher coefficient using 1-NN classifier (crit_2),
3. number of misclassified samples described by 10 best features according to clustering-based method using 1-NN classifier (crit_3) and
4. number of misclassified samples described be 10 best features according to clustering-based method using hierarchical clustering algorithm (crit_4).

Results obtained in calculations are summarized in Table 2

Table 2. Results of verification test

Verification criteria	2–cluster sets min / max / average	3–cluster sets min / max / average
crit_1	8 / 4 / 6.27	6 / 4 / 4.23
crit_2	1 / 0 / 0.07	2 / 0 / 0.18
crit_3	0 / 0 / 0.00	3 / 0 / 0.27
crit_4	0 / 0 / 0.00	2 / 0 / 0.15

As it can be seen, our feature selection method proved effective. Numbers of features simultaneously found by unsupervised and supervised algorithms are very high (five in average). When more clusters are included in the data set, numbers of common features drop, but the accuracy of classification remains satisfactory. Both tested algorithms perform well when describing samples with features selected either by the supervised or unsupervised methods.

In addition, notice that performance of a classifier can be improved, if we disregard superfluous features in the 10–best subset. It often happens that the value of evaluation function is greater for a lower than desired number of features. Thus, the result of our method is triple:

1. feature subset of the size specified by the user,
2. feature subset of the minimal size ($c-1$, if c denotes number of clusters),
3. feature subset of the optimal size (maximizes evaluation function).

From our experiments it arises that in such optimum-sized subsets, majority are the features simultaneously found by the help of Fisher coefficient.

5 Conclusions

We have presented novel feature selection method for the use in unsupervised context. It is based on clustering quality measures and utilizes SFFS algorithm to find the best feature subset. The uniqueness of our approach is constituted by the fact, that it embodies three different aspects of clustering validation: quality of clusters, quality of separation of credibility of partitioning.

Using compactness coefficient as the main factor of quality evaluation enhances objectivism of the method: estimated quality is verifiable not only in terms of specific clustering algorithm used to separate data. Moreover, the method is straightforward and robust. The underlying principle is intuitive and all calculations involve basic mathematical operations. This also facilitates implementation issues.

Experiments that we performed on textured images confirmed effectiveness of our technique in comparison to supervised feature selection algorithms and by testing performance of different classification procedures. Possible fields of the method application are not restricted to unsupervised learning tasks. It can introduce objectivism into data vectors discrimination. Feature reduction methods which make use of the a priori information about class–membership of samples are inclined to find features that best match this separation. If selection is performed in unsupervised manner, selected features will most probably reflect some more general rule that forced partitioning of the given data set into classes.

References

1. Law M, Figueiredo M, Jain A (2004) IEEE Trans. on Pattern Analysis and Machine Intelligence 26:1154–1166
2. Materka A, Strzelecki M, Szczypinski P (2000) Texture Analysis Software MaZda. Mazda User's Manual. On: http://www.eletel.p.lodz.pl/cost/progr_mazda.html
3. The MathWorks, Inc. (2002) MATLAB Documentation: Statistics Toolbox 4.0 User's Guide
4. Mao J, Jain A (1995) IEEE Trans. on Neural Networks 6: 296–317
5. Oh I, Lee J, Moon B (2004) IEEE Trans. on Pattern Analysis and Machine Intelligence 26:1424–1437
6. Wolf L, Shashua A (2003) submitted to Journal of Machine Learning Research
7. Zaït M, Messatfa H (1997) Future Generation Computer Systems 13:149–159

Conceptual Ontological Object Knowledge Base and Language

Marek Krótkiewicz[1] and Krystian Wojtkiewicz[2]

[1] Institute of Mathematics and Informatics, University of Opole
`mkrotki@math.uni.opole.pl`

[2] Institute of Mathematics and Informatics, University of Opole
`kwojt@math.uni.opole.pl`

Summary. This paper deals with AI in aspect of knowledge acquisition and ontology base structure. The core of the system was designed in an object model to optimize it for further processing. Direct concept linking was used to assure fast semantic network processing. Predefined attributes used in the core minimize the number of basic connections within the ontology and help in inference. The system is assumed to generate questions and to specify the knowledge. The AI system defined in this way opens a possibility for better understanding of such basic human mind mechanisms as learning or analyzing.

1 Ambiguity

Ambiguity is one of the main concepts of Artificial Intelligence studies and it is mainly linked with text understanding, which is identified with an interpretation i.e. with a reference to ontology. An ambiguous term is the one with many meanings. Therefore, it is possible to assign many different meanings to one identifier, or simply one word. For example, we can take the word 'date' which has at least three different meanings. However, there are also some concepts that have more than one word they can refer to. We call them synonymic expressions e.g. man, person, human. The information gaps in communication resulting from the the lack of determination of all the connections between concepts cause a necessity to fill them in with guess words. Take the sentence 'John is at school', which could mean 'John is at his school' and more precisely 'John is at the school he studies in'. To be unequivocal we should say that this sentence refers to the following statement: the object 'John' is in the relation 'to be at' with the object 'school' with whom John is in the relation 'to study in'. However, this might not be precise enough, so we should say: the object 'John' is in the relation 'to be at' with the object 'school'. As we can see, the ambiguity is the main factor affecting the process of understanding. The methods that are currently in use are focused on an analysis of direct and indirect contexts. Although they are effective, they are not designed to understand the text. They just assume that some terms are assigned to certain concepts. The assumption is based on the topic of the whole text. Totally different is an approach set in the direction of understanding the text, which is represented as linking of concepts derived from this text with the ontology[1]. In this way, we have an opportunity to excel in interpretation based on the knowledge saved in ontology. This aspect of building the ontology refers to the process of human mind development[2]. With the use of the AI System configured in this way, we can examine its behavior in the situation where the main factors are a changeable structure of information or the inference module. Taking this as a leading point of studies, we can try to find the source of 'stupid questions' asked by children.

2 Active Knowledge Acquisition

Taking previously presented theory as the leading one, an obvious conclusion is that one of the main features of AI system is the knowledge acquisition stored in the attributes of the semantic network elements and in the relations between those elements. This activity requires high accuracy in choosing questions. Already collected knowledge should create the base for determining unambiguous and correct interpretation of statements. This can start further processing, e.g. deduction. Processing ambiguous information may lead to wrong conclusions. For example, we can take a sentence 'John has a date with Ms Johnson'. During the analysis, we find that 'date' has three different meanings in the ontology. However, semantic structure of 'date' as the one in calendar excludes it from further processing. The other two are equally possible to use, so it is a classical problem of ambiguity. The next step is to determine which of these meanings should be used in this case. The problem of ambiguity can be easily solved with the questions about detailed attributes of connections in the semantic network. The knowledge gathering action cannot create a wrongful conclusion. Children have their own mechanism of understanding ambiguous statements that generates such questions as: Who? What? Where? When? How?[2] Trying to imitate evolving mind, AI System should behave exactly in the same way. It is expected that the number of questions declines together with the growth of knowledge base, analogically to the human behavior.

There are a few well known methods of knowledge representation such as: rules, semantic networks, logical notations[3]. All of them have positive and negative aspects. Although essential is that the chosen one would in the best possible way ensure obtaining required functionalities in the scope of storing ontological knowledge, facts, rules as well as in fast and easy processing in aspects of inference, question generating and complex facts matching.

3 Main Features of COOS

The subject area of this study includes the structure of ontological base shown on Fig. 1. During its construction, the main assumption was to ensure maximal capacity of knowledge directly accessible through the connections between specified elements: CONCEPT, ASSOCIATION, PREPOSITION, CLASS, FEATURE, VALUE, DATATYPE, being a part of a given microtheory, which contains rules. The implemented semantic network between concepts ensures

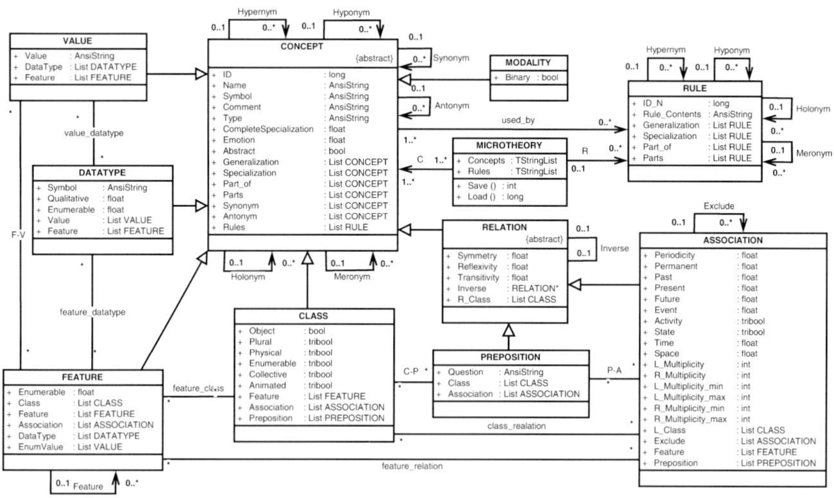

Fig. 1. The object model of the ontological base structure

a possibility of direct search for essential elements. This is very important for question generation. For example, CLASS has a structural connection with concepts ASSOCIATION, FEATURE and PREPOSITION, but it does not have a straight link with VALUE. It is derived from the assumption that VALUE cannot refer to CLASS. The statement 'red car' (VALUE:='red' and CLASS:='car') within the presented ontology has to be filled out with the information connecting them i.e. FEATURE. This statement should be presented as 'car of red color' or 'red color car' i.e. `C#car [F#color V#red]`. The studies of the ontological base structure show the following categories of information gaps:

- a lack of concepts in the ontological database,
- a lack of information e.g. attribute values of already known concepts,
- a lack of information on concept linking in the database.

If the system does not have a concept in its base, it will ask for it. We should interpret it as an attempt to create a new object and to fill in its attributes with appropriate values. The lack of information about specific values of essential attributes generates a question that enables filling the values. In case there is no connection in the ontological database between certain concepts that are connected in the statement, the system question will refer to the correctness of a given connection. The concepts `C#car` and `F#color` may not be linked in the ontology. That means the system does not know whether the class 'car' is connected with the feature 'color'. If the system obtains the information that this connection is allowed, its future uses will be clear. Moreover, the system will use this information in any other statement that contains the class 'car' to generate questions about this feature. The rule applies to all the other connections within the ontological database. The presented structure is the core of the ontology. It allows for an instant verification of concepts connection admissibility. This core contains all the information about essential attributes for each of the concepts. A large number of basic attributes decreases considerably the number of connections. For each of the concepts, there are specified attributes needed for the correct interpretation. Not only do they make the system operate faster, but they are also used to distinguish between different categories. Basic connections and object model proprieties are implemented directly into the structure. Most of them are described in the next section.

The class CONCEPT has a number of technical attributes such as ID, Name, Symbol, Comment, Type, whose meaning is obvious or is not essential in this matter. Therefore, we do not characterize them. The attribute CompleteSpecialization is bound with Specialization containing the list of concepts being specialization of a given concept according to the scheme specialization-generalization. If CompleteSpecialization is set to value one, it means that the list Specialization is complete, and it is not possible to add any more concepts to it. Generalization contains the list of all the concepts being generalization of a given concept. The attributes Part_of and Parts are used for a description of the dependency - 'whole-part'. The lists of synonyms and antonyms of the concept are located in the attributes Synonym and Antonym.

The additional attribute Rules is a list of rules, in which the concept is used. Its task is to accelerate considerably the context search for concepts, which is of high importance for its future uses in the scope of ontological database analysis.

The class named CLASS is a description of the objects representing classes and objects presented in knowledge base. The attribute Object differs objects from classes. An example of the class is `C#car`, and of the object is `C#Alan_Turing`.

The meaning of the attributes: Plural, Physical, Enumerable, Collective, Animated is compliant with their intuitive meanings. The lists FEATURE, ASSOCIATION, PREPOSITION describe possibilities of linking elements of those lists with a given class.

The objects of class FEATURE contain information about the features that can be related to CLASS, FEATURE, ASSOCIATION, DATATYPE or

VALUE. The limitations of those links are within the values of the attributes: Class, Feature, Association, DataType, EnumValue.

VALUE is a class with only three attributes whereas Value is a string representing a value of this class e.g. `V#5` or `V#red`. The lists DataType and Feature are analogical to the previously described ones. They set the scope of possible connections between the mentioned concepts.

The class RELATION is an abstract class, which is the generalization of class PREPOSITION and ASSOCIATION. The attributes Symmetry, Reflexivity, Transitivity are the basic features of relation. The attribute Inverse shows an inversed relation to a given one and R_Class is a list of admissible right operands of relation.

The class PREPOSITION contains the lists - Class and Association - and the attribute Question that contains question expressions appropriate for given preposition.

The class ASSOCIATION describes the key element for the majority of statements. The attribute Periodicity is linked with the periodicity of association. Permanent is the information about permanency of association. The attributes Past, Present, Future are used for fact positioning on the time axis. Event gives information whether the duration of association is important or not. Activity and State are used for telling the difference between states e.g. `A#sleep` and activities `A#go`. Attributes Time and Space contain the information about invariability in time and space. The attributes within the Multiplicity group determine the multiplicity of a given association or the scope within which the multiplicity may be set. L_Class is a list of allowed left operands. A list of concepts excluded from the simultaneous use is stored in the attribute Exclude. Feature and Preposition are the lists of features and prepositions that are suitable for a use with a given association.

The class MODALITY contains the description of objects making up modalities of statements. If the attribute Binary has a value True, then the modality is bipolar. The class RULE has the main attribute Rule_Contents being the text representation of a rule. The lists Generalization, Specialization, Part_of, Parts have pointers to the rules being in relation 'generalization-specialization' or 'whole-part' with a given rule.

MICROTHEORY is a class describing fields that are aggregations of concepts and rules. Thanks to it, it is possible to move quickly through the inheritance hierarchy, which results in the verification efficiency of related features or of creatable relations, and in other aspects of the object model. The logical attribute type is used in the form of a tribool to enable the operation in a three-valued logic. The issue of answering the problem questions such as 'why?', 'what for?' is moved to the competence of the inference module. The general system scheme of the module orientated structure is presented on Fig. 2. It presents a transition of information from the outside of the system to the knowledge base and in the opposite direction. One of the most important features of this construction is a possibility of multilingual text interpretation and translation.

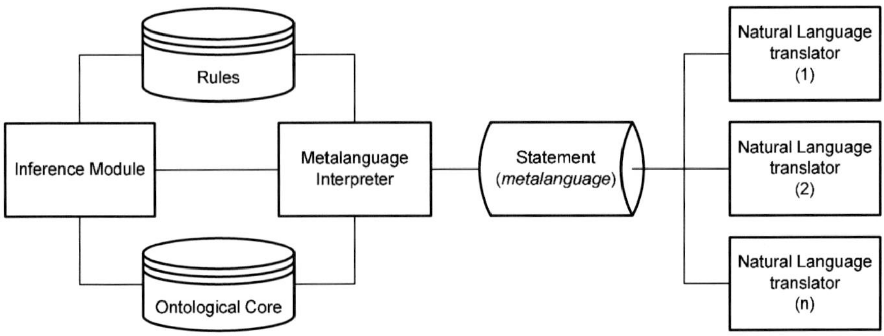

Fig. 2. The general diagram of the connections between the main system elements

4 Metalanguage

Metalanguage was created to obtain a possibility of knowledge storing and processing. It performs an interface function between the knowledge base and natural languages[4][5]. A series of assumptions was made while developing grammar and semantic rules. Metalanguage has to be accurate. It means that the possibility of multi-interpretation has to be limited as well as the structure has to force the use of some specifications. It is also very important for this language to be as compact as possible, but at the same time easy to interpret. The latter statement is in opposition to the assumption of simplicity of automatic processing to make different forms of inference. The basic features of the metalanguage are presented below on the examples drawn from the paper[6]. To provide a comparison with other languages of the type, examples are provided in: English (E), Formalized-English (FE), Frame Conceptual Graphs (FCG), Conceptual Graphs Linear Form (CGLF), Conceptual Graphs Interchange Format (CGIF), Knowledge Interchange Format (KIF) and Conceptual Ontological Object Language (COOL) along with its simplified grammar tree.

Example: 1

```
E:    Tom owns a dog that is not Snoopy.
FE:   Tom is owner of a dog different_from Snoopy.
FCG:  [Tom,owner of: (a dog != Snoopy)]
CGLF: [T:Tom]<-(owner)<-[dog:*x!=Snoopy]
CGIF: [dog:*x] (owner ?x Tom) (different_from ?x Snoopy)
KIF:  (exists ((?x dog)) (and(owner ?x Tom) (/= ?x Snoopy)))
COOL: {C#Tom} A#own {C#dog} [F#name D#text ~V#Snoopy]
Tree: A#own
        C#Tom
        C#dog [F#name D#text ~V#Snoopy]
```

The basic element of the statement is a connection between the association and the left and right operands in the form of: `{X#...} A#... {X#...}`. `C#Tom` means a class, but actually it is an object. The statement `C#dog` equals to the class, but the further modification of the feature `F#name D#text ~V#Snoopy` narrows the class only to the elements whose feature "name" is different from the text 'Snoopy'. The symbol ~ is used as a negation, e.g. if it precedes the association `~#go` it means 'don't go'. The symbol ~ before a feature means 'different from', so `~V#Snoopy` has to be interpreted as 'different from Snoopy'. The features of concepts should be given in square brackets as a list, e.g `C#dog [F#name D#text ~V#Snoopy, F#color D#color V#red]`. The elements of the list separated with a comma stand for a conjunction. To create an alternative for the elements we need to separate them with a semicolon. The statement `C#car [F#size V#big; F#speed V#fast]` means a big car or a fast car, or big or fast cars. The attribute Plural may be used to deermine the above mentioned equivocality, e.g. `C#car [Plural=1] [F#size V#big; F#speed V#fast]`.

Example: 2

```
E:    Tom believes Mary now likes him (in 2002) and before she did
      not.
FE:   Tom is believer of '*p' Mary is liking Tom' at time 2002' and
      is believer of '!*p is before 2002'.
FCG:  [Tom,believer of:[*p [Mary, agent of:(a liking,object: Tom)],
      time:2002], believer of: [!*p, before: 2002] ]
CGLF: [proposition: *p [T:Mary]<-(agent)<-[liking]->(object)
      ->[T:Tom] ][T:Tom]- { (believer)<-[[situation: ?p]->
      (time)->[time:"2002"], (believer)<-[ [situation:~?p]->
      (before)->[time:"2002"] ] }
CGIF: [proposition *p: (agent [liking *1] Mary) (object ?1
      Tom) ] (believer [situation: (time[situation: ?p]
      "2002")] Tom) (believer [situation: (before [situation:
      ~[?p]] "2002")] Tom)
KIF:  (exists (?p) (and (= ?p '(exists((?x liking)) (and (agent
      *1 Mary)(object ?1 Tom))) (believer ^(time ,?p 2002)
      Tom) //',?p'->the value of ?p is quoted (believer ^(before
      (not ,?p) 2002) Tom)))
COOL: {C#Tom} A#believe ({C#Mary} A#like[Present=1][F#time
      D#year V#2002] {C#Tom}, {C#Mary} ~A#like[Past=1] {C#Tom})
Tree: A#believe
         C#Tom
         A#like[Present=1]
            C#Mary
            C#Tom
      A#believe
         C#Tom
         ~A#like[Past=1]
            C#Mary
            C#Tom
```

The example shows the use of time attributes `Present=1` and `Past=0`. It also presents the representation of nested associations. The association `A#believe` has the right operand in the form of a list because Tom's belief refers to two facts, which may be written down separately. This rule could be explained as follows: `{X#a} A#r (expr_1, expr_2)` which means: `{X#a} A#r expr_1 && {X#a} A#r expr_2`, and `{X#a} A#r (expr_1; expr_2)` which equals to `{X#a} A#r expr_1 || {X#a} A#r expr_2`. In the above example `&&` stands for a conjunction and `||` for an alternative.

5 Conclusion

The optimal ontology development is still the main concern of the studies over AI Systems. The part of AI systems is being designed to imitate the human intelligence. They often just try to fulfill Turing criteria but the only rational way is by building the system, which can understand any given information. Understanding is a possibility of writing knowledge within ontology base and a further verification and inference. The value of Turing criteria is disputable and there are other conditions to fulfill such as efficiency in the knowledge base search, the verification of the statement semantic correctness, complete answers providing detailed features, the relations possible for concepts, generalizations or specializations etc. All of these can be easily fulfilled by the structure presented in this article. The article describes mainly the core of the system and is a short introduction to the metalanguage. The complete and compact description of the metalanguage and the realization of the semantic network will be presented in the next articles.

References

1. Dyachko A.G. (1997) Text Processing and Cognitive Technologies, Moscow, Pushchino, ONTI PNTS RAN
2. Włodarski Z. (1996) Psychology of learning, PWN Warszawa (in Polish)
3. Fuchs, N.E., Schwertel, U., Torge, S. (1999) Controlled Natural Language Can Replace First-Order Logic. 14th IEEE International Conference on Automated Software Engineering, Cocoa Beach, Florida
4. Landauer T. K., Littman M. L. (1990) Fully automatic cross-language document retrieval using latent semantic indexing. Sixth Annual Conference of the UW Centre for the New Oxford English Dictionary and Text Research. Waterloo Ontario: UW Centre for the New OED and Text Research 31-38
5. Frederking R., Mitamura T., Nyberg E., Carbonell, J. (1997) Translingual information access, AAAI Symposium on Cross-Language Text and Speech Retrieval. American Association for Artificial Intelligence
6. Philippe M. (2002) Knowledge representation in CGLF, CGIF, KIF, Frame-CG and Formalized-English, 10th International Conference on Conceptual Structures, Springer Verlag, LNAI 2393, pp. 77-91) Borovets, Bulgaria

A Method of Supervised Discrimination of Textures Based on Serial Statistical Tests

Juliusz L. Kulikowski[1], Malgorzata Przytulska[1], and Diana Wierzbicka[1]

Institute of Biocybernetics and Biomedical Engineering PAS, 4 Ks. Trojdena Str., 02-109 Warsaw, Poland jlkulik@ibib.waw.pl

Summary. It is presented a new type of learning textures recognition algorithms based on serial statistical tests. It is assumed that a texture can be formally represented by a multi-component random vector whose probabilistic characteristics are, in general, a priori unknown. Discrimination of textures is equivalent to a discrimination of random vectors of different but a priori unknown statistical properties. For this purpose non-parametric statistical tests based on serial statistics are used. Construction of serial statistics needs a linear ordering of multi-dimensional observation space. The method is illustrated by numerical examples.

1 Introduction

Recognition of textures is one of important and the most difficult pattern recognition problems. A texture can be defined as a set of locally homogenous photo-optical, morphological, spectral, fractal and/or statistical features visible in images of real objects and caused by their physical or morphological structure. Some authors define it also as "all that is not a boundary in an image" [1]. This pragmatic definition suits well in the case when texture recognition in a fixed area of an image is the aim of image analysis. However, it leads to a logical vicious circle if textures are to be discriminated in order to detect the separating them contours. It is thus reasonable to distinguish between texture recognition, when we ask about the type of texture under observation and discrimination of textures when one should decide whether two (or more) samples of image represent the same or different textures. In the last case a set of such decisions leads to image segmentation, i.e. image partition into mutually disjoint sub-areas, each one being covered by a homogenous texture, different than the ones covering the adjacent sub-areas. Our interest will be focused below on textures discrimination rather than on texture recognition. However, it should be remarked that textures discrimination may be a preliminary step to texture recognition, as well.

There will be considered below 2D images only. The approach used here is based on the assumption that visible parts of textures are instances of 2D random fields. However, probabilistic characteristics of textures under observations are a priori unknown and it is assumed that one should distinguish between textures whose properties are significantly different in statistical sense. For this purpose non-parametric statistical tests can be used. In particular, using serial statistical tests to texture recognition in [2] has been proposed. In the mentioned works textures discrimination without learning was considered. This means that a statistical test used to textures discrimination within a given experiment remained unchanged and the discrimination was performed on a fixed statistical quality level. However, it was also observed that this quality depends on some test features. In particular, in the case of serial statistical tests it depends on the dimensionality and on the way of linear ordering of a vector observation space.

Textures, in general, describe multi-level morphological structures of observed objects. For example, in satellite Earth surface images on the highest level textures corresponding to clouds, continents and oceans can be distinguished. On a lower level of continents imaging the textures of deserts, glaciers and other types of continental earth covering are visible. In the last ones, on a lower level the textures of forests, towns, agricultural fields, etc. can be observed. In aerial images of agricultural fields the textures corresponding to various types of crops can be recognized, etc. A similar hierarchy of textures occurs in image analysis of biological specimens, natural or artificial materials, etc. In all the above mentioned cases the results of texture analysis depend strongly on the spatial resolution of imaging tracts on one, and on the size of a basic observation window on the other hand. Such macro-parameters of image processing algorithms are usually fixed by an observer and they cannot be changed by a self-learning textures discrimination algorithm. However, the last is able to choose, within certain limits, its inner parameters or properties so that a decision quality indicator is improved while the set of observations is extended. Our attention will be focused below on the way of a vector observation space linear ordering.

2 Linear Ordering of Observation Space

Before going to the details several basic notions will be introduced.

2.1 Basic notions

It is assumed here that an observed image is given in thew form of a finite, $I \times J$ matrix $\boldsymbol{u}$ of elements called pixel values, I and J being natural numbers denoting, correspondingly, the number of rows and of columns of the matrix.

A pixel is given by a triple of elements $u_{ij} = [i, j, v_{ij}]$, $1 \leq i \leq I$, $1 \leq j \leq J$, $v_{ij} \in V$, V being a finite, 2^k -element linearly ordered set of elements called pixel values, where k, $k = 1, 2, \ldots$, is a fixed natural number (the number of bits per a pixel value).

A basic window is a rectangular, $a \times b$-size sub-matrix $\boldsymbol{w} \subset \boldsymbol{u}$ such that its elements are considered as an instance of a random vector $\boldsymbol{W}$ representing a texture. In fact, a basic window is also characterized by its localization in $\boldsymbol{u}$, for example, by the address $[\alpha, \beta]$ of its upper-left element, and as such, it should be denoted by $\boldsymbol{w}^{(\alpha,\beta)}$. Taking into account that all such basic windows are mutually isomorphic, for the sake of simplicity the upper indices will be omitted, everywhere it is possible.

The set $V^{a\times b}$ of all possible instances of the random vector $\boldsymbol{W}$ represented by the given basic window $\boldsymbol{w}$ is called a basic observation space. Of course, it is $V^{a\times b} \in V^{I\times J}$ where $V^{I\times J}$ is an observation space of the total image.

For testing hypotheses of statistical identity (vs. difference) of random vectors representing the textures finite samples of vectors assigned to some collections of basic windows will be considered. Such collections, called sample windows, will be, generally, denoted by $\boldsymbol{s}$. The form and size of sample windows are also subjects of arbitrary decisions. For the sake of simplicity it is here assumed that all sample windows under consideration have the same form and size and they consist of compact, rectangular $c \times d$ compositions of mutually adjacent basic windows. The natural number $c \cdot d$ describes thus the size of a sample of vectors taken into account in a decision about its statistical identity or difference to another sample of vectors.

A collection of sample windows whose statistical identity has been established and, thus, representing the same texture, is called a textural segment. Partition of image into mutually disjoint textural segments is called a texture-based image segmentation.

2.2 Linear ordering of observations - a new pattern recognition paradigm

Early pattern recognition methods were based on observation space partition into decision regions corresponding to the classes of similar objects. For many years a concept of similarity of observed objects played thus the role of a basic paradigm of pattern recognition. Such situation still exists besides the fact that the paradigm occurs ineffective in the case of composite patterns recognition. In such case it can be removed by a paradigm of structural description of composite patterns using formal linguistic or relational tools. Linear ordering of observation space is a new paradigm in pattern recognition which suits well to the case of pattern recognition under extremely poor primary information about the patterns being to be discriminated and/or recognized. Its idea is illustrated in Fig. 1. Instead of partition of observation space into decision regions by hyper-planes (Fig. 1a) its elements are ordered linearly so that the elements belonging to different classes are grouped in separate segments of a curve tracing the order in the observation space (Fig. 1b).

Linear ordering of observation space is a new paradigm arising in pattern recognition. It follows from the above-given definition that the basic observation space $V^{a\times b}$ is a finite set consisting of $a \cdot b \cdot 2^k$ elements which, in general, can be linearly ordered in $(a \cdot b \cdot 2^k)!$ ways. We shall denote by Π the set of all possible permutations of the elements of observation space; Π is thus isomorphic to the set Λ of all possible ways of the observation space linear orderings. Let us denote by S_1, S_2 two finite, mutually disjoint subsets of $V^{a \times b}$ representing a pair of different classes of elements (i.e. a pair of textures). We can take into consideration the set $S_1 \cup S_2$ and arrange its elements according to a fixed way λ, $\lambda \in \Lambda$, of linear ordering introduced into $V^{a \times b}$. Then, the linearly ordered set $S_1 \cup S_2$ can be divided into a number of homogenous subsets of consecutive elements belonging to S_1 or to S_2. Such subset of elements is called a series and we are interested, in general, in the number and in the

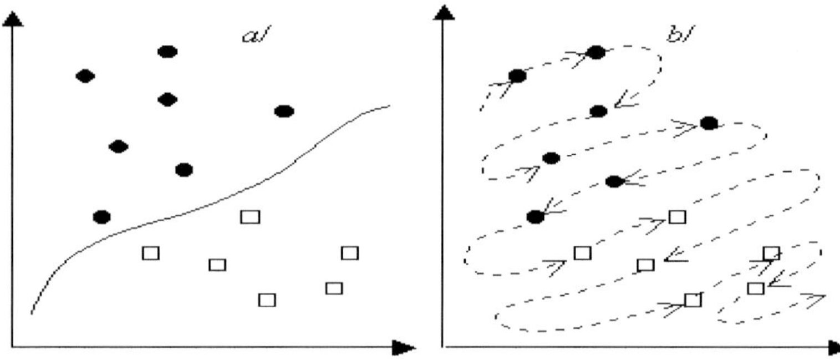

Fig. 1. Observation space partition a) and linear ordering b).

lengths of such series. For any given pair S_1, S_2 the number ν of series may be, at least, $\nu_{min} = 2$ and, at most, $\nu_{max} = 2min\{|S_1|, |S_2|\} + 1$, where $|S|$ denotes the number of elements of S. For the best discrimination between the elements of S_1 and S_2 a linear order leading to the minimum number of series is desired. Therefore, the ratio:

$$\rho = \frac{\nu_{max} - \nu}{\nu_{max} - \nu_{min}} \tag{1}$$

can be used as a quality measure of the given way of observation space linear ordering. Only few permutations satisfy the requirement $\nu = \nu_{min}$ of the given pair of subsets exact separation.

2.3 A multi-level linear lexicographical ordering of observation space

There is no general methodology of the observation space optimum linear ordering. Our attention is paid below to some classes of orderings making possible their more effective improving than by a random searching procedure. In particular, in this section it will be shown how a step-wise procedure can be applied to a multi-level linear lexicographical order improving.

The multi-level linear lexicographical order will be introduced into the observation space $V^{a \times b}$ in the following way:

a) $V^{a \times b}$ will be subjected to a first-level partition into mutually disjoint sub-areas $Q_{\kappa}^{(1)}$, $1 \leq \kappa \leq K$:

$$V^{(a,b)} = \bigcup_{\kappa=1}^{K} Q_{\kappa}^{(1)} \tag{2}$$

b) Selected sub-areas $Q_{\kappa}^{(1)}$, of the first-level partition will be subjected to a second-level partition into sub-sub-areas $Q_{\kappa,\lambda}^{(2)}$, $1 \leq \kappa \leq K$, $1 \leq \kappa \leq L$:

$$Q_{\kappa}^{(1)} = \bigcup_{\lambda=1}^{L} Q_{\kappa,\lambda}^{(2)} \tag{3}$$

etc. , up to the highest, Rth-level partition.

c) At a given partition level the indices (κ,λ,...) of the corresponding sub-areas establish their linear order within the higher-order area.

d) The elements of any sub-area being at its highest sub-partition level (i.e. .not subjected to further, higher-order partition) become lexicographically linearly ordered according to:
   - the values of vector-components,
   - the indices of vector-components

   as it has been described in [3].

e) The elements (observation vectors) of $V^{a \times b}$ are then lexicographically linearly ordered in the following steps:
   1) according to the first-level partition index (point a));
   2) according to the rule of linear ordering described in point d) if the highest partition level has been reached or according to the next partition level index (points b),c)).

The following example will illustrate the above-given rules of linear ordering.

Example 1. It will be considered a two-dimensional discrete observation space V^2 shown in Fig. 2 The co-ordinates in the square-area denote values of pixels; for example, a pair 2.3 corresponds to a vector $\boldsymbol{u}$ = [2,3]. The vectors are lexicographically linearly ordered according to the position of row of the first vector component and of column of the second component. However, the position of vectors can be permuted by permutations of. rows, columns, and/or some blocks obtained due to the observation space partition. The tables a), b), c) and d) show three consecutive steps of observation space partition and re-ordering. In all tables a first-level partition into four 4×4 sub-areas (denoted by bold numbers) has been introduced. In table b) the positions of sub-areas **(2)** and **(4)** have been interchanged. In table c) the sub-area **(2)** has been sub-divided into four second-level 2×2 squares. In table d) the position of the second-order sub-areas **(2.1)** and **(2.3)** once more has been interchanged. The above-described operations changes the original lexicographical order of the observation space. For example, a series of vectors presented in its original order as (see table a)):

[2.2], [2.7], [3.2], [3.7], [6.2], [6.7], [7.2], [7.7],

after reordering takes the following form (see table d)):

[2.2], [6.7], [3.2], [7.7], [6.2], [2.7], [7.2], [3.7],

The above-presented permutations can be shortly denoted as follows:

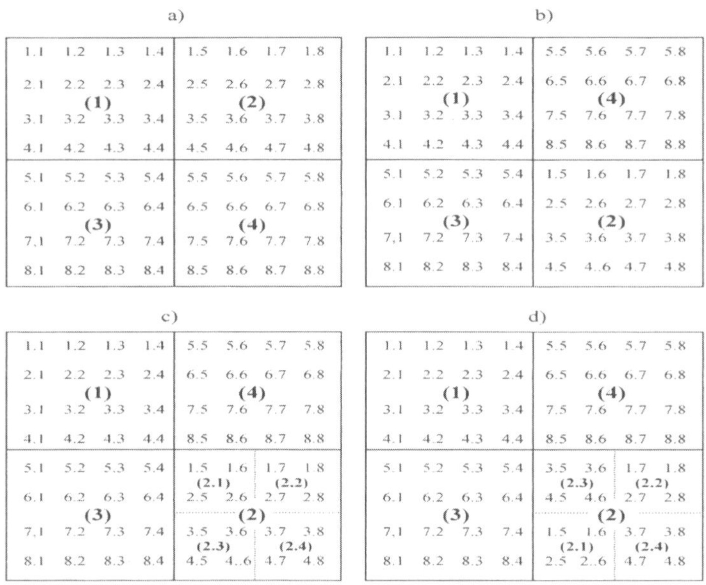

Fig. 2. Linear ordering of observation space by sub-partitions and permutations.

$$[(2), (4)] \Rightarrow [(4), (2)] * [(2.1), (2.3)] \Rightarrow [(2.3)(2.1)]$$

where symbols in the square brackets stand for the tags of the square blocks, $\Rightarrow$ indicates a logical direction of changes (from - to), and $*$ denotes a (non-commutative) conjunction of permutations. Such a notation contains, in general, all information about the modified linear order of observation space used to making decisions concerning discrimination of textures.

3 Improving the Algorithm of Separation of Classes

Let us assume that there are given two finite and mutually disjoint learning sets, S_1, S_2 of the observation space $V^{a \times b}$ (see sec. 2.2). It will be considered the set $S_1 \cup S_2$ linearly ordered according to a primary ordering rule $\lambda^{(0)}$ (the upper-case index denotes the step of the observation space reordering process). Without loss of generality it may be assumed that $\lambda^{(0)}$ is a natural lexicographical order of vectors. Then the number ν of series and the measure $\rho^{(0)}$ of quality of the linear order $\lambda^{(0)}$ according to the method described in sec. 2.2 can be calculated. If $\nu = \nu_{min}$ then the classes are perfectly separated, $\rho^{(0)} = 1$ and no observation space reordering is necessary. Let us thus assume that $\nu = \nu_{max}$. Then this means that the elements of S_1 and S_2 are perfectly mixed and the corresponding statistical classes, even if being different can not be discriminated on the basis of the available data. Therefore, let us assume that $\nu_{min} < \nu < \nu_{max}$ and, consequently, $0 < \rho^{(0)} < 1$. In such case, taking into account that $V^{a\times b}$ is a finite set, there is always a possibility to reorder the observation space so that $\nu = \nu_{min}$ and $\rho^{(1)} = 1$. For example, for this purpose one could put: 1st all elements of S_1 taken in any linear order, 2nd join to them, in any linear order, all elements of $V^{a\times b} \backslash (S_1 \cup S_2)$, and 3rd join to them, also in any linear order, all elements of S_2. However, such method of observation space reordering from the learning process point of view would be rather useless. Taking into account that in practice the cardinal number of $V^{a\times b}$ is much higher than this of $S_1 \cup S_2$, it can be expected that next elements of $V^{a\times b}$ with great probability will belong to $V^{a\times b} \backslash (S_1 \cup S_2)$ where the linear order has been chosen arbitrarily. Such method of reordering would thus have rather poor ability of experience generalization. However, it can be improved if the new-introduced linear order in $V^{a\times b}$ localizes, if possible, any two its elements the closer each to the other one on the order tracing curve (see Fig. 1b) the closer they were in the original space. This can be reached if the class Π of permutations at each step of reordering is limited to permutations of sub-areas at a given level of observation space partition into fixed-size rectangular parallelepipeds, as it has been shown above. Let us assume that the linearly ordered set $S_1 \cup S_2$ is presented in the form of a sequence of pairs $\Sigma = [(\nu_\mu, s_\mu)]$, $\mu = 1, 2, \ldots, M$, where ν_μ is the μ-th element (vector) in the linearly ordered set $S_1 \cup S_2$, $s_\mu \in \{1, 2\}$, s_μ is the index of the learning set (S_1 or S_2) from which ν_μ has been drawn out, M is the cardinal number of $S_1 \cup S_2$. Let us also assume that at a given, l-th partition level the observation space is presented by the formula (2). Then Σ can be also partitioned into K sub-sequences (for the sake of simplicity K is used here without the upper index indicating its dependence on the partition level l):

$$\Sigma = [\Sigma_1, \Sigma_2, \ldots, \Sigma_\kappa, \ldots, \Sigma_K] \tag{4}$$

preserving the linear order of Σ and such that Σ_κ consists of all pairs (ν_μ, s_μ) belonging to the sub-area $Q^{(}l)_\kappa$. Of course, some sub-sequences in (4)may be empty if no elements of the learning sets S_1, S_2 belong to the corresponding sub-area $Q^{(}l)_\kappa$. For any given non-empty sub-sequence Σ_κ there will be evaluated the numbers $\beta_\kappa^{(1)}$ of its elements (ordered pairs) drawn out from S_1 and $\beta_\kappa^{(2)}$ of those ones drawn out from S_2. Then, there will be calculated the sub-area indicators:

$$r_\kappa = \frac{1}{2} \cdot \frac{\beta_\kappa^{(1)} + 2\beta_\kappa^{(2)}}{\beta_\kappa^{(1)} + \beta_\kappa^{(2)}} \tag{5}$$

The indicators take value $1//2$ if $\beta_\kappa^{(2)} = 0$, $3//4$ if $\beta_\kappa^{(1)} = \beta_\kappa^{(2)}$ and 1 if $\beta_\kappa^{(1)} = 0$. In addition, we put $r_\kappa = 0$ if $\beta_\kappa^{(1)} = \beta_\kappa^{(2)} \equiv 0$. Then the order of sub-areas $Q^(l)_\kappa$ will be changed in the following way:

1. all sub-areas are arranged in the order of increasing values of r_κ ;
2. the sub-areas for which $r_\kappa = 0$ preserve their original mutual order;
3. two or more sub-areas having the same value $r_\kappa (r_\kappa > 0)$ are permuted so that the number of sub-series in $S_1 \cup S_2$ is minimized.

It can be easily shown that this reordering of observation space reduces the number of sub-series in $S_1 \cup S_2$ or keeps it unchanged. However, the number of sub-series can be reduced if the procedure is repeated on the next level of partition of the observation space. Then a statistical decision concerning similarity or difference of a pair of texture samples can be based on the statistics of the number of sub-series, as proposed in [2] or on this of the number of sub-series described in [3]. In both cases additional data included into the learning sets S_1 and S_2 and a consecutive observation space reordering improves the quality of the next decisions concerning discrimination of textures.

4 Conclusions

It has been shown that the method of discrimination of textures based on serial statistics can be modified so that the cumulated results of former decisions are used to a supervised improving of the discrimination algorithm by reordering of observation space. For effective decision making at a given optimization step it is enough to take into account only the information about observation space reordering, instead of the elements of learning sets.

References

1. Bruno A., Collorec R., Bezy-Wendling J., et al. (1997). Texture Analysis in Medical Imaging. In: Roux C., Coatrieux J.-L. (eds) Contemporary Perspectives in Three-Dimensional Biomedical Imaging. IOS Press, Amsterdam: 133-164.
2. Kulikowski J.L., Wierzbicka D. (2004). Texture Analysis Based on Application of Non-Parametric Serial Statistical Tests. Biocybernetics and Biomedical Engineering vol. 24 No 2: 27-39.
3. Runyon R.P. (1977). Nonparametric Statistics. A Contemporary Approach. Addison-Wesley Publishing Company, Reading, Mass., USA.

Approximation Algorithm for the Argument Reduction Problem

Piotr Kułaga[1], Piotr Sapiecha[2], and Krzysztof Sęp[3] *

Warsaw School of Information Technology Newelska 6, 01-447 Warsaw, Poland
Department of Electronics and Information Technology Warsaw University of Technology , ul. Nowowiejska 15/19, 00-665 Warsaw, Poland
System Research Institute Newelska 6, 01-447 Warsaw, Poland

Summary. This paper proposes a new method of solving the argument reduction problem. Our method is different to the classical approach using the greedy algorithm, independently invented by Lovasz, Johnson, and Chvatal. However, sometimes the classical method does not produce minimal sets in the sense of cardinality. According to the results of computer tests, better results can be achieved by application of our method in combination with the classical method. Therefore, improvements are found in the quality of solutions when it is applied as a post-processing method.

1 Argument Reduction Problem, Basic Concepts

Argument reduction methods of decision tables based on combinatorial algorithms are described in this paper. With the help of these methods, reduction in the size of decision tables can be achieved. In this section we will introduce some basic concepts of information systems [16]. The information system contains data about *objects*, characterized by certain *attributes* (often two classes of attributes are distinguished: condition and decision attributes). Such an information system is called a decision table. The decision table describes conditions that must be satisfied in order to carry out the decisions specified for them. With every decision table we can associate a decision algorithm which is a set of *if... then...* rules. This decision algorithm can be simplified, which results in an optimal description of the data in the information system. More formally, *an information system* is a pair $S = (U, A)$, where U is a nonempty set of *objects* called the universe, and A is a nonempty set of *attributes*. If a certain information system $S = (U, A)$ is given, often it turns out that some of its condition attributes are redundant. An *argument reduction* problem is an algorithmic problem

*Research partly supported by the grant from the State Committee for Scientific Research. Decision no. 55/E-82/SPB/5.PR UE/DZ 385/2003-2005 of 16.07.2003

of removing as many condition attributes (input variables) from a given information system (truth table) so that it still remains consistent. In this problem, two notions play an important role, namely the *discernibility matrix and discernibility function* [16]. A discernibility matrix of a decision table S is a matrix $n \times n$ (where n is the number of rows in the decision table) whose elements are denoted as follows: $m_{ij} = \{x \in X : x(i) \neq x(j)\}$ iff $\exists_{y \in Y} y(i) \neq y(j)$, otherwise $m_{ij} = \varnothing$ (where $x(i)$ denotes "a value of the variable x in the ith row of the truth table). The meaning of this definition is: the element m_{ij} is an empty set if values of all output variables in rows i and j are compatible, but otherwise it is a set of all input variables that have incompatible values in both rows. A discernibility function f_s is a boolean function with boolean variables v_i corresponding to attributes a_i, and is define as follows: $f_s(v_1, v_2, \ldots, v_m) = \bigwedge\{\bigvee m_{ij} : 1 \leq j < i \leq n, m_{ij} \neq \varnothing\}$, $\bigvee m_{ij}$ where is the disjunction of all variables v_k such that $x_k \in m_{ij}$. When writing discernibility functions, we will give the variables of f_s the same names as the variables from X when no confusion can arise.
The introduction of discernibility matrix is very useful for the process of argument reduction. It is easy to check that from the definition of its elements follows the property: if m_{ij} is nonempty, at least one from the variables from m_{ij} cannot be removed. The discernibility function is a formalization of this property: the minimum set of variables that have to be left in the output function is equal to $MIN(f_s)$.

Note that minimization of the discernibility function is equivalent to transforming it from the *CNF -Conjunctive Normal Form* (in which it is originally constructed) to the *DNF - Disjunctive Normal Form* and finding a minimum implicant. Such transformation is usually time-consuming. Therefore, it is important to ask about the overall time complexity of the argument reduction problem. The following theorem gives the desired information:

Theorem 1. *The argument reduction problem is NP-hard. The NP-complete minimum k-test collection problem can be reduced in a polynomial number of steps to the argument reduction problem.*

Some combinatorial properties of the argument reduction problem, notably transversal problem are presented in the following paper. (blocking sets) [2], [3], [6], [11], [13]
One of the most common ways of solving the argument reduction problem is to reduce it to another problem, a well-known combinatorial minimum transversal problem. Since this approach will be used in one of the algorithms proposed for the problem, it is necessary to describe the details of this construction.
A *hypergraph* is a pair $H = (V, E)$, where V is a nonempty set of *vertices* and E is a set of *edges*. Each edge is a subset of vertices, i.e. $E \subseteq 2^V$. Note that an edge can contain any number of vertices (even one), which makes a difference between hypergraphs and graphs. In fact, a hypergraph can be viewed as a direct generalization of graphs. A *transversal* $T \subseteq V$ is a subset of vertices

which has the property $\forall_{e \in E}\, e \cap T \neq \varnothing$. In other words, a transversal is a set of vertices that covers (or blocks, hence the other name: *blocking set*) all the edges. A transversal is called minimum if there exists no other transversal having fewer elements.
As we recall, the idea of argument reduction is to find a minimum implicant in the discernibility function. For doing this, the discernibility function can be expressed as a hypergraph, where vertices correspond to variables, and edges correspond to elements from the discernibility matrix.
We presented some practical proposals for solving argument reduction problems. One of the possible algorithms is an exact one, using a backtracking approach. But the worst-case number of steps to find a minimum transversal is $O(2^{|V|})$, and an expected number of steps is: $E(X) = \sum_{k=0}^{|V|} \binom{n}{k} 2^{-2^k}$.
Thus it is practically useless to solve real problems.
The other one is a much faster, but usually not as good, approximation algorithm often called a greedy algorithm. It operates directly on a hypergraph constructed from a given information system. Because of this, essentially it is the algorithm for solving the set-covering problem. Therefore, the hypergraph must be constructed first and it must be constructed effectively. Algorithm 1 describes such a procedure.
It can be easily noted that the number of steps required to construct the hypergraph is $O(|U|^2|A|)$

Algorithm 1 Creation of a hypergraph from a decision table

```
Create a hypergraph H with no edges and in which each vertex
corresponds to one condition attribute of the
initial information system;
for all pairs of rows in the decision table that have incompatible values
of decision attributes do
        Check which condition attributes have different
        values in these rows;
        Create an edge that joins all vertices corresponding to
        these attributes;
end for
```

2 Greedy Algorithms

The first, well-known greedy algorithm for the minimum transversal problem is based on a procedure proposed by Lovasz, Johnson and Chvatal [6], [11], [13]. It is capable of finding a transversal not bigger than $1 + lg_2|V|$ times the size of the optimal solution, always in polynomial time.

Algorithm 2 Greedy algorithm for set-cover

while H has any vertices **do**
  begin
    v :=the vertex that covers the most edges;
    $S_{min} := S_{min} \cup \{v\}$;
    $V = V - \{v\}$;
    $E := E - \{e : v \in e\}$;
end;

The quantity of the approximation is $t \leq t_a \leq t(1 + lg_2 n)$ where t is a minimal transversal in the sense of cardinality, t_a is a transversal achieved by the greedy algorithm.

On the other hand, the second greedy strategy, which can to compute the minimal transversal-in the sense of inclusion relation for a given hypergraph, could be taken. under consideration. This approach may application of the solve the transversal problem, or what would be more interesting, be used as a post- processing method, after *Lovasz-Johnson-Chvatal* algorithm application.

Let $H = (X, F)$ is a hypergraph
X -set of vertices
F -set of edges
$m(X)$ -vertex from X which belong to the least number of edges
$Z(X, F)$ -set of vertices in the reduct of $H = (X, F)$ not belong to any edge
$F(x)$ -set of all edges with vertex x.

Algorithm 3 New Greedy algorithm for set-cover

Input: $Tr := \varnothing; V := X;\ Q := X;\ E := F;$
Output: Tr
while $V \neq \varnothing$ and $E \neq \varnothing$ **do**
  begin
    $k := m(V)$;
    $V := V\ \{k\}$;
    if V is not a transversal of hypergraph (Q, E) **do**
      begin
        $Tr := Tr \cup \{k\}$;
        $E := E \setminus F(x)$
        $V := V \setminus Z(V, E)$;
      end;
    $Q := V$;
  end;

3 Results of computer simulations

The presented algorithms were implemented as a software package with application of a gcc compiler. We used random hypergraphs from a uniform distribution of cover. According to computer tests it could be claimed that the *Lovasz-Johnson-Chvatal* algorithm achieved better results than the second algorithm for given data sets. However, usually the solutions generated by the *Lovasz-Johnson-Chvatal* algorithm were not minimal in the sense of cardinality. The second heuristic could improve the quality of these solutions as a post-processing method. In little hypergraphs we didn't see great improvement. For big, infrequent hypergraphs, we saw much better improvement. For hypergraphs with 600 vertices and 24 000 edges we saw an average improvement of about 6 vertices, e.g. the *Lovasz-Johnson-Chvatal* algorithm generated a transversal with 348 vertices and our algorithm decreased it to 338 vertices. In 50 hypergraphs with 200 vertices and 100 000 edges we saw an average improvement of about 2 vertices, eg. the *Lovasz-Johnson-Chvatal* algorithm generated a transversal with 75 vertices and our algorithm decreased the transversal to 71 vertices. In 1 200 tests our algorithm reduced the transversals by about 2%.

4 Conclusions

In this paper we have shown that for the argument reduction problem our new algorithm which can be applied as a post-processing method after the classical greedy algorithm achieves better results, in the sense of cardinality, but minimal in the sense of inclusion.

References

1. Aussiello G, Crescenzi P, Gambosi G, Kann V, Marchetti-Spaccamela A, Protasi M, (1999) "Complexity and Approximation", Springer
2. Ballobas B (1986) "Combinatorics, Set Systems, Hypergraphs, Families of Vectors and Combinatorial Probability", Cambridge University Press
3. Berge C (1989) "Hypergraphs, Combinatorics of Finite Sets", North-Holland
4. Berger B, Rompel J (1994) "Efficient NC Algorithms for Set Cover with Applications to Learning and Geometry", Journal of Computer and System Sciences 49, 454–477
5. Blot J, Hemandez de la Vega W, Paschos V Th, Saad R (1995) "Average case analysis of greedy algorithms for optimisation problems on set problems", Theoretical Computer Science 147, 267–288
6. Buciak P, Łuba T, Niewiadomski H, Pleban M, Sapiecha P, Selvaraj H (2002) : Decomposition and Argument Reduction of Neural Networks, IEEE Sixth International Conference on Neural Networks and Soft Computing (ICNNSC'02), Poland, June 11-15, 2002

7. Chvatal V (1979) "A greedy heuristic for the set-covering problem", Mathematics and Operations Research, 4, 233–235
8. Feige U (1998), "A Threshold of ln(n) for Approximating Set Cover", Journal of the ACM
9. Hromkovic J (2001) "Algorithmics for Hard Problems", Springer
10. Hochbaum D S (Ed.)(1997) "Approximation Algorithms for NP-hard Problems", PWS Publishing Company
11. Korte B, Vygen J (2000) "Combinatorial Optimization, Theory and Algorithms", Springer
12. Johnson D S (1974) "Approximation Algorithms for Combinatorial Problems", Journal of Computer and System Sciences, 9, 256–278
13. Lovasz L (1975) "On the ratio of optimal integral and fractional covers", Discrete Mathematics 13 383–390
14. Lund C, Yannakakis M (1994) "On the Hardness of Approximating Minimization Problems", Journal of the ACM
15. de Micheli G (1994) "Synthesis and Optimization of Digital Circuits", McGraw-Hill Inc.
16. Motvani R (1992) "Lecture Notes on Approximation Algorithms - Volume F, Stanford University
17. Niewiadomski H, Buciak P, Pleban M, Selvaraj H, Sapiecha P, Łuba T (2002) Decomposition of Large Neural Networks, Proceedings of the IASTED International Conference, Applied Informatics, International Symposium on Artificial Intelligence and Applications, pp. 165–170, Insbruck, Austria, February 18-21, 2002
18. Pleban M, Niewiadomski H, Buciak P, Sapiecha P, Yanushkevich S, Shmerko V (2002) Argument reduction algorithms for multi-valued relations, IASTED International Conference on Artificial Intelligence and Soft Computing (ASC 2002), pp. 609-614, Banff, Canada, July 17-19, 2002
19. Skowron A, Rauszer C (1992) "The discemibility matrices and functions in information systems", in: R. Slowinski (Ed.), "Intelligent Decision Support. Handbook of Applications and Advances on the Rough Set Theory", Kluwer
20. Slavik P(1996) "A Tight Analysis of the Greedy Algorithm for Set Cover", STOC'96, USA
21. Srinivasan A (1995) "Improved Approximation Guamatees for Packing and Covering Integer Programs", DIMACS TR.
22. Vazirani V (1997), "Approximation Algorithms", Lecture Notes 1997, Georgia Institute of Technology
23. Williamson D (1999) "Lecture Notes on Approximation Algorithms", IBM Research Report, RC 21409,02/17/1999

Spread Histogram – A Method for Calculating Spatial Relations Between Objects

Halina Kwasnicka[1] and Mariusz Paradowski[1]

Wroclaw University of Technology, Institute of Applied Informatics
[halina.kwasnicka, mariusz.paradowski]@pwr.wroc.pl

Summary. This paper presents a novel approach called Spread Histogram for calculation of spatial relations between objects. It allows to determine such relations as *INSIDE*, *OUTSIDE*, *ENCOMPASS*. Additionally, the method cooperates very well with standard histogram methods like Histogram of Angles for determining the directional spatial relations.

1 Introduction

The need of spatial relations determination is very high in research areas such as image understanding. It is used for the image content description and is one of the most important sources of knowledge about image content. There are many approaches used for spatial relations determination [1][2][3][4].

Histogram of angles, a method for extracting spatial relation between two objects is able to represent *RIGHT OF*, *LEFT OF*, *BELOW*, *ABOVE* relations [3]. It takes in the account an angle between vector connecting two points, one from the first object and second from the other, and one of the axes. Such angle is calculated for every pair of points. Sets of angles are labeled by proper spatial relations. The dominant relations between the objects can be read from the histogram. Force Histogram [2][5], also called F–Histogram, uses the value of scalar of forces associated with certain angles between two objects. R–Histogram [3][6], the extension of Histogram of Angles method, uses the concept of labeled distance to handle *INSIDE* and *OUTSIDE* spatial relations. Labeled distance calculation is partially based on the Euclidian distance calculation. Euler Histogram [4][7][8] and its extensions are based on division of image into cells and assigning a bucket to each cell. Values of buckets are modified based on presence of objects in relevant cells.

This paper presents a novel approach called Spread Histogram for representing spatial relations between two objects. It allows, with cooperation of Histogram of Angles, to determine the relations *RIGHT OF*, *LEFT OF*, *ABOVE*, *BELOW*, *INSIDE*, *OUTSIDE*, and, what seems to be important, the

ENCOMPASS relation. Results are presented for exemplary pairs of objects. The last part of the paper contains a short discussion of the S–Histogram method and further research directions.

2 The method

Let us assume X and Y to be sets of points representing two shapes:

$$X = \{x_1, x_2, ..., x_m\} \tag{1}$$

$$Y = \{y_1, y_2, ..., y_n\} \tag{2}$$

For every x_i there can be calculated a vector of angles between x_i and every point in Y:

$$A_i = [\alpha_1, \alpha_2, ..., \alpha_n] \tag{3}$$

Let us order the angles in the vector of angles A_i in an ascending manner:

$$\forall_{a \in \{1,2,...,n-1\}} \forall_{b \in \{a+1,...,n\}} \quad \alpha_a \leq \alpha_b \tag{4}$$

Let us calculate the *coefficient of spread* for the given point x_i:

$$\beta_i = max(\alpha_2 - \alpha_1, \alpha_3 - \alpha_2, ..., \alpha_n - \alpha_{n-1}, 2\pi - \alpha_n + \alpha_1) \tag{5}$$

If (x_i is inside Y and x_i is not on the border of Y) or x_i is encompassed by Y then

$$(n \to \infty) \Rightarrow (\beta_i \to 0) \tag{6}$$

If (x_i is not inside Y or x_i is on the border of Y) and x_i is not encompassed by Y then

$$(n \to \infty) \Rightarrow (\beta_i \gg 0) \tag{7}$$

The difference between *INSIDE*, *ENCOMPASS* and a partial *ENCOMPASS* has been shown in fig. 1.

Fig. 1. Examples of *INSIDE*, *ENCOMPASS* and partial *ENCOMPASS* spatial relations

The histogram of β angles is calculated for every $x \in X$. Based on this statistic, the spatial relations between X and Y can be determined. Calculation of β value require angle sorting. Standard sorting algorithms take $O(n \log n)$, but in this case bucket sorting ($O(n)$) is enough to calculate β value with a given precision. Let us denote $S'(X,Y,\beta)$ as an unnormalized Spread Histogram (eq. 8).

$$\forall_{i\in\{1,2,\ldots,m-1\}} \quad S'(X,Y,\beta) = \begin{cases} S'(X,Y,\beta)+1 & if \quad \beta_i = \beta \\ S'(X,Y,\beta) & otherwise \end{cases} \tag{8}$$

After the calculation of histogram it is required to normalize it. Let us denote $S(X,Y,\beta)$ as a normalized Spread Histogram (eq. 9).

$$S(X,Y,\beta) = \frac{S'(X,Y,\beta)}{\sum_{\beta\in<0,2\pi)} S'(X,Y,\beta)} \tag{9}$$

Spread Histogram (also called S–Histogram) is both rotation and scale invariant. Rotating and Scaling objects do not introduce any changes in the S–Histogram.

Spread Histogram can be very easily calculated when the Histogram of Angles is calculated. It takes $O(nm + mk)$ to calculate the histogram, where m is the number of points in X, n is the number of points in Y and k is the number of "bins" in bucket sorting used to calculate the β values. Exemplary vectors with calculated β angle have been shown in fig. 2.

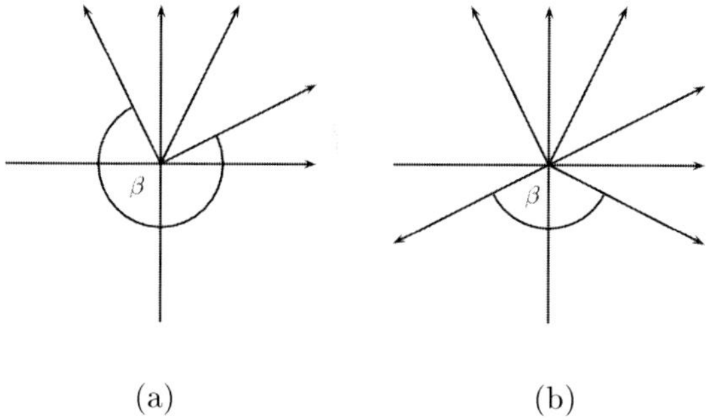

Fig. 2. Exemplary β angles, point outside object (a), point inside object (b)

Classification rules

The classification of spatial relation is based on a set of functions. These functions represent the intensity of each of the spatial relations. Values of every function are normalized. Let us denote $A(X,Y,\alpha)$ as a Histogram of Angles value for α angle.

$$Left(X,Y) = \frac{\sum_{\alpha \in <3/2\pi, 2\pi) \cup <0, \pi/2)} A(X,Y,\alpha)}{\sum_{\alpha \in <0,2\pi)} A(X,Y,\alpha)} \tag{10}$$

$$Right(X,Y) = \frac{\sum_{\alpha \in <\pi/2, 3/2\pi)} A(X,Y,\alpha)}{\sum_{\alpha \in <0,2\pi)} A(X,Y,\alpha)} \tag{11}$$

$$Below(X,Y) = \frac{\sum_{\alpha \in <0,\pi)} A(X,Y,\alpha)}{\sum_{\alpha \in <0,2\pi)} A(X,Y,\alpha)} \tag{12}$$

$$Above(X,Y) = \frac{\sum_{\alpha \in <\pi,2\pi)} A(X,Y,\alpha)}{\sum_{\alpha \in <0,2\pi)} A(X,Y,\alpha)} \tag{13}$$

$$Inside(X,Y) = \frac{\sum_{\beta \in <0,\pi)} S(X,Y,\beta)}{\sum_{\beta \in <0,2\pi)} S(X,Y,\beta)} \tag{14}$$

$$Outside(X,Y) = \frac{\sum_{\beta \in <\pi,2\pi)} S(X,Y,\beta)}{\sum_{\beta \in <0,2\pi)} S(X,Y,\beta)} \tag{15}$$

Equations 10, 11, 12 and 13 are based only on Histogram of Angles. Equations 14 and 15 are based on the Spread Histogram. Additionally the Spread Histogram allows to determine *ENCOMPASS* and *IS ENCOMPASSED* spatial relations. For both of these relations Histogram of Angles is almost identical. It is required to have additional information and such information is provided by Spread Histogram.

Let us define a function $f^A_{cont}(X,Y)$ based only on Histogram of Angles. Function $f^A_{cont}(X,Y)$ has normalized values and gives the information how much *X encompasses Y* or *Y encompasses X*. All experiments in this paper for *ENCOMPASS* and *IS ENCOMPASSED* have been performed using heuristic function f^A_{cont} given by eq. 17.

$$Encompass(X,Y) = f^A_{cont}(X,Y) \frac{\sum_{\beta \in <\pi,2\pi)} S(X,Y,\beta)}{\sum_{\beta \in <0,2\pi)} S(X,Y,\beta)} \tag{16}$$

$$f^A_{cont}(X,Y) = 16 Left(X,Y) Right(X,Y) Below(X,Y) Above(X,Y) \tag{17}$$

Another important spatial relation is *IS ENCOMPASSED*. This method is not able to differentiate *INSIDE* and *IS ENCOMPASSED*, but is able to differentiate *OUTSIDE* and *ENCOMPASS*. The differentiation between *INSIDE* and *IS ENCOMPASSED* can be very easy performed by calculating $A(Y,X,\alpha)$, $S(Y,X,\beta)$ and $Encompass(Y,X)$ as given by eq. 18.

$$IsEncompassed(X,Y) = Encompass(Y,X) \tag{18}$$

For an object placed in the center of *encompassed area* values of all directional functions are close to 0.5. To normalize for an object placed in the center

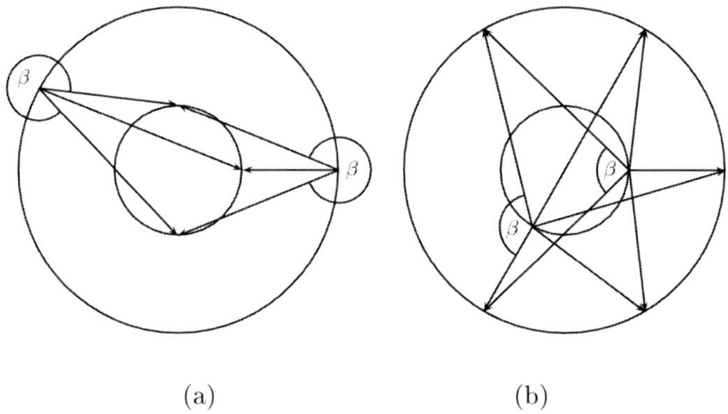

Fig. 3. β angles for *ENCOMPASS* (a) and *IS ENCOMPASSED* (b)

the value of the function has to be multiplied by 16. S–Histogram enables differentiation between *ENCOMPASS* and *IS ENCOMPASSED*. For each point of object encompassing another object β angles have large values (fig. 3a). All vectors outgoing from one point have similar direction. For each point, the directions of vectors are completely different, but the β angle still is large. For each point of object encompassed by another object β angles have small values (fig. 3b). All vectors outgoing from one point are distributed along the whole angle domain. Such situation occurs for every point in the object.

3 Results of experiments

All experiments have been conducted for spatial relations between two objects. The black–dotted object is referred as object X, the grey–crossed object is referred as object Y.

Figure 4 shows object X encompassing object Y. Spread Histogram has higher than 0 values for $\beta \in < \pi, 2\pi)$. Histogram of Angles shows uniform distribution of all angles. All directional spatial relations are present between these two objects.

Figure 5 shows object X partially inside object Y. Spatial Histogram allows for approximate calculation how much of the object is *INSIDE* or *IS ENCOMPASSED* by another object. As shown in fig. 5, $Inside(X, Y) = 0.29$ of object X is inside object Y. Spread Histogram has greater than 0 values for both $\beta \in < 0, \pi)$ and $\beta \in < \pi, 2\pi)$. It is impossible to determine *INSIDE* and *OUTSIDE* spatial relations only with Histogram of Angles. Situation with very similar results from Histogram of Angles as in fig. 5 have been shown in fig. 6. It is required to use information from the Spread Histogram to calculate the difference.

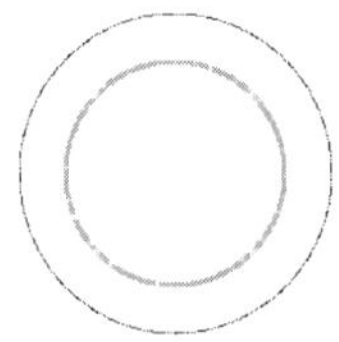

Relation	Result
X right of Y	0.50
X left of Y	0.49
X below Y	0.49
X above Y	0.50
X inside/enc. by Y	0.00
X outside Y	1.00
X encompass Y	0.95

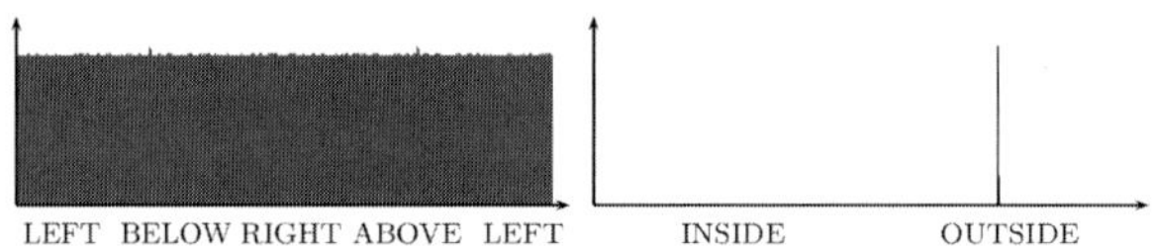

Fig. 4. Object X encompasses object Y

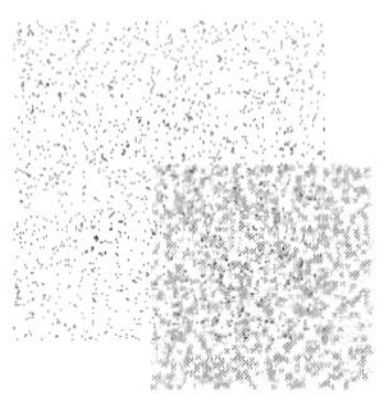

Relation	Result
X right of Y	0.21
X left of Y	0.78
X below Y	0.21
X above Y	0.78
X inside/enc. by Y	0.29
X outside Y	0.70
X encompass Y	0.30

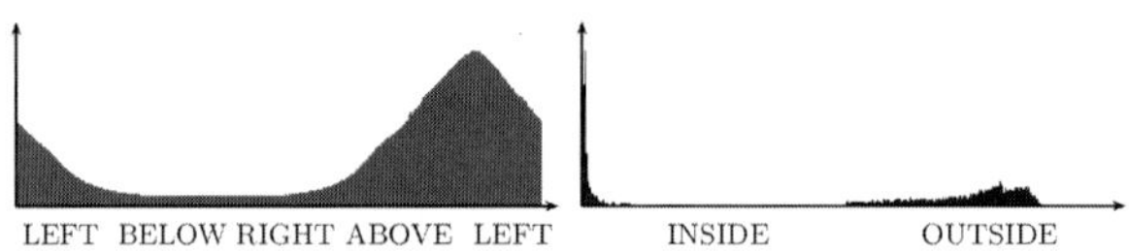

Fig. 5. Object X partially inside object Y

Figure 7 shows object X inside object Y. Spread Histogram has greater than 0 values only for $\beta \in < 0, \pi)$. For every point of X β angles are small, because points of Y are around it.

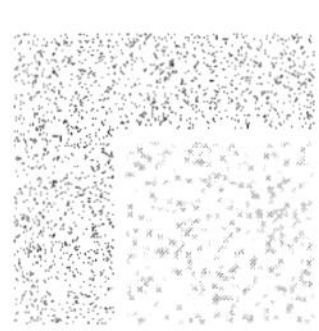

Relation	Result
X right of Y	0.17
X left of Y	0.82
X below Y	0.18
X above Y	0.81
X inside/enc. by Y	0.00
X outside Y	1.00
X encompass Y	0.32

Fig. 6. Object X outside object Y, X partially encompasses Y

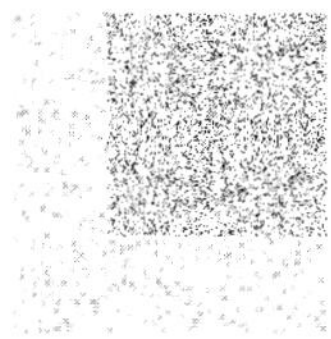

Relation	Result
X right of Y	0.65
X left of Y	0.34
X below Y	0.34
X above Y	0.65
X inside/enc. by Y	1.00
X outside Y	0.00
X encompass Y	0.00

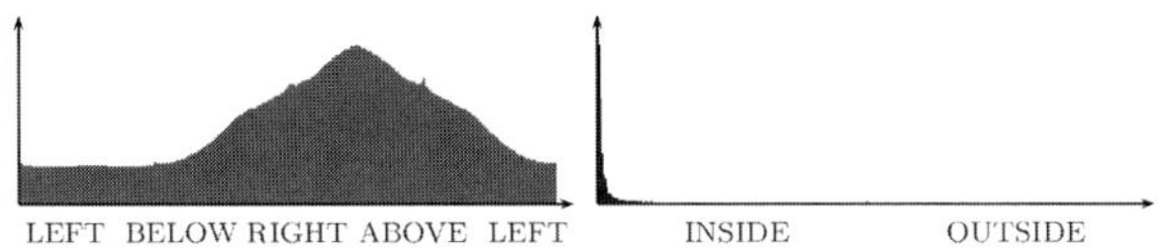

Fig. 7. Object X inside object Y

Figure 8 shows object X outside object Y. Spread Histogram has greater than 0 values only for $\beta \in < \pi, 2\pi)$. For every point of X β angles are large, because all points of Y are placed only in one direction from it.

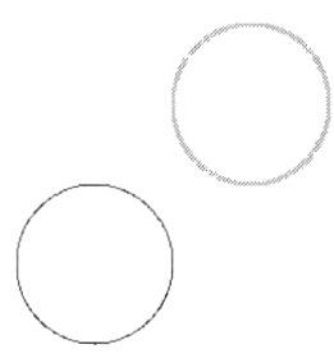

Relation	Result
X right of Y	0.00
X left of Y	1.00
X below Y	1.00
X above Y	0.00
X inside/enc. by Y	0.00
X outside Y	1.00
X encompass Y	0.00

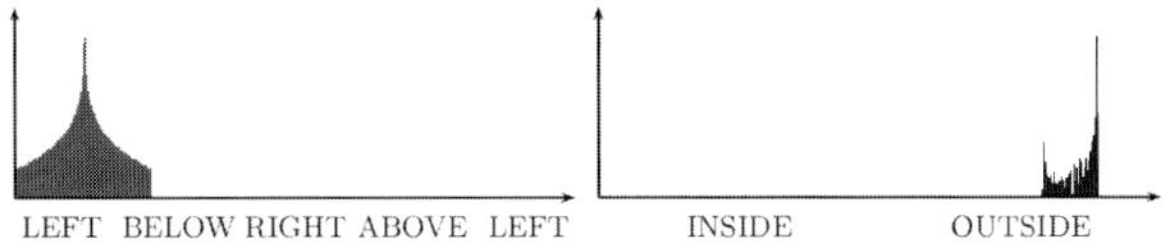

Fig. 8. Object X outside Y

Spread Histogram allows in certain cases to distinguish between *NEAR* and *FAR* spatial relations. It is working very well for rather circular objects, but gives false result for very long and flat objects. *FAR* spatial relation is present if Histogram of Angles has detected 1 or 2 of non–opposite directional spatial relations. Additionally Spread Histogram can have larger than 0 values for only very large β values. *NEAR* spatial relation is the opposite, it is present only if one *FAR* spatial relation conditions is not met.

4 Discussion and further research

Presented method allows in a very simple and precise way to distinguish between spatial relations like *INSIDE* and *OUTSIDE*. Additionally it presents a way to distinguish between *OUTSIDE*, *ENCOMPASS* and *IS ENCOMPASSED*, which is often needed in image analysis. In some degree it also handles *FAR* and *NEAR* spatial relations. The advantage of the Spread Histogram is that it is based only on angles calculation, which makes very easy to incorporate it into Histogram of Angles.

Further research will focus on the following aspects:

- Replacing functions given by eqs 10 to 18 with a classifier with supervised learning. This will allow to adjust the method to user preferences.
- Using S–Histogram method for determining spatial relations in real-life images. The method is plan to be used on skin cancer images for determining the relations between melanocytic lesions [9].

References

1. Dimitris Papadias, Marinos Kavouras (1994) "Acquiring, Representing and Processing Spatial Relations", Proceedings of the 6th International Symposium on Spatial Data Handling, Advances in GIS
2. Pascal Matasakis, James M. Keller, Ozy Sjahputera, and Jonathon Marjamaa (2004) "The Use of Force Histograms for Affine-Invariant Relative Position Description", IEEE Transactions on Pattern Analysis and Machine Intelligence
3. Yuhang Wang, Fillia Makedon (2003) "R-Histogram: Quantitative Representation of Spatial Relations for Similarity-Based Image Retrieval", The 11th Annual ACM International Conference on Multimedia
4. Xuemin Lin, Qing Liu, Yidong Yuan, Xiaofang Zhou (2003) "Multiscale Histograms: Summarizing Topological Relations in Large Spatial Datasets", Proceedings of the 29th VLDB Conference, Berlin, Germany
5. R. Bondugula, P. Matsakis, J. Keller (2004) "Force Histograms and Neural Networks for Human-Based Spatial Relationship Generalization", Proceedings of IASTED Int. Conf. on Neural Networks and Computational Intelligence
6. Yuhang Wang, Fillia Makedon, James Ford, Li Shen, Dina Goldin (2004) "Generating Fuzzy Semantic Metadata Describing Spatial Relations from Images using the R-Histogram", Proceedings of the 4th ACM and IEEE-CS joint conference on Digital libraries
7. Chengyu Sun, Divyakant Agrawal, Amr El Abbadi (2002) "Selectivity Estimation for Spatial Joins with Geometric Selections", Proceeding of the EDBT 2002, LNCS 2287, pp. 609–626
8. Qing Liu, Yidong Yuan, Xuemin Lin (2003) "Multi-resolution Algorithms for Building Spatial Histograms", Conferences in Research and Practice in Information Technology, Vol. 16
9. Jerzy W. Grzymala-Busse, Jay Hamilton, Zdzislaw S. Hippe (2004) "Diagnosis of Melanoma Using IRIM, a Data Mining System", Artificial Intelligence and Soft Computing - ICAISC 2004, LNAI 3070

Comparing Modification Operators Used in Clustering Algorithm Based on a Sequence of Discriminant Rules

Dariusz Mazur

Silesian University of Technology, Faculty of Organisation and Management, ul. Roosevelta 26, 41-800 Zabrze, Poland
e-mail: dmazur@polsl.gliwice.pl

Summary. Clustering as a data exploration technique is very widely applied. It is based on clustering algorithms whose usefulness depends strictly on the form and style of the incoming data. The following article comparing operator in evolutionary algorithms used to clustering of symbolic data. Clustering methods is based on list of decision rules.

1 Introduction

Clustering problem has been addressed in many context and by researchers in many disciplines like statistics [4], machine learning [5, 2], data bases [8]. Recently, the problem of data clustering has been redefined in the data mining area. The concept of *cluster mining* [9] is used to represent a method which analyzes very large data sets to efficiently identify a small sets of high-quality clusters of data items. Cluster mining does not aim at partitioning of all the data items – instead, less frequent noise and outliers are ignored. In other words, cluster mining finds only the highest density areas in the data space.

Clustering is generally considered as unsupervised learning where the learner is given a training set including non-categorised (unlabelled) patterns. The proper categories must be then proposed by the learner on the basis of clustering rule implement in the algorithm [2].

2 Proposed approach with genetic algorithm

This paper is devoted to problems of improving clustering methods based on genetic algorithms.

In this research, an application of genetic algorithm to clustering using decision lists was developed. One important issue has been extensively studied: to choose proper modification operator which has the best performance.

3 Clustering based on a sequence of discriminant rules

In hierarchical algorithms the solution is represented by a graph in form of an upturned tree (dendogram). In such a tree the leaves represent particular objects and the branches reflect particular groups. The number of groups depends on the height of the tree. The groups determine all branches at a particular level. The leaves belonging to different branches represent objects belonging to different groups.

In the described method a significant alteration is suggested: the graph will be created on the basis of a list of discriminant descriptions, analogously as in CN2 algorithm [3]. The same quality measure based on entropy function from information theory is also applied. As mentioned above, the obtained solution has the form of an ordered decision rules list. The notion of rules is based on the theory of *symbolic classification*[6]. The form of the rules is:

$$D_j ::> K_i, \tag{1}$$

where D_j is the descriptive rule, K_i represents the cluster, and ::> the assignation operator. The rules can be interpreted as follows: *If a given object fulfills the D_j condition it then belongs to K_i cluster.* In order to adapt the notion of rules in document set clustering the following assumption was made. The rule will have the form of the occurrence of a particular term in the text, that is, if the term used in rule D_j occurs in the document X_n it is then assigned to K_i cluster. The rules are represented in the form of an ordered list. The process of assigning objects to cluster has the following procedure.

A *decision list* consists of nodes and branches to partition a set of objects info a set of covering decision rules. In each node, a single test or decision is made to obtain a partition. Each node has one leaf, which a decision if made of the class assignment and one branch to next node. The starting node is usually referred as the head node. In each node, the main task is to select an attribute that makes test.

The list is searched until the appropriate rule is found which means that the list should contain a proper number of rules so that each object gets assigned. The list can be treated as a gene being a subject to processing in genetic algorithms. The received structure of clustering is a graph in the form of a degenerated tree in which, out of each node, only one branch or one or more leaves stem out. The leaves represent the objects present in one or more node. This fact is a significant element in comparison with the hierarchical methods.

4 Application of genetic algorithm for clustering

The clustering problem can be solved by genetic algorithm since clustering can be converted to an searching problems. In order to apply genetics algorithms

(GA) to clustering problem, is needed to select internal representation of the space to be searched. The traditional representation of GA involves using fixed-length strings to represent points in the space to be searched. However, such representations do not appear well-suited for representing the space of concept descriptions that are generally symbolic in nature, that have both syntactic and semantic constraints, and that can be widely varying length and complexity. Besides internal representation, for resolve issue of symbolic clustering, genetics algorithms need also changing the fundamental GA operators (crossover and mutation) to work effectively with complex objects. One of method clustering has been proposed in [7]. In this paper new modification operator for evolutionary algorithm of clustering based on decision list is intruduced.

4.1 Chromosomal representation

GAs represent the partitioning of the given set of objects by the chromosome, which contain encoded list of rules e.q. decision lists. Each gene contain one encoded rule as in equation (1). The chromosome is obtained by concatenating all the gene. Each chromosome, therefore, contains a possible solution to the problem. The population processed by evolutionary algorithm consist of N chromosome (individuals).

4.2 Crossover

Proposed form of chromosome is very special, each gene should be in chromosome only one time, and with crossover we may lost this. Crossover cause that some kind of allele (representing one rule) are get off, and some are duplicated. Presented methods do not accept this case of modification.

4.3 Mutation

In suggested chromosome mutation should be form of permutation: we may change only place of each gene, we don't may add or delete any of gene. The adequate method of mutation was worked out for the way of representation of decision list. All of them are kind of permutation:

- we choose randomly two gene and swap them,(fig. 4.3), rest of list stay unchanged, name of operator is M_{swap},
- we choose randomly two gene, second we insert in front of first (move it up toward head) (fig. 4.3), all rules goes down by one place, name of operator if M_{ins}.

During evaluations, we notice, that we can do small improvement of these operators. Because only small part of list is used to build clusters we need, that at least one gene belong to this part. If we don't do it, in that case new build clusters are identical to previous, we lost able to connect some of object to particular clusters.

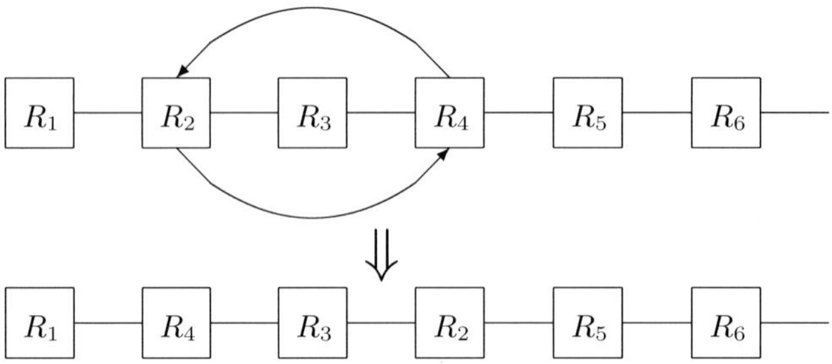

Fig. 1. Random mutation by swap

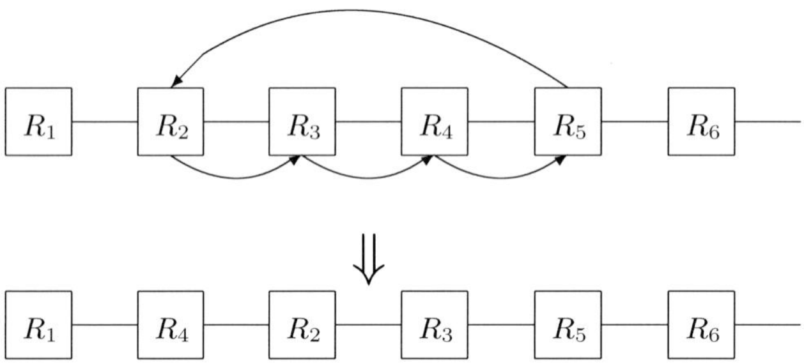

Fig. 2. Insert rule from tail

4.4 The criterion function

The fitness function is the same, as criterion function choose often as represented by formulas taken from the theory of information. It is assumed that clustering which provides the highest information gain is optimal for it corresponds to the minor differentiation of categories in the subsets. Information included in the set of labeled pattern P can be represented as follows:

$$I(P) = \sum_{d \in \mathcal{D}} -\frac{|P^d|}{P} \log \frac{|P^d|}{P}. \tag{2}$$

5 Experimental Results

For an empirical evaluation of presented system, conducted experiment on artificial domains has been done. Main goal was checks, that genetics clustering based on rules works on, at least, one data base. First experiment used small data base introduced in [6], results are shown in Table 1. We also performed clustering of transaction of Congressional Voting Record Database [1]. We then evaluated these transaction clusters to see to what extent they represent voting records of congressmen belonging to the same party. In Table 5 we see results of using above operator. We can compare count of iterations and time of computing. We can also compare quality of clustering done by each of operators. The precision each of operators is presented in Table 3. Third experiment has been performed on Mushroom DataBase [1]. We try to discover edible and poisonous mushroom. In Table 4 we see results of using compared operator.

Table 1. Compare modification operators for *small database.*

name of operator	informations gain	time	count
operator SWAP	0,3312	0: 0:04	19
operator INSERT	0,3312	0: 0:05	73

Table 2. Compare modification operators for *Votes database*

name of operator	informations gain	time	count
operator SWAP	1,0245	0: 1:08	4430
operator INSERT	1,1208	0: 0:26	360

Table 3. Precision for *Votes database.*

operator	democrat YES	democrat NO	republican YES	republican NO	Precision %
operator SWAP	212	55	161	7	85,75
operator INSERT	249	18	140	28	89,43

Table 4. Compare modification operators for for *Mushroom Database*

name of operator	informations gain	time	count
operator SWAP	1,0805	0: 2:21	2288
operator INSERT	1,1290	0: 1:20	1193

6 Concluding Remarks

We consider the problem of clustering data base with categorical values. We introduce two variants of the algorithm using different modification operator

Table 5. Precision for *Mushroom Database*.

operator	edible YES	edible NO	poisonous YES	poisonous NO	Precision %
operator SWAP	3102	814	3424	784	80,33
operator INSERT	3100	816	4096	112	88,58

in Genetics Algorithm used in clustering. Genetics algorithms are providing themselves in solving real problems in data mining, especially in cases where data are noisy, requires the solution of multi-objective optimization problem or data are too ill-behaved (such as multimodal and/or non-differentiable) for more conventional hill-climbing and derivative based techniques. The traditional neighborhood clustering algorithm usually needs the user to provide distance d for the clustering. But a unique d for a set of objects often cause problems because there may be some natural cluster in which the objects are not close to one another within the distance d. Proposed algorithm avoids this kind of problem by processing the data in a global view. We have shown, that the operator *insert* used as modification operator in clustering algorithm are more effective, received better results in shorter time (need less iterations). Proposed operators are very simple, only one small improvement has been done. In future work we can try to find more robust operators, that will done more effective clustering.

References

1. C. L. Blake and C. J. Merz. UCI repository of machine learning databases, 1998
2. P. Cichosz. Systemy uczace sie. WNT, Warszawa, 2000
3. P. Clark and T. Niblett. The CN2 induction algorithm. Machine Learning, 3:261–283, 1989
4. R. O. Duda and P. E. Hart. Pattern Classification and Scene Analysis. Wiley, New York, 1973
5. D. Fisher. Iterative optimization and simplification of hierarchical clusterings. Journal of Artificial Intelligence Research, 4:147–180, 1996
6. D. Mazur. Clustering algorithm based on a sequence of discriminant rules. In T. Burczyński, W. Cholewa, and W. Moczulski, editors, Methods of Artificial Intelligence, Gliwice, 2003. AI-METH Series.
7. D. Mazur. Clustering based on genetics algorithm. In T. Burczyński, W. Cholewa, and W. Moczulski, editors, Methods of Artificial Intelligence, Gliwice, 2004. AI-METH Series
8. R. T. Ng and J. Han. Efficient and effective clustering methods for spatial data mining. In J. Bocca, M. Jarke, and C. Zaniolo, editors, 20th International Conference on Very Large Data Bases, pages 144–155, Los Altos, USA, 1994. Morgan Kaufmann Publishers
9. M. Perkowitz and O. Etzioni. Towards adaptive Web sites: conceptual framework and case study. Computer Networks (Amsterdam, Netherlands: 1999), 31(11–16):1245–1258, 1999

Multi-Channels Time-Domain-Constrained Fuzzy c-Regression Models

Aleksander Owczarek[1], Adam Gacek[1], and Jacek M. Leski[12]

[1]Institute of Medical Technology and Equipment.
Roosevelt St. 118A, 41-800 Zabrze. Poland. aleck@itam.zabrze.pl
[2]Institute of Electronics. Silesian University of Technology.
Akademicka 16, 44-100 Gliwice. Poland. jleski@polsl.pl

Summary. The paper presents a new fuzzy clustering method with constraint in time domain which may be used to multi-channels biomedical signal analysis. Proposed method makes it possible: 1) to include a natural constraint for signal analysis using fuzzy clustering, that is, the neighbouring samples of signal belong to the same cluster, 2) to incorporate some domain knowledge which yields to a hybrid clustering environment based on simultaneous usage of numerical data and domain knowledge. The paper shows the results of using the multi-channels time-domain-constrained fuzzy c-regression models in analysis of artificial signals and the real noised ECG signals.

1 Introduction

The Fuzzy C-Means (FCM) method is one of the most popular clustering method based on the minimization of a criterion function [2] and plays an important role in many engineering fields such as: pattern recognition and data mining, computer vision, image analysis and so on [4]. Many modifications of the FCM method are described in the literature. These modifications may be treated as inclusion of an additional information into the process of clustering [4]. However, a few applications to the signal analysis may be find in the literature [1], [5]. If we assume that clustered objects represent the consecutive in time domain samples of a signal, then it is a great possibility that the neighbouring samples of signal belong to the same cluster. Such constraint may be called a time-domain-constraint. The new clustering method based on such constraint is presented in [4] and [5]. Additionally, if we incorporate some domain knowledge elicitation into the clustering process [6], [7] we obtain a new clustering method which can be called the multi-channels time-domain-constrained clustering.

The goal of this work is twofold: first, to introduce multi-channels time-domain-constrained fuzzy c-regression models clustering method. Then, to show obtained results for the artificial as well as real ECG signals.

2 A new clustering method

Let us assume that we have a set $\mathcal{Z}^{(N)} = \{(\mathbf{x}_k, y_k)\}_{k=1}^{N}$, where each independent datum $\mathbf{x}_k \in \mathbb{R}^t$ has a corresponding dependent datum $y_k \in \mathbb{R}$ and N is data cardinality. In fact, data pairs $(\mathbf{x}_k, y_k)$ are unlabeled. We assume that data pairs from $\mathcal{Z}^{(N)}$ set are drawn from switching regression model, that consist of c models in the following linear form: $y_k = w_0^{(i)} + \widetilde{\mathbf{w}}^{(i)T}\mathbf{x}_k + e_k$ for $k = 1, 2, \cdots, N$ where e_k represents uncertainty (with zero mean) for kth pair, $\mathbf{w}^{(i)} = \left[w_0^{(i)}, \widetilde{\mathbf{w}}^{(i)T}\right]^T \in \mathbb{R}^{t+1}$ are parameters of ith model. Any information about membership degree of kth data pair to the ith model $u_{i,k}$ and parameters of these models $\mathbf{w}^{(i)}$ are not given. So, there is a necessity of simultaneous estimation of c-partition of the data pairs set and parameters of models. This problem, called **F**uzzy **C-R**egression **M**odels (FCRM), is solved by Hathaway and Bezdek in [3].

Using a loss function L the criterion function of the fuzzy c-regression models is given in the form [3]:

$$J_m(\mathbf{U}, \mathbf{W}) = \sum_{i=1}^{c}\sum_{k=1}^{N} (u_{i,k})^m \; L\left(\mathbf{x}_k, y_k; \mathbf{w}^{(i)}\right), \tag{1}$$

where $\mathbf{W} = [\mathbf{w}^{(1)}, \mathbf{w}^{(2)}, \cdots, \mathbf{w}^{(c)}] \in \mathbb{R}^{(t+1)\times c}$. Due to simplicity and low computational burden, usually the squared error is used as a loss function:

$$L\left(\mathbf{x}_k, y_k; \mathbf{w}^{(i)}\right) = \left(y_k - \mathbf{w}^{(i)T}\mathbf{x}'_k\right)^2, \tag{2}$$

where $\mathbf{x}'_k \triangleq \left[1, \mathbf{x}_k^T\right]^T$ is the augmented input vector.

A fuzziness of the regression models runs on the parameter m. The larger m the fuzzier models are obtained. For $m \longrightarrow 1^+$, the fuzzy c-regression models solution becomes a hard one. For $m \longrightarrow \infty$ the solution is as fuzzy as possible. Usually $m = 2$ is used.

Now, let us assume that we have M sets $\mathcal{Z}^{(n,N)} = \left\{\left(\mathbf{x}_k^{(n)}, y_k^{(n)}\right)\right\}_{k=1}^{N}$, $n = 1, 2, \cdots, M$. If the nth channel of analyzed discrete-time signal is denoted by $s^{(n)}(k\Delta)$, where k is discrete time and Δ is sampling period, then, in this case, $y_k^{(n)}$ denotes an amplitude of signal for kth sample ($y_k^{(n)} = s^{(n)}(k\Delta)$) and $\mathbf{x}_k^{(n)}$ stands for a discrete time index ($\mathbf{x}_k^{(n)} = x_k = k\Delta$) or amplitudes of previous ξ samples ($\mathbf{x}_k^{(n)} \in \left[s^{(n)}((k-1)\Delta), s^{(n)}((k-2)\Delta), \cdots, s^{(n)}((k-\xi)\Delta)\right]^T$). Thus, the nth set $\mathcal{Z}^{(n,N)}$ consists of data from nth channel of biomedical data. If we define the following vector and matrices in nth channel:

$$\mathbf{y}^{(n)} = \left[y_1^{(n)}, y_2^{(n)}, \cdots, y_N^{(n)} \right]^T = \left[\left(\widetilde{\mathbf{y}}^{(n)} \right)^T, y_N^{(n)} \right]^T, \tag{3}$$

$$\mathbf{X}^{(n)} = \begin{bmatrix} \widetilde{\mathbf{X}}^{(n)} \\ \left(\mathbf{x}_N^{(n)\prime} \right)^T \end{bmatrix} = \begin{bmatrix} 1, & \left(\mathbf{x}_1^{(n)} \right)^T \\ 1, & \left(\mathbf{x}_2^{(n)} \right)^T \\ \vdots & \vdots \\ 1, & \left(\mathbf{x}_N^{(n)} \right)^T \end{bmatrix} \in \mathrm{I\!R}^{N \times (t+1)}, \tag{4}$$

$\mathbf{D}_1^{(n,i)} = \mathrm{diag}\left(\left(u_{i,1}^{(n)} \right)^m, \left(u_{i,2}^{(n)} \right)^m, \cdots, \left(u_{i,N}^{(n)} \right)^m \right) \in \mathrm{I\!R}^{N \times N}$,
$\mathbf{D}_2^{(n,i)} = \mathrm{diag}\left(\left(u_{i,1}^{(n)} - u_{i,2}^{(n)} \right)^m, \left(u_{i,2}^{(n)} - u_{i,3}^{(n)} \right)^m, \cdots, \left(u_{i,N-1}^{(n)} - u_{i,N}^{(n)} \right)^m \right) \in \mathrm{I\!R}^{(N-1)\times(N-1)}$ and
$\mathbf{D}_3^{(n,l,i)} = \mathrm{diag}\left(\left(u_{i,1}^{(n)} - u_{i,1}^{(l)} \right)^m, \left(u_{i,2}^{(n)} - u_{i,2}^{(l)} \right)^m, \cdots, \left(u_{i,N}^{(n)} - u_{i,N}^{(l)} \right)^m \right) \in \mathrm{I\!R}^{N \times N}$ for $i = 1, 2, \cdots, c$; $n, l = 1, 2, \cdots, M$ then new criterion function of the multi-channels time-domain-constrained fuzzy c-regression models for the nth channel with loss function (2) may be written in the following matrix form

$$\begin{aligned} J_m' \left(\mathbf{U}^{(n)}, \mathbf{W}^{(n)} \right) = \sum_{i=1}^{c} \Big\{ & \left(\mathbf{y}^{(n)} - \mathbf{X}^{(n)} \mathbf{w}^{(n,i)} \right)^T \mathbf{D}_1^{(n,i)} \left(\mathbf{y}^{(n)} - \mathbf{X}^{(n)} \mathbf{w}^{(n,i)} \right) \\ & + \gamma \left(\widetilde{\mathbf{y}}^{(n)} - \widetilde{\mathbf{X}}^{(n)} \mathbf{w}^{(n,i)} \right)^T \mathbf{D}_2^{(n,i)} \left(\widetilde{\mathbf{y}}^{(n)} - \widetilde{\mathbf{X}}^{(n)} \mathbf{w}^{(n,i)} \right) \\ & + \sum_{l=1}^{M} \xi_{n \leftarrow l} \left(\mathbf{y}^{(n)} - \mathbf{X}^{(n)} \mathbf{w}^{(n,i)} \right)^T \mathbf{D}_3^{(n,l,i)} \left(\mathbf{y}^{(n)} - \mathbf{X}^{(n)} \mathbf{w}^{(n,i)} \right) \Big\}. \end{aligned} \tag{5}$$

where

$$\boldsymbol{\Upsilon} = \begin{bmatrix} 0, & \xi_{1 \leftarrow 2}, & \cdots & \xi_{1 \leftarrow M}, \\ \xi_{2 \leftarrow 1}, & 0, & \cdots & \xi_{2 \leftarrow M}, \\ \vdots & \vdots & \ddots & \vdots \\ \xi_{M \leftarrow 1}, & \xi_{M \leftarrow 2}, & \cdots & 0, \end{bmatrix} \in \mathrm{I\!R}^{M \times M}, \tag{6}$$

and M denotes the number of channels and $\xi_{n \leftarrow l}$ stands for the influence of lth channel to nth channel.

In the case of signal analysis the second term in (5) imposes the time-domain-constraints (the neighbouring samples of a signal belong to the same model) and the third term in (5) imposes the domain knowledge elicitation (interaction by exchanging information about partition matrix of each channel). Therefore, in (5) we have three terms: a standard error term from the FCRM method, regularizing term depending on smoothness of membership degrees $u_{i,k}$ and regularizing term depending on collaboration between the corresponding channels. Parameter $\gamma \geq 0$ controls the trade-off between the smoothness of membership degrees and the amount up to which clustering errors are tolerated. A larger γ results in more smoothness membership degrees and greater clustering errors. For $\gamma \to \infty$ the solution is as smoothness as possible. Matrix $\boldsymbol{\Upsilon}$ which contains interactions coefficients $\xi_{n\leftarrow l}$ controls amount of collaboration between channels. The higher value of the interaction coefficient $\xi_{n\leftarrow l}$ the stronger influence of lth channel to nth channel. For $\gamma = 0$ and $\boldsymbol{\Upsilon} = \mathbf{0}$ (where $\mathbf{0}$ denotes a zero matrix of dimension $M \times M$) the original FCRM method in each channel is obtained.

The following theorem may be formulated for the weighting exponent equal to 2. If $m = 2$, $1 < c < N$, $\boldsymbol{\Upsilon} \neq \mathbf{0}$ and $\gamma \geq 0$ are fixed parameters and $I_k^{(n)}$, $\widetilde{I}_k^{(n)}$ are sets defined as:

$$\underset{\substack{1\leq k\leq N \\ 1\leq n\leq M}}{\forall} \begin{cases} I_k^{(n)} = \left\{ i \,\middle|\, 1 \leq i \leq c;\ L\left(\mathbf{x}_k^{(n)}, y_k^{(n)}; \mathbf{w}^{(n,i)}\right) = 0 \right\}, \\ \widetilde{I}_k^{(n)} = \{1, 2, \cdots, c\} \setminus I_k^{(n)}, \end{cases} \tag{7}$$

and

$$\mathcal{M}_{fc} = \left\{ \mathbf{U} = [0,1]^{c\times N} \,\middle|\, \underset{\substack{1\leq i\leq c \\ 1\leq k\leq N}}{\forall} \sum_{i=1}^{c} u_{i,k} = 1;\ 0 < \sum_{k=1}^{N} u_{i,k} < N \right\}, \tag{8}$$

then $(\mathbf{U}^{(n)}, \mathbf{W}^{(n)}) \in (\mathcal{M}_{fc} \times \mathrm{I\!R}^{(t+1)\times c})$ may be globally minimal for $J_2'\left(\mathbf{U}^{(n)}, \mathbf{W}^{(n)}\right)$ only if:

for $I_k^{(n)} = \emptyset$ and $k = N$:

$$\underset{\substack{1\leq i\leq c \\ 1\leq k\leq N \\ 1\leq n\leq M}}{\forall} u_{i,k}^{(n)} = \frac{1 + \sum\limits_{l=1}^{M} \xi_{n\leftarrow l} u_{i,k}^{(l)} \sum\limits_{j=1}^{c} \frac{L\left(\mathbf{x}_k^{(n)}, y_k^{(n)}; \mathbf{w}^{(n,i)}\right)}{L\left(\mathbf{x}_k^{(n)}, y_k^{(n)}; \mathbf{w}^{(n,j)}\right)}}{\left(1 + \sum\limits_{l=1}^{M} \xi_{q\leftarrow l}\right) \sum\limits_{j=1}^{c} \frac{L\left(\mathbf{x}_k^{(n)}, y_k^{(n)}; \mathbf{w}^{(n,i)}\right)}{L\left(\mathbf{x}_k^{(n)}, y_k^{(n)}; \mathbf{w}^{(n,j)}\right)}}, \tag{9a}$$

for $I_k^{(n)} = \emptyset$ and $k = N-1, \cdots, 1$:

$$\underset{\substack{1\leq i\leq c \\ 1\leq k\leq N \\ 1\leq n\leq M}}{\forall} u_{i,k}^{(n)} = \frac{1 + \left(\gamma u_{i,k+1}^{(n)} + \sum\limits_{l=1}^{M} \xi_{n\leftarrow l} u_{i,k}^{(l)}\right) \sum\limits_{j=1}^{c} \frac{L\left(\mathbf{x}_k^{(n)}, y_k^{(n)}; \mathbf{w}^{(n,i)}\right)}{L\left(\mathbf{x}_k^{(n)}, y_k^{(n)}; \mathbf{w}^{(n,j)}\right)}}{\left(1 + \gamma + \sum\limits_{l=1}^{M} \xi_{n\leftarrow l}\right) \sum\limits_{j=1}^{c} \frac{L\left(\mathbf{x}_k^{(n)}, y_k^{(n)}; \mathbf{w}^{(n,i)}\right)}{L\left(\mathbf{x}_k^{(n)}, y_k^{(n)}; \mathbf{w}^{(n,j)}\right)}}, \tag{9b}$$

for $I_k^{(n)} \neq \emptyset$:

$$\underset{\substack{1\le i\le c\\1\le k\le N\\1\le n\le M}}{\forall}\quad u_{i,k}^{(n)} = \begin{cases} 0, & i \in \widetilde{I}_k^{(n)} \\ \sum\limits_{i\in I_k} u_{i,k}^{(n)} = 1, & i \in I_k^{(n)} \end{cases} \tag{9c}$$

and for all $i = 1, 2, \cdots, c$; $n = 1, 2, \cdots, M$

$$\mathbf{w}^{(n,i)} = \left(\mathbf{X}^{(n)T}\mathbf{D}_1^{(n,i)}\mathbf{X}^{(n)} + \gamma\widetilde{\mathbf{X}}^{(n)T}\mathbf{D}_2^{(n,i)}\widetilde{\mathbf{X}}^{(n)} + \sum_{l=1}^{M}\xi_{n\leftarrow l}\mathbf{X}^{(n)T}\mathbf{D}_3^{(n,l,i)}\mathbf{X}^{(n)}\right)^{-1}$$
$$\cdot\left(\mathbf{X}^{(n)T}\mathbf{D}_1^{(n,i)}\mathbf{y} + \gamma\widetilde{\mathbf{X}}^{(n)T}\mathbf{D}_2^{(n,i)}\widetilde{\mathbf{y}} + \sum_{l=1}^{M}\xi_{n\leftarrow l}\mathbf{X}^{(n)T}\mathbf{D}_3^{(n,l,i)}\mathbf{y}\right). \tag{10}$$

The optimal partition in each channel is obtained by iterating through: (9a)–(9c) and (10). The **M**ulti-**C**hannels **T**ime-**D**omain-**C**onstrained **F**uzzy **C**-**R**egression **M**odels (MCTDCFCRM) method may be summarized in the following steps:

1. Fix $c \in (1, N)$, $m = 2$, $\boldsymbol{\Upsilon} \neq \mathbf{0}$ and $\gamma \geq 0$. Initialize $\mathbf{U}^{(n)[0]} \in \mathcal{M}_{fc}$ for each channel. Set the iteration index $j = 1$.
2. For each channel, calculate the parameters matrix $\mathbf{W}^{(n)[j]}$ for jth iteration using (10) and $\mathbf{U}^{(n)[j-1]}$.
3. For each channel, update the fuzzy partition matrix $\mathbf{U}^{(n)[j]}$ for jth iteration using (9a) and $\mathbf{W}^{(n)[j]}$.
4. If $\underset{1\le n\le M}{\forall} \left\|\mathbf{U}^{(n)[j]} - \mathbf{U}^{(n)[j-1]}\right\|_F > \xi$ then $j \leftarrow j+1$ and go to 2. else stop.

$\|\cdot\|_F$ denotes the Frobenius norm ($\|\mathbf{U}\|_F^2 = Tr\left(\mathbf{U}\mathbf{U}^T\right) = \sum_i\sum_k u_{i,k}^2$) and ξ is a pre-set parameter.

3 Numerical experiments

In all experiments for the FCRM [3], TDCFCRM [4] and MCTDCFCRM the weighted exponent $m = 2$ was used. The iterations were stopped as soon as the Frobenius norm in a successive pair of $\mathbf{U}^{(n)}$ matrices in all of M channels was less than 10^{-3}. In all experiments a one characteristic point in each signal is searched, therefore only two regression models are used ($c = 2$). All experiments were performed in the MATLAB environment. The uniform and Gaussian random numbers were generated using the MATLAB "rand" and "randn" functions.

The performances of the clustering methods in the presence of noise were evaluated for different Signal-to-Noise Ratio (SNR) defined as

$$\mathrm{SNR} = 20\log\frac{\sigma_S}{\sigma_N}, \tag{11}$$

where σ_S, σ_N denote the signal and noise standard deviations, respectively.

3.1 Synthetic signals

The purpose of this experiment was to compare the performance of the TDCFCRM and MCTDCFCRM methods for simple synthetic signals with different kinds of noise level. Three-dimensional (one input – time $x_k^{(1)} = x_k^{(2)}$ and two outputs – signal value in two channels, $y_n^{(1)}$ and $y_n^{(2)}$) dataset consists of a pair of true linear functions with Gaussian random noise. The true but unknown (for algorithm) models are: $y_k^{(1)} = 1 \cdot x_k + e_k$; $y_k^{(2)} = 0 \cdot x_k + e_k$ for $k = 1, 2, \cdots, 40$ and $y_k^{(1)} = 41 + e_k$; $y_k^{(2)} = 1 \cdot x_k - 40 + e_k$ for $k = 41, 42, \cdots, 100$, where e_k represents a realization of a random noise. The dataset consists of 100 samples. Each datum pair $(x_k, \mathbf{y}_k)$ was generated by the following technique: for each value of $x_k^{(1)} = x_k^{(2)} = k$ the value of $y_k^{(n)}$ was obtained using appropriate linear model for each channel and Gaussian random noise with the zero mean value.

For each kind of noise level 100 different realization of signals were generated. For each realization of signal the characteristic point x_P in each of two channels (point for which models are switched) was determined in the following way. First, the parameters of models are obtained by a one of the clustering method, $\mathbf{w}^{(n,1)} = \left[w_0^{(n,1)}, w_1^{(n,1)}\right]^T$ and $\mathbf{w}^{(n,2)} = \left[w_0^{(n,2)}, w_1^{(n,2)}\right]^T$. Next, the characteristic point is determined as a solution of linear equation system:

$$\begin{cases} y^{(n)} = w_0^{(n,1)} + w_1^{(n,1)} x, \\ y^{(n)} = w_0^{(n,2)} + w_1^{(n,2)} x, \end{cases} \tag{12}$$

i.e. $x_P^{(n)} = \left(w_0^{(n,2)} - w_0^{(n,1)}\right) / \left(w_1^{(n,1)} - w_1^{(n,2)}\right)$ for $w_0^{(n,2)} \neq w_0^{(n,1)}$ and $w_1^{(n,1)} \neq w_1^{(n,2)}$. For $w_1^{(n,1)} = w_1^{(n,2)}$ and $w_0^{(n,2)} \neq w_0^{(n,1)}$ there is no solution of (12) — the lines are parallel. For $w_1^{(n,1)} = w_1^{(n,2)}$ and $w_0^{(n,2)} = w_0^{(n,1)}$ there are infinite many solutions of (12) — the lines are overlap. In the last two cases the localization of the characteristic point is assumed in the midpoint of analyzed interval, $x_P = x_{N/2}$ for even N and $x_P = x_{(N-1)/2}$ for odd N. For each kind of noise level the mean $\overline{x}_P$ and standard deviation σ_P of characteristic points' localization in each channel are calculated over 100 experiments. The parameter $\gamma = 5$ were used based on previous works [4], [5] and $\xi_{2 \leftarrow 1} = 1$, $\xi_{1 \leftarrow 2} = 0$. The result obtained for both algorithms are presented in Table 1. The expected localization of characteristic point was in 40 sample.

As we can see from Table 1, the smaller value of SNR the worst results of characteristic point detection in the second channel are obtained by the TDCFCRM method (each channel analyzed separately). If we incorporate the influence of the first channel (with higher SNR) into the second channel (with smaller SNR) during clustering process we obtain an improvement of characteristic point detection in spite of the poor SNR.

Table 1. The results of traditional and the new clustering method obtained for synthetic signals. Mean and standard deviation of x_P for Gaussian noise.

	1st ECG channel			2nd ECG channel	
SNR	TDCFCRM	MCTDCFCRM	SNR	TDCFCRM	MCTDCFCRM
7.3	38.86 ± 1.78	38.86 ± 1.78	3.2	63.22 ± 47.63	44.26 ± 6.70
6.1	39.13 ± 2.00	39.13 ± 2.00	2.0	45.86 ± 68.22	44.22 ± 7.87
4.2	39.33 ± 8.91	39.33 ± 8.91	0	41.09 ± 88.10	43.07 ± 12.00

3.2 Real ECG signals

The purpose of this experiment was to compare the performance of the TDCFCRM and MCTDCFCRM methods for the J point detection in a real noised ECG signal from QTDB database. The expected J point occurrence was at $P = 30$ sample for both channels. Obtained result for both algorithms are presented in Figure 1. The dashed lines denote models of the TDCFCRM algorithm ($P = 25$ in I channel, $P = 78$ in II channel), while the dashed lines denotes models of the MCTDCFCRM ($P = 25$ in I channel, $P = 16$ in II channel). The parameter $\gamma = 5$ were used and $\xi_{2\leftarrow 1} = 5$, $\xi_{1\leftarrow 2} = 0$.

As we can see in Fig. 1, the results obtained by the TDCFCRM method in the second channel are wander much more then it can be accepted. Using of the MCTDCFCRM method yields a great improvement of J point detection.

4 Conclusions

In the paper the new method of clustering called multi-channels time-domain-constrained fuzzy c-regression models is presented. The numerical examples are presented to illustrate the validity of the proposed method in signal analysis. Numerical examples show the usefulness of the method in characteristic point detection for Gaussian noise and real ECG signal (e.g. J point detection).

References

1. Bargiela A, Pedrycz W (2003) Recursive Information Granulation: Aggregation and Interpretation Issues. IEEE Transaction on Systems, Man and Cybernetics – Part B: Cybernetics, vol.33, no.1, pp.96–112.
2. Bezdek JC (1982) Pattern Recognition with Fuzzy Objective Function Algorithms. Plenum Press, New York.
3. Hathaway RJ, Bezdek JC (1993) Switching Regression Models and Fuzzy Clustering. IEEE Transaction on Fuzzy Systems, vol.1, no.3, pp. 195–204.
4. Leski JM, Owczarek AJ (submitted) A time-domain-constrained fuzzy clustering method and its application to signal analysis. Fuzzy Sets and Systems.

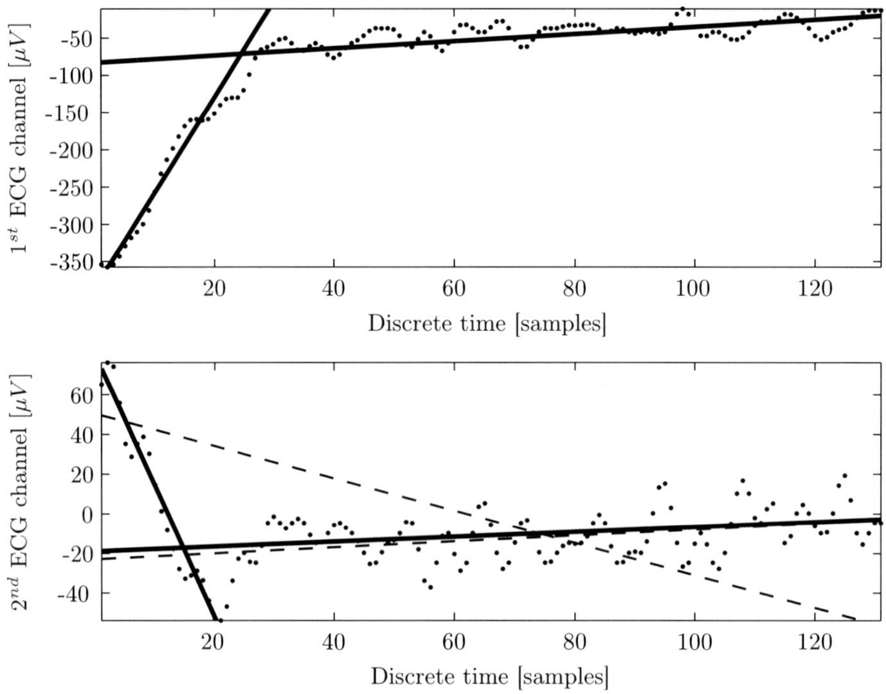

Fig. 1. The results of J point detection obtained by algorithms in channel I and II.

5. Owczarek AJ (2004) A new method of fuzzy clustering and its application to ECG signal analysis. PhD Thesis. Silesian Technical University, Gliwice, 2004.
6. Pedrycz W (2002) Distributed Collaborative Knowledge Elicitation. Computer Assisted Mechanics and Engineering Sciences, vol.9, pp.87–104.
7. Pedrycz W, Gacek A (2004) Knowledge-Based Clustering as a Conceptual and Algorithmic Environment of Biomedical Data Analysis. Journal of Medical Informatics and Technologies, vol.7, pp.13–21.

Pairwise Selection of Features and Prototypes

Elżbieta Pekalska, Artsiom Harol, Carmen Lai and Robert P.W. Duin

Information and Communication Theory group
Faculty of Electrical Engineering, Mathematics and Computer Science
Delft University of Technology, The Netherlands
{e.pekalska,a.harol,c.lai,r.p.w.duin}@ewi.tudelft.nl

Summary. Learning from given patterns is realized by learning from their appropriate representations. This is usually practiced either by defining a set of features or by measuring proximities between pairs of objects. Both approaches are problem dependent and aim at the construction of some representation space, where discrimination functions can be defined.

In most situations, some feature reduction or prototype selection is mandatory. In this paper, a pairwise selection for creating a suitable representation space is proposed. To determine an informative set of features (or prototypes), the correlations between feature pairs are taken into account. By this, some dependencies are detected, while overtraining is avoided as the criterion is evaluated in two-dimensional feature spaces. Several experiments show that for small sample size problems, the proposed algorithm can outperform traditional selection methods.

1 Introduction

The construction of a proper representation space is essential for designing successful learning procedures. Concerning both the computational efficiency and performance of a recognition system, one is usually interested in a space of a low dimensionality. Since the initial space may be large, some reduction methods are necessary to either detect or create informative features. An ideal technique is capable of reducing the dimensionality effectively, while preserving the class separability in the data. As some information is unavoidably lost in this process, it is, therefore, desirable to formulate a method that significantly reduces the dimensionality, but still preserves the information. In this paper, we focus on feature selection approaches.

Feature selection methods rely on a quantitative criterion that measures their performance. This criterion is used in some optimization process to determine a subset of informative features. Selection methods are usually divided into filters and wrappers [10]. Filters evaluate the relevance of features based on a feature capacity to discriminate between the classes. Wrappers employ a classification algorithm, used later to build the final classifier, to judge the quality of a feature. Both approaches involve a combinatorial search through the constructed space of possible feature subsets. Usually, greedy procedures such as forward or backward eliminations are employed due to their computational attractiveness. More complex procedures such as floating searches and genetic algorithms can also be applied [5, 10, 14, 11].

Concerning the evaluation of the criterion, selection techniques are either univariate or multivariate. Univariate approaches are simple and fast. Multivariate approaches evaluate the relevance of features in a group, taking their interdependencies into account. When features are correlated, these techniques are able to construct good feature subsets, while univariate techniques may fail. A disadvantage of multivariate approaches, however, is that they evaluate features in a multidimensional space, not only demanding a considerable computational effort, but also resulting in a loss of accuracy in case of a limited training set. Due to overfitting, feature subsets that do not ensure a good discrimination may be still judged as 'good'. The more features have to be selected, the worse this problem becomes.

In this paper, a pairwise feature selection procedure is investigated. Some ideas in this direction can be found in [2], where a particular pairwise selection algorithm was proposed for gene expression data. Since pairs of features are considered, second order dependencies are taken into account. Multidimensional spaces are now restricted to two dimensions, hence this method does not suffer from overfitting as other multivariate approaches do.

The problem of feature selection is similar to the selection of prototypes used to define a linear embedding of proximity data. In this case, a set of objects is represented by a dissimilarity matrix, where each entry describes a degree of commonality between pairs of objects. The chosen prototypes determine a vector space, in which all objects are represented as points and the corresponding dissimilarities are preserved as well as possible. Pairwise prototype selection is an appealing alternative to random, individual and multivariate selections [7, 12, 13], especially for low-dimensional embedded spaces.

2 Feature selection techniques

Feature selection techniques try to determine a small subset of features, which are sufficient for a good discrimination. Usually, some type of a combinatorial search, in a forward (an incremental addition of features, starting from a single one), backward (an incremental removal of features, starting from the entire set) or floating manner is employed to find this feature subset. This optimization relies on some specified criterion, usually related to the class separability, and the way the relevance of a feature to be added (or removed) is evaluated.

Three incremental selection methods are considered here. These are individual, forward and pairwise strategies. Assume that $F = \{f_1, f_2, \ldots, f_m\}$ is a set of m features. Denote by $\tilde{F} \subset F$ a subset of the selected features. In each step, a feature or a pair of features is chosen according to some criterion J and added to $\tilde{F}$. Note that $\tilde{F} = \emptyset$ in the beginning.

Individual (univariate) selection. In this approach, the informativeness of each feature is evaluated individually according to the criterion J. In each step, a single best feature is chosen. This can be formally written as:

$$\begin{aligned} \tilde{F} &:= \tilde{F} \cup f, \quad \text{where } f \colon \max_{f_i \in F} J(f_i) \\ F &:= F \backslash f \end{aligned} \tag{1}$$

In this procedure, features are ranked from the most to the least informative according to the criterion J and the most indicative features are finally selected.

Forward selection. Forward feature selection starts with the single most informative feature and adds next most informative features in a greedy fashion. Each step can be formalized as follows:

$$\begin{aligned} \tilde{F} &:= \tilde{F} \cup f, \quad \text{where } f \colon \max_{f_i \in F} J(F \cup f_i) \\ F &:= F \backslash f \end{aligned} \tag{2}$$

Pairwise selection. The relevance of features is judged by evaluating pairs of features. In each step, the best feature pair is detected. Two approaches are here possible. Either both features are chosen from the current unselected feature set F or only one of them, as the other one comes from $\tilde{F}$. In each step, one has:

$$\begin{aligned} \tilde{F} &:= \tilde{F} \cup \{f \cup f'\}, \quad \text{where } \{f \cup f'\} \colon \max_{f_i \notin \tilde{F} \vee (f_j \neq f_i \wedge f_j \notin \tilde{F})} J(f_i \cup f_j) \\ F &:= F \backslash \{f \cup f'\} \end{aligned} \tag{3}$$

Criterion. In our experiments, the inter-intra criterion is used [8]. It is applied in some representation space, where the between-scatter S_b and within-scatter S_w matrices are computed. S_w measures the average dispersion of a class sample around its mean, while S_b describes the scattering of the class means around the overall average. Given n samples, K classes and n_k samples per class, the inter-intra criterion is given as

$$J = \text{trace}(S_w^{-1} S_b), \tag{4}$$

where $S_w = \frac{1}{n}\sum_{k=1}^{K} n_k S_k$, $S_b = \frac{1}{n}\sum_{k=1}^{K} n_k(\mathbf{m}_k - \mathbf{m})(\mathbf{m}_k - \mathbf{m})^T$, and $\mathbf{m}$ is the estimated overall mean and $\mathbf{m}_k$ and S_k are the estimated mean and covariance matrix of the k-th class, respectively. The higher value of the criterion, the more informative the corresponding feature. For a single feature and two-class problems, this criterion is equivalent to the Fisher criterion [5] $J_{FC} = \frac{|m_1 - m_2|}{\sqrt{s_1^2 + s_2^2}}$, where s_1 and s_2 are the class standard deviations.

3 Examples

The potential of pairwise feature selection is illustrated by three examples.

3.1 Artificial example

An artificial data set with some correlated feature pairs is generated to investigate the behavior of feature selection methods in a controlled environment. Assume that n samples and m features are given, where only q features, generated in correlated pairs, are informative. The samples for each correlated feature pair are drawn from a Gaussian distribution with the class means $\mu_1 = \begin{bmatrix}0 & 0\end{bmatrix}^T$ and $\mu_2 = \frac{\sqrt{2}}{2}\begin{bmatrix}r & -r\end{bmatrix}^T$ for some $r > 0$. The covariance matrix, identical for both classes, is $\Sigma_1 = \Sigma_2 = \left[\begin{smallmatrix} v+1 & v-1 \\ v-1 & v+1 \end{smallmatrix}\right]$. The remaining $m-q$ features are uninformative, i.e. the two classes are drawn from a spherical Gaussian distribution $\mathcal{N}(\mathbf{0}, \frac{v}{\sqrt{2}}I)$, where I is the identity matrix. We set $m = 300$, $q = 20$, and, in order to have a class overlap, $r = 3$ and $v = \sqrt{40}$. Since we want to simulate a small sample size problem, we chose $n = 100$ for the training set, while the size of the test set is set to $n = 10000$.

For each selection method, a Fisher linear discriminant (FLD) [5] is trained on a training set with a growing number of features (starting from the best two features) and tested on an independent test set. All selection methods rely on the criterion (4). As a result, the classification error can be plotted versus the number of features used. The error is estimated based on 50 repetitions of the experiments with different generations of the training set.

Figure 1, left plot, shows the behavior of different feature selection methods as judged by the average classification error. The peaking phenomenon visible in the plot occurs when the number of samples is comparable to the number of features. This is due to the use of a pseudo-inverse instead of the usual inverse of the sample covariance matrix on which the FLD relies [15]. Some solutions to avoid this problem can be found e.g. in [16].

The univariate approach performs the worst and the pairwise selection performs the best. This is expected, since pairs of features are strongly correlated. Although the forward search reaches a higher accuracy than the univariate technique, it is limited by the greedy procedure it is based on.

Figure 1, right plot, shows the number of detected informative features versus the number of selected features. The results are the average of 50 experiments. The pairwise approach determines all informative features perfectly. In this small sample sizes problem, the forward search retrieves more uninformative features than the univariate approach.

3.2 Feature selection example

A feature selection example on a more general data set is here presented. The Waveform data, as described in [4], is chosen as it clearly shows what can be gained by the pairwise selection. This three-class problem is based on sampling triangle shaped waves and, thereby, it really needs a significant subset of the 21 original features in order to reach a proper class separation. There are 5000 objects in total, approximately equally distributed over the three classes.

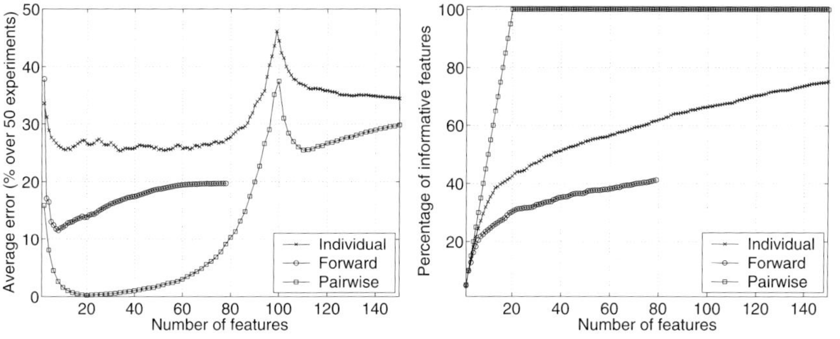

Fig. 1. Artificial data. Left: average classification error over 50 repetitions for the three feature selection procedures. Right: percentage of relevant features retrieved by the selection techniques.

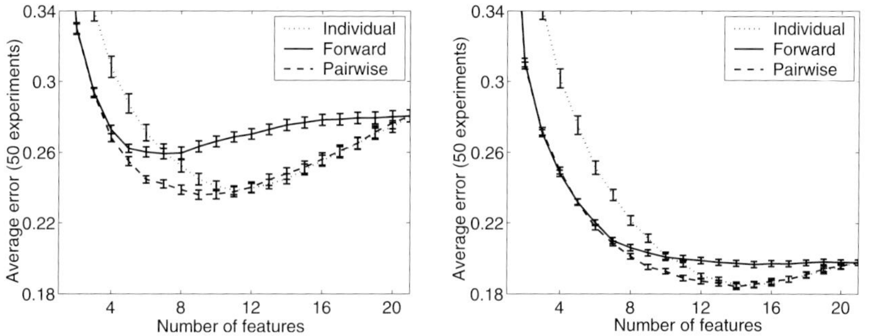

Fig. 2. Waveform data. Average classification error over 50 repetitions of the NLC for three feature selection procedures using 35 (left) and 100 (right) objects per class for feature selection and classifier training.

Figure 2, left plot, shows the average classification error over 50 experiments, in which 35 objects per class have been chosen at random. Using these objects, feature selections are performed based on the inter-intra criterion, formula (4). In the resulting feature spaces, a Bayes normal-density based linear classifier (NLC), assuming class normal distributions with equal covariance matrices [5], is trained on the same training set. The classifiers are tested on the remaining objects. The resulting error rates are averaged out and the standard deviations of the means are computed and shown in the plot. This is a clear example, where a pairwise selection behaves equal or better than the forward selection as well as the individual selection. For larger feature sizes, the forward procedure cannot compute the criterion values in a sufficiently accurate way, which leads to suboptimal feature subsets. The pairwise procedure shows a continuous improvement.

Figure 2, right plot, presents the results for 100 objects per class used for the feature selection and training. The same phenomena can be observed as

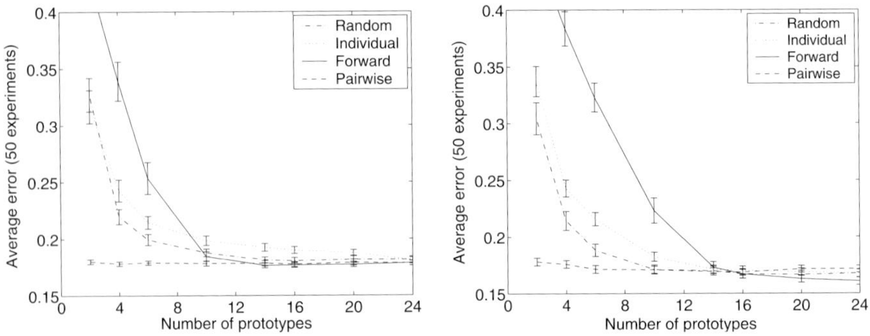

Fig. 3. Heart disease data. Average classification error over 50 repetitions based on 50 (left) and 150 (right) objects in total used for prototype selection, embedding and classifier training. The NLC is trained in embedded spaces.

for 35 objects per class, but less pronounced, as the forward procedure now suffers less from overtraining.

3.3 Prototype selection example

The heart disease data set [1] is considered for the prototype selection example. There are 303 cases, where 139 and 164 refer to ill and healthy patients, respectively. A subset of 13 attributes of mixed types is used to compute a Gower's distance representation [6], which is known to be Euclidean.

The data are split into a training set T and test set S with the same prior probabilities (either 50 or 150 objects in total for training). The distance matrix $D(T,T)$ is used for prototype selection, embedding and training the NLC. For testing, the distance matrix $D(S,T)$ is used. In the individual and forward selection methods, the prototypes are determined by using the inter-intra criterion (4) applied to $D(T,T)$. This means that a distance representation is interpreted in a vector space, where each dimension describes a distance to a particular object from T [12, 13]. Inspired by [17], the pairwise prototype selection is realized by evaluating the criterion in two-dimensional spaces determined by isometric embeddings of $D(\cdot,[p_i\ p_j])$, where p_i and p_j are different objects. The details can be found in [17, 7]. Having found a prototype set R, the classical scaling [3] is used for a linear isometric embedding of $D(R,R)$. The remaining $D(T\backslash R,R)$ are then projected to the embedded space and the NLC is trained [3, 13]. Testing is realized by projecting $D(S,T)$ to the same space and applying the trained NLC. The experiments are performed for growing prototype sets and repeated 50 times for various splits into the training and test sets. The average classification error is plotted as a function of the prototype set size; see Figure 3.

Note that the number of prototypes m defines the embedding and describes the dimensionality p of an embedded space ($m=p+1$). The prototypes should be significantly different (i.e. vectors of distances to them should differ) to preserve the most of the distance information in the data. For a small number of prototypes, this holds for the pairwise selection and can be observed in Figure 3. The forward selection performs then the worst, since the embedding is defined by prototypes p_i and p_j which are characterized by correlated vectors of distances $D(\cdot, p_i)$ and $D(\cdot, p_j)$. In this case, this does not ensure yet that the resulting embedded space will be good for discrimination. The random selection is better here than the forward and individual selections as it tends to choose objects that differ with respect to distance information.

4 Discussion and conclusions

The need for dimension reduction holds in a similar way for traditional feature spaces as for embedded spaces defined on the dissimilarities to a set of prototype objects. In this paper, we presented a new procedure for dimension reduction by selection. There are several reasons to lower the dimensionality of a representation space in which classifiers have to be trained. First, less dimensions implies less computational effort to represent new objects to be classified: less features to be measured or less proximities to be computed. Secondly, in low-dimensional spaces the accuracy of trained classifiers is higher than in spaces with more dimensions. The trade-off is, however, that by removing dimensions (features) the class separability is deteriorated. So, feature selection should be done carefully.

An issue often neglected in previous studies on feature selection is that the accuracy of the criterion itself, like the classifier, also may suffer from small training set sizes. Procedures like backward elimination, branch and bound, forward selection [9] and floating search [14] evaluate the criteria in a multi-dimensional space. The estimation of the criterion values suffers from noise. Many criteria used for judging class separability are biased for small sample sizes: classes seem to be better separable than they are, in fact. Even when corrections for such a bias are made, e.g. by using the F-statistics, there is still a bias caused by the selection mechanism itself. This is for high-dimensional spaces more severe than for low-dimensional spaces as the variance in the criterion estimate is larger in the former case.

The individual evaluation and ranking of features suffers the least from this problem. It is, however, entirely unable to take into account the dependency between features in estimating the separability. The pairwise selection procedure studied in this paper makes some trade-off. Feature spaces are judged just in various combinations of feature pairs. So, whenever the dependency between two features is of importance, it is detected and can be used. This procedure is expected to be almost always better than individual ranking, except for very small sample sizes or for very large feature sets, as in these cases also pairwise evaluation will cause an overtraining. The proposed procedure may be better than the multivariate techniques when more than just a few features are needed and the training set size is small. For large training sets multivariate approaches do not suffer from overtraining and may detect higher order useful dependencies between features. If the problem can be solved by a small set of features, multivariate techniques may find them as well.

In conclusion, the pairwise procedure for the selection of features or prototypes may be a useful strategy in case of small sample size problems. Some examples are presented to support this claim.

References

1. Blake CL and Merz CJ (1998) UCI Repository of machine learning databases, http://www.ics.uci.edu/ mlearn/MLRepository.html.
2. Bo T and Jonassen I (2002) New feature subset selection procedures for classification of expression profiles, Genome biology 3.
3. Borg I and Groenen P (1997) Modern Multidimensional Scaling. Springer-Verlag.
4. Breiman L, Friedman JH, Olshen RA and Stone CJ (1984) Classification and regression trees, Wadsworth, California.
5. Duda RO, Hart PE and Stork DG (2001) Pattern Classification 2nd. edition, John Wiley & Sons.
6. Gower JC, A general coefficient of similarity and some of its properties, Biometrics vol. 27, 25-33, 1971.
7. Harol A, Pekalska E and Duin RPW (2005), Pairwise prototype selection on distance data, submitted.
8. van der Heijden F, Duin RPW, de Ridder D and Tax DMJ (2004) Classification, Parameter Estimation and State Estimation. An Engineering Approach using Matlab. John Wiley & Sons Ltd.
9. Jain AK, Duin RPW and Mao J (2000) Statistical Pattern Recognition: A Review. IEEE Trans on PAMI 22:4-37.
10. Kohavi R and John GH (1997) Wrappers for feature subset selection. Artificial Intelligence 97:273-324.
11. Li L, Weinberg CR, Darden TA and Pedersen LG (2001) Gene selection for sample classification based on gene expression data: study of sensitivity to choice of parameters of the GA/KNN method. Bioinformatics 17:1131-1142.
12. Pekalska E, Duin RPW and Paclík P (2005), Prototype Selection for Dissimilarity-based Classifiers. accepted to Pattern Recognition.
13. Pekalska E (2005) Dissimilarity representations in pattern recognition. Concepts, theory and applications. PhD thesis. Delft University of Technology.
14. Pudil P, Novovicova J, and Kittler J (1994) Floating search methods in feature selection. Pattern Recognition Letters 15:1119-1125.
15. Raudys S, and Duin RPW (1998) On expected classification error of the Fisher linear classifier with pseudo-inverse covariance matrix. Pattern Recognition Letters 19:385-392.
16. Skurichina M (2001), Stabilizing weak classifiers. PhD thesis. Delft University of Technology.
17. Somorjai RL, Dolenko B, Demko A, Mandelzweig M, Nikulin AE, Baumgartner R, Pizzi NJ (2004) Mapping high-dimensional data onto a relative distance plane an exact method for visualizing and characterizing high-dimensional patterns, Journal of Biomedical Informatics 37:366-379.

Evolutionary Method in Grouping of Units

Henryk Potrzebowski, Jarosław Stańczak, and Krzysztof Sęp*

Systems Research Institute, Polish Academy of Science
Newelska 6, 01-447 Warsaw, Poland
potrzeb@ibspan.waw.pl, stanczak@ibspan.waw.pl, sep@ibspan.waw.pl

Summary. This paper deals with the clustering problem, where an order of elements plays a pivotal role. This formulation is very usable for wide range of Decision Support System (DSS) applications. The proposed clustering method consists of two stages. The first is a stage of data matrix reorganization, using a specialized evolutionary algorithm. The second stage is a final clustering step and is performed using a simple clustering method.

1 Introduction

Many decision problems deals with the task of appropriate design of a system structure, where it is important to take into account many interactions among its elements. The Design Structure Matrix (DSM) is a good tool for representing and solving such problems.
The DSM contains elements a_{ij} (where $i \in \{1..|R|\}, j \in \{1..|S|\}$, $|.|$ denotes a cardinality of set, R and S are sets of elements of the system). The element a_{ij} is a measure of the relationship strength between elements of sets R and S. It can show some kind of connections between units (data or goods flow, communication connections, etc.). By proper permuting of rows and/or columns of such array it is possible to obtain clusters, which are subsets of R strongly related to corresponding subsets of S [3]. The widely used method of DSM clustering is to maximize interactions among elements of the cluster while minimizing interactions among clusters [2].
The DSM clustering problem can be easily transformed to a well-known TSP (Traveling Salesman Problem) problem and in practical cases only approximate (greedy, 2-opt, 3-opt [3], [10] or BEA - Bounded Energy Algorithm [4]) or heuristic methods (genetic algorithm [1], [5], [7], [8], [11]) can be used. In our approach an adjusted evolutionary algorithm has been used to solve the clustering problem with an aid of simple clustering method described in

*Research partly supported by the grant from the State Committee for Scientific Research. Decision no. 55/E-82/SPB/5.PR UE/DZ 385/2003-2005 of 16.07.2003

section 3.
The adjustment of the genetic algorithm to the solved problem requires a proper encoding of individuals, an invention of specialized genetic operators and designing a fitness function to be optimized by the algorithm. The problem's quality function is closely connected with the fitness function that values the members of the population. The classification problem is not a typical optimization task and its quality function is some artificial formula tuned to the problem. There are probably many possible fitness functions that can be used there.
The accepted member of the population structure requires specialized genetic operators, which modify the population of solutions. Simple random operators are easy to think out, and similar to the widely used mutation and crossover (or exchange of parts of solutions). Also a set of heuristic operators was worked out and successfully tested: 2-opt and intelligent exchange.
Evolutionary method with the simple clustering algorithm and obtained results are discussed in the following sections.

2 Evolutionary Algorithm To Solve The Clustering Problem

Standard evolutionary algorithm (EA) works in the manner as it is shown in the Algorithm 1, but this simple scheme requires many problem specific improvements to work efficiently.
The adjustment of the genetic algorithm to the solved problem requires a proper encoding of solutions, an invention of specialized genetic operators for the problem, accepted data structure and a fitness function to be optimized by the algorithm.

1. Random initialization of the population of solutions.
2. Reproduction and modification of solutions using genetic operators.
3. Valuation of the obtained solutions.
4. Selection of individuals for the next generation.
5. If a stop condition is not satisfied, go to 2.

Algorithm 1. The evolutionary algorithm.

2.1 Individual Representation

The whole information about the problem is stored in an array of data (square and symmetric for the symmetric case or rectangular for non-symmetric case) that describes all data connections. There is only one global data table in described approach. Members of the population (Fig.1) contain their own solutions of the problem as vectors of indexes to rows or rows and columns in the non-symmetric case. Vectors of indexes are permutations of rows or columns of the global data array. This method of encoding saves memory, especially for big data arrays

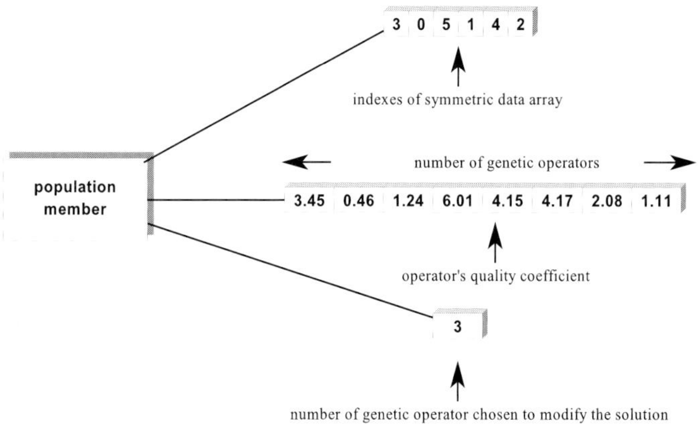

Fig. 1. A structure of the population member (symmetric case).

Beside it, the member of the population contains several more data including: a vector of real numbers, which describe its knowledge about genetic operators and a number of the operator chosen for current iteration - more details abut them will be given later in this chapter.

2.2 Fitness Function

The problem's quality function is closely connected with the fitness function, which evaluates the members of the population. In the problem of data clustering, the maximized fitness function is the problem's quality function 1.

$$Q = \sum_{i=1}^{n} \sum_{j=1}^{m} a[i,j] \left(a\left[i,j+1\right] + a[i+1,j]\right) \tag{1}$$

where:
n, m- numbers of rows and columns, for symmetric case $n = m$;
$a[i,j]$ - element of the data array.
It is assumed that elements $a[0,j]$, $a[i,0]$, $a[n+1,j]$, $a[i,m+1]$ equal 0.

This function is maximized using all possible permutations of rows and columns of DSM matrix (for non-symmetric case sequence of rows and columns can be changed separately, for symmetric one rows and columns must be changed in the same manner).

Maximalization of this function make the DSM matrix more close around diagonal (preprocessing) and thus it is easier to divide it into compact clusters.

2.3 Specialized Operators

The described data structure requires specialized genetic operators, which modify the population of solutions. Only operators permuting indexes to rows and/or columns are allowed in that problem:

- mutation - an exchange of randomly chosen subset of indexes;
- multiple mutation - the mutation operator performed several times;
- intelligent exchange - one randomly selected index is exchanged with some others, solution with the best value of quality function is an effect of this operator appearance;
- a multiple version of intelligent exchange is also applied;
- 2-opt operator - indexes are exchanged in pairs, best found modification is stored as a new solution.

2.4 Evolutionary Algorithm Used To Solve The Problem

Using of specialized genetic operators requires applying some method of sampling them in all iterations of the algorithm. In the used approach [6], [9]it is assumed that an operator that generates good results should have bigger probability and more frequently effect the population. But it is very likely that the operator, that is good for one individual, gives worse effects for another, for instance because of its location in the domain of possible solutions. Thus every individual may have its own preferences. Every individual has a vector of floating point numbers, beside encoded solution. Each number corresponds to one genetic operation. It is a measure of quality of the genetic operator. The higher the number is, the higher is the probability of the operator.
The ranking of qualities becomes a base to compute the probabilities of appearance and execution of genetic operators. This set of probabilities is also a base of experience of every individual and according to it, an operator is chosen in each epoch of the algorithm. Due to the gathered experience one can maximize chances of its offspring to survive.
The applied selection method consists of two methods with different properties: a histogram selection (increases the diversity of the population) and a deterministic roulette (strongly promotes best individuals) [9], which are selected in random during the execution of the algorithm. The probability of executing of the selection method is obtained from the formula 2.

$$p_{his}(t+1) = \begin{cases} p_{his}(t) * (1-a) & \text{for } \max(F_{av}(t) - F_{\min}(t), \\ & F_{\max}(t) - F_{av}(t)) > 3 * \sigma(F(t)), \\ p_{his}(t) * (1-a) + 0.5 * a & \text{for } (\max(F_{av}(t) - F_{\min}(t), \\ & F_{\max}(t) - F_{av}(t)) \geq 0.5 * \sigma(F(t)) \\ & \wedge(F_{av}(t) - F_{min}(t), \\ & F_{max}(t) - F_{av}(t)) \leq 3 * \sigma(F(t))) \\ p_{his}(t) * (1-a) + a & \text{for } \max F_{av}(t) - F_{\min}(t), \\ & F_{\max}(t) - F_{av}(t)) < 0.5 * \sigma(F(t)) \end{cases} \tag{2}$$

where:
$p_{his}(t+1), p_{his}(t)$ - probability of histogram selection appearance in following iterations ($1-p_{his}(t)$ is a probability of deterministic roulette method $p_{det}(t)$);
$F_{av}(t), F_{min}(t), F_{max}(t)$ - average, minimal and maximal values of fitness function in the population;
$\sigma(F(t))$ - standard deviation of fitness function $(F(t))$ in the population of solutions;
a - a small value to change probability $p_{his}(t)$, in simulations $a = 0.05$.
If individuals in the population are described by too small standard deviation of the fitness function $\sigma(F(t)))$ with respect to the extent of this function $(max(F_{av}(t) - F_{min}(t), F_{max}(t) - F_{av}(t)))$, then it is desirable to increase the probability of appearance of the histogram selection. On the contrary the probability of the deterministic roulette selection is increased. As far as parameters of the population are located in some range, considered as profitable we may keep approximately the same probabilities of appearance for both methods of selection. It is important that always $p_{his}(t) + p_{det}(t) = 1$- it means that some method of selection must be executed.

3 Clustering Method

After EA preprocessing of DSM a following method is performed to obtain a final clustering. Let elements of vector G take values obtained from the following formula:

$$G[j] = \sum_{i=1}^{n} a[i,j] \left(a\,[i,j-1] + a\,[i,j+1] + a\,[i-1,j] + a\,[i+1,j] \right) \tag{3}$$

where:
j - number of row, $j = 1, 2, \ldots, m$;
i - number of column, $i = 1, 2, \ldots, n$;
$a[i,j]$ - element of the data array.

A simple clustering algorithm:

1.Calculate value of G for each row of the data array;
2.Find all local minimums in vector G;
3.Each local minimum is a point of partitioning for new cluster;
4.Add each partitioning element to closest class.

Algorithm 2. Clustering method.

4 Computer Simulation Results

A testing data set describes a symmetric 50 x 50 problem with a non-symmetric matrix (data array), where 50 data attributes are considered. This problem is called "Import-export" example and has been taken from [3]. It describes economical connections among regions of Indonesia. The aim is to maximize the quality criterion 1 or roughly s peaking to "thicken" non-zero elements near a diagonal of the data array, to simplify proper clustering of data. Solving this problem can help to find better model of connections among regions and to develop them in larger areas. The best-found solution of the problem with explicit clustering is shown below.
As it can be seen, the initial data array is very scattered (Fig.2) and difficult to interpret. Hence, the clustering of this form of data array is useless and results are rather poor (Fig 3 and Fig 2). After EA preprocessing, the data array is much more concentrated (Fig 5) and clustering can be much more helpful to find closely related regions of Indonesia (Fig 4).

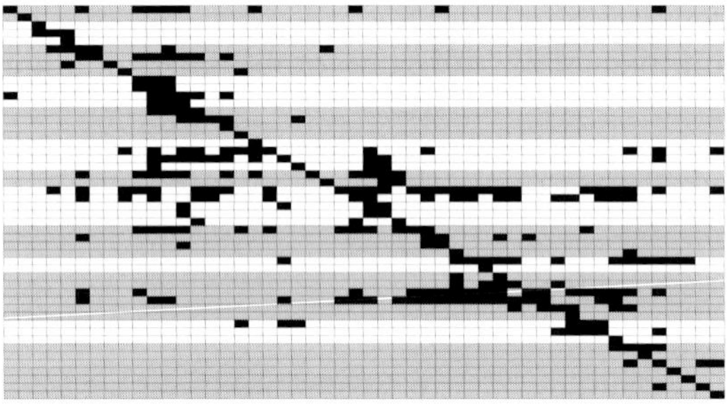

Fig. 2. Original data

Conclusions

The evolutionary algorithm is a very powerful tool to solve combinatorial optimization problems. It has been successfully applied for reordering of DSM matrix (its rows and columns) to obtain groups of closely connected elements. The simple clustering method, added at the end of the EA procedure, helps

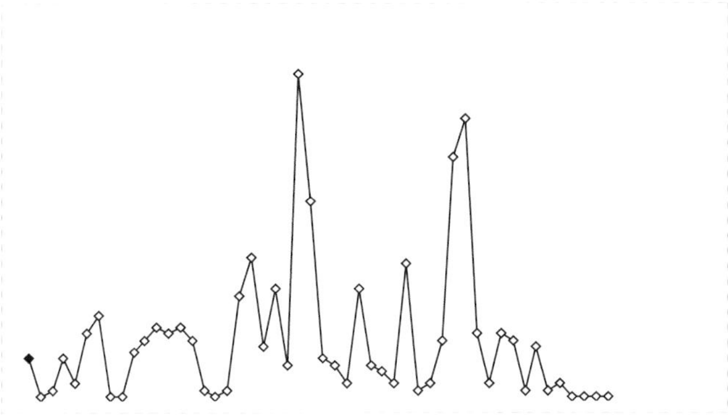

Fig. 3. Results of clustering without EA preprocessing ($Q = 223$, 13 clusters)

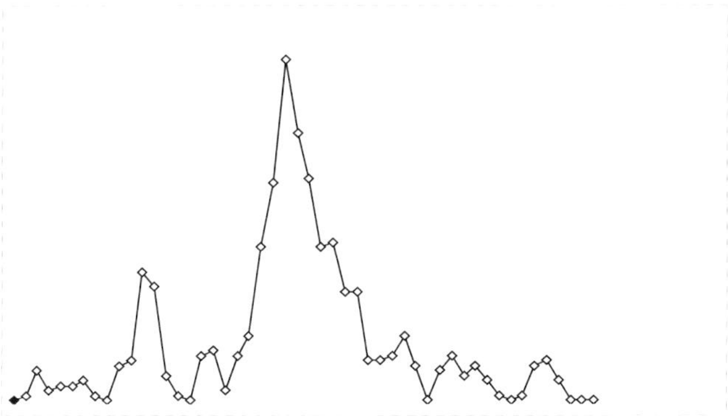

Fig. 4. Results of clustering with EA preprocessing ($Q = 290$, 9 clusters)

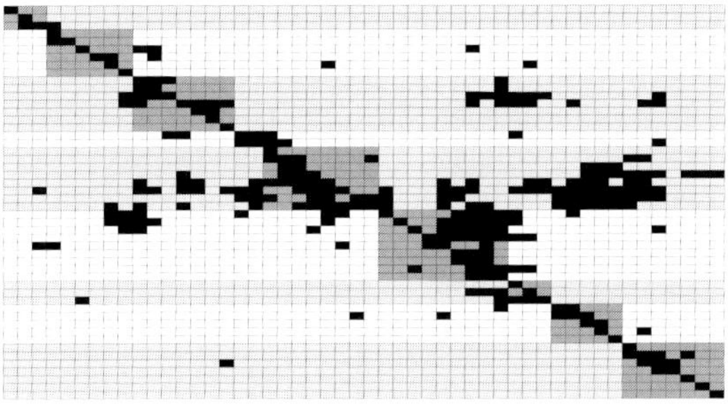

Fig. 5. Results of clustering with EA preprocessing, corresponding to Fig 4

to separate precisely elements into clusters and finally find closely related regions.
Described method can be helpful with finding connections between different groups of objects and may be used to optimize some processes of communication, data flow, design and similar fields of activity.

References

1. Altus S S, Kroo I M, Gage P J (1996) "A Genetic Algorithm for Scheduling and Decomposition of Multidisciplinary Design Problems", Journal of Mechanical Design, Vol. 118, No. 4, 486–489
2. Browning T R (2001) "Applying the Design Structure Matrix to System Decomposition and Integration Problems: A Review and New Directions", IEEE Transactions on Engineering Management, Vol. 48, No 3
3. Lenkstra J K (1977) "Sequencing by Enumerative Methods", Matematisch Centrum, Amsterdam
4. McCormick W T , Schweitzer P J, White T W "Problem decomposition and data reorganization by a clustering technique", Operations Res. 20, 1972, 993–1009
5. McCulley C, Bloebaum C (1996) A Genetic Tool for Optimal Design Sequencing in Complex Engineering Systems, Structural Optimization, Vol. 12, No. 2-3, 186–201
6. Mulawka J, Stańczak J (1999) "Genetic Algorithms with Adaptive Probabilities of Operators Selection", Proceedings of ICCIMA'99, New Delhi, India, pp. 464--468.
7. Potrzebowski H, Stańczak J , Sęp K (2004) "Evolutionary method in grouping of units with argument reduction", ICSS, Wrocław
8. Rogers J L (1997) "Reducing Design Cycle Time and Cost Thorough Process Resequencing", International Conference on Engineering Design ICED 997, Tampere, Finland
9. Stańczak J (1999) "Rozwój koncepcji i algorytmów dla samodoskonalących się systemów ewolucyjnych", Ph.D. Dissertation, Politechnika Warszawska
10. Sysło M M, Deo N, Kowalik J S (1983) "Algorithms of discrete optimization", Prentice-Hall
11. Yu T L, Goldberg D E, Yassine A, Yassine C (2003) "A Genetic Algorithm Design Inspired by Organizational Theory", Genetic and Evolutionary Computation Conference (GECCO) 2003, Chicago, Illinois, USA, Publ. Springer-Verlag, Heidelberg, Lecture Notes in Computer Science, Vol 2724/2003, 1620–1621.

Interpretation of Medical Symptoms Using Fuzzy Focal Elements

Ewa Straszecka[1] and Joanna Straszecka[2]

[1] Institute of Electronics, Silesian University of Technology, 16 Akademicka St. estraszecka@polsl.pl
[2] Department of Internal Diseases, School of Health Care and Education, Medical University of Silesia, 102 Edukacji St.

Summary. In medical diagnosis symptoms often differ in nature varying form linguistic information trough numerical values to crisp statements. Hence, unification of their interpretation during diagnosis support is difficult. The authors propose the diagnosis support using an extension of the Dempster-Shafer theory for fuzzy focal elements. Performance of the method depends on the membership function shapes. The paper provides indications for imprecise symptom representation and its membership function determination. The problems are discussed for thyroid gland diseases. Conclusions helpful in diagnosis support are formulated.

1 Introduction

The Dempster-Shafer theory of evidence (DST) can be useful for a medical diagnosis support model building as it makes it possible to assign symptoms to diagnoses. In the model, symptoms are focal elements for which belief (Bel) and plausibility (Pl) of diagnoses can be calculated. Thus, certainty of a diagnosis is described by Bel and Pl measures. However, it is also inevitable to express imprecision of symptoms. The symptoms often differ in nature varying form linguistic information to images and from sociological knowledge to signal features. Therefore, it is difficult to find a uniform interpretation for them. Such an interpretation is possible if they are focal elements, as the latter are defined in DST as predicates, which can be formulated using strict or imprecise items. The imprecise symptoms can be represented by fuzzy sets. This involves extending DST for fuzzy focal elements and defining membership functions for the symptoms. The paper aim at providing indications for imprecise symptoms representation and their membership functions determination. Following problems of symptom's representation for medical diagnosis support must be solved:

- interpretation of linguistic descriptions, e.g.: "exacerbated", or "significant", used to characterize patient's disorders;

- flexible laboratory test results handling around norm limits;

- possibility of fuzzy evidence input, for instance "rapid disease progress".

The problems are discussed for the diagnosis support in thyroid gland diseases. Authors' experiences of the DST diagnosis model building as well as calculations for a database provided in the Internet make it possible to formulate conclusions helpful in diagnosis support.

2 Model of diagnosis support in the Dempster-Shafer theory with fuzzy focal elements

The Dempster-Shafer theory of evidence (DST) defines focal elements as predicates that we have information about [1]. The information consists in a basic probability assignment (BPA). In medical knowledge representation the set of focal elements consists of single or complex symptoms. The f element that describes a symptom that is always false can be also defined to complete assumptions of DST. Let's denote the set of focal elements as S_l. The set is determined for the l-th diagnostic conclusion and comprises its symptoms and usually $S_j \cap S_k \neq \varnothing$. BPA for the l-th diagnosis can be defined as [2]:

$$m_l(f) = 0, \sum_{s_i^l \subset S_l} m_l(s_i^l) = 1, \tag{1}$$

where f can be understood as "none of symptoms of the l-th diagnosis is present" and s_i^l can be either a single symptom or a collection of symptoms (e.g. both systolic and diastolic blood pressure).

Once BPA is determined, certainty measures of belief (Bel) and plausibility (Pl) are defined in the following way [2]:

$$Bel(d_l) = \sum_{s_i^{l*} \subset S_l} m_l(s_i^{l*}), Pl(d_l) = \sum_{S_l \setminus s_i^{l\circ} \subset S_l} m_l(S_l \setminus s_i^{l\circ}), \tag{2}$$

when present (s_i^{l*}) and non-absent ($S_l \setminus s_i^{l\circ}$) symptoms of the l-th diagnosis are considered.

The concept of a non-absent symptom regards complex symptoms. When at least one of composing symptoms is present, then the complex symptom is non-absent On the other hand, a symptom is present if its membership is greater or equal a threshold. Obviously, membership of crisp symptoms takes values out of the $\{0, 1\}$ set and membership of fuzzy symptoms belongs to the $[0, 1]$ interval. If the threshold (let's denote it η_{BPA}) is just greater than 0, then classical BPA definition holds true. Nothing is changed with the definition (1) if we elevate η_{BPA} too, as only interpretation of inclusion, not the sum condition, is different. In such a case a symptom is included in S_l when its membership is greater or equal η_{BPA}. We are able to determine η_i, i.e. an individual membership of a data case. Thus, the membership for a single symptom equals $\eta_i = \mu(x^*)$, while for a complex symptom e.g. $\eta_i = \min(\mu_{i1}(x^*), \mu_{i2}(y^*))$ where x^* and y^*denote observed values (i.e. data case elements) and $\mu_{i1}(x)$, $\mu_{i2}(y)$ knowledge about the symptoms given in the form of fuzzy sets. Frequencies of occurrence of data cases that fulfill $\eta_i \geq \eta_{BPA}$, for chosen learning data and η_{BPA} may be normalized and so m_l can be determined. BPA and symptoms' membership functions compose general diagnostic knowledge. In the similar way, belief and plausibility can be calculated using a threshold. Since the threshold value may be different, let's denote it as η_T. Individual patient's diagnosis can be deduced by analyzing belief and plausibility values for present ($\eta_i \geq \eta_T$) and non-absent ($\eta_{iOR} = \max(\mu_{i1}(x^*), \mu_{i2}(y^*)) \geq \eta_T$) symptoms. The $[Bel(d_l), Pl(d_l)]$ interval determines credibility of the diagnosis. Belief is a measure of the most certain information, while plausibility characterizes available information about the patient under consideration. Medical diagnosis must be cautious, so final diagnosis can be stated by comparing beliefs of all possible diagnoses. The beliefs were previously determined for the maximum $Pl(l)$. Single $Bel(d_l)$ interpretation (without the comparison) can be deceptive, particularly in case of many symptoms absence.

The η_{BPA} threshold determines quality of knowledge, the higher it is, the more sure manifestations are considered during diagnosis support. Since it is not obvious whether only most distinct symptoms should influence diagnosis, lowering η_{BPA} values down to $[0.8, 0.9]$ should be regarded. The η_T threshold decides if we want to interpret only the most clear symptoms (high η_T values) or we think over ambiguous manifestations, too (low η_T values). Membership functions shape influences both knowledge representation and interpretation of symptoms during a consultation session. Therefore, they should be carefully defined.

3 Defining membership functions for fuzzy focal elements

Exact shape of membership functions is not crucial, hence most often trapezoidal or bell-shaped curves are used. The former are convenient for numerical procedures, the latter make their differentiation (which is sometimes necessary) possible. We propose trapezoidal shapes, thus 4 points must be indicated to determine a membership function. Roughly speaking, their x-coordinates create division of a domain for intervals related to individual diagnoses, while y-coordinates show certainty of considering parameter's value as a symptom of selected diagnosis. Fuzzy sets characterized by the membership functions cannot create an empty intersection, otherwise gaps in a parameter interpretation appear. In the absence of an expert, membership functions must be designed using learning data. The membership functions should intersect at the same level for each variable. The level regards certainty with which a symptom evidences only one diagnosis Explicit value of the crossing level is necessary for η_T and η_{BPA} thresholds choice. Let's assume the crossing level is 0.5. Thus, y-coordinates for two points of the trapezoidal function are stated. It is reasonable to set their x-coordinates at the points of theoretical distributions crossing for two diagnoses. Hence, two points of the membership function are determined. The points are indicated by experts, whenever they are available. They suggest the interval of values significantly implying a diagnosis. The points are limits of the interval. Norms given for a laboratory test naturally create such an interval.

For the other two points y-coordinates equal 1. In the present investigation their x-coordinates were primary assumed as lower and upper quartiles of the empirical distribution of learning data. This was not a bad choice, still it could be improved, as experiments showed.

It is worth noticing that not only crisp, but also fuzzy evidences can be interpreted using the thresholds. Maximum of conjunction of a fuzzy focal element and a fuzzy observation can be compared with the threshold. In such a way, an observation like "rapid disease progress" can be represented and easily included in diagnostic inference.

4 Example of membership functions tuning

The proposed method of diagnosis support using DST with fuzzy focal elements was verified for thyroid gland diseases. Three diagnostic categories: euthyroidism (health)- d_1, hyperthyroidism - d_2, and hypothyroidism - d_3, were considered. The database *ftp.ics.uci.-edu/pub/machine-learning-databases/thyroid-disease*, files *new-thyr.** were used to test the performance of the proposed method. The data included values of 5 laboratory test results for which a diagnosis about a thyroid gland disease was stated. The data were divided for learning/test sets with the following number of cases: 75/75 (d_1), 15/20 (d_2), 15/15 (d_3). Membership functions were determined using the procedure described in Section 3, and learning data sets. However, it turned out that empirical quartiles not always corresponded points of theoretical distribution intersections. In such situations, membership function slope was assumed as "sufficiently steep", i.e. with 99% gradient.

Global error, i.e. the percentage of wrong or not stated diagnoses was calculated for test data. If two belief values were the same, a conclusion was not stated (an error occurred). Otherwise, the final conclusion was the hypothesis with the greatest belief. The global error depended on threshold choice. Its lowest value of 2.67% occured for $\eta_{BPA} \in [0.1, 0.4] \cup [0.8, 0.9]$ and $\eta_T \in [0.2, 0.4]$. This satisfactory result could still be improved by membership functions tuning. To this end, shapes of membership functions were changed by modifying x-coordinates of points with y-coordinate value equal 1. The modification consisted in moving the points toward less and more steep slopes.

If the original point was $(x_o, 1)$, then it was changed following the pattern: $(x_o + \varepsilon \left|\bar{x} - x_{0.5}\right|, 1)$, where: $\bar{x}$ denotes mean value of learning data, $x_{0.5}$ - stated point of theoretical distribution intersection, $\varepsilon = -0.8 \div 2.0$ with the step 0.1 Wider change was inconsistent due to lack of conformity with $\bar{x}$ and $x_{0.5}$. Inconsistency occurred for some membership functions for smaller than $\varepsilon = 2.0$ coefficients too, for instance it turned out that $x_o + \varepsilon \left|\bar{x} - x_{0.5}\right| > \bar{x}$. In such situations, the slope of the membership function was not changed. "Sufficiently steep" slopes were not changed, too. Fig.1 shows the modification for the diagnosis "healthy" and the v_1 variable of the Internet data. Right slope of the membership function is "sufficiently steep".

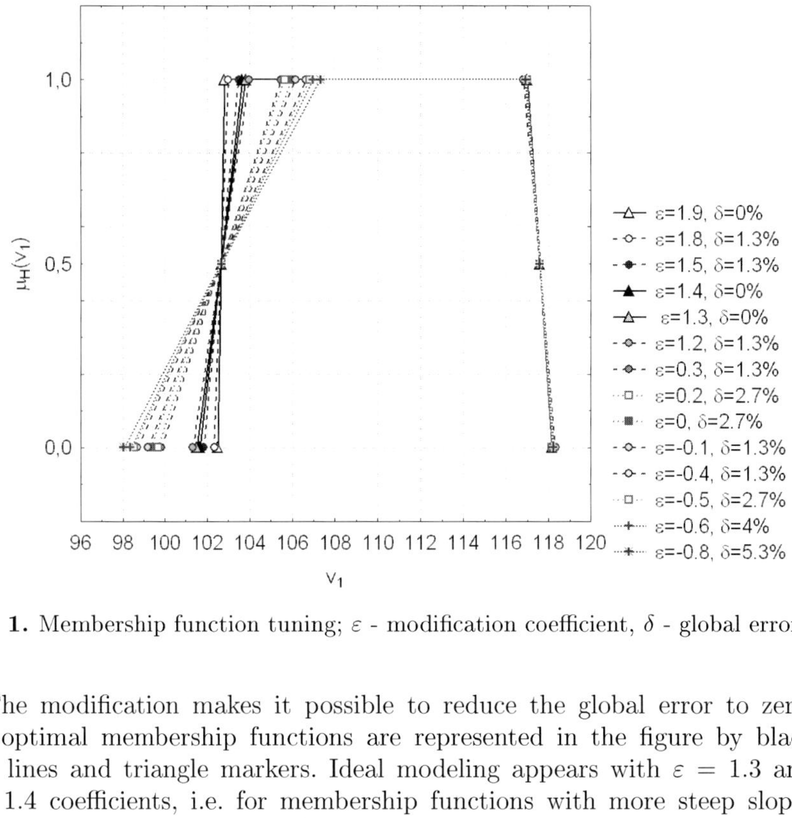

Fig. 1. Membership function tuning; ε - modification coefficient, δ - global error.

The modification makes it possible to reduce the global error to zero. The optimal membership functions are represented in the figure by black solid lines and triangle markers. Ideal modeling appears with $\varepsilon = 1.3$ and $\varepsilon = 1.4$ coefficients, i.e. for membership functions with more steep slopes than originally proposed ($\varepsilon = 0$). As ε values of zero error are surrounded by the coefficient values that result in small error (equal 1.3%), it can be stated that membership functions must have rather steep slopes to minimize error. Additionally, the zero error occurs for majority of threshold values: it is enough to assume η_{BPA}, $\eta_T \geq 0.2$. Still, extremely steep slopes are not recommended as for the $\varepsilon = 2.0$ again the error reaches 2.67%. At that stage almost all slopes are already the most steep. Zero error also appeared for

$\varepsilon = 1.9$, yet it was probably caused by an accidental simultaneous influence of unchanged (because of inconsistent values) and very steep membership functions. Anyway, this was a local minimum and it could not be used to formulate indications for the optimal membership function shape.

The proposed method of diagnosis support using DST with fuzzy focal elements is effective, in comparison with classical ISODATA and fuzzy ISODATA algorithms that result in global classification errors of 30% [3] for the Internet data. As the example shows, particular shape of membership function could significantly improve efficiency of the method.

5 Conclusions

The proposed method of diagnosis support using DST with fuzzy focal elements is a new application for fuzzy sets. It can be errorless, providing right threshold choice and membership function tuning. Hence, it is much better than classical ISODATA and fuzzy ISODATA algorithms. Still, both thresholds assumed while basic probability assignment calculation and diagnosis inference, as well as membership function shapes, influence the method's performance. The thresholds should be greater or equal 0.2, which ensures rejection of the most dubious symptoms. They also should not be higher than 0.9, as not only the most sure, but also some uncertain symptoms must take part in diagnosis inferring. Generally, the threshold for the basic probability assignment should be higher than that for diagnosis reasoning. It is intuitively understandable, as the quality of knowledge about a domain should be better than precision of symptoms. The latter can be unclear but must be analyzed, as they are the only source of information for a diagnostician. Fuzzy observations can be included in diagnostic reasoning trough their matching with focal elements using the thresholds. This makes the diagnosis support flexible.

The membership functions could be trapezoidal, with range of maximal values that covers an interval wider than a distance between empirical quartiles of learning data distribution. The example shows that the global error may be reduced by membership function change. It is not possible to ensure that the error will be equal to zero in every application, still it is possible to indicate the way of membership function modeling. We can start from quartiles of empirical distribution and wider the maximal value interval until minimum of error for learning data is reached.

The proposed method makes it possible to reason about a diagnosis using exclusively learning data. Still, it is desirable to work with an expert during diagnostic rules formulation and membership function building. Particularly norms for medical tests are valuable information for membership function defining.

Acknowledgements

This work has been partly supported by the Polish Ministry of Scientific Research and Information Technology grant nr 4 T11E 037 25

References

1. Gordon J, Shortliffe EH (1984) The Dempster-Shafer Theory of Evidence,. In: Rule-BasedExpertSystems Buchanan BG, Shortliffe EH (eds), Addison Wesley, 272-292
2. Kacprzyk J, Fedrizzi M (eds) (1994) Advances in Dempster-Shafer Theory of Evidence.
3. Straszecka E (2000) An interpretation of focal elements as fuzzy sets, Int J Inteligent Systems 18: 821-835

User Model for Conceptual and Personalized Search

D. Thenmozhi, G. Annapoorani, K. Baskaran, Senthil Kumar

Department of Computer Science and Engineering,
SSN College of Engineering, (*)Research Scholar(Anna University)
Affiliated to Anna University,
SSN Nagar-603110, Tamil Nadu, India
theni_d@yahoo.com

Summary. Search Engines of today have evolved from generic ones that find matches based on keywords, to those that provide personalized search based on concepts specified explicitly by the users. In this paper we present an approach to build dynamic user model for personalized search based on the user's browsing history using Open Directory Project Concept hierarchy that not only learns user's interest implicitly, but also tracks the temporal evolution and digression of their interests and modifies their profile accordingly.

1 Introduction

The huge amount of data on the Web means that there is information on almost any topic available online. Search engines are a very popular way to locate information, but the information provided is too general to efficiently solve individual user's information needs. Although users differ in their specific interests, they tend to provide very short queries that do not adequately describe the specific information they seek. Thus, the same results are shown to all users who may be looking for entirely different types of information. A typical search engine matches the words in the user's query with the words in its document database, and returns the documents that contain those words. However, a simple word can have many meanings depending on the context and the typical word matching mechanisms are unable to distinguish between meanings, causing the search engines to retrieve documents that are not relevant to the user. For example, suppose a person is looking for information about wild animals and queries a search engine with the keyword "wildcat". This query will retrieve information about the animal, but it will also bring information about universities and sports teams. The user could refine the query to get more accurate results, but this is a time-consuming task. Many users simply get discouraged using the search service after trying a few queries and getting a huge amount of unwanted information. Devising a way to constrain the search results to the user's current interests may allow the search engine to retrieve a higher proportion of relevant documents without requiring the user to refine the query. One way to do this is to summarize information about the user's interests in a user model. This information may be used later to re-rank search engine results in such a way that the pages that promise to be more interesting to the user appear first. Alternatively, the model can be used to filter the results so that the uninteresting pages will not be shown at all. Although many studies have been conducted on incorporating user profile technology into web search, little has been done about studying the way profiles evolve over an extended period of time as the user interests change. For example, suppose a musician uses the Web for his/her daily work, and one day he/she decides to go on vacation and search for information about travel packages on the Internet. Being able to track user interests with a high degree of accuracy may increase the precision of retrieved results in a typical web search and the user's satisfaction with the system. In this paper an approach to personalizing web search engines using a user model is presented. The user model is created based on the user's search history through proxy servers by implicitly learning the concept of visited documents using Open Directory Project(ODP) concept hierarchy. In contrast to long-term user interest, the short-term interest what the user is working on at the time they conduct a search is also tackled.

2 Related Works

2.1 Ontologies and Semantic Web

"An ontology is a specification of a conceptualization". Ontologies can be defined in different ways but they all represent taxonomy of concepts along with the relations between them. OntoSeek [7] is an example of system based on ontologies. Utilizing information sources such as product catalogs and yellow pages it applies conceptual graphs to represent both queries and resources. However, concept hierarchies can also be used as simple ontologies. The OBIWAN project [9] uses the ODP's [12] concept hierarchy as ontology. This ontology has been used to represent the content of Web sites and as the basis of user profiles [6] for personalized search and browsing. "Semantic Web" describes the extension of the Web to deal with the meaning of available content rather than just its syntactic form. One way to provide conceptual search is to explicitly state the meaning of the content in a Web page. Research in this area tries to address the problem by having the creators of the content explicitly specify the meaning associated with a page using a knowledge representation language. One such knowledge representation languages is Ontobroker [13]. Our approach shares many of the same goals as the Semantic Web, however we use ODP concept hierarchy as ontology as the basis for User Model creation for personalized search.

2.2 Personalization

Web personalization is the process of selecting, preparing and delivering Web contents for a given user, by taking into account his specific needs and preferences. Personalization means delivering to the user the most relevant contents, in the most adequate way, and at the most appropriate time. User browsing histories are the most frequently used source of information about user interests. Trajkova and Gauch [15] use this information to create user profiles represented as weighted concept hierarchies. The user profiles are created by classifying the collected Web pages with respect to a reference ontology. In the OBIWAN project [6], search results from a conventional search engine are classified with respect to a reference ontology based upon the snippets summarizing the retrieved documents. Documents are re-ranked based upon how well their concepts match those that appear highly weighted in the user profile. Spretta [14] explored the use of a less-invasive means of gathering user information for personalized search. The user profile was build based on activity at the search site itself and was used to provide personalized search. The user profiles were created by classifying the information into concepts from the ODProject concept hierarchy and then were used to re-rank the search results. This system just reorganizes the existing search results but not providing the relevant result to the user. The system KeyConcept [2] indexed documents with their automatically identified concepts from the ODP in addition to the keywords. The concepts for each query were provided manually or by running a text related to the query through the classifier. This system performs conceptual retrieval with explicitly specifying concepts. Our approach builds a user model for both conceptual and personalized search by implicitly learning the concepts based on Open Directory Project concept hierarchy that overcomes the above issue for user interested web pages.

2.3 Text Classification

Text classification organizes information by associating a document with the best possible concept(s) from a predefined set of concepts. Several methods for text classification have been developed based on different models for comparing the new documents to the reference set. These including comparison between vector representations of the documents (Support Vector Machines, k-Nearest Neighbor, Linear Least-Squares Fit), use of the joint probabilities of words being in the same document (Naïve Bayesian), decision trees, and neural networks. Some of these approaches include implementing unsupervised learning algorithms like Latent Semantic Analysis and using AI rule-base trees to compute the conceptual relevancy of search results. Approaches to hierarchical text classification include the use of enhanced clustering techniques and the use of Support Vector Machines. A thorough survey and comparison of such methods is presented in [5]. In particular, [16] examines the complexities involved in different methods of text categorization and especially in hierarchical categorization. In the OBIWAN project [6], user profiles are represented by weighted concept hierarchies where the weight of a concept represents the user's interest in that topic. The concept weights are determined implicitly by classifying web pages browsed by the user. Queries are submitted to a traditional search engine and the results are classified into the ontology concepts based on their summaries. The documents in the result set are then re-ranked based on matches between the summary concepts and the highly weighted concepts in the user's profile. This system takes this approach and classifies the visited pages from the web log using vector space model (TF-IDF) and concept weights are found implicitly. Highly weighted concepts are stored in the user model.

2.4 Construction of User Model

Widyantoro, Ioerger and Yen [4] have developed a three-descriptor representation to monitor user interest dynamics. This model maintains a long-term interest descriptor to capture user's general interests and a short-term interest descriptor to keep track of user's more recent faster changing interests. Goecks and Shavlik [10] learn user's interests by looking at more than just the pages themselves. They also observe and measure user mouse and scrolling activity in addition to user browsing activity. V.Challam [1] developed an approach for building a user profile as ontology-based contextual profiles that captures what the user is working on at the time they conduct a search. The profiles were used to personalize the search results to suit the information needs of the user at a particular instance of time. Jason [8] developed ChatProfile, a system that uses text classification based on the vector space model to create a profile of chat data, in essence creating a summary of the chat. In our system the user model is constructed based on the browsing history of the user. The concepts are stored in the user model. The user model is updated dynamically by the concepts of current browsing context.

2.5 Contextual Search

Rather than building long-term user profiles, contextual systems try to adapt to the user's current task. Watson [11] monitors users' tasks, anticipates task-based information needs, and proactively provide users with relevant information. The user's tasks are monitored by capturing content from Internet Explorer and Microsoft Word applications. Stuff I've Seen [3], developed at Microsoft Research indexes the content seen by a user and uses the index to provide easier access to information already seen by the user and also to provide rich contextual information for Web searches. V. Challam[1] developed a system for contextual search using ontology-based user profiles by capturing what the user is working on at the time they conduct a search. The profiles were used to personalize the search results to suit the information needs of the user at a particular instance of time. Contextual information was extracted from open word documents and web pages. Our system gathers the contextual information from the current session's browsing history. Information is classified to concepts and maintained continuously in the user model.

3 Approach

The main components of our system are as follows: 1. A Proxy Server that captures the visited web pages of the users. 2. A Classifier that classifies the web document into a concept based on the Open Directory Project concept hierarchy. 3. User Model that maintains the concept in which the user is interested along with his/her identification. 4. A system that tracks the current browsing information of the users. The System Architecture consists of the following modules and is shown in Fig.1.

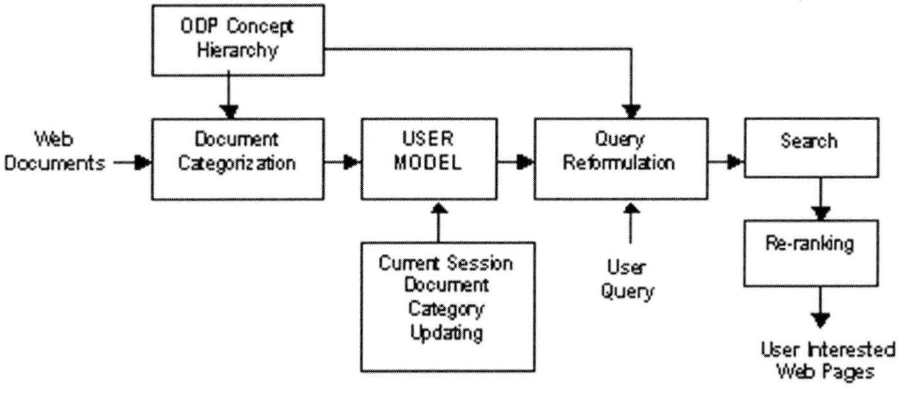

Fig. 1. System Architecture

3.1 Document Categorization

The browsing activity of N number of users is captured through the proxy server for a period of time. The access log information from the web log is filtered to extract the visited URL's of the user. The content of each URL is retrieved and the document is classified to a concept. Before classifying a document to a category, it should be preprocessed to get the most important words of a document. This preprocessing is done with the following steps. * Lexical analysis of the text:: Tokenization is the process taking place during the lexical analysis phase of the text. It is the extraction of plainwords and terms from a document, stripping out administrative metadata and structural or formatting elements, e.g. removing HTML tags from the HTML source for a Web page. This operation needs to be performed prior to indexing or before categorizing documents to a concept. * Stop word removal:: Filtering out the useless words such as prepositions, etc, which does not give any special meaning to the document are eliminated in the document. * Stemming:: Stemming algorithms are used in natural language systems (e.g., IR systems) for stripping the endings from words, so that related words such as "oceaneering", "oceanic", "oceanics", "oceanization", "oceans" can be matched to their stem "ocean". There are three main different types of Stemmer: Lovins, Porter and Paice/Husk. Each of these three stemmers have different properties and stemming rules, which result in varying output and their stemming performance can vary depending on the given input. Porter's stemming algorithm yields better result when it is implemented in Java. This paper proposes to implement the Porter's algorithm for stemming. * Index term selection:: Not all words are equally significant for representing the semantics of a document and most of the semantics is carried by the noun words. Hence index terms are selected from the nouns of the document. During this phase the document is being categorized to concepts based on the vector-space model(TF-IDF) which is found as follows. Term Frequency(TF) = (tf1, tf2,......tfn) where tfi if the frequency of ith term in the document Inverse Document Frequency = log(N / dfi) where N is the total number of documents in the collection and dfi is the number of documents that contain the ith term TF-IDF representation = (tf1 log(N/df1), tf2 log(N/df2), ... tfn log(N/dfn)) With this model the term frequency for each term selected as index term is determined. The term with high term frequency is taken as a concept and is generalized to a concept based on the Open Directory Project concept hierarchy. This phase includes the ranking of user's category. The category for each of the document visited by the user is found. The weight for each category is found. The category with a weight higher than a threshold value will be ranked higher.

3.2 User Model Construction

The category with higher rank is stored in a User Model along with the user identification. When the user gives query for searching a page the current session concepts are identified and updated dynamically in the user model.

3.3 Current Session Tracking

When the user gives query for searching web pages, the browsed pages of the current session is tracked and the pages has to categorized as per the document categorization procedure and are maintained as current session categories and updated periodically in the user model

3.4 Query Reformulation

Query given by the user is reformulated as following * The query concept is identified based on the ODP concept hierarchy. * The user interest is identified either from the current session category or from the user model. * Transform the query by adding the concept along with query to make the query more specific.

3.5 Search

The reformulated query is given to the search engine to retrieve the web pages. But the search engines retrieve pages based on their own ranking methodology. Hence the search results are re-ranked using the user model.

3.6 Search Result Re-ranking

This part is used to improve the performance of the search result and hence the retrieved pages are most relevant to the user query. There are several methods to perform re-ranking. The following techniques can be used for the search results from a single search engine. * Anchor text matching * Click rate analysis * Keyword matching * Link analysis Since this system searches the information from the single search engine say Google, the Keyword matching techniques is used. The final rank for a document is calculated by based on the concepts of user model as follows. Final Rank = a * Conceptual Rank + (1-a) * Keyword Rank a Has the value between 0 and 1 a - 0 : Keyword based ranking a - 1 : Conceptual ranking

4 Conclusion

Section 2 reviews the different approaches that have been adopted so far for the conceptualization, personalization, classification, user model construction and contextual search for personalized search engines and outlines our approach to implement each of these processes. Detailed description of the main components of our system is given in section 3 for building a user model that implicitly learns the user's interest for personalizing search based on the users browsing history and the model is used to retrieve most relevant web pages for the given query. The approach helps to improve the performance of the search result by re-ranking the search result using the user model.

References

1. Challam, V.: Contextual information retrieval using ontology based user profiles. University of Kansas (2004)

2. Devanand, R.: KeyConcept : Exploiting hierarchical relationships for conceptually indexed data. Master's thesis. Department of Electrical Engineering and Computer Science, University of Kansas, Kansas, USA (2004)
3. Dumais, S.T., Cutrell, E., Cadiz, J.J., Jancke, G., Sarin, R., Robbins, D.C.: A system for personal information retrieval and re-use. Proceedings of SIGIR (2003)
4. Widyantoro, D.H., Ioerger, T.R., Yen, J.: Learning User Interest Dynamics with a Three-Descriptor Representation. Journal of the American Society for Information Science, 52(3) (2000) 212- 225
5. Sebastiani, F.: Machine Learning in Automated Text Categorization. In ACM Computing Surveys 34(1) (2002) pp. 1-47
6. Gauch, S., Chafee, J., Pretschner, A.: Ontology Based Personalized Search and Browsing, Web Intelligence and Agent Systems. Vol. 1 No. 3-4, April (2004) pp.219-234.
7. Guarino, N., Masolo, C., Vetere, G.: OntoSeek: Content Based Access to the web. IEEE Intelligent System. Volume 14, no. 3, (1999) pp. 70 - 80.
8. Bengel, J., Gauch, S., Mittur, E., Vijayaraghavan, R. : ChatTrack: Chat Room Topic Detection Using Classification. University of Kansas (2004)
9. Chaffee, J., Gauch, S.: Personal Ontologies for Web Navigation. In proceedings of the 9th International Conference on Information and Knowledge Management (CIKM), 2000, pp 227-234.
10. Shavlik, J., Calcari, S., Eliassi-Rad, T., Solock, J.: An Instructable, Adaptive Interface for Discovering and Monitoring Information on the World Wide Web. Proceedings of the International Conference on Intelligent User Interfaces (1999) pp. 157 - 160.
11. Leake, D., Scherle, R., Budzik, J., Hammond, K. Selecting Task Relevant Sources for Just-in-Time Retrieval. In Proceedings of the AAAI-99 Workshop on Intelligent Information Systems. AAAI Press, Menlo Park, CA (1999)
12. The Open Directory Project (ODP). http://dmoz.org.
13. Decker, S., Erdmann, M., Fensel, D., Studer, R.: Ontobroker: Ontology based Access to Distributed and Semi Structured Information. Proceedings of W3C Query Language Workshop QL'98. (1998)
14. Speretta, M.: Personalizing Search Based on User Search Histories. Master's thesis, The University of Kansas, Lawrence, KS (2004).
15. Trajkova, J., Gauch, S.: Improving Ontology Based User Profiles. RIAO, Vaucluse, France, April 26-28 (2004) pp. 380-389.
16. Yang, Y., Zhang, J., Kisiel, B.: A scalability analysis of classifiers in text categorization . In Proceedings of 26th Annual International ACM SIGIR Conference,July to Auguest (2003), pp 96-103.

On the Relationship Between Active Contours and Contextual Classification

Arkadiusz Tomczyk[1] and Piotr S. Szczepaniak[1,2]

[1] Institute of Computer Science, Technical University of Lodz Wolczanska 215, 93-005, Lodz, Poland tomczyk@ics.p.lodz.pl

[2] Systems Research Institute, Polish Academy of Sciences Newelska 6, 01-447 Warsaw, Poland

Summary. To discuss the relationship between *active contours* and *contextual classification*, a formal definition of the *contour* as well as a uniform approach to the all *active contour* methods are proposed first, and then a *contextual classification* problem is introduced and formalized. The equivalence relationship between *contours* and *classifiers*, thoroughly considered and illustrated by examples, proves to allow incorporation of the methods and techniques specific for the *active contour* approach to the *contextual classification* and vice versa.

1 Introduction

Being the first step of image understanding, the image analysis usually aims at the precise identification of the image content. This identification implicates two tasks: *classification* of the image elements (usually pixels) and localization of the object *contours*. Because of differences in the two approaches, they were originally developed separately. That is why they differ e.g. in the utilization of *a priori* knowledge about the problem domain (this knowledge is necessary for correct object identification and consequently for proper image understanding). This article reveals a relationship between the considered approaches leading in consequence to an easy transfer of concepts between the two techniques.

The article is organized as follows: section 2 describes the idea of *active contours* and introduces a formal *contour* definition, section 3 focuses on the problem of *classification* with and without context, *feature spaces* and class boundaries, section 4 presents two examples that show the relationship between *contours* and *classifiers*, while section 5 formalizes this relationship. Finally, the last section presents the main ideas for the future research directions.

2 Active Contours

2.1 Contours

The term *contour* coming from the image analysis, is used to define an outline (boundary) of objects and thus divides an image (in this article an image is treated as a function I with domain in $\mathbf{R}^2$) into two parts: object and its background (of course *contour* uniquely defines an object and vice versa). There are many ways to describe closed *contours* in an image. It can be defined as a parametric curve ([1, 14]), as a polygon, as a spline, it can be encoded by means of Freeman codes ([5]) etc. However, all these descriptions have one common drawback. Using them it is hard to descibe objects of different topologies (e.g. having holes). This is why another description (geodesic *contour*) overcoming that problem (Fig. 1) was proposed in [2]. The definition below formalizes this concept:

Definition 1. *Let ρ denote any metric (e.g. Euclidean metric) in $\mathbf{R}^2$ and let $K(\mathbf{x_0}, \varepsilon) = \{\mathbf{x} \in \mathbf{R}^2 : \rho(\mathbf{x_0}, \mathbf{x}) < \varepsilon\}$ denote the sphere with centre $\mathbf{x_0} \in \mathbf{R}^2$ and radius $\varepsilon > 0$. The set $c \subseteq \mathbf{R}^2$ is called a contour iff there exists a function $f : \mathbf{R}^2 \to \mathbf{R}$ such that:*

$$c = \{\mathbf{x} \in \mathbf{R}^2 : \forall_{\varepsilon>0}\ \exists_{\mathbf{x_1},\mathbf{x_2} \in K(\mathbf{x},\varepsilon)}\ f(\mathbf{x_1}) \geq 0 \wedge f(\mathbf{x_2}) < 0\}$$

Using this description, an object can be defined as $\{\mathbf{x} \in \mathbf{R}^2 : f(\mathbf{x}) \geq 0\}$ and its background as $\{\mathbf{x} \in \mathbf{R}^2 : f(\mathbf{x}) < 0\}$. The symbol $\mathcal{C}$ will be used to denote the space of all *contours*.

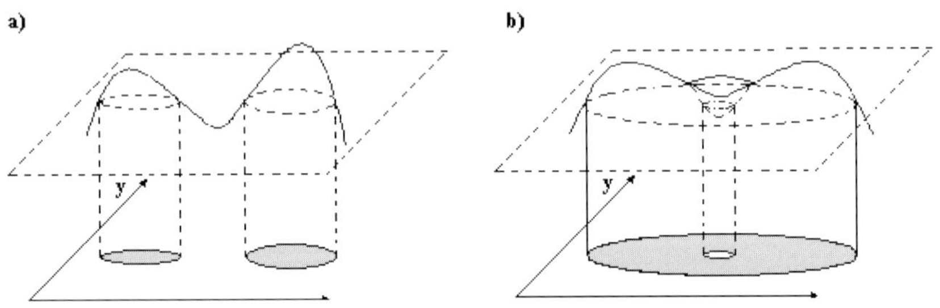

Fig. 1. Sample *contours* and objects with different topologies that can be described using proposed definition.

2.2 Active contours

The aim of *active contour* methods is to find an object in a given image. Obviously, there can exist many *contours* in $\mathcal{C}$ and only one of them represents the boundary of the searched object. In the *active contour* approach this search is formulated as an optimization problem of function $E : \mathcal{C} \rightarrow \mathbf{R}$ called *energy*. Function E is used for evaluation of the *contour* and should assign the optimal value to the *contour* that actually describes a boundary of the searched object. When evaluating *contours* this function can take under consideration the *contour* itself (*internal energy*), the image (*external energy*) or both ([1, 2, 3, 4, 5, 9, 14, 15]).

3 Classification

3.1 Classification

In all *classification* problems, which are the *pattern recognition* problems, there exists a set of objects $\mathcal{O}$ that are to be classified i.e. assigned a proper label from the finite set of labels $\mathcal{L}$ (where e.g. $\mathcal{L} = \{1, \ldots, L\}$ and L is a number of classes). Such an assignment can formally be described as a classification function (*classifier*) $k : \mathcal{O} \rightarrow \mathcal{L}$ (each object $o \in \mathcal{O}$ receives a unique label $l \in \mathcal{L}$). Because there are many functions $k \in \mathcal{K}$ ($\mathcal{K}$ - set of all possible *classifiers* in a given problem) that map $\mathcal{O}$ into $\mathcal{L}$, there must be any *a priori* knowledge that enables a choice of the best *classifier* (e.g. a *training set* with correctly labeled objects). This knowledge can always be formulated in the form of *performance index* $Q : \mathcal{K} \rightarrow \mathbf{R}$ capable of evaluation of the usefullness of each function k. Thus the problem of classification can be expressed as optimization of *performance index* Q ([6, 7, 10, 13]).

The same problem can also be described in a different though equivalent way ([6, 11]). Instead of looking for an optimal function k, the optimal set of L *decision functions* $d_l : \mathcal{O} \rightarrow \mathbf{R}$ is sought ($d_l \in \mathcal{D}$). Those functions should have the property that for a given object $o \in \mathcal{O}$ function d_l possesses the largest value iff $k(o) = l$. Thus the values of the functions can identify uniquely the class of the object.

3.2 Feature space

In the real world no *classifier* is able to classify objects directly. It can operate only on the data extracted from an object (i.e. sensed, measured etc.). Those data are usually coded as real numbers and called *features* of the object. The extraction process can be described as function $e : \mathcal{O} \rightarrow \mathcal{X}$ that assigns *features* to an object. The *features* compose *feature vector* $\mathbf{x} \in \mathcal{X} \subseteq \mathbf{R}^n$ (n is a number of extracted *features*) and so called *feature space* $\mathcal{X}$. With no loss of generality it can be assumed that $\mathcal{X} = \mathbf{R}^n$ because the *classifier* which is optimal for a larger space will also be optimal in the subspace.

In order not to complicate the notation in the further part of this article the objects and their *feature vectors* will be used interchangeably when the *feature* extraction method e is either precisely known or insignificant.

3.3 Boundaries

Classifier $k : \mathcal{X} \to \mathcal{L}$ divides *feature space* $\mathcal{X}$ (along with set of objects $\mathcal{O}$) into L subsets. This division unambigiously determines boundaries between classes in the *feature space.*

Definition 2. *Let ρ denote any metric (e.g. Euclidean metric) in $\mathcal{X} = \mathbf{R}^n$ and let $K(\mathbf{x_0}, \varepsilon) = \{\mathbf{x} \in \mathcal{X} : \rho(\mathbf{x_0}, \mathbf{x}) < \varepsilon\}$ denote the sphere with centre $\mathbf{x_0} \in \mathcal{X}$ and radius $\varepsilon > 0$. The boundary between two classes $l_1 \in \mathcal{L}$ and $l_2 \in \mathcal{L}$ is a set of points*

$$b_{l_1 l_2} = \left\{\mathbf{x} \in \mathcal{X} : \forall_{\varepsilon>0}\ \exists_{\mathbf{x_1},\mathbf{x_2} \in K(\mathbf{x},\varepsilon)}\ k(\mathbf{x_1}) = l_1 \wedge k(\mathbf{x_2}) = l_2\right\}.$$

3.4 Contextual classification

Let $\mathcal{S}$ denote the whole set of objects that can be considered in the *classification* problem (this differs from the set $\mathcal{O}$ described above because not necessarily all objects are supposed to be labeled).

Definition 3. *Let $s \in \mathcal{S}$. Then $\mathcal{U}_s$ is defined as a set of all subsets of $\mathcal{S}$ that does not contain s. In other words $\mathcal{U}_s = \left\{\mathcal{U} \in 2^{\mathcal{S}} : s \notin \mathcal{U}\right\}$.*

In *contextual classification*, the decision on the label that should be assigned to object $s \in \mathcal{S}$ must take into account not only the given object but also some other objects from space $\mathcal{S}$ that compose a context of s, and therefore the *classification* problem has to be formulated differently. Using the formalism of section 3.1 the objects under classification are now defined as follows: $o = \langle s, \mathcal{U} \rangle \in \mathcal{S} \times 2^{\mathcal{S}}$ where $\mathcal{U} \in \mathcal{U}_s$.

In *contextual classification* viewed as presented above, one fact needs to be strongly emphasized. For classification purposes, the context $\mathcal{U} \in \mathcal{U}_s$, apart from its assumed sufficiency for correct classification, must be precisely defined for each object.

3.5 Classification and contextual classification

As stated in the previous section, *contextual classification* can be treated as a usual *classification.* That means that all considerations and techniques described so far can be applied to *contextual classification* in the unchanged form. Thus for example, with object $o = \langle s, \mathcal{U} \rangle$ one can associate a *feature vector.* Though usually different from the *feature vectors* extracted for s and all the objects from $\mathcal{U}$, this vector can be calculated basing only on the *features* of context components since no other information is available. Similarily, the *feature space* can be constructed on those *features* and boundaries of the classes can be found.

It is worth mentioning that this formulation of *contextual classification* can be viewed as a generalization of the *classification* problem presented in section 3.1 with $\mathcal{O} = \{\langle s, \emptyset \rangle : s \in \mathcal{S}\}$.

4 Examples

4.1 Example without context

Let us consider the image presented in Fig. 2a and let W and H denote its width and height respectively. Pixels are the only elements $o \in \mathcal{O}$ which need to be classified. Because this is a grey scale image (with 256 grey levels) the *features* that can be extracted are pixel coordinates (x, y) and intensity $I(x, y)$. The *features* which will be actually considered in this example are y and $I(x, y)$ i.e. $e(o) = \mathbf{x} = \langle x, y, I(x, y)\rangle$ and $\mathbf{x} \in \mathcal{X} = \mathbf{R}^2$. A sample *classifier* can be defined as:

$$k(\mathbf{x}) = \begin{cases} 1 \text{ if } y \geq H/2 \text{ and } I(x,y) \geq 128 \\ 2 \text{ if } y < H/2 \text{ or } I(x,y) < 128 \end{cases}$$

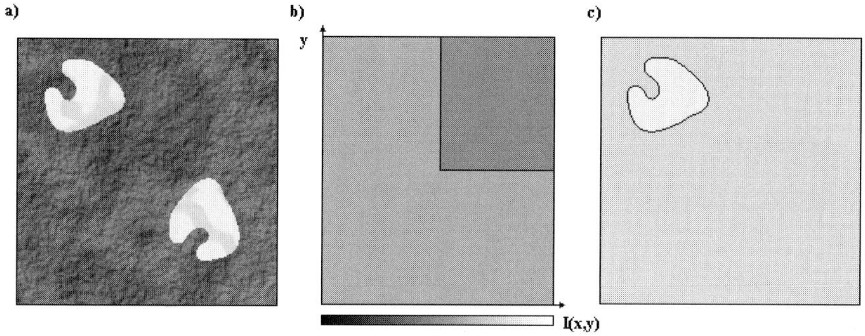

Fig. 2. An example of *classification*: (a) the analysed image, (b) *feature space* and class boundaries, (c) the result of *classification* with *contour* of the object.

In this case each object (pixel) $o \in \mathcal{O}$ can be classified separately basing on the *features* which can be extraced from it. Thus it does not matter in which context it appears. Fig. 2b presents the *feature space* $\mathcal{X}$ in the considered problem and the boundary b_{12} in this space. In Fig. 2c the classification results are presented. It is worth mentioning that the boundary between classes created in the space of pixel coordinates actually represents a *contour*.

The same problem can be addressed differently. Object searching can be expressed in the formalism of *active contours*. The *energy* of the *contour* c in this case can be defined as follows: $E(c) = \int_I g(x, y)\, dxdy$ where $g : \mathbf{R}^2 \to \mathbf{R}$:

$$g(x,y) = \begin{cases} 1 & \text{if } ((y < H/2 \text{ or } I(x,y) < 128) \text{ and } f(x,y) \geq 0) \\ & \text{or } ((y \geq H/2 \text{ and } I(x,y) \geq 128) \text{ and } f(x,y) < 0) \\ 0 & \text{if } ((y \geq H/2 \text{ and } I(x,y) \geq 128) \text{ and } f(x,y) \geq 0) \\ & \text{or } ((y < H/2 \text{ or } I(x,y) < 128) \text{ and } f(x,y) < 0) \end{cases}$$

4.2 Example with context

Let us now consider the image presented in Fig. 3a. In this example of the *contextual classification* pixels compose the whole space $\mathcal{S}$ of objects. The context for each pixel s_1 with coordinates (x, y) is pixel s_2 with coordinates $(x, y-1)$. Thus the objects that need to be classified can be described as $o = \langle s_1, \{s_2\} \rangle$. Let $e(o) = \mathbf{x} = \langle I(x, y), I(x, y-1) \rangle$ and $\mathbf{x} \in \mathcal{X} = \mathbf{R}^2$. A *classifier* can be defined as:

$$k(\mathbf{x}) = \begin{cases} 1 \text{ if } |I(x,y) - I(x,y-1)| \geq 128 \\ 2 \text{ if } |I(x,y) - I(x,y-1)| < 128 \end{cases}$$

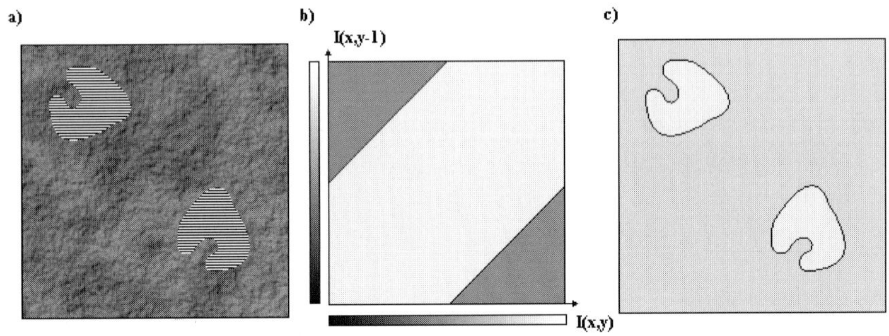

Fig. 3. An example of *contextual classification*: (a) the analysed image, (b) *feature space* and class boundaries, (c) the result of *classification* with *contour* of the object.

Like in section 4.1 Fig. 3b presents the *feature space* and class boundaries and Fig. 3b is the result of *classification* with *contour* of the object. Again, the problem can be expressed as minimization of the *energy* E where $E(c) = \int_I g(x, y)\, dxdy$ and:

$$g(x,y) = \begin{cases} 1 & \text{if } ((|I(x,y) - I(x,y-1)| < 128) \text{ and } f(x,y) \geq 0) \\ & \text{or } ((|I(x,y) - I(x,y-1)| \geq 128) \text{ and } f(x,y) < 0) \\ 0 & \text{if } ((|I(x,y) - I(x,y-1)| \geq 128) \text{ and } f(x,y) \geq 0) \\ & \text{or } ((|I(x,y) - I(x,y-1)| < 128) \text{ and } f(x,y) < 0) \end{cases}$$

5 Relationship between active contours and contextual classification

5.1 Image analysis

In the image analysis, the labels assigned to pixels provide information on what kind of objects are represented by particular pixels. The *features* that can be extracted from the labels are pixel coordinates *(x,y)* and information about the pixel's color (for grey scale images it is intensity, for color images it can be R, G, B components etc.). What is crucial is the fact that coordinates are the *features* that describe the pixels uniquely i.e. there are no two pixels having the same coordinates (other *features* not necessarily must possess the uniqueness property).

5.2 From classifiers to contours

Let us consider *classifier* k with $L = 2$ which assigns labels to pixels. It does not matter whether it is a contextual classification or not (in other words, objects of what kind are considered) and what kind of *features* (what extraction function e) is used beacuse pixels are always uniquely labeled. Consequently, the set of objects is divided into two subsets (with labels 1 and 2 respectively) and moreover, as each pixel has unique coordinates, *feature space* $\mathcal{X}$ constructed on these coordinates is also divided. The boundaries created by this division are *contours* in the image plane. Formally this *contour* can be defined as $c = b_{12}$ in space $\mathcal{X}$.

5.3 From contours to classifiers

Similarily to the reasoning presented in section 5.2 let us consider *contour* c associated with function f. Such a *contour* divides the image into two parts that can be labeled by 1 and 2. It is worth mentioning that pixels of the same color not necessarily must be assigned the same label, which means that the context can be significant. Like before, because the coordinates characterize the pixels uniquely such a *contour* is equivalent to a *classifier* which formally can be defined e.g. by means of *decision functions* $d_1 = f$ and $d_2 = -f$.

5.4 Other relationships

In *active contour* techniques, the final *contour* can be searched in an iterative process of optimization (e.g. simulated annealing) of the *energy* function E. Similarily, the search for the optimal *classifier* is sometimes an iterative process (e.g. the neural network training) for the specified *performance index* Q. Both approaches are equivalent in view of the relationships presented in sections 5.2 and 5.3.

The relationship presented above can also be generalized on any arbitrary chosen number of classes L. In such a situation, a separate *contour* must be defined for each class $c_1, \ldots, c_L$. The relationship can then be described as $c_i = \bigcup_{j=1}^{L} b_{ij}$ and $d_i = f_i$.

6 Conclusions

The presented relationship between *active contours* and *contextual classification* reveals that the two techniques can be considered equivalent in the image analysis. Both *classifiers* and *contours* are capable of the unique identification of an object. So far these methods have been considered and developed separately. The relationship described in the paper allows the transfer of concepts between them (e.g. the idea of *internal energy* can be used in *classification*). Moreover, the presented theoretical results can be easily generalized e.g. the same idea can be applied to 3-D recognition problems where *contours* represent surfaces (*active surfaces*). All these aspects are presently under further investigation.

References

1. Kass M., Witkin W., Terzopoulos D., (1988) *Snakes: Active Contour Models*, International Journal of Computer Vision, 321–331
2. Caselles V., Kimmel R., Sapiro G., (1997) *Geodesic Active Contours*, International Journal of Computer Vision 22(1) 61–79
3. Xu Ch., Yezzi A., Prince J., (2000) *On the Relationship between Parametric and Geometric Active Contours*, in Proc. of 34th Asilomar Conference on Signals, Systems and Computers 483–489
4. Cootes T., Taylor C., Cooper D., Graham J., (1994) *Active Shape Model - Their Training and Application*, CVGIP Image Understanding, 61(1) 38–59 Janvier
5. Grzeszczuk R., Levin D., (1997) *Brownian Strings: Segmenting Images with Stochastically Deformable Models*, IEEE Transactions on Pattern Analysis and Machine Intelligence vol. 19 no. 10 1100-1013
6. Tadeusiewicz R., Flasinski M., (1991) *Pattern Recognition*, PWN, Warsaw (in Polish)
7. Kwiatkowski W., (2001) *Methods of Automatic Pattern Recognition*, WAT, Warsaw (in Polish)
8. Sobczak W., Malina W., (1985) *Methods of Information Selection and Reduction*, WNT, Warsaw (in Polish)
9. Nikolaidis N., Pitas I., (2001) *3-D Image Processing Algorithms*, John Wiley and Sons Inc., New York
10. Bishop Ch., (1993) *Neural Networks for Pattern Recognition*, Clarendon Press, Oxford
11. Pal S., Mitra S., (1999) *Neuro-fuzzy Pattern Pecognition, Methods in Soft Computing*, John Wiley and Sons Inc., New York
12. Bennamoun M., Mamic G., (2002) *Object Recognition, Fundamental and Case Studies*, Springer-Verlag, London
13. Looney C., (1997) *Pattern Recognition Using Neural Networks*, Theory and Algorithms for Engineers and Scientists, Oxford University Press, New York
14. Sonka M., Hlavec V., Boyle R., (1994) *Image Processing, Analysis and Machine Vision*, Chapman and Hall, Cambridge
15. Gonzalez R., Woods R., (2002) *Digital Image Processing*, Prentice-Hall Inc., New Jersey

An Improvement on LDA Algorithm for Face Recognition

Vo Dinh Minh Nhat[1] and Sungyoung Lee[2]

[1] Kyung Hee University, South of Korea vdmnhat@oslab.khu.ac.kr
[2] Kyung Hee University, South of Korea sylee@oslab.khu.ac.kr

Summary. Linear discrimination analysis (LDA) technique is an important and well-developed area of image recognition and to date many linear discrimination methods have been put forward. Despite these efforts, there persist in the traditional LDA some weaknesses. In this paper, we propose a new LDA-based method that can overcome the drawback existed in the traditional LDA method. It redefines the between-class scatter by adding a weight function according to the between-class distance, which helps to separate the classes as much as possible. At the same time, in this method, we firstly remove the null space of total scatter matrix which has been proved to be the common null space of both between-class and within-class scatter matrix, and useless for discrimination. Then in the lower-dimensional projected space, the null space of the resulting within-class scatter matrix is calculated. This lower-dimensional null space, combined with the previous projection, represents a subspace of the whole null space of within-class scatter matrix, and is really useful for discrimination. The optimal discriminant vectors of LDA are derived from it. Experiment results show our method achieves better performance in comparison with the traditional LDA methods.

1 Introduction

PRINCIPAL component analysis (PCA), also known as Karhunen-Loeve expansion, is a classical feature extraction and data representation technique widely used in the areas of pattern recognition and computer vision. Within this context, Turk and Pentland [1] presented the well-known Eigenfaces method for face recognition in 1991. Since then, PCA has been widely investigated and has become one of the most successful approaches in face recognition. However, PCA could not capture even the simplest invariance unless this information is explicitly provided in the training data. It also cannot make full use of pattern separability information like the Fisher criterion, and its recognition effect is not ideal when the size of the sample set is large.

The Fisherface method [4] combines PCA and the Fisher criterion [9] to extract the information that discriminates between the classes of a sample set. It is a most representative method of LDA. Nevertheless, Martinez *et al.* demonstrated that when the training data set is small, the Eigenface method outperforms the Fisherface method [7]. Should the latter be outperformed by the former? This provoked a variety of explanations. Liu *et al.* thought that it might have been because the Fisherface method uses all the principal components, but the components with the small eigenvalues correspond to high-frequency components and usually encode noise [11], leading to recognition results that are less than ideal. In line with this theory, they presented two enhanced Fisher linear discrimination (FLD) models (EFMs) [11] and an enhanced Fisher classifier [12] for face recognition. Their experiential explanation lacks sufficient theoretical demonstration, however, and EFM does not provide an automatic strategy for selecting the components. Chen *et al.* proved that the null space of the within-class scatter matrix contains the most discriminative information when a small sample size problem takes place [13]. Their method is also inadequate, however, as it does not use any of the information outside the null space. In [5], Yu *et al.* propose a direct LDA (DLDA) approach to solve this problem. It removes the null space of the between-class scatter matrix firstly by doing eigen-analysis. Then a simultaneous diagonalization procedure is used to seek the optimal discriminant vectors in the subspace of the between-class scatter matrix. However, in this method, removing the null space of the between-class scatter matrix by dimensionality reduction would indirectly lead to the losing of the null space of the within-class scatter matrix which contains considerable discriminative information. Rui Huang [10] proposed the method in which the null space of total scatter matrix which has been proved to be the common null space of both between-class and within-class scatter matrix, and useless for discrimination, is firstly removed. Then in the lower-dimensional projected space, the null space of the resulting within-class scatter matrix is calculated. This lower-dimensional null space, combined with the previous projection, represents a subspace of the whole null space of within-class scatter matrix, and is really useful for discrimination. The optimal discriminant vectors of LDA are derived from it.

In general, for a K-class classification problem where $K > 2$, traditional LDA methods are not optimal, because they overemphasize the larger distance between classes and cause large overlaps of neighbouring classes. So, Loog et al. [6] redefine the between-class scatter to solve this overemphasizing problem. And in this paper we improve the LDA algorithm in [10] by redefining the between-class scatter [6] by adding a weight function according to the between-class distance, which helps to separate the classes as much as possible. Our new method takes the advantages of both Rui Huang's method [10] for dealing with high dimensional data to avoid singularity and a redefinition of the between-class scatter matrix proposed by Loog et al. [6]. The remainder of this paper is organized as follows: In Section 2, the traditional LDA method is reviewed. The idea of the proposed method and its algorithm are described in Section 3. In Section 4, experimental results are presented for the ORL face image databases to demonstrate the effectiveness of our method. Finally, conclusions are presented in Section 5.

2 Linear disciminant analysis

Let us consider a set of N sample images taking values in an n-dimensional image space, and assume that each image belongs to one of c classes $\{X_1, X_2, ..., X_c\}$. Let N_i be the number of the samples in class $X_i (i = 1, 2, ..., c)$, $\mu_i = \frac{1}{N_i} \sum_{x \in X_i} x$ be the mean of the samples in class X_i, $\mu = \frac{1}{N} \sum_{i=1}^{N} x_i$ be the mean of all samples. Then the between-class scatter matrix S_b is defined as

$$S_b = \frac{1}{N} \sum_{i=1}^{c} N_i (\mu_i - \mu)(\mu_i - \mu)^T = \frac{1}{N} \Phi_b \Phi_b^T \tag{1}$$

and the within-class scatter matrix S_w is defined as

$$S_w = \frac{1}{N} \sum_{i=1}^{c} \sum_{x_k \in X_i} (x_k - \mu_i)(x_k - \mu_i)^T = \frac{1}{N} \Phi_w \Phi_w^T \tag{2}$$

Also, the total scatter matrix or mixture scatter matrix S_t is defined as

$$S_t = S_b + S_w = \frac{1}{N} \sum_{i=1}^{N} (x_i - \mu)(x_i - \mu)^T = \frac{1}{N} \Phi_t \Phi_t^T \tag{3}$$

which is also the covariance matrix of all the samples.

In LDA, the projection W_{opt} is chosen to maximize the ratio of the determinant of the between-class scatter matrix of the projected samples to the determinant of the within-class scatter matrix of the projected samples, i.e.,

$$W_{opt} = \arg\max_{W} \frac{|W^T S_b W|}{|W^T S_w W|} = [w_1 w_2 ... w_m] \tag{4}$$

where $\{w_i \,| i = 1, 2, ..., m\}$ is the set of generalized eigenvectors of S_b and S_w corresponding to the m largest generalized eigenvalues $\{\lambda_i \,| i = 1, 2, ..., m\}$, i.e.,

$$S_b w_i = \lambda_i S_w w_i \; i = 1, 2, ..., m \tag{5}$$

3 New LDA-based method

It is easy to prove that the upper bounds of the rank of S_b, S_w, and S_t are respectively $c - 1$,$N - c$, and $N - 1$, which are all much less than n

in many practical problems, i.e., S_b, S_w, and S_t are all usually singular in practice. To deal with the singularity of S_w, a regularization method was mentioned in [2]. S_w can be slightly modified to $S_w + KI$,where K is a very small (relative to the eigenvalues of S_w)positive number such that $S_w + KI$ is strictly positive definite. This is a pure LDA method without dimensionality reduction. However, the computational complexity is very high to handle such a high-dimensional S_w. Another kind of method to solve the small sample size problem is projecting the original samples to a lower-dimensional space to make the resulting within-class scatter matrix full-rank. Various subspaces have been used previously. The most widely used subspace method [2, 3, 4] performs Principle Component Analysis (PCA) firstly to reduce the dimension of the samples which must be not more than the rank of S_w so as to make the resulting within-class scatter matrix full-rank. Another novel method called Direct LDA [5] removes the null space of S_b firstly by doing eigen-analysis. Then a simultaneous diagonalization procedure is used to seek the optimal discriminant vectors in the subspace of S_w.It is notable that Direct LDA appears to avoid removing the null space of S_w, but cannot substantially avoid it. In fact, the rank of S_b is usually smaller than that of S_w, so the subspace that guarantees the full-rank of S_b also guarantees the full rank of S_w. Therefore, removing the null space of S_b by dimensionality reduction would indirectly lead to the losing of the null space of S_w which contains considerable discriminative information. In fact, if $q^T S_w q = 0$ and $q^T S_b q \neq 0$, then q is very useful for discrimination. But if $q^T S_w q = 0$ and $q^T S_b q = 0$ too, then q is not useful for discrimination. This means that not the whole null space of S_w is useful for discrimination. In an effort to remove the null space of S_b while keeping useful information for discrimination in the null space of S_w, Rui Huang [10] proposed the method in which the null space of S_t (actually null space of S_t is the common null space of both S_b and S_w), can be removed firstly by eigen-analysis without losing useful discriminative information. Only in the lower-dimensional projected space does the null space of the resulting within-class scatter matrix need to be determined. Through above procedure, a much smaller and equally useful subspace of the null space of S_w is found, which is then used to derive the optimal discriminant vectors of LDA.

However, we notice a drawback in traditional LDA. The definition of the between-class scatter matrix only makes the variations between the class mean and the sample mean as much as large, instead of the variations among the mean of each class. Therefore, it might cause large overlaps of neigh-bouring classes. As a consequence, there is a large overlap among the remaining classes, leading to an overall low and suboptimal classification rate. Hence, in general, LDA is not optimal with respect to minimizing the classification error rate in the lower-dimensional space [6]. To overcome this problem, in our proposed method, which is different from [10], the between-class scatter matrix , S_b is redefined according to Loog's approach [6]

$$S_b = \sum_{i=1}^{c-1} \sum_{j=i+1}^{c} P_i P_j \omega(\Delta_{ij})(\mu_i - \mu_j)(\mu_i - \mu_j)^T \tag{6}$$

where P_i, P_j is the prior probability, Δ_{ij} is the Euclidean distance, $\omega(.)$ is the weighting function. Clearly, choosing $\omega(.)$ to be the constant function will result in the conventional Fisher Criteria. According the analysis in the last paragraph, since we want to alleviate the dominant role of the outlier classes in the eigen-analysis, we tend to assign them with smaller weights. Therefore, we propose normalizing all the vectors $\mu_i - \mu_j$ by $\frac{\mu_i - \mu_j}{\Delta_{ij}}$. After doing that, all the vectors have the same length. In this case, all of classes will be fairly treated.

According to this normalization rule, the weighting function is defined as

$$\omega(\Delta_{ij}) = \frac{1}{\Delta_{ij}^2} \tag{7}$$

And now, we can describe our new LDA-based method as follows :

1. According to (1) and (2)(6)(7) compute the within-class scatter and the between-class scatter respectively.

2. Remove the null space of S_t. This can be done by doing eigen-analysis on the NxN matrix $\frac{1}{N}\Phi_b^T \Phi_b$, instead of the nxn matrix S_t. Let U be the matrix whose columns are all the eigenvectors of S_t, corresponding to the nonzero eigenvalues, then we get :

$$S'_w = U^T S_w U \tag{8}$$

$$S'_b = U^T S_b U \tag{9}$$

3. Calculate the null space of S'_w. After step 1, the dimension of S'_w is at most $N-1$, for the rank of S_t is at most $N-1$. It is now quite manageable to calculate the null space of S'_w by doing eigen- analysis again. The dimension of this null space (nullity of S'_w) is usually $c-1$, because the rank of S'_w is usually equal to that of S_w, which is usually $N-c$. Let Q be the null space of S'_w, then we get :

$$S''_w = Q^T S'_w Q = (UQ)^T S_w (UQ) = 0 \tag{10}$$

$$S''_b = Q^T S'_b Q = (UQ)^T S_b (UQ) \tag{11}$$

UQ is a subspace of the whole null space of S_w, and is really useful for discrimination.

4. Remove the null space of S''_b if it exists, and reduce dimension further if necessary.

Do eigen-analysis on S''_b. Let V be the matrix whose columns are all the eigenvectors of S''_b corresponding to the nonzero eigenvalues or part of them associated with the largest eigenvalues (for further dimensionality reduction), then the final LDA projection is : $W = UQV$.

The last step is optional, because S''_b is usually full-rank. So the number of the optimal discriminant vectors was $c-1$, which coincide with the number of ideal features for classification.

4 Experimental results

This section evaluates the performance of our propoped algorithm compared with that of the original Fisherface algorithm,Direct LDA algorithm, and Rui Huang's algorithm [10] based on using ORL face database. In the ORL database, there are ten different images of each of 40 distinct subjects. For some subjects, the images were taken at different times, varying the lighting, facial expressions (open / closed eyes, smiling / not smiling) and facial details (glasses / no glasses). All the images were taken against a dark homogeneous background with the subjects in an upright, frontal position (with tolerance for some side movement). All the images are used in the experiments without any pre-processing steps.

Firstly, we tested the recognition rates with different number of training samples. $k(k = 2, 3, 4, 5, 6)$ images of each subject are randomly selected from the database for training and the remaining images of each subject for testing. For each value of k, 50 runs are performed with different random partition between training set and testing set, and *Table 1* shows the average recognition rates (%) with ORL database. We choose 39 (i.e. $c-1$) as the final dimension.

Table 1. The recognition rates on ORL databases

k	Fisherface	Direct LDA	Rui Huang[10]	Our method
2	77.83	79.63	82.06	84.73
3	86.09	86.33	88.61	92.78
4	91.49	92.10	92.67	94.80
5	93.19	93.68	94.06	96.51
6	94.96	95.60	96.34	97.67

Next, we tested the recognition rates with different number of dimensions. In this case, we use 3 sample images of each subject for training and the other images for testing. The dimensions which are from 10 to 39 are tested during this experiment. In Fig. 1, we can see the result of recognition rate vs. the number of dimensions and our method achieves the best recognition rate compared to the other methods.

5 Conclusions

A new LDA-based method for face recognition has been proposed in this paper. In order to enhance the classification rate, we use the new definition of the between-class scatter, which aims to separate different classes as much as possible and alleviate the dominant role of the outlier classes in the eigendecomposition. When applying LDA, if the within-class scatter is singular, we firstly remove the null space of total scatter matrix which has been proved to

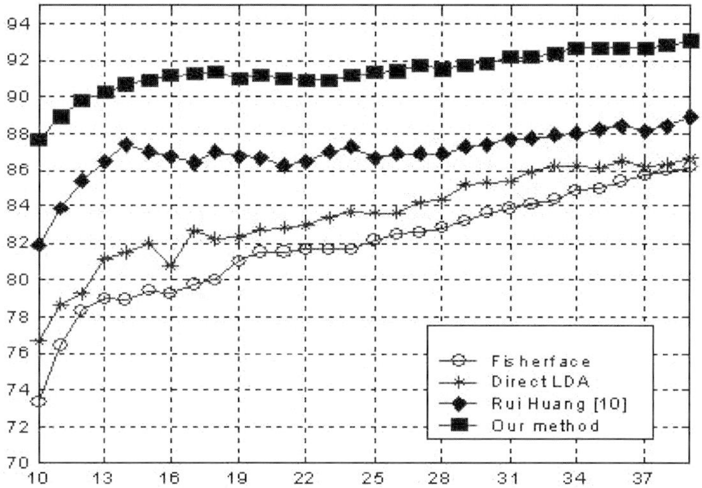

Fig. 1. The recognition rate vs. the number of dimensions.

be the common null space of both between-class and within-class scatter matrix, and useless for discrimination. Then in the lower-dimensional projected space, the null space of the resulting within-class scatter matrix is calculated. This lower-dimensional null space, combined with the previous projection, represents a subspace of the whole null space of within-class scatter matrix, and is really useful for discrimination. The optimal discriminant vectors of LDA are derived from it. By solving the small sample size problem, this paper proposes a practical algorithm for applying LDA on image recognition applications, and shows the efficiency in face recognition application. It has the advantage of easy training, efficient testing, and good performance compared to other linear classifiers.

References

1. M. Turk and A. Pentland, "Eigenfaces for recognition," *Int. J. Cog. Neurosci.*, vol. 3, no. 1, pp. 71–86, 1991.
2. W. Zhao, R. Chellappa, and P.J. Phillips, "Subspace Linear Discriminant Analysis for Face Recognition", Technical Report CAR-TR-914, Center for Automation Research, University of Maryland, 1999.
3. D. L. Swets and J. J. Weng, "Using discrimination eigenfeatures for image retrieval," IEEE Trans. Pattern Anal. Machine Intell., vol. 18, pp. 831–836, Aug. 1996.
4. P. N. Belhumeur, J. P. Hespanha, and D. J. Kriegman, "Eigenfaces vs. fisherface: Recognition using class specific linear projection," *IEEE Trans. Pattern Anal. Machine Intell.*, vol. 19, pp. 711–720, July 1997.

5. H. Yu and J. Yang, "A direct LDA algorithm for high-dimensional data with application to face recognition," *Pattern Recognit.*, vol. 34, no. 12, pp. 2067–2070, 2001.
6. M. Loog, R. P. W. Duin, and R. Haeb-Umbach, "Multiclass linear dimension reduction by weighted pairwise fisher criteria," *IEEE Trans. Pattern Anal. Machine Intell.*, vol. 23, pp. 762–766, July 2001.
7. A. M. Martinez and A. C. Kak, "PCA versus LDA," *IEEE Trans. Pattern Anal. Machine Intell.*, vol. 23, pp. 228–233, Feb. 2001.
8. D. H. Foley and J. W. Sammon, "An optimal set of discrimination vectors," *IEEE Trans. Comput.*, vol. C-24, pp. 281–289, Mar. 1975.
9. R. A. Fisher, "The use of multiple measurements in taxonomic problems," *Ann. Eugenics*, vol. 7, pp. 178–188, 1936.
10. Rui Huang; Qingshan Liu; Hanqing Lu; Songde Ma, "Solving the small sample size problem of LDA" Pattern Recognition, 2002. Proceedings. 16th International Conference on , Volume: 3 , 11-15 Aug. 2002 Pages:29 - 32 vol.3
11. C. Liu and H. Wechsler, "Robust coding scheme for indexing and retrieval from large face databases," *IEEE Trans. Image Processing*, vol. 9, pp. 132–137, Jan. 2000.
12. Chengjun Liu; Wechsler, H. "A shape- and texture-based enhanced Fisher classifier for face recognition," *IEEE Trans. Image Processing*, vol. 10, pp. 598–608, Apr. 2001.
13. L. Chen, H. M. Liao, M. Ko, J. Lin, and G. Yu, "A new LDA-based face recognition system which can solve the small sample size problem," *Pattern Recognit.*, vol. 33, no. 10, pp. 1713–1726, 2000.
14. "ORL face database" : www.uk.research.att.com/facedatabase.html

Estimating and Calculating Consensus with Simple Dependencies of Attributes

Michal Zgrzywa[1] and Ngoc Thanh Nguyen[1]

Institute of Control and Systems Engineering, Wroclaw University of Technology, Poland {mzgrzywa,thanh}@pwr.wroc.pl

Summary. In this paper we consider some problems related to attribute dependencies in consensus determining. These problems concern the dependencies of attributes representing the content of conflicts, which cause that one may not treat the attributes independently in consensus determining. We show how to cope with situations when consensus values calculated for the attributes do not fulfill the dependency function: an error estimation method and an algorithm for calculating a consensus for dependent attribute are presented.

1 Introduction

Conflict resolution is one of the most important aspects in distributed systems and multi-agent systems. The resources of conflicts in these kinds of systems come from the autonomy feature of their sites (nodes). This feature means that each site of a distributed or multi-agent system processes a task independently. There are several reasons to organize a system in such an architecture [1]. First of all, information collected in the system is easier to obtain – some sites may be nearer or not as busy as others. Although the reliability of such systems is better – the failure of one node may be compensated by using others. Finally, the trustworthiness of the system may be increased when several agents are investigating the same issue. Unfortunately, there may arise such a situation that for the same task, different sites may generate different solutions. Thus, one deals with a conflict.

In distributed and multi-agent systems three origins of conflicts can be found: insufficient resources, differences of data models and differences of data semantic [2]. For a semantic conflict one can distinguish the following three components: conflict body, conflict subject and conflict content. Consensus models, among others, seem to be useful in semantic conflict solving [3]. The oldest consensus model was introduced by such authors as Arrow [4], Condorcet and Kemeny. This model serves to solve such conflicts in which the content may be represented by orders or rankings. The models of Barthelemy and Janowitz [5], Barthelemy and Leclerc [6] and Day [7] enable such conflicts for which the structures of the conflict contents are n-trees, semillatices, partitions etc to be solved. The common characteristic of these models is that they are one-attribute, which means that conflicts are considered only referring to one feature. Multi-feature conflicts have not been investigated.

In works [8] and [9] the author presents a consensus model, in which multi-attribute conflicts may be represented. Furthermore, in this model attributes are multi-valued, which means that for representing an opinion on some issue an agent may use not only one elementary value (such as +, -, or 0) [10] but a set of elementary values. This model enables the processing of multi-feature conflicts, but attributes are mainly treated as independent. However, in many practical conflict situations some attributes are dependent on others. For example, in a meteorological system the attribute *Wind power* (with values: *weak*, *medium*, *strong*) is dependent on the attribute *Wind speed*, the values of which are measured in unit m/s. This dependency follows that if the value of the first attribute is known, then the value of the second attribute is also known. It is natural that if a conflict includes these attributes then in the consensus the dependency should also take place. Unfortunately, in [11] the authors have shown that it is not enough to determine the consensus for the conflict referring to the attribute *Wind speed*. In other words, it is not always true that if some value is a consensus for the conflict referring to the attribute *Wind speed*, then its corresponding value of the attribute *Wind power* is also a consensus for the conflict referring to this attribute. In this article we will show how to estimate the maximum error of using the corresponding value instead of the proper consensus. Also an algorithm for calculating the consensus for a dependent attribute will be presented.

2 The Outline of Consensus Model

The consensus model which enables processing multi-attribute and multi-valued conflicts has been discussed in detail in works [8] and [9]. In this section we present only some of its elements with extensions needed for the consideration of attribute dependencies. We assume that a real world situation is commonly considered by a set of agents that are placed in different sites of a distributed system. The interest of the agents consists of events which occur (or have to occur) in this world. Their task is based on determining the values of the attributes describing these events. If several agents consider the same event then they may generate different descriptions (which consist of, for example, scenarios, timestamps etc.) for this event. Thus we say that a conflict takes place. Let us consider a simple example. Let the meteorological system introduced in Section 1 give information about the temperature forecast for the next day. Every agent - meteorological station - predicts the weather for a few regions of the country. Consequently, many stations made a forecast for the central region. The following predictions were gathered: two stations claimed that the temperature will be 24 Celsius degrees and one forecast 25 Celsius degrees. Which prediction should be presented in the official forecast? Intuition suggests that 24 is the correct value, but how could it be formally chosen? In other words, how to calculate the consensus?

For representing ontologies of potential conflicts we use a finite set $\boldsymbol{A}$ of attributes and a set $\boldsymbol{V}$ of attribute elementary values, where $\boldsymbol{V} = \bigcup_{a \in A} V_a (V_a$ is the finite domain of attribute a). Let $\prod(V_a)$ denote the set of subsets of set V_a and $\prod(V_B) = \bigcup_{b \in B} \prod(V_b)$. Let B$\subseteq \boldsymbol{A}$, a tuple r_B of type B is a function $r_B : B \rightarrow \prod(V_B)$ where $r_B(b) \subseteq V_b$ for each $b \in B$. Empty tuples are denoted by symbol ϕ. The set of all tuples of type B is denoted by $TYPE(B)$. The conflict ontology is defined as the quadruple: $\langle \boldsymbol{A}, \boldsymbol{X}, \boldsymbol{P}, \boldsymbol{F} \rangle$, where:

- A is a finite set of attributes, which includes a special attribute *Agent*; a value of attribute a where $a \neq Agent$ is a subset of V_a; values of attribute *Agent* are singletons which identify the agents;
- $\boldsymbol{X} = \{\prod(V_a) : a \in A\}$ is a finite set of conflict carriers;
- $\boldsymbol{P}$ is a finite set of relations on carriers from $\boldsymbol{X}$, each relation $P \in \boldsymbol{P}$ is of some type T_P (for $T_P \subseteq \boldsymbol{A}$ and $Agent \in T_P$). Relations belonging to set $\boldsymbol{P}$ are classified into two groups identified by the upper indices '+' and '−'. For example, if R is the name of a group, then relation R^+ is called the positive relation (contains positive knowledge – possible events) and R^- is the negative relation (contains negative knowledge – impossible events).
- Lastly, $\boldsymbol{F}$ is a set of function dependencies between sets of attributes.

The structures of the conflict carriers are defined by means of a distance function between tuples of the same type. In the article, distance functions will be represented by the symbols δ and σ. A consensus is considered within a conflict situation, which is defined as a pair$s = <\{P^+, P^-\}, A \rightarrow B>$ where$A, B \subseteq \boldsymbol{A}$, $A \cap B = \emptyset$ and $r_A \neq \phi$ holds for any tuple $r \in P^+ \cup P^-$. The first element of a conflict situation (i.e. set of relations $\{P^+, P^-\}$) includes the domain from which consensus should be chosen, and the second element (i.e. relationship $A \rightarrow B$) presents the schemas of consensus subjects of the consensus content, such that for a subject e (as a tuple of type A, included in P^+ or P^-) there should be assigned only one tuple of type B. A conflict situation yields a set $Subject(s)$ of conflict subjects which are represented by tuples of type A. For each subject e two conflict profiles, i.e. $profile(e)^+$ and $profile(e)^-$, as relations of $TYPE(\{Agent\} \cup B)$ should be determined. Profile $profile(e)^+$ contains the positive opinions of the agents on the subject e, while profile $profile(e)^-$ contains agents' negative opinions on this subject.

Definition 1. *Consensus on a subject* $e \in Subject(s)$ *is a pair* $(C(s,e)^+, C(s,e)^-)$ *of 2 tuples of type* $A \cup B$ *which fulfill the following conditions:*

a) $C(s,e)^+_A = C(s,e)^-_A = e$ *and* $C(s,e)^+_B \cap C(s,e)^-_B = \phi$,

b) The sums $\sum\limits_{r \in profile(e)^+} \sigma(r_B, C(s,e)^+_B)$ *and* $\sum\limits_{r \in profile(e)^-} \sigma(r_B, C(s,e)^-_B)$

are minimal.

Any tuples $C(s,e)^+$ and $C(s,e)^-$ satisfying the conditions of Definition 1 are called consensus of profiles *profile*$(e)^+$ and *profile*$(e)^-$, respectively. Let us consider the example from the beginning of this section. The ontology of that conflict is the following four: <{Region, Temperature}, {$\prod$(Silesia, Great Poland, Little Poland, Pomerania), $\prod$(-30,-29,...,34,35)}, {Weather$^+$}, ϕ >. We can distinguish one conflict situation: <{Weather$^+$,ϕ}, {Region}→{Temperature}>. The information about the conflict for the subject Silesia is gathered in Table 1.

Table 1. The relation Weather$^+$.

Meteorological station	Region	Temperature
S_1	Silesia	24
S_2	Silesia	25
S_3	Silesia	24

The column Temperature is in fact the *profile*(Silesia)$^+$. To compare the values of different votes for the Temperature attribute the function δ_{Temp} will be used: $\delta_{Temp}(t_1, t_2) = |t_1 - t_2|$. Now, after calculating all the necessary distances, we can use Definition 1b to determine the consensus for this subject (24 Celsius degrees).

3 Some Aspects of Attribute Dependencies

In Definition 1, condition b) is the most important. It requires the tuples $C(s,e)^+_B$and $C(s,e)^-_B$ to be determined in such a way thus the sums $\sum_{r \in profile(e)^+} \sigma(r_B, C(s,e)^+_B)$ and $\sum_{r \in profile(e)^-} \sigma(r_B, C(s,e)^-_B)$ are minimal. These tuples could be calculated in the following way: for each attribute $b \in B$ one can determine sets $C(s,e)^+_b$and $C(s,e)^-_b$, which minimize sums $\sum_{r \in profile(e)^+} \sigma(r_b, C(s,e)^+_b)$ and $\sum_{r \in profile(e)^-} \sigma(r_b, C(s,e)^-_b)$ respectively. This way is an effective one, but it is correct only if the attributes from set B are independent ($\boldsymbol{F}=\phi$). In this section we consider consensus choice assuming that some attributes from set B are dependent on some others. The definition of attribute dependency given below is consistent with those given in the information system model ([12]):

Definition 2. *Attribute b is dependent on attribute a if and only if there exists a function f^a_b:$V_a \to V_b$ such that in every conflict ontology $\langle \boldsymbol{A}, \boldsymbol{X}, \boldsymbol{P}, \boldsymbol{F}\rangle$ ($f^a_b \in \boldsymbol{F}$) for each relation $P \in \boldsymbol{P}$of type T_P and a,b$\in T_P$ the formula $(\forall r \in P)(r_b = \bigcup_{x \in r_a} \{f^a_b(x)\})$ is true.*

The dependency of attribute b on attribute a means that in the real world if for some object the value of a is known, then the value of $b$is also known. In practice, owing to this property for determining the values of attribute b, it is enough to know the value of attribute a. Instead of $\bigcup_{x\in Y}\{f_b^a(x)\}$ we will abbreviate it to $f_b^a(Y)$.

Consider now a conflict situation s=<$\{P^+, P^-\}, A \rightarrow B$ >, in which attribute b is dependent on attribute a where $a, b \in B$. Let *profile*$(e)^+$ be the positive profile for given conflict subject $e \in$*Subject*(s). The problem relies on determining the consensus for this profile. We can solve this problem using two approaches:

1. Notice that *profile*$(e)^+$ is a relation of type $B\cup\{Agent\}$. The dependency of attribute b on attribute a implies that there exists a function from set $TYPE(B\cup\{Agent\})$ to set $TYPE(B\cup\{Agent\}\backslash\{b\})$ such that for each profile *profile*$(e)^+$ one can assign one set *profile*'$(e)^+$= $\{r_{B\cup\{Agent\}\backslash\{b\}}$: $r \in$*profile*$(e)^+\}$. Set *profile*'$(e)^+$ can be treated as a profile for subject e in the following conflict situation s' = <$\{P^+, P^-\}$, $A \rightarrow B\backslash\{b\}$>.

Notice that the difference between the profiles *profile*$(e)^+$ and *profile*'$(e)^+$ relies only on the lack of attribute b and its values in profile *profile*$(e)^+$. Thus one can expect that the consensus $C(s,e)^+$for profile *profile*$(e)^+$ can be determined from the consensus $C(s,e)'^+$for profile *profile*$(e)'^+$ by adding to tuple $C(s,e)'^+$ attribute b and its value which is equal to $f_b^a(C(s,e)'^+_a)$. In a similar way one can determine consensus for the *profile*$(e)^-$.

2. In the second approach attributes a and b are treated equivalently. That means they play the same role in consensus determining for the profiles *profile*$(e)^+$ and $^-$. The consensus for the profiles *profile*$(e)^+$ and *profile*$(e)^-$ are defined as follows:

Definition 3. *The consensus for subject* $e\in$ *Subject*(s) *in situation* s=<$\{P^+, P^-\}$, $A \rightarrow B$ > *is a pair of tuples* $(C(s,e)^+, C(s,e)^-)$ *of type* $A\cup B$, *which satisfy the conditions:*

a) $C(s,e)^+_A = C(s,e)^-_A = e$ *and* $C(s,e)^+_B \cap C(s,e)^-_B = \phi$,

b) $C(s,e)^+_b = f_b^a(C(s,e)^+_a)$ *and* $C(s,e)^-_b = f_b^a(C(s,e)^-_a)$,

c) The sums $\sum_{r\in profile(e)^+} \sigma(r_B, C(s,e)^+_B)$ *and* $\sum_{r\in profile(e)^-} \sigma(r_B, C(s,e)^-_B)$ *are minimal.*

We are interested in the cases when conditions b) and c) of Definition 3 can be satisfied simultaneously. The question is: is it true that if set $C(s,e)^+_a$ is a consensus for profile *profile*$(e)^+_a$ (as the projection of profile *profile*$(e)^+$ on attribute a) then set $f_b^a(C(s,e)^+_a)$ will be a consensus for profile *profile*$(e)^+_b$ (as the projection of profile *profile*$(e)^+$ on attribute b)? Let us suppose that the distance between the values of attribute a and between the values of attribute b may be measured by functions δ_a and δ_b (the minimal costs of the operation which transforms a one-element set into an one-element set is the distance between their elements), respectively. Let $q : R \rightarrow R$ be a function such that: $\delta_b(f_b^a(a_1), f_b^a(a_2)) = q(\delta_a(a_1, a_2))$, where a_1, a_2 are values of the attribute a. In [11] authors considered which forms of q and δ guarantee that all dependencies between attributes will be fulfilled in consensus. The following theorem was proved:

Theorem 1. *If function q is monotonic and both distance functions are metrics then the dependencies between the attributes are not always fulfilled in calculated consensus.*

So $f_b^a(C(s,e)_a^+)$ is not always $C(s,e)_b^+$. But the first value is much easier to calculate. In many cases the difference between them may be so small that calculating $C(s,e)_b^+$ would not be sensible. In the next section we will show the method of estimating the maximum error of using the first value.

4 Estimating the Error of Using a Dependency Function

Sometimes it is not required to find the best solution for the conflict – a good solution may be enough if it is easier (quicker) to obtain. The quickest method of determining a candidate for consensus for the dependent attribute (b) is to use the dependency function f_b^a on consensus on attribute a. But how to estimate how good such a candidate is? Let us define a few useful functions:

Definition 4. *Value $maxError(e, f_b^a)$ is the maximum difference between a sum of distances from consensus for attribute b to a whole $profile(e)_b$ and a sum of distances from the image of the consensus for attribute a to a whole $profile(e)_b$:*

$$maxError(e, f_b^a) = \sum_{r \in profile(e)} \delta_b(r_b, C(s,e)_b) - \sum_{r \in profile(e)} \delta_b(r_b, f_b^a(C(s,e)_a)), \tag{1}$$

Definition 5. *Value $convertCost(a_i, f_b^a)$ is the difference between a distance from candidate i to a whole $profile(e)_a$ and a distance from the image of candidate i to a whole $profile(e)_b$:*

$$convert\_cost(a_i, f_b^a) = \sum_{r \in profile(e)} \delta_b(r_b, f_b^a(a_i)) - \sum_{r \in profile(e)} \delta_a(r_a, a_i), \tag{2}$$

Definition 6. *Value $minConvert(f_b^a)$ is the minimal difference between a distance and its image:*

$$minConvert(f_b^a) = \min_{d \in \delta_a}(q(d) - d). \tag{3}$$

If $minConvert(f_b^a)$ is known for the dependency, the maximum error may be calculated using Theorem 2. For many forms of q the $minConvert$ value is easy to find, for example when a transformation may only enlarge the distances, the $minConvert = 0$ may be used.

Theorem 2. *For every dependence f_b^a and every subject $e \in Subject(s)$ in situation $s=<\{P\},A \rightarrow B>$, $a,b \in B$, the maximum error of using $f_b^a(C(s,e)_a)$ instead of $C(s,e)_b$ satisfies the condition:*

$$maxError(e, f_b^a) < convertCost(C(s,e)_a, f_b^a) - (\overline{profile(e)} - 1) * minConvert(f_b^a). \quad (4)$$

In other words, the maximum error is less than the cost of converting the consensus for a minus the smallest cost of converting any proposition. To better illustrate Theorem 2 let us consider the following example.

Example 1. An attribute a may have four values: a_1, a_2, a_3 and a_4. Also attribute b may have four values: b_1, b_2, b_3 and b_4. There is a dependency function f_b^a such that $f_b^a(a_1) = b_1$, $f_b^a(a_2) = b_2$, $f_b^a(a_3) = b_3$ and $f_b^a(a_4) = b_4$. All four pairs (a_i, b_i) were suggested as votes. The distances for attributes a and b equal respectively:

Table 2. The distance functions δ_a and δ_b.

$\delta_a(a_1,a_2) = 30$	$\delta_b(b_1,b_2) = 30$
$\delta_a(a_2,a_3) = 50$	$\delta_b(b_2,b_3) = 50$
$\delta_a(a_1,a_3) = 60$	$\delta_b(b_1,b_3) = 69$
$\delta_a(a_1,a_4) = 70$	$\delta_b(b_1,b_4) = 79$
$\delta_a(a_3,a_4) = 80$	$\delta_b(b_3,b_4) = 80$
$\delta_a(a_2,a_4) = 90$	$\delta_b(b_2,b_4) = 90$

After determining the consensus for both attributes it appeared that value a_1 was chosen for attribute a and value b_2 was chosen for attribute b (the distance from *profile*$(e)_b$ to b_1 is 178 and to b_2 is 170). Thus, dependency f_b^a is not fulfilled in such a determined consensus and we can not calculate the consensus for b using only f_b^a.

Now, let us use Theorem 2. As one can see, the *minConvert* for the dependency is 0. The *convertCost* of the consensus for a is 18 because the distance from *profile*$(e)_a$ to a_1 is 160 and the distance from *profile*$(e)_b$ to b_1 is 178. Finally, if we used the theorem we could calculate that $maxError(e,f_b^a) < 18$ (the real error is 8) and decide that such a small deviation is not a problem in our application.

In the next section we will consider the conflict with conditions described in Theorem 1 and try to find a consensus for a dependent attribute in such a situation.

5 Calculating the Consensus for a Dependant Attribute

The conditions of Theorem 1 are very intuitive. Let function q be monotonic, for example non-decreasing. This means that: $\delta_a(a_1,a_2) \geq \delta_a(a_3,a_4) \Rightarrow$

$\delta_b(b_1,b_2) \geq \delta_b(b_3,b_4)$. Consequently, all the possible distances for functions δ_a and δ_b will be in the same order. Such functions are usually called dependent (in this case δ_b depends on δ_a). Surprisingly, as it was shown in [11], dependencies between attributes are not always fulfilled in consensus calculated in such conditions. Let us consider Example 1 again. As one can see, function δ_b depends on function δ_a and q is increasing (monotonic). Additionally, both distance functions are metrics, so all the conditions for the Theorem 1 are fulfilled. Indeed, the consensus for b is not equal to the image of the consensus for a. Does this mean that, in such conditions, a consensus for a dependant attribute must always be calculated separately? No. The algorithm presented in this section facilitates the fact that q is monotonic and allows us not to calculate all the distances for all the candidates for the consensus (which is required in conventional consensus determining).

Algorithm 1. *Input:* dependency function f_b^a, candidates for a consensus for attribute a: a_x, a_y, the consensus for attribute a: $C(s,e)_a$.

Output: the consensus for attribute b: $C(s,e)_b$.

STEP 1: Sort all the distances $\delta_a(a_x,a_y)$ in an ascending order D, SET i = 1,

SET CONS = $f_b^a(C(s,e)_a)$.

STEP 2: IF $D_i = \max(\delta_a(C(s,e)_a,a_x))$ GOTO STEP 7.

STEP 3: IF D_i IS LIKE $\delta_a(C(s,e)_a,a_x)$ OR $\delta_a(a_x,C(s,e)_a)$ GOTO STEP 5.

STEP 4: D_i IS LIKE $\delta_a(a_x,a_y)$,

IF $\sum\limits_{r\in profile(e)} \delta_b(r_b, f_b^a(a_x)) <$ CONS SET CONS $= f_b^a(a_x)$,

IF $\sum\limits_{r\in profile(e)} \delta_b(r_b, f_b^a(a_y)) <$ CONS SET CONS $= f_b^a(a_y)$.

STEP 5: SET i = i + 1.

STEP 6: GOTO STEP 2.

STEP 7: RETURN CONS.

Additionally, if it is possible to estimate a *maxError* value for the case, we may omit all the candidates a_x that satisfy the condition:

$$maxError(e, f_b^a) < \sum_{r\in profile(e)} \delta_a(r_a, a_x) - \sum_{r\in profile(e)} \delta_a(r_a, C(s,e)_a). \quad (5)$$

The above sums are already known for all candidates, because they are required to calculate the consensus for attribute a.

Let us use Algorithm 1 for the conflict from Example 1. We know that the consensus for a is a_1 and its worst distance is $\delta_a(a_1,a_4)$. Now, the first candidate that does not satisfy the conditions from steps 2 and 3 is a_2. In this conflict the *maxError* value equals 18, so the candidate a_2 passes the test (sum of all $\delta_a(a_2,a_x)$ is 170). When we add all the distances $\delta_b(b_1,b_x)$ (also 170) we will find that the result is better than the sum for b_1 (178). We have the new value of CONS. Then, the next (and last) candidate that does not satisfy the conditions from steps 2 and 3 is a_3. Fortunately, the sum of all

$\delta_a(a_3, a_x)$ is 190, so the candidate may be omitted. Thus, the algorithm ends with candidate b_2 without calculating all the sums for b_3 and b_4.

6 Summary

In this article the problem of consensus determining in conflicts with attribute dependencies was considered. It appears that in the majority of cases the consensus for dependent attributes cannot be calculated by counting the dependency function of the source attribute. To avoid calculating the consensus for both attributes (which may be very time and memory-consuming) the authors provided a method of estimating the maximum error of using only the dependency function. In many real problems, determining the best result is not strictly required. When the maximum error is relatively small, calculating a consensus for a dependant attribute is not practical.

For conflicts where the consensus for a dependant attribute must be determined, the authors proposed an algorithm. Imposing some requirements on the form of the dependency, the algorithm calculates the consensus very efficiently by considering only a subset of candidates. In future research the algorithm will be tested in a multi-agent system to evaluate its efficiency.

References

1. Coulouris G, Dollimore J, Kindberg T (1996) Distributed systems, Concepts and design.
2. Pawlak Z (1998) An Inquiry into Anatomy of Conflicts. In: Information Sciences 108: 65–78.
3. Tessier C, Chaudron L, Müller HJ (2001) Conflicting agents: conflict management in multi-agent systems. Kluwer Academic Publishers, Boston.
4. Arrow, KJ (1963) Social Choice and Individual Values. Wiley, New York.
5. Barthelemy JP, Janowitz MF (1991) A Formal Theory of Consensus. In: Siam J Discrete Math 4: 305–322.
6. Barthelemy JP, Leclerc B (1995) The Median Procedure for Partitions. In: DIMACS Series in Discrete Mathematics and Theoretical Computer Science 19: 3–33.
7. Day WHE (1988) Consensus Methods as Tools for Data Analysis. In: Bock HH (ed) Classification and Related Methods for Data Analysis: 312–324. North Holland.
8. Nguyen NT (2002) Consensus System for Solving Conflicts in Distributed Systems. In: Information Sciences – An International Journal 147: 91–122.
9. Nguyen NT (2002) Methods for Consensus Choice and their Applications in Conflict Resolving in Distributed Systems. Wroclaw University of Technology Press (in Polish).
10. Pawlak Z (1998) An Inquiry into Anatomy of Conflicts. In: Information Sciences 108: 65–78.

11. Zgrzywa M, Nguyen NT (2004) Determining Consensus with Simple Dependencies of Attributes. In: Nguyen NT (ed) Intelligent Technologies for Inconsistent Knowledge Processing: 57–72. Advanced Knowledge International, Australia.
12. Pawlak Z, Skowron A (1993) A Rough Set Approach to Decision Rules Generation. In: Reports of the Institute of Computer Science. Warsaw University of Technology.

The Empirical Study of the Naive Bayes Classifier in the Case of Markov Chain Recognition Task

Andrzej Zolnierek and Bartlomiej Rubacha

Chair of Systems and Computer Networks, Wroclaw University of Technology, Wybrzeze Wyspianskiego 27, 50-370 Wroclaw, Poland
andrzej.zolnierek@pwr.wroc.pl

Summary. In this paper the problems of sequential pattern recognition are considered. As a statistical model of dependence, in the sequences of patterns, the first-order Markov chain is assumed. Additionally, the assumption about independence between the attributes in the feature vector is made. The pattern recognition algorithms with such assumption are called in the literature "naive Bayes algorithm". In this paper such approach is made to the pattern recognition algorithm for first-order Markov chain and some results of numerical investigation are presented. The main goal of these investigations was to verify if it is reasonable to make such assumption in the real recognition tasks.

1 Introduction

In many pattern recognition problems there are exists dependencies among the patterns to be recognized. Additionally there are exists dependencies in the feature vectors of recognized patterns. For instance, this situation is typical for recognition of state in technological processes or in the computer-aided medical diagnosis [10], [11] to name only a few. In such cases, we should recognize the sequence of patterns, for example the sequence of states of technological processes or the sequence of patient's states in medical diagnosis. In this paper, similar as in the [6], [9], [12] the assumption of Markov dependence in the sequence of classes of recognized patterns is made. Furthermore, it is assumed that the features vectors are conditional independent. Based on this model and using Bayes' approach in the case of complete probabilistic information the pattern recognition algorithms for first-order Markov chains can be presented in recurrent form [5], [9].

In the practice the pattern recognition algorithms in the case of complete statistical information are the base on which we can propose the algorithms with learning, because usually instead of knowing needed probabilities and class density functions we have so called set of learning sequences. One conceptually simple and effective method of building the algorithms with learning is to replace these probabilities by their estimators and to calculate the needed values of density functions using nonparametric estimators like Parzen estimator for example[4]. In such cases, for calculation reasons and to simplify the learning procedures, the assumption about the conditional independency among the attributes in the feature vector is usually made. Such classifiers are known in the literature as naive Bayes classifiers [1], [2], [8]. There are plenty of papers, in which the naive Bayes classifier is examined. Some interesting results, concerning text recognition, can be found in [7].

However, in this paper some results of simulation investigations of naive Bayes classifier in the case of Markov chain recognition task are presented. In these investigations, the sequences of patterns with dependency in two dimensional features vector, which form the first-order Markov chains, were generated by simulation and next in the case of complete probabilistic information the pattern recognition algorithm for Markov chain was tested. In this paper we want to investigate if taking into account the dependencies in the feature vectors can really improve the quality of recognition in comparison with the attempt in which assumption of class conditional independence in the feature vector is made. Additionally, results of Markov chain algorithms were compared with results, which were obtained using very well known Bayes' decision rule for independent patterns.

2 Statement of the problem

Let us consider the classical problem of pattern recognition that is concerned with the assignment of a given pattern to one of m possible classes from the set of classes $M = \{1, \ldots, m\}$. Let x_n, which takes values in $r-$dimensional Euclidean space $\mathbb{E}^r$, $x_n = \left[x_n^1, x_n^2, \ldots, x_n^r\right]^T$ denote the vector of measured features of the n-th pattern to be recognized and let j_n denote the label of the class to which the pattern in question belongs. Thus: $\bar{x}_n = \{x_1, x_2, ..., x_n\}$, $\bar{j}_n = \{j_1, j_2, ..., j_n\}$ denote the sequence of feature vectors and true identities, respectively. In this paper it is assumed that x_n, j_n are observed values of random variables X_n, J_n for $n = 1, 2, \ldots$, while random variablesX_n are multidimensional. It is also assumed that the sequence of classifications forms first-order Markov chain:

$$P\left(J_n = j_n \,|J_{n-1} = j_{n-1}, J_{n-2} = j_{n-2}, \ldots, J_1 = j_1\right) = \tag{1}$$
$$= P\left(J_n = j_n \,|J_{n-1} = j_{n-1}\right).$$

Note that the first-order Markov chain is described by the initial probabilities:

$$P\left(J_1 = j_1\right) = p_{j1}\ , \qquad j_1 \in M, \tag{2}$$

and the set of transition probabilities:

$$p^{n}_{j_n,\,j_{n-1}} = P\left(J_n = j_n \,|\, J_{n-1} = j_{n-1}\right), \quad j_n, j_{n-1} \in M, \; n = 2, 3, \ldots . \quad (3)$$

Let $f\,(x_n\,|j\,)$ be the conditional density function of random variable X_n given that $J_n = j$, $j \in M$, identically for all natural n. For simplicity we shall assume conditional independence among variables X_n, n = 1, 2, ..., which implies that

$$\bar{f}_n\,(\bar{x}_n\,|\,\bar{j}_n\,) = \prod_{\alpha=1}^{n} f\,(x_\alpha\,|\,j_\alpha\,), \quad n = 1, 2, \ldots . \quad (4)$$

This assumption states that, given the true identity of a pattern, the distribution of measurement vector is independent of the features and true identities of previous and future patterns, but it is dependent only on the true identity of the pattern in question.

In the naive Bayes pattern recognition algorithms we also assume that the attributes in the features vectors x_n, $n = 1, 2, \ldots$given the class $j \in M$are conditional independent :

$$f\,(x_n\,|\,j\,) = f\left(x_n^1, x_n^2, \ldots, x_n^r\,|j\right) = \prod_{\beta=1}^{r} f_\beta\left(x_n^\beta\,|j\right) \quad (5)$$

This assumption greatly simplify the learning procedures, but we lose the part of complete probabilistic information.

3 Pattern recognition algorithm

In this part, under the assumption of complete statistical information and using Bayes' approach, the decision rules of pattern recognition algorithm for second-order Markov chains are presented. In the papers [5], [6] it is shown, that for the special case of a 0-1 loss function, i.e.:

$$L\,(i_n, j_n\,) = 0 \; if \; i_n = j_n \; or \; 1 \; otherwise \; for \; n = 1, 2, \ldots, \quad (6)$$

where $L\,(i_n, j_n\,)$ is the loss incurred by the classifier if a pattern from the class j_n is assigned to the class i_n at the moment n, the Bayes' decision rule assigns the n-th recognized pattern to the class i_n with the highest a posterior probability after observing $\bar{x}_n$ for all natural n :

$$i_n = \Psi_n^*\,(\bar{x}_n\,), \quad n = 1, 2, \ldots, \quad (7)$$

if for every $s \in M$, $s \neq i_n$,

$$p_{1,n}\,(i_n\,|\,\bar{x}_n) > p_{1,n}\,(s\,|\,\bar{x}_n\,) \quad (8)$$

where: $p_{1,n}(j_n \mid \bar{x}_n) = P\left(J_n = j_n \mid \bar{X}_n = \bar{x}_n\right)$, and $f_{1,n}(\bar{x}_n)$ denotes the joint conditional density function of the sequence of random variables $\bar{X}_n$. Instead of calculate the probabilities (8) it is sufficient to maximize only the following discriminate functions:

$$d(j_n, \bar{x}_n) = f_{2,n}(\bar{x}_n \mid j_n) \cdot p_{2,n}(j_n) = p_{1,n}(j_n \mid \bar{x}_n) \cdot f_{1,n}(\bar{x}_n) \qquad (9)$$

$$n = 1, 2, \ldots \qquad j_n \in M$$

where $p_{2,n}(j_n) = P(J_n = j_n)$ and $f_{2,n}(\bar{x}_n \mid j_n)$ denotes the joint conditional density function of the sequence of random variables $\bar{X}_n$ given that $J_n = j_n$. In the papers [5], [6] it is shown, that the discriminate functions can be calculated recursively as follows:

$$d(j_n, \bar{x}_n) = f(x_n \mid j_n) \cdot \sum_{j_{n-1}=1}^{m} p^n_{j_n, j_{n-1}} \cdot d(j_{n-1}, \bar{x}_{n-1}) \qquad (10)$$

for all natural $n \geq 2$ and for every $j_n, j_{n-1} \in M$. At the first step of classification the discriminate functions can be obtained immediately:

$$d(j_1, x_1) = f(x_1 \mid j_1)\, p_{j_1} \qquad n = 1,\ j_1 \in M \qquad (11)$$

In order to classify the n-th pattern we take into account in our recognition task the whole sequence of patterns to this moment ("context") calculating the discriminant functions (9) according to the recursive formulas (10), (11). Then the pattern recognition algorithm for first-order Markov chain classifies the n-th recognized pattern to this class for which the discriminant function is maximal. The difference between the naive Bayes classifier and complete algorithm in the case of Markov chain recognition consists in the way in which the needed values of density functions in the discriminant functions are obtained. In the first case we simplify calculation using assumption (5) and in the second case we take complex class density functions.

4 Simulation investigations

During the simulation investigations we considered the set of three classes $M = \{1, 2, 3\}$. The parameters of Markov chain were constant i.e. $p_{j1} = [0.25\ 0.5\ 0.25]^T$ and the transition probabilities (3) formed constant transition matrix:

$$P = \begin{bmatrix} 0.6 & 0.6 & 0.1 \\ 0.1 & 0.1 & 0.8 \\ 0.3 & 0.3 & 0.1 \end{bmatrix}$$

where for example $P(J_n = 2 \mid J_{n-1} = 2) = 0.8$ for every n. In each class two-dimensional Gaussian density functions were assumed. This density function was described by the vector of mean values $m_j = [m_{j1}, m_{j2}]^T$ and the covariance matrix:

$$\Sigma_j = \begin{bmatrix} \sigma_{j1}^2 & \rho\,\sigma_{j1}\sigma_{j2} \\ \rho\,\sigma_{j1}\sigma_{j2} & \sigma_{j2}^2 \end{bmatrix},$$

with the same parameter of correlation ρ for j=1,2,3.

In every experiment two hundred testing sequences of length equal to ten were generated using program Matlab 6.5. The vectors of mean values were constant and equal:

$$m_1 = [70,\ 110]^T,\ m_2 = [80,\ 190]^T,\ m_3 = [90,\ 270]^T.$$

The variances of each attribute in every class were the same i.e. $\sigma_{j1}^2 = \sigma_{j2}^2 = 900$ for j = 1, 2, 3. In this way according to the changes of correlation parameter the covariance matrix for every class was changing in the same way. In these simulation investigations three different coefficients of correlation were taken into account : $\rho = 0.3$, $\rho = 0.5$, $\rho = 0.7$ so the covariance matrix for each class were:

$$\Sigma_j = \begin{bmatrix} 900 & 270 \\ 270 & 900 \end{bmatrix},\ \Sigma_j = \begin{bmatrix} 900 & 450 \\ 450 & 900 \end{bmatrix},\ \Sigma_j = \begin{bmatrix} 900 & 630 \\ 630 & 900 \end{bmatrix}$$

respectively. Of course in the case of Gaussian two dimensional density function the correlation parameter ρ is the measure of dependence between attributes in the feature vectors. Let us notice, that treating our attributes as independent the covariance matrix become diagonal and the value of joint density function in needed point x_n can be calculated for every class as follows:

$$f(x_n \mid j_n) = f(x_n^1, x_n^2 \mid j_n) = f_1(x_n^1 \mid j_n) \cdot f_2(x_n^2 \mid j_n) \tag{12}$$
$$n = 1, 2, \ldots, \quad j_n \in M$$

where:

$$f_1(x_n^1 \mid 1) = N(70, 30),\ f_2(x_n^2 \mid 1) = N(110, 30)$$
$$f_1(x_n^1 \mid 2) = N(80, 30),\ f_2(x_n^2 \mid 2) = N(190, 30)$$
$$f_1(x_n^1 \mid 3) = N(90, 30),\ f_2(x_n^2 \mid 3) = N(270, 30)$$

Then calculating the values of density functions either according to the joint distribution $f(x_n^1, x_n^2 \mid j_n)$ or to the formula $f_1(x_n^1 \mid j_n) \cdot f_2(x_n^2 \mid j_n)$ we can investigate, in the case of complete probabilistic information, how important in the recognition task is taking into account the dependencies among attributes in the feature vectors. The results of pattern recognition algorithms for first-order Markov chain in these two above mentioned methods of calculating the values of density functions were compared with the results obtained using algorithm which does not take into account any dependencies in the sequences of patterns. The decision functions of this algorithm are as follows:

$$d\,(j_n, x_n) = f\,(x_n\,|\,j_n)\;p_{j_n} \quad , n = 1, 2, \ldots , j_n \in M \tag{13}$$

The probabilities:

$$P\,(J_n = j_n) = p_{j_n} , \quad j_n \in M,\; n = 1, 2, \ldots \tag{14}$$

we can assume as constant and equal to the initial probabilities of Markov chain $P\,(\,J_1 = j_1\,) = p_{j_1}\,,\quad j_1 \in M$, or we can calculate them according to the step of classification and to the recursive formula:

$$p_{j_n} = \sum_{j_{n-1}}^{m} p^{n}_{j_n\,,j_{n-1}} p_{j_{n-1}}\,, \quad j_n\,,j_{n-1} \in M\,, \quad n = 2, 3, \ldots \tag{15}$$

with the initial condition for n =1:

$$P\,(\,J_1 = j_1\,) = p_{j_1}\,, \quad j_1 \in M. \tag{16}$$

In this way during simulation investigations the following pattern recognition algorithms were compared by calculating the percentage of correctly classified patterns:

A - algorithm for Markov chain with dependent attributes,
B - algorithm for Markov chain with independent attributes,
C - modified Bayes algorithm with dependent attributes,
D - modified Bayes algorithm with independent attributes,
E - classical Bayes algorithm with dependent attributes,
F - classical Bayes algorithm with independent attributes.

The results of numerical example, in which the correlation parameter ρ was changed, are presented in the table 1:

Table 1. The results of simulation

Algorithm	A	B	C	D	E	F
$\rho = 0.3$	92,4	91,1	89,5	88,4	87,4	86,7
$\rho = 0.5$	93,7	90,4	90,4	87,1	89,6	85,7
$\rho = 0.7$	95,8	89,6	94,6	86,7	93,8	83,6

Looking at these results we can see the improvement of the quality of classification, taking into account the dependencies among the attributes in the feature vector. In addition, as it can be expected, the improvement is bigger if the correlation parameter is growing up. In this example of Markov chain we can notice, that the difference between the algorithm A and B for every investigated degree of correlation is less significant that the difference between algorithms C and D, or between E and F. This difference for Markov chain recognition is less than 6 % in the case of complete statistical information. In the real problems we have to solve the problem of learning, which can be very complicated if we want to use nonparametric technique (like Parzen method for example) and if we want to take into account the dependency in the feature vector. Additionally,in the case of discrete attribute the number of needed samples in the learning sequences must be greater because we have to estimate all needed joint probabilities.

5 Conclusions

In this paper some results of numerical investigation for pattern recognition algorithm for first-order Markov chain with the "naive" assumption concerning the feature vector are presented. This assumption state, that in every class the attributes in the feature vector are conditionally independent. From theoretical point of view, i.e. in the case of complete probabilistic information in which we know the class conditional joint distribution of feature vector or the class conditional joint density function, there is no problem in calculating the values of decision functions. However the problem remains if have to consider the algorithms with learning. In such case the simplifying assumption about the independencies among the attributes in the feature vector is very useful. First assuming the independency in the feature vector, the calculation of estimates of needed joint probabilities requires less training samples. Second, using nonparametric estimators of values of density function we have the problem of finding the proper kernel function for the kind of dependence in the feature vector. That's way such assumption is usual made. Concluding, in the case of Markov chain recognition, if we can accept less quality of pattern recognition algorithm (about 6 %) the assumption of independence of attribute in the feature vector can be made. Similar investigations should be done for another cases of Markov chain recognition tasks, i.e. for higher-order or controlled Markov chains.

References

1. Duda R, Hart P (1973) Pattern classification and scene analysis. John Wiley, New York
2. Friedman N, Geiger D, Goldszmidt M (1997) Bayesian network classfiers. Machine learning 29:131-163
3. Fu K (1974) Syntactic methods in pattern recognition. New York Academic Press
4. Greblicki W (1978) Pattern recognition procedures with nonparametric density estimates. In: IEEE Trans. on SMC 8:809–812
5. Kurzynski M (1997) Pattern recognition-statistical approach. Publishers of Wroclaw University of Technology.

6. Kurzynski M, Zolnierek A (1980) A recursive classifying decision rule for second-order Markov chain. Control and Cybernetics 9:141–147
7. McCallum, Nigam K (1998) A comparison of event models for naive Bayes text classification. In AAAI-98 Workshop on Learning and Text Categorization, Madison, WI, USA:41–48
8. Mitchel T (1997) Machine learning. McGraw Hill, New York
9. Raviv J (1967) Decision making in Markov chain applied to the problem of pattern recognition. In: IEEE Trans. on IT 21:536–551
10. Zolnierek A (1982) Computer-aided recognition of the human acid-base state. In Proc. of 6-th Int. Conf. on Pattern Recognition:1219
11. Zolnierek A (1983) Pattern recognition algorithms for controlled Markov chains and their application to medical diagnosis. Pattern Recognition Letters 1:299–303
12. Zolnierek A (2003) The simulation investigations of pattern recognition algorithm for second-order Markov chains. In: Proc. of the 37-th conference, Brno, Czech Republic, Acta MOSIS 92:29–35

Using Diversity Measure in Building Classifier Ensembles for Combination Method Analysis

Héla Zouari, Laurent Heutte, and Yves Lecourtier

Laboratoire Perception, Systèmes, Information (PSI), Université de Rouen, Place Emile Blondel, 76821 Mont-Saint-Aignan Cedex, France
{Hela.Khoufi,Laurent.Heutte,Yves.Lecourtier}@univ-rouen.fr

Summary. We propose a method for the random generation of classifier outputs with specified individual accuracies and fixed pairwise agreement. A diversity measure (kappa) is used to control the agreement among classifiers for building the classifier teams. The generated team outputs can be used to study the behaviour of class-type combination methods such as voting rules over multiple dependent classifiers.

1 Introduction

In recent years, classifier combination has been used extensively for improving the performance of recognition systems. Diversity among classifiers is considered as a desired characteristic to achieve this improvement. Though still an ill-defined concept, the diversity has been studied recently in many multiple classifier problems such as identifying the minimal subset of classifiers that achieves the best prediction accuracy [1] or building ensembles of diverse classifiers [9, 2]. Several techniques have been investigated to enforce the diversity between the classifiers. According to Duin [3], the principal one is to use different data representations adapted to the classifiers [10]. The diversity can also be implicitly encouraged either by varying the classifier topology [5] or by training each classifier on different parts of the data which is done for example by Bagging [6]. On the contrary, there do exist some experimental studies which design a set of different classifiers by asserting the diversity measure in the process of building ensembles [9, 2]. The advantage of incorporating these measures is to control a priori the diversity between the classifier outputs and then to facilitate the analysis of the combination behaviour. In [9] for example, the authors propose a comprehensive study of the random generation of dependent classifiers to evaluate class type combination methods especially the majority voting. Formulas are derived according to how two classifiers can be generated with specified accuracies and dependencies between them. Based on these formulas, the authors propose an algorithm for generating

multiple dependent classifiers. Lecce et al. have also studied the influence of correlation among classifiers to evaluate combination methods of class type [2]. The performance of a combination method is measured on several sets of classifiers, each set differing from the others in terms of recognition rate and level of correlation. Each classifier is simulated at output level and a similarity index is used to estimate the stochastic correlation among the classifiers of each set.
In this paper, we are interested in the use of diversity for designing classifier teams. The idea behind the proposed method is to generate classifier outputs with desired accuracies and fixed agreement. A statistic measure, chosen in advance, is used to determine the pairwise agreement between the classifiers. The paper is thus organized as follows. Section 2 presents the diversity measure used to generate the pairwise agreement. Section 3 describes the algorithm for generating two classifiers according to the predefined agreement measure and fixed accuracies. The case of generating more than two classifiers is addressed in section 4. Simulation results are reported in section 5 and the conclusions are finally reported in section 6.

2 Measuring diversity

To evaluate the diversity, an appropriate measure is needed [7]. In this study, we use the kappa measure κ to estimate the level of agreement between the classifier outputs [8]. We slightly favoured this measure for the following reasons: κ depends on the individual accuracies of the classifiers, and has a specific value 0 for statistically independent classifiers. κ varies between -1 and +1. κ close to 1 indicates that the classifiers tend to recognize the same objects correctly. κ=-1 means that the responses of the two classifiers are different for the same objects. In this paper, we assess a global agreement on correct and incorrect classification. Precisely, let A_i and A_j be two classifiers providing S outputs for a N-class classification problem. Each output of the classifier A_i (respectively A_j) is made up of an input label b_s (s=1, ..., S) and one class label $o_{i,s}$ (respectively $o_{j,s}$). We consider that when A_i and A_j propose the true class label, they agree. If they both propose incorrect labels, they also agree. In all the other cases, they disagree. Let p_i be the accuracy of A_i representing the number of outputs in which the true class appears. We can represent the outputs of the classifier A_i as a S-dimensional binary vector $V_i = [v_{i,1}, ..., v_{i,S}]$ such that $v_{i,s}$ =c (correct), if the output s contains the true class label b_s, and $v_{i,s}$=w (wrong) otherwise. So that S*p_i elements of V_i are c and S*(1-p_i) elements are w. Thus, the N-class problem is transformed into a 2-class problem depending on the presence or absence of the true class in the classifier outputs. The pairwise agreement between A_i and A_j having accuracies p_i and p_j can be obtained by the following function [7]:

$$\kappa_{i,j} = 1 - \frac{P_{cw} + P_{wc}}{2\overline{p}(1-\overline{p})} \tag{1}$$

where $\overline{p}$ is the mean accuracy of the two classifiers, $\overline{p} = \frac{p_i+p_j}{2}$; P_{xy} is the probability that $v_{i,s}$ =x and $v_{j,s}$ =y (see Table 1). $P_{cc} = \frac{n_{cc}}{S}$, $P_{cw} = \frac{n_{cw}}{S}$, $P_{wc} = \frac{n_{wc}}{S}$, $P_{ww} = \frac{n_{ww}}{S}$ with $n_{cc} + n_{cw} + n_{wc} + n_{ww} = S$; n_{ab} is the number of outputs in which the classifiers A_i and A_j propose or not the true class b_s ($v_{i,s}$ =a and $v_{j,s}$ =b).

3 The generation algorithm for two classifiers

Recently, Kuncheva et al. [9] proposed a method for generating multiple classifier outputs (binary). They derive formulas according to which two classifiers can be generated with desired accuracies and dependency Q between them. Inspired by this study, we present here another possible solution based on the measure of agreement. The idea is to determine a priori the relation (agreement or disagreement) between the two classifiers through the diversity matrix DM according to fixed accuracies and pre-specified agreement level and then to generate the classifier outputs respecting this matrix. The generation of the first classifier outputs is beyond the scope of this paper and has been discussed earlier in a separate paper [4]. Here we concentrate on the generation of the second classifier based on the outputs of the first classifier and the diversity matrix. Thus, when two classifiers A_1 and A_2 are needed, the procedure of generation follows 3 steps:

1. calculus of the diversity matrix DM_{12} between A_1 and A_2 according to pre-specified agreement level $\kappa_{1,2}$,
2. generation of the outputs of the basic classifier A_1 according to the fixed accuracy p_1
3. generation of the outputs of the second classifier A_2 from A_1 respecting the fixed accuracy p_2 and the diversity matrix DM_{12} .

Now, we detail the three main steps of this algorithm. The goal of the first step is to determine the values of the elements of the diversity matrix DM_{12} (i.e. P_{cc} , P_{cw} , P_{wc} and P_{ww}) with fixed parameters p_1 , p_2 and $\kappa_{1,2}$. As presented above, the pair-wise agreement, in the case of two classifiers A_1 and A_2 having accuracies p_1 and p_2, is:

$$\kappa_{1,2} = 1 - \frac{P_{cw} + P_{wc}}{2\overline{p}(1 - \overline{p})} \qquad (2)$$

Table 1. Diversity matrix DM_{ij} representing the percentage of agreement and disagreement between two classifiers A_i and A_j

	A_j (c)	A_j (w)
A_i (c)	P_{cc}	P_{cw}
A_i (w)	P_{wc}	P_{ww}

$P_{cc} + P_{cw} + P_{wc} + P_{ww}$=1

After simple algebraic manipulations using (2), we can express the entries P_{cc}, P_{cw}, P_{wc} and P_{ww} using p_1, p_2, $\overline{p}$ and $\kappa_{1,2}$ as:

$$\begin{aligned} P_{cw} &= \frac{(p_1-p_2)+(1-\kappa_{1,2})2\overline{p}(1-\overline{p})}{2} \\ P_{wc} &= P_{cw} - (p_1 - p_2) \\ P_{ww} &= \frac{[(1-p_1)+(1-p_2)]-(P_{cw}+P_{wc})}{2} \\ P_{cc} &= 1 - (P_{cw} + P_{wc} + P_{ww}) \end{aligned} \quad (3)$$

We can therefore calculate each probability from P_{cw} to P_{cc}. After computing the diversity matrix DM_{12}, we can generate the outputs $O_1 = [o_{1,1}, ..., o_{1,s}]$ of the basic classifier A_1 according to the desired accuracy p_1 and we can transform it into the binary vector V_1. Based on this vector and DM_{12}, we generate the vector V_2 of the classifier A_2 as follows. For each element $v_{1,s}$ of V_1 (s=1 to S), we determine the value of $v_{2,s}$. To do that, we draw a value X in the range $[I_1, I_2]$ where $I_1 = [1, n_{v_{1,s}c}]$ and $I_2 =]\, n_{v_{1,s}c}, n_{v_{1,s}c} + n_{v_{1,s}w}]$ (X $\in I_1$ means that $v_{2,s}$=c and X $\in I_2$ means that $v_{2,s}$=w) and we decrement the number $n_{v_{1,s}v_{2,s}}$. Now we generate the vector O_2 of the classifier A_2 according to O_1, V_1 and V_2. For each label b_s, if the two classifiers agree (i.e. $v_{1,s}=v_{2,s}$) then we rewrite $o_{1,s}$ in $o_{2,s}$. Otherwise (the two classifiers disagree), two cases must be managed:

- either $v_{2,s}$=c (i.e. we assume that the classifier A_2 provides the true class), then we write b_s in $o_{2,s}$.
- or $v_{2,s}$=w, then we choose randomly the class label different to b_s and we place it in $o_{2,s}$.

Note that there are cases where the generation of classifier ensembles with fixed diversity and accuracy is not possible (especially for negative diversity). This can be explained by the fact that the kappa measure depends on the individual classifier accuracy. Shown in Table 2 are the possibilities to obtain a perfect generation of pairs of classifiers. We have obtained these results by simulating two classifiers with recognition rate p_1 and p_2 in {0.5, 0.6, 0.7, 0.8, 0.9} for each of different values of diversity κ={-1, -0.9, ..., 1}. For each combination of p_1, p_2 and κ, we compute the values of P_{cc}, P_{cw}, P_{wc} and P_{ww}. We consider the perfect generation when these 4 values are positive. As we can seen from Table 2, for two classifiers having performance equal to 0.5, the diversity can vary between -1 and +1. But, as the accuracy increases, the range of the diversity decreases. There are more possibilities to generate positively dependent classifiers than negatively dependent ones.

4 The case of more than two classifiers

The aim of the procedure described in this section is to generate automatically, for a N class problem, S outputs of L classifiers to be used for analysing the performance of abstract-level combination methods. The input is the number

Table 2. Possible cases for producing the outputs of two classifiers A_1 and A_2 with fixed accuracies and diversity

	0.5	0.6	0.7	0.8	0.9
0.5	[-1, 1]	[-0.8, 0.8]	[-0.6, 0.5]	[-0.5, 0.3]	[-0.4, 0]
0.6		[-0.6, 1]	[-0.5, 0.7]	[-0.4, 0.5]	[-0.3, 0.2]
0.7			[-0.4, 1]	[-0.3, 0.7]	[-0.2, 0.3]
0.8				[-0.2, 1]	[-0.1, 0.6]
0.9					[-0.1, 1]

of classes N, the number of outputs S, a vector with desired individual accuracies p=$[p_1, ...p_L]^T$, and a Kappa matrix $\kappa = [\kappa_{i,j}]$, where $\kappa_{i,j}$ is the desired agreement between classifiers A_i and A_j (i=1, ...,L-1; j=i+1, ...,L). In the first step, the diversity matrices DM_{ij} are determined for each pair of classifiers A_i and A_j using the relations defined in the previous section. With L classifiers, we have to calculate $\frac{L(L-1)}{2}$ diversity matrices DM_{ij}. The outputs of the first classifier are next generated according to the desired accuracy p_1. For each label b_s, a random permutation of the number of classifiers from 2 to L is chosen to indicate the order in which the classifiers will be picked. Starting from the outputs of the first classifier, new vector outputs are generated according to this permutation. For example, take the permutation (1, 3, 2) for 3 classifiers. First, A_1 is used to set the output label of A_3 for b_s using the diversity matrix DM_{13}. Further, the output label of A_2 (for the same b_s) is generated according to the diversity matrices DM_{12} and DM_{23}. This generation process is repeated S times. At the end of this process, we us obtain the outputs of L-1 classifiers.

5 Simulation results

In this section, we report two experiments aimed at demonstrating the effectiveness of the proposed method in generating dependent classifiers according to pre-defined accuracies and diversity, and the difficulty to respect in the same time these two parameters when more than two classifiers have to be simulated. In the first experiment, we have chosen to simulate L=3 classifiers having the same recognition rates p_1=p_2= p_3={0.6, 0.7, 0.8, 0.9} with S = N*1000 outputs. For each recognition rate, we vary the desired diversity agreement (κ_{target}) from -1 to +1 by 0.1 step (when it is possible). For each combination (p_i, κ_{target}), we generate 100 ensembles and we measure the obtained agreement in the ensemble ($\kappa_{obtained}$). Figure 1 shows the relation between κ_{target} (desired value of kappa) and $\kappa_{obtained}$ (kappa value measured on the generated ensemble) for p=0.6 and 0.9. One can first remark that the generation is perfect for values of diversity between -0.1 and 1 (most of the points are on the diagonal) especially for p=0.9. For p=0.6 and $\kappa_{target} <$ -0.2, the target values of kappa are not always respected. This is due to the addi-

tion of the third classifier in the ensemble. Indeed, with two classifiers there is no problem, the desired kappa is well respected. However, when the number of classifiers increases, it becomes more difficult to respect both the specified accuracy of each classifier and the desired level of diversity between each pair of classifier in the ensemble.
We have also studied the obtained values of p ($p_{obtained}$) versus the target values of p (p_{target}) for desired values of kappa. Figure 2 shows the relation between $p_{obtained}$-p_{target} and κ_{target} for p_{target}=0.6 and 0.9. We can see that the individual recognition rates are less respected when the values of the desired kappa are negative. In this case, if we want to generate the outputs with specified kappa and recognition rate, a selection procedure could be applied as proposed in [9] to accept only the outputs whose kappa and p are sufficiently close to the desired values.
In the second experiment, the proposed method is run 100 times with L=5 classifiers with different values of diversity and recognition rates. For each team, we determine the best and the averaged approximation. Tables 3 and 4 show some examples of the averaged and best matrices obtained for 5 classifiers. To determine the best matrices, we compute the distance d between the desired and the obtained matrices and we keep the matrice with the smallest distance defined by:

$$d = \sum_{i=1}^{L-1} \sum_{j=i+1}^{L} |M_{ij}^{target} - M_{ij}^{obtained}|$$

The mean and the standard deviations of the recognition rates of the classifiers have been also calculated (see table 5). As mentioned in the previous results (for 3 classifiers), it is not always possible to keep precise values of both kappa and p especially for negative values of kappa. Generating negatively correlated ensembles is not straightforward when the number of classifiers increases. Values of kappa in [-0.1, 1] can be easily obtained whereas values inferior to -0.1 may not always be respected especially for p=0.6 and 0.7.

6 Conclusion

In this paper, we have proposed a simulation method for randomly generating dependent classifier ensembles. An algorithm for building two classifiers with specified accuracies and a pairwise agreement between them has been presented. The algorithm for generating more than 2 classifiers has also been proposed. The interest of such a simulator is its usefulness for studying the behaviour of abstract-level combination methods especially in the case of combining positively correlated classifiers which is in practice the most encountered situation. The simulation results show however that it is not straightforward to respect precisely both the desired kappa and the desired classifier performance.

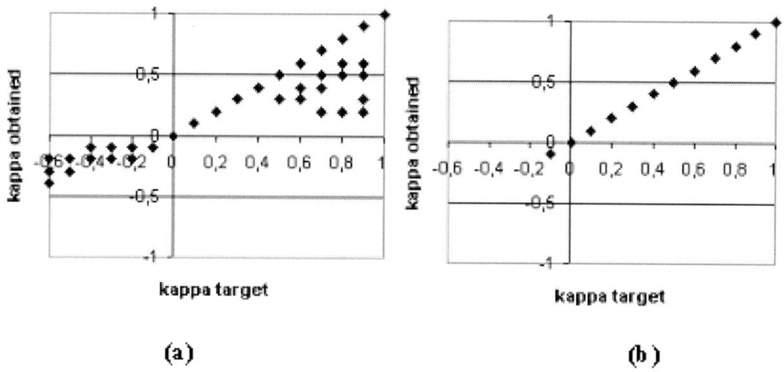

Fig. 1. Relation between κ_{target} and $\kappa_{obtained}$ for ensembles of 3 classifiers (a) p=0.6 (b) p=0.9

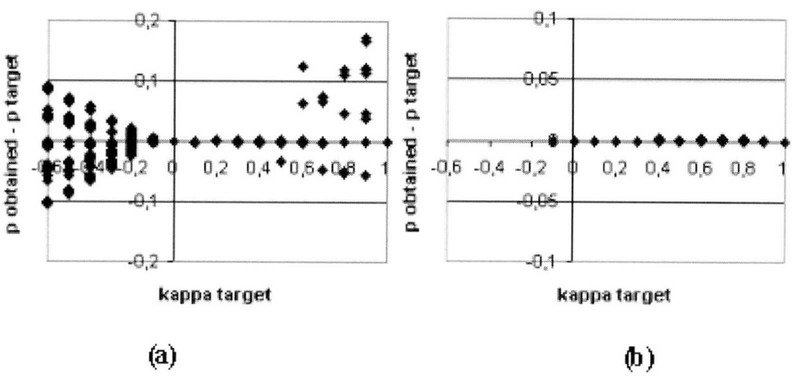

Fig. 2. Relation between $p_{obtained}$-p_{target} and the desired values of kappa for ensembles of 3 classifiers (a) p=0.6 (b) p=0.9

Table 3. The diversity matrices: the best matrix found and the averaged over 100 experiments for p=0.6 and Kappa = -0.6

Best kappa				Averaged kappa			
-0.3000	-0.3000	-0.4000	-0.4000	-0.2750	-0.3100	-0.2650	-0.2980
	-0.1000	-0.1000	-0.2000		0.1320	0.1110	0.1090
		0.0000	0.0000			0.1250	0.1430
			0.1000				0.1240

Table 4. The diversity matrices: the best matrix found and the averaged over 100 experiments for p=0.6 and Kappa = 0.4

Best kappa				Averaged kappa			
0.4000	0.4000	0.4000	0.4000	0.3700	0.3790	0.3650	0.3760
	0.4000	0.4000	0.4000		0.3550	0.3560	0.3580
		0.4000	0.4000			0.3550	0.3620
			0.4000				0.3540

Table 5. The mean and the standard deviations of the recognition rates of the classifiers

	A_1	A_2	A_3	A_4	A_5
p=0.6, κ=-0.6	0.600±0.000	0.585±0.138	0.589±0.136	0.582±0.136	0.591±0.139
p=0.6, κ=0.4	0.600±0.000	0.611±0.022	0.606±0.018	0.613±0.023	0.608±0.019

References

1. Aksela M (2003) Comparison of Classifier Selection Methods for Improving Committee Performance. 4th International Workshop on Multiple Classifier Systems, Guildford, UK, LNCS 2709, 84–93
2. Lecce VD, Dimauro G, Guerrierro A, Impedovo S, Pirlo G, Salzo A (2000) Classifier Combination: The Role of AŋPriori Knowledge. In: 7th International Workshop On Frontiers In Handwriting Recognition. Amsterdam 143–152
3. Duin RPW (2002) The Combining Classifier: To Train or not to Train?. ICPR16, Proceedings 16th International Conference on Pattern Recognition, Quebec City, Canada, Vol. II, IEEE Computer Society Press, Los Alamitos, 765ŋ770
4. Zouari H, Heutte L, Lecourtier Y, Alimi A (2003) Simulating Classifier Outputs for Evaluating Parallel Combination Method. 4th International Workshop on Multiple Classifier Systems, Guildford, UK, LNCS 2709, 296–305
5. Tax DMJ, Breukelen MV, Duin RPW, Kittler J (2000) Combining Multiple Classifiers by Averaging or by Multiplying?. Pattern Recognition 33:1475–1485
6. Breiman L (1996) Bagging Predictors. Machine Learning 24:123–140
7. Kuncheva LI and Whitaker CJ (2003) Measures of Diversity in Classifier Enŋ sembles and their Relationship with the Ensemble Accuracy. Machine Learning 51:181–207
8. Fleiss J (1981) Statistical Methods for Rates and Proportions. John Wiley and sons
9. Kuncheva LI and Kountchev RK (2002) Generating Classifier Outputs of Fixed Accuracy and Diversity. Pattern Recognition Letters 23:593–600
10. Kittler J, Hatef M, Duin RPW and Matas J (1998) On Combining Classifiers. IEEE Transactions on Pattern Analysis and Machine Intelligence 20:226–239

Part III

IMAGE PROCESSING AND COMPUTER VISION

An Evolutionary Algorithm for Solving the Inverse Problem for Iterated Function Systems for a Two Dimensional Image

Andrzej Bielecki and Barbara Strug

Institute of Computer Science, Jagiellonian University,
Nawojki 11, Cracow, Poland
{bielecki, strug}@softlab.ii.uj.edu.pl

Summary. This paper presents an approach based on evolutionary computations to the IFS inverse problem. Having a bitmap image we look for a set of functions that can reproduce a good approximation if a given image. A method using variable number of mappings is proposed. A number of different crossover operators is described and tested. Different parameters for fitness functions are also tested. The paper ends with some experimental results showing images we were able to generate with our method

Introduction

This paper is a continuation of our previous research [2] concerning finding an Iterated Function System (IFS) that can be used to generate a given two dimensional target image - so called inverse problem. Such a problem may be considered to be a search problem, where the search space consists of all possible IFSs. The main problem lies in the size of such a search space.

There is a number of well known search algorithms [9]. It has been shown that genetic algorithms (GA) and their generalizations offer a nearly optimal compromise between exploitation and exploration as it keeps a large population of different solutions that carry out a parallel search of the space [5, 6, 8].

The evolutionary methods has been applied to the inverse problem by Giles [4], who described the use of GA but he used the Hausdorff metric as a fitness function what resulted in very long running time of his implementation. The evolutionary programming has also been used in [7] where a predefined set of contraction mappings is used. The GAs has also been applied to a similar problem of image compression with the IFS [12]. But this approach does not actually try to solve an inverse problem but it divides image into smaller parts and tries to find possible mappings between this parts and bigger ones.

In this paper an approach based on evolutionary search is presented. The main difference to the majority of other approaches based on genetic algorithms the real numbers representation is used rather then a binary strings. Moreover the number of mappings a given IFS consists of is not fixed but can be changed during the evolutionary search. The solution presented in this paper differs from the one presented in [2] by using modified form of the fitness function and by the introduction of a new evolutionary operator. Some new experiments have also been carried out.

1 Theoretical Basis

Iterated Function Systems. Iterated Function Systems were developed by Barnsley [1, 3]. The theory is based on contractive mappings in metric spaces. Here we briefly recall definitions and results that are essential to our method.

Let (X, d) be a metric space. A transformation $f : X \to X$ is called *contractive mapping* if there exists a constant $s \in [0, 1)$ such that $d(f(x), f(y)) \leq s \cdot d(x, y)$ for each $x, y \in X$. The number s is called the contractive factor of f. If f is a mapping then the point $x \in X$ such that $f(x) = x$ is called the *fixed point* of the mapping. Let $(\mathcal{H}(X), h)$ be a space of nonempty compact subsets with the Hausdorff metric. Let $w_i : X \to X, (i = 1 \ldots n)$ be a set of contractive mappings on (X, d). Let's define $W_i : \mathcal{H}(X) \to \mathcal{H}(X), (i = 1 \ldots n)$ as $W_i(A) = \{w_i(p) : p \in A\}$ for each $A \in \mathcal{H}(X)$. Then $W : \mathcal{H}(X) \to \mathcal{H}(X)$ defined as $W(A) = \bigcup_{i=1}^{n} W_i(A)$ is a contractive mapping and the corresponding IFS is defined as $\{X, W_i : i = 1 \ldots n\}$. The mappings used on Euclidean plane are usually affine and can be written as $W_i(p) = w_i(x, y) = (a_i \cdot x + b_i \cdot y + e_i, c_i \cdot x + d_i \cdot y + f_i) = A_i \cdot p + t_i$.

The mapping W can be defined recursively as $W^0(A) = A, W^n(A) = W(W^{n-1}(A))$. Thus the fixed point of W is defined by $\mathcal{A} = \lim_{n\to\infty} W^n(A)$, for each $A \in \mathcal{H}(X)$. This fixed point is the attractor of W.Its existence and uniqueness is guaranteed by Banach Fixed Point Theorem.

The important property of any IFS is the fact that attractor is independent from the image to which W is applied. It means that the attractor $\mathcal{A}$ is fully defined by the set of coefficients of W and can be generated be repeatedly applying the IFS to any starting point.

Moreover for the continuous transformations of the IFS its attractor is also continuous. It means in turn that small changes to coefficients of transformations result in small changes of the attractor. Thus by adjusting parameters we can move closer to the IFS whose attractor is similar enough to a desired image. The existence of such IFS is guaranteed by the Collage Theorem [1].

Evolutionary Computations. Genetic algorithms were first researched by Holland [6]. He named a number of elements that must be present in such a system: a search space (simulating the environment), a population of solutions (individuals), a method for changing the individuals in the population and the measure of the quality of each solution (fitness function).

In case of genetic algorithms each solution has been represented as a binary code of a fixed length. Later this method was extended by using other representations (evolutionary strategies) and hierarchization [10, 11]. Goldberg and Michalewicz [5, 8] proposed a hybrid evolutionary algorithm where a representation is taken from the problem and it is then evolved using appropriately modified genetic operators.

2 Representation, Algorithm and Experiments

Representation. An IFS can most naturally be represented as a number of vectors; each consisting of 6 parameters (coefficients of affine functions). Thus each IFS consists of a number NF defining the number of IFS functions used and the functions themselves i.e each IFS is represented as $\{n_i, v_1, \ldots v_k\}$, where n_i is the number of affine mappings and $v_i = (a_i, b_i, c_i, d_i, e_i, f_i)$ is a vector of coefficients of w_i. For example the following sequence represents a pattern based on square: (3, (0.5,0.0,0.0,0.5,-2.5638,-0.0000003) (0.5,0.0,0.0,0.5,2.43555,-0.0000003) (0.0,-0.5,0.5,0.0,4.8731,7.3635)) shown in fig 1a.

Fig. 1. Target image

Genetic Operators and Selection. The individuals in each population have to be updated iteratively to form the next population. The updating process must take into account the individuals quality in respect to a problem being solved and the search space constraints (if there are any).

For the representation based on real values a number of different types of crossover can be defined. One of them is the so called arithmetic crossover which takes two selected IFSs - parents p_1 and p_2 and produces two offsprings $c1 = p1 \cdot (1-a) + p2 \cdot a$, $c2 = p1 \cdot a + p2 \cdot (1-a)$, where $a \in [0, 1]$). The second recombination operator is a vector one-point crossover. It takes two vectors from parent population $v_1 = (a_1, b_1, c_1, d_1, e_1, f_1)$ and $v_2 = (a_2, b_2, c_2, d_2, e_2, f_2)$ and exchanges the parts divided at a random point in a way based on standard one-point crossover in GA. The number of offsprings generated by each of these operators is controlled by global parameters tuned experimentally.

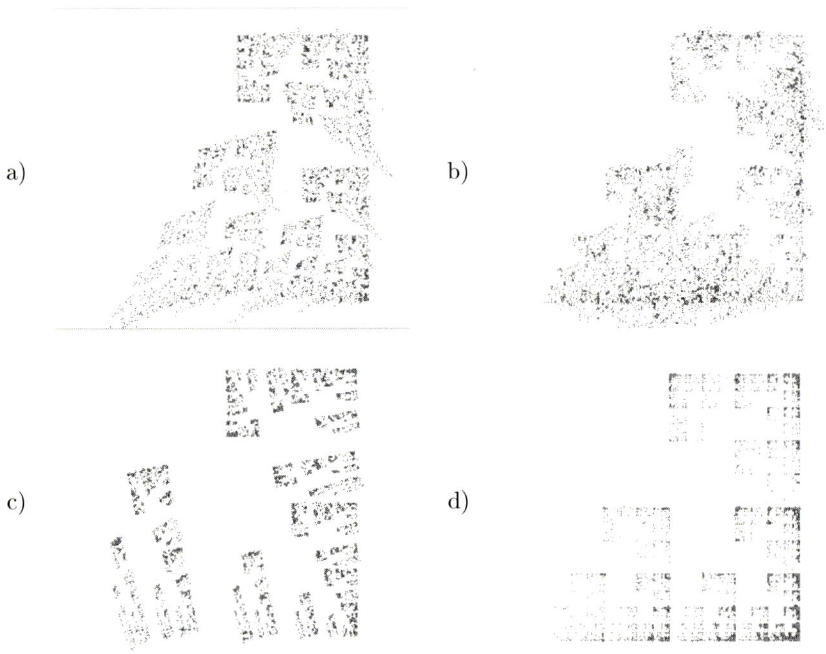

Fig. 2. The best elements in the first experiment after a) 1000 generations, b) 1500 generations, c) 2000 generations and d) 4000 generations

In this papers two types of mutation are applied. One of them is a random mutation where a number $x \in [a, b]$ is mapped into $x' \in [a, b]$ using a uniform distribution. The second type of mutation is based on Gaussian distribution and a number $x \in [a, b]$ is mapped into $x' = x + N(0, d)$ (and clipped to [a,b]) where $d = r \cdot (b - a)$. The bigger the value of r the greater neighbourhood of a mutated solution is reached. The probability of applying a given type of mutation changes over time with random mutation being most likely to occur at the beginning to prevent the premature convergence and more controlled type of mutation is more likely to dominate in later stages of the evolution. Parameter r is also changed over time to ensure smaller mutations to take place in more advanced stage of the search. The total probability of mutation is not influenced by the relative probabilities of different types of mutation.

Selection is based on a standard fitness proportional procedure in which the fittest individual is most likely to be selected as a parent. Moreover the elitism is used in which a set percentage of fittest individuals always survive to the next generation.

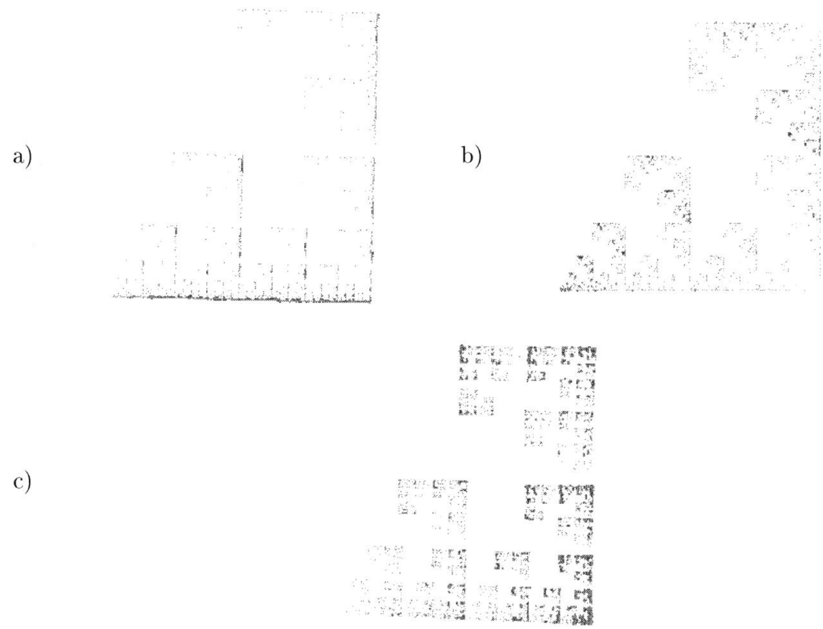

Fig. 3. The best elements in the second experiment after a) 500 generations, b) 2500 generations and c) 4000 generations

Fitness Function. Although in a number of approaches a Hausdorff distance of a target image and an attractor of IFS being evaluated is considered, its computational time makes it unpractical. In this paper a fitness function of an IFS is based on the relative points coverage of the target image. To evaluate how well the attractor covers an image we calculate the number of differences: the number of points that are in the image but not in the attractor - N_{ND} (not drawn points) and the number of points present in the attractor but not in the image -N_{NN} (points not needed). If we denote by N_A the number of points in the attractor and by N_I - in the image then $RC = N_{ND}/N_I$ is the relative coverage of the attractor and $RO = N_{NN}/N_A$ calculates how many points of an attractor are outside the image. Thus the smaller each of this values the better the solution. Two combine both values we use the fitness function defined as $(1 - RC) + (1 - RO)$. This function has to be maximized. The points produced by an IFS outside the image rectangle are also classified as points not needed thus leading to low fitness and fast elimination of such functions.

In such a fitness function an equal importance is given to the number of points drawn correctly and points drawn outside. It seems useful to first eliminate IFSs drawing too many points outside the image and then concentrating on improving point coverage. This requires replacing this fitness function with

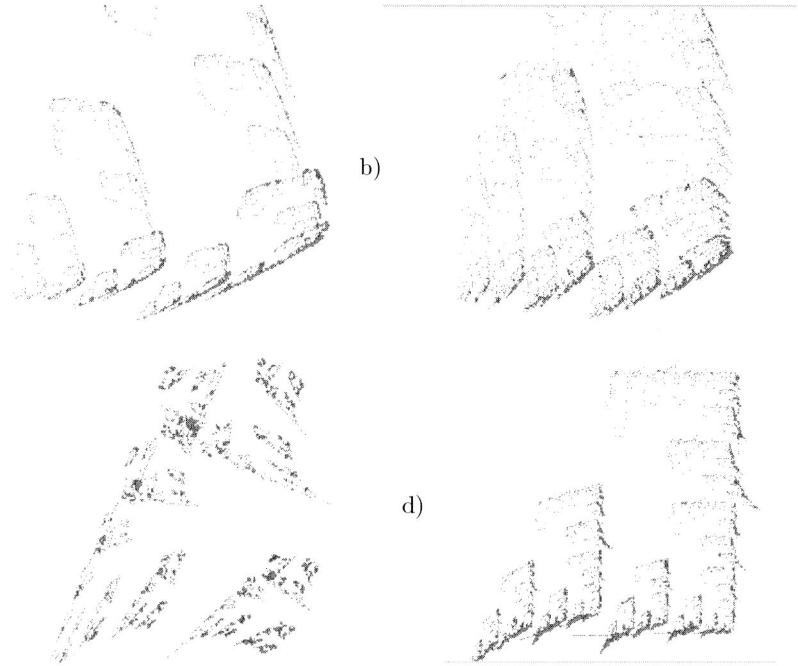

a) b) c) d)

Fig. 4. The best elements in the third experiment after a) 1000 generations, b) 1500 generations, c) 2500 generations and d) 4000 generations

its modified version $p_{rc} \cdot (1 - RC) + p_{ro} \cdot (1 - RO)$ and changing parameters p_{rc} and p_{ro}.

Growth. As it is not possible to establish the number of mappings needed for an IFS to produce an attractor well approximating a target 2D image number of mappings is usually set. In this paper we propose a different approach. We start with the minimal number of mappings (2) and search the space. If no satisfactory approximation of target image is attained IFSs are allowed to "grow" i.e. number of mappings they consist of is increased by 1. This increase may occur if a population stabilizes (i.e. there are small changes in fitness) but the fitness of the best individual is still very low.

Experiments. We carried out a number of experiments. Figs. 2 to 5 depict some of the results for the pattern based on square. In the first experiment the fitness function without any parameters was used and mainly arithmetic crossover was applied. For the second experiment we used the following parameters of the fitness function: $p_{rc} = 1$ and $p_{ro} = 5$ and arithmetic crossover. In the third experiment we set $p_{rc} = 1$ and $p_{ro} = 4$ while in the fourth one - $p_{rc} = 2$ and $p_{ro} = 3$.

In all experiments we used a population of 200 elements. It took usually over 20 hours to compute 2000 generations on a desktop PC. Although the best images in each experiment are not absolutely equal to the target image

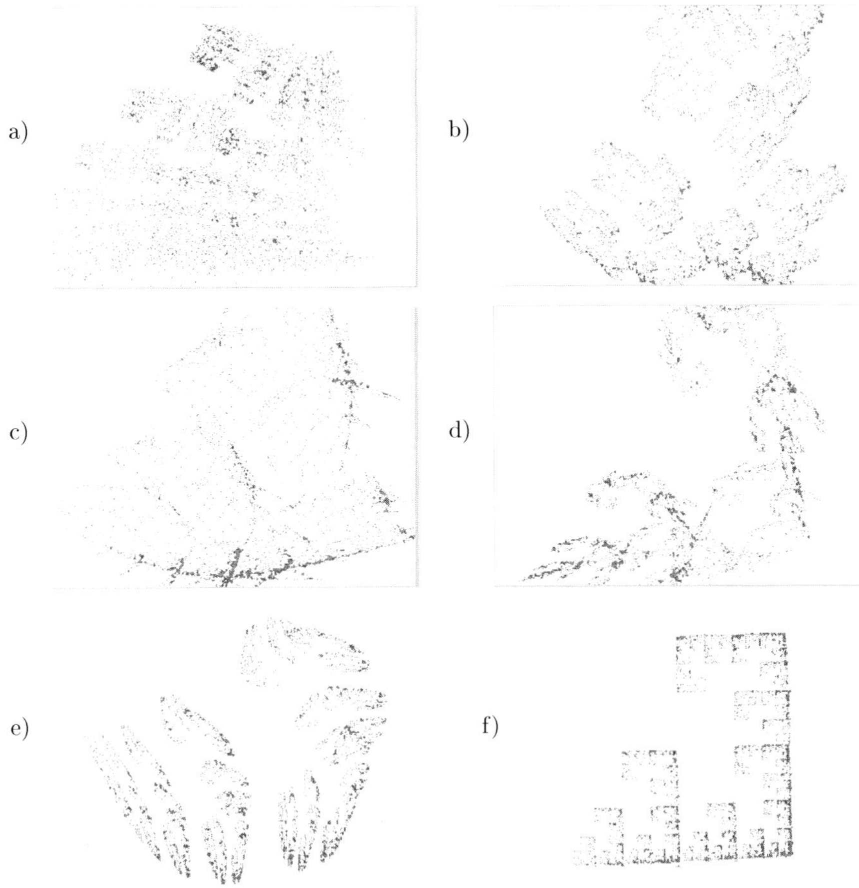

Fig. 5. The best elements in the fourth experiment after a) 500 generations, b) 1000 generations, c) 2500 generations, d) 3000 generations, e) 3500 generations and f) 4500 generations

in majority of the cases they may be considered to be satisfactory approximations.

3 Concluding Remarks

In this paper a method of finding IFS for a given image based on variable number of mappings and evolutionary computation is presented. We have tested the fitness function where different importance is given to the number of points drawn correctly and points drawn outside. As it could be seen from the results it seems very useful to first eliminate IFSs drawing too many points outside the image and then concentrating on improving point coverage. The parameters p_{rc} and p_{ro} were set for each experiment and were constant during the evolutionary process. In the further exploration of the problem we plan to allow for the parameters to be changed during the run of the evolutionary algorithm. Thus it should be possible to start with high value of the parameter responsible for removing outside points and then lower this value and focus on quality of coverage of the image.

In the current implementation the standard roulette wheel selection is used. To avoid loosing good solution it is combined with elitism. Other selection methods that may prove better for this problem are currently tested. The structural analysis of the image could also be used to set the starting number of mappings.

References

1. Barnsley,M.F. (1988) Fractals Everywhere,Academic Press.
2. Bielecki, A., Strug, B. (to apear) Evolutionary Approach to Finding Itarated Function Systems for a Two Dimensional Image in Computational Imaging and Vision, Kluver,
3. Falconer, K., (1990) Fractal Geometry: Mathematical Foundations and Applications, Wiley.
4. Giles,P.A. (1990) Iterated Function Systems and Shape Representation. PhD thesis, University of Durham, UK.
5. Goldberg,D. E.,(1989) Genetic Algorithms in Search, Optimization and Machine Learning, Reading, MA, Addison-Wesley,
6. Holland, J. H. (1975) Adaptation in Natural and Artificial Systems, Ann Arbor.
7. D.E. Hoskins and J. Vagners. (1992) Image compression using Iterated Function Systems and Evolutionary Programming: Image compression without image metrics. In Proceedings of the 26th Asilomar Conference on Signals, Systems and Computers.
8. Michalewicz, Z.: (1996) Genetic Algorithms + Data Structures = Evolution Programs. Springer,Berlin.
9. Michalewicz, Z. Fogel, D. B.: (2000) How to Solve It: Modern Heuristics. Springer.
10. Schaefer, R. (2002) Foundations of Global Genetic Optimization, UJ, Krakow, (in Polish)
11. Schaefer, R., Kolodziej, J.(2003) Genetic Search Reinforced by the Population Hierarchy in Foundations of Genetics Algorithms 7, De Jong, K.A., Poli, R. , Rowe, J.E. - eds.,pp 383-399, Morgan Kaufmann.
12. Vences, L. and Rudomin, I. (1994) Fractal compression of single images and image sequences using genetic algorithms. Manuscript, Institute of Technology, University of Monterrey, 1994. available from ftp://ftp.informatik.uni-freiburg.de/papers/fractal/VeRu94.ps.gz.

Specification of the Evidence Accumulation-Based Line Detection Algorithm

Towards Finding Blood Vessels in Mammograms

Leszek J. Chmielewski

Institute of Fundamental Technological Research, PAS, Świętokrzyska 21, 00-049 Warsaw, Poland lchmiel@ippt.gov.pl

Summary. The recently proposed algorithm, using the evidence accumulation principle, for finding lines (ridges) having shape which can be neither parameterized nor tabularized is described in detail. This fuzzy, multi-scale algorithm stores the evidence in the accumulator congruent with the image domain. The primary application was finding blood vessels in mammograms.

1 Introduction

The useful information contained in images is dispersed. By this statement it is meant that, besides the extremal cases, no useful information is contained in a single pixel; on the contrary, it is always contained in groups of pixels, not necessarily neighbouring each other. Before the analysis is completed it is not possible, as a rule, to distinguish between pixels carrying relevant information and those which contain irrelevant, misleading or erroneous information, or from which the information is missing. Although this statement is straightforward, it has large practical significance in the design of robust image analysis methods: the information should be collected from regions having location and size consistent with the features sought, and the dependence of the result on non-relevant data should be minimized.

The important class of methods which have good properties in the above described sense are the evidence accumulation methods. In the image analysis and feature detection settlement, the Hough transform (HT) was historically the first such method ([1], reviewed, among others, in [2, 3]). The term *evidence accumulation* was used in [4] and the notion of *evidence gathering* in [5]. The necessary condition for using these methods is the possibility of representing the shape of the object of interest either in a parametric form – standard HT, or with the use of a template or table – generalized HT [6, 7], or with shape descriptors [5].

In this paper, the method for detecting lines (ridges) having shapes which can be neither parameterized nor tabularized, proposed recently [8], is described in more detail. The lines detected are of various and slowly changing widths and can have a slowly, but unpredictably changing direction. The feature of the bright line, chosen for the implementation, is that its edges have large gradients directed towards each other. Pairs of pixels, lying each on the other side of the line, are considered in an elementary accumulation. The consistence of line presence and its slowly changing direction among numerous pairs of pixels in the considered region is what makes the accumulation result meaningful. The measure of presence, or the *intensity* of the line, or simply *lineness*, emerges in the accumulator which is congruent with the image domain. The accumulation process is fuzzified in several simple ways (see [9] for an in-depth analysis of the fuzzy accumulation). The accumulator can be analyzed according to the requirements related to the application considered.

The method described has relatively high complexity, inherited both from the m-to-1 HT [2] (here, $m=2$) and from the Symmetry Transform [10], to which it is related in respect of using pairs of pixels and consistence conditions. However, with the equipment currently available, images of moderate size can be analyzed in acceptable time. It should be stressed that the method is inherently multi-scale, that is, the analysis is performed for the specified range of expected line widths simultaneously, in a single run. The complexity is high only under the assumption that the bounds of the range of line widths are related to the image size.

The primary motivation of the previous and the presented work was the detection of blood vessels in mammographic images. The location of the features of malignancy, like microcalcifications, neoplastic masses and spiculated patterns, with respect to the elongated normal structures, like blood vessels and milk ducts, is important for the diagnosis.

Recently, the detection of elongated objects specifically in mammograms was studied in [11], being a continuation of [12]. From a number of detectors, the line operator proposed in [13] and the orientated bins detector proposed in [11] occurred to be the best. It seems that the strength of these principally simple methods was in that raw grey-level data were used rather than their derivatives. One of the compared methods, proposed in [14], which was claimed in [11] to be widely recognized as a benchmark for ridge and line detectors, used second-order derivatives and failed in comparisons. The methods using scale-orientation signatures [15] appeared to be even more reliable.

The quality of the mammograms, which is inherently limited due to small differences in the density of the visualized tissues, makes it impractical to use the methods typically applied to the detection of blood vessels in angiography, where contrast enhancing agents are added to blood.

The algorithm introduced in [8] and studied here in more detail is generally different from those described in the cited papers. Although a thorough comparative study is still to be done, it can be said that the proposed algorithm seems to work well in images of the quality typical for mammograms (see Fig. 7). Therefore it is suggested that it can also be applied to other images having comparably unfavourable quality.

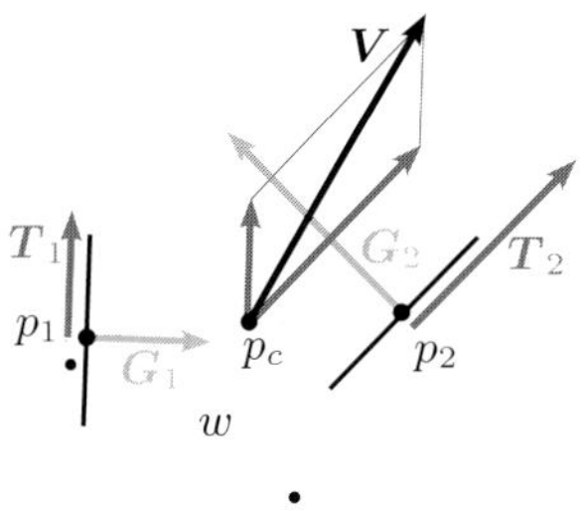

Fig. 1. Derivation of the accumulated value in pixel p_c from gradients in two pixels p_1, p_2 (see text).

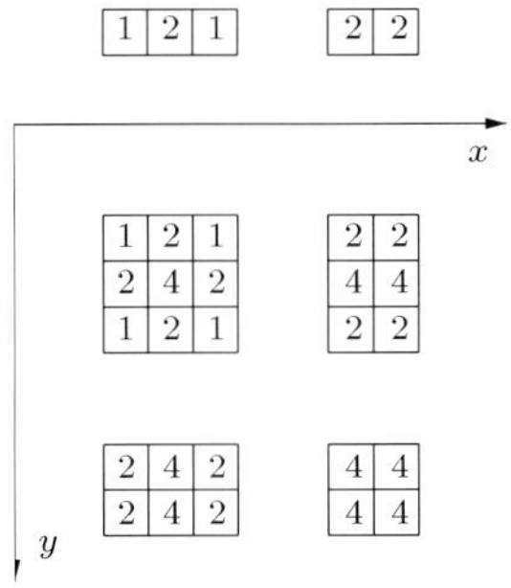

Fig. 2. Masks defining the membership function in fuzzy voting in a central pixel p_c (see text, p. 358).

2 Line detection as evidence accumulation

Model and notations It is assumed that a bright line is an object having two nearly parallel edges characterized by large image brightness gradients directed towards each other. Pairs of pixels $p_i, i = 1, 2$, are considered in an elementary accumulation. Pixels in a pair are located at distances within a range corresponding to the expected range of line widths. Notations are shown in Fig. 1. Letters in bold denote vectors. Each pixel is related to an edgel, shown as black line going through the pixel, having the *edgeness* (edge intensity) equal to the gradient modulus $|\boldsymbol{G}_i|$. Vectors $\boldsymbol{T}_i$ tangent to the edgels are the gradients rotated by $\pi/2$ in two opposite directions. The accumulation for each pair is performed in its central pixel p_c.

Necessary conditions For each pixel p_1, the second pixel p_2 is considered within the specified region (explained further). The gradient in each pixel must point towards the second pixel; for example, $\boldsymbol{G}_1$ must form an acute angle with $\overrightarrow{p_1 p_2}$. Vectors $\boldsymbol{T}_1, \boldsymbol{T}_2$ must form an acute angle. Image intensities in all pixels of $\overline{p_1 p_2}$ must not be smaller than that in its darker end.

Direction, width and intensity Line direction, defined by angle $\measuredangle(\overrightarrow{Ox}, \boldsymbol{V})$, is the direction of the sum $\boldsymbol{V}$ of the tangent vectors $\boldsymbol{T}_i$. The length $|\boldsymbol{V}|$ is the basis for calculation of the line intensity to be accumulated. The line width w is the length of the projection of $\overline{p_1 p_2}$ onto the normal to $\boldsymbol{V}$.

Penalty functions and $\cos^2$ *function* Before accumulation, $|\boldsymbol{V}|$ is multiplied by the penalty functions of directional consistence c_d and edgeness consistence c_e:

$$c_d = \cos^2[\measuredangle(\boldsymbol{T}_1, \boldsymbol{T}_2)] , \tag{1}$$

$$c_e = \cos^2[\pi \; (1 - \min(e_1, e_2)/\max(e_1, e_2)) \; / \; 2] , \tag{2}$$

where e_i is the edgeness, $e_i = |\boldsymbol{G}_i|$. There is no penalty for the acuteness of angles formed by $\overline{p_1 p_2}$ and the line, so the evidence is gathered from short as well as longer distances from p_c with equal influence on the result.

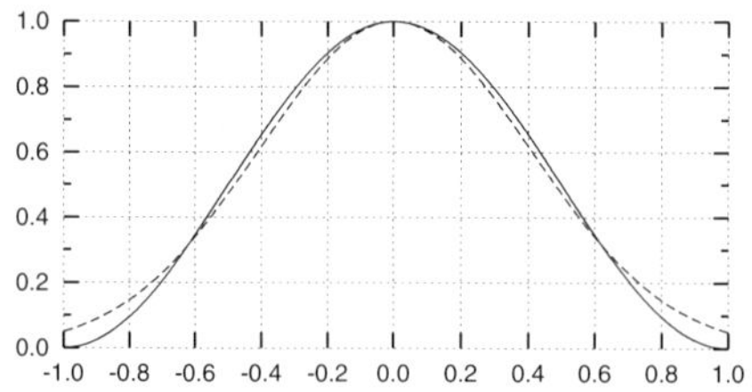

Fig. 3. Cosine square with zeros in $\{-1, 1\}$ (*solid line*) and Gaussian with the $\pm 2\sigma$ range $[-1, 1]$ (*dashed line*).

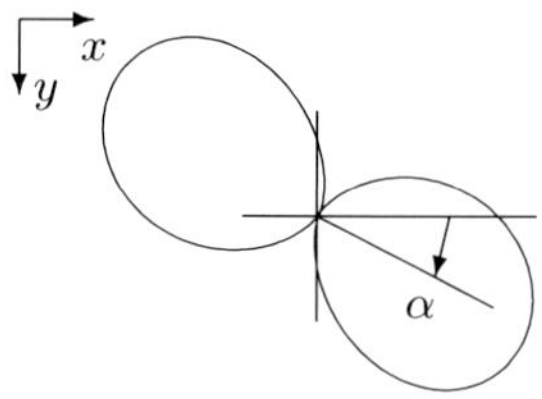

Fig. 4. Lineness fuzzified with the $\cos^2$ function (see text, p. 360).

Note that the cosine square function, with a proper frequency, has been used in both cases. This function has been used as the weighting function, or the membership function, throughout this paper. Its shape is similar to that of the Gaussian function, frequently used to model uncertainty (Fig. 3), but has some valuable feature which will become apparent further.

Accumulated value and accumulator The accumulated value – line intensity or *lineness* l – is calculated as

$$l = c_d \, c_e \, |\boldsymbol{V}| \, . \tag{3}$$

The accumulator $L_{xyw\alpha}$ is four-dimensional, real-valued. The x, y are the image coordinates of the central pixel p_c of the pair and $x \in \langle 1, x_u \rangle$, $y \in \langle 1, y_u \rangle$; further, $w \in \langle w_w, w_u \rangle$ is the edge width, and α is its direction, $\alpha \in \langle 0, \pi \rangle$. Indexes w and u denote the lo<u>w</u>er (min) and <u>u</u>pper (max) value, respectively. The final result of the detector in a pixel (x, y) is the lineness equal to the maximum over w and α for this pixel, and the w and α at the maximum are the local characteristics of the line. The accumulator can be analysed according to the requirements related to the application considerd.

Odd and even distances between p_1 and p_2 If the distance between the pixels in the pair is odd in x or y direction, then the central point has non-integer coordinate in this direction and does not fall into any pixel and into any element of the accumulator. Rounding would introduce a systematic error of half a pixel in locations of lines having odd widths. One remedy for this is to introduce appropriate rows and columns into the accumulator, but this would (nearly) double its size. Another remedy, applied here, is to fuzzify the votes between the neighbouring accumulator elements so as to reflect the non-integer location of the central point. In this way, the final decision on the location of the ridge of lineness, that is, the maximum of the so obtained membership function, is postponed until the accumulation is finished.

In Fig. 2 the masks giving rise to the membership functions used in fuzzy accumulation are shown. Above the coordinate system the basic masks for integer (*left*) and non-integer (*right*) coordinates of the central point are displayed. They both have four votes. By transposing one of the masks and con-

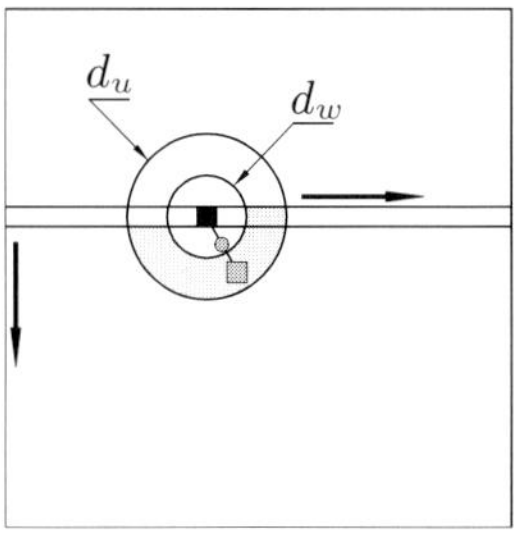

Fig. 5. Region (*grey*) swept by pixel p_2 (*grey square*) for a given pixel p_1 (*black square*). Central pixel p_c (*grey circle*) in between (see text).

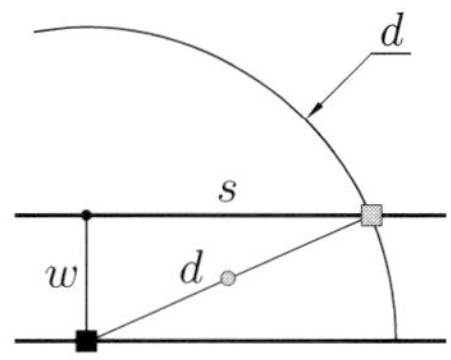

Fig. 6. Limits on distance d of pixels of a pair: p_1 (*black square*) and p_2 (*grey square*) and on line width w (see text).

volving it with the other one, the masks for four combinations of integer and non-integer x and y are shown inside the coordinate system: x integer (*left*), x non-integer (*right*); y integer (*upper*), y non-integer (*lower*). The membership function values are equal to mask elements divided by 4.

Range of accumulation The region to which the second pixel p_2 can belong for a given first pixel p_1 is restricted by the lower and upper bounds on the expected widths of lines in the image: w_w and w_u. This is explained in Figs. 5 and 6. The lower bound on distance d_w is simply w_w. The region swept by the pixel p_2 is inside the ring $R(p_1, d_w, d_u)$ and in front of and below p_1.

The region to which the voting pixels belong should have similar proportions for narrow as well as for wide lines. Therefore, the maximum projection of distance of the voting pixels on the line direction is limited (Fig. 6):

$$s \leq f_d \, w \;, \tag{4}$$

where f_d is the maximum voting pixel distance factor. This condition is applied in two ways. First, when the range of line widths expected in the image is specified, the corresponding global range of distances is established once:

$$d_w = w_w \;, \tag{5}$$

$$d_u = \sqrt{w_u^2 + (f_d w_u)^2} \;. \tag{6}$$

Second, for each elementary accumulation, as soon as w is found for the given pair, the condition corresponding to (6) is checked

$$d \leq \sqrt{w_u^2 (1 + f_d)^2} \tag{7}$$

and the current elementary accumulation is rejected in case it is false. In the examples, $f_d = 3$ was assumed.

As a result of these formulations, more pixels vote for a wide line than for a narrow one. Consequently, in place of Eq. (3), the following is used:

$$l = c_d \, c_e \, |\boldsymbol{V}| \,/\, w \;. \tag{8}$$

This issue related to the scale invariance of the detector is extended in [16].

Fuzzification along the angle and 3D accumulator The cosine square function (Fig. 4) was used for fuzzifying the voting for the angle. The fuzzified function can be rewritten as $l\cos^2(\varphi-\alpha) = 0.5l\cos(2\varphi-2\alpha)+0.5$. In the accumulation process, the systematic multiplicative and additive constants are insignificant. Hence, instead of adding the values resulting from the fuzzification to all the adequate accumulator elements, it is possible to add two harmonic functions having the same frequency, but differing by amplitude l and phase 2α. Now, the size of the accumulator can be reduced so that it stores only two reals, l and α, in each L_{xyw}, which can be written as L_{xywv_2}, where $v_2 = l$ or α. It is more practical to store three reals: l, $\sin 2\alpha$ and $\cos 2\alpha$ during the calculations, and to find α for each (x, y, w) after the accumulation is completed.

Fuzzification along the direction To further enhance the elongated objects and omit the short ones, after the accumulation the accumulator elements are fuzzified along the line direction. As the membership function, once more the $\cos^2$ function is used, scaled to span the range $\pm 2w$.

Postprocessing The condition of local homogeneity of line direction is used. This is calculated as the opposite of the standard deviation of angle in a circular neighbourhood of a pixel, of diameter w. After this result is calculated for the whole image, it is mapped into an interval $\langle 0, 1\rangle$.

In the end, a simple ridge following algorithm is run on the accumulator L_{xywv_2}, starting from the strong local maxima of l – larger than the lineness threshold $f_l \max(L)$. The next pixel is chosen as the one having the largest value from the six pixels found as follows. For the width w of the current element of L and its two neighbouring widths, two pixels are considered, which determine the directions according to two roundings (up and down) of the local direction α. If any of these directions is in contradiction with the previous move, the respective pixel is rejected. The ridge following stops if another ridge or image edge is reached, or if the accumulated value of the next pixel is too low: less than f_a-th part of the average accumulated value for the currently analysed ridge. The coefficient f_a is fixed to 0.25, and the lineness threshold coefficient f_l is the parameter of the algorithm.

3 Parameters, complexity and storage

The significant parameters of the algorithm are the lower and upper edge widths w_w, w_u and the lineness threshold coefficient f_l. Other parameters: the size of the mask for the calculation of the gradient, set to 5×5, the range of fuzzification along the line direction, set to $\pm 2w$, the maximum voting pixel distance factor $f_d = 3$ and the average value coefficient for a ridge $f_a = 0.25$ are the 'hidden' parameters which practically do not need tuning.

As explained above, for a given pixel p_1, a second pixel p_2 is taken from the front-and-lower part of ring of radii d_w, d_u (Fig. 5), and the radii depend linearly on w_w, w_u (Eqs. (5), (6)). The number of pairs to analyse is then proportional to $x_u\, y_u\, (w_u^2 - w_w^2)$ (in practice, less than 10% of pairs meet the conditions). The fuzzification along the line direction is performed for $x_u\, y_u\, (w_u - w_w + 1)$ accumulator elements and spans along $(4w_u + 4w_w)/2$ pixels average for each cell, which leads to the same order as for the accumulation: $O(x_u\, y_u\, (w_u^2 - w_w^2))$. The storage requirements are of the order $O(x_u\, y_u\, (w_u - w_w))$. Under the assumption that the image is square: $x_u = y_u = n$, and that the expected line widths are proportional to the image size n, we receive $O(n^4)$ for complexity and $O(n^3)$ for storage. If the expected line widths do not depend of the image size, it is $O(n^2)$ for complexity and $O(n^2)$ for storage. This makes it possible to use the method effectively for images of moderate size, even with the computers with typical processors.

4 Example

In the following example, the version of the algorithm with the modifications discussed in [16] was used. The image is one of the phantom images used by Zwiggelaar et al. in [11, 12] (the *Dense Mammographic Background* image) available at `http://www2.cmp.uea.ac.uk/~rz/miu/miuLinear.html` . The image contains short straight lines, 8 pixel wide. The parameters used in the result presented in (Fig. 7) were: $w_w = 6, w_u = 10, f_l = 0.25$. Time was 4 min 18 s (accumulation 195 s, fuzzification along direction 45 s, angle homogeneity 13 s, ridge following 5 s; Pentium 1 GHz). One spurious line was detected – upper right corner (for $f_l = 0.29$ neither lines are omitted nor false lines appear). Shapes of detected lines are not equal but close to those expected. The detector elongates the lines by the fuzzification along the direction. As a side-effect, some of the lines were elongated beyond their actual lengths – especially the line in the second column from left, fourth row from top.

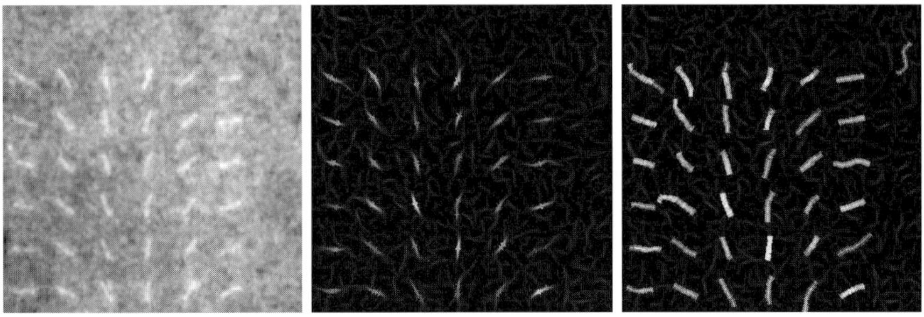

a b c

Fig. 7. Phantom of a mammographic image from [11, 12], 512 × 512, 8 bits/pixel. (**a**) source; (**b**) line intensity (*grey* proportional to lineness) with line widths marked (*grey* equal to that in the maximum) in local line intensity maxima larger than $f_l L$ (*black dots*); (**c**) widths of detected lines (*grey*) and their ridges (*white lines*).

5 Conclusion and suggested future work

A new detector of lines having variable orientation and width, recently proposed [8], has been described in detail. Although it has been designed primarily for the detection of blood vessels in mammograms, it is a generic ridge detector. It seems that the power of the algorithm lies in the use of the fuzzy accumulation principle. The quality of the methods selected as the most reliable in [11, 12] stems from the use of raw grey levels, rather than the gradients, as in the present paper. This suggests that merging the accumulation principle and fuzziness, with the use of raw grey levels instead of derivatives, or with the principles used in [15], should be a good direction of future work.

References

1. Hough PVC (1959) In: *Proc Int Conf High Energy Accelerators and Instrumentation.* CERN.
2. Maître H (1985) Un panorama de la transformation de Hough. *Traitement du Signal,* 2(4):305–317
3. Leavers VF (1993) Which Hough transform? *CVGIP-IU* 58:250–264
4. Lam WCY, Lam MTS et al. (1994) A general evidence accumulation technique for Hough transformation. In: *Proc IEEE Int Conf SMC,* vol 3, 2414–2419
5. Aguado AS, Nixon MS, Montiel EM (1998) Parameterizing arbitrary shapes via Fourier descriptors for evidence-gathering extraction. *CVIU* 69(2):202–211
6. Merlin PM, Farber DJ (1975) A parallel mechanism for detecting curves in pictures. *IEEE Trans Comp* 24:96–98
7. Ballard DH (1981) Generalizing the Hough transform to detect arbitrary shapes. *Pat Rec* 13:111–122
8. Chmielewski L (2004) Detection of non-parametric lines by evidence accumulation: Finding blood vessels in mammograms. In: *Proc. ICCVG 2004,* vol of Computational Imaging and Vision. Springer. In print.
9. Strauss O (1999) Use the Fuzzy Hough Transform towards reduction of the precision-uncertainty duality. *Pat Rec* 32:1911–1922
10. Reisfeld D, Wolfson H, Yeshurun Y (1995) Context-free attentional operators: the Generalised Symmetry Transform. *Int J Comput Vis* 14:119–130
11. Zwiggelaar R, Astley SM et al. (2004) Linear structures in mammographic images: detection and classification. *IEEE Trans Med Imag* 23(9):1077–1086
12. Zwiggelaar R, Parr TC, Taylor CJ (1996) Finding orientated line patterns in digital mammographic images. In: *Proc 7th BMVC 96* 715–724
13. Dixon RN, Taylor CJ (1979) Automated asbestos fibre counting. In: *Proc Inst Phys Conf Series* vol 44, 178–185
14. Lindeberg T (1998) Edge detection and ridge detection with automatic scale selection. *Int J Comput Vis* 30(2):117–156
15. Zwiggelaar R, Boggis CRM (2001) Classification of linear structures in mammographic images. In: *Proc Conf Med Image Underst Analysis 2001*
16. Chmielewski L (2005) Scale and direction invariance of the evidence accumulation-based line detection algorithm. In: *Proc. CORES 2005,* vol of *Advances in Soft Computing.* Springer. (In the same volume.)

Scale and Rotation Invariance of the Evidence Accumulation-Based Line Detection Algorithm

Leszek J. Chmielewski

Institute of Fundamental Technological Research, PAS, Świętokrzyska 21, 00-049 Warsaw, Poland lchmiel@ippt.gov.pl

Summary. The scale and rotation invariance properties of a recently proposed algorithm, using the fuzzy evidence accumulation principle, for finding lines (ridges) of non-parametric shapes is analysed. The proposed modifications consist in scaling the accumulated value with the inverse of the line width and further fuzzifying the accumulation process – along the line width. Good invariance properties received are tested on artificial images and confirmed on real-life mammographic images.

1 Introduction

The algorithm analysed in the present paper has been recently proposed in [1] and is described in detail in [2]. Two properties of this algorithm will be investigated: its sensitivity to the scale and to the direction of the detected objects. It is assumed that a multiscale line detector which might be described as *good* should detect narrow lines as well as wide ones, and lines inclined at arbitrary angles, and that the only feature which should differentiate between *strong* and *weak* lines should be their contrast with respect to the background.

The algorithm has originally been designed with the application to the analysis of elongated objects in mammographic images. This demanding task (see [3]) necessitates for superior properties of the algorithm, if the decisions made on the grounds of its results are expected to be reliable.

2 Scale invariance

In the previous papers [1, 2], the line intensity was primarily defined using the vector sum of tangents to the line at its two edges (see [2], in the same volume, Eq. (3)):

$$l = c_d \, c_e \, |\boldsymbol{V}|. \tag{1}$$

The coefficients – directional consistence c_d and edgeness consistence c_e of the tangent vectors – have been defined in [2].

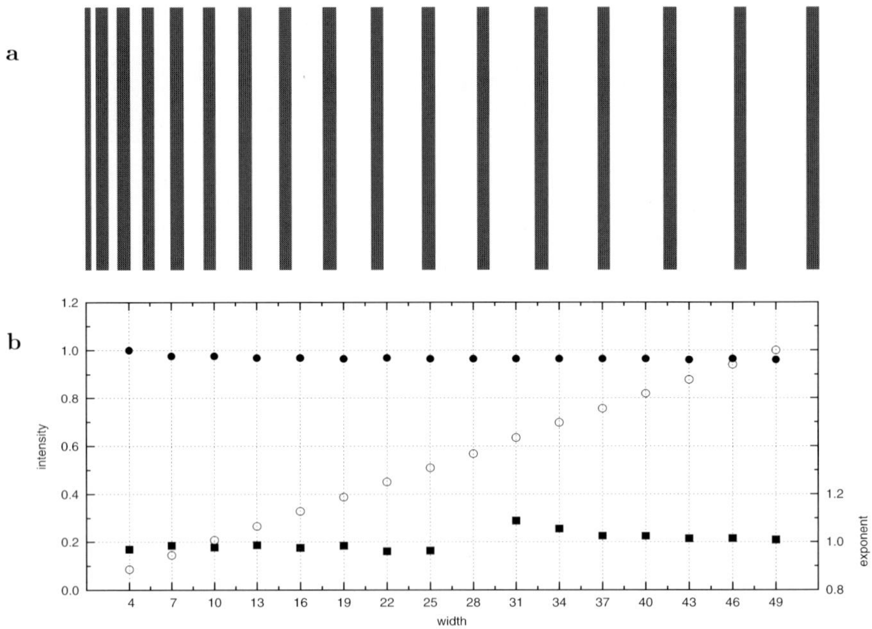

Fig. 1. Introducing scale invariance. (**a**) test image; (**b**) graphs of the accumulated value in centres of lines: lineness l according to Eq. (1) (*empty circles*) and scale invariant lineness l_s according to (5) (*full circles*); exponents found from (4) for subsequent widths (*full squares*, except reference 28) are all close to one.

We shall understand the *scale invariance* of the line detector as the independence of the line intensity of the line width. Let us denote such scale invariant line intensity as l_s. Assume the following relation of this intensity to l as defined in Eq. (1):

$$l_s = l/w^p \ . \tag{2}$$

The exponent function is introduced to allow for possible non-linearities, potentially stemming from the relation of pixel size to the line width. Let us check for which exponent p the intensity l_s is actually invariant to the line width. Let w_1 and w_2 be the widths of two lines. It should be then $l_s(w_1) = l_s(w_2)$ which gives

$$l_1/w_1^p = l_2/w_2^p \ . \tag{3}$$

A test image containing 16 bright lines of width from 4 to 49 pixels every 3 pixels (Fig. 1**a**) was used to find the exponent p. Intensities were taken in maxima at line centres, in the middle row of the image. Line of width 28 pixels was used as a reference and 15 values of p were calculated from

$$p_i = \frac{ln(l_i/l_{28})}{ln(w_i/w_{28})} \ , \quad i = 4, 7, ..., 22, 25, 31, 34, ..., 49 \ . \tag{4}$$

In the graph of p_i in Fig. 1**b** (*squares*) it can be seen that all the exponents are close to one, which leads to a simple equation for the scale-invariant accumulated line intensity:

$$l_s = c_d\, c_e\, |\boldsymbol{V}| \;/\; w\;. \tag{5}$$

In the graph of p_i in Fig. 1**b** (*full circles*) it can be seen that the line intensity according to this equation is indeed close to constant for all the widths.

An example of results for a real-life image is shown in Fig. 2. The input image is a window 300×300 starting from pixel $(850, 170)$ of the mammographic image `MDB007LL.TIF` from the MIAS database [4], originally 4320×2400. Calculations were made for line widths from 7 to 35 and lineness threshold coefficient 0.25. Time of calculations was 23 min (Pentium 1 Ghz).

a b c

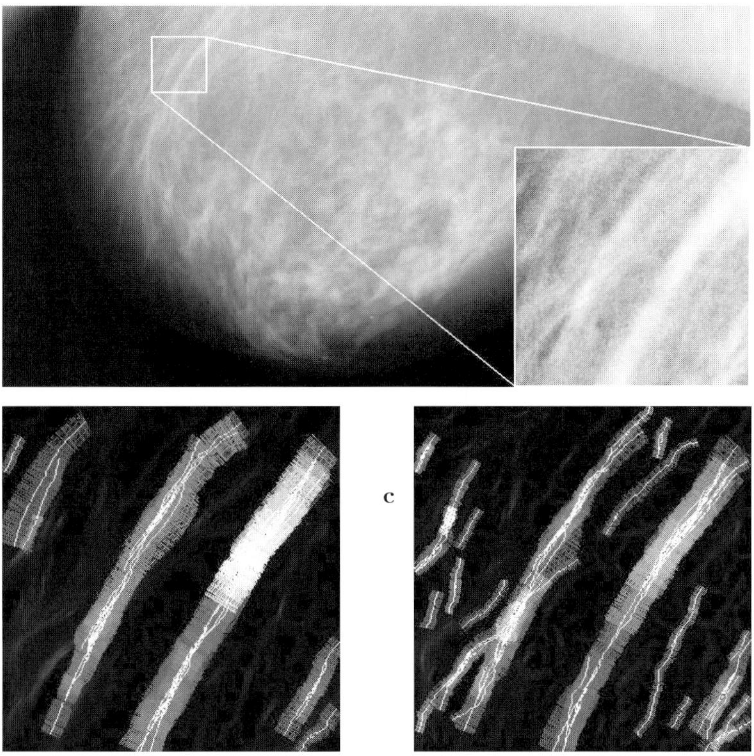

Fig. 2. Example of results for a mammographic image. (**a**) input: window of image `MDB007LL.TIF` from the MIAS database; (**b**) accumulated lineness according to Eq. (1) with lineness ridges and line widths – overestimated line widths and narrow lines overlooked; (**c**) accumulated scale-invariant lineness according to Eq. (5) with lineness ridges and line widths – the number of relevant details is larger.

3 Rotation invariance

The rotation invariance was tested on seven series of images containing lines of various inclination and widths. Experiments on four series of test images, sufficiently demonstrating the phenomenon of interest, will be reported. In each series there were images of a line 11 pix wide, inclined at angles growing from 0 to 90° by 5°, and by 0.1° near 45° (Fig. 3). Line brightness was 250 and background was 50. Each of the series corresponds to a class of typical line edges which can appear in an image: a line plotted with standard anti-aliasing, possibly with noise (additive, Gaussian); a line without anti-aliasing, and a line with blurred edges in the form of gentle slopes. The size of the images was 127×127 pixels, so that the line was narrow with respect to the image. For images with the line at subsequent angles, the line intensity according to Eq. (5) and the line width in the central pixel (63, 63) were recorded.

The results are shown in Fig. 4. For easier comparisons, the relative line intensity in the graphs is plotted, with the unit corresponding to 120000. For all the cases, the discrepancies of the lineness from the constant function are substantial, except for the normally anti-aliased lines with noise. The errors of width, which should be 11 in all the cases, are moderate to large, except for the normally anti-aliased lines.

The largest discrepancies appeared at 'round' angles: 0°, 45° and 90°. This phenomenon can be investigated by visualizing the influences of specific pixels on the line intensity. For angle 44.2° (Fig. 5) some pixels have much smaller influence than the others, which results in smaller line intensity. This is due to that the rate of errors made in the calculation of width at angles close to the 'round' angles is much different from that for other angles (Fig. 6).

The erroneous calculations are closely related to the square tessellation of the image. The angular distribution of image properties is far from isotropic, which can not be avoided. However, the unevenness of the distribution of errors across the edge widths can be reduced by means of fuzzification introduced in storing the results of elementary accumulation for pairs of pixels.

a b c d

Fig. 3. Central windows (63×63) of examples of test images used in the tests. Here, line normal inclination angle is 5°. (**a**) with normal anti-aliasing; (**b**) anti-aliasing and noise $\sigma = 15$; (**c**) no anti-aliasing; (**d**) slope edges.

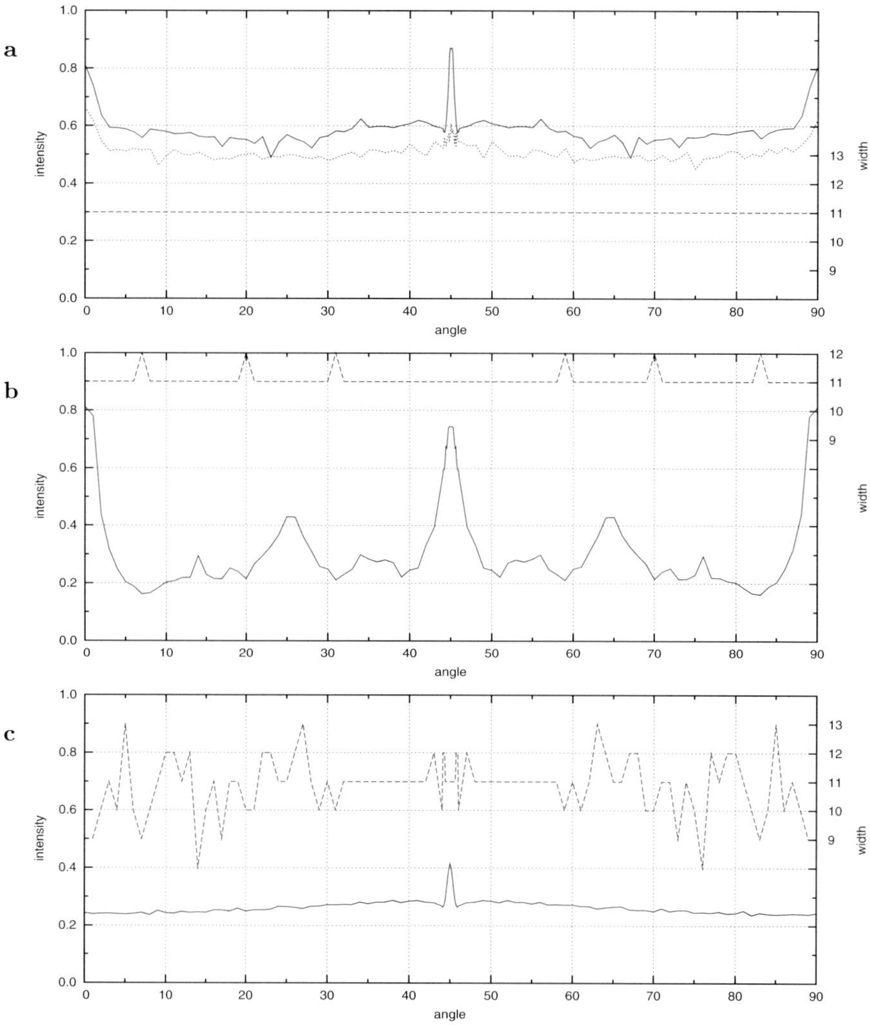

Fig. 4. Results in the central pixel of the algorithm before the modification: line intensity (*solid line*) and width (*dashed line*); (**a**) normal anti-aliasing, clear (*solid line*) and with noise (*dotted line*); (**b**) no anti-aliasing; (**c**) slope edges.

It is proposed to introduce a fuzzy membership function of width defined on triplets of neighbouring widths. Each elementary accumulation is performed for a given width and its two neighbouring ones. The tested membership functions were: flat function $(1, 1, 1)$, function $(1/2, 1, 1/2)$ and $(2/3, 1, 2/3)$. Results for the normally anti-aliased lines (Fig. 3**a**) are shown in Fig. 7. The results for the function $(2/3, 1, 2/3)$ appeared to be the most uniform and this function has been finally chosen.

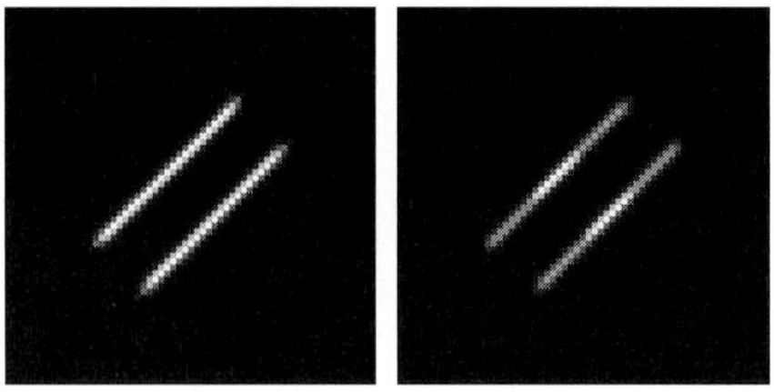

a **b**

Fig. 5. Maps of influence of pixels on the accumulated value. Brighter pixels have larger influence. (**a**) for angle 45.0°; (**b**) for angle 44.2° – the minimum in the graph from Fig. 4**a**, solid line.

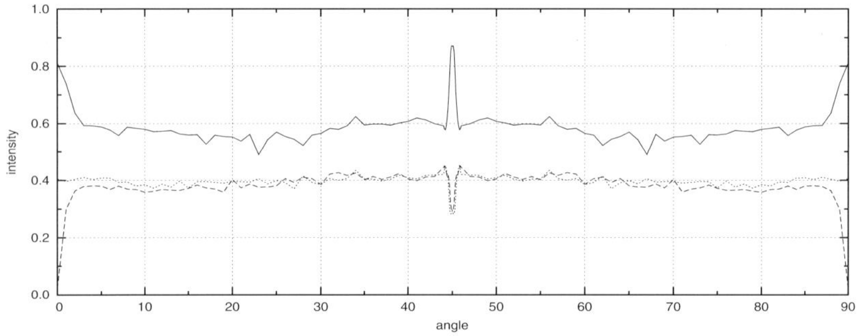

Fig. 6. Result of accumulation for three neighbouring widths: the expected width 11 (*solid line*), width 10 (*dashed line*) and width 12 (*dotted line*). For angles close to 0° and 45° less votes than usual go to widths other than that expected.

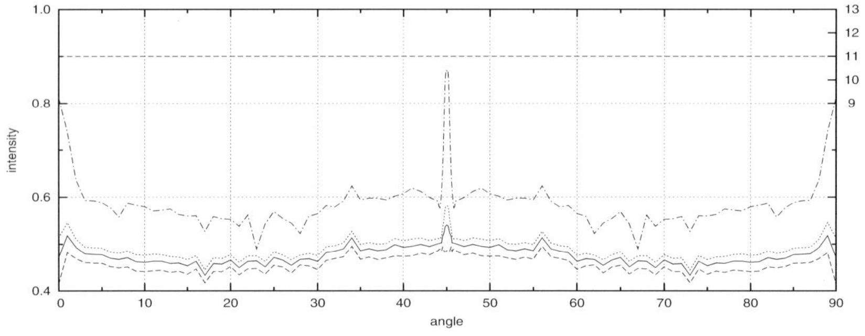

Fig. 7. Result of accumulation before (*dash-dot line*) and after fuzzification: with flat membership function (1, 1, 1) (*dashed line*), with function (1/2, 1, 1/2) (*dotted line*) and with function (2/3, 1, 2/3) (*solid line*).

An example of results with and without the rotation invariance received for a test image can be seen in Fig. 8. The test image is a rosette with lines 7 pixel wide at 5° angular intervals. Lines have maximum intensity 165 and have the longitudinal shape of a $\cos^2$ function having maxima at a distance of

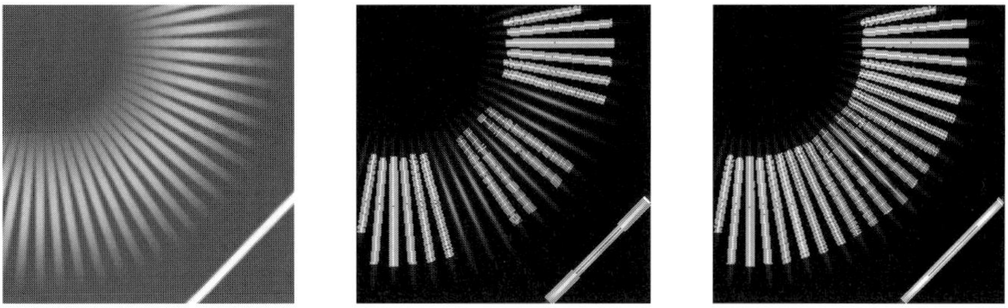

a b c

Fig. 8. Results for the rosette test image. (**a**) input image; (**b**) result of the directionally non-invariant algorithm; (**c**) result of the directionally invariant one.

140 from the centre in pixel (32, 32) and the period of 200 pixels. Transversal profile of lines is a result of dithering the intensity into slopes 3 pixel wide. The line in the lower right corner has intensity of 250. This line is necessary to receive the maximum accumulated value so that the intensities of the lines in the rosette can be thresholded with respect to it.

The result of the algorithm with the rotation invariance enhanced by the fuzzification in width does not exhibit the effect of variable sensitivity to lines depending on their inclination, as the primary algorithm does.

An example for a real-life image is shown in Fig. 9. The input image is a window 250×250 starting from pixel (3250, 650) of the mammographic image `MDB059LS.TIF` from the MIAS database [4], originally 4320×1600. Calculations were made for line widths from 15 to 25 and lineness threshold coefficient 0.67. Time of calculation was 6 and 7 min for the two algorithms, respectively (Pentium 1 Ghz).

In the result of the algorithm with the rotation invariance enhanced by the fuzzification in width the spurious lines stemming as extensions from the line in the centre were eliminated.

4 Conclusion and suggested future work

The scale and rotation invariance of the recently proposed line detection algorithm using the evidence accumulation principle was studied. Each of the considered features was tested on specially designed images. The modifications of the algorithm aiming at improving its invariance properties were introduced and tested on selected real-life images from the data base of mammographic images. As a result, good scale and rotation invariance properties were received. This seems to confirm the suitability of the algorithm of interest to be used in the analysis of mammographic images. The future work should go towards comparative tests of the detector against the existing detectors which use other principles [3, 5], with the use of real-life images.

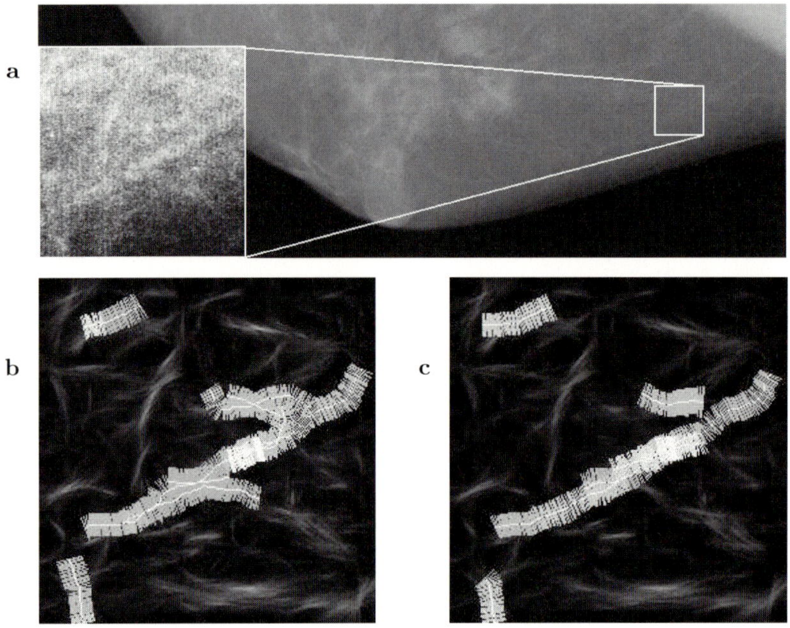

Fig. 9. Example of results for rotation invariance for a mammographic image. (a) input: window of image MDB059LS.TIF from the MIAS database; (b) accumulated lineness and line widths, directionally non-invariant algorithm; (c) the same for the directionally invariant algorithm – number of spurious lines smaller.

References

1. Chmielewski L (2004) Detection of non-parametric lines by evidence accumulation: Finding blood vessels in mammograms. In: *Proc ICCVG 2004*, vol of Computational Imaging and Vision. Springer. In print.
2. Chmielewski L (2005) Specification of the evidence accumulation-based line detection algorithm. In: *Proc CORES 2005*, vol of *Advances in Soft Computing*. Springer. (In the same volume.)
3. Zwiggelaar R, Astley SM et al. (2004) Linear structures in mammographic images: detection and classification. *IEEE Trans Med Imag* 23(9):1077–1086
4. Suckling J, Parker J et al. (1994) The Mammographic Images Analysis Society digital mammogram database. In: Gale AG, Astley SM et al. (eds) *Digital Mammography*, vol 1069 of *Excerpta Medica International Congress Series* 375–378 (`http://www.wiau.man.ac.uk/services/MIAS/MIASweb.html`).
5. Zwiggelaar R, Boggis CRM (2001) Classification of linear structures in mammographic images. In: *Proc Conf Med Image Underst Analysis 2001*

Content Based Image Retrieval Technique

Ryszard S. Choraś[1], Tomasz Andrysiak[1], and Michał Choraś[1]

Faculty of Telecommunications
University of Technology & Agriculture
85-796 Bydgoszcz, S. Kaliskiego 7, Poland
choras@mail.atr.bydgoszcz.pl

1 Introduction

Conventional image databases are text-annotated, where image retrieval is based on keyword searching. Such approach has many disadvantages. Computer retrieval systems, based on image content, would be more desirable for large image databases. Color, texture, local shape and spatial information, in a variety of forms, are the most widely used features in such systems. In response to a user's query, the system returns images that are similar in some user-defined sense.
This paper describes content based image retrieval method which uses Gabor filtration for determining a number of Regions-of-Interest (ROI), in which fast and effective feature extraction is performed.

2 Points of Interest Extraction Using Gabor Filters

The points of interest detection is based on decomposition of Gabor wavelets transform. In our work we use a bank of filters built from these Gabor functions for texture feature extraction. Before filtration, we normalize an image to remove the effects of sensor noise and gray level deformation.

The general functional of the two-dimensional Gabor filter family can be represented as a Gaussian function modulated by a complex sinusoidal signal. Specifically, a two dimensional Gabor filter $\psi(x, y; \sigma, \lambda, \theta_k)$ can be formulated as:

$$\psi(x, y; \sigma, \lambda, \theta_k) = \exp\left(-\frac{x_\theta^2 + \gamma^2 y_\theta^2}{2\sigma^2}\right) \exp\left(\frac{2\pi x_{\theta_k}}{\lambda} i\right) \tag{1}$$

where $x_{\theta_k} = x\cos\theta_k + ysix\theta_k$; $y_{\theta_k} = -x\sin\theta_k + y\cos\theta_k$, σ is the standard deviation of the Gaussian envelope along the x- and y-dimensions, and λ and θ_k are the wavelength and orientation, respectively.

The parameter γ is usually equal to 0.5. Since the spatial aspect ratio γ is constant, we do not use it as Gabor filter parameter. Rotation of the $x - y$ plane by an angle θ_k will result in a Gabor filter at orientation θ_k. The $\theta_k = \frac{\pi}{n}(k-1)$ for $k = 1, 2, ..., n$ and $n \in N$, where n denotes the number of orientations.
The parameter λ is the wavelength and $\frac{1}{\lambda}$ the spatial frequency of the harmonic factor $\cos(2\pi x_\theta/\lambda)$ or $\sin(2\pi x_\theta/\lambda)$. The ratio $\frac{\sigma}{\lambda}$ determines the spatial frequency bandwidth of the Gabor filters. The ratio $\frac{\sigma}{\lambda}$, which is used, is constant ($\frac{\sigma}{\lambda} = 0,56$) for all filters in the bank and corresponds to a half-response spatial frequency bandwidth of one octave.
Input image $f(x, y)$ is convolved with a 2-D Gabor odd filters $\psi_o(x, y; \sigma, \lambda, \theta_k)$ to obtain a Gabor image responses $\Phi(x, y)$ as follows:

$$\Phi_o(x, y; \sigma, \lambda, \theta_k) = \sum \sum f(\eta, \zeta)\psi_o(x - \eta, y - \zeta; \sigma, \lambda, \theta_k)d\eta d\zeta. \quad (2)$$

In order to time-effective and quality-effective comparison of query and database images, we search for image features only within the Regions of Interest, which are created around points of interest.
The consecutive steps of the points of interest extraction algorithm are as follows:

- Step 1. We divide images $\Phi(.,.;.,.,\theta_k)$ onto not overlapping blocks $b^k(i,j)$ of $a \times a$ size for $k = 1, 2, 3, 4$.
- Step 2. For each of those blocks we compute variance $V^k(i,j)$.
- Step 3. We search for p blocks $b_p^k(i,j)$ of the maximum variance, which unambiguously characterize such blocks. Blocks of maximum variance contain considerable illumination changes, usually corresponding to contours of the objects within the image.
- Step 4. Then in each block $b_p^k(i,j)$ we search for the point with the maximum value of the the filter response. Next, coordinates of the found points are presented on PoI image (points of interest) $PoI = \sum_k \Phi(.,.;.,.,\theta_k)$.

The results of points of interest extraction algorithm are presented in Figures 1.

3 Texture Feature Based on Gabor Filters

Each texture in the image is characterized by a given localized spatial frequency or a narrow range of dominant localized spatial frequency that differ significantly from dominant frequencies of other textures. The Gabor filters encode the textured images into multiple narrow frequency and orientation channels. The filter responses that result from the application of a filter bank of Gabor filters can be used directly as texture features, but we used the most simple idea to obtain features by applying threshold on Gabor filter results. The thresholded Gabor features are computed as follows:

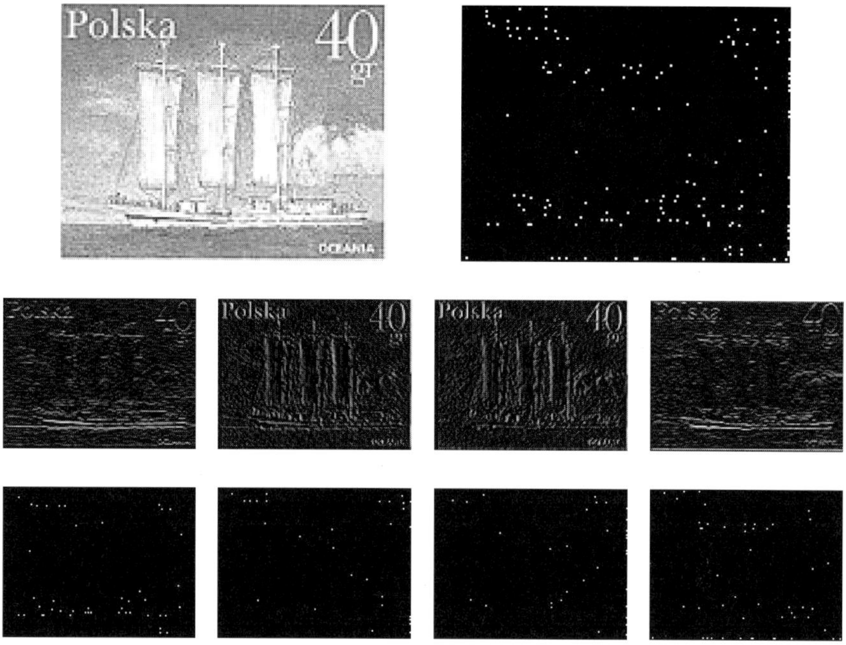

a)
b)
c)

Fig. 1. Original image and extracted points of interest (a); Gabor filters responses for orientations 0°,45°,90°,135° respectively b) and extracted points of interest for Gabor filters responses images c)

$$
\begin{aligned}
T_o(x, y; \sigma, \lambda, \theta_k) &= \chi(\Phi_o(x, y; \sigma, \lambda, \theta_k)) \\
T_e\,(x, y; \sigma, \lambda, \theta_k) &= \chi\,(\Phi_e\,(x, y; \sigma, \lambda, \theta_k))
\end{aligned}
\tag{3}
$$

where:

$$
\chi\,(z) = \begin{cases} 0 \; for \; z < 0 \\ z \; for \; z \geq 0 \end{cases}
\tag{4}
$$

and $\Phi_o\,(x, y; \sigma, \lambda, \theta_k)$ and $\Phi_e\,(x, y; \sigma, \lambda, \theta_k)$ are the odd and even components of Gabor filter responses.

The method of feature extraction is based on decomposing the images and calculating the entropy and energy in each of the ROI. In an $m \cdot a \times m \cdot a$ size ROI, normalized energy and entropy are computed accordingly to the following relations:

$$
E_t = \frac{\sum_i \sum_j T_o^2(x, y; \sigma, \lambda, \theta_k) + T_e^2(x, y; \sigma, \lambda, \theta_k)}{(m \cdot a)^2}
\tag{5}
$$

$$Entropy = -\frac{\sum_i \sum_j [\frac{T_o^2(x,y;\sigma,\lambda,\theta_k)+T_e^2(x,y;\sigma,\lambda,\theta_k))}{sr^2}] \log_2 [\frac{T_o^2(x,y;\sigma,\lambda,\theta_k)+T_e^2(x,y;\sigma,\lambda,\theta_k))}{sr^2}]}{\log_2(m \cdot a)^2} \tag{6}$$

and

$$sr^2 = \sum_i \sum_j (T_o^2(x,y;\sigma,\lambda,\theta_k) + T_e^2(x,y;\sigma,\lambda,\theta_k)) \tag{7}$$

Finally, we obtain texture feature vector:

$$F_{texture} = \{E_{t1}, E_{t2}, \ldots, E_{tp}, Entropy_1, Entropy_2, \ldots, Entropy_p\} \tag{8}$$

where p is the number of ROI.

The texture similarity of a query image Q and an image D in the database is defined as:

$$d^{(Q)(D)}(E_t, Entropy) = \sum_p d_p^{(Q)(D)} \tag{9}$$

where:

$$d_p^{(Q)(D)} = \left| E_{tp}^{(Q)} - E_{tp}^{(D)} \right| + \left| Entropy_p^{(Q)} - Entropy_p^{(D)} \right| . \tag{10}$$

4 Color Feature Extraction

Color models global content of images and provides a good basis for similarity measure. The YUV space is widely used in image compression and image applications. Y represents the luminance of a color, while U and V represent the chromaticity of a color.

$$Y = \lfloor \frac{R + 2G + B}{4} \rfloor \quad ; \quad U = R - G \quad ; \quad V = B - G \tag{11}$$

This method is based on the assumption that visual system is more sensitive to Y component than to two chrominance components, and that y component typically uses higher density (eg. JPEG and MPEG standards). The extraction of color features includes two steps. Firstly, we convert RGB value of pixels from image to an YUV value. Secondly, we construct the Y (luminance) component histogram (Figure 2).
The histogram of each ROI is created. The ROI contains $m \cdot a \times m \cdot a$ pixels, which gives $(m \cdot a)^2$ bins (Figure 3).

$$H_x(j) = \frac{number\ of\ pixels\ with\ luminance\ j}{number\ of pixels\ in\ x} \tag{12}$$

Histogram similarity metric is used to compare histograms of a given image with histograms of images from the database. Let's denote the j histogram bin value of query image as $H_Q(j)$, and the j histogram bin value in the database as $H_D(j)$. Then for $H_Q(j)$ and $H_D(j)$ which are given by equation (12):

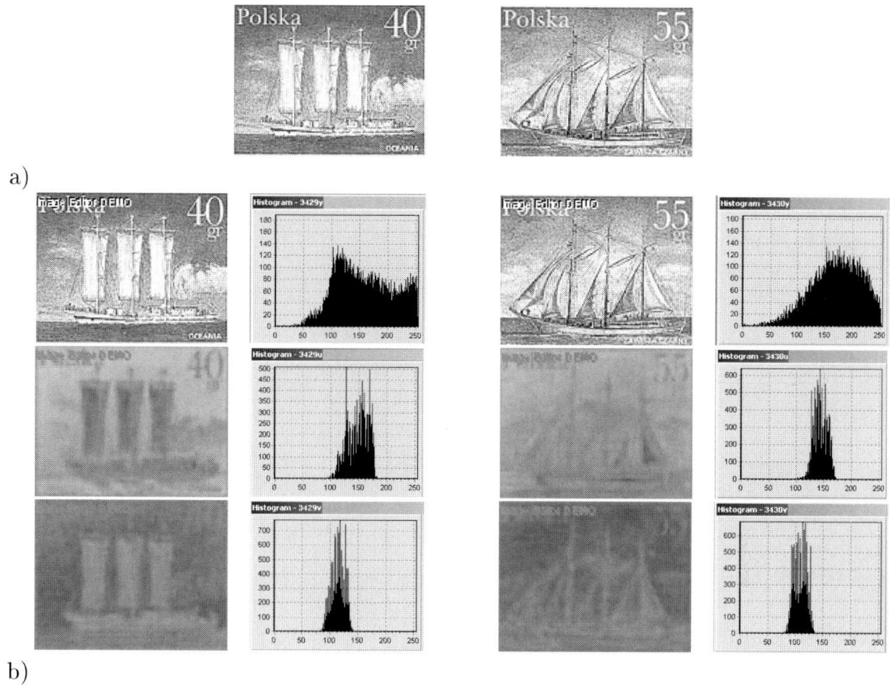

Fig. 2. Original images(a) and YUV components with appropriate histograms

$$D = \sum_{j=1}^{N} |H_Q(j) - H_D(j)| \qquad (13)$$

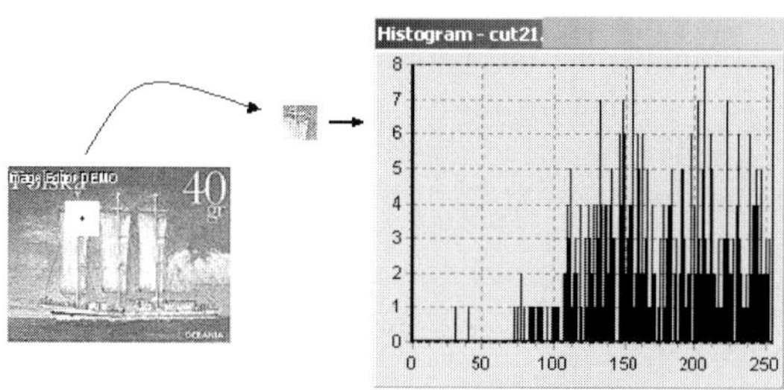

Fig. 3. Histogram of the extracted ROI

Next, we present the color content of an image also as the color moment values.

$$E_p = \frac{1}{m \cdot a} \sum_{s=1}^{m \cdot a} Y_{p,s} \quad ; \quad \sigma_p = \sqrt{\frac{1}{m \cdot a} \sum_{s=1}^{m \cdot a} (y_{ps} - E_p)^2} \tag{14}$$

Finally, we obtain color feature vector:

$$F_{color} = \{H_{Q_1}, H_{Q_2}, \ldots, H_{Q_p}, E_1, E_2, \ldots, E_p, \sigma_1, \sigma_2, \ldots, \sigma_p\} \tag{15}$$

where p is the number of ROI.

The color similarity of a query image Q and a image D in the database is defined as:

$$d^{(Q)(D)}(D_p, E_p, \sigma_p) = \sum_p d_p^{(Q)(D)} \tag{16}$$

where:

$$d_p^{(Q)(D)} = \left| D_p^{(Q)} - D_p^{(D)} \right| + \left| E_p^{(Q)} - E_p^{(D)} \right| + \left| \sigma_p^{(Q)} - \sigma_p^{(D)} \right|. \tag{17}$$

5 Shape features

Basically, shape based image retrieval is the measuring of similarity between shapes represented by their features. Two steps are essential in shape based image retrieval, and those are, feature extraction and similarity measurement between the extracted features.

The region-based shape descriptor belongs to the broad class of shape analysis techniques based on moments [7]. Zernike moment (ZM) sequence, Z_{nm} is uniquely determined by the image $f(x,y)$ and conversely, $f(x,y)$ is uniquely described by Z_{nm}. Orthogonality property of the ZM enables redundancy reduction among their respective description, and thus it helps to improve the computation efficiency.

The kernel of Zernike moments is a set of orthogonal Zernike polynomials defined over the polar coordinate space inside a unit circle.

We have p ROI's and around every ROI we build the boxes on a 8-neighborhood eg. representative shape box is $(m \cdot a) \times (m \cdot a)$ pixels. In this box we extract the objects and then we compute local axial moments.

5.1 Zernike moments

Zernike moment of order n and repetition m is defined as [7], [8]:

$$Z_{nm} = \frac{n+1}{\pi} \iint_{x^2+y^2 \leq 1} V_{nm}(\rho, \theta) f(x,y) dx dy \tag{18}$$

where:

- $f(x,y)$ is the image intensity at (x,y) in Cartesian coordinates,
- $V_{nm}(\rho, \theta)$ is a complex conjugate of $V_{nm}(\rho, \theta) = R_{nm}(\rho) e^{-jm\theta}$ in polar

coordinates (ρ, θ) and $j = \sqrt{-1}$,
- $n \geq 0$, and $n - |m|$ is even positive integer.

The polar coordinates (ρ, θ) in the image domain are related to the Cartesian coordinates (x, y) as $x = \rho cos(\theta)$ and $y = \rho sin(\theta)$.
$R_{nm}(\rho)$ is a radial defined by [8], as follows:

$$R_{nm}(\rho) = \sum_{s=0}^{\frac{n-m}{2}} \frac{(-1)^s[(n-s)!]\rho^{n-2s}}{s!(\frac{n+|m|}{2} - s)!(\frac{n-|m|}{2} - s)!} \tag{19}$$

The discrete approximation of equation 18 is given as:

$$Z_{nm} = \frac{4(n+1)}{(N-1)^2\pi} \sum_{k=0}^{N-1} \sum_{l=0}^{N-1} f(k,l) R_{nm}(\rho_{k,l}) e^{-jm\theta_{kl}} \quad ; \quad 0 \leq \rho_{k,l} \leq 1 \tag{20}$$

where the discrete polar coordinates:

$$\rho_{k,l} = \sqrt{x_k^2 + y_l^2} \quad ; \quad \theta_{kl} = arctan(\frac{y_l}{x_k}) \tag{21}$$

are transformed by:

$$x_k = \frac{\sqrt{2}}{N-1} k + \frac{-1}{\sqrt{2}} \quad ; \quad y_l = \frac{\sqrt{2}}{N-1} l + \frac{-1}{\sqrt{2}} \tag{22}$$

for $k = 0, \ldots, N-1$ and $l = 0, \ldots, N-1$, as shown in Figure 4.

To calculate the Zernike moments of an image $f(x, y)$, the image is first mapped onto the unit disk using polar coordinates, where the center of the image is the origin of the unit disk. Pixels falling outside the unit disk are not used in the calculation.

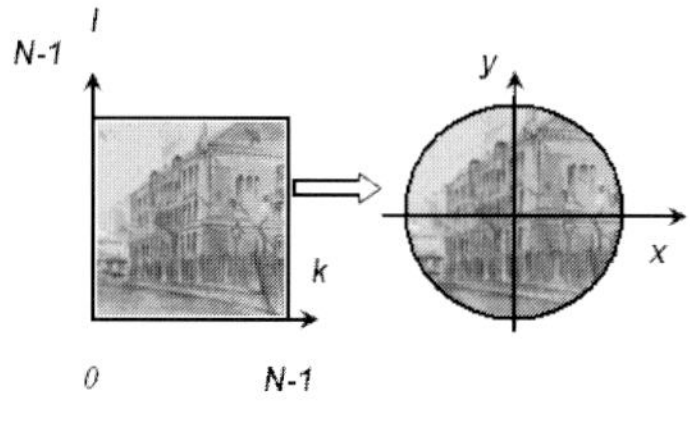

Fig. 4. The square - to - circular transformation

Because Z_{mn} is complex, we often use the Zernike moments modules $|Z_{mn}|$ as the features of shape in the recognition of pattern.
The magnitude of Zernike moments has rotational invariance property. An

image can be better described by a small set of its Zernike moments than any other type of moments such as geometric moments, Legendre moments, rotational moments, and complex moments in terms of mean-square error. Zernike moments do not have the properties of translation invariance and scaling invariance. The way to achieve such invariance is image translation and image normalization before calculation of Zernike moments.

To characterize the shape we used a feature vector:

$$SFV = (Z_{1m}, Z_{2m}, \ldots, Z_{nm}) \tag{23}$$

consisting of the Zernike moments. This vector is used to index each shape in the database. The distance between two feature vectors is determined by city block distance measure.

6 Conclusion and future work

A retrieval methodology which integrates color, texture and shape information is presented in this paper. Consequently, the overall image similarity is developed through the similarity based on all the feature components. Alternatively to known CBIR systems, we compute features only in the finite number of extracted ROIs. There are some other known methods of determining ROIs, but our method of extracting ROI based on points of interest detection and Gabor filtration, enables to use filter responses also to describe texture parameters. The described method was tested on a small post stamps database (130 stamps), for which we achieved comparable results as for *Blobworld* system. Presented method is further developed in postal image analysis and retrieval system.

References

1. Teh C C, Chin R T (1988) On image analysis by the methods of moments, IEEE Trans. Pattern Anal. Machine Intell., vol. 10, pp. 496-513
2. Haralick R, Shanmugam K, Dinstein I (1973) Textural features for image classification, IEEE Trans. on Systems, Man, and Cybernetics, SMC-3(6), pp.610-621
3. Khotanzad A, Hong Y H (1990) Invariant image recognition by Zernike moments, IEEE Trans. Pattern Anal. Machine Intell., 12 (5) , 489-498
4. Andrysiak T, Choraś M (2003) Hierarchical Object Recognition Using Gabor Wavelets, Proc of KOSYR, 271-278
5. Choraś R (2003) Content-Based Retrieval Using Color, Texture, and Shape Information. In Sanfeliu A, Ruiz-Shulcloper J (eds): Progress in Pattern Recognition, Speech and Image Analysis, Springer, Berlin Heidelberg New York
6. Fogel I, Sagi D (1989) Gabor filters as texture discriminator, Biological Cybernetics, 61: 103-113
7. Jain A , Farrokhnia F (1991) Unsupervised texture segmentation using Gabor filters, Pattern Recognition, 24(12):1167-1186

Scanning Faces Surroudings - New Concept in 3D Exact Multiviews of Nonconvex Polyhedron Generation

Maciej Frydler[1] and Wojciech S. Mokrzycki[2]

[1] Institute of Mathematical Machines, BI, Krzywickiego 34, Warsaw, Poland; m.frydler@imm.org.pl
[2] Institute of Informatics, University of Podlasie, Sienkiewicza 51, Siedlce, Poland; mokrzycki@ii3.ap.siedlce.pl

Summary. This article concerns generating of 3D multiview exact models that are a complete representation of polyhedron, according to viewing sphere with perspective projection. Those models are going to be used for visual identification based on them and a scene depth map. We give a new concept and an algorithm for face-depended generation of multi-face views. It does not require any preprocessing nor auxiliary mechanisms or complex calculations connected with them

1 Introduction

Method of generating multiview representation of polyhedron for object visual identification described in many papers, e.g. [1]-[7]. Method described in [2] concerns of 3D views and is based on the following idea: Centrally generate views relative object features chosen for identification, calculate single-view areas on viewing sphere which correspond to earlier generated views, check if whole viewing sphere is covered with single-view areas. If this cover is complete, generation of viewing representation is finished. If not, than generate additional views corresponding to uncovered areas of viewing sphere and again check if this cover is complete. Continue until complete viewing sphere cover is done. Complete viewing sphere cover with single-view areas means that generated representation is complete.

Methods from [4, 5, 6] are better. To achieve complete representation one don't need to work in a loop. Complete representation is obtained by strict covering viewing sphere by single-view areas in spiral way and controlling "edge" register (of no covered area).When register is set to "empty" generation of multiview representation is done. Generated representation is complete which follows from the generation method. However to achieve complete representation we have to calculate single-view areas on viewing sphere and operate them in a given order. Without their help it is not possible to get a complete set of views of virtual polyhedron model. On the top of that described methods are for convex polyhedron only.

In the sec.4 of this article we present a method for generating a complete viewing representation polyhedron more computational efficient then described above, and in the sec. 5 - a complete viewing representation for nonconvex plyhedron.

2 Research assumptions

This research focuses on developing a method and an algorithm for generation of multiview, polyhedron representation. For representation generation we use viewing sphere with perspective projection, [2]. For this following conditions have to be met:

1. Models are accurate - every model is equivalent to B_{rep} model.
2. Models are viewing models - it is possible to identify object from any view.

We consider polyhedrons that's non transparent and monotonous, and contains (if) only convex concavities. Use of a viewing sphere with projection as a projection space allows simple view standardization.

Uses: Recognition of objects not bigger then a few meters and distant (from the system) not more then 10 - 20 meters.

Mentioned above uses allow to make certain assumptions about recognition system strategies. We assume following steps of recognition processes:

1. Determining recognizable object types.
2. Definition of identification task.
3. Generation of viewing models for each object system should identify.
4. Creation of database containing all views of all models.
5. Acquisition of scene space data and visual data.
6. Isolation of scene elements and their transformation to model structures stored in the database.
7. Identification of objects by comparing them with database models.

3 View generation space - viewing sphere with perspective projection. Basic concepts.

Let object be a convex, non transparent polyhedron without holes or pits. Let's consider its faces as features areas, those areas will be used as a foundation for accurate multiview model determining. This model is a set of accurate views, acquired through perspective projection from viewing sphere, according to the model from [2]. This model is best for 3D scene data acquisition as gives identification system reliability.

Complete set of views for a given polyhedron is obtained by completely covering of viewing sphere with corresponding single-view areas. Algorithm makes this approach complete. Changing one view to the other is a **visual event**. This event occurs as a result of point VP movement. This event is manifested by appearance of a new feature in a view, disappearance of a feature or both.

4 Using views by scanning of faces normals surrounding

Idea: In [7] we described an original algorithm $FK^{\cap V}$ of 3D multiview exact representation. Main idea of the algorithm is to rotate complementary cone around translated faces normals to obtain all views: for each vector v_i from vector representation V_rep of polyhedron rotate around it (vector) complementary cone and note every visual event. This event is manifested by appearance of a new versor inside complementary cone or by disappearance of versor. Result of such routine is set of vectors faces that can be seen by observer from view sphere.

If the polyhedron is convex all the faces inside view contour are visible. Finally create set of different sets of vectors faces that can be seen. It is possible to achieve this by adding to the set all obtained sets and by removing sets that repeats.

For convex polyhedron fact that some faces normals translated to center of small view sphere contains in complementary cone mean that there are in the same view. By rotating complementary cone around all normals all combination of faces normals that are in some view are obtained. That is why it is true to assume that by this algorithm all views will by collect.

5 Mathematical approach

At the beginning lets describe the algorithm for computing orientation of complementary cone (around vector v_i) in which it intersect. It is useful to use idea of coordinates transformation to simplify this task. For vector v_i described in cartesian coordinates, that is in base B that contains three vectors $x(1,0,0)$, $y(0,1,0)$, $z(0,0,1)$ or simply $B((1,0,0),(0,1,0),(0,0,1))$, build base B'. $B'(z',y',z')$ is such base that can be created by multiplying by (transformation) matrix $[B']$. B' is selected in such way that coordinates of vector v_i in this base equals $v_i'(1,0,0)$

$$v_i' = [B'] * v_i.$$

Set of matrix's that fulfils mentioned condition is infinite. It is possible to use following approach to acquire one of them. From base vectors select this one that angle distance between it and v_i is closest to 90°. That is select this one that dot product is closest to 0. Mark it as t, than:

$$x' = v_i$$

$$y' = v_i \wedge t$$

$$z' = v_i \wedge (v_i \wedge t),$$

$B'(x', y', z')$ fulfils our condition. Now lets represent complementary cone that is rotated around vector v_i' as vector $r(s)$ that lies in center of symmetry of cone and is function of rotation angle around v_i' (if consider the base B'). Compute $r(0)$ as rotation vector $(1, 0, 0)$ around z' axis at $\beta//2$ angle (β is complementary cone angle), than:

$$r(s) = (r(0).x, r(0).y * cos(s), r(0).y * sin(s)).$$

Complementary cone intersect with h' (that is vector h transformed from base B to B') only if there is such s that:

$$r(s) * h' = cos(\beta//2).$$

It is true because h' intersect with complementary cone only if angle distance to center of symmetry of cone equals half of cone angle. It leads to following equation

$$r(0).x * h'.x + r(0).y * cos(s) * h'.y + r(0).y * sin(s) * h'.z = cos(\beta//2).$$

If mark known values as

$$A = r(0).y * h'.y$$

$$B = r(0).y * h'.z$$

$$C = r(0).x * h'.x - cos(\beta//2)$$

than equation simplify to

$$A * cos(s) + B * sin(s) - C = 0.$$

For particular data this equation can have 0 solutions when vector h' do not intersect with cone for all s, one solution when cone "touch" h' for some s and 2 solutions ($s0 < s1$) when h' enters and leaves cone during its rotation. Only third case h is considered as to "be in one view" with vector v. Computational results achieved in base B' are also true in base B because of transformation approach .It is important to check and note if h belongs to cone if $s0 < s < s1$ or ($0 < s < s1$ or $s2 < s < 2\pi$) fulfill task conditions. Let's mark this result as Sh. Next step consider computing set of sets of vectors that contains vectors that for some s interval lies inside complementary cone. Create set H that contains all „cone intersecting" solutions $Sh(i)$ for all $h(i)$ vectors from vector representation of polyhedron that are not v vector. Now by moving from 0 to 2π check and note appearance and disappearance of previously computed solution ($Sh(j)$).This is equals to computation of sets of vectors that are in one view. Notice, that each set contain vector v.

6 Using views by scanning of faces surrounding

The idea of presented algorithm $FM^{\cup}NV$ is a extension and generalization of $FM^{\cap V}$ algorithm from [7] of generation of view representation for convex polyhedrons based on V_{rep} representation. This generalization consider generating of full representation of polyhedrons using scanning normal faces by complementary cone nad tracing aperances and disaperances of faces from view.

This is short description of previous algorithm.

- For each face build its vector representation Vp_{rep}. Vp_{rep} representation it is set (for each face) of sets: normal vector of face (that is element v_i of $V_{rep} representation$) together with edges vector of side faces of pyramid that is element of natural representation N_{rep} of polyhedron.
- Scan pace around each face by rotation complementary cone along its normal vector and tracing of aperances and dsiaperances of others normals from Vp_{rep}.

The surrounding scanning performs as follows:

- Let's rotate complementary cone around vectors designated by lateral edges of NRP.
- Let's make it in such way that complementary cone surface is tangent to vector designed by one of these edges. To obtain optimal result rotate complementary cone in such range that pyramid, that selected edge belong to, do not leave cone. During rotation (space scanning) collect information about polyhedron faces that enters and leaves complementary cone.

Fig. 1. Fig. 4. a) Represent pyramid designated by face and three vectors from geometric center. b) Complementary cone rotates around blue edge of pyramid - view from selected edge of pyramid of normal representation.

The following events may occur:

- Complementary cone include faces together their normals; they create set of visible faces (at some view).
- Complementary cone includes faces but does not its normals; they are not visible in the view; some of the potentially visible faces can be occluded by the invisible ones.
- Faces with some edges outside cone are invisible; no matter is the state of its normal vector; they belong to opposite site of polyhedron.

During scanning edge's vectors will leaves end enters complementary/scanning cone. Such events may results as new **visual events** (new view appears). Several different events may appear. Any of them may leads to different situation.

- Covering (by rotating/scanning space by complementary cone) last edge vector of some face causes potential extension of view. However face could remain invisible or became invisible.
- Covering normals of faces, that are currently in cone causes these faces to became potentially visible, however it could still remain occluded.
- Covering faces that have their normals outside cone causes these faces to be invisible. These faces may occlude other faces that currently are inside immersion.
- Face may enter to cone but it normal does not. It causes occlusion.
- Leaving complementary cone by any edge vector of some face cause this face to leaves actual view.

If during scanning some event occurs it must by carefully check if actual view include immersion and if it is true than information about this view must be enrich with information about shadow boarders.

Occlusion study: When complementary cone covers all edge vectors of some face except its normal that means this face belong to view contour (potentially belong to view) but it is inclined backwards to observer therefore it is invisible and occludes other faces of immersion (potentially visible) and is called area of occlusion. Area of occlusion is designated by viewpoint and edges of contour of inclusion that belong to face that causes occlusion. When cone, during its rotation, approaches immersion occlusion get smaller and finally disappears (when normal enter the cone).

Data that needs to be collect during scanning to makes view representation complete (for identification reason):

- Information about view contour (neighborhoods information, coordinate of vertices edges and faces.
- Information about invisible faces that lay inside view contour.
- Information about faces that are partially or completely occluded and lay inside view contour (information about border of shadow).

All this elements defines view content. Any differences between such sets make them different.

Inclusion to contour study: Faces that are inside cone but its normals does not - belong to immersion. Faces that are visible (all vectors edges and normal belong to cone) may also belong to immersion. Condition that we use to designate faces that belong to immersion are: normals of faces that belongs to immersion cross each other. Immersion contour is designated by border between immersion faces and other faces that belongs to view.

Preceding analysis makes possible to construct following algorithm of generation view representation of monotonic polyhedrons from its natural representation.

Algorithm $FM^{\cup NV}$

1. Create Vp_{rep} representation for given polyhedron, select face n_1.
2. Perform cone scanning of surroundings for any v_i from V_{rep} and any face of polyhedron and control appearance of visual events caused b v_j and n_k.
3. Select face n_1 and perform scanning around this face. Let's consider situation when only one immersion is inside view and is completely visible. Registered view contains only visible faces.
4. Start rotating complementary cone and note:
   a) If some face from convex part of polyhedron appears or disappears (with its normal) let's consider it as casual view event, new view and register it.
   b) If faces leaves cone abut its normal stays it mean that view lost one of its element; it is olso casual event. Register it.
   c) If only normal of some faces leaves cone that's mean immersion is now in cone, and one face disappeared from view (element of immersion contour) and new immersion faces are going to be occluded. Lets start algorithm of designating shadow plane for face edge; Start algorithm of tracing changes in visibility of immersion faces:
      i. Register new view (face is invisible but it is in view contour.
      ii. Designate immersion elements, faces and contour:
         A. Find faces that adjoin with faces f and check how they interfere with each other (if they normals cross)
            If normal vector of some face from neighborhood cross with f normal that's mean that this face is inside immersion
            If not it face doesn't belong to immersion.
         B. For each face from immersion perform same action as for face that disapear (f).
         C. Find view contour as a set of edges that are between faces from immersion and from other faces from actual view.
      iii. Find shadow plane SP for immersion contour that is the edge of face that disarmer.
      iv. Continue scanning and trace covering of vertices of faces that lies inside immersion by Shadow plane SP:
         A. If only fragment of some face became occluded (by SP) designate shadow border and register it as new visual event.
         B. If whole face became occluded consider it as visual event and register new view.
         C. If another face becomes invisible move to upper level of algorithm and found its shadow plane.

      Perform these actions until all immersion faces became visible.
   d) If inside cone appear face without its normal that mean it is part of immersion and its only one immersion face that is invisible. Other immersion faces enters cone at same time but they normal enters earlier. Those faces are invisible. At this time find immersion faces and contour, register new view and start routine of finding SP for invisible

face (plane that intersect face edge, from contour, and viewpoint) . Continue movment of scanning cone until PC intersect new vertex taht belong to immersion faces. Consider it as new view. Designate edge of shadow, that is border between parts of faces that are in shadow and rest of immersion. Continue rotation until invisible faces become visible (this cause whole immersion to be visible).

e) If inside scanning cone appear only normal vector of some face it annonce of soon aperance new immersion in view.

5. Perform full rotation scan.
6. Perform scanning for all faces.
7. Finish.

Summary

This latest approach to generating 3D multiview models of polyhedron $FM^{\cup NV}$ is in phase of programming.

Algorithm is going to be carefully checked and verified on set of test polyhedrons. Positive results are expected because problem has been precisely analyzed. Result will be presented soon.

References

1. Arbel T., Ferrie F.P. 1996: Informative views and sequential recognition. Proc. ECCV'96, Cambridge, UK, April,469-481.
2. Dźbkowska M., Mokrzycki W.S. 1997: Multi-view models of convex polyhedron. MGV, 6(4), 419-450.
3. Dabkowska M., Mokrzycki W.S. 1998: A new view model of convex polyhedron with feature dependent view.MGV, 7(1//2), (Proc. GKPO'98, Borki, Poland, 18-22 May), 325-334.
4. Kowalczyk M., Mokrzycki W.S. 2002: A new method of finding one-view areas and tight view sphere covering. Proc. ICCVG'02,Zakopane, Poland, Sept. 25-29, 443-449.
5. Kowalczyk M., Mokrzycki W.S. 2003: Obtaining complete 2 1/2D view representation of polyhedron using concept of seedling single-view area. CV&IU 91,208-301.
6. Kowalczyk M., Mokrzycki W.S.: Methods of generation $3D$ exact views of convex polyhedron for visual identification. Part II: Noniterative methods, implementation and tests results. MG&V, 12(4), 435-452.
7. Frydler M., Mokrzycki W.S.: New, fast algorithm of 3D multiview polyhedron representation generation on view sphere with perspective. Proc. CCV(ACV)'04, Prague, May, 15-16.

Image Decomposition by Grade Analysis - an Illustration

Maria Grzegorek

Institute of Computer Science, Polish Academy of Sciences,
Ordona 21, 01-237 Warsaw
mary@ipipan.waw.pl

Summary. Two test images are decomposed into sequences of ten ordered images which result from a clustering of pixels. The first image is supposed to contain pixels belonging to *edge*, the tenth image — pixels belonging to *region interior*. The remaining images gradually change from edge related to interior related. The clustering is provided by the so called Grade Correspondence — Cluster Analysis (GCCA), described in lastly published book on grade models and methods for data analysis. The GCCA is applied to the data matrices formed by a set of 12 variables which include gradient module, gray level, and ten variables describing the nearest neighborhood of each pixel according to the increasing level of module differentiation. Data matrices are visualized in form of the so called "ordered overrepresentation maps" and "grade stripcharts".

1 Introduction

Grade data analysis is a new branch of intelligent data analysis. Grade models and methods are described in [1]. Last three chapters of this book deal with exploration of multivariate data matrices. The idea of the present paper is to apply such grade exploration to matrices in which rows correspond to image pixels, while columns correspond to a selected set of variables describing pixels. The paper exemplifies possible applications, providing a decomposition of two test images called *"blocks"* and *"Lena"* (Fig. 1). Both images and their decompositions are presented in Section 2, while Section 3 introduces the set of 12 variables forming the data matrix. Two of the variables, gray level and gradient module, are generally used to describe pixels. Ten remaining variables introduced in the paper count, for each pixel separately, how many pixels in the pixel's neighborhood have a gradient module differing from that of the pixel under consideration by less than a certain percentage of the maximal gradient module in the whole image; the percentages increase from 1 to 10.

A short outline of the grade analysis is given in Sections 4 and 5. The main method used, named the *Grade Correspondence – Cluster Analysis* (GCCA), provides a so called overrepresentation matrix with pixels and variables suitably ordered and clustered. The matrix is then visualized as an overrepresentation map (given in Fig. 4 for *blocks*) and it is usually supplemented by a sequence of grade stripcharts (Fig. 5), i.e. by a sequence of line charts of the values of variables (the orderings of variables in the sequence and also of the pixels are established by GCCA). Next, in Section 6, the clusters of pixels indicated at the line charts are used to construct the sequence of images which forms the ordered image decomposition (Fig. 2).

2 Results obtained for two selected test images

Two typical test gray images, *"blocks"* and *"Lena"*, selected for illustrating the grade decomposition of their pixel sets into ordered meaningful subsets, are shown in Fig. 1. Table 1 provides information on their resolution, number of pixels and bits per pixel.

Table 1. General information on test images

	resolution	number of pixels	bits per pixel
blocks	384 × 288	107920	8
Lena	512 × 512	258064	8

Fig. 1. Two selected test images: blocks (left) and Lena (right)

The grade decomposition of the test images is partly shown in Fig. 2. It contains four of ten images of disjoint ordered subsets of pixels (more precisely, four pairs of images, one for *blocks* and one for *Lena* in each pair). The first pair of images contains edges (contours), the second presents details surrounding the edges (e.g., Lena's eyes, nose and lips, as well as grains at the wooden surface of one of the blocks). The last (fourth) pair of images extracts interior regions with uniform texture and so is in opposition to the first pair. The whole sequence of ten subsets is ordered from most strongly "edge related" to most strongly "interior related". The third pair of images in Fig. 2 is taken from the middle of the sequence: it represents this stage at which "edge related" subsets just turn to be "interior related".

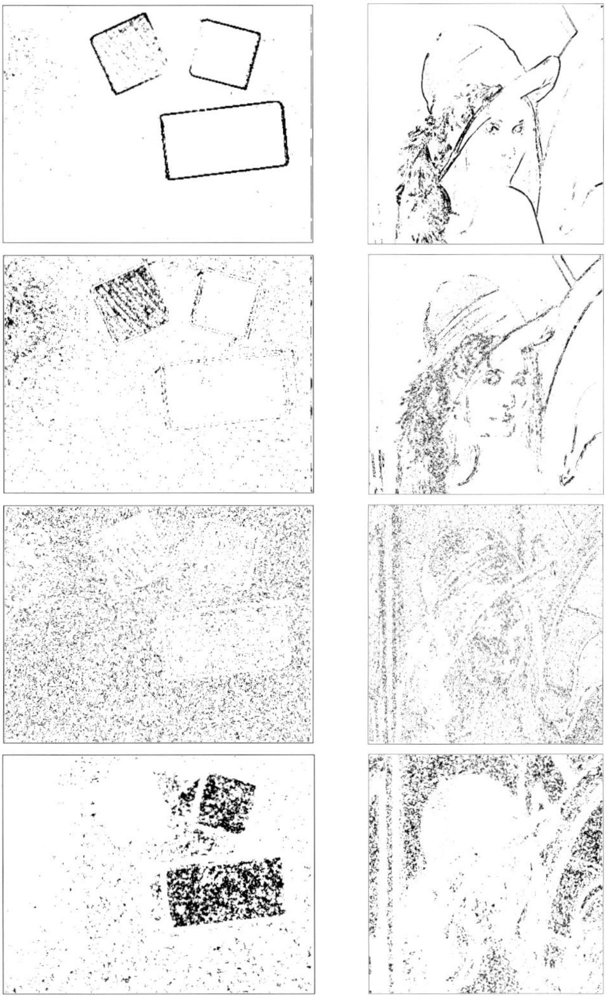

Fig. 2. Grade decomposition of *blocks* (left) and *Lena* (right): four pairs of ordered images.

3 Variables used for describing pixels

The decomposition of pixel sets presented in Section 2 was formed from the data matrix corresponding to specially chosen sequence of variables, including:

- gray level (gl)
- gradient module (gm)
- variables n_j ($j = 1, \ldots, 10$) describing pixel's neighborhood, specified by their threshold parameters t_j.

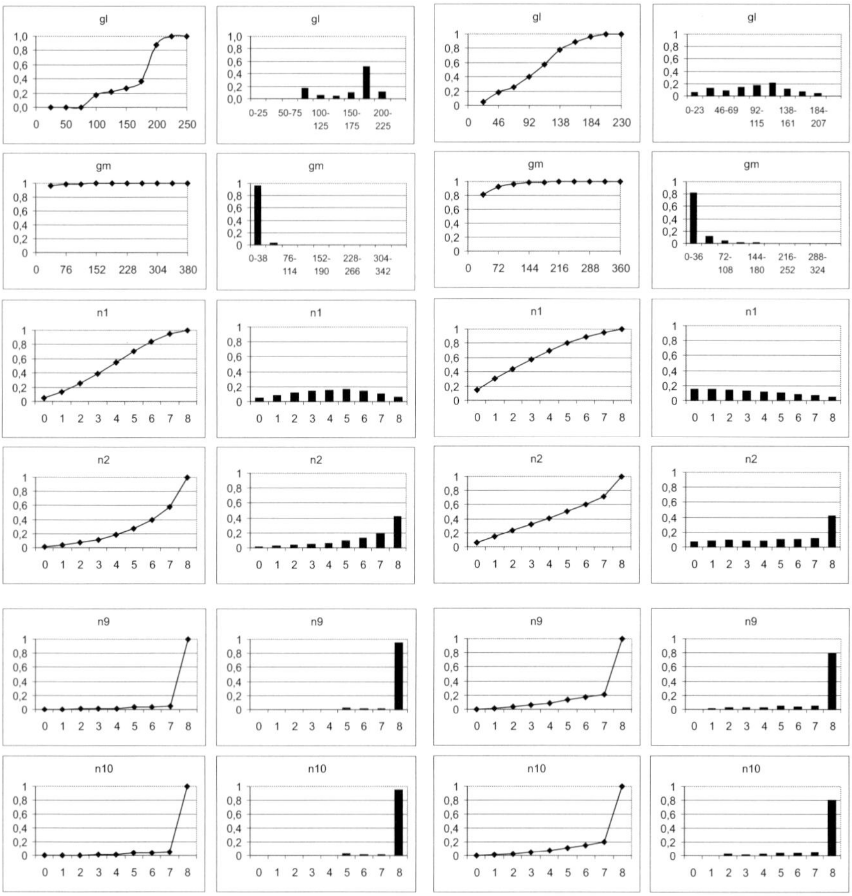

Fig. 3. Histograms and cdf's of variables gl, gm, n_1, n_2, n_3, n_8, n_9, n_{10} for *blocks* (left) and *Lena* (right).

The pixel's neighborhood is defined as a 3×3 pixels window. Thresholds t_j ($j = 1, \ldots, 10$) are calculated for each image as j% of the maximal value of gm appearing in this image. For *blocks*, $t_j = \frac{j*462}{100}$, $j = 1, \ldots, 10$; for *Lena*, $t_j = \frac{j*355}{100}$, $j = 1, \ldots, 10$. The value of n_j for a particular pixel shows how many pixels in the neighborhood have a gradient module differing from that of the considered pixel by less than t_j. Evidently, $n_j \geq n_{j-1}$, $j = 2, \ldots, 10$. The histograms and cdf's (*cumulative distribution functions*) of selected variables are given in Fig. 3. An insight into the suitably transformed joint distribution of (gl, gm, $n_1, \ldots, n_{10}$) will be given in two next sections in the form of the overrepresentation maps and sequences of stripcharts.

4 Ordered overrepresentation maps (with clustering)

The raw data matrices are starting points of the grade procedure providing image decompositions shown in Fig. 2. The raw matrix for *blocks* contains 107 920 pixels (rows) and 12 variables $gl, \ldots, n_{10}$ (columns), the matrix for *Lena* is of size 250 064×12. Each variable is then normalized: the values of the variable in the respective column of the raw data matrix are divided by the column total. The resulting matrix will be called normalized. Next, for each pixel (row) in turn, the normalized values in this row are divided by their sum and multiplied by twelve. The result is called the *overrepresentation index* related to the variable and pixel under consideration. The index is 1 if the non-normalized value is strictly *proportional* to the product of the row and column sums in the initial data matrix (which means "fair" representation); the index *exceeds* 1 if the initial value is larger than expected under the assumption of strict proportionality to the product of marginal sums (i.e. if the initial value shows "overrepresentation" with respect to the expected proportional value); the index is smaller than 1 if the initial value is smaller than expected under proportionality (i.e. if the initial value is "underrepresented"). Underrepresentation is opposite to overrepresentation: one is the reciprocal of the other. The resulting matrix is called the overrepresentation matrix.

A probabilistically oriented reader should observe that the overrepresentation index might be treated as the density of a continuous bivariate distribution defined on the unit square. The square is divided by vertical and horizontal lines into rectangles, the density is constant in each rectangle and equal to the respective overrepresentation index. The width of each stripe between adjacent horizontal lines is equal to the ratio of the sum of elements in the respective row of the normalized matrix to the total sum of elements of this matrix; the width of each stripe between adjacent vertical lines is equal to 1/12.

Each rectangle is then classified into one of five categories: white ("strongly underrepresented") if the overrepresentation index is smaller than 2/3; light grey ("weakly underrepresented") if this index is between 2/3 and 0.95; grey ("fair represented") if this index is between 0.95 and 1.05; dark grey ("weakly overrepresented") if the index is between 1.05 and 1.5; black ("strongly overrepresented") if the index exceeds 1.5. So light rectangles show underrepresentation, dark - overrepresentation. The unit square with suitably colored rectangles is called the overrepresentation map of the normalized data matrix.

A grade procedure called GCA (from *Grade Correspondence Analysis*) reorders rows and columns of the overrepresentation matrix so that darker rectangles are placed as close as possible on two opposite corners of the unit square and also close to a decreasing curve which joins these two corners. The criterion used maximizes rows-columns positive dependence measured by the Spearman's correlation (i.e. by the *grade correlation coefficient*). Another grade procedure clusters rows and columns so that the resulting aggregated matrix is possibly strongly positive dependent. GCA supplemented by clustering is called the GCCA (*Grade Correspondence Cluster Analysis*). The clusters are shown on the map called GCCA overrepresentation map (see Fig. 4 in the case of *blocks*).

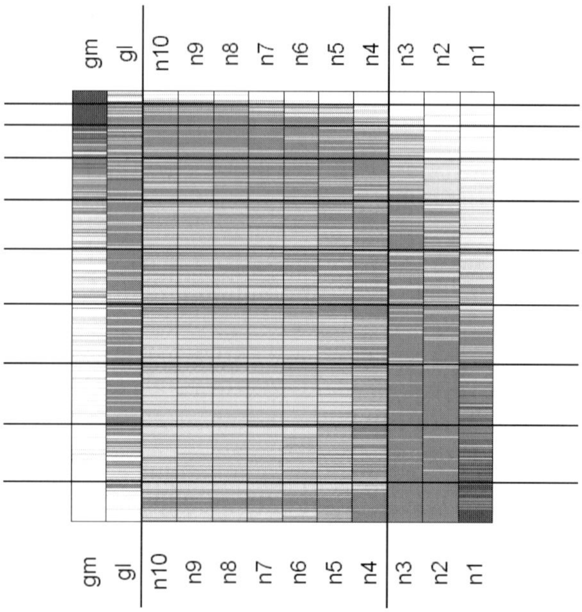

Fig. 4. The GCCA overrepresentation map for the data matrix formed for *blocks*; ten row clusters, three column clusters.

Regularly diminishing positive dependence of variables ensures regular differentiation of variables and also regular differentiation of records. In the case of ideally regular data matrix one expects that in each row and in each column the darkness intensity will have just one peak (not necessarily black) and will gradually diminish on both sides of the peak; moreover, the peaks will concentrare near a curve decreasing from the upper left corner to the lower right one. The GCCA map in Fig. 4 is irregular due to *gl* which distinctly outlies from the regularly diminishing positive dependence pattern; but (gm, n_7, n_6, n_5, n_4, n_3, n_2, n_1) tend to follow that pattern. Variables $n_7 - n_{10}$ are very close one to another and seem to be superfluous. Influential subsets of variables on both ends, (gm, gl) and (n_3, n_2, n_1), form separate extreme clusters.

The GCCA overrepresentation map for *Lena* occurred to be amazingly similar to that for *blocks* (this is not shown because of the restricted length of the paper).

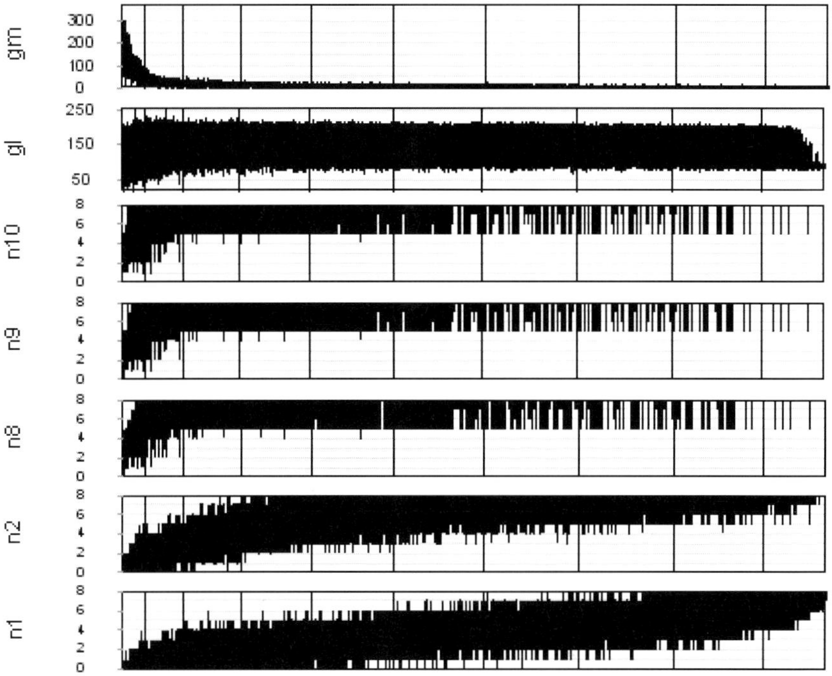

Fig. 5. Grade stripcharts for *blocks* (seven selected variables).

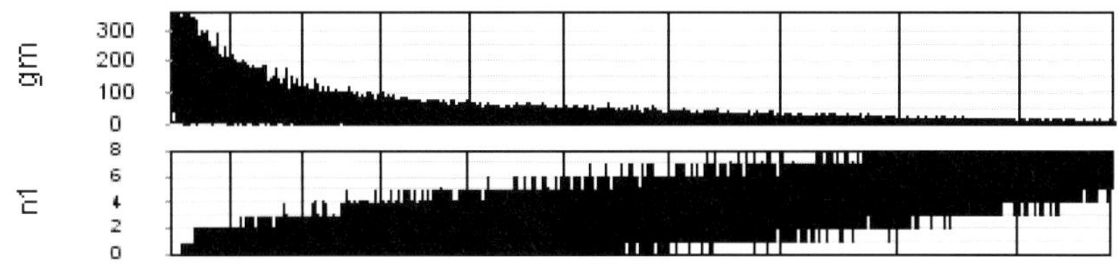

Fig. 6. Grade stripcharts for *Lena* (two selected variables).

5 Sequences of stripcharts ordered and clustered according to GCCA

Let us repeat that the overrepresentation maps directly visualize overrepresentation matrices and *not* raw data matrices or normalized matrices. Therefore it is advised to supplement the maps by the sequence of GCCA stripcharts, i.e. the charts of initial raw values of particular variables obtained for pixels which are ordered according to GCA, while variables in the sequence are also ordered according to GCA. Although the widths of stripcharts are the same, the original raw scales are shown at the vertical axes. A sequence of GCCA stripcharts in Fig. 5 corresponds to *blocks*; its analog for *Lena* in Fig. 6 takes into account only two selected variables which show that the differences between twin strips in Figs. 5 and 6 are relatively small.

6 Restoration of the spatial structure for the GCCA clusters of pixels

Pixels from each of the 10 clusters in Fig. 5 and Fig. 6 possess their coordinates (x, y) in the initial images in Fig. 1. Therefore, spatial structure is restored when these pixels are reintroduced - for each cluster separately - to the image. The result is given in Fig. 2.

7 Closing remarks

A comparison of image decompositions provided by various sets of variables is outside the scope of this paper, although the author is engaged in that line of research. Special attention will be given to outlying variables and clusters. An investigation on applications of the ordered image decomposition based on GCCA is also thought of.

The paper seems to be the first application of grade data analysis to image processing. The grade methods are specially useful as data mining (exploration) techniques. The mathematical background of grade methods is quite simple; what should be noted is that they introduce a new consistent infrastructure of statistical concepts. Book [1] is the only existing full presentation of this direction of intelligent data analysis, with a list of case studies and research papers in the bibliography. The set of grade procedures is implemented by application GradeStat (`http://gradestat.ipipan.waw.pl/`) prepared by Olaf Matyja. GradeStat page will soon have a list of papers on grade methods.

References

1. Kowalczyk T., Pleszczynska E., Ruland F. (eds.) (2004) Grade Models and Methods for Data Analysis, With Applications for the Analysis of Data Populations. Series: Studies in Fuzziness and Soft Computing, vol. 151, 477 p., Springer Verlag Berlin Heidelberg New York.

Detection of Rectangular Landmarks

Ireneusz Hallmann[1]

Institute of Fundamental Technological Research, 21 Swietokrzyska Str., 00-049 Warsaw, Poland irhal@ippt.gov.pl

Summary. In this paprer a method of automated detection of rectangular landmarks is presented. A landmark can be found if it has a monochromatic and characteristic color. The landmark doesn't need to be visible as a whole, it can be partially obstructed by other objects.

1 Introduction

A mobile robot needs to know its location and orientation. It computes them basing on number of various sensors. Among them is a camera, recently more often used because it needs a lot of computational power of computer. The use of a camera allows simultanous detection of many objects for various purposes, like localization or objects tracking.
In this paper a method of detection of rectangular objects is presented. They can be natural object that are usually present in the environment, or artificial landmarks, introduced for easy orientation and localization. Because in indoor environments objects usually have rectangular shapes, the chance that such objects will be found is big.
The block schema of the method is presented in the figure 1. An example of image with a searched landmark is presented in figure 2. The green paper card on the desk is the searched landmark.

2 Image preprocessing

Image of the environment captured from a camera usually cannot be directly used for detection of objects. The most common reasons are: geometrical distortions which cause that real stright lines are curved, and high frequency noise which moves single pixels in detected edges. Before a picture can be used, the captured image must be *preprocessed*.
The nonlinear distortions are introduced when the spherical surface of the

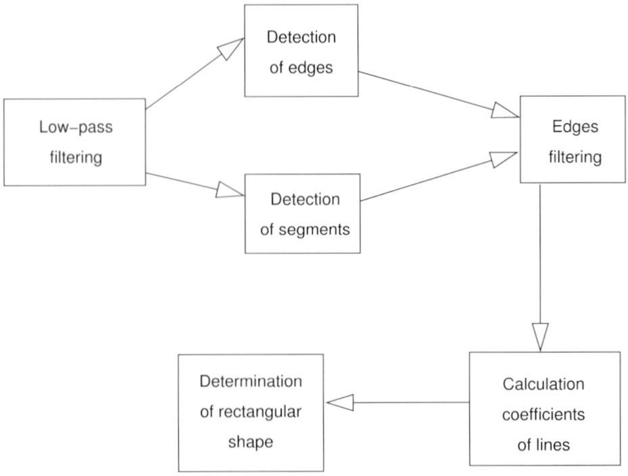

Fig. 1. Schema of rectangular landmarks detection

Fig. 2. An example image with the rectangular landmark

lens is projected to the rectangular CCD sensor or CMOS sensor. Short focal lenses cause larger distortions. In addition, the optical axis of the camera often isn't projected to the image centre. For correction of the distorted image is needed the distortion function. In various papers, eg [5], are described methods of calculation of these distortion functions. In a large part of them radial distortions and considered, and more rarely other distortions, like distance between the optical axis of the camera and the image centre.

In the method used by the author, during the camera calibration a map of distortions is created based on measurements. For calibration a rectangular grid printed on paper is used. I assume that in undistorted image a rectangular grid should be visible. But in real image it is distorted, what is presented in the figure 3. From such an image lines are extracted. Figure 4 shows the extracted lines. Next, each grid cell is mapped to the position and size, that it should have in the undistorted image. Each cell is mapped independently using the bilinear transformation [6]. The size of undistorted cells is assumed to be equal to the largest cell in the image. After calibration, each image is corrected basing on the map made during the calibration. The nonlinear distortion removal is described in [3].

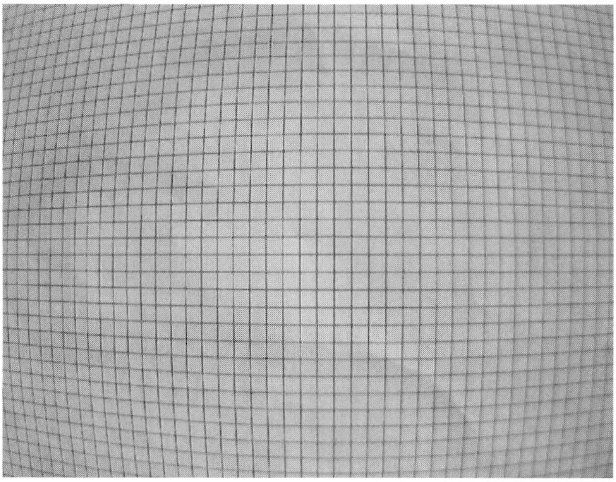

Fig. 3. The source of the calibration image

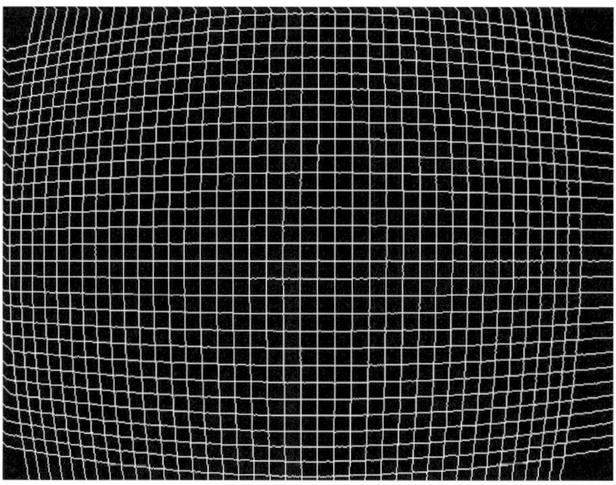

Fig. 4. The distortion map

3 Detection of characteristic segments

In the next stage monochromatic segments are found. Each of those segments may be the searched object, but it isn't yet known at this stage. When the source image is in grayscale, the only segment extracting criterion is the brightness of pixels. When the searched object is black, image should be thresholded with value about 80 (the pixels brightness can vary in range $0 \div 255$). When the source image has colors, segments of pixels having a characteristic hue and sauration can also be extracted. Those colors should be fixed earlier, when the robot learns its work environment. After that stage the image has only 2 colors, white and black. Pixels, which have searched hue, saturation and brightness become white, and other pixels become black. For HSV conversion the method decribed in [1] is used. Examples of source and output images at this stage are shown in figures 5 and 6.

Fig. 5. The source image before the detection of segments

After thresholding, the median filter is used to remove the high-frequency noise. It is really a part of the preprocessing stage, but it is performed after threshold because a faster algorithm can be used. The median filter is better than the linear filters, because the median filter doesn't introduce new colors to the image.

The next stage is edge detection. To save the time, before edge detection the image is divided into parts with segments and the border of size about 20 pixels.It is done for improving speed, because the alogithms which are used are slow on larger images.

First, the Sobel method is used. It detects edges very well, but they are too wide to calculate precisely their equations. For edge detection with maximal precision the method described in [4] is used. After performing the Sobel

Fig. 6. The image after the detection of segments

method, three tables of the second derivatives of the image are calculated. They store the $\frac{\partial^2 Img}{\partial x^2}$, $\frac{\partial^2 Img}{\partial y^2}$ and $\frac{\partial^2 Img}{\partial x \partial y}$ derivatives. The Sobel detector produces few pixel wide lines, and the second derivatives allow to check which among them is the closest pixel to the point, where the second derivative is 0 and the sign of the first derivative changes. Those points are selected as edges. The results of the Sobel detector and the results of edge detector improved by Marr and Hildreth are shown in the figure 7.

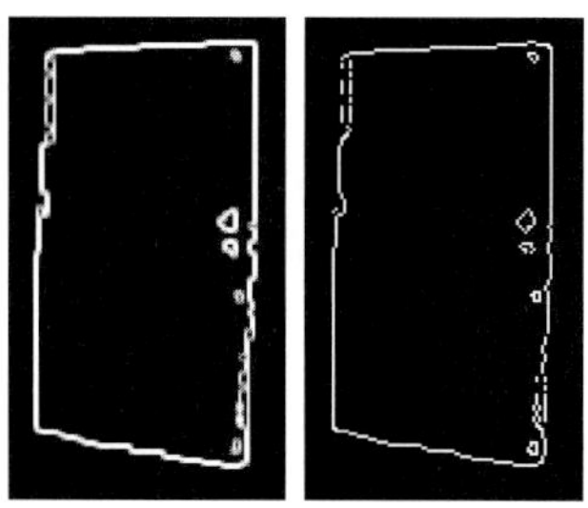

Fig. 7. Line detected by the classical Sobel method and the enhanced Sobel method

When the edges are precisely extracted, some pixels are close to the borders of the segments and some are far from them. Because only borders of the segments are needed for further processing, the number of edge pixels can be reduced by selecting only the pixels near borders of the segments. The proper edge pixels are not further than 5 pixels from the border of the segment inside the segment and not further than 10 pixels outside the segment. Figure 8 shows edge pixels filtering.

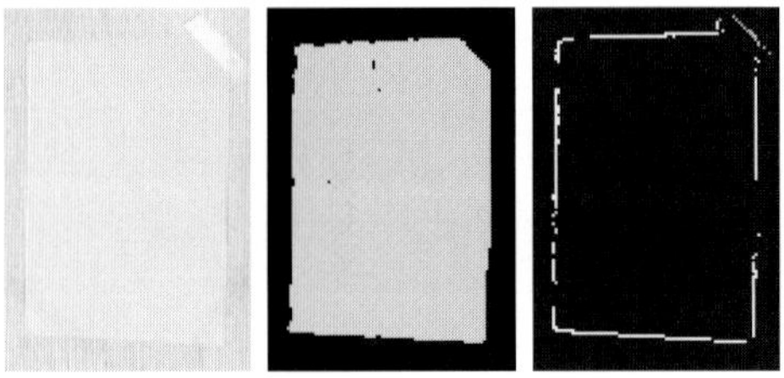

Fig. 8. Removal of edge pixels far from the segment borders

4 Calculations of lines coefficients

At the beginning of this stage the input data are sets of edge pixels near the borders of the detected segments. In this stage straight lines equations are calculated. Those lines are the longest lines that can be build from the input set of pixels. This is performed in two stages:

1. Approximation of straight lines using the Hough transform;
2. Precise calculation of coefficients using the linear regression.

The Hough transform calculation takes a lot of time when the resolution of its coordinates is good. This causes that the author used it only to approximate the coefficients of lines. To calculate the precise coefficients the linear regression is used. Points belong to lines only when their neighbours also belong, what avoids false and very short lines detection. After this stage the data are set of equations of lines, among them are the borders of detected segments.

4.1 The modified Hough transform

The Hough transform calculates for each point $P = (x, y)$ a set of straight lines, that include this point. Those lines are described by their coefficients α and c. α is an angle between the line and the OX line, and c is the distance between the line and the point $(0, 0)$. The lines are described by equations like eq. 1:

$$x \cos \alpha + y \sin \alpha = c. \tag{1}$$

The Hough transform is a function in the coordinates α and c. The value of this function is equal to the number of points (edge pixels) that this line can include. This value describes each line length. The higher resolution of coordinates α and c can improve the precision of coefficients, but considerably increases the time needed to calculations.
The edges detection using the Hough transform allows to calculate the lines coefficients with low precision. This method detects the longest lines, that don't need to be continuous, what is useful when not all edge points are detected.

The Hough transform is sensitive to false points that don't belong to the detected line, but are close to it or are on that line, but at a place where the line doesn't exist in the image. To avoid such effects is used modified Hough transform, described in [2]. The point is considered to be on the line only when that line includes also neighbours of that point. This makes calculations faster and improves precision.

4.2 Linear regression

The linear regression allows to compute the best coefficients of equation of line that should include all points given. The computed line has the minimal distance to each point. It uses the smallest-square technique. The line equation is $y = ax + b$, and the coefficients are computed using the formule 2 and 3:

$$a = \frac{n \sum x_i y_i - (\sum x_i)^2}{n \sum x_i^2 - (\sum x_i)^2}, \tag{2}$$

$$b = \frac{n \sum x_i^2 \sum y_i - \sum x_i y_i \sum x_i}{n \sum x_i^2 - (\sum x_i)^2}, \tag{3}$$

where n is the number of used points.
The set of points used to compute each line is the set of points that indicate one line in the Hough transform.

5 Rectangular landmarks determination

All previous stages lead to the set of monochromatic segments with a set of lines near the borders of each segment. In this stage, all that information is used to determine whether the segment is rectangular or not. The segment is considered to be rectangular, if the following conditions are met:

1. there must exist at least one line on the left side and one on the right side of the segment; those lines must be vertical or almost vertical,
2. there must exist at least one line on the bottom and on one the top of the segment, those lines don't need to be horizontal (because of the perspective), but their coefficient a must meet the condition $-1 < a < 1$.

If at one or more sides of the segment are no such lines, the segment isn't rectangular or is too obscured. If there are more than one line at one side of the segment, the border line is the longest of them (all points in those lines are near borders). In such a way a partially obscured objects can be recognized. During the experiments it turned out that if over 60% of their border lines were visible, the object was recognized as rectangular. Figure 9 shows the rectangular landmark with one corner obscured by the light effects. It was succesfully recognized with precisely detected borders.

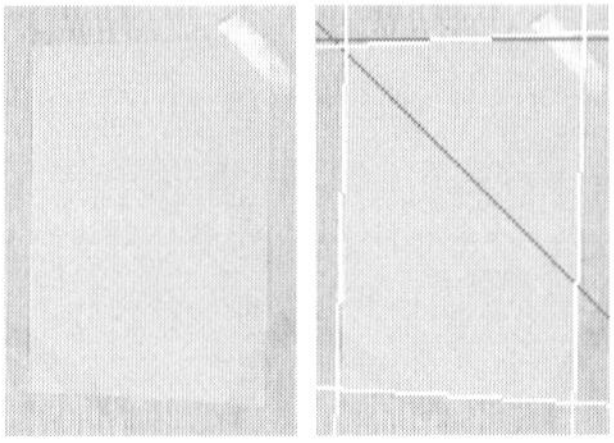

Fig. 9. Example of recognition of a obscured landmark

References

1. Wolczyk P. Bal A., Palus H. Selected properties of perceptual colour spaces. In *Proceeedings of* 2nd *Conference on Computer Recognition Systems KOSYR'01*. Milkow, 2001.
2. Siemiatkowska Barbara. The use of hough transform for localization of the mobile robot. *Pomiar Automatyka Kontrola*, (4/2004), 2004.
3. Borkowski Adam Hallmann Ireneusz. Correction of non-linear distortions of images. In *Automation 2002*. PIAP, 2002.
4. D. Marr and E. Hildreth. Theory of edge detection. *Proc. of the Royal Society*, B-207:187–217, 1980.
5. Rahul Swaminathan and Shree K. Nayar. Non-metric calibration of wide-angle lenses and polycameras.
6. Christopher D. Watkins, Alberto Sadun, and Stephen Marenka. *Nowoczesne metody przetwarzania obrazu*. Wydawnictwa Naukowo-Techniczne, 1995.

Q-shift Complex Wavelet-based Image Registration Algorithm

Hala S. Own and Aboul Ella Hassanien

[1] National Research Institute of Astronomy and Geophysics, Helwan, Cairo Egypt hown@hotmail.com
[2] Kuwait University, Faculty of Business Administration, Quantitative Methods and Information Systems Department P.O..Box 5969 Safat, code no. 13060 Kuwait Abo@cba.edu.kw

Summary. This paper presents an efficient image registration technique using the Q-shift complex wavelet transform (Q-shift CWT). It is chosen for its key advantages compared to other wavelet transforms; such as shift invariance, directional selectivity, perfect reconstruction, limited redundancy and efficient computation. The experiments show that the proposed algorithm improves the computational efficiency and yields robust and consistent image registration compared with the classical wavelet transform.

1 Introduction

Today, images are collected much faster and in significantly greater quantities compared to a few years ago. Accurate registration of these images are vital for comparing the similarities and differences between multiple images. Since human analysis is tedious and error prone for large data sets, we require an automatic, efficient, robust, and accurate method to register images. Image registration is defined as the process that determines the most accurate matching between two or more images acquired at the same or at different times by different or identical sensors [2,3,5,13,14].

Wavelet transforms have proven useful for a variety of signal and image processing tasks, including image registration. The advantages of wavelet transform is that it can analyze signal in time domain and frequency domain respectively and the multi-resolution analysis is similar with Human Vision System (HVS). The Discrete Wavelet Transform (DWT) established by Mallat [15] is widely used in image processing now. Unfortunately, discrete wavelet suffer from the two following problems:

∘ Lack of shift invariance. This means that small shifts in the input signal can cause major variations in the distribution of energy between DWT coefficients at different scales.

∘ Poor directional selectivity for diagonal features, because the wavelet features are separable and real.

Complex wavelets [1,4,6,7,8,9] are shown to overcome these two key problems by introducing limited redundancy into the transform. Recently, Q-shift complex wavelet transform has been developed, which allows perfect reconstruction while still providing the other advantages of complex wavelets.

The objective of this paper is to introduce an automatic registration algorithm to correct the geometric errors of the astronomical distorted images with respect to the reference image. It describes a form of discrete wavelet transform, which generates complex coefficients by using the extended version of the dual tree complex wavelet called Q-shift complex wavelet [7,8]. In addition, a new approach for automatic control point extraction based on multiresolution local contrast entropy is introduced. It extracts more significant control points according to their contribution weight to the multiresolution local contrast entropy.

A similarity is often used for matching process to define the objective functions; therefore, appropriate similarity metrics are crucial to the extraction of high quality control points. Common similarity metrics used in image registration include sum of absolute difference, normalized cross correlation, correlation coefficient, and mutual information. Mutual information measure [16,17] become one of the favorite similarity measures for many researchers because of its superior performance. The matching process was performed using mutual information metric, which is an accurate and robust measure for the degree of similarity between the selected control points.

The rest of this paper is organized as follows. A brief description of the complex wavelet transform and the Q-shift CWT are introduced in section 2. Details of the proposed image registration algorithm is discussed in section 3. The performance of the proposed algorithm as well as some results are presented in Section 4. The conclusion is included in Section 5

2 Complex Wavelet Transform

Because of its desirable multiresolution properties, the two-dimensional wavelet transform happens to be highly applicable to many areas, very notably to the field of image processing. However, its lack of shift-invariance tends to be a major inconvenience, and a transform that provides multiresolution as well as shift-invariance would be highly useful almost everywhere wavelets are used. Complex wavelets are an answer to this problem, and a solid mathematical foundation that allowed practical use of complex wavelets in image processing was originally set up in 1997 by Nick Kingsbury [7,8,9].

The complex two-dimensional wavelet transform provides all of the advantages that the separable discrete wavelet transform provides: multiresolution, sparse representation, and useful characterization of the structure of an image. What makes the complex wavelet basis exceptionally useful for our purposes is that it provides a high degree of shift-invariance in its magnitude. Complex coefficient is obtained by interpreting the wavelet coefficient from one DWT as the real part of complex coefficient, while the corresponding wavelet coefficient from the other tree is interpreted as the imaginary part. The complex wavelet function $\Psi(x)$ is defined as:

$$\Psi(x) = \Psi(x) + jH\Psi(x) \tag{1}$$

The wavelets coefficients calculated as follows:

$$C_{j,k} = W^{r}_{j,k} + iW^{r}_{j,k} \tag{2}$$

Filters are chosen to be linear phase so odd length high pass filters have even symmetry and even length high pass filters have odd symmetry about their midpoints. So the impulse responses of these filters are like the real and imaginary parts of a complex wavelet. The reader can refer to [7,8,9,12] for more illustration. For 2-D signals, we can filter separately along columns and then rows by the way like 1-D.

2.1 Q-Shift Complex Wavelet Transform

The Q-shift complex wavelet transform retain the good shift invariance and directionality properties of the original while also improving the sampling structure. There are two sets of filters has been used; the filters at level 1, and the filters at all higher levels. The filters beyond level 1 have even length but are no longer strictly linear phase. Extension of Q-shift dual in 2D produces three sub images in each of spectral quadrants 1 and 2 giving six band pass sub images of complex coefficients at each level, which are strongly oriented at angles of $\pm 15^o, \pm 45^o, and \pm 75^o$. The strong orientation occurs because complex filter can separate positive from negative frequencies vertically and horizontally. The result image from transformation shows good rotational invariance, because each image is using coefficients from all six directional subbands at the given wavelet level.

3 Registration Algorithm

An automatic registration algorithm for image which globally distorted by 2D affine transformation is discussed here. The search strategy utilize the multiresolution property of Q-shift complex wavelet transform by starting the search from the higher level to the lowest level resulting in reducing the computation complexity and operation time. The algorithm can be summarized in the following steps:

Input
distorted and reference image A and B.

Processing

Step-1: Decomposition process and control point extraction
1.1 For i=1 to k (at coarsest level) do
1.2 Decompose A and B by Q-shift wavelet transform
1.3 Compute the magnitude of the wavelet coefficient at point (i,j) using the following form:

$$M(i,j) = \sqrt{W_r(i,j)^2 + W_i(i,j)^2} \tag{3}$$

Where $W_r(i,j)$ and $W_i(i,j)$ are the real and imaginary parts of the of A and B.
1.4 Divide the Q-shift wavelet matrix into blocks with same size
1.5 For each Q-shift wavelet coefficient block Do
1.5.1 Calculate the probability of the multiresolution local contrast entropy (MLCE) using the following equation:

$$MLCE(x,y) = \frac{ME(x,y)}{\sum_{i=1}^{N}\sum_{j=1}^{N} ME(i,j)} \tag{4}$$

Where $ME(x,y)$ is the multiresolution entropy calculated by the following form:

$$ME(x,y) = \sum_{j=1}^{l} \frac{M_j(x,y)^2}{2\sigma_j^2} \tag{5}$$

1.5.2. The control point is coefficients within the block if the following condition is satisfied:

$$MLCE(x,y) = max_{\acute{x},\acute{y} \in NP} MLCE(\acute{x},\acute{y}) \tag{6}$$

where, NP is the neighborhood of $MLCE(\acute{x},\acute{y})$.

Step-2 Similarity measure
2.1 Compute the mutual information as follows:

$$I(A,B) = \sum P_{AB}(a,b) log \frac{P_{AB}(a,b)}{P_A(a).P_B(b)} \tag{7}$$

Where $P_A(a)$ and $P_B(b)$ are the marginal probability distributions, and $P_{AB}(a,b)$ is the joint probability distribution of A and B. To estimate MI between A and B based on the last equation, we only need to estimate the joint histogram between A and B, $h_{A,B}(i,j)$, where $P_A(i) = \frac{1}{N}h_{A,B}(i,j)$, $P_B(j) = \frac{1}{N}h_{A,B}(i,j)$, and $P_{A,B}(i,j) = \frac{1}{N}h_{A,B}(i,j)$. N is the total number of pixels within each sub-images.

Step-3: Matching

3.1 if MI is maximum over small size window surrounding each control point within the coarsest level in wavelet transform then the control points of the A and B is matched.

Step-4: Deformation model

4.1 Define the deformation parameters using the affine transformation with rotation, translation and scaling parameters $(c, \alpha, \Delta a, \Delta b)$.

4.2 For all control points Do

4.2.1 estimate the deformation parameters using least square method [10,18]. 4.3 Evaluate the performance of the least square fitting and the accuracy of the registration using the root mean square error (RMSE). It is defined as follows:

$$RMSE = (\sum_{1}^{N}((\beta a_i + \delta b_i + \Delta a - A_i)^2 + (\beta a_i + \delta b_i + \Delta b - B_i)^2)^{\frac{1}{N}} \quad (8)$$

$\beta = cCos\alpha$, $\delta = cSin\alpha$, and N is the number of matched control points.

Step-5:Warping process

5.1 Warp A using the estimated parameters.

5.2 Doubled the window size to keep computation complexity the same at higher resolution.

5.3 Repeat al steps for the next finer resolution, using the current corrected image as the initial input.

Step-6:Reconstruction process

6.1 Apply the inverse transform to reconstruct A.

Output

The registered image.

4 Results and discussion

In this section, the performance of the algorithm is introduced. A 256 x 256 gray scale of Horsehead Nebula and Galaxy NGC266 test input images are shown in Figure (1a) and Figure (2a), respectively. The input image is globally distorted by 2-D affine transformation as shown in Figure 1(b) and Figure 2(b). There are obvious large scale geometric distortion ($\Delta a = 20, \Delta b = 9, c = 5, \alpha = 20$). The registration result is shown in Figure 1(c) and Figure 2(c). The result image shows good rotational invariance, because each image is using coefficients from all six directional subband at the given Q-shift CWT level. Table (1) shows the estimation of the simulated parameters using the proposed algorithm in different levels and the residual error ξ^2 expressed in dB as SNR between the aligned simulated source and reference image.

(a) Reference image (b)Warped image (c)Registered image

Fig 1: Registration results (Horsehead Nebula image)

(a) Reference image (b)Warped image (c)Registered image

Fig 2: Registration results (Galaxy NGC266 image)

Different complex wavelet transform can be thought of to quantify the accuracy of the proposed registration algorithm in different wavelet transforms levels. Table (2) shows the root mean square error (RMSE)values at different

Table 1. Registration parameters

Levels	Δ a=20	Δ b=9	α =20	c=5	ξ^2
1	19.9	8.9	5.00	4.9	44.17
2	19.9	9.00	20.002	4.00	44.26
3	20.00	9.00	20.001	4.00	44.32
4	20.00	9.00	20.00	5.00	44.43

level of decomposition applied on the gray scale of Horsehead Nebula image as a sample of the implementation. It is shown that the RMSE for Q-shift is a good performance compared with other methods [10,11,18].

Table 2. RMSE on levels 4,3 and 1

wavelet	RMSE-l4	RMSE-l3	RMSE-l1
DWT	0.1	0.233	0.235
Steerable	0.05	0.0916	0.236
DT-CWT	0	0	0.135
Q-Shift CWT	0	0.05	0.089

5 Conclusion

The approximate shift-invariant property of the complex wavelet coefficients allows us to infer pixel shifts, which can be used to invert geometric distortion. An image registration algorithm based on this idea is described in this paper. In addition, a new approach for control point extraction based on the multiresolution entropy within a Q-Shift wavelet transform is also presented. Experimental results show that the quality of image registration based on Q-Shift complex wavelet transform is better than registration image based DWT, Dual-tree and Steerable wavelet.

References

1. P. Bao (1998) Panoramic image Mosaics via Complex Wavelet Pyramid. In proceedings of IEEE Int. Conf. in systems, Man, and Cyberetics, 5, 4614-4619.
2. L. Brown (1992) A Survey of Image Registration Techniques. ACM Comput. Surv., 24(4), 325-376.
3. J. P. Djamdli, A. Bijaoui and R. Maniere (1993)Geometrical Registration of Remotely Sensed Images With The Use of The Wavelet Transform. SPIE International Symposium on optical Engineering and photonics, Vol. 1938, Orlando, FL, USA, 421-422.

4. F. C. A. Fernades, R. L. Spaendonck and C. S. Burrus (2001) A New Directional, Low-Redundancy, Complex-Wavelet Transform. In proceedings of IEEE Int. Conf. In Acoustics, Speech, and signal processing, 6, 3653-3656.
5. L. M. G. Fonseca, and M. H. M. Costa (1997) Automatic Registration of Satellite Images. Brazilian Symposium on Graphic Computation and Image Processing, IEEE Computer Society, 219-226.
6. M. A. Woodward , J. J. Rowland and D. B. Kell (2004) Fast automatic registration of images using the phase of a complex wavelet transform: application to proteome gels. The Analyst journal,129(6), 542 - 552.
7. N.G. Kingsbury (1998)Dual-tree Complex Wavelet Transform : a new technique for shift invariance and directional filters. IEEE Digital Signal Processing workshop, DSP 98, Bryce Canyon.
8. N. Kingsbury (2002) Complex Wavelet for Shift Invariant Analysis and Filtering of Signals. http://www-sigproc.eng.cam.ac.uk/ ngk/.
9. N. Kingsbury (2003)Design of Q-Shift Complex Wavelets for Image Processing Using Frequency Domain Energy Minimization. The international confernce in image processing, ICIP2003, Septamber 14-167, 2003, Barcelona,Spain.
10. I. Koren, A. Laine, and F. Taylor (1995) Image Fusion Using Steerable Dyadic Wavelet Transform. In proceedings of IEEE Int. Conf. In image processing, 3, 132-135.
11. K. Kozlov, E. Myasnikova, M. Samsonova, J. Reinitz, and D. Kosman (2000) Fast Redundant Dyadic Wavelet Transform in Application to Spatial Registration of the Experssion Patterns of Drosophila Segmentation Genes. 15th IEEE Inter. Conf. In Pattern Recognation, 459-462.
12. Hala S. Own. (2005) Image Registration Algorithm Based on Complex Wavelet Transform.ICGST International Journal on Graphics, Vision and Image Processing, vol.2, Jan.9-15.
13. J. Le Moigne, W. Xia, J. C. Tilton, T. EL-Ghazawi, M. Mareboyana, N. Netanyahu, W. J. Campbell, and R. E. Cronp (1998) First Evaluation of Automatic Image Registration Methods. IGARSSŠ98, July, 6-10.
14. J. Le Moigne, N. S. Netanyahu, J. G. Masek, D. M. Mount, S. Goward, and Honzak (2000) Geo-Registration of Landsat Data by Robust Matching of Wavelet Features. IEEE Inter. Conf., in Geoscience and Remote Sensing Symposium, 4, 1610-1612.
15. S. Mallat (1989) A Theory for Multiresolution Signal Decomposition. IEEE PAMI, 11(7), 674-693.
16. B. Likar, F. Pernus (2001)Hierarchical Approach to Elastic Registration Based on Mutual Information. Image and Vision Computing, 19, 2001.
17. J. P. W. Pluim, J. B. A. Maintz, and M. A. Viergever (2001) Mutual Information Matching in Multiresolution Contexts. Image and Vision Computing, 19, 45-52.
18. E. Simoncelli, and A. Karasaridis (1996) A Filter Design Technique for Steerable Pyramid ImageTransforms. Int'l Conf. Acoustics Speech and Signal Processing. Atlanta GA, http://www.cis.upenn.edu/ eero/steerpyr.html

Concept of 3D View Model of Non-Convex Polyhedra Generation Using View Sphere Partitioning into Single-View Areas

Monika Kowalczyk[1] and Wojciech S. Mokrzycki[2]

[1] Institute of Fundamental Technological Research PAS, Swietokrzyska 21, Warsaw, Poland; mkowalczyk@ippt.gov.pl

[2] Institute of Informatics, University of Podlasie, Sienkiewicza 51, Siedlce, Poland; mokrzycki@ii3.ap.siedlce.pl

Summary. The paper concerns the construction of the solids representation for a model-based visual identification system. The chosen representation is a complete 3D view model and the modelable solids are convex polyhedra and some class of non-convex polyhedra. A described algorithm uses the concepts of the view sphere with perspective projection and partitioning the sphere into the single-view areas.

1 Introduction

The 3D view models described in this paper are destined for object visual identification based on the models and the object data educed from a depth map. The depth map can be obtained by any technic of receiving 3D data about a scene. The model-based identification consists in matching of a real object image or view with the model views generated automatically and stored in a system memory.

2 View models of objects

There are many kinds of object representations described in the literature which can be used in different applications. It seems that for visual identification the 2D and 3D view models are the most suitable. The 2D view models for visual identification have to be ordered in a graph structure called **an aspect graph**, [1, 2, 3].

The methods of generating 3D view models for convex polyhedra are described in i.e. [4, 5, 6]. In the view models a polyhedron is represented by a set of views. Each view is **a relative arrangement of the distinct object features** seen from a certain viewpoint and forming so-called **an aspect**.

Identification of objects can be conducted by comparing **an object data** from a scene with **an object models** from a database and finding identity. Information about the physical object is derived from a depth map and stored in a mathematical structure. The same structure is used for object view models in the database. The database consists of a certain object models number and the system can identify just these objects which are modelled. The basic elements of the structure are the pre-selected distinct features of the objects (faces, edges and vertices) which form an aspect.

3 Generalization of KM^{zWW} algorithm to KM^{zWN} form

In the papers mentioned before the algorithms generating 3D view representation of opaque convex polyhedra (WW) for the viewpoints lying on the view sphere (in a fixed distance from an object center) with perspective projection are described. In particular an algorithm KM^{zWW} ([5]) is presented. The algorithm consists in determining of a so-called seedling single-view area on the view sphere and then in spiral finding the adjacent single-view areas until they cover the view sphere completely, fig. 1. For each found single-view area a 3D view is generated. The complete set of such views is a right view representation of an object.

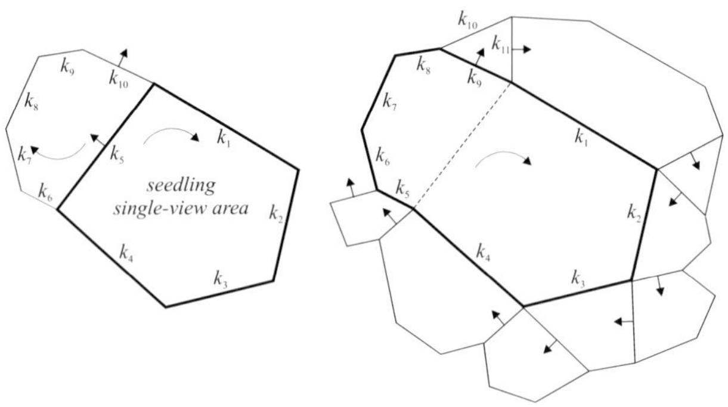

Fig. 1. The method of spiral covering the view sphere by determining and joining the adjacent single-view areas

3.1 Primary assumptions

In this paper we propose an extension of the KM^{zWW} algorithm for a certain group of the opaque non-convex polyhedra (WN) – KM^{zWN} algorithm. Our intention is to use the previous tried and true concepts in a maximal range, especially the algorithm for WW based on the seedling single-view area.

The presented concept of automatic generation of a view model embraces a class of opaque convex polyhedra and also a class of **opaque non-convex polyhedra without lodges, holes, humps, etc.** More precisely, it is assumed that each presumptive concavity in a solid (it may be many of them but separated) considered as it is (individually, without a solid) has a form of convex polyhedron, and also the whole solid considered without concavities is a convex polyhedron. For this kind of solids it is possible to choose the viewpoints (on the view sphere) in such a way that each little recess of a concavity can be seen. The solids have also a property that all self-occlusions and all invisibilities in the visible part of the solid appears only inside a concavity (i.e. they can be brought on only by the faces of the concavity and they can embrace only the faces of the concavity). It is assumed also that in a phase of real image acquiring a polyhedron is completely visible.

3.2 Terms

- **view** – a set of polyhedron elements (faces, edges and vertices) which are inside a view contour; faces of the view can be visible, partly or completely occluded (faces of concavity only) or invisible (invisible faces of concavity); Hence, there are three kind of views:
  1. ordinary view – when all faces inside the view contour are visible;
  2. view with invisibility – when certain faces inside the view contour are invisible because they are inverted for the viewpoint back;
  3. view with self-occlusion – it is strictly related with a view with invisibility when an edge of invisible concavity face causes occlusion of other concavity faces; if all vertices of an occluded face are invisible the face is completely occluded;
- **view contour** – it is a sequence of edges common for view faces and out-of-view faces;
- **concavity contour** – it is a sequence of edges common for concavity faces and adjoining faces of convex part of the solid;
- **visual event** – it causes a change of a view; we assume that a new view appears when visibility of at least one solid feature (a face or a vertex) changes; hence, a new view appears if:
  - at the border of the view contour one face appeared or disappeared,
  - at the border of the view contour one face appeared and another one disappeared,
  - some concavity face became invisible,
  - some vertex inside a concavity contour appeared or disappeared,
  - at the same time a change at the border of the view contour and a change inside the concavity contour appeared;
- **view cone** – a solid angle placed inside the view sphere which has the vertex on the view sphere (a viewpoint VP) and the cone angle width is connected with a visual head parameter and is equal to 2α; this cone includes a sphere circumscribing the object which has its middlepoint in the (0,0,0) point and is tangent to the cone along its circumference;
- **complementary cone** – it is coupled with the view cone in such way, that they have the same view axis direction (but inversed the turns), its vertex is in the (0,0,0) point (i.e. in the middle of the solid and in the

center of the sphere circumscribing the object and also in the center of the view sphere), the cone angle width is $\pi - 2\alpha$, and both the cones intersect at the right angle;

3.3 Analysis of problems

Let's try to generate a 3D multiview representation of a single polyhedron with concavities (without lodges, holes, humps etc.) using the view sphere with perspective projection and the seedling single-view area concept, [5].

Comparing with convex polyhedra the concavities are the most important reason of difficulties. A concavity may cause that some faces in the view are not visible because they are inverted for the viewpoint back. It may happen that at first some faces are qualified as the view faces which are in fact out-of-the-view but they are turned towards the viewpoint. These faces shouldn't remain in the view (because we assume that the solid is opaque).

In a case of convex polyhedron crossing any face plane causes a change in a view (a visual event) which is disappearing of a face from the view or appearing a new face in the view. A change in the set of view elements is connected with passing into another single-view area. For convex polyhedra the planes of solid faces are the only ones which limit the single-view areas.

In a case of opaque non-convex polyhedra some planes of solid faces doesn't bound the single-view areas. The planes of concavity faces delimit only the single-view areas lying in the part of the view sphere from which the concavity is visible. On the other hand, the face planes aren't enough to indicate all the visual events since an occlusion of concavity faces by a concavity contour edge (belonging to an invisible concavity face) is also the visual event. Such event (**a self-occlusion event**) is not connected with crossing any face plane but it is caused by moving away from the crossed face plane (which became invisible)

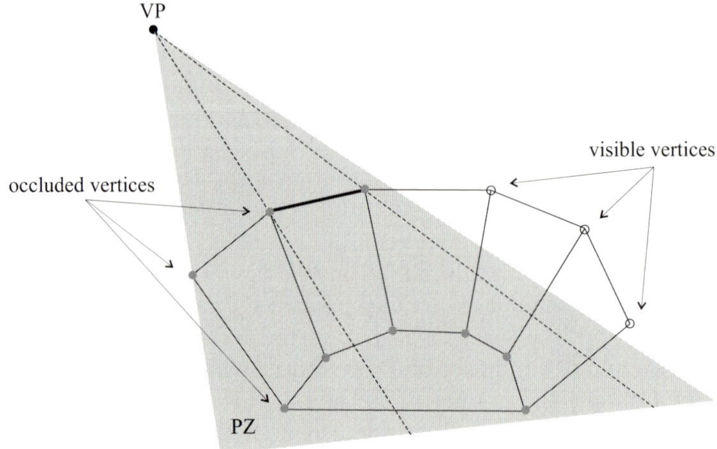

Fig. 2. Occlusion plane PZ

until meeting another boundary plane. An additional **occlusion plane** PZ (fig. 2) helps in pointing out a place of the occlusion event. The occlusion plane PZ is determined by a concavity contour edge belonging to the invisible concavity face and the viewpoint VP. When PZ crosses any vertex inside the concavity then an occlusion of this vertex causes. If a concavity face has all its vertices occluded then it is recognised as a completely occluded face and in this place on the view sphere should appear a boundary of the single-view area. This boundary is formed by PZ plane. Hence, the boundaries of the single-view areas are set up by the polyhedron face planes (all of them but the concavity face planes set up boundaries on the part of the view sphere only) and some additional planes – the occlusion planes PZ (which are defined in each concavity by a concavity contour edge and a vertex inside the concavity) which also set up boundaries on the part of the view sphere only.

3.4 Scheme of algorithm

The input data for the algorithm of of multiview representation generation an exact mathematical solid model (B_{rep}), i.e. 3D coordinates of each solid vertex and all mutual relation between vertices, edges and faces of the polyhedron.

The first, preliminary step of the algorithm is **a solid analysis**. In this step a solid is checked if it meets the assumptions and all concavities are pointed out. After such preparing of data the algorithm starts from choosing a viewpoint VP on the view sphere. **The first view** can be calculated by checking if the normal versors of the faces are inside the complementary cone. After calculating the view the seedling single-view area is beeing determined. **The view contour** is formed by the edges common for the faces belonging to the view and for the faces out-of-the view. For non-convex solid we check if any invisible or occluded faces are in the view. All possible occlusions caused by an invisible face in the view should be calculated.

Then we proceed with determining a neighbouring single-view area. The single-view areas are being calculated in such a sequence that they cover the whole view sphere in a spiral way. Each new area is joined with the previous ones and the edges of the covered part of the view sphere are inserted into **the boundary register**. So, the edges gathered in the boundary register divide the view sphere into known and unknown part. When the boundary register is empty it means that the whole sphere became explored. And this is a stop condition of the algorithm.

As a result of this algorithm we receive a complete set of the solid views which is the smallest and enough to identify the solid basing on only one shot of it when few constraints on placing the camera are satisfied.

Solid analysis

The aim of this part of the algorithm is to recognise if a given solid satisfies the assumptions (i.e. if it is an opaque polyhedron without or with concavities which are convex) and to point out all concavity faces if they exist. For this purpose all the vertices (or edges) in the solid can be checked, the angles between the normal versors of neighbouring faces can be calculated and all neighbours forming the concavities can be localised. After these operations each solid face is qualified as "a concavity face" or "an out-of-concavity face" and each solid edge is qualified as "a concavity contour edge", "an inside concavity edge" or "an out-of-concavity edge".

In the next step of the algorithm all the occlusion planes which causes complete occlusion of some concavity face should be found because passing this kind of plane is connected with a change in the view and with stepping into a new single-view area.

Determining of view faces and view contour

At first the visibility of all faces in a view has to be specified. In the case of convex polyhedra a view differs from previous one in visibility of a boundary face sg_1 which is just crossed – in the actual view this face is added (if it becomes visible) or subtract (if it disappears) from the previous view. In the case of non-convex polyhedra this operation (of determining a new view) is more complicated. **An algorithm of generating a view** proceeds as follow:

1. check the normal versor of all faces:
   - if the normal versor of a face lies *inside* the complementary cone then label the face with +1 (as a potential face of the view),
   - if the normal versor of a face lies *outside* the complementary cone then label the face with -1 (as a potential face out of the view),
2. conduct additional selection of **the concavity faces**:
   - find faces, among concavity faces which are labelled with -1, which are in the view but inverted back to the viewpoint – they are surrounded by the faces of the view (labelled with a positive number); label these faces with +5 (as invisible faces of the view);
   - find faces, among concavity faces which are labelled with +1, which are out of the view but turned towards the viewpoint – they are surrounded by the faces out of the view (labelled with a negative number); label these faces with -5 (as concavity faces out of the view);
   - for the rest of concavity faces (labelled still with +1) check the visibility of all its vertices and change the label of a face properly:
     - if all vertices are visible label the face with +2 (as a visible face),
     - if some face vertices are visible and the others are occluded label the face with +3 (as a partly occluded face),
     - if all the face vertices are occluded label the face with +4 (as a completely occluded face);

At the end **the edges of the view contour** must be selected – they are edges connecting faces of the view (labelled with positive number) and faces out of the view (labelled with negative number).

Determining of seedling single-view area

About some solid faces we know in advance that they don't limit any single-view area and they shouldn't be taken into consideration. Therefore we create **a set of possibly boundary faces** of a single-view area. They are the faces forming a view contour and the concavity faces in the view.

Now, we point out the faces which really limit the seedling area:

1. take the first face s from the possibly boundary faces set and create a halving plane PP, [5], going through the viewpoint VP and the normal vector of the face s;
2. going along the plane PP in a fixed direction the viewpoint VP meets a plane of another possibly boundary face; the first met plane sg_1 is the boundary plane and the intersection point of the plane PP and the plane sg_1 is the first vertex v_1 of the seedling single-view area;
3. going further along the plane sg_1 starting from the vertex v_1 in chosen direction (e.g. clockwise around VP) VP meets a plane of another possibly boundary face – this is the second boundary plane sg_2 and the intersection point is the second vertex v_2 of the area;
4. and so on until the met plane is sg_1.

After this calculations the seedling single-view area is set by an ordered list of boundary faces sg_1, sg_2, sg_3, ...

Calculating of self-occlusions

A change of the viewpoint VP position (the view cone and the complementary cone also) may cause disappearing of some (concavity) face s_i because of going out from the complementary cone of its normal vector (although the face remains inside the view and it is registered as a part of the view). It is a visual event of starting a new view (view with invisibility) followed by a possibility of occluding other concavity faces (if VP moves further).

An occlusion plane PZ helps in establishing which faces will be occluded (partially or completely). The occlusion plane is determined by the viewpoint VP and a concavity contour edge k_j of the face s_i and it contains the face s_i which just disappears ($PZ_{s_i k_j}$). The viewpoint VP moving further slants the occlusion plane (which is now only PZ_{k_j} because it doesn't contain the face s_i, it contains only the edge k_j) which occludes progressively the other concavity faces. Checking the position of vertices in relation to the plane PZ (the vertices which are under the plane are occluded and the vertices which are above are visible) we can say if any part of a face is still visible.

3.5 Generation of view model of non-convex polyhedra algorithm

After all these considerations we may now present a scheme of full version of the algorithm KM^{zWN} generating 3D multiview model of non-convex polyhedra with applying the seedling single-viewa area concept.

ALGORITHM KM^{zWN}

1. Having a mathematical model (B_{rep}) of a polyhedron analyse the solid.
2. Compute the radius r of a sphere circumscribed around the object and the radius R of a view sphere.
3. Choose a viewpoint VP, generate and register a view, mark the view contour edges.
4. Determine a seedling single-view area; put its edges into the boundary register RG.
5. Find an adjoining single-view area; put new edges of the area into the boundary register instead of repeated ones.
6. Choose a viewpoint VP in this single-view area, generate and register a view, mark the view contour edges.
7. Check the boundary register RG:
   7.1. if it is not empty find an adjoining single-view area; back to step 6.
   7.2. if it is empty then go to step 8.
8. End the generation.

4 Future work

At present particular functions of the algorithm are being programmed. After implementing the whole algorithm the tests on a group of solids are going to be made. After presented analysis of the problem and the solution we strongly expect positive results. The tests results will be published in our next papers.

References

1. Stewman J., Bowyer K. (1988) Creating the Perspective Projection Aspect Graph of Polyhedral Objects. Proc. 2nd IEEE Int. Conf. on Computer Vision (ICCV88), 494–500
2. Stapor K., Skabek K., Tomaka A. (1999) Model-Based Recognition of Polyhedral Objects from Single Intensity Image Using Aspect Graph. Machine Graphics & Vision, 8(2), 249–264
3. Cyr C.M., Kimia B.B. (2001) 3D Object Recognition Using Shape Similarity-Based Aspect Graph. Proc. 8th IEEE Int. Conf. on Computer Vision (ICCV01), 1, 254–261
4. Kowalczyk M., Mokrzycki W.S. (2003) Obtaining Complete $2\frac{1}{2}$D View Representation of Polyhedra Using Concept of Seedling Single-View Area. Computer Vision & Image Understanding, 91(3), 280–301
5. Kowalczyk M., Mokrzycki W.S. (2003) Methods of Generation 3D Exact Views of Convex Polyhedron for Visual Identification. Part II: Noniterative Methods, Implementation and Tests Results. Machine Graphics & Vision, 12(4), 435–452
6. Frydler M., Mokrzycki W.S. (2004) New, Fast Algorithm of 3D Multiview Polyhedron Representation Generation on View Sphere with Perspective. ECCV – Workshop: Applications of Comp. Vision (ECCV-ACV04)

Using Modified Hough Transform for Grouping of Image Features

Leszek Przybylski

Institute of Control and Information Engineering, Poznan University of Technology 3a Piotrowo Street, 60 - 965 Poznan leszek.przybylski@delphi.com

Summary. A modified Hough transform has been proposed for grouping of image feature-carriers. The method has adjustable parameters, which are used in grouping and adding of missing image feature-carriers (due to registration noise). The adding of missing image features is based on performing of the secondary Hough-transform over the small window, in order to increase of transform resolution. The article contains of the research results addressing the influence of the method parameters on grouping of image features of real images.

1 Introduction

As a result of extraction image feature-carriers, i.e. some separated elements (or groups of elements), which represents spatial units (primitives) are detected. Due to the noise, the detected primitives do not create connected units in most cases and do not let to establish the elements of geometrical description of objects on the scene. Usually interrupted, disconnected segments are extracted, which are parts of contours, or isolated pixels. Scene interpretation requires grouping of the appropriate image features into some larger units. For grouping process have been proposed [12,15] a lot of rules (for example Gestalt rules) and methods (for example clustering, fitting a model). Grouping methods should use information coded by image features-carriers. In the article, some grouping rules has been proposed, which are based on feature-class denominators (semantic cues) and on the context - the bit-map neighborhood of appropriate image feature carriers. Linear segments are of primary interest for the recognition of the large class of objects visible on the scenes. This is the reason why as a method of grouping the directly extracted image features, the Hough transform has been chosen. The classical Hough transform has been extended with additional algorithmic mechanism enabling to connect some detected disconnected pieces of contours (gap filling process), and to join the obtained continuous segments into larger parts of curvilinear contours (shape primitives).

2 The modified Hough transform

The essence of the classical Hough transform [7] is the remark, that each image point may potentially belong to some parametric line, its image coordinates (x_i, y_i) satisfying a parametric equation of some model. So by searching in the parameter space, it is possible to find the number of pixels supporting the hypothesis about the particular contour model. The simplest linear model it is $y_i = a * x_i + b$ and the parameter space is 2D. Strong evidence supporting a hypothesis on the appearance of the model-line on the original image is represented by the local maxima in the parameter space. In the case of linear-model hypothesis, for each value of a_i, b is calculated as $b = -x_k * a_i + y_k$ and further it is quantized to some value b_j. As a result one gets the table of accumulators holding a number of "votes" for a particular pair of (a_i, b_i). To overcome the problems with discontinuity in parameter space - for case of $a = 0$, the following parametric equation is used: $x * cos\theta + y * sin\theta = \rho$, where θ denotes the angle between the Ox axis and the semi-line from the origin, orthogonal to the model-line. ρ is here the distance of the model-line to the origin of the parameter space coordinates system.

After detecting particular image features in the Hough parameter space, the analysis is performed along a detected line to establish the length of gaps separating the linear domains of image feature-carriers (in classical setting these are edgels clusters). Grouping process is based on joining of these linear domains of image feature-carriers as a function of the distance of adjacent groups (l_gap parameter of the method) and of the number of feature-carriers, which constitutes these groups (l_joint parameter of the method). Modification of grouping mechanisms (which use preliminary results of the Hough transform) is based on introducing the possibility of adjusting the l_gap and l_joint parameters. Neighbor pixels - carrying features, which are sought for, are treated henceforth as elementary segments of some contour. Whenever $d < l\_gap$, where d is the distance separating segments, the gap is interpolated with pixels (image feature-carriers) at higher resolution. When the number of pixels (image feature-carriers, which forms segments) is equal or higher to value of the l_joint parameter, this segment is treated as a part of some contour and co-ordinates of this segment is added to the list of detected segments. Short, compact segments (or separated pixels), for which the mutual distance is larger than l_gap parameter are ignored. Gap filling process uses a window (Fig. 1) of dimensions depending upon the end-points coordinates of closest segments and consists of the linear interpolation over the image raster.

This gap filling process step consists of a secondary Hough transformation over upper defined window, in order to detect potential pixels supporting already detected model-line, but skipped in the first step as a result of primary rough quantization of the Hough-parameter space. In this step, the finest (integer value) resolution in the parameter-space, θ and ρ is used. Newly-found pixels (image feature-carriers) are associated to the primary line and small

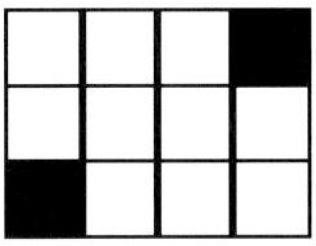

Fig. 1. The window of gap filling process

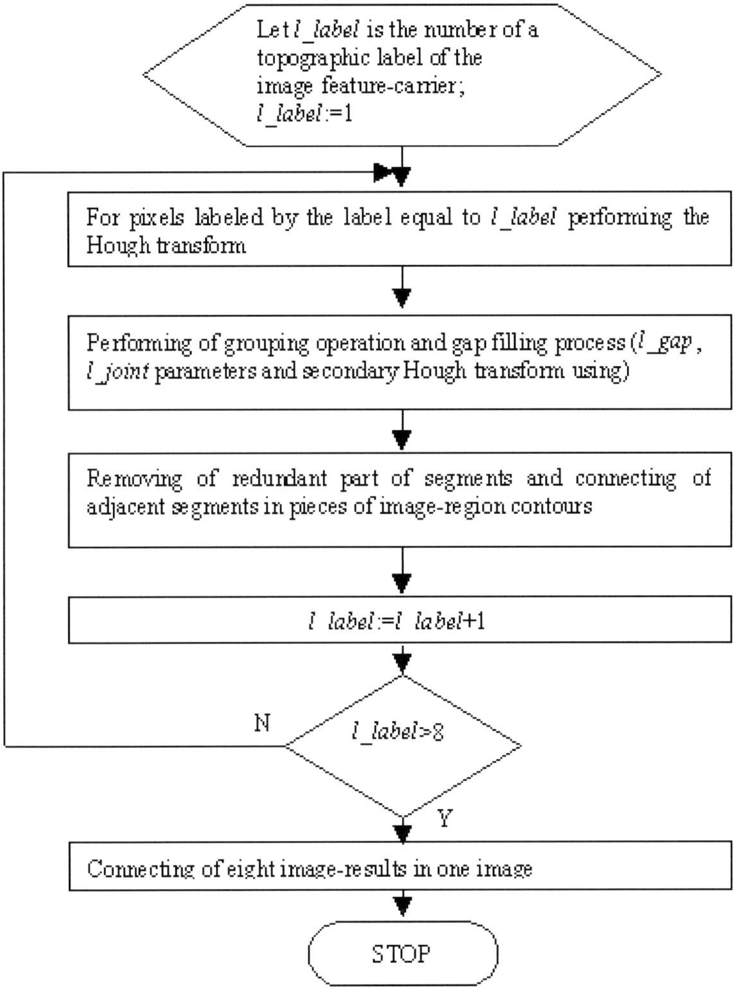

Fig. 2. The algorithm of the grouping by using modified Hough transform

enough gaps are filled with new image feature-carriers.

As the obtained segments may belong to some larger curvilinear structures, the next step consists of checking for the geometrical relationships between the adjacent segments with the line aggregation as a target. In this way connected larger pieces of image-region contours can be obtained. Due to using the modified Hough transform to grouping of topographic feature-carriers (image feature-carriers are labeled with one from eight topographic labels - coded on images by grey-levels: the valley, the ridge, the foot, the bluff, the fault up, the fault down, the soft valley and the soft ridge), the grouping process is done separately for each topographical category. As a result eight images are received corresponding to particular features categories. The next step is the fusion of eight labeled images in one-end image.

3 Demonstration of results

In next part, there have been presented results of the described algorithm. Input images in grouping process are the images, which include extracted and labeled image topographic feature-carriers. The following grey-images, representing physical objects (automotive parts - accumulator-dehydrator and geometrical figures) have been use for the extraction of image feature-carriers (Fig. 3).

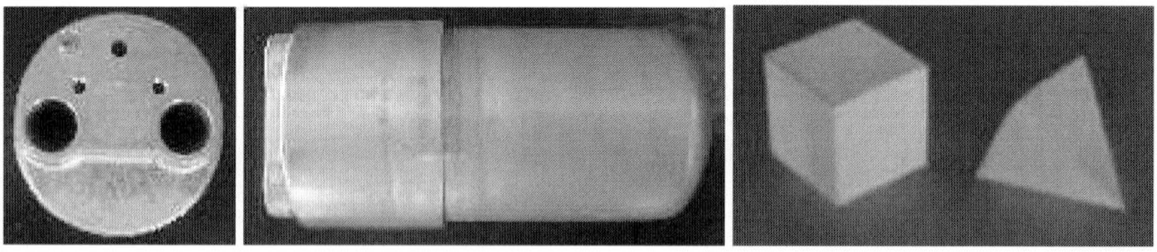

Fig. 3. Test images

On the graph in Fig. 4 it have been presented the influence of l_gap and l_joint parameters of the modified Hough transform as a function of the number of segments (forming contours), obtained as the result of grouping (accumulator-dehydrator upper-view image). On this figure, values on the X axis correspond to the length ranges (expressed in the image feature-carriers units) of grouped segments: $1 \cong 5-9$, $2 \cong 10-19$, $3 \cong 20-29$, $4 \cong 30-39$, $5 \cong 40-49$, $6 \cong 50-59$, $7 \cong 60-69$, $8 \cong 70-79$, $9 \cong 80-89$, $10 \cong 90-99$, $11 \cong 100-149$, $12 \cong 150-199$; the Y axis corresponds to the number of segments within each range; the parameters have value: $l\_gap = 1$, $l\_joint = 10$ (black line); $l\_gap = 1$, $l\_joint = 5$; $l\_gap = 2$, $l\_joint = 10$; $l\_gap = 2$, $l\_joint = 5$ (white line).

It is possible to see, that the most segments are connected, whenever the l_gap parameter has a high value and the l_joint parameter has a low value.

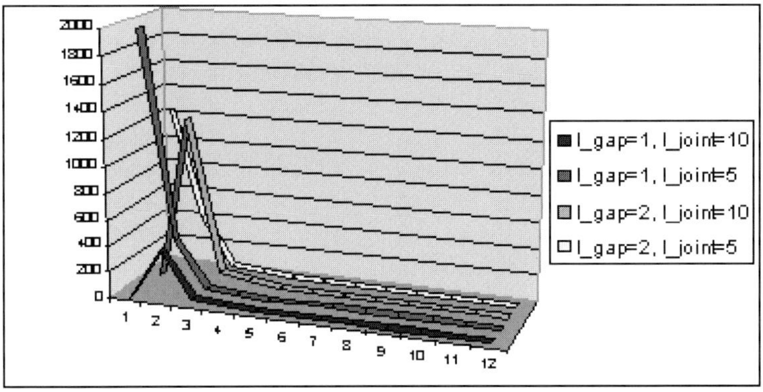

Fig. 4. Number of segments as a function of segment length for different value of l_gap and l_joint parameters

For natural images, as a result of grouping process, mainly short segment has been a received: $5 - 19$ image feature-carriers (and shorter segments).

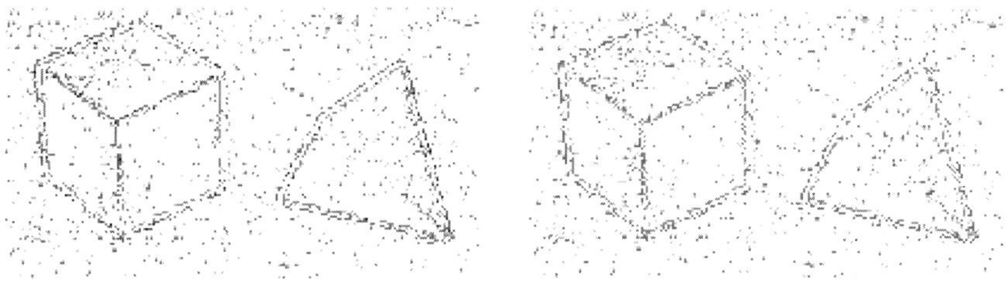

Fig. 5. Source (gradient) image with geometrical figures (left) and grouping result of this image by using the classical Hough transform (right)

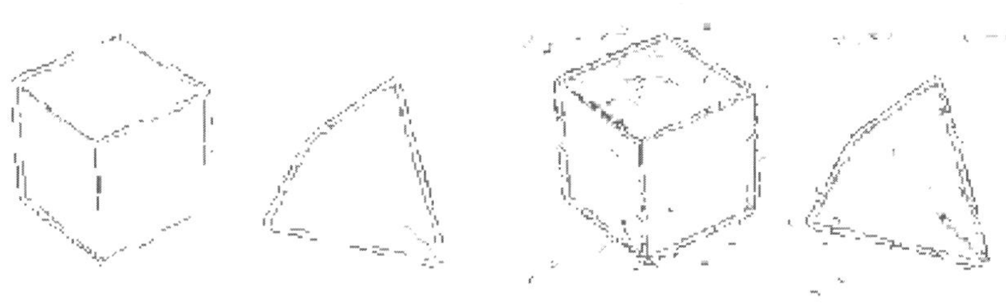

Fig. 6. Results of grouping of image feature-carriers by using the modified Hough transform for $l\_gap = 1$, $l\_joint = 10$ (left); $l\_gap = 1$, $l\_joint = 5$ (right)

Hough transform has been calculated for $\theta = 1$, $\rho = 5$. On resulting images, is possible to see the influence of the l_gap and l_joint parameters

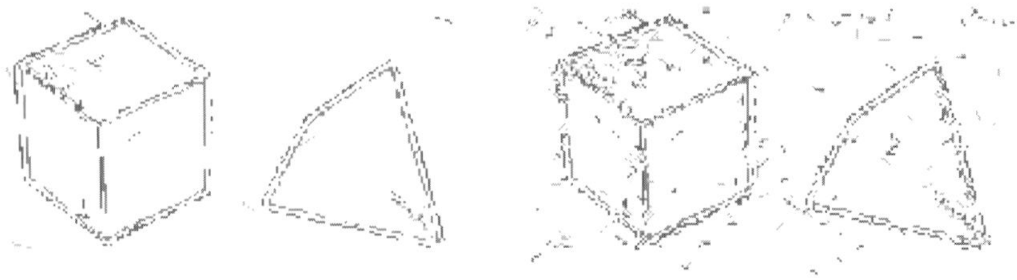

Fig. 7. Results of grouping of image feature-carriers by using the modified Hough transform for $l\_gap = 2$, $l\_joint = 10$ (left); $l\_gap = 2$, $l\_joint = 5$ (right)

on the effects of grouping process. When the l_gap has a high value, image feature-carriers are often connected into single segment, although they do not represent the same contour. Thus the augmentation of the l_gap value leads to the presence of artefacts. High value of the l_joint parameter leads to the elimination of short edges. Values of those parameters should be adjusted to the type of the image under study. Images with relevant smooth straight line contours of objects can be processed with higher values of l_gap and l_joint. In this case, significant parts of contours will be detected. Images with curved line contours (and for cases when the level of noise is high) must be treated with parameter l_joint kept low to avoid the discontinuities of the resulting curvilinear contour. It is also necessary to upper-bound the value of the l_gap, in order to avoid artefact edge-arcs, which are obtained by connecting the adjacent segments being part of close object's contours, representing neighbor objects on the scene. Comparing to results of using the classical Hough transform (presented on Fig. 5, 8, 10) is visible how the modified Hough transform connects interrupted segments and removes separated image feature-carriers. The introduced modification proved to be efficient in extracting curvilinear contours of limited curvature from images, by grouping short straight-segments from the input image into larger contour-arcs.

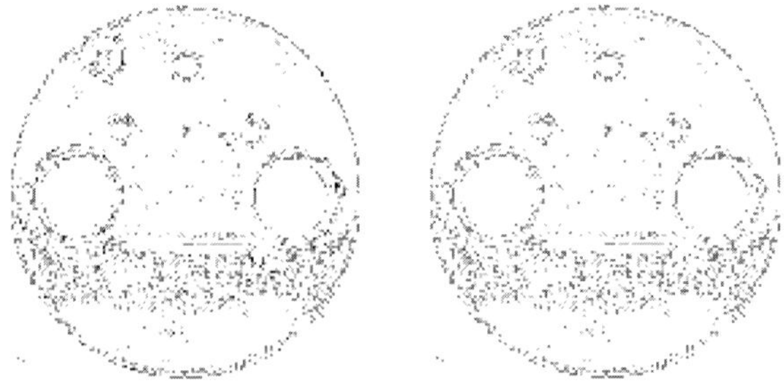

Fig. 8. Source (gradient) image with accumulator-dehydrator upper-view (left) and grouping result of this image by using the classical Hough transform (right)

Fig. 9. Results of grouping of image feature-carriers by using the modified Hough transformation for $l\_gap = 1$, $l\_joint = 10$ (left); $l\_gap = 1$, $l\_joint = 5$ (middle); $l\_gap = 2$, $l\_joint = 10$ (right)

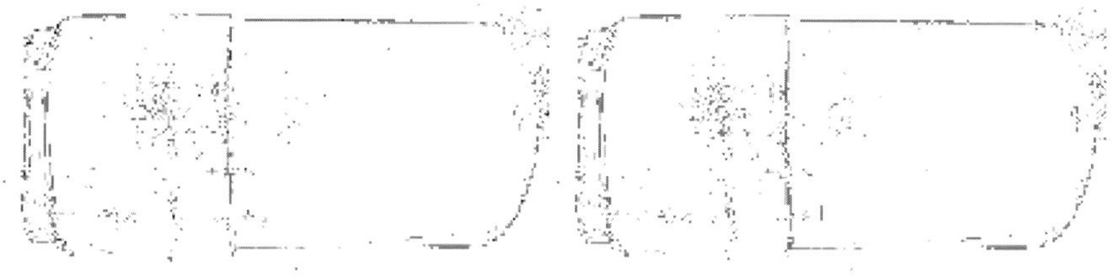

Fig. 10. Source (gradient) image with accumulator-dehydrator side-view (left) and grouping result of this image by using classical Hough transform (right)

Fig. 11. Results of grouping of image feature-carriers by using the modified Hough transform for $l\_gap = 1$, $l\_joint = 10$ (left); $l\_gap = 2$, $l\_joint = 10$ (right)

4 Conclusions

The modified Hough transform, with gap filling process and adjustable l_gap and l_joint parameters allows for grouping of image feature-carriers with straight linear and curvilinear disposition on the source image. The optimal selection value of l_gap and l_joint should be related to the type of image under study (dependent whether the image under study is with smooth straight-line-contours or with strongly curvilinear contours or whether it is a noisy image). Grouping process is performed on gradient images which provide specific topographic feature-carriers. These feature-carriers are extracted with image by KAS method [8]. In this method, it has been assumed, that in most practical cases, meaningful contours on images are mutually separated by the distance of 2-3 pixels (this refers to objects with surfaces having no fine and/or strong texture). This assumption prevents from the false gap-filling process (for topographic feature-carriers which are close, but does not create the same straight-linear or curvilinear contour). Grouping based on detected topographic feature-carriers results with image function contours having richer geometrical interpretation than in the case of ordinary edge-detection operators (as also the type of the edge is determined). Grouped contours, with clear topographic description can provide spatial cues, which can support the automatic interpretation of the image contents.

References

1. Ballard DH, Brown CM (1982) Computer Vision. Prentice Hall, New Jersey
2. Chutatape O, Guo L (1999) A modified Hough Transform for Line Detection and Its Performance. Pattern Recognition 32:181–192
3. Conker R (1988) A Dual Plane Variation of the Hough Transform for Detecting Non-concentric Circles of Different Radii. Computer Vision, Graphics and Image Processing 43:115–132
4. Gonzales RC, Woods RE (1992) Digital Image Processing. Addison Wesley
5. Guil N, Zapata EL (1997) Lower Order Circle and Ellipse Hough Transform. Pattern Recognition 30:10:1729–1744
6. Haralick R (1983) Ridges and Valleys on Digital Images. Computer Vision, Graphics and Image Processing 22:28–38
7. Hough PVC (1962) Method and Means for Recognizing Complex Patterns. US Patent 3069654
8. Kasinski A (1997) Smoothing Noisy Images without Destroying Predefined Feature Carriers. Computer Analysis of Images and Patterns, Proc. 7th International Conference Computer Analysis of Images and Patterns CAIP '97 Kiel, Germany, Springer, Berlin Heidelberg 519–526
9. Kasinski A, Przybylski L (1999) Segmention of Topographical Image Features. Control and Information Engineering Research 24:105–120 (in Polish)
10. Kasinski A, Przybylski L (2001) Using Perceptual Labels for Topographical Features in Scene Recognition. National Conference KOSYR 2001:395–400
11. Kasinski A, Przybylski L (2003) Grouping Based on Perceptual Labels in Topographical Analysis of the Image Function. Journal of Applied Computer Science 11:2:41–53
12. Medioni G, Lee M, Tang Ch (2000) A computational Framework for Segmentation and Grouping, Elsevier
13. Xu L, Oja E (1993) Randomized Hough Transform (RHT): Basic Mechanisms, Algorithms, and Computational Complexities. Computer Vision, Graphics and Image Processing: Image Understandig 57:2:131–154
14. Zhang Y, Webber R (1996) A Windowing Approach to Detecting Line Segments Using Hough Transform. Pattern Recognition 29:2:255–265
15. Forsyth DA, Ponce J (2003) Computer Vision. A Modern Approach. Prentice Hall, New Jersey

Planning Positioning Actions of a Mobile Robot Cooperating with Distributed Sensors

Piotr Skrzypczyński

Poznań University of Technology, Institute of Control and Information Engineering
ul. Piotrowo 3A, PL-60-965 Poznań, Poland, e-mail: ps@ar-kari.put.poznan.pl

Summary. Localization procedures for a mobile robot cooperating with external cameras and artificial navigation aids (landmarks) are discussed. An action planning method, taking into account in an exact way both the action cost and positioning uncertainty is presented. Its performance is illustrated by results of simulations.

1 Introduction

The pose (position and orientation) of a mobile robot has to be corrected from time to time by using measurements from exteroceptive sensors. Currently, CCD cameras are the most compact and low cost sensors for mobile robots. Unfortunately, most of the general-purpose, vision-based localization methods fail under common environmental conditions, due to occlusions, shadows, etc. A solution for limited environments such as warehouses or factories is to develop an external infrastructure [9] providing pose estimates to the robots. Operational characteristics of the on-board vision sensors can be also improved by deploying unobtrusive artificial landmarks in the environment [2].

In the previous work [6] we have proposed a negotiation framework used by the robots to choose best positioning data from the available external sources. A robot either uses its on-board vision to localize artificial landmarks or asks for positioning service from the stationary cameras treated as Perception Agents (PA). The robot compares proposals from particular PAs and awards the contract to the one, which offers the best pose estimate. Some of the results [7] pointed out a certain weakness of this method. We have observed, that at some points the positional uncertainty resulting from the negotiations is higher than the threshold used (a robot requests positioning service whenever the positional uncertainty exceeds this threshold). This is caused by local nature of the negotiations – the robot uses the best positioning data available at the given point of the path, but the choice of the point is not optimized. To cope with this problem, and improve robustness of localization using the external infrastructure, this article contributes a method, which plans a global sequence of the positioning actions undertaken by a robot. If a robot has complete knowledge about the external cameras available in the system, and it knows also where the artificial landmarks are placed, it can compute in advance an optimal positioning strategy for the path it has to follow. The aim of the optimization is to minimize the time spent by the robot at communication and sensing actions (requests to PAs and observations of landmarks), that are necessary to keep the pose uncertainty within bounds.

There are path planning methods, known from the literature, which take into account the localization uncertainty. However, most of these works assume continuous sensing by means of some range sensors [10] and a complete environment model available to the robot.

We take a different approach in which the sensing is opportunistic (i.e. the robot updates its pose only when it sees some landmarks or it is seen by an external camera) and the robot knows only locations of elements of the external navigation infrastructure: cameras and landmarks. The robot does not plan the path (known in advance) but the sequence of positioning actions executed in order to get to the goal. This sequence minimizes the overall cost of the positioning actions, ensuring that the positional uncertainty at any point of the path is lower than the given threshold.

2 Uncertainty in Vision-Based Positioning

2.1 Spatial Uncertainty Model

Localization based on the external infrastructure uses both fixed and on-board cameras, and exploits artificial visual cues in the form of passive, printed landmarks and active LED markers on the robots. Due to these visual cues, simple and fast image processing methods could be employed, resulting in reliable and accurate positioning of the mobile robot with regard to (w.r.t.) the global reference frame [7]. The robot pose $\mathbf{X}_R = [x_r \; y_r \; \theta_r]^T$ uncertainty is described by the covariance matrix $\mathbf{C}_R$ [3]. We obtain closed-form formulas expressing this matrix as function of the robot configuration w.r.t. the given external navigation aid.

The uncertainty analysis uses first order covariance propagation [4], and is focused on influence of the relative position and orientation between the robot and the elements of the external infrastructure to the uncertainty of the pose estimate. The uncertainty caused by the quantization error is considered. Errors due to electronic noise in the image are not taken into account, because they largely do not depend on the spatial configuration of the robot w.r.t. the landmark or camera. The analysis enables to predict the pose uncertainty before taking and processing an image. To construct an uncertainty map for the given external sensor or landmark, we have adopted the equiprobability ellipsoid computed from the $\mathbf{C}_R$ matrix. The ellipse obtained by projecting this ellipsoid on the floor plane shows the area which contains the robot position $\mathbf{X}_{R_{xy}} = [x_r\ y_r]^T$ with the given level of probability [8]. We employ the area of the predicted ellipse (for the 95% probability) as the positioning goodness value in the uncertainty maps.

2.2 Distributed Overhead Cameras

The distributed vision system uses B/W cameras mounted to the ceiling. The cameras are equipped with wide angle (fish-eye) lenses, their optical axes are orthogonal to the ground plane.

The *Labmate* wheeled robot has been equipped with active LED markers attached symmetrically at the corners. Detection of the robot is performed on a difference image, which is computed from a pair of images taken when the LEDs are on, and then off. Three LEDs must be visible to form a minimal detectable pattern. More details about the image processing procedure can be found in [7].

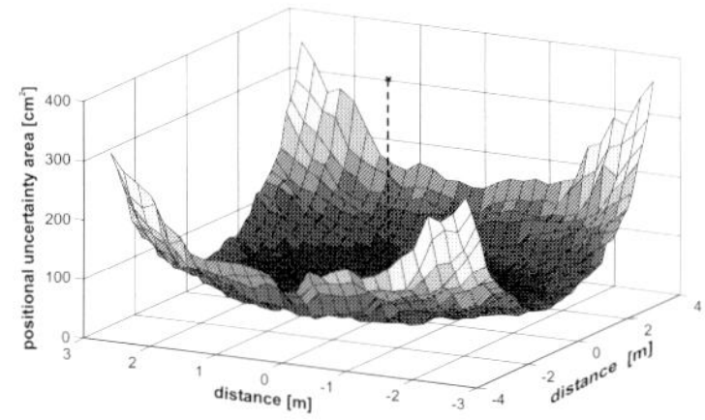

Fig. 1. Positional uncertainty as a function of the robot configuration w.r.t. the overhead camera

The spatial uncertainty of a robot localized by the overhead camera depends mainly on the uncertainty of the location of the points of the LED-pattern in the camera image. Correction of the fish-eye distortion results in shifting pixels from their original positions. Errors arise also because the assumption of the orthogonality of the optical axis to the floor plane is not perfectly satisfied. Spatial distribution of the errors in pixel location along the x and y axes (after correction) has been evaluated by comparing the image of a calibration pattern with the ground truth [3]. This assessment of the errors in the overhead camera images provides the primary uncertainty for the calculation of the estimated spatial uncertainty of the robot. It is computed as the covariance matrix $\mathbf{C}_{pix}$, taking into account the errors introduced by the camera mounting and the correction procedure for the given pixel location $[u\ v]^T$ in the image coordinates: $\mathbf{C}_{pix}(u,v) = \mathbf{C}_{mount}(u,v) + \mathbf{C}_{fish\_eye}(u,v)$, where:

$$\mathbf{C}_{pix}(u,v) = \begin{bmatrix} \sigma_u^2 & \sigma_{uv} \\ \sigma_{vu} & \sigma_v^2 \end{bmatrix}. \tag{1}$$

This uncertainty is propagated to the uncertainty of a single LED marker location $\mathbf{D} = [x_d\ y_d]^T$ in the global frame, using the first order approximation [7]. The overhead camera pose in the global frame $\mathbf{X}_K = [x_k\ y_k\ \theta_k]^T$ is assumed to be certain. Finally, spatial uncertainty of the LED-pattern centre position is computed, which coincides with the centre of the robot. Fig. 1 shows the positional uncertainty map for the overhead camera.

2.3 On-board Vision with Artificial Landmarks

Artificial landmarks are made of A4 paper sheets. A chessboard-like pattern placed inside the landmark defines an unique code.

Landmark recognition leads to determination of the image-coordinates of its reference points (see [2] for the image processing details). Then, the vector $\mathbf{P} = [x_1\ y_1\ x_2\ y_2]^T$ of coordinates of the centres of left and right frame edges are calculated. These data are used to calculate the vector $\mathbf{L} = [l\ \varphi_1\ \varphi_2]^T$ determining robot pose w.r.t. the landmark.

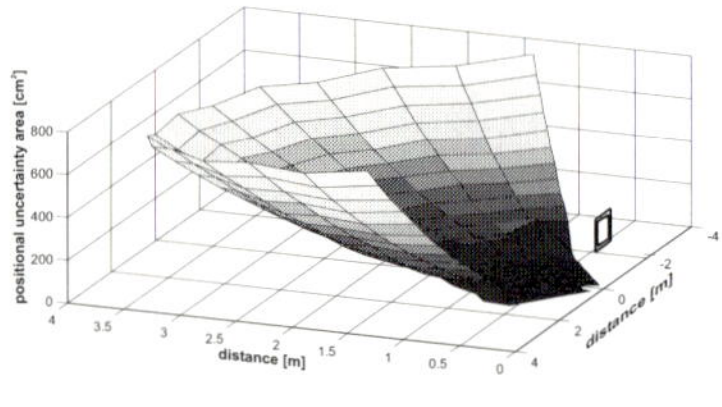

Fig. 2. Positional uncertainty as a function of the robot configuration w.r.t. the landmark

We assume, that positions of the landmark reference points are corrupted by zero-mean Gaussian noise of standard deviation equal to one pixel. This uncertainty, caused by quantization, is represented by the 4×4 diagonal covariance matrix $\mathbf{C}_P$ [2]. It is then propagated to the uncertainty of the vector $\mathbf{L}$ parameters. The uncertainty of $\mathbf{L}$ is described by the 3×3 covariance matrix $\mathbf{C}_L$ computed as:

$$\mathbf{C}_L = \mathbf{J}_P \mathbf{C}_P \mathbf{J}_P^T, \tag{2}$$

where $\mathbf{J}_P$ is Jacobian matrix of the nonlinear transformation between $\mathbf{P}$ and $\mathbf{L}$. The uncertainties in the angles and distance to the landmark cause uncertainty of the robot pose $\mathbf{X}_R$ in the global frame:

$$\mathbf{X}_R = f_R(\mathbf{X}_{L_i}, \mathbf{L}), \tag{3}$$

where $\mathbf{X}_{L_i} = [x_{l_i}\ y_{l_i}\ \theta_{l_i}]^T$ are coordinates of the i-th landmark in the global frame. We assume, that the landmark coordinates are certain. Thus, the uncertainty of $\mathbf{X}_R$ is described by the covariance matrix $\mathbf{C}_R$, which is a result of uncertainty propagation from the $\mathbf{C}_L$ matrix. Because the relation (3) is nonlinear, the matrix $\mathbf{C}_R$ is computed from a first order approximation [2]. Fig. 2 shows the positional uncertainty map for the artificial landmark.

3 Planning the Positioning Actions

The action planning framework we propose is based on the classic approach to search of the shortest path in a graph. The first step is to generate a discrete action space (Fig. 3A). The nominal path of the robot is sampled uniformly, and the possible positioning actions, which can be undertaken by the robot along this path are generated. An action at the i-th path point is described by the pose of the robot $\mathbf{X}_{R_i}$, the action type, the covariance matrix $\mathbf{C}_{S_i}$ describing the pose uncertainty resulting from the given type of positioning action performed in this particular configuration, and the cost of the positioning T_i.

The covariance matrix is predicted using the closed-form formulas. The cost is an integer value generated upon a simple look-up table of the experimentally determined time (in seconds), the robot spends at positioning depending on the type of sensing and/or communication action and the position w.r.t. the external navigation aid. Actions are nodes of the graph $G(V,E)$ (Fig. 3B). The node v_i is connected to the node v_j by an edge $e_{i,j}$ if it is possible for the robot to move from v_i to v_j, keeping the pose uncertainty below a given threshold defined by the uncertainty ellipse area C_{max} (scalar value). Because we assume, that the robot can update its pose only at the action nodes, we use the odometry model of the differential drive robot to compute the maximum admissible distance between two connected nodes.

Because more than one positioning action may be available at the given (x,y) point, several nodes having the same $\mathbf{X}_R$ but different $\mathbf{C}_S$ and T can be generated. The edges of the graph are labelled with the costs of the positioning actions. The edge $e_{i,j}$ has the traversal uncertainty $\mathbf{C}_{i,j}$ (from odometry), and the cost T_j, as we assume that progressing to the particular node means execution of the positioning action associated to this node. The resulting action space is a directed graph. Because a positioning action can be performed only once by the robot travelling along a given path, the graph is acyclic.

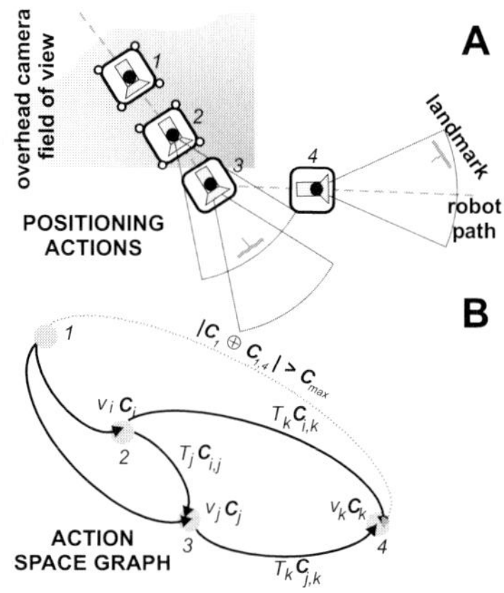

Fig. 3. Action space concept for the positioning sequence

Although simple search in the action space will return the shortest path in the sense of minimal action cost (minimal time), it cannot guarantee that the positional uncertainty will be kept all the time below the given threshold. We cast the positioning action planning as a constrained optimization problem. An obvious solution is to construct the action space in such a way, that any path in the graph guarantees the required positioning precision from the start node v_s to the goal v_g. The positional uncertainty at the node v_k of a particular edge $e_{k,l}$ depends on the previous positioning actions executed along the path from v_s to v_k. However, assuming a conservative initial uncertainty of an edge, which is yielded by the positioning action at v_k (known in advance), permits to build a *safe* graph $G_C(V,E_C)$. Two given nodes in this graph are connected by an edge $e_{k,l} \in E_C$ only if the merged uncertainty of the edge traversal (from odometry), and the positioning action undertaken at v_k is below C_{max}. The Kalman filter used to merge the pose estimates [8] guarantees, that the result is not worse than the best estimate taken as input, while we know that one of the input estimates at v_k has the uncertainty of $\mathbf{C}_{S_k}$. A search in $G_C(V,E_C)$ by means of the Dijkstra algorithm w.r.t the positioning cost T yields an optimal sequence of actions and guarantees the pose uncertainty within the given bounds.

However, a robot can traverse between two action nodes keeping the uncertainty under C_{max} even if these nodes are not connected by an edge according to the above-described conservative approach. The robot can achieve this by acquiring pose information in other nodes on its path, thus having the pose estimate at v_k (after merging) better than $\mathbf{C}_{S_k}$. As a result, search in the safe graph may result in a failure (safe strategy not found), even if a sequence of actions keeping the uncertainty below the given threshold does exist. An action space, which permits a search taking into account the actual accumulated uncertainty can be constructed by using at the initial node v_k of a given edge uncertainty value smaller than $\mathbf{C}_{S_k}$. The smallest pose uncertainty the robot can ever achieve is the uncertainty of the most effective positioning action known to the system, i.e. the smallest $\mathbf{C}_{S_i}$ in the whole action space. When this uncertainty is used, the resulting graph $G_D(V, E_D)$ has more edges, because starting with smaller $\mathbf{C}_S$ permits the odometry to take the robot further without violating the C_{max} constraint.

From the theoretical point of view, we are facing the restricted shortest path problem (RSP), which is known to be NP-complete [1]. However, regarding practical importance of the RSP problem (e.g. to the quality of service routing in communication networks [11]) approximate, but efficient algorithms to solve it have been published. There are also publicly available implementations and software packages, such as SAMCRA [11] and CNOP [12]. However, to compare results of the RSP-based search in the G_D graph to the simple shortest-path search in the G_C graph we developed a much simpler dynamic programming solution, provided below. The computational complexity of this algorithm is pseudo-polynomial, but it can be turned into FPAS (fully polynomial approximation scheme) by using the approximation from [5]. In the following algorithm, $\mathbf{C}[v,t]$ denotes a vector associated with each node v, which stores the minimum uncertainty on any path from v_s to v, that has a total cost of t. T_{max} is the maximum cost of a path from v_s to v_g in the graph, obtained by search w.r.t the cost information only. It is a stop condition for the dynamic program. When the FPAS is used, T_{max} makes sure that the scaling error is not too large. $\mathbf{C}_{u,v}$ represents the uncertainty evolved by odometry during traverse of the edge $e_{u,v}$, $\mathbf{C}_u$ is uncertainty of the positioning action at node u. The notation $|\mathbf{C}|$ means computation of a scalar value (ellipse area) from the given covariance matrix. Pose uncertainty along the given path required to run the algorithm is achieved by compounding (denoted by $\oplus$) and merging (denoted by $\otimes$) [8] the uncertainty matrices of the consecutive edges and nodes.

procedure ActionPlanningAsRSP($G_D(V,E), v_s, v_g, C_{max}, T_{max}$)
for each $v \in V - \{v_s\}$ $\mathbf{C}[v,0]{:=}\infty$
$\mathbf{C}[v_s,0]{:=}0$
for $t := 1$ **to** T_{max} **do**
  for each $v \in V$ **do**
    $\mathbf{C}[v,t]{:=}\mathbf{C}[v,t-1]$
    $P[v,t]{:=}nil$
  for each $e_{u,v} \in E$ **do**
    $\mathbf{C}_{temp}{:=}\mathbf{C}[u,t-T_{u,v}] \otimes \mathbf{C}_u \oplus \mathbf{C}_{u,v}$

        if $|\mathbf{C}_{temp}| < |\mathbf{C}[v,t]|$ **and** $|\mathbf{C}_{temp}| \leq C_{max}$ **then**
            $\mathbf{C}[v,t]{:=}\mathbf{C}_{temp}$
            $P[v,t]{:=}u$ *{update sequence}*
    if $C[v_g,t] \leq C_{max}$ **then** RETURNSEQUENCE($P[v_g,t]$)
RETURNFAILURE $\{T_{max}$ *has been exceeded*$\}$

4 Results

The two approaches to positioning action planning described in the previous section have been compared in simulation, to investigate if the more complex RSP-based search yields better results than simple search in the safe graph. The simulated environment shown in Fig. 4A contains four overhead cameras

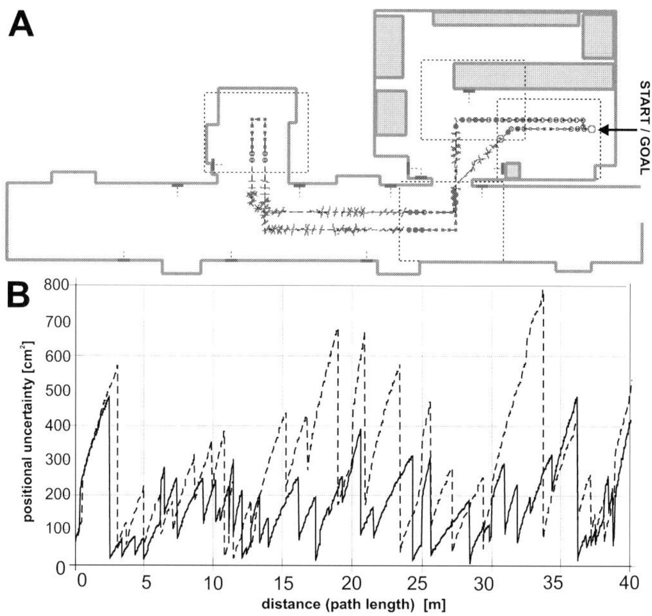

Fig. 4. Comparison of results for the two action planning methods

(fields of view indicated with dotted lines), and twelve artificial landmarks attached to the walls (indicated with small bars). Fig. 4A shows also the predicted positional uncertainty ellipses for all possible positioning actions along the nominal path. The robot followed a pre-planned path (dashed line) executing the positioning actions according to a plan obtained from the G_C graph, and then from the G_D graph by using the RSP algorithm. From the plots in Fig. 4B it can be seen, that the safe graph approach (dashed line) enables to reduce the positional uncertainty to a 800 cm^2 ellipse, while with the RSP algorithm (solid line) a uncertainty threshold below 500 cm^2 is achievable, using the same action space.

The safe graph approach has been also validated practically, results of the experiments with the *Labmate* robot and two monitoring cameras have been reported in [7].

5 Conclusions

We have presented a method for planning the positioning actions of a mobile robot, which cooperates with external navigation infrastructure containing distributed cameras and artificial landmarks enhancing capabilities of the on-board vision. Formulation of the positioning action planning as the RSP problem enables to use efficient search algorithms, developed recently in the field of computer/communication networks. Although the RSP search has higher computational cost, it enables to obtain better sequences of actions, as it has been shown in simulation.

The action planning algorithm can be easily generalized to any other type of stationary sensor if the uncertainty related to positioning w.r.t this sensor is described by closed-form formulas.

References

1. Ahuja R., Magnanti T., Orlin J. (1993) Network Flows: Theory, Algorithms and Applications, Prentice Hall.
2. Bączyk R., Kasiński A., Skrzypczyński P. (2003) Vision-Based Mobile Robot Localization with Simple Artificial Landmarks, Prepr. 7th IFAC Symp. on Robot Control, Wroclaw, 217–222.
3. Bączyk R., Skrzypczyński P. (2003) A Framework for Vision-Based Positioning in a Distributed Robotic System, Proc. European Conf. on Mobile Robots, Warsaw, 153–158.
4. Haralick R. M. (1996) Propagating Covariance in Computer Vision, Int. J. Pattern Recog. and Artif. Intell., 10:561–572.
5. Lorenz D. H., Raz D. (2001) A Simple Efficient Approximation Scheme for the Restricted Shortest Path Problem, Oper. Res. Lett., 28:213–219.
6. Skrzypczyński P. (2004) A Team of Mobile Robots and Monitoring Sensors – From Concept to Experiment, Advanced Robotics, 18(6):583–610.
7. Skrzypczyński P. (2005) Uncertainty Models of the Vision Sensors in Mobile Robot Positioning, Int. J. of Appl. Mathematics and Comp. Sci., 15(1), in press
8. Smith R., Cheeseman P. (1987) On the Estimation and Representation of Spatial Uncertainty, Int. J. of Robotics Res., 5(4):56–68.
9. Sogo T., Ishiguro H., Ishida T. (1999) Mobile Robot Navigation by Distributed Vision Agents, In: Approaches to Intelligent Agents, (H. Nakashima and C. Zhang, eds.) Springer-Verlag, Berlin, 96–110.
10. Takeda H., Facchinetti C., Latombe J.-C. (1994) Planning the Motions of a Mobile Robot in a Sensory Uncertainty Field, IEEE Trans. on Pattern Anal. and Machine Intell., 16(10):1002–1017.
11. Van Mieghem P., Kuipers F. A. (2004) Concepts of Exact Quality of Service Algorithms, IEEE/ACM Trans. on Networking, 12(5):851–864.
12. Ziegelmann M. (2001) Constrained shortest paths and related problems, PhD Thesis, Universität des Saarlandes, Saarbrücken.

Merging Probabilistic and Fuzzy Frameworks for Uncertain Spatial Knowledge Modelling

Piotr Skrzypczyński

Poznań University of Technology, Institute of Control and Information Engineering
ul. Piotrowo 3A, PL-60-965 Poznań, Poland, ps@ar-kari.put.poznan.pl

Summary. The issues of spatial knowledge representation for mobile robots are considered. Two types of maps, grid and feature based, and two uncertainty representations, probabilistic and fuzzy are merged in one framework to obtain accurate and consistent geometric maps of the environment from range sensor readings.

1 Introduction

Automated building of spatial knowledge from sensory data is one of the central problems for autonomous robots. Many particular environment representations have been proposed in the literature [10]. The grid-based maps [7] represent space as an array of equal cells. They can easily be updated with range sensor readings, tolerating data uncertainty and ambiguity, but require a large amount of memory to cover bigger areas with a dense grid. Feature-based maps contain concise and explicit representations of the geometric entities [2, 8], but are less popular because of difficulties with the direct interpretation of raw sensory data.

An important aspect of the spatial knowledge representations is the mathematical framework used to handle the uncertainty. The grid maps represent uncertainty by estimating the confidence, that the given cell is empty or occupied. Most popular mathematical formalism is the Bayesian theory, which has well established foundations, but does not have ability to represent the lack of information. The Dempster-Shafer theory of evidence provides a better representation for the ignorance [1]. Also the fuzzy-set-based method proposed in [5] provides a good representation of different forms of uncertainty and incompleteness of information. The feature-based maps are built in terms of primitives represented by vectors of parameters and their covariance matrices modelling spatial uncertainty. Coordinates of the features are assumed to be disturbed by unimodal Gaussian noise. Many authors employed feature-based maps in the self-localization methods based on the Extended Kalman Filter (EKF) formalism (e.g. [2]). Although good results in self-localization are achieved, in most cases the built maps consist of a high number of short segments, and lack the ability to represent corners or connections between objects. Looking closer at such a map (Fig. 1) one finds many multiplied segments crossing each other, what makes a map inconsistent with the common sense properties of the observed objects.

This paper presents a spatial knowledge representation for a mobile robot, that combines the most desirable features of both the grid-based and the geometric maps. From the other hand, we combine two frameworks for the uncertainty representation: the classical probabilistic methods to explicitly propagate those uncertainties, that can be expressed by statistical measures computed upon the series of measurements (i.e. standard deviations), and the fuzzy set theory to support the decisions, that are necessary during the map building process (e.g. to merge or not two segments).

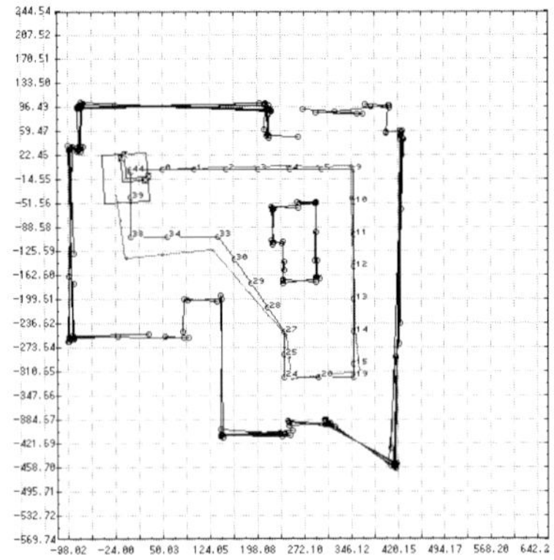

Fig. 1. Example of a segment-based map obtained with EKF from raw laser range data. Scale is in [cm]

2 Local Fuzzy Grids for Data Fusion and Filtering

The grid maps are able to accumulate data taken from several poses (positions and orientations) of the robot filtering out unreliable measurements, and reinforcing the good ones. Because of that, the grid maps are utilized here as the first-step, local representation of the spatial knowledge in the mobile robot. The grid map is a two-dimensional tessellation of the environment, consisting of cells, being either occupied ($\mathcal{O}$) or empty ($\mathcal{E}$).

To find the mathematical framework most appropriate for the grid map updating, we have examined the Bayesian, Dempster-Shafer (D-S), and fuzzy methods. A comparison of these three frameworks by using sonar data is presented in [7], but to find the method that best fits to the requirements of the multi-sensor robotic system, we have performed experiments with the sonars, and two different types of 2D laser scanners in a controlled environment. Figure 2A depicts three snapshots of the data obtained from the sensors used to update the grid map. The half-ring of eight sonars provides only very coarse range measurements (left), and exhibits all the well-known problems due to the wide beam and multiple reflections. In contrary, the laser scanners have better angular resolution and small beam extent. The home-built scanner, utilizing an AMCW[1] laser rangefinder, and mounted on top of the robot, has a 360° field of view (middle), but exhibits significant systematic errors and

[1] Amplitude Modulated Continuous Wave

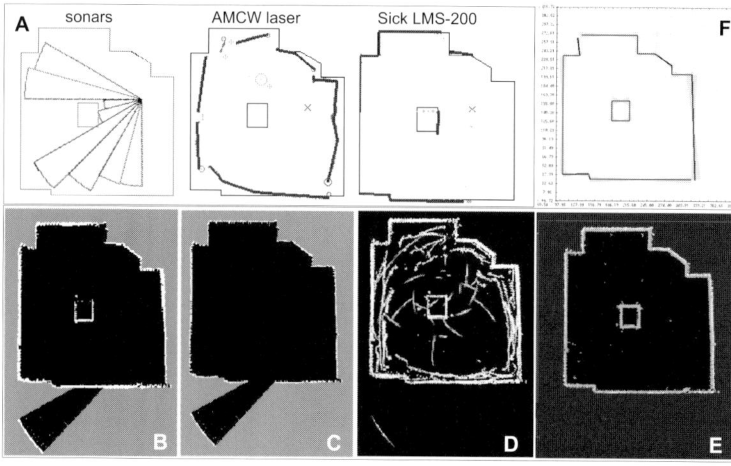

Fig. 2. Examples of the sensory data (A), and the grid maps computed according to three different methods: Bayesian (B), D-S (C), and fuzzy (D,E)

produces many outliers [8]. The *SICK LMS-200* time-of-flight laser scanner mounted at the front of the robot provides most accurate range data, but has a limited field of view (right). The resulting maps of occupied cells, obtained from twelve robot poses are shown in Fig. 2B,C,D, for the Bayesian, D-S, and fuzzy method, respectively. The cell size is 2×2 [cm]. The Bayesian and D-S maps have been implemented according to the methods known from the literature [7], where more details can be found.

Fuzzy grid updating is described in more detail, because this framework is used further in the paper. The sets of occupied $\mathcal{O}_i^k$ and empty cells $\mathcal{E}_i^k$ are determined, by computing their membership functions according to the sensor beam models, for every measurement s_i^k taken from the k-th robot pose. The sensor models represent the degree of membership of a given cell to the sets of occupied and empty cells, according to the range sensor reading s, and its uncertainty Δs (defined here as three times the standard deviation):

$$\mu_o\left(r,\alpha,s,\Delta s\right)=\begin{cases}0, & 0\le r<s-\Delta s\\ \lambda(\alpha)k_o\left(1-\left(\frac{s-r}{\Delta s}\right)^2\right), & s-\Delta s\le r<s+\Delta s\\ 0, & r\ge s+\Delta s\end{cases}, \tag{1}$$

$$\mu_e\left(r,\alpha,s,\Delta s\right)=\begin{cases}\lambda(\alpha)k_e, & 0\le r<s-\Delta s\\ \lambda(\alpha)k_e\left(\frac{s-r}{\Delta s}\right)^2, & s-\Delta s\le r<s\\ 0, & r\ge s\end{cases},$$

where: r is the distance from the sensor to the center of the given cell, α is the angle between the beam axis and the bearing to the cell, and k_o, k_e are the limits of μ_o and μ_e, such that $k_o + k_e \le 1$. New beam models have been defined for the laser scanners. The essential difference from the known sonar beam patterns [5, 7] is in the function $\lambda(\alpha)$, which defines much smaller

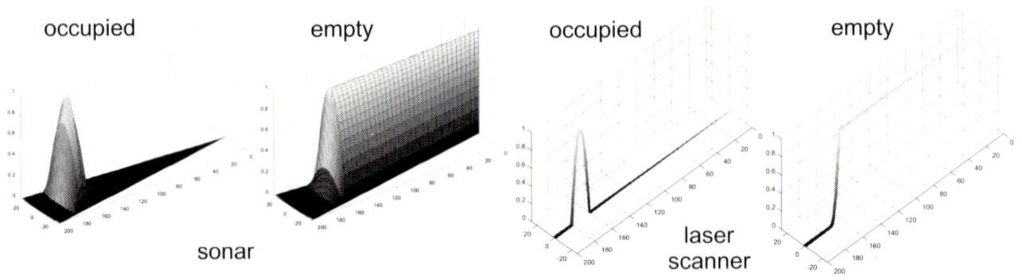

Fig. 3. Sensor beam models of the sonar (left) and the laser scanners (right)

angular width of the beam and constant value of being occupied/empty:

$$\lambda_{son}(\alpha) = \begin{cases} 1 - \left(\frac{\alpha}{12.5^o}\right)^2, & 0 \leq |\alpha| \leq 12.5^o \\ 0, & |\alpha| \geq 12.5^o \end{cases}, \quad \lambda_{las}(\alpha) = \begin{cases} 1, & 0 \leq |\alpha| \leq \varphi_{las} \\ 0, & |\alpha| \geq \varphi_{las} \end{cases}, \tag{2}$$

where the half-width of the beam φ_{las} is 0.5^o for the LMS-200, and 1.5^o for the AMCW laser. The 3D profiles of these functions are given in Fig. 3, for a distance reading of 1.7 [m].

The data gathered from a single robot pose are aggregated to the sets $\mathcal{O}^k$ and $\mathcal{E}^k$, representing the occupied and empty cells, respectively:

$$\mathcal{O}^k = \bigcup_i \mathcal{O}_i^k, \quad \mathcal{E}^k = \bigcup_i \mathcal{E}_i^k. \tag{3}$$

The sets $\mathcal{E}^k$ and $\mathcal{O}^k$ generated at the k-th pose are aggregated with the previously available information. Also two sets describing the lack of knowledge, by identifying the cells being ambiguous ($\mathcal{A}$) or indeterminate ($\mathcal{I}$) are computed:

$$\mathcal{O} = \mathcal{O} \cup \mathcal{O}^k, \quad \mathcal{E} = \mathcal{E} \cup \mathcal{E}^k, \quad \mathcal{A} = \mathcal{E} \cap \mathcal{O}, \quad \mathcal{I} = \bar{\mathcal{E}} \cap \bar{\mathcal{O}}. \tag{4}$$

Then, the sets of cells, that are useful for particular robot tasks can be defined upon the known values of $\mathcal{O}$, $\mathcal{E}$, $\mathcal{A}$, and $\mathcal{I}$ [5].

In Fig. 2B,C,D the grid maps of the *occupied* areas are presented. From these maps, one can see, that the Bayesian and D-S methods produced quite similar maps, however the Bayesian one shows more outliers due to the AMCW laser measurements, while the D-S map does not include the small box in the center, which is below the beam plane of the scanner mounted on top of the robot. In contrary, the fuzzy grid map, displaying the $\mathcal{O}$ fuzzy set (i.e. all occupied cells) is very conservative, indicating many empty areas as occupied. However, the occupied and empty sets are not complementary, thus partial membership to $\mathcal{O}$ and $\mathcal{E}$ is possible, what enables to identify the level of contradiction between the measurements.

Because the aim of this work is to build a segment-based model of the environment, we are looking for cells providing reliable support for line extraction. To this end, we define a set of *support* cells, being very occupied and unambiguous. The term "very occupied" emphasizes the difference between low and high values of occupancy, and is implemented by squaring the value of the membership function:

$$\mathcal{S} = \mathcal{O}^2 \cap \bar{\mathcal{A}} \cap \bar{\mathcal{I}}. \tag{5}$$

The resulting map of the fuzzy support cells is shown in Fig. 2E. It describes boundaries of the objects (including the small box in the center) quite precisely, suppressing almost all outliers, that gave rise to the contradictory values of $\mathcal{O}$ and $\mathcal{E}$.

Finally, each cell in the map is a tuple: $c_{ij} = \{\mathcal{O}, \mathcal{E}, \mathcal{I}, \mathcal{A}, \mathcal{S}\}$. In each cell the n_p raw laser data points $\mathbf{P}_i = [x_i \; y_i]^T$, $i = 1 \ldots n_p$, that contributed to the occupancy of this cell, are stored. The sonar measurements are discarded, once they have updated the grid.

3 Extracting Structured Local Maps

The geometric, segment-based map is appropriate to represent the entire working area of the robot. The problems of data association and filtering of unreliable measurements are solved by extracting segments from the local fuzzy grids. Extraction of the lines is based on the Hough transform [3]. Object contours are aligned cells of high membership degree to the $\mathcal{S}$ set. The found lines are represented by the equation:

$$x \cos(\phi) + y \sin(\phi) - \rho = 0, \tag{6}$$

where ρ is the perpendicular distance to the line, and ϕ is the line orientation in the coordinates of the local map. They are also the coordinates of the Hough parameter space. The Non Maxima Suppression is done in the parameter space to eliminate false positives. Next, the segments are determined by examining local connections between those cells, that are members of the found lines. The Hough transform does not take uncertainty into account when estimating the line parameters, and introduces itself uncertainty due to the discrete parameter space. The loss of precision is also inherent to the grid representation, because the individual range readings are incorporated into the cells of given size, hence the information about their precise position is lost.

To overcome these problems, we refine the line segment representation. The weighted line fitting via a maximum likelihood formalism is applied to find the vectors $\mathbf{L} = [\rho \; \phi]^T$ and covariance matrices $\mathbf{C}_L$ of the support lines, using the coordinates of the laser range measurements stored in the segment-member cells of the grid. The sonar data have to low spatial density to contribute to the object contours with the required precision. However, they contribute to the fuzzy line support measure computed from Eq. (5), reinforcing those laser measurements that are consonant with the sonars. The line fitting method we use (more details about this procedure, and the closed-form formulas for the covariance of the line fit can be found in [6]) considers the accuracy of the range measurements when updating the line model. Finally, the extremities of the segments are computed by projecting the data points into the infinite

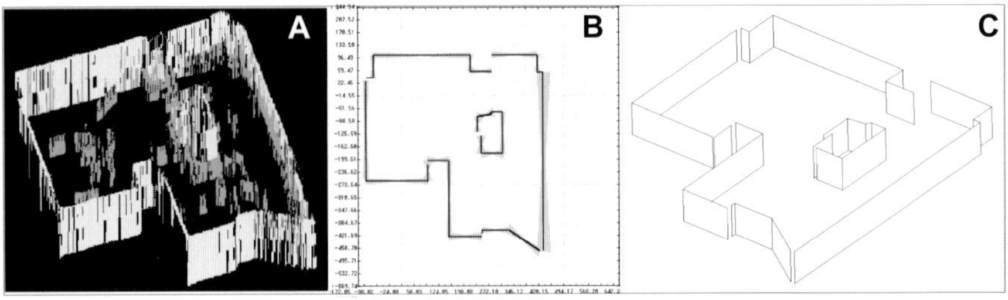

Fig. 4. Fuzzy grid $\mathcal{S}$ (A), the resulting structured map with the fuzzy representation of spatial uncertainty (B), and its 3D view (C)

line and trimming the line at the extreme endpoints. For each line segment its center point $\mathbf{P}_c = [x_c \; y_c]^T$, and the half-length l are determined. Hence, the i-th line segment is a tuple $L_i = \{\mathbf{L}_i, \mathbf{C}_{L_i}, \mathbf{P}_{c_i}, l_i\}$.

We have found useful to define a fuzzy uncertainty representation upon the covariance matrix $\mathbf{C}_L$. This is inspired by the fuzzy segment model [4], that employs trapezoidal fuzzy sets to represent the spatial uncertainty of a segment built from sonar readouts. In [4] the uncertainty of the location of a segment is due to the distance between the object and the sensor, scattering of the measurement around the segment, and the uncertainty of the robot pose. The uncertainty due to these three factors is treated separately, by introducing constants depending on the particular sensor and environment type, what makes the model rather subjective. We take a different approach, by defining the fuzzy sets on the $\mathbf{L} = [\rho \; \phi]^T$ parameters, using the matrix:

$$\mathbf{C}_L = \begin{bmatrix} \sigma_\rho^2 & \sigma_{\rho\phi} \\ \sigma_{\phi\rho} & \sigma_\phi^2 \end{bmatrix} . \tag{7}$$

The trapezoidal fuzzy sets $\mathcal{R}$ and $\mathcal{P}$ are built by using the normal distributions on ρ and ϕ, respectively. The intervals of one standard deviation (confidence level 0.68) are assigned to the α-cut in 1, and the intervals of two times the standard deviation (0.95) are assigned to the α-cut in 0 :

$$\mathcal{R} = \{- \mid t_{0.025} \mid \sigma_\rho, - \mid t_{0.16} \mid \sigma_\rho, \mid t_{0.16} \mid \sigma_\rho, \mid t_{0.025} \mid \sigma_\rho\}, \tag{8}$$

$$\mathcal{P} = \{- \mid t_{0.025} \mid \sigma_\phi, - \mid t_{0.16} \mid \sigma_\phi, \mid t_{0.16} \mid \sigma_\phi, \mid t_{0.025} \mid \sigma_\phi\}, \tag{9}$$

where $t_{\frac{\beta}{2}}$ are the $\frac{\beta}{2}$ percentage points defining the 1-β symmetrical confidence limits [4]. Hence, we obtain a structured local map, being a list of fuzzy segments defined as: $FL_i = \{\mathbf{L}_i, \mathbf{C}_{L_i}, \mathbf{P}_{c_i}, l_i, \mathcal{R}_i, \mathcal{P}_i\}$. The fuzzy sets, together with the known center and length, allow to draw for each segment its uncertainty boundaries.

In Fig. 2F the structured map extracted from the fuzzy grid of Fig. 2E is shown. The grey areas around the segments are the fuzzy uncertainty boundaries, for sake of clarity they are omitted on very short segments. Figure 4 shows results of another experiment, in which only the *LMS-200* was used. The robot took data from 46 poses in a laboratory environment. In the 5

$\times$ 5 meters room only one local fuzzy grid was built. Its 3D view is shown in Fig. 4A, where the higher peaks mean better support for line extraction, while the darker objects are the outliers, mostly due to people walking by. The extracted geometric map with the fuzzy representation of the spatial uncertainty is shown in Fig. 4B, while Fig. 4C depicts the 3D (extruded) view of this map, that cleary shows its geometric consistency. The final structured map consists of only 25 line segments, while the map obtained from direct interpretation of the same range data (used for self-localization, cf. Fig. 1) has 72 partially overlapping segments.

4 Structured Global Map

In larger environments several fuzzy grids are built, and the global map is constructed by integrating the segments from local structured maps. At the integration stage, the segments have to be transformed to the global coordinates. The estimate of the robot pose (being the origin of the local map coordinates) $\mathbf{X}_R = [x_r\ y_r\ \theta_r]^T$ is obtained from the overhead cameras mounted in the laboratory, or from the EKF-based self-localization. The uncertainty of a line in the global map $\mathbf{C}_{L_g}$ depends on both its uncertainty in the local frame $\mathbf{C}_L$ and the uncertainty of the local map coordinates in the global frame $\mathbf{C}_R$:

$$\mathbf{L}_G = f(\mathbf{L}, \mathbf{X}_R), \quad \mathbf{C}_{L_g} = \mathbf{J}_L \mathbf{C}_L \mathbf{J}_L^T + \mathbf{J}_R \mathbf{C}_R \mathbf{J}_R^T, \tag{10}$$

where $\mathbf{J}_L, \mathbf{J}_R$ are the Jacobians of the relation f w.r.t. $\mathbf{L}$ and $\mathbf{X}_R$, respectively [9]. The matching procedure determines whether the candidate segments are similar enough to merge. The standard way of matching line segments is to employ the squared Mahalanobis distance test [3], to check whether the infinite lines are similar taking into account their uncertainty, and then to use some geometric tests to check if the segments are overlapping. The geometric tests usually do not take the spatial uncertainty into account, using ad hoc defined tolerance values [2, 8]. This approach often leads to false matches and multiple segments in the map.

To remedy this problem, we use the fuzzy sets defined by (9) to determine if the uncertain segments are actually overlapping. This is done according to the procedure proposed in [4], by checking their colinearity and intersection. Since at the matching stage the fuzzy sets $\mathcal{R}$ and $\mathcal{P}$ must reflect the spatial uncertainty of the local segment in the global frame, they are computed from the updated covariance matrix $\mathbf{C}_{L_g}$. In [4] the fuzzy uncertainty in ϕ is only inferred from the uncertainty in ρ by a simple geometric computation. If the candidate segments pass these fuzzy tests, and the Mahalanobis distance test, they can be fused into one. A Kalman filter algorithm [2] determines the best estimate of the infinite line. The end points of the two segments are projected onto the estimated line, they replace the old end points if they make the segment longer. The new center point and half-length are computed as

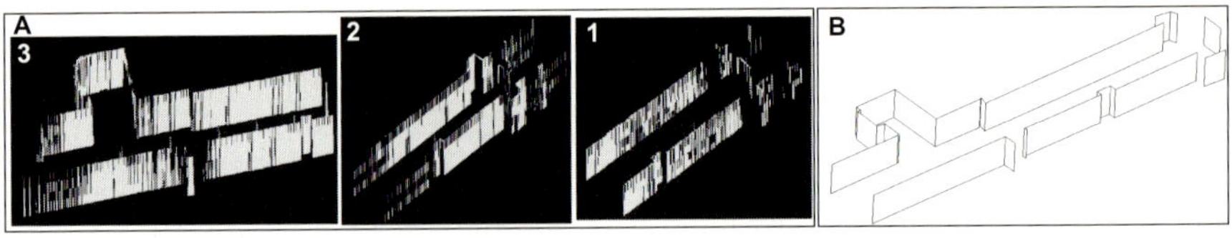

Fig. 5. Local fuzzy grids (A) and a 3D view of the structured global map (B)

well. Again, the fuzzy sets $\mathcal{R}$ and $\mathcal{P}$ are computed from the updated covariance matrix. Figure 5 shows results of an experiment, in which the robot took measurements (both lasers only) from 16 poses along a hallway, with some people walking by. Three fuzzy grids have been built (Fig. 5A), then interpreted as local structured maps, and fused into the global map depicted in Fig. 5B, that consists of 23 segments.

5 Conclusions

We have proposed to merge two types of spatial knowledge representation – grid and feature based, and two frameworks for the uncertainty description: probabilistic and fuzzy. The experimental results show, that this approach is superior to the classical direct segment reconstruction, in terms of map accuracy and geometric self-consistency. The fuzzy uncertainty representation serves at two levels: being useful in dealing with outliers in the grids, and then providing support for decisions at the global integration stage.

References

1. Bolc L., Borodziewicz W., Wójcik M. (1991) Fundamentals of Uncertain and Incomplete Information Processing (in Polish), PWN, Warsaw.
2. Castellanos J., Tardòs J. (1999) Mobile Robot Localization and Map Building. A Multisensor Fusion Approach, Kluwer, London.
3. Duda R., Hart P., Stork D. (2001) Pattern Classification, J. Wiley & Sons, New York.
4. Gasós J., Rosetti A. (1999) Uncertainty Representation for Mobile Robots: Perception, Modeling and Navigation in Unknown Environments, Fuzzy Sets and Syst., 107:1–24.
5. Oriolo G., Ulivi G., Vendittelli M. (1997) Fuzzy Maps: A New Tool for Mobile Robot Perception and Planning, J. of Robotic Syst., 14(3):179–197.
6. Pfister S. (2002) Weighted Line Fitting and Merging, Tech. Rep., California Inst. of Tech.
7. Ribo M., Pinz A. (2001) A Comparison of Three Uncertainty Calculi for Building Sonar-based Occupancy Grids, Robotics and Autonom. Syst., 35:201–209.
8. Skrzypczyński P. (1997) 2D and 3D World Modelling Using Optical Scanner Data, in: Intelligent Robots: Sensing, Modeling and Planning, (R. Bolles *et al.*, eds.), World Scientific, Singapore, 211–228.
9. Smith R., Cheeseman P. (1987) On the Estimation and Representation of Spatial Uncertainty, Int. J. of Robotics Research, 5(4):56–68.
10. Thrun S. (2002) Robotic Mapping: A Survey, Tech. Rep., CMU-CS-02-111.

Detection of Elliptical Shapes Using Contour Grouping

Marcin Smereka

Institute of Engineering Cybernetics, Wroclaw University of Technology,
Janiszewski Str. 11/17, 50–372 Wroclaw, Poland
martin@ict.pwr.wroc.pl

Summary. In this paper the ellipse–specific contour grouping algorithm is introduced. It is a technique for detecting elliptical shapes from images. The algorithm uses ellipse–specific direct least square fitting and elliptic variance descriptor to assess an error of fit. A survey of other methods for detecting elliptical shapes is performed. Contour grouping techniques are discussed in details. The algorithm was illustrated on real images obtained from cytological phase contrast microscopy.

1 Introduction

Detection of circular and elliptical shapes is a common task in computer vision and image recognition. The join research conducted in Wroclaw University of Technology and Gynecological Clinic GMW in Opole is aimed at recognition and classification of objects present in the phase contrast (PC) cytological images (Fig. 1a). The PC microscopy is a new technology to conduct gynecological examinations. With this technology the diagnosis can be made immediately. Computer aided processing of the PC images can additionally improve performance and quality of these examinations. The most significant objects in PC images are cell nuclei. The size and the shape of a nucleus bring a lot of information about a precancerous lesions in progress [1, 2]. Large, irregularly outlined and non-uniformly filled nuclei are suspected to be pathological ones.

The cell nuclei detection algorithm is based on the following assumptions:

- PC microscopy emphasizes edges of objects, thus searching for nuclei is substituted with searching for their edges;
- the shape of nucleus is circular or elliptic, so oval patterns are of particular interest;
- the image magnification is known in advance, thus only objects within the specific range of radii (r_{min}, r_{max}) are considered.

The problem of detection of oval shapes has been widely studied in literature. Methods solving the problem are generally divided into two categories:

global and local ones. Global methods apply mathematical modeling techniques to describe boundaries of objects in images. The examples of global methods follow:

- Simple Shape Descriptors – Many numerical descriptors have been formulated to classify shapes in binary images. Peura and Ilvarinen studied some of them [3]. The descriptor known as elliptic variance is especially useful for detecting ellipses. Rosin proposed other simple descriptors (moment invariants, Euclidean distances) that can be adapted to measure ellipticity of shapes [4].
- Direct Least–Square Fitting – Pilu and Fitzgibbon were first who presented a direct method for fitting ellipses to scattered data in the least squares sense [5]. Previous methods used a generic conic fitting or an iterative approach to recover elliptic solutions. A variety of error of fit (EOF) functions have been discussed by Rosin [6].
- Hough Transform (HT) – The Hough Transform has been recognized as a very powerful method to detect parametric curves in images [7]. It relies on voting process that maps image edge points into manifolds in an appropriately defined parameter space. Peaks in the parameter space correspond to detected curves. The direct HT method for detecting ellipses is computationally expensive due to high number of dimensions in the parameter space. Many improvements have been proposed to make these methods more efficient [8, 10] and more robust to irregularities [9, 11].

Local methods attempt to close the edge gaps by seeking those edge pixels that are most likely connected in the neighbourhood of initial edge points. A few examples of local methods are:

- Active Contours – Active Contours (snakes) are designed to detect objects whose boundaries are not necessarily defined by gradients [12]. The basic idea is to evolve a curve, subject to constraints from a given image. The initial curve moves governed by appropriately designed energy function until it stops at the local optimum. The final curve is the object boundary. Ray and Acton showed that active contours can also be employed for tracking of moving objects [13].
- Contour Grouping – Low level edge detection operators do not guarantee the generation of continuous boundaries of objects. This makes many image analysis tasks difficult, especially for noisy images. The aim of contour grouping algorithms is to connect edges that are suspected to be parts of the same object. Known contour grouping techniques were concentrated on detecting salient curves without any specific limitations. Following sections present a new contour grouping algorithm that detects curves subject to elliptic constraints.

2 Contour Grouping Techniques

In contour grouping process, pixels placed in and near an edge segment are analysed for their degrees of similarity to the neighbouring edges. Newly identified edge segments are added to a set of edges already found, subject to connectivity criterion. Zhu and Payne [14] defined Directional Potential Function (DPF) to form criteria for edge linking. An edge image is modeled as a potential field with energy depositions at the detected edge points. The DPF evaluates the potential forces generated by the energy depositions. When the potential force value exceeds a given threshold, a conversion of non-edge pixels to edge pixels is done at an edge broken points. This iterative process gradually connects discontinuous edge segments into continual boundary curves. Shashua and Ullman [15] introduced a saliency measure for edge segments that favours long and smooth contours. The goal of grouping is to find a disjoint set of groups of edges that maximizes the overall saliency measure. Unoptimal but satisfactory and computationally acceptable solution is obtained in a two stage process. In the first stage, the saliency measure is computed for each pixel. In the second stage, an iterative process is executed to connect pairs of edge elements and to update the saliency measure for newly formed pairs. Alter and Basri [16] discussed strengths and weaknesses of this method in details. Elder and Zucker [17] tried to impose some constraints on the grouping process. They noticed that the grouping improves perceptually when closed contours are preferred. They claimed that closure computations can potentially complement region grouping methods by extending the class of structures segmented to include heterogeneous structures.

3 Ellipse–Specific Contour Grouping

For detecting nuclei from cytological images it is advisable to search for contours that form ellipses. Therefore, designed algorithm should take into account elliptic constraints. The constraints apply to both the regularity and the size of ellipse. Let us denote the size constraints a_{min} and a_{max} as the lower and the upper limit for the major semiaxis of an elliptical contour, respectively. The algorithm consists of the following stages:

1. Image preprocessing (acquisition, thresholding, labeling, selecting of contour points for each edge element).
2. Building a graph of neighbouring edge elements.
3. Finding the optimal set of paths in the graph that form disjoint elliptical groups.

3.1 Image Preprocessing

The image acquisition requires photographic camera, CCD camera or another device that enables acquiring a source image from the microscope. An ac-

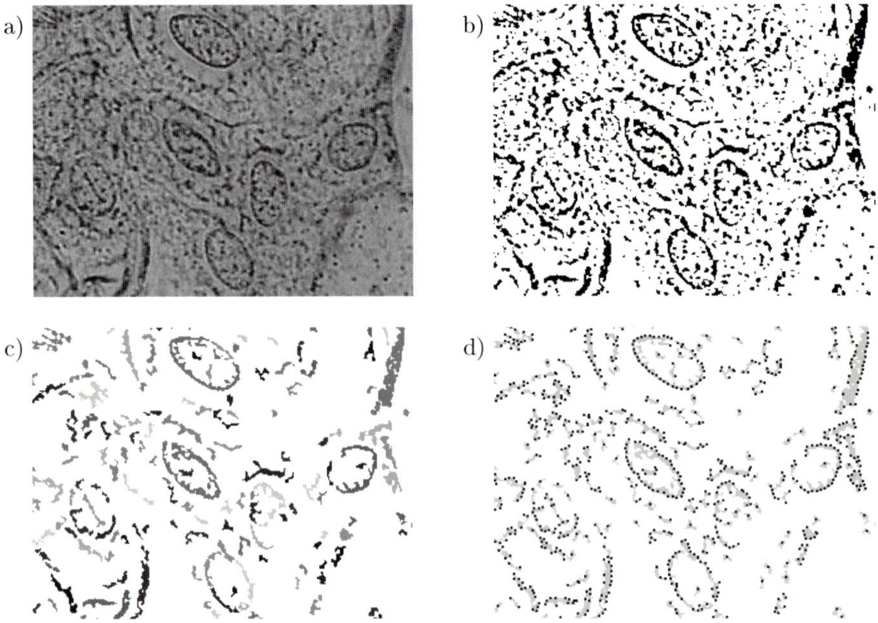

Fig. 1. Preprocessing of PC image. (**a**) initial image, (**b**) result of thresholding, (**c**) after labeling and removing small edgels, (**d**) edgels with marked contour points

quired image should be initially filtered in order to ignore chrominance (color information) and to equalize the background level (Fig. 1a).

The next step is the image thresholding to extract boundaries of objects. For PC images a threshold was empirically set to the value $\frac{9}{8}b$, where b is the equalized background level of image (Fig. 1b). The resulting binary image $I = \{\mathbf{p}\}$ is a set of pixels that exceed the threshold. Symbol $\mathbf{p}$ (and other bold lowercase symbols) means a column vector of the pixel coordinates $\mathbf{p} = (p_x, p_y)^T$. The image $\mathbf{I}$ is then labeled to extract connected segments of edges (edgels) $\mathbf{E_j}$. The label j is uniquely assigned to each edgel:

$$\mathbf{E_j} = \{\mathbf{p} | \mathbf{p} \in \mathbf{I}; Label(\mathbf{p}) = j\}. \tag{1}$$

Small edgels (containing less than 20 pixels) are removed from the image (Fig. 1c). They are too small to be segments of nuclei boundaries. Moreover, if the small edgels are not removed, the computational complexity of the algorithm increases rapidly.

Most of contour grouping techniques require thin contours to measure curvature and perform grouping. Edgels obtained from PC microscopy are not thin, so for each edgel a subset of points (contour points) must be selected that represents this edgel in a grouping process. Let $\mathbf{c_j}$ be a centroid of the edgel $\mathbf{E_j}$. For each angle parameter $\alpha \in [0, 2\pi)$ a ray given by the expression:

$$\mathbf{r_j}(\alpha, k) = \mathbf{c_j} + k \begin{bmatrix} \cos \alpha \\ \sin \alpha \end{bmatrix}; \quad k > 0; \quad \alpha \in [0, 2\pi), \tag{2}$$

is sourced form the centroid. The farthest ray point from centroid that intersects the edgel $\mathbf{E_j}$ is a contour point $\mathbf{r_j^*}(\alpha)$. Contour $\mathbf{Q_j}$ of edgel $\mathbf{E_j}$ is a set of contour points for all values of α:

$$\mathbf{r_j^*}(\alpha) = \mathbf{r_j}(\alpha, \max_{\mathbf{r_j}(\alpha,k)\in\mathbf{E_j}} k); \quad \mathbf{Q_j} = \{\mathbf{r_j^*}(\alpha) | \alpha \in [0, 2\pi)\}. \tag{3}$$

In practice, the uncountable set $\mathbf{Q_j}$ is discretized to keep the constant distance δ between consecutive points of $\mathbf{Q_j}$, if possible. The parameter δ was empirically set to three pixels. A reasonable number of contour points keeps the computational load at a moderate level. Selected contour points are marked by black dots in Fig. 1d.

3.2 Building a Graph of Edgels

For the purpose of grouping, it is convenient to consider edgels as a graph $\mathbf{G} = \langle \mathbf{V}, \mathbf{A} \rangle$, whose vertices $\mathbf{V}$ correspond to edgels $\mathbf{E}$ and arcs connect these pairs of edgels that lay close enough each other:

$$\mathbf{A} = \{(\mathbf{E_i}, \mathbf{E_j}) : \min_{\substack{\mathbf{p}\in\mathbf{E_i} \\ \mathbf{q}\in\mathbf{E_j}}} \|\mathbf{p} - \mathbf{q}\| < \epsilon\}. \tag{4}$$

Symbol $\|\cdot\|$ introduces the Euclidean norm of a vector. For PC images the parameter ϵ was empirically set to five pixels. Let r_{xj} be the width and r_{yj} be the height of the bounding rectangle of edgel $\mathbf{E_j}$. Let $\mathbf{f_j}, a_j, b_j, \Theta_j$ denote, respectively: a center, major and minor semiaxes, and the orientation angle of an ellipse that fits to the contour points $\mathbf{Q_j}$, due to ellipse–specific direct least square fitting algorithm [5]. Let us determine an elliptic variance $evar_j$ as a proportional mean–squared error between the ellipse and contour points for each edgel [3]. For ideal ellipses $evar_j$ equals 0 and increases due to irregularities of the contour. Let us also define a measure $d_j = \|\mathbf{f_j} - \mathbf{c_j}\|/(a_j + b_j)$ for the distance rate between the center of ellipse $\mathbf{f_j}$ and the centroid $\mathbf{c_j}$ of the edgel. For full ellipses d_j tends to 0. For elliptical arcs d_j increases, because larger displacement between the center of ellipse and the centroid of arc can be observed.

Each vertex of the graph is initially classified to one among the following classes:

- SPOT – when $r_{xj} < \lambda$ and $r_{yj} < \lambda$ or the number of contour points $\mathbf{Q_j}$ is smaller than 6;
- HUGE – when $r_{xj} > 2a_{max} + \lambda$ or $r_{yj} > 2a_{max} + \lambda$;
- ELLIPSE – when $a_j > a_{min}$ and $a_j < a_{max}$ and $b_j > a_j/2$ and $evar_j < \zeta$ and $d_j < \eta$;
- ARC – when $a_j > a_{min}$ and $a_j < a_{max}$ and $b_j > a_j/2$;

- NOTHING – otherwise;

To isolate SPOTS and HUGE objects the parameter λ is set to 10. Parameters ζ, η specify the required quality of detected ellipses. For PC images they are empirically set to $\zeta = 0.1, \eta = 0.1$. All HUGE edgels are excluded from the graph, because they cannot form any ellipse that meets the specified size constraints.

3.3 Finding Optimal Groups

Usually, boundary of a nucleus consists of one or more edgels. The goal of contour grouping algorithm is to establish groups of edgels that form the boundary of a nucleus. A path in the graph (group of adjacent edgels) corresponds to a contour in the image. A path is defined as an ordered set of edgels (multiedgel) $\mathbf{M} = (\mathbf{E_1}, \mathbf{E_2}, \ldots, \mathbf{E_n})$. Initially each edgel can be considered as a multiedgel composed of only one edgel.

For all vertices of the graph **G** perform the search according to the following steps:

1. Set all vertices as multiedgels and put them in the queue **L**.
2. Take the consecutive multiedgel **M** from the queue **L**.
3. Concatenate **M** with all the vertices that are adjacent to the last edgel of the considered multiedgel **M** and put the newly created multiedgels to the end of the queue **L**. New multiedgels are again classified to one of the previously mentioned classes. Concatenation succeeds only when the number of edgels in the multiedgel **M** is smaller than 6 and the new multiedgel is classified as neither SPOT nor HUGE. If concatenation fails, the new multiedgel is not added to the queue.
4. Repeat steps 2–3 until no new multiedgel can be created.
5. Remove from the queue **L** all multiedgels not classified as ELLIPSE.
6. Select the best multiedgel **M** from the queue **L**. The best is the multiedgel that has the smallest value of elliptic variance $evar$.
7. Add **M** to the list of the final groups of contours.
8. Remove from the queue **L** all multiedgels that contain at least one of edgels that belong to **M**.
9. Repeat steps 6–8 until **L** is empty.

The presented algorithm outputs the list of disjoint set of edgels. Each set of edgels forms an elliptical contour in the image **I**.

4 Experiments

The algorithm was tested on a set of 54 real PC cytological images of size 640×480. Nuclei of small and great intermediate cells were searched for, thus parameters (a_{min}, a_{max}) were set to $(20, 40)$. Examples of output images

Table 1. Statistical results of nuclei detection

	True Positive	False Negative	False Positive
Samples	136	31	19
Percentage	81.4%	18.6%	11.4%

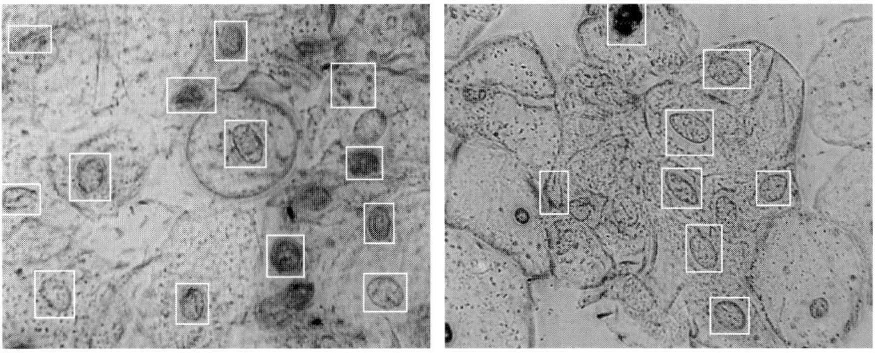

Fig. 2. Examples of detection of elliptical objects

are given in Fig. 2 and statistical results are collected in Table 1. The total number of true objects indicated by a medical expert in all tested images was 167. These were the cell nuclei and a little number of other elliptical objects (granulocytes, particles of dust). These objects were also treated as true objects because the ellipticity criterion does not distinguish them from the nuclei. 136 objects (81.4%) were correctly detected. Additionaly the algorithm detected 19 false objects (11.4%). These objects were mainly bends of cell walls that form elliptical arcs. The overall time of computations per one image is about 1 up to 4 seconds (for very complicated images). The performance was measured on AMD Athlon 1.8GHz. The computational complexity depends mainly on the number of possible paths in the graph **G**. To increase the computational efficiency, wrong paths should be eliminated as soon as possible and right paths should be followed as deep as needed.

5 Conclusions

In this paper the new efficient algorithm for detection of elliptical objects was presented. The algorithm joins features of local detection algorithms (contour grouping) and global detection algorithms (by imposing elliptic constrains on objects being detected). General contour grouping techniques are designed to find long smooth edges and cannot be used to this particular task. Other global methods (simple shape descriptors, Hough Transform) does not deal with this task effectively due to high level of noise and disturbances (edges of cell walls, other objects) present in images. Application of Hough Transform methods [11] works properly for circular objects, but the more irregular the objects are, the worse the results of detection become. Further research will be concentrated on incorporating other features of detected objects to distinguish between normal nuclei, pathological nuclei and other nuclei–like objects.

References

1. Glab G., Florczak K., Jaronski J., Licznerski T., Diagnostic cytology and phase contrast microscopy, Blackhorse, Warszawa 2001 (in Polish)
2. Smereka M., Towards computer aided phase contrast cytology. Reports ICT PWr. (54), 2003.
3. Peura M., Iivarinen J., Efficiency of simple shape descriptors, Aspects of Visual Form, pp. 443–451, World Scientific, 1997
4. Rosin P. L., Measuring Shape: Ellipticity, Rectangularity, and Triangularity, citeseer.ist.psu.edu/553726.html
5. Pilu M., Fitzgibbon A.W., Fisher R.B., Ellipse–Specific Direct Least–Square Fitting, IEEE Trans. PAMI, 21(5), pp. 477–480, 1999
6. Rosin P. L., Assessing Error of Fit Functions for Ellipses, Graphical models and image processing: GMIP, 58(5), pp. 494–502, 1996
7. Duda R.O., Hart P.E., Use of the Hough Transform to Detect Lines and Curves in Pictures, Communications of the ACM 15, pp. 11–15, 1972
8. Guil N., Zapata E.L., Low Order Circle and Ellipse Hough Transform, J. Pattern Recognition, vol. 30(10), pp. 1729–1744, 1997
9. Atherton T. J., Kerbyson D. J., Size invariant circle detection, Image and Vision Computing, 17, pp. 795–803, 1999
10. Atiquzzaman M., Coarse–to–Fine Search Technique to Detect Circles in Images, Int. Journal of Advanced Manufacture Technologies, 15, pp. 96–102, 1999
11. Smereka M., Nuclei recognition in phase contrast microscopy images, Conf. on Computer Recognition Systems, Wroclaw Univ. of Technology Publ., pp. 35–40, 2003
12. Chan T.F.,Vese L.A., Active contours without edges, IEEE Trans. Image Processing, 10(2), pp. 266–277, 2001
13. Ray N., Acton S. T.,Ley K. F., Tracking leukocytes in vivo with shape and size constrained active contours, IEEE Trans. Med. Imag. (Special Issue on Image Analysis in Drug Discovery and Clinical Trials), 21, pp. 1222–1235, 2002
14. Zhu Q., Payne M., Riordan V., Edge linking by a directional potential functions (DPF), Image and Vision Computing, 14, pp. 59–70, 1996
15. Shashua A., Ullman S., Grouping contours by iterated pairing network, Neural Info, 3, pp. 335–341, 1991
16. Alter T., Basri R., Extracting salient curves from images: An analysis of the saliency network, Int. Journal on Computer Vision, 27(1), pp. 51–69, 1998
17. Elder J. H., Zucker S. W., Computing Contour Closure, ECCV, 1, pp. 399–412, 1996

Active Shape Models in Practice

Maciej Smiatacz and Witold Malina

Faculty of Electronics, Telecommunications and Informatics, Gdansk University of Technology, ul. Narutowicza 11/12, 80-952 Gdansk, Poland
{slowhand, malwit}@eti.pg.gda.pl

Summary. Active Shape Models (ASM) were proposed in the last decade of the 20th century as a versatile method of object localization and recognition. The theoretical concept on which the algorithm is based seems very attractive but the practical value of this technique still needs to be verified. The authors developed a multi-purpose object locating system containing an implementation of the ASMs and the experiments performed with the help of the system revealed serious drawbacks of the method. The discovered practical problems related to the use of the ASMs are presented in the paper.

1 Introduction

The general task of automatic segmentation of images appears exceptionally difficult because the number of different types of objects that we deal with in real life is unlimited and nearly every object can vary considerably in size and shape. The possible changes of illumination, perspective or background make the situation even more complicated. Many specialized algorithms capable of locating the objects of specific type have been proposed but they are useful only in particular situations and the lack of versatility is their main drawback. The more advanced methods are usually based on some *model* of the object that we are looking for. The very important feature of such model is the *flexibility* that allows it to adapt to the changes in object's appearance. The well-known example of a flexible structure that can stretch and fit to the image features is the Active Contour Model described by Kass et al [1]. This approach utilizes the iterative energy minimization technique, which is only capable of local adaptation and ignores the global constraints. Another example of a flexible model is the Deformable Template proposed by Yuille [2]. It is a hand built structure that consists of several geometric parts that can deform and move to fit an image but it must be prepared individually for each application. The Active Shape Models introduced by Cootes et al [3] are free from the disadvantages of the previous methods. This is why we have used this technique to construct our object locating system called ASMTEST [4, 5]. It contains an implementation of the original method as well as the version based on geometric histograms. The system can serve as a part of a preprocessing module helpful in solving the pattern recognition problems and may also be used for educational purposes. Here we describe its application for medical image segmentation, face localization and vehicle tracking.

2 Active Shape Models

The detailed description of the Active Shape Models is beyond the scope of this article. In the following sections we introduce the basic concepts of this technique and encourage the reader to analyze the original publications [3, 6].

2.1 Classic ASM algorithm

Active Shape Model (ASM) is a structure containing information about an average shape of the object of some type and the data concerning the most characteristic shape deviations observed in the training set. The models can be modified by algorithms that try to match them to the real shape while preventing unnatural deformations. First we have to define a set of labeled points that represent the shape. This is called the Point Distribution Model (PDM). The landmark points must be placed at equivalent locations on each of the M training examples. This way we obtain a set of sample shapes stored as vectors containing coordinates of the landmark points. The information concerning scale and position is removed from these vectors by the aligning procedure [6] and only the shape description remains. At the next stage we extract the statistical characteristics of the training set by using the principal components analysis. We define the $\mathbf{\Gamma}_i$ vectors that describe n points of each shape, then we compute the mean shape $\bar{\mathbf{\Gamma}}$ and the deviation from the mean for each element of the training set, i.e. $\mathbf{\Phi}_i = \mathbf{\Gamma}_i - \bar{\mathbf{\Gamma}}$. Now we must construct the appropriate covariance matrix:

$$\mathbf{C} = \frac{1}{M} \sum_{i=1}^{M} \mathbf{\Phi}_i \mathbf{\Phi}_i^{\mathrm{T}} \tag{1}$$

and calculate its eigenvectors $\mathbf{u}_i$. A new instance of the model $\mathbf{X}$ can be generated by using the mean shape and the weighted sum of the most characteristic shape deviations represented by the eigenvectors $\mathbf{u}_i$ i.e.:

$$\mathbf{X} = \bar{\mathbf{\Gamma}} + \mathbf{Pb} \tag{2}$$

where $\mathbf{P} = (\mathbf{u}_1 \mathbf{u}_2 ... \mathbf{u}_t)$ is the matrix containing the first t eigenvectors of $\mathbf{C}$ and $\mathbf{b} = (b_1 b_2 ... b_t)^{\mathrm{T}}$ is the vector of their weights.

The matching algorithm consists of the following steps:

1. Research a small neighborhood of each landmark point in order to get the desired deformation of the model.
2. Calculate translation, scale and rotation parameters that allow to put the unchanged model as close to the real shape as possible.
3. Update the model parameters to deform the template shape within acceptable limits.

In the 1st step we have to calculate the suggested movement for each landmark point. This can be achieved by looking for the strongest edge along the line perpendicular to the model boundary at a given point. If the information about the gray scale profile around each point is included in the model definition then the search involves finding a nearby region that better matches this profile.

In the 2nd step we utilize the vector describing suggested point movements $\mathbf{d_x}$ to find the translation $\mathbf{t_x}$, rotation $d\Theta$ and scaling $(1 + ds)$ which best map the current set of points $\mathbf{x}$ onto the set of points given by $(\mathbf{x} + \mathbf{d_x})$ [3].

Having adjusted the pose variables we calculate the residual $\mathbf{\Delta_x}$ adjustments that can only be satisfied by deforming the shape in the 3rd step. We do not want, however, to disturb the consistency of the model so we have to determine the adjustments to shape parameters $\mathbf{\Delta_b}$ that would satisfy the equation:

$$\mathbf{x} + \mathbf{\Delta_x} = \mathbf{x} + \mathbf{P}(\mathbf{b} + \mathbf{\Delta_b}) \tag{3}$$

It can be proved [3] that

$$\mathbf{\Delta_b} \approx \mathbf{P}^{\mathrm{T}} \mathbf{\Delta_x} \tag{4}$$

2.2 Pairwise Geometric Histograms

The Pairwise Geometric Histograms (PGH) were introduced by Evans et al in [7]. In this method every straight line of the model is labeled with its local geometrical context so, theoretically, the changes in illumination, shapes in the background or even occlusions does not cause so much trouble as in the case of grey level distributions.

If we want to use the PGH method, every shape in the scene must be represented by a set of straight lines approximating its contour. Each segment of the contour is called a *baseline* and all other segments that appear in the circle with the radius r drawn around the baseline, form the set of its local features. The geometrical relation between the baseline and a segment can be described by the three parameters: the angle Θ and the distances from the baseline to the beginning and the end of the segment (Fig. 1a). The distribution of values of such parameters is stored in the so-called geometrical histogram (Fig. 1b).

Each entry of the histogram receives a value equal to the product of the length of a baseline and the length of the segment. If we treat each histogram as a vector, then the degree of similarity between the $\mathbf{M}_j$ histogram of the

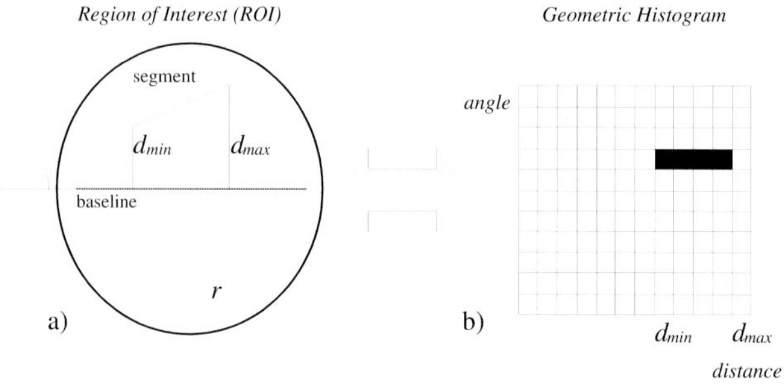

Fig. 1. (**a**) The angle and the distances between the baseline and the segment, (**b**) appropriate entry in a geometric histogram

j-th model segment and the **I** histogram created for some line in the image can be described by the following correlation:

$$D_j = \sum_{i=1}^{R} \sqrt{I_i M_{ij}} \qquad j = 1, ..., s \tag{5}$$

where R – number of histogram entries, s – number of line segments in the model. The exhaustive description of the PGH algorithm can be found in [5].

3 Experimental results

3.1 The versatile object locating system

We have implemented the theory of Active Shape Models in an object locating system called ASMTEST that can be run on any standard PC. The object-searching engine is encapsulated in a dynamic link library so the pattern recognition systems are able to employ it as a part of the preprocessing stage. Fig. 2 shows the simplified block diagram of the system.

3.2 Face localization experiments

The model that we used in this experiment consisted of 84 landmark points and was trained on 8 images. The generalization ability of the model was impressive, for example changing the b_1 weight of the 1st eigenvector we could obtain rotations around vertical axis (Fig. 3). During the search the algorithm coped with scale changes, rotations, different face expressions (smiling, surprised etc.) or even occlusions of the face (Fig. 4) but only if the image showed the person whose photographs had been included in the training set. Localization of the faces on images showing other persons, photographed in different lighting conditions appeared much more problematic and the results

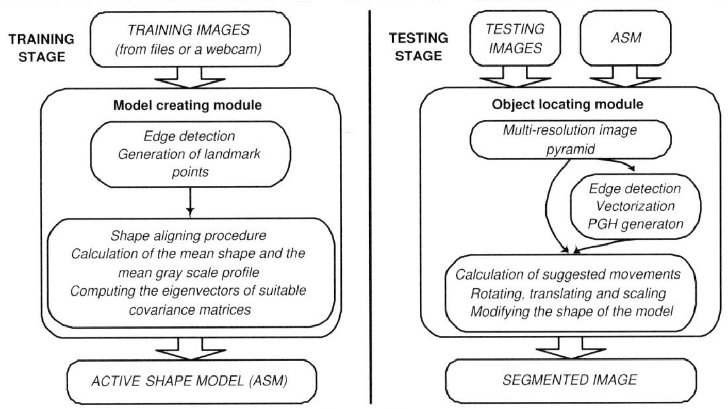

Fig. 2. The simplified block diagram of the object locating system

were often unsatisfactory. We discovered, however, that slight changes in image brightness and contrast could easily help to solve the problem in many cases. Generally saying, although the shape statistics are perfectly extracted from the training set, the mechanism of finding the suggested landmark movements (based on the analysis of grey levels) is the weakest point of the method.

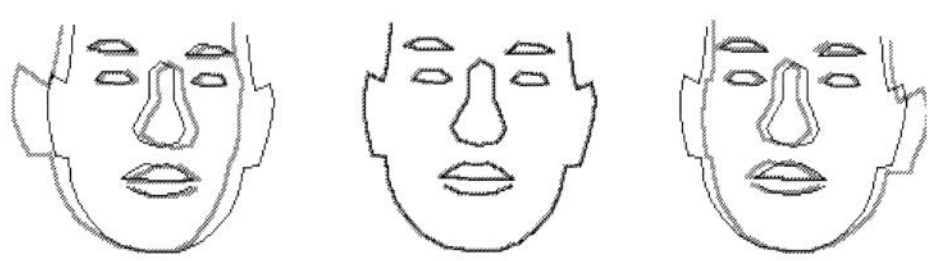

Fig. 3. Information contained in the 1st eigenvector of the face model

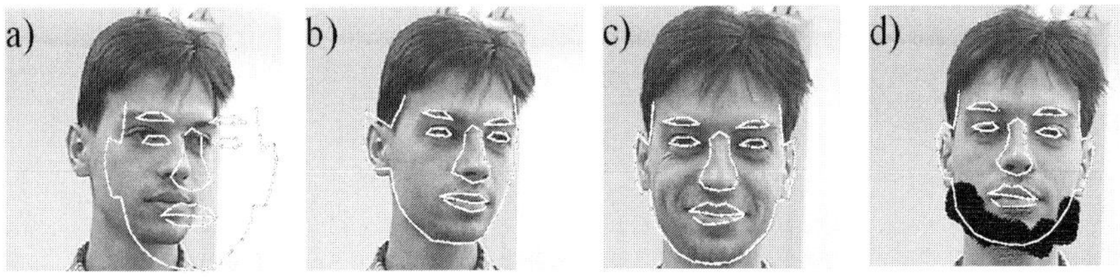

Fig. 4. Localization process: (**a**) starting position, (**b**) rotated face, (**c**) smile, (**d**) partial occlusion

The method performed much worse when the background became more complex, i.e. when it contained numerous distinct edges that tended to attract the landmark points. Blurring the image improved the results but it couldn't eliminate the intrinsic drawbacks of the grey level profile analysis.

3.3 Medical image segmentation

The aim of our experiments with medical image segmentation was to locate the brain ventricles on MR image. The model consisted of 20 points and was trained on only 3 images. Such a small number of samples should be enough in this case, because the shape itself is very characteristic and does not change dramatically from image to image. Although the task seemed to be relatively easy and we managed to obtain some acceptable results, the overall evaluation of the ASM performance must be critical. In most of the cases the model was unable to find even the approximate location of the object because it had a tendency to adapt to the artifacts it found during the search.

The poor results of this experiment were mostly caused by the specificity of MR images, i.e. the very high contrast and presence of large areas with the same intensity value. If the search was performed by the number of models starting from different positions at the same time, the probability of correct localization would be certainly higher but then the difficult problem of choosing the best-fitted model would have to be solved.

3.4 Vehicle localization

This was the most exhaustive series of our experiments and it allowed us to evaluate the method of Pairwise Geometric Histograms. Primary tests were performed on artificial images showing simple 3D car-like objects. This helped us to choose the best parameters of the method, such as number of landmark points and edges, the radius of ROI, the number of the training samples or the resolution of geometric histograms. Careful selection of edge detection and line approximation algorithms along with their parameters was also extremely important.

The localization of cars in real life scenes is much more challenging, because of the changes in illumination and existence of background objects. Moreover, the shapes of today's vehicles are highly complex and it is often difficult to approximate them with straight lines. In the first set of tests we used the database containing photographs of car miniatures, available from the Weizmann Institute [8]. The model used for artificial image segmentation had to be replaced with a more complicated one. The results we obtained were very good (Fig. 5) but only for the type of car that had been chosen for training (Golf). Other brands of cars were localized less precisely.

The localization error was defined as the mean distance (in pixels) between the position of the landmark after the search was completed and the actual position of the point represented by the landmark. The error rates obtained for experiments performed on three car miniatures are summarized in Fig. 6a. The score of 0-10 points means a very good localization with exact fitment

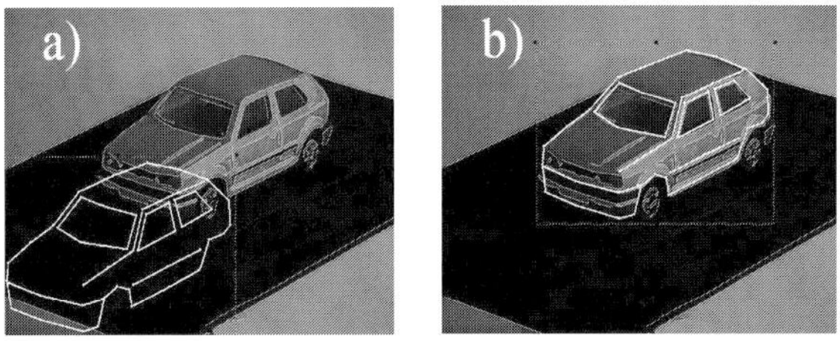

Fig. 5. An example of car miniature localization: (**a**) starting position, (**b**) search result

to the object's shape, 10-20 points – small part of the model does not fit the object, 20-30 points – only object's position and angle were found correctly, 30-40 points – accurate position was found, 40-60 – the object's position was roughly estimated, more than 60 points – a total failure.

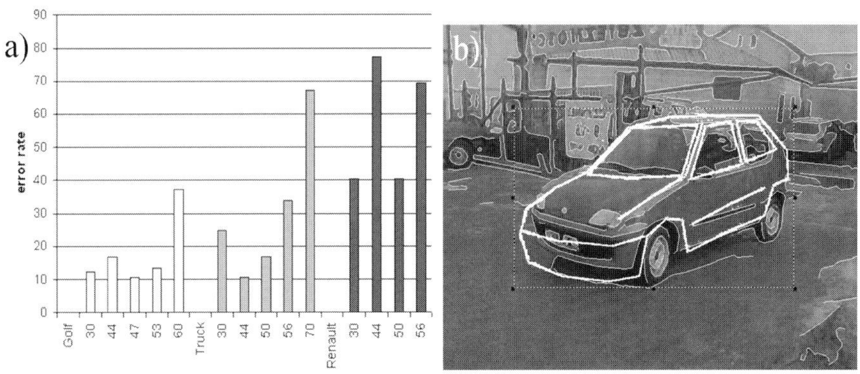

Fig. 6. (**a**) Error rates obtained during experiments with car miniature localization, (**b**) one of the best results of the segmentation of real-life images

The results of the segmentation of the outdoor images showing authentic cars in realistic surroundings were very poor. One of the best-fitted models is shown on Fig. 6b but this success was a matter of chance rather than a typical outcome of the algorithm. Even in this case the localization errors are clearly visible but even very slight change of the starting position of the model in the same image usually led to the absolute failure. The dark painted cars were much more difficult to localize than the bright ones because it was tricky to detect the edges of the vehicle correctly even if the lighting conditions were good. On the other hand, large background objects with distinct straight edges (e.g. buildings) pulled the model segments towards themselves so strongly that the algorithm was not able to find even the approximate location of the car. Even if the starting position of the model was very close to the correct location, the method performed disappointingly because quite often two different model segments tried to fit to the same straight line in the image. In many cases these conflicts were responsible for localization failures.

Generally, the method appeared to be very unstable. Although in most of the problematic situations we were able to achieve good results finally, the proper segmentation depended on the correct choice of numerous parameters that had to be fine tuned individually for every image.

4 Conclusion

We have performed exhaustive practical tests of the Active Shape Models. We confirm that this method is capable of finding the objects that show significant shape variations and it can generalize the information contained in the training set or localize the shapes that were not included in it. On the other hand, the search process can be easily disturbed by background items or artifacts and in our opinion it is not possible to construct a fully automatic, stable and robust object localization system for real-life applications on the basis of ASMs. Although the theory of deformable models seems to be promising, the fundamental problems of creating correct edge maps or finding appropriate suggested movements of model points still haven't been solved. The limited size of this publication does not allow us to present all the practical observations and findings that we made during our experiments. Nevertheless, we encourage the readers to perform their own tests with the help of the ASMTEST system available from our web site [9].

References

1. Kass M., Witkin A., Terzopoulos D. (1987), Snakes: Active Contour Models. Int. J. Computer Vision, vol. 1, no. 4, pp. 321-331
2. Yuille A. (1991), Deformable Templates for Face Recognition. J. Cognitive Neuroscience, vol. 3, no. 1 pp. 59-70
3. Cootes T., Taylor C. (1992), Active Shape Models – "Smart Snakes". British Mach. Vision Conf., pp. 266-275
4. Dudzinski J. (1999), Active Shape Models in Object Localization. MSc dissertation, Gdansk University of Technology, Gdansk (in Polish)
5. Glowinski D. (2004), Automatic Object Locating System. MSc dissertation, Gdansk University of Technology, Gdansk (in Polish)
6. Cootes T., Taylor C., Cooper D., Graham J. (1992), Training Models of Shape from Sets of Examples. British Mach. Vision Conf., pp. 9-18
7. Evans A., Thacker N., Mayhew J. (1993), The Use of Geometric Histograms for Model Based Object Recognition. British Mach. Vision Conf., pp. 429-438
8. Car Model Database, http://www.wisdom.weizmann.ac.il/~cars/
9. ASMTEST, http://www.eti.pg.gda.pl/katedry/kiw/software/ASM/index.html

Fast Detection and Impulsive Noise Attenuation in Color Images

Bogdan Smolka, Andrzej Chydzinski, and Konrad Wojciechowski

Silesian University of Technology, Faculty of Automatic Control, Electronics and Computer Science, Akademicka 16, 44-100, Gliwice, Poland,
{Bogdan.Smolka;Andrzej.Chydzinski;KWojciechowski}@polsl.pl

Summary. In this paper a novel approach to the impulsive noise removal in color images is presented. The proposed technique employs the switching scheme based on the impulse detection mechanism using the so called *peer group* concept. Compared to the vector median filter, the proposed technique consistently yields better results in suppressing both the random-valued and fixed-valued impulsive noise. The main advantage of the proposed noise detection framework is its enormous computational speed, which enables efficient filtering of large images in real-time applications.

1 Introduction

Nonlinear image processing methods continue to grow in popularity and the advances in computing performance have accelerated the process of moving from theoretical explorations to practical implementations. The nonstationarity of images, the significance of visual cues such as edges and the nonlinearity of human visual system, all contribute to the importance of nonlinear methods in imaging applications.

In this paper a novel approach to the detection and removal of impulsive noise in color images is presented. The main advantage of the described technique is its simplicity and enormous computational speed. The proposed method is using the well known vector median filter for the suppression of the detected noise, however different techniques can be used for the denoising of the previously detected impulses.

2 Impulsive Noise Removal

The majority of the nonlinear, multichannel filters are based on the ordering of vectors in a sliding filter window. The output of these filters is defined as the lowest ranked vector according to a specific vector ordering technique.

Let the color images be represented in the RGB color space and let $\mathbf{x}_1$, $\mathbf{x}_2$, ..., $\mathbf{x}_n$ be n samples from the sliding filter window W. Each of the $\mathbf{x}_i$ is an μ-dimensional multichannel vector, (in our case $\mu = 3$). The goal of the vector ordering is to arrange the set of n vectors $\{\mathbf{x}_1, \mathbf{x}_2, \ldots, \mathbf{x}_n\}$ belonging to W using some sorting criterion.

In [12, 18, 6] the ordering based on the cumulative distance function $R(\mathbf{x}_i)$ has been proposed: $R(\mathbf{x}_i) = \sum_{j=1}^{n} \rho(\mathbf{x}_i, \mathbf{x}_j)$, where $\rho(\mathbf{x}_i, \mathbf{x}_j)$ is a function of the distance among $\mathbf{x}_i$ and $\mathbf{x}_j$. The ordering of the scalar quantities according to $R(\mathbf{x}_i)$ generates the ordered set of vectors.

One of the most important noise reduction filter is the vector median. In the case of gray scale images, given a set W containing n samples, the median of the set is defined as $x_{(1)} \in W$ such that

$$\sum_j \left| x_{(1)} - x_j \right| \leq \sum_j \left| x_i - x_j \right| , \quad \forall \; x_i, x_j \in W. \tag{1}$$

Median filters exhibit good noise reduction capabilities, and outperform simple nonadaptive linear filters in preserving signal discontinuities. As in many applications the signal is multidimensional, in [2] the *Vector Median Filter* (VMF) was introduced, by generalizing the definition (1) using a suitable vector norm. Given a set W of n vectors, the vector median of the set is defined as $\mathbf{x}_{(1)} \in W$ satisfying

$$\sum_j \left\| \mathbf{x}_{(1)} - \mathbf{x}_j \right\| \leq \sum_j \left\| \mathbf{x}_i - \mathbf{x}_j \right\| , \quad \forall \; \mathbf{x}_i, \mathbf{x}_j \in W \, . \tag{2}$$

The orientation difference between vectors can also be used as their distance measure. This so-called vector angle criterion is used by the *Vector Directional Filters* (VDF), to remove vectors with atypical directions, [19]. The *Basic Vector Directional Filter* (BVDF) is a ranked-order, nonlinear filter which parallelizes the VMF operation, minimizing the sum of the angles with the other vectors. To improve the efficiency of the directional filters, another method called *Directional-Distance Filter* (DDF) was proposed. This filter retains the structure of the BVDF but utilizes the combined distance criterions to order the vectors inside the processing window, [19, 13, 9, 10].

3 Proposed Noise Detection Algorithm

The main objective of the noise reduction algorithms is to suppress noise while preserving important image features like edges, corners or texture.

Over the years various impulsive noise reduction algorithms have been proposed, [13]. The main drawback of many popular and extensively used filters is the fact that they fail to distinguish between the original uncorrupted pixels and pixels affected by the noise process, which leads to poor visual quality of the restored image.

This is also a serious drawback of the very popular vector median filter. It is quite easy to notice that the VMF offers good performance in the removal of impulsive noise, but at the same time it introduces unnecessary changes to the pixels not corrupted by the impulsive noise, which leads to image blurring, destruction of image texture and even artifacts like artificial blotches. This behavior of the VMF can be easily observed in Fig. 1 d,e, in which the the black pixels indicate those image pixels that were changed by the VMF algorithm. The test image was distorted by 5% random valued impulsive noise and the VMF replaced 80.7 % of the image pixels.

Let us now modify the concept of the *peer group* introduced in [7] and extensively used in various filtering designs, mostly under the name of extended spatial neighborhood, [8, 3, 20].

The *peer group* $\mathcal{P}(\mathbf{x}_i, m, d)$, in this paper will denote the set of m pixels belonging to the filtering window W centered at the pixel $\mathbf{x}_i$, which satisfy he following condition: $\|\mathbf{x}_i - \mathbf{x}_j\| \leq d$, $\mathbf{x}_j \in W, j = 1, \ldots, m$. In other words the peer group $\mathcal{P}$ associated with the central pixel of W is a set of pixels belonging to W whose distance to the central pixel is not exceeding d.

The proposed impulsive noise detection algorithm works as follows: if the central pixel $\mathbf{x}_i$ belongs to the peer group $\mathcal{P}(\mathbf{x}_i, m, d)$, then the pixel $\mathbf{x}_i$ is treated as not corrupted by noise, otherwise it is declared as noisy and can be filtered with any efficient noise reduction algorithm.

If $\mathbf{y}_1$ denotes the output of the filtering operation, x_1 the central pixel in W and $\mathbf{x}_{(1)}$ the output of the VMF, then we can construct the following simple filtering algorithm.

$$\mathbf{y}_1 = \begin{cases} \mathbf{x}_1, & \text{if} \quad \mathbf{x}_1 \in \mathcal{P}(\mathbf{x}_1, m, d), \\ \mathbf{x}_{(1)}, & \qquad \text{otherwise}. \end{cases} \tag{3}$$

In other words, the central pixel is retained if there are $m - 1$ neighbors in W, which are 'close' enough, where the closeness is defined by parameter d.

As the output is switched between the identity and a filtering operation, various filtering designs can be used instead of the VMF, [4, 5, 17, 1, 11]. In this paper we have chosen the VMF mainly to demonstrate the efficiency and extremely low computational effort of the proposed noise detection framework.

The low computational complexity stems from the fact that when the peer group parameter m is low, for example $m = 2$, then if the algorithm finds two pixels, which are close enough to the central pixel under consideration, then $\mathbf{x}_i$ is declared as noise-free and the sliding window moves to the adjacent pixel. Very often only a few calculations of the distances $\|\mathbf{x}_i - \mathbf{x}_j\|$, $\mathbf{x}_j \in W$ are needed to classify the pixel as undisturbed by noise. The minimal number of calculation of the distances needed to classify the pixel is thus equal to m and the maximal number of distances is $n - m - 1$, where n is the number of pixels contained in the filtering window W. The number of distances needed for the detection of impulses is extremely low when compared with the number of distances needed by the VMF algorithm which is equal to $n(n-1)/2$.

4 Simulation Results

In many practical situations, images are corrupted by noise caused either by faulty image sensors or due to transmission errors resulting from man-made phenomena such as ignition transients in the vicinity of the receivers or even natural phenomena such as lightning in the atmosphere.

The impulsive noise is often generated by bit errors, especially during the scanning or transmission over noisy information channels, [3]. In this paper the noisy signal is modelled as $\mathbf{x}_i = \{x_{i1}, x_{i2}, x_{i3}\}$, where

$$x_{ik} = \begin{cases} v_{ik} & \text{with probability } \pi\,, \\ o_{ik} & \text{with probability } 1-\pi\,, \end{cases} \tag{4}$$

and the contamination component v_{ik} is a random variable. We will assume two models, which will be called impulsive *salt & pepper* or *fixed-valued* noise, when $v_{ik} = \{0, 255\}$ and impulsive *uniform* or *random-valued* noise, when $v_{ik} \in [0, 255]$. It can be noticed that the first model is a special case of the *uniform* noise, as this noise can take on only two values 0 or 255 with the same probability, assuming 8-Bit per channel color image representation.

In both noise models the contamination of the color image components is uncorrelated, and the overall contamination rate is $p = 1 - (1-\pi)^3$.

For the measurement of the restoration quality the commonly used *Root Mean Squared Error* (RMSE) expressed through the *Peak Signal to Noise Ratio* (PSNR) was used as the RMSE is a good measure of the efficiency of impulsive noise suppression. The PSNR is defined as

$$PSNR = 20 \log_{10}\left(\frac{255}{\sqrt{MSE}}\right), \quad MSE = \frac{\sum\limits_{i=1}^{Q}\sum\limits_{k=1}^{\mu}(x_{ik} - o_{ik})^2}{Nm}, \tag{5}$$

where N is the total number of image pixels, and x_{ik}, o_{ik} denote the k-th component of the noisy image pixel channel and its original, undisturbed value at a pixel position i, respectively.

The parameters m and d provide control over the performance of the impulsive noise detection process. For its assessment a series of simulations on natural images was performed.

With regard to the parameter m of the peer group $\mathcal{P}$ the simulation results show that when the contamination intensity is low, good results are achieved for $m = 2$ in case of both the fixed valued and impulsive noise, (Figs. 2 a,d). For higher noise probability p, the images contaminated by fixed valued impulsive noise require $m = 3$, (Figs. 2 b,c). Surprisingly, good results are achieved for $m = 2$ when the images are contaminated by random valued noise, (Figs. 2 e,f). As the filtering results are not very sensitive to the choice of m we used $m = 3$ for the comparisons with the VMF.

The experiments conducted on a broad variety of natural color images have shown, [15, 16, 14] that the parameter d should be equal to about 50,

(Fig. 3) and as such a setting guarantees good performance of the proposed switching scheme independently on the image characteristics, noise model and contamination intensity.

The main advantage of the proposed noise detection technique is its enormous computational speed. The comparison with the VMF, presented in Tab. 1 shows that the new technique is for low contamination intensities 2-4 times faster than the VMF. The computational efficiency is decreasing with increasing noise intensity because with increasing p, the proposed filter converges to the pure VMF.

The good performance of the proposed technique can be also observed in Fig. 4, in which zoomed parts of the test color images were distorted by uniform impulsive noise and restored with VMF and with the new filter. As can be observed the incorporated switching scheme enables the preservation of edges and fine image details. This behavior is also confirmed in Fig. 1 f,g, which shows that the new filter rejects the impulses and replaces only a small fraction of the undisturbed pixes, (in this example the contamination intensity was $p = 5\%$ and only 6.7% of the pixels were replaced by the VMF).

5 Conclusion

In this paper a new approach to the problem of impulsive noise detection and removal in color images has been presented. The main advantage of the proposed technique is its extraordinary high computational speed, which makes it attractive for real-time applications. The noise detection scheme has been coupled in this paper with the vector median filter, however the computational speed can be further increased when employing a less computationally demanding noise removal algorithm.

Table 1. Filtering efficiency of the proposed noise removal algorithm in comparison with the VMF for *salt & pepper* **a**) and *uniform* noise for LENA with $d = 50$ and $m = 3$.

(a)

p [%]	PSNR NEW	PSNR VMF	speed gain
0	40.43	33.51	4.57
5	38.41	33.33	3.20
10	37.01	33.18	2.67
15	35.79	32.99	2.14
20	34.67	32.76	1.88
30	32.46	32.02	1.52
40	29.88	30.55	1.23
50	27.17	27.99	1.03
52	26.75	27.52	1.00

(b)

p [%]	PSNR NEW	PSNR VMF	speed gain
0	40.43	33.51	4.57
5	40.05	33.00	4.00
10	37.89	32.48	3.20
15	35.72	31.79	2.46
20	33.90	30.88	2.00
30	29.70	28.00	1.60
40	25.49	24.41	1.33
50	21.37	24.41	1.14
56	19.33	18.82	1.00

References

1. Abreu E, Lightstone M, Mitra SK, Karakawa A (1996) A new efficient approach for the removal of impulsive noise from higly corrupted images. IEEE Trans Image Processing 5:1012–1025
2. Astola J, Haavisto P, Neuvo Y (1990) Vector median filters. Proceedings of IEEE 78:678–689
3. Astola J, Kuosmanen P (1997) Fundamentals of nonlinear digital filtrering. CRC Press, Roca Baton, New York
4. Chen T, Wu HR (2001) Adaptive impulse detection using center-weighted median filters. IEEE Signal Processing Letters 8,1:1–3
5. Chen T, Ma KK, Chen LH (1999) Tri-state median filter for image denoising. IEEE Trans Image Processing 8:1834–1838
6. Dougherty E, Astola J (1999) Nonlinear filters for image processing. SPIE/IEEE Series on Imaging Science & Engineering, IEEE Press, New York
7. Kenney C, Deng Y, Manjunath BS, Hewer G (2001) Peer group image enhancement. IEEE Trans Image Processing 10,2:326–334
8. Kober V, Mozerov M, Alvarez-Borrego J (2001) Nonlinear filters with spatially-connected neighborhoods. Optical Engineering 40,6:971–983
9. Lukac R (2002) Color image filtering by vector directional order-statistics. Pattern Recognition and Image Analysis 12,3:279–285
10. Lukac R, Smolka B, Plataniotis KN, Venetsanopulos AN (2004) Selection weighted vector directional filters. Computer Vision and Image Understanding 94:140–167
11. Lukac R, Smolka B, Plataniotis KN, Venetsanopoulos AN (2003) Entropy vector median filter. Lecture Notes in Computer Science 2652:1117–1125
12. Pitas I, Tsakalides P (1991) Multivariate ordering in color image processing. IEEE Trans on Circuits and Systems for Video Technology 1,3:247–256
13. Plataniotis KN, Venetsanopoulos AN (2000) Color image processing and applications. Springer Verlag
14. Smolka B, Lukac R, Chydzinski A, Plataniotis KN, Wojciechowski K (2003) Fast adaptive similarity based impulsive noise reduction filter. Real-Time Imaging 9,4:261–276
15. Smolka B, Plataniotis KN, Chydzinski A, Szczepanski M, Venetsanopulos AN, Wojciechowski K (2002) Self-adaptive algorithm of impulsive noise reduction in color images. Pattern Recognition 35:1771–1784
16. Smolka B (2002) Adaptive modification of the vector median filter. Machine Graphics & Vision 11,2-3:327–350
17. Sun T, Nuevo Y (1994) Detail preserving median based filters in image processing. Pattern Recognition Letters 15:341–347
18. Tang K, Astola J, Neuovo Y (1995) Nonlinear multivariate image filtering techniques. IEEE Trans on Image Processing 4,6:788–797
19. Trahanias PE, Venetsanopoulos AN (1993) Vector directional filters: a new class of multichannel image processing filters. IEEE Trans on Image Processing 2,4:528–534
20. Yaroslavsky L, Eden M (1996) Fundamentals of digital optics. Birkhauser, Boston

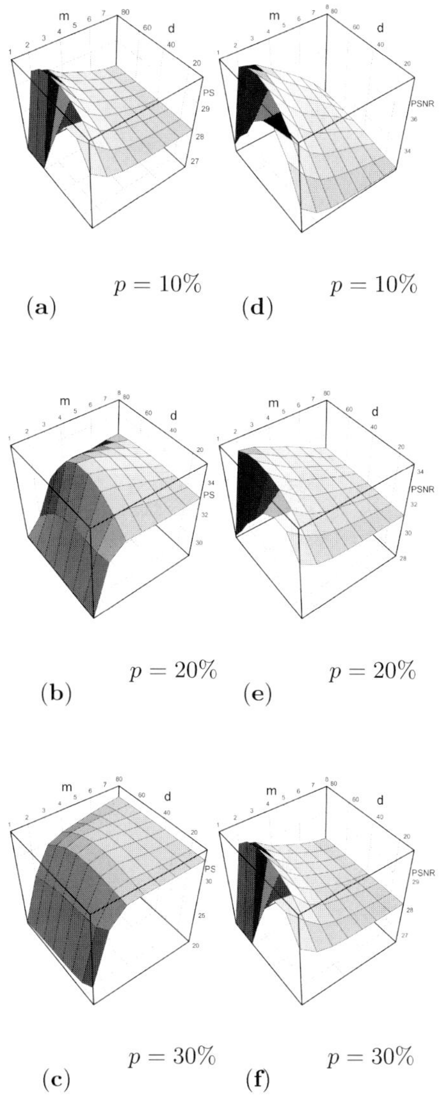

Fig. 1. Illustration of the efficiency of the new filtering design: a) image GOLDHILL, b) contaminated by $p = 5\%$, c) black dots show the pixels disturbed by noise, d) image restored with VMF, e) difference between original and VMF, f) new filter output, $m = 3, d = 50$ and below the corresponding residual image g).

Fig. 2. Dependence of the PSNR on the parameters m and d for the LENA image contaminated by *salt & pepper* (**a**-**c**) and *uniform* (**d**-**f**) impulsive noise for $p = 10\%, 20\%$ and 30%.

(a)

(b)

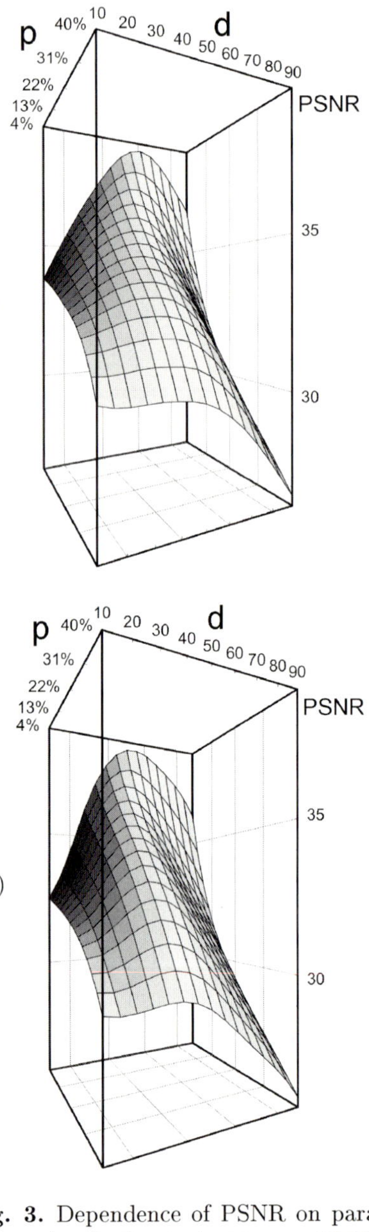

Fig. 3. Dependence of PSNR on parameters d and p for the test images LENA corrupted by *salt & pepper* (**a**) and *uniform* (**b**) impulsive noise for $m = 3$.

(a)

(b)

(c)

(d)

Fig. 4. Illustrative examples of the filtering efficiency: **a**) zoomed parts of the color test images, **b**) images contaminated by 5% *uniform* noise, **c**) restoration achieved with the VMF, **d**) filtering results achieved using the new noise detection technique.

Part IV

MEDICAL APPLICATIONS

Adaptive Wearable Vital Signs Monitor for Home Care and Sports

Piotr Augustyniak[1]

Department of Automatics, Biocybernetic Laboratory AGH University of Science and Technology Krakow, Poland august@agh.edu.pl

Summary. This paper presents new idea of remote control over a wearable general purpose health monitor based on biosignals. The autonomy of wearable devices is usually an opposite requisite to the interpretation intelligence requiring computation power. The idea of device programmability is applied to continuous optimization of resources use towards the best diagnosis quality. The proposed concept of programmable recorder assumes high flexibility of the remote device, however only certain aspects of adaptation were implemented up to today. The application of such monitors extends beyond the traditional long term ECG recording and covers the area of sports, endurance, stress, pregnancy and elderly surveillance in various combinations.

1 Introduction

The remote monitoring of physiological signals is currently one of the hottest topics [2], [8] and includes not only specialized monitors for clinical but also home care devices accessible for everyone [3], [7]. Several networks aiming at continuous monitoring of cardiac risk people are already matured in US and Europe. Those approaches assume the capturing device to interpret the electrocardiogram and to issue an alert message in case of abnormalities. Although the spread interpretation intelligence limits the communication costs, due to resources limitation typical to a wearable computer, the percentage of false alarms is rather high. An alternative approach uses triggered acquisition method typical for the ECG event recorders. However, a manually operated independent device risks to miss an electrocardiogram when the patient in pain is unable to start the capture session.

At first glance, the advantage of remotely controlled device is thus twofold:

- the signal is interpreted on-line and if necessary transmitted without delay and the reanimation may start immediately if necessary
- the acquisition is controlled by the experienced staff with support of technically unlimited knowledge base and with consideration of previous results.

Considering further advantages of remote programmability two dimensions should be point out: the levels and the aspects of adaptation. They are discussed in details throughout chapter 2. Chapter 3 presents an experimental biosignal recording device partially meeting the challenge of assumed adaptability and the preliminary result of the in-field tests. Conclusions, future plans and final remarks are included in chapter 4.

2 Adaptivity Concept

In a typical topology of surveillance network (fig. 1) remote wearable recorders are supervised and controlled by a node archiving the captured information. Assuming both device types are equipped with signal interpretation software, the analysis of other constraints leads to the following remarks:

- higher interpretation performance of the wearable device results in higher power consumption and in shorter autonomy time,
- lower interpretation performance of the wearable device augments the data stream and increases the costs of digital communication,
- the interpretation needs and priorities vary with time and patient, they depend on many factors known before the examination starts, but also on previous examination results,
- the node has not to be mobile, therefore it benefits from a world wide knowledge resources and can be supported by human experts.

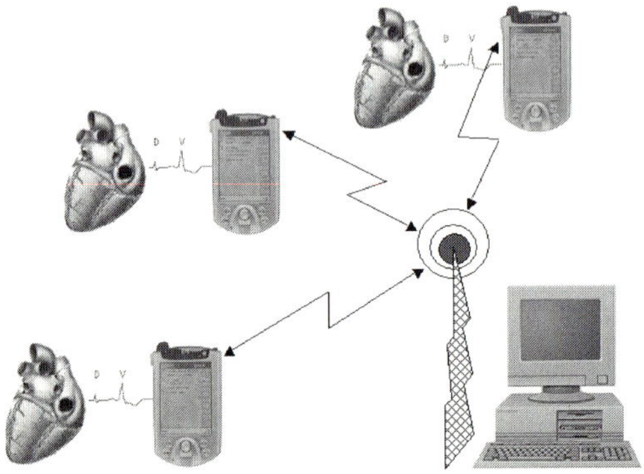

Fig. 1. Typical topology of surveillance network using wireless digital communication

With consideration of these remarks a new concept of adaptive wearable vital signs monitor was worked out in our laboratory. This concept joins the artificial intelligence approach to both device types in the network and the generalized division of tasks practiced by human medics. Main assumptions of our concept are the following:

- The interpretation is done partially in the remote device and partially by a complementary software thread running on the node host computer. The results are prioritized following the changes of diagnostic goals and current patient state.
- The actual data contents and format results from negotiations between the node and the remote monitor. The negotiations process may be driven by distributed optimizations of power consumption, transmission channel use and diagnosis quality.

High flexibility of a vital signs monitor may be achieved remotely in real time on various levels of adaptation:

- modification of the hardware structure and functionality with use of analog and digital reprogrammable circuits
- modification of the software structure and functionality by means of dynamically linked libraries,
- modification of interpretation parameters.

Combining these adaptation levels opens the possibility of deep changes of device functionality and purpose. That is called in this paper aspects of adaptation:

- acquisition and interpretation of many different vital signs (ECG, EMG, EOG, blood pressure, phonocardiography, uterine contraction and other signals from the surrounding) up to the number of channels available in the hardware and with their proper sampling characteristics.
- cooperation with a node as a 'transparent' recorder, in partial autonomy with optimized interpretation task share or as an independent remote device with signal interpretation.
- operation in a continuous surveillance mode or as an event monitor triggered manually or by a given physiological event with an optional pre-trigger,
- continuous adaptation of interpretation depth following the patient state and the diagnosis goals.

3 Experimental Device Design and Performance

The portable ECG recorder was developed in our Laboratory on demand from cardiology researchers and comply with the following criteria:

- three simultaneous channels sampled at up to 200 Hz,
- co-operation with a GSM modem for on-line wireless transmission of recorded signal or with the PDA as the source of data for interpretation,
- autonomy for at least 24 hours of operating.

The recorder was designed for medical experiments, the lack of build-in interpretive algorithm was intended making the research on the adaptive software fully independent. The recorder does not contain reprogrammable hardware, but the basic set of remote configuration commands, offers high flexibility of acquisition parameters.

The recorder was developed around the popular circuit from the MicroConverter family integrating analog to digital converters, serial communication interfaces, internal flash memory and a '51-type processing kernel running at 2,7V. The on-chip PLL-adjustable oscillator with the Fast Interrupt Response feature is very useful in a portable device where the power management is critical. The diagram of the recorder's circuitry is displayed in figure 2.

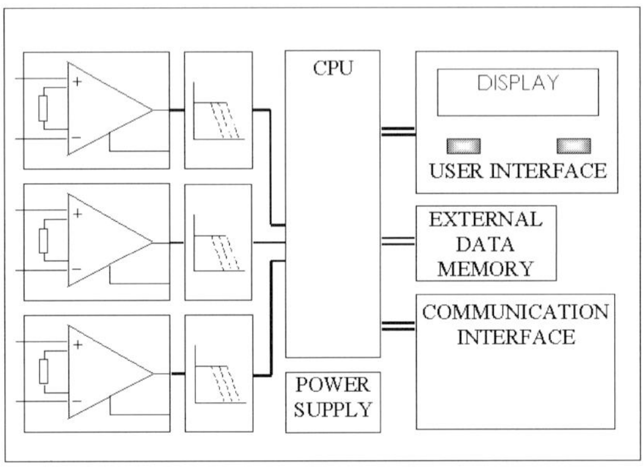

Fig. 2. The block diagram of the recorder's circuitry

The analog circuitry repeats the same architecture in each recording channel and uses micropower (230 μW) instrumental amplifiers with rail-to-rail input and output signal swing. This approach maximizes the usable dynamic range of the a/d converter even for low supply voltage. The analog inputs are FET-based and protected by 10mA ESD diodes against the overvoltages up to 20V. The effective input voltage range can be set by the software from ś2mV to ś16mV in order to cover all the area of applications. The digitizers provided in a MicroConverter [1] chip guarantee 12-bits resolution (4096 levels) and 1LSB nonlinearity over expected temperature range.

The internal non-volatile memory is used to store the recorder configuration. Therefore, the device status, the data organization and other settings are not lost on power failure.

The external memory is used for data storage in the recording mode and acts as a data buffer in the interactive mode. This function prevents from the loss of data when the transmission channel is temporarily not available. Third function of the external memory is the closed loop buffer used for defined length of pre-trigger data. The external memory capacity lasts for ca. 12

minutes of data, and is easily extensible with use of MMC or SD memory card thanks to the use of SPI interface.

The communication interface uses the bi-directional UART that could be directly connected to a PDA computer or to a mobile telephone for independent wireless connection. In this case the modem initialization sequence is stored in the configuration area of the internal flash memory. Except for data transmission, the communication channel is used for sending the textual messages to the user and the configuration data to the device.

The user interface is as simple as possible and consists of an alphanumerical LCD module intended for displaying messages and two programmable push buttons. For the applications in animals, the buttons are ignored and the LCD is remotely switched off for the sake of energy saving.

The recorder was designed using surface mount components (SMD) and four-layer printed circuit board. The top layer contains digital components and voltage controllers while the bottom layer is reserved for analog circuits. The high immunity to the electromagnetic interference was achieved thanks to the minimum length of signal wires. Although, the recorder size was taken into the consideration, further reduction by up to a half of current dimensions is possible. During all the development process the requirements of international standards for medical devices and electromagnetic compatibility [4], [5], [6] were carefully observed.

Since the signal quality is of crucial importance, extensive tests were performed in order to confirm the recorder's ability to deliver a medically meaningful signal representation. The electrical tests were made in a specialized laboratory complying with TUV/ISO measurements standards. The parameter set and testing procedure used were typical for ECG long-term recorders. For the transmission channel the electrical tests were limited to the electromagnetic compatibility (EMC) and interference immunity issues. Main results of these measurements are displayed in table 1. The following procedures necessary for independent transmission were thoroughly tested: scheduled acquisition to the memory; scheduled acquisition and transmission over a GSM telephone in various conditions; acquisition and transmission initiated remotely over a GSM telephone; displaying of a textual message sent over a GSM telephone; changing of the configuration memory contents over a GSM telephone.

Table 1. Selected results of recorders electrical tests

parameter	value	conditions
1 LSB linearity range	-1.87 $\div$ 1.83 mV	2mV range
CMRR	92 dB	DC $\div$ 100 Hz (worst case)
bandwidth	0.03 $\div$ 100 Hz	-3 dB
voltage noise (ref. to input)	8.3 μV	0.1 $\div$ 10 Hz
channel crosstalk	-77 dB	DC $\div$ 100 Hz (worst case)

The tests were performed on a single recording device connected to the Siemens S55 mobile telephone. A PC computer equipped with a GSM modem hosted the signal storage and the device control software. The test results confirm the correct support of transmission break, multiple connection retries, data stream redirection etc. The power supply monitoring enables the data-safe shutdown and wake-up with reporting to the supervising remote station. Unfortunately, sudden power failure (battery disconnection) is too fast to be serviced correctly.

Separated challenge is the recorder test in a configuration with the interpretation unit based on a PDA computer (HP). This choice was justified by easy software development and interfacing with standard peripherals: wireless transceiver (GSM), signal acquisition module and extensible memory buffer. In this configuration the adaptability includes adjustment of processing parameters, on-line modification of communication protocol and processing routines. The PDA uses Pocket Windows operating system that is compatible with Microsoft Windows platform for desktop PCs. The software architecture consists of a process management and communication control kernel and of a set of basic interpretation routines linked upon request. Each routine is implemented as a dynamic function library and can be adjusted remotely with a vector of interpretation parameters or replaced by an alternative routine from the basic set or by the code provided by the supervising node (fig. 3).

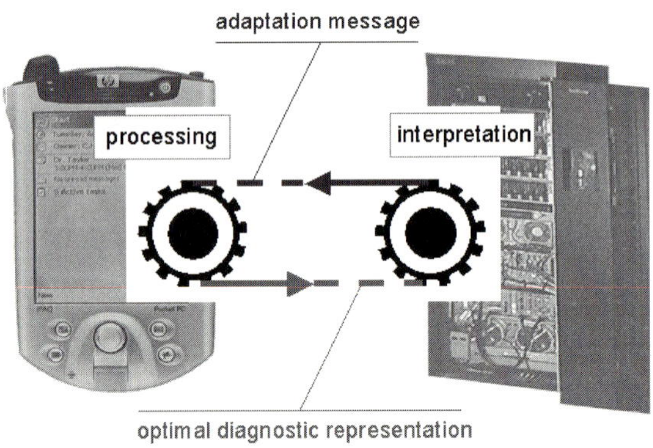

Fig. 3. Cooperation of the remote monitor and the supervising node aiming at optimizations of power consumption, transmission channel use and diagnosis quality

The consequence of interpretation programmability is the multitude of output signal formats ranging from raw electrocardiogram to the sparse data (e.g. heart rate). The modifiable transmission protocol is very useful for optimization of wireless channel use aiming at keeping the monitoring costs at the acceptable level. The general rule assumes the transmission of basic interpretation results for all the monitoring time and more detailed reports for short time intervals. Every occurrence or suspicion of any event results in a more detailed report including up to the corresponding strip of raw signal. This approach was proposed as a result of cardiologist's behavior analysis, but it can be remotely programmed upon request.

At this stage the compatibility of the interpretive software on the remote device and on the supervising node was problematic. For the reported tests the node is running the same procedures as the remote monitor. Consequently, because of no influence of task sharing to the diagnosis quality, the transmission channel use factor dominated over the battery consumption factor and the task sharing found as optimal favored interpretation done by the PDA.

4 Discussion

Except for satisfying the requirements of technical specifications for electrocardiographs, the recorder was evaluated in several applications including daily activity and intensive training of sportsmen. The medical research already completed with use of the recorder include:

- muscle fatigue assessment during training of downhill ski competitors,
- uterine contraction detection based on abdominal potentials in patients at risk of premature delivery,
- the investigation of influence from environmental stress to the physiology of domestic animals.

The concept of wireless physiological monitors proposed in this paper may be extended to open networks providing various medical services and having a considerable impact to the health care in the future. Its principal advantage is the flexibility of automated interpretation very close to the human medics. I is manifested by:

- adaptive patient description level varying from a general to a detailed report dependent on the result severity,
- adjustability of monitoring and auto-alerting parameters accordingly to the patient-specific signal; during the initial recording phase and anytime thereafter the device may be remotely taught what is correct and what is wrong in the acquired signal,
- possibility of following of any unexpected event and if the remote interpretation is not flexible enough, the uncommon signal is interpreted in the network node with intervention of human supervisors.

The audiovisual communication with the patient or his attendee provides an interactive channel for instructions necessary in case of technical troubles (e.g. electrode replacement), medical risk (e.g. physical overload), medication intake or modification of wiring up to the monitor's function.

Although main scientific goal was achieved, several problems emerged during the design and testing of the vital signs recorder. A majority of them could be resolved by increasing the processing power. Next version of acquisition part should be fit into a standard PCMCIA case after elimination of LCD and charge-pump voltage converter. The user interface provided by the PDA must be programmed to support easy and unambiguous operating by the patient in any condition. The built-in speaker may also be used to generate voice message guiding the operation of the monitor. Another future considerations include:

- deep and medically justified investigation of the interpretation process and reporting format changes implied by the dependency of previous interpretation results.
- development of multi-threaded software for the cardiology centre in order to perform independent supervising of several remote monitors and for the management of patient's data archive.

5 Acknowledgment

This work was supported by Polish State Committee for Scientific Research grant no: 3T11E 001 27

References

1. ANALOG DEVICES http://www.analog.com/ microconverters
2. Chiarugi F. et al. (2002). Real-time Cardiac Monitoring over a Regional Health Network: Preliminary Results from Initial Field Testing Computers in Cardiology 29, 347-350.
3. Gouaux F., et al. (2002). Ambient Intelligence and Pervasive Systems for the Monitoring of Citizens at Cardiac Risk: New Solutions from the EPI-MEDICS Project Computers in Cardiology 29, 289-292.
4. IEC 60601-2-25 (ed. 1999). Medical electrical equipment: Particular requirements for the safety of electrocardiographs.
5. IEC 60601-2-27 (ed. 1994). Medical electrical equipment: Particular requirements for the safety electrocardiographic monitoring equipments.
6. IEC 60601-2-47 (ed. 2001). Medical electrical equipment: Particular requirements for the safety, including essential performance, of ambulatory electrocardiographic systems.
7. Maglaveras N., et al. (2002). Using Contact Centers in Telemanagement and Home Care of Congestive Heart Failure Patients: The CHS Experience Computers in Cardiology 29, 281-284.
8. Nelwan S.P., van Dam T.B., Klootwijk P. and Meil SH. (2002). Ubiquitous Mobile Access to Real-time Patient Monitoring Data Computers in Cardiology 29, pp. 557-560.

Fractal Based Approaches to Morphological Analysis of Fundus Eye Images

Maria Berndt-Schreiber

Faculty of Mathematics and Computer Science,
Nicolaus Copernicus University,
87-100 Torun, ul. Chopina 12/18, Poland
berndt@mat.uni.torun.pl

Summary. Methodological aspects of quantitative morphological analysis of digital fundus eye images are described, both in the spatial and frequency domains. They refer to the notion of fractal dimension as a measure of image texture complexity thus providing an additional descriptor of the image, which may be useful for medical data classifications.

1 Introduction

It is commonly accepted that many irregular complex biological systems - resulting both from random and non-linear processes - may be described as fractals [1-2]. Fractal geometry has been frequently applied as a tool in quantifying the structures of naturally occurring objects [3-7], recently also in advanced nuclear medicine imaging of different types [8-15].

In particular, the textures of digital fundus eye images have been proved to represent fractal patterns [16-20]. Fundus images are collected in routine ophthalmologic non-invasive examinations, usually with only the pupils dilated in order to reveal the entire optic disc areas and the surfaces of retinal vessel patterns. Below in Figure 1 sample fundus images are presented for a normal case (a) and pathological ones (b) - (e), respectively.

Quantitative bias-free analysis of digital fundus eye images is a crucial problem nowadays since it may be relevant to various pathologies and appears fundamental in early diagnosis. Accurate description of the fundus eye images becomes essential not only for ophthalmologists - the observations and assessments of these images for detection and monitoring morphological changes in retinal microvascular patterns are also of importance for nephrology and neurology experts currently [21-22].

Regardless of recent technological advances proper diagnosing procedures for fundus images remain extremely complicated and very often uncertain.

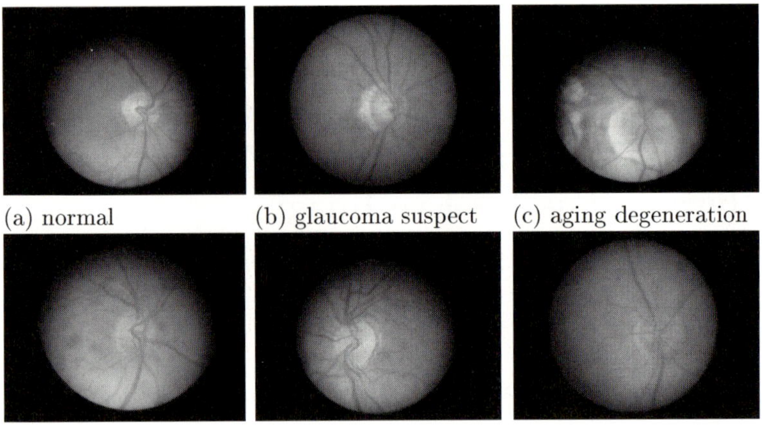

(a) normal (b) glaucoma suspect (c) aging degeneration

(d) arteriosclerosis (e) diabetic retinopathy (f) hypertension effects

Fig. 1. Sample fundus images: normal (a) and pathological cases (b)-(e) [39].

Therefore methodological investigations are carried on regularly in many clinical centers all over the world, also due to their potential significance in telemedical applications [23-34].

Two complementary approaches to morphological fractal analysis of digital fundus images are described below. They refer to the analysis in both the spatial and frequency domains, respectively. The methods, providing additional descriptors for the images in the form of fractal dimensions being the measure of the presented structure complexity, have already been applied for medical data classification and some obtained preliminary results seem encouraging [35-38]. Since both the approaches have not been described extensively before, here a special attention is paid to the methodological aspects of the analysis and the perspectives for future improvement when facing more sophisticated digital technologies, on the one hand, and availability of extended medical database systems on the other.

2 Spatial-domain Analysis

2.1 Method Outline

The assumptions is made that the original digital fundus image may be treated as a three dimensional 3D surface: $z = f(x, y)$, with x, y referring to pixel positions coordinates and z representing the gray level values. It is illustrated in Figure 2 for two sample fundus images.

Eventually, the observed texture of the fundus image is considered as the spatial 3D representation of the gray level values and a quantitative measure of the roughness of such a surface, representing vessel patterns and their surrounding, should be found. The estimation of fractal dimension - as a measure

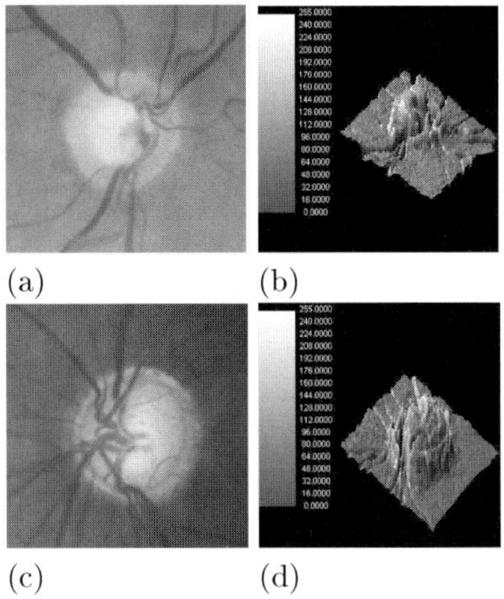

Fig. 2. Original fundus images (a),(c); and their 3D spatial gray levels representations (b),(d).

of the surface complexity - is performed with referring to the definition due to Minkowski and Bouligand (MB) and implemented for the discrete surface in a specific variation scheme (comp. [1] and [35] and references therein), as described shortly below.

Let E be a fractal set embedded in a d-dimensional space $\mathbf{R}^d$ and $E(\epsilon)$ a set of points in $\mathbf{R}^d$ at a distance less than ϵ from E. Then $E(\epsilon)$ is defined as the union $E(\epsilon) = \bigcup_{x \in E} \mathbf{B}_\epsilon(x)$, where $\mathbf{B}_\epsilon(x)$ is a ball in a d-dimensional space centered at x and of radius ϵ (it is often called a Minkowski sausage, a thickening or dilation of E). By definition the Minkowski-Bouligand (MB) dimension is expressed as:

$$\Delta(E) = \lim_{\epsilon \to 0} \frac{d - \log \mathrm{Vol}_d[E(\epsilon)]}{\log \epsilon}, \tag{1}$$

where Vol_d is the volume in d dimensions. To approximate the MB fractal dimension value according to (1) for surfaces $z = f(x, y)$ in discrete domains one assumes the following: For a surface $z = f(x, y)$ the so called ϵ-oscillation at a given point (x', y') is defined as the difference between extreme values admitted by function f in the ϵ-neighbourhood of the point (x', y'):

$$V_f(x', y', \epsilon) = \max_{\mathbf{B}_\epsilon(x', y')} f(x, y) - \min_{\mathbf{B}_\epsilon(x', y')} f(x, y). \tag{2}$$

The average of $V_f(x', y', \epsilon)$ taken over all (x', y') yields the so called ϵ-variation of f denoted by $V_f(\epsilon)$. Eventually, the MB fractal dimension in the three dimensional space is estimated as:

$$\Delta(G) = \lim_{\epsilon \to 0} \frac{3 - \log V_f(\epsilon)}{\log \epsilon}, \tag{3}$$

with G denoting the considered surface.

2.2 Algorithm Description

The function f is given as digitized data representing images. The data are grouped into R^2 bins, where R is an integer. The value of ϵ_n is chosen as $\epsilon_n = k_n/R$, where $\{k_n\}$ is a sequence of increasing integers satisfying the condition: $k_1 = 1, k_n \leq 2k_{n-1}$. For all x, y inside an elementary box $[i/R, (i+1)/R] \times [j/R, (j+1)/R]$, where $i, j \in \{0, 1, ...R\}$, the ϵ-oscillation for $\epsilon = k_n/R$ is estimated:

$$V_f(x', y', \epsilon) \approx \max f(i,j) - \min f(i,j), \tag{4}$$

where $\max f(i,j)$ and $\min f(i,j)$, respectively denote the maximum and minimum values of the function f admitted on the considered elementary box at (i,j). Finally, the so called ϵ-variation of the function f is calculated as:

$$V_f(\epsilon) \approx \frac{1}{(R+1)^2} \sum_{i=0}^{R} \sum_{j=0}^{R} (\max f(i,j) - \min f(i,j)), \tag{5}$$

and the corresponding log-log plot can be used to evaluate $\Delta(G)$ as in equation (3). It is worth to notice here, that the implemented procedure for $\Delta(G)$ evaluation assumes scanning the entire discrete surface of the image to check the changes of $f(x,y)$ values, i.e. gray scale levels of the image, in all directions - thus providing a detailed quantitative analysis of the image texture. Prior to extended medical applications[37] the algorithm, described above, was tested on mathematically defined fractal surfaces, representing superposition of pyramids of various frequencies attenuated by a scaling factor. It has been shown that the various surfaces may be reasonably classified by MB fractal dimension values [35].

3 Frequency-domain Analysis

3.1 Method Outline

Fundus eye images transformed to frequency domain may reveal specific individual features and therefore their quantitative description - as a complementary one - may be essential for diagnoses. Two samples, presenting original fundus images with the optic disc areas, their Fast Fourier Transform power spectra in gray scale and also in the 3D representations and appropriate cross-sections, are shown in Figures 3 and 4, below, for glaucoma and normal cases, respectively. One can easily observe differences appearing in the two sets of images.

In the frequency domain we are usually interested in the amplitude at each frequency - which is precisely the information given by the power spectrum of the image. The central idea of the approach presented here is to view an image in the frequency domain as a collection of iso-intensity contours, as it is schematically illustrated in Figure 5, below, for a given gray level set

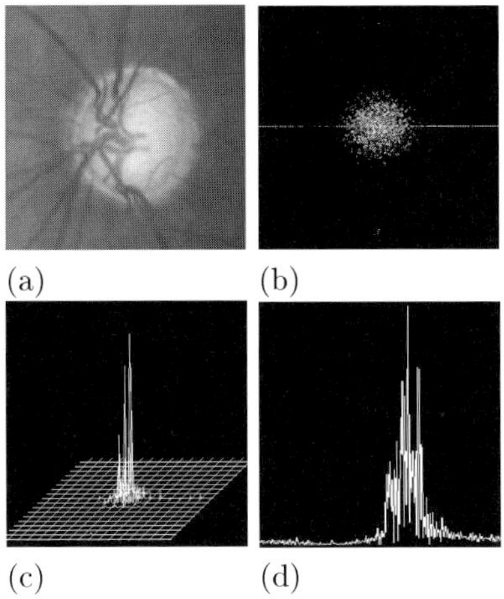

Fig. 3. Original fundus image for glaucoma case: (a), its power spectrum projection: (b), power spectrum of (b) in 3D representation:(c), and its cross-section: (d).

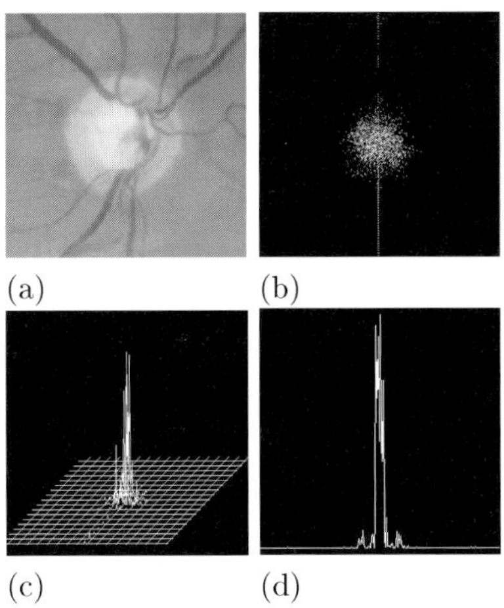

Fig. 4. Original fundus image for normal case: (a), its power spectrum projection: (b), power spectrum of (b) in 3D representation: (c), and its cross-section: (d).

cut-off. Such a model approach allows to estimate the fractal dimension value from the relationship between the total number of pixels and the cut-off levels transformed appropriately into logarithms as described below.

3.2 Algorithm Description

The assumption is made that the power spectrum F may be described as:

$$F(x, y) = Cr^{-\beta}, \tag{6}$$

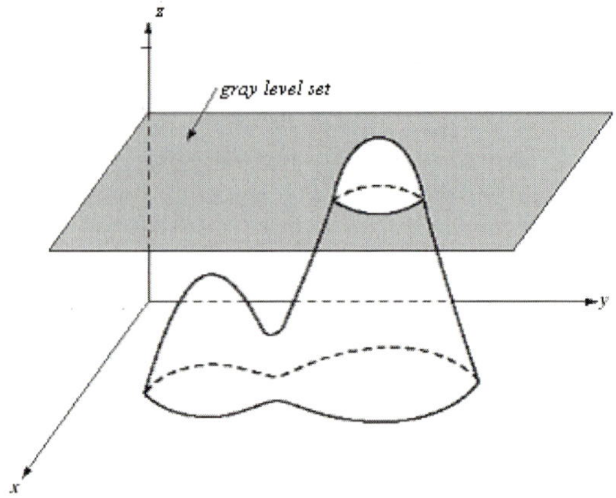

Fig. 5. A model of the image power spectrum in frequency domain in 3D gray scale.

where $r^2 = x^2 + y^2$. The relation is treated in a flexible way, assuming in practice that the graph of the power spectrum may be represented as well by a cone surface with a basis being not of a regular circular shape, and with the center being possibly shifted from the zero position. Notice, that at a given height h the level set of points (x, y) for which $\{F(x, y) \geq h\}$ will be a disc of radius r satisfying the relations below:

$$Cr^{-\beta} = h \Rightarrow r = C^{1/\beta} h^{-1/\beta}. \tag{7}$$

Eventually, the number of pixels in the level set of height h will be of the order of the area S of the disk of radius r, i.e. $\pi C^{2/\beta} h^{-2/\beta}$. For a more general case (for different shapes of the cone bases, and different values of the constant C) it may be stated that the number of pixels at the h level set is given by the following formula:

$$S(h) = C' h^{-2/\beta}, \tag{8}$$

where C' is unknown. Taking logarithms of both sides of the equations we get a linear relation:

$$\log S(h) = \log C' - [2/\beta] \log h. \tag{9}$$

The term $\log C'$ can be ignored as providing no information and the calculation of the coefficient $a = -2/\beta$ is performed using the least squares method to find beta. Finally, the approach refers to the notion of spectral fractal dimension Ds as related to a quantitative description of power spectrum and estimated as follows [36]:

$$\mathrm{Ds} = (7 - \beta)/2. \tag{10}$$

Detailed numerical analysis of the method, as well as its preliminary application to ophthalmology data classification has been recently described in [38].

4 Conclusions and Further Work

The fractal type approaches to the morphological analysis of fundus images, presented above, are more refined that those applied earlier based on the simple box-counting method [18-20]. Estimated fractal dimensions both in the spatial and frequency domains may be treated as additional descriptors of the images useful for medical data classification. The spatial method provides global information about the image texture; while the discrete image Fourier transform may be used for texture description for selected regions of interest (e.g. optic nerve disc or macula area for a fundus image). A joint spatial/frequency approach is generally recommended for the analysis of complex texture structure [40].

From the practical point of view the proposed algorithms seem reasonable also due to their rather simple implementation in the routine computer processing. They have already been applied to the analysis of gray scaled pictures using the database system of fundus images [39]. With the advanced digital imaging technologies available the fractal analysis of fundus images performed separately in the three color channels seems a special challenge currently. There are some observations confirming a possibility of extracting different features of the images, both in the spatial and frequency domains, for separated color channels. Preliminary works are in progress in the area.

Direct implementations of the proposed spatial/frequency algorithms to various radiological digital imaging procedures might be also an interesting task for the future.

References

1. Losa G A, Merlini D, Nonnemacher T F, Weibel T R (eds) (1998) Fractals in Biology and Medicine, Birkhäuser Verlag.
2. Klonowski W (2000) Machine Graphics & Vision 9: 403-432
3. Tourassi G, Frederick E D, Vittitoe N F, Coleman R E, (2000) Computers and Biomedical Res 33: 161-171
4. Morfill G, Bunk W (2001) Europhysics News 32: 123-136
5. Ahammer H, DeVaneyT T J, Tritthart H A (2001) Eur Biophys J 30: 494-499
6. Nagao M, Murase K, Yasuhara Y, Ikezoe J (1998) Am J Roentgenol 171: 1657-1663
7. Chung H-W, Huang Y-H (2000) Am J Roentgenol 174:1055-1059
8. Chung H-W, Nagao M (2001) J Nucl Med 42:177-178
9. Nagao M, Murase K, Kikuchi T, Ikeda M, Nebu A, Fukuhara R, Sugawara Y, Miki H, Ikezoe J (2001) J Nucl Med 42:1446-1450
10. Kuikka J T, Nagao M (2002) J Nucl Med 43:1727-1728
11. Chung H-W, Nagao M (2003) J Nucl Med 44:316-317
12. Chung H-W (2003) J Nucl Med 44:1874-1880
13. Yosikawa T, Murase K, Oku N, Imaizumi M, Takasawa M, Rishu P, Kimura Y, Ikejiri Y, Kitagawa K, Hori M, Hatazawa J (2003) Am J Neuroradiol 24:1341-1347

14. Lee J S , Lee D S, Park K S, Chung J-K, Lee M C (2004) J Neuroimaging 14:350-356
15. Thie J A (2004) J Nucl Med 45: 724-730
16. Family F, Masters B R , Platt D E (1989) Physica D 38: 98-102
17. Kyriacos S, Nekka F, Vicco P, Cartilier L (1997) The Retinal Vasculature: Towards an Understanding of the Formation Process [in] Fractals in Engineering, (eds.) Levy Vehel J E, Lutton E, Tricot C, Springer
18. Landini G, Mission G P, Murray P I (1993) Curr Eye Res 12: 23-27
19. Daxter A(1993) Grafe's Arch Clin Ophthalmol, 231: 681-68
20. Daxter A (1993) Curr Eye Res 12: 1103-1109
21. Hubbard L D, Brothers R J, King W N, Clegg L X, Klein R, Cooper L S, Sharrett A R, Davis M D, Clai J (1999) Ophthalmology 6: 59-80
22. Wong T Y, Klein R, Klein B E, Tielsch J M, Hubbard L, Nieto F J (2001) Survey of Ophthalmology 46: 59-80
23. Porta M, Bandello F (2002) Diabetologia, 45:.1617-1634
24. Lamminen H, Ruohonen K (2002) J. Telemed Telecare 8:.255-258
25. Lee S C, Lee E T, Kingsley R M, Wang Y, Russell D, Klein R, Warn A (2001) Arch. Ophthalmol 119: 509-515
26. Sinathanayothin C, Boyce J F, Williamson T H, Cook H L., Mensah E, Lal S, Usher D (2002) Diabetic Medicine 19: 105-112
27. Teng T, Lefley M, Claremont D (2002) Med. .Biol Eng Comput 40: 2-13
28. Yogesan K, Constable I J, Barry C J, Eikelboom R H, McAllister I L, Tay-Kearney M L (2000) Telemed J 6: 219-223,
29. Liesenfeld B, Kohner E, Piehlmeier W, Kluthe S, Aldington S, Porta M, Bek T, Obermaeir M , Mayer H, Mann G, Holle R , Hepp K D (2000) Diabets 23: 345-348
30. Harper R, Reeves B, Smith G (2000) Ophthal Physiol Opt 20:.265-273
31. Gilchrist J (2000) Ophthal Physiol Opt 20: 452-463
32. Maberley D, Cruess A F, Barile G, Slakter J (2002) Ophthalmic Epidemiol 9: 169-178
33. Tennant M T , Greeve M D, Rudnisky C J, Hillson T R, Hinz B J (2001) Can J Ophthlamol 36:187-196
34. Ege B M, Hejlesen O K, Larsen OV, Moller K, Jennings B, Kerr D, Cavan D A (2000) Computer Methods and Programs in Medicine 62: 165-175
35. Berndt-Schreiber M, Bieganowski L (2000) Machine Graphics & Vision 9: 433-438
36. Berndt-Schreiber M, Mikolajczak I (2001) Fourier Based Analysis of Fundus Eye Images, In: Computer Recognition Systems, KOSYR 2001 (Kurzynski M., Puchala E., Wozniak M., eds.) Division of Systems and Computer Networks, Wroclaw University of Technology, Wroclaw
37. Berndt-Schreiber M, Bieganowski L, Kowalczyk A, Maciejewski K (2002) Polish J Med Phys & Eng 8: 89-98
38. Berndt-Schreiber M, Baczkowska A (2004) J Med Informatics & Technology 7:15-22
39. Arlukowicz M, Berndt-Schreiber M, Bieganowski L, Brozek M, Jazowiecka A, Kazmierska H, Kowalczyk A, Mutrynowska J (2004) Acta Medica (accepted for publication)
40. Gonzales RC, Woods R E, Eddins S I (2004) Digital Image Processing Using Matlab, Prentice Hall Inc.

Recognition of Subtle Microcalcifications in High-Resolution Mammograms

Krzysztof Boryczko[1] and Marcin Kurdziel[1]

AGH University of Science and Technology, Institute of Computer Science,
al. Mickiewicza 30, 30-059 Kraków, Poland.
boryczko@agh.edu.pl, kurdziel@icsr.agh.edu.pl

Summary. In this article we present a new method for recognition of subtle microcalcifications in high-resolution digital mammograms. The identification of suspicious regions in mammogram images is carried out using a method based on the 2D discrete wavelet transform. For classification of regions we use the SVC algorithm. Initially, a number of statistical features of mammogram regions was employed to achieve high classification accuracy. Using a novel clustering technique we performed feature selection and identified 18 highly descriptive features. Our method can achieve sensitivity as high as 0.97 while maintaining specificity above 0.90.

1 Introduction

Breast cancer is the most frequently occurring tumor among women. According to [1] in 2004 around 40,100 women died from the breast cancer. Furthermore, over 215,000 new cases of the invasive breast cancer have been diagnosed. For the localized breast cancer the 5-year survival rate is close to 97%. However, it drops to 79% in case of the regionally advanced disease and 23% for the metastatic cancer. This shows that the effectiveness of the therapy relies on the early detection of the cancer.

In this article we propose a method for recognition of subtle microcalcifications in high-resolution digital mammograms. We focus mainly on the classification of occlusions that frequently occur in those images. The goal of our work is to design a classification scheme that is highly sensitive to microcalcifications while yielding a low ratio of false-positive detections. We decided to use Support Vector Machine [3] algorithm as the underlying classifier. For the identification of regions in mammogram image that contain suspicious occlusions we developed a simple, wavelet-based algorithm.

2 Identification of regions of interest

In order to identify suspicious regions in a mammogram image, further called regions of interest (ROIs), we employ an algorithm based on the 2D discrete wavelet transform. The pseudo-code of the algorithm is provided in the Algorithm. 1.

Algorithm 1 Wavelet-based algorithm for detection of Regions of Interest in mammogram image

INPUT: mammogram image I of size $n \times m$. Black color is coded by 0 whereas white by 1
OUTPUT: a set of ROIs with size 31×31 pixels each
1. Perform 4-level 2D discrete wavelet transform of I using spline bi-orthogonal wavelets of order 2
2. I_r = perform image reconstruction with approximate coefficients set to 0
3. I_f = filter I_r with a 20×20 averaging filter;
4. $I_d = I_r - I_f$; Rescale I_d to $\langle 0, 1 \rangle$ interval
5. Compute histogram of I_d; Store bin values in H_v and bin positions in H_p
6. H_i = find positions of bins whose value is below $\frac{1}{7}$ of maximal histogram value
7. $t = \frac{\sum_i H_p(i) \cdot H_v(i)}{\sum_i H_v(i)} + [H_p(\max_k H_i(k)) - H_p(\min_k H_i(k))]$
8. M = treshold I_d with t and find connected components in the resultant image
9. Remove from M components whose area is above 200 pixels or below 6 pixels
10. $\forall c \in M$: find the center of gravity of c and using it select a ROI from I

Microcalcifications appear on a mammogram image as small, spot-like protrusions. Consequently we assume that a wavelet decomposition will preserve the microcalcifications within the detailed coefficients. This assumption lies at the basis of the proposed method. First, the algorithm performs a 4-level wavelet decomposition of the mammogram image and reconstructs it using only the detailed coefficients. Next, in steps 3 and 4, a *difference image* I_d is constructed that emphasizes abrupt changes in the pixel intensities. In steps 5-7 the histogram of difference image is employed to estimate the threshold value t. Afterwards, in step 8, the image I_r is tresholded to segmentate the microcalcifications. In Fig. 1 we presented an example result of this segmentation. Finally in steps 8-10 the algorithm constructs the ROIs.

The proposed algorithm is highly sensitive to abrupt changes in pixel intensities. After visual inspection of results produced by the method on several mammograms, we concluded that it detects most of the microcalcifications and a number of other occlusions presented in the images. As our focus is on the classification of mammogram ROIs the detailed evaluation of sensitivity of the algorithm is beyond the scope of this study.

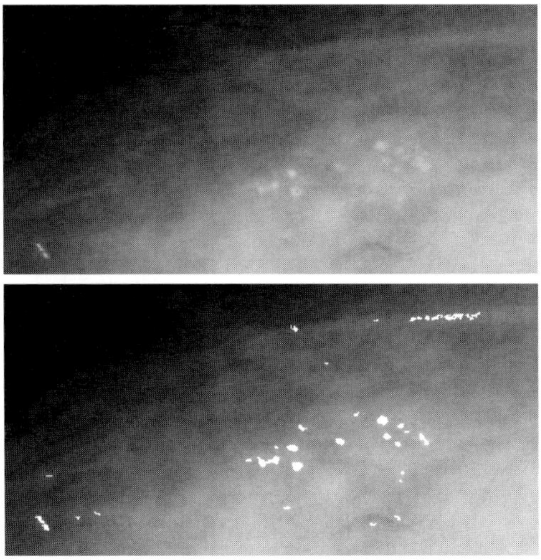

Fig. 1. Result of microcalcifications segmentation using the Algorithm 1. A mammogram region containing microcalcifications is depicted above. Below, the same region is presented on which we overlayed the result of segmentation.

3 Feature extraction from Regions of Interest

In order to recognize ROIs that contain microcalcifications we employed a set of statistical features. The detailed listing is provided in table 1. We use the following notation:

- R – mammogram ROI of size $N \times N$,
- H – histogram of the ROI,
- μ_h, σ_h^2 – mean value and variance of the histogram,
- C – second order histogram of ROI (co-occurrence matrix). The size of matrix C is $G \times G$.
- μ_{Cx}, μ_{Cy} – mean values of marginal distributions of histogram C.
- σ_{Cx}, σ_{Cy} – standard deviations of marginal distributions of histogram C.

4 Experimental results

In this section we evaluate the accuracy of recognition of microcalcifications with the Support Vector Classifier. Recently a number of successful applications of this method was reported. Moreover, learning of the SVC is carried out as a convex optimization problem and thus, cannot stick in local minimum [2]. Furthermore, heuristical methods have been developed that can solve this optimization problem efficiently [7]. A detailed description of the SVC algorithm can be found in [5].

Mean pixels intensity	$\mu = \frac{1}{N^2}\sum_{i,j=1}^{N} R_{ij}$
Variance of pixels intensities	$\sigma^2 = \frac{1}{N^2}\sum_{i,j=1}^{N}(R_{ij}-\mu)^2$
Range of pixels intensities	$r = \max_{i,j=1}^{N} R_{ij} - \min_{i,j=1}^{N} R_{ij}$
Kurtosis of pixels intensities	$\mu_4 = \frac{1}{N^2\sigma^4}\sum_{i,j=1}^{N}(R_{ij}-\mu)^4$
Skewness of pixels intensities	$\mu_3 = \frac{1}{N^2\sigma^3}\sum_{i,j=1}^{N}(R_{ij}-\mu)^3$
Fifth and sixth central moments of pixels intensities	$\mu_5 = \frac{1}{N^2\sigma^5}\sum_{i,j=1}^{N}(R_{ij}-\mu)^5$ $\mu_6 = \frac{1}{N^2\sigma^6}\sum_{i,j=1}^{N}(R_{ij}-\mu)^6$
Mean absolute deviation of pixels intensities	$m = \frac{1}{N^2}\sum_{i,j=1}^{N}\lvert R_{ij}-\mu\rvert$
Percentiles of pixels intensities	p_l = value greater than l% of pixel intensities and smaller than (100-l)% of pixel intensities $l = 10\%,\ 50\%,\ 90\%,\ 95\%$
Interquartile range of pixels intensities	$iq = p_{75} - p_{25}$
Skewness of a histogram	$\mu_{h3} = \frac{1}{N\sigma_h^3}\sum_{i=1}^{N}(H_i-\mu_h)^3$
Kurtosis of a histogram	$\mu_{h4} = \frac{1}{N\sigma_h^4}\sum_{i=1}^{N}(H_i-\mu_h)^4$
Entropy of pixels intensities	$S = -\sum_{i=1}^{N} H_i \cdot \log_2 H_i$
Energy of a histogram	$E = \sum_{i=1}^{N} H_i^2$
Entropy of a co-occurrence matrix	$S_C = -\sum_{i,j=1}^{G} C_{ij} \cdot \log_2 C_{ij}$
Energy of a co-occurrence matrix	$E_C = \sum_{i,j=1}^{G} C_{ij}^2$
Autocorrelation a co-occurrence matrix	$\rho = \sum_{i,j=1}^{G} \frac{(i-\mu_{Cx})(j-\mu_{Cy})C_{ij}}{\sigma_{Cx}\sigma_{Cy}}$
Contrast	$v = \sum i,j = 1^G (i-j)^2 C_{ij}$
Inverse difference of a co-occurrence matrix	$I = \sum_{i,j=1}^{G} \frac{C_{ij}}{1+(i-j)^2}$
Maximal value of a co-occurrence matrix	$M_c = \max_{i,j=1}^{G} C_{ij}$

Table 1. Statistical features for mammogram regions of interest.

4.1 Dataset

The evaluation was carried out on 200 high-resolution digital mammograms from the DDSM database [6]. Using the algorithm described in section 2 we extracted from these images 93277 ROIs (31×31 pixels each). Next, we selected a random subset of 1500 ROIs and visually classified them according to the presence of microcalcifications. 312 ROIs contained microcalcifications whereas remaining 1188 depicted other types of occlusions. These sets were split into *training dataset* and *validation dataset*, each containing 156 ROIs with microcalcifications and 594 ROIs with other occlusions. Remaining 91777 ROIs were assigned to *bulk dataset* and used to perform feature selection.

4.2 Feature selection

For each of the ROIs the set of numerical features from table 1 was computed. Furthermore these features were computed on the central, rectangular part of ROI with size 11 × 11 pixels. Consequently, 46 numerical features were assigned to each region. Each feature was normalized to the zero mean and unit variance.

The initial set of features was composed of commonly used statistical textures characteristics. However, no additional attention was put to select features appropriate for classification of mammogram ROIs. Consequently, some of the features are useless for this task. This increase the computational cost of the system and might decrease tits accuracy. Therefore, a feature selection need to be performed.

To perform a coarse evaluation of features, we carried out K-Means clustering of the *bulk dataset*. The result is presented in the Fig. 2a. The projection into 2D space was done using the Principal Component Analysis (PCA) technique. Next, for each feature we computed 6 histograms - one inside each of the discovered clusters. Visual investigation of these histograms allowed for identification of 19 features whose values do not differ among clusters. Example histograms of such, useless, feature are presented in Fig 3b. Fig. 3a presents histograms of a feature that shows significant variability between the clusters and therefore is useful for classification purposes.

To further investigate the remaining 27 features, we performed clustering of the *bulk dataset* with a novel clustering method proposed in [4] i.e. the SNN algorithm. This algorithm not only discovers clusters, but can also remove noise points from dataset. We used it to identify compact clusters, that should corresponds to different appearances of microcalcifications or other occlusions. The algorithm discovered 5 clusters presented in Fig. 2b. The projection into 3D space was done using the PCA technique. For each of the clusters we computed the center of gravity C_i and afterwards the following score value:

$$S_n = \sum_{i,j=1...5} \sqrt{\sum_{k=0}^{27} [C_i(k) - C_j(k)]^2} - \sum_{i,j=1...5} \sqrt{\sum_{k=0}^{27} \left[C_i^{(n)}(k) - C_j^{(n)}(k)\right]^2}$$

Here, $C_i^{(n)}$ denotes the center of gravity of the i-th cluster when the n-th coordinate is set to 0. As it can be seen, the score value S_n reflects the contribution of the n-th feature to the sum of distances between the centers of clusters. Therefore, it can be used as a measure of usefulness of the feature in classification of ROIs. Experimental results revealed, that the 18 features with the biggest score values provides classification accuracy close to the one given by all 27 features. These 18 features are:

- *from ROIs of size* 11 × 11 *pixels*: 95-th percentile of pixels intensities, entropy of pixels intensities, energy of a histogram, entropy of a co-occurrence matrix, inverse difference of a co-occurrence matrix.

- *from ROIs of size* 31×31 *pixels*: mean pixels intensity, skewness of pixels intensities, 10th, 50th, 90th and 95th-percentile of pixels intensities, skewness of a histogram, kurtosis of a histogram, entropy of pixels intensities, energy of a histogram, entropy of a co-occurrence matrix, contrast, inverse difference of a co-occurrence matrix.

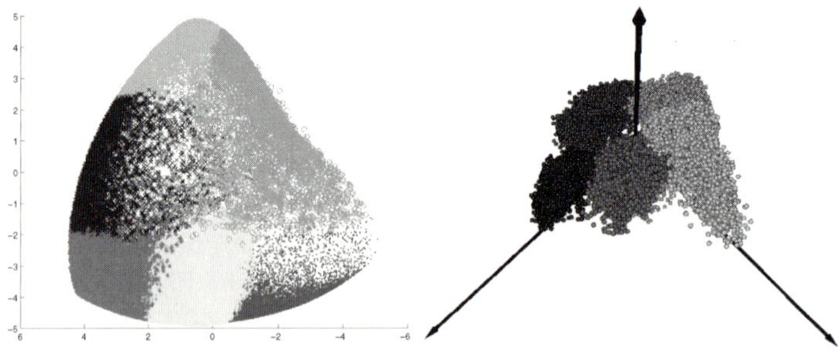

(a) (b)

Fig. 2. The Result of clustering of the feature vectors with the K-Means algorithm (a) and the SNN method (b)

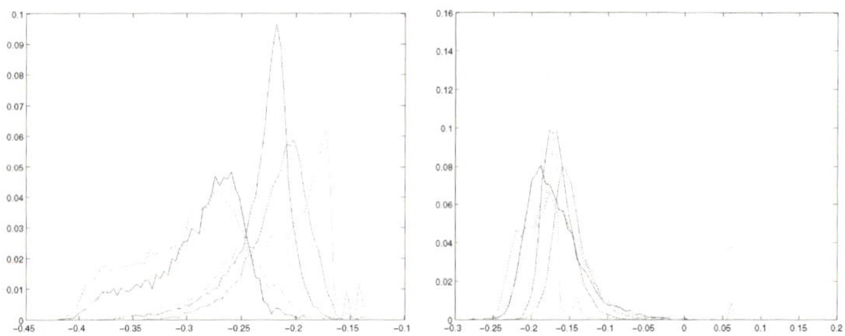

(a) (b)

Fig. 3. Histograms of two ROI features computed separately for each of the clusters found by K-Means. Feature from the diagram (a) differs significantly between the clusters. Values of feature from diagram (b) are similar inside each of the clusters.

4.3 Classification performance

The classification was carried out using the Radial Basis Function kernel:

$$K(\mathbf{x}_i, \mathbf{x}_j) = \mathrm{e}^{-\sigma \|\mathbf{x}_i - \mathbf{x}_j\|^2}$$

Initial values for the parameter σ and the SVC misclassification penalty C were set manually. Afterwards we evaluated the classification accuracy for each value $C \in \{1, 2, \dots 200\}$. The training was done on the *training dataset* whereas classification accuracy was measured for *validation dataset*. The results are depicted in diagram 4a. The value $C = 60$ yielded the best classification. Similarly, we evaluated the classifier performance for each value $\sigma \in \{0.01, 0.02, \dots 2.00\}$. The parameter C was set to 60 during this study. The results are presented in diagram 4b. The best classification accuracy was achieved for $\sigma = 0.80$. ROC curves for the best values of parameters C and

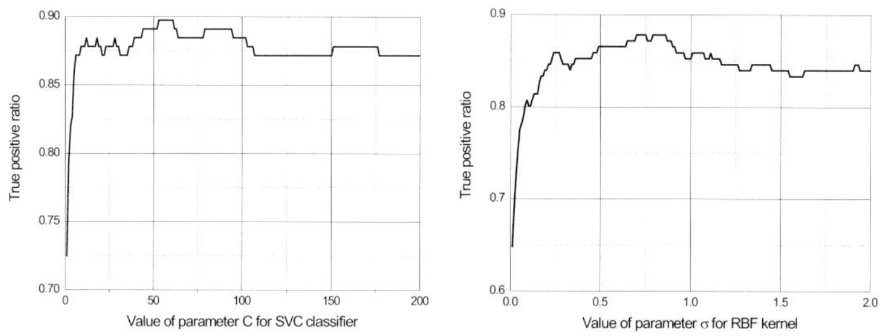

(a) (b)

Fig. 4. Evaluation of the SVC misclassification penalty and the kernel parameter σ.

σ are presented in Fig. 5. The red curve correspond to classification accuracy on all 46 features. The green and blue curves correspond to the accuracy for the subset of 27 and 18 features respectively. As we can see, in all three cases the classifier achieved similar accuracy. This is reflected by the areas under the ROC curves which are equal to 0.98 in each case. Therefore the feature selection, which allow for significant reduction of computational cost, do not impair the classifier accuracy. Using the 18 best features SVC can achieve sensitivity as high as 0.97 while maintaining specificity slightly above 0.90.

5 Conclusions and future work

We have developed a new method for recognition of microcalcifications in digital mammograms. Our algorithm was tested on 200 high resolution mammograms. The analysis of ROC curves revealed that it can achieve high recognition accuracy. Using a novel clustering algorithm we design a set of 18 highly descriptive features of mammogram regions. The results showed that further increase in number of features does not improve classification accuracy. Our future work will focus on development of more advanced method for identifica-

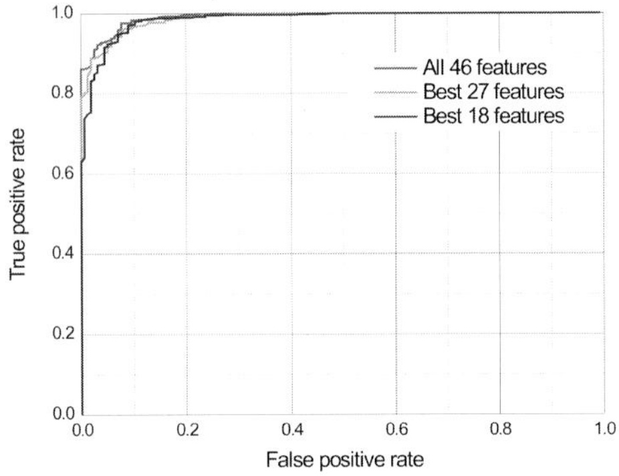

Fig. 5. The ROC curves for the SVC. The classifier was trained on *training dataset*. The performance was evaluated on the *validation dataset*

tion of suspected regions in mammograms. We will also work on an algorithm for detection of clusters of microcalcifications in mammogram images.

Acknowledgements This work was partially founded by the Polish Committee for Scientific Research (KBN) Grant no. 3T11C05926.

References

1. Cancer fast and figures. The American Society, 2004
2. C. J. C. Burges and D. Crisp. Uniques of the SVM soution. In *Advances in Neural Information Processing Systems 12*, pages 223-229, Cambridge, MA, 1999. MIT Press
3. C. Cortes and V. Vapnik. Support vector networks. *Machine Learning*, 20:273-297, 1995
4. L. Ertoz, M. Steinach and V. Kumar. Finding clusters of different size, shapes and densities in noisy, high dimensional data. In *Proc. of SIAM International Conference on Data Mining*, 2003
5. S. Gunn. Support vector machines for classification and regression. Technical Report, Dept. of Electronics and Computer Science, University of Southampton, U.K., 1998.
6. M. Heath, K. Bowyer, D. Kopans, R. Moore and P.K. Jr. The digital database for scrennig mammography. In *The Preceedensig of the 5th International Workshop on Digital Mamography*, pages 662-671, 2000.
7. J. Platt. Fast training of support vector machines using sequential minimal optimization. In B. Schölkopf, C. J. C. Burges, and A. J. Smola, editors, Advances in Kernel Methods – Support Vector Learning, pages 185-20, Cambridge, MA, 1999. MIT Press.

The Gait Characteristic Data Spectrum Extraction

Slawomir Chandzlik[1] and Jan Piecha[1]

[1]University of Silesia, Institute of Informatics, Dept. of Computer Systems, Sosnowiec, Poland, piecha@us.edu.pl

Summary. The gait characteristic features can be used for various diseases source and level diagnosis. The interferences of a physiological data record caused by any characteristic abnormal data spectrum (of the disease) corresponds to the disease source and level. The disease source can easy be classified by the relevant periodical functions. The interferences characteristics were analysed then matched to the well defined mathematical expressions for doing several manipulations on the data set. Knowing a range of the data records disorders the physiological data record can be modified due to multiply the data set of the defined disease class. This way the efficient amount of the data for the neural network training can be obtained. The paper shows several aspects of the gait data record analysis concerning two neurological diseases.

1 Introduction

The present paper shows authors recent investigations that allow avoiding very difficult condition for neural network training process - shortage of clinical records that are used for: first, a neural network structure selection, then training the neural network by big set of well defined records.

The large number of well classified records are used for running the artificial conclusion making system. While, the effectiveness of medical diagnosis depends on equipment quality that a doctor has for his disposal.

Various computer systems provide the operator with precise measures, not only for the disease classification but also (the most important condition) for an automatic conclusion making system training.

Numbers of works presented bellow concern the gait characteristics analysis of patients suffering from two representative neurological diseases [1, 2, 3, 5]. These works were based on Parotec System for Windows (PSW) metrological equipment [4, 5, 8, 9]. The PSW software package main goal concerns the pressure distribution analysis describing a foot shape for orthopaedic purposes [4, 8, 9].

These elementary options were widely described in several earlier works [2, 4, 5, 5, 6, 8, 9]. The works discussed in the paper concern another field of the PSW application - neurological diseases recognition and classification.

Also various extensions for neurological diseases diagnosis are needed, partly described in several papers [1, 2, 5, 6]. The given illustration concern in majority to diseases: group one left- and right- lateral hemiparesis, group two Parkinson disease.

Analysing the control group several regularities were distinguished and extracted. These regularities were expressed by several math-formulas, defining the disturbance of gait functions. They were used as an interference spectrum of the physiological gait record.

The defined formulas allow producing the virtual data records on a basis of the data record determined as physiological one. The virtual data record generator produces thousand of virtual data sets on a basis of characteristic clinical measurements.

Although the PSW data concerns orthopaedic diseases big number of additional data can be extracted from this basic data spectrum.

2 The Interferences Data Spectrum

An aim of neural network training is finding a global minimum of the cost function. This minimum of the function is found as faster as the training is better controlled, not running chaotically [6, 7]. This goal is found as fast as the training is more carefully driven. The N-dimension training sequence has to be given on the inputs of the neural network in case the weights of vector W were determined correctly with the training factor η [2, 5, 6].

Various experiments with the neural network selection were carried out on the basis of four sets of records: the control group of patients, the group with left-lateral hemiparesis, the right-lateral hemiparesis and records describing as Parkinson's disease.

Almost one hundred cases for putting the diagnosis has been assigned. These clinical records were divided into a following groups: 25 cases of the control group, 29 records classified by medical experts as the left-lateral hemiparesis, 28 cases for the right-lateral hemiparesis and 10 records concerning the Parkinson's disease.

This number of records is not sufficient for any satisfying optimisation result of neural network training process [1, 6, 7]. For increasing the size N of the training data set additional virtual records have been produced.

Below one can find several formalities explaining how the virtual records are produced.

Definition 1. *A step is understood as a period of time while a patient's foot (left or right) touches a floor.*

Definition 2. *A steps-cycle is a time between a left step (or right) beginning and a right step (or left) ends if the left step (or right) is finished. It defines the observation cycle for only one left-foot step and only one right-foot step within the single steps-cycle.*

Definition 3. *The overlap phase is a time period of parallel floor contact of both feet in the dynamic part of the data, during a walking cycle. It is a time when the body weight is moved from one foot to the other one.*

Let us assume that the value of a defines an active foot:

- for left foot $\alpha = \text{l}$
- for right foot $\alpha = \text{r}$.

Let us also determine the force measured on an α-foot in a current i steps-cycle as:

$$F(\alpha, t) = \sum_{i=1}^{n_\alpha} F_i(t) = \sum_{i=1}^{n_\alpha} P_i(t) \cdot S_i \tag{1}$$

where: η_a is a number of sensors installed on an insole (of the α-foot), $F_i(t)$ determines the force recorded in a time t on each sensor i, $P_i(t)$ determines a pressure value in a time t on each sensor i, S_i describes the hydrocell surface of every sensor i.

The time distribution of these forces (on a footprint) has been presented by functions F in Fig. 1a. Similarly the gait function W (Fig. 1b) can be defined as:

$$W(t) = F(l, t) - F(r, t) \tag{2}$$

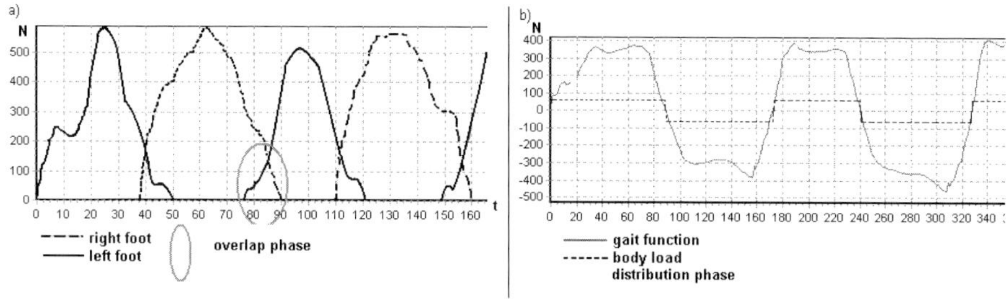

Fig. 1. The forces time distribution: a) the functions of forces F, b) the gait function W.

The positive values of the gait function W determine a gravity centre of the patient's body movement into the left side of the body. The negative values of W determine the overload on a right foot where the gravity centre moved into the right side of the body.

A dynamic part of the data record contains pressure samples recorded during the gait time. For this data part the functions F_i (1) registered by sensors set is defined by a spline interpolation algorithm producing all continuous values of functions F_i - widely described in the paper [3].

Thanks to the given approximation, we obtained continuous functions F_i:

$$F_i : [0, T_D] \rightarrow \Re^+ \cup \{0\} \tag{3}$$

where: T_D is a time period of a dynamic part of the measurement.

The virtual data records can be produced in the case the interferences of the pathological records are recognised. Then multiplying the clinical cases into well-defined classes can cover the needs of large number of training data set. The function F representing distribution of forces at a patient's foot was determined on the basis of control group.

Let A denotes the gait function of the virtual record, as:

$$A(t) = W(t) + E(t) \tag{4}$$

where: W is a gait function given by formula (2) for a record of the control group, E represents a gait disturbances function defined as:

$$E(t) = W_1(t) + W_2(t) \tag{5}$$

where: W_i defines the gait functions obtained from formula (2) from two clinical data record R_1 and R_2.

In the case the numbers of steps-cycles m_1, m_2 are different then they have to be reduced into the same size - into smaller number of steps-cycles $m = \min\{m_1, m_2\}$. Moreover functions W_1 and W_2 operate on these sets:

$$\begin{array}{l} W_1 : [0, T_{D_1}] \rightarrow \Re \\ W_2 : [0, T_{D_2}] \rightarrow \Re \end{array} \tag{6}$$

where: T_{D_1}, T_{D_2} concern the time markers of the dynamic units for R_1 and R_2 data records respectively for m number of steps-cycles.

The values of T_{D_1} and T_{D_2} should be additionally equal. This condition enable that the $W_i(t)$ functions are defined for every time unit $t \in T_{DD}$ and the T_{DD} set is defined as:

$$T_{DD} = [0, T_{D_1}] \tag{7}$$

Then the time T_{D_2} has to be redefined by k_T factor, as:

$$k_T = \frac{T_{D_1}}{T_{D_2}} \tag{8}$$

for m steps-cycles.

After this operations $T_{D_1} = T'_{D_2}$ are equal and both functions W_1 and W_2 are defined for a whole T_{DD} time period. What is more the values of disturbances functions E respond to the real data of the data record.

3 The Interference Extraction

Let us compare two data records R_w and R_p. Let us also assume that R_w record represents clinical case defined as physiological (the control group record) and R_p record is a pathological case (outside the control group).

For these cases the gait functions W_w and W_p are determined in accordance with the formulas presented above (illustrated in Fig. 2a,b). The interferences of the gait physiology are obtained as:

$$E(t) = W_p(t) - W_w(t) \tag{9}$$

The above E function shows the character of the gait data spectrum interferences. This function of interferences is then used for multiplying the records with the extracted character of interferences; here with the left-lateral hemiparesis.

The example of the E expression for a left-lateral hemiparesis is presented in Fig. 2c.

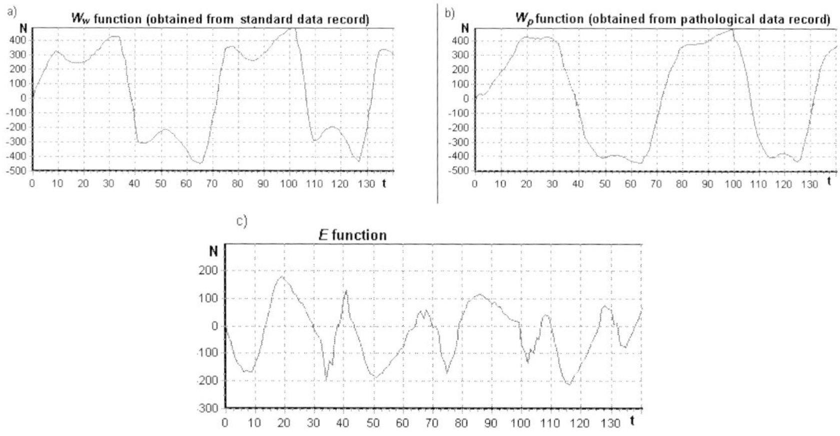

Fig. 2. Interferences of physiological records: a) the gait function W_w of a physiological record, b) the gait function W_p of a pathological record, c) interferences E.

4 The Experiments Analysis

The gait interferences defined by function E determined for each sensor of the insole. The same operation is done for every data record. In tab. 1 four classes of a gait abnormality have been determined - defined as *.efz files, with virtual products of the data record.

Table 1. Number of virtual *.efz files

	Total records			
	of control	of hemiparesis		of Parinson's
	group	left lateral	right lateral	disease
	25	29	28	10
The virtual data products; **.efz files for four disease classes	625	725	700	250

The function E of the gait interferences are used to produce the *.efz files, however the gait functions W are determined from any clinical data record of the control group.

4.1 An Overlap Phase

An overlap phase is one of the characteristic features of the patient's gait that has to be determined in the set of virtual records (def. 1).

The overlap phase number p is defined by formula:

$$p = 2m - 1 \tag{10}$$

where: m is a number of steps-cycles.

There is possible to obtain approximations of distribution of the force function F_O, that appears at a patient's foot during the overlap phase in the virtual data record, directly from the gait function A given by formula (4). The following formula determines discussed approximation of force distribution that has to be assigned to every of overlap phases i:

$$F_{O_i}(\alpha, t) = A(t) - \beta_\alpha(t) \tag{11}$$

where: A is a gait function generated for the virtual data record, β is an approximation factor defined by:

$$\begin{aligned} \beta_l(t) &= m_o - m_o\,(1 - f_2(t))\, f_1(t) \\ \beta_r(t) &= M_o - M_o\,(1 - f_2(t))\,(1 - f_1(t)) \end{aligned} \tag{12}$$

where:

$m_o = \min\limits_{t=[0,t_{O_i}]} (A(t))$ is a minimal value of the function A of i overlap phase,

$M_o = \max\limits_{t=[0,t_{O_i}]} (A(t))$ is a maximum value of the function A of i of the overlap phase, f_1 is a linear function determining monotonic character of the function E, given by formula:

$$f_1(t) = \begin{cases} g_1(t), & if A(t_{SO_i}) < A(t_{SO_i} + t_{O_i}) \\ 1 - g_1(t) if A(t_{SOi}) > A(t_{SO_i} + t_{O_i}) \end{cases} \tag{13}$$

where: $g_1(t) = \frac{t}{t_{Oi}}$,
t_{O_i} is a time duration of overlap phase i,
t_{SO_i} is a time overlap phase i marker, where the overlap phase begin,
f_2 is a function of correction defined by formula: $f_2(t) = c_G \sin(\Pi g_1(t))$ where: c_G is a scaling constant.

The F_{O_i} functions are assigned at ranges of $[t_{SO_i}, t_{SO_i} + t_{O_i}]$, i.e. only for time markers where the overlap phase exists. The example forces distribution on feet in time period of the overlap phase and its approximations by F_{O_i} functions are presented in Fig. 3a.

4.2 The Overlap Approximation Error

For the virtual overlap estimation the error analysis has been carried out first. This makes the optimisation of the function β coefficient possible. For all experiments the relative error has been defined:

$$R_E(t) = \frac{|F_O(l,t) - F_O(r,t) - (F(l,t) + F(r,t))|}{F(l,t) + F(r,t)} \tag{14}$$

where: F_O concerns estimated values of approximated forces on a foot distribution, F concerns real values of the forces on a foot.

With this error definition its measure can be defined, as:

$$R_i = \int_{t_{SOi}}^{t_{SOi}+t_{Oi}} R_E(t)dt \tag{15}$$

where: t_{O_i}, t_{SO_i} are described as in the above formula (14).

The 50 clinical records have been discussed carefully, where the estimation error was the analysis subject. The smallest value of the measure R was obtained for coefficient $c_G = 0.22$. With the average value of the R measure for this coefficient c_G was $\overline{R} = 1.7704$ (Fig. 3b).

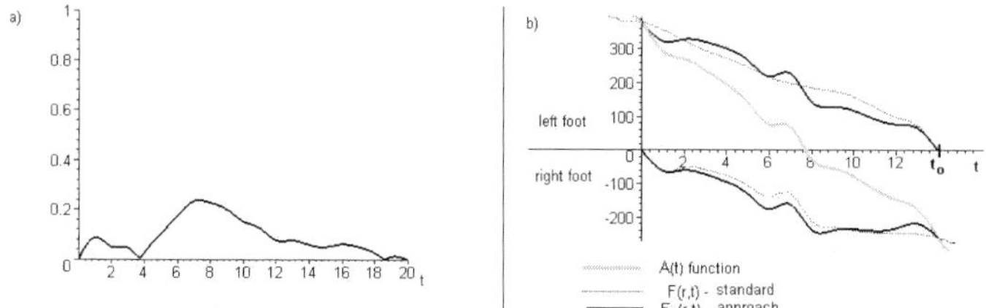

Fig. 3. a) An error R_E, b) the overlap functions A.

The Distribution of Forces in Virtual Records

An algorithm describes the forces distribution for the virtual record describing the patient's foot load in the dynamic part of the data:

- the gait function W extraction from a pattern record as in Fig. 2a,
- normalisation of a time duration T_{D_w} of the dynamic part of the pattern record using the scaling coefficient kT,
- the interference function E (Fig. 2c) reading from an *.efz file,
- the gait function A extraction from the virtual record as a product of functions concatenation E with W (equ. 4 and Fig. 1b),
- extraction of the forces F_O occur on an overlap of the feet using the gait function A,
- absolute values of forces, representing pressure recorded on a right foot.

For virtual product definition one file *.efz and one pattern file is needed. This way a large number of addition records can be obtained (tab. 2).

Table 2. The virtual record database structure

	Total records			
	of control	of hemiparesis		of Parinson's
	group	left lateral	right lateral	disease
Number of clinical records	25	29	28	10
Number of virual records	15 625	18 125	17 500	6 250

5 Conclusions

Although the discussed method of forces distribution was evaluated for the dynamic part of the record, the same methodology can be used for static part of the record, with several modifications. The interferences concern duration of the gait time, the amplitude of interference functions E and the amplitude of the gait function W for a pattern clinical record.

These virtual records (first properly classified) have been used for the conclusion-making unit training [2].

In the conclusion making unit examination of the automatic classification of current records have been done for: left- and right- lateral hemiparesis, Parkinson's disease and the control group.

Acknowledgements

The work is supported by Committee for Scientific Research funds, KBN Grant No. 3 T11E 055 26 in 2004/2005.

References

1. Chandzlik S, Kopicera K (2000) Experiments with neural network parameters - selection for foot abnormalities recognition. Journal of Medical Informatics & Technologies MIT 5:CS-71–CS-78
2. Chandzlik S, Piecha J (2003) The body balance measures for neurological disease estimation and classification. Journal of Medical Informatics & Technologies MIT 6:IT-87–IT-94
3. Chandzlik S, Piecha J (2002) A patient walk-data-record modelling using a spline interpolation method. Journal of Medical Informatics & Technologies MIT 3:MIT-153–MIT-160
4. Kopicera K, Piecha J (2001) The fuzzy estimation unit of foot-print abnormality recognition. Journal of Medical Informatics & Technologies MIT 2:MI183–MI188
5. Kopicera K, Piecha J, Zygula J (1999) The neural networks in diagnostics support for PSW system. Proc. of Int. Conference ASIS'99:113–118
6. Piecha J (2000) The neural network conclusion-making system for foot abnormality recognition. Proceedings of IMACS World Congress, Lausanne
7. Pieha J (2001) The neutral network selection for a medical diagnostic system using an artificial data set. Journal of Computing and Information Technology CIT 9:123–132
8. Piecha J, Zygula J, Lyczak J, Gazdzik T, Proksa J (1996) The advanced measuring system for orthopaedic pathologies diagnostics using a static and dynamic footprints. Chirurgia narzadow ruchu i ortopedia polska LXI, suplement 3B:119–124 (in Polish)
9. Zbrojkiewicz J, Piecha J, Jarzabek-Stepniak A (2001) The gait pattern detection in PSW records of the acute sciatic neuralgia. Proc. on KOSYR'01:29–36

On The Construction of the Syntactic Pattern Recognition-Based Expert System for Auditory Brainstem Response Analysis *

Mariusz Flasiński[1], Elżbieta Reroń[2], Janusz Jurek[1], Piotr Wójtowicz[1], and Krzysztof Atłasiewicz[1]

[1] Chair of Applied Computer Science, Institute of Computer Science, Jagiellonian University Nawojki 11, 30-072 Cracow, Poland
[2] Otolaryngological Clinic, Jagiellonian University ul. Śniadeckich 2, 31-501 Cracow, Poland

Summary. Recent developments of the construction of the syntactic pattern recognition-based expert System for Auditory Brainstem Response Analysis (SABRA) are described. The paper includes a brief characterization of problems related to the the use of computer science in ABR analysis, an outline of the concept of the SABRA system and a short description of GDPLL(k) and ACLL(k) grammars as the solution for fundamental problems that arose during the design of the system.

1 Introduction

The research conducted by Jewett and Wilson [9] laid foundation to the wide range of an organ of hearing testing methods called Electric Response Audiometry (ERA). In these tests, an ear is stimulated by an artificial sound. The electric signals generated in the process of hearing are measured by electrodes placed on the body of a patient and recorded by a special equipment. From the medical point of view response in the first 10 ms after the application of the stimulus is particulary important. Such tests are called Auditory Brainstem Response (ABR).

Theoretically, in ABR chart one can identify 5 waves[3] denoted as I, II, III, IV and V (see: Figure 1). Each wave corresponds to a certain organ involved in hearing. However, the identification of these waves is difficult in many cases[4]

*This work was supported by the Polish State Committee for Scientific Research (KBN) under Grant No. 3 T11C 054 26.

[3]Actually, in some cases up to 7 waves can be identified.

[4]Waves II and IV are especially difficult to observe as in some cases they overlap each other.

and involves an expertise of a skilled physician[5]. The identification of the V wave is easiest and serves as the basis of assessment of so called *hearing threshold level*[6]. The necessity of medical expertise in ABR tests causes high cost and organizational overhead, especially when the results are not satisfactory and the test should be repeated. There have been many attempts to develop an automated system of the level of hearing detection utilizing various techniques of Pattern Recognition. The range of methods used to solve this problem include: signal processing and filtering (eg. [1]), recognition based upon calculation of certain parameters of the ABR chart (eg. [3]), syntactic pattern recognition (eg. [2]), artificial neural networks (eg. [8]). Although some attempts have been quite successful, the task of constructing reliable, widely recognized system for ABR analysis is still far from being completed.

In the paper we present the recent results of the research done by the teams from Chair of Applied Computer Science, Jagiellonian University, and Otolaryngological Clinic, Jagiellonian University. The research has been aimed at the application of syntactic pattern recognition methods in the expert System for Auditory Brainstem Response Analysis (SABRA), responsible for evaluating of an organ of hearing in neonates.

In case of neonates the shape of the ABR charts differ from each other to the large extend and depends on many factors like age, mothers diseases (eg. diabetes) or child diseases (eg. Down or Pierre-Robin syndromes) [12]. The shape of ABR is rarely the subject of medical investigation, although it can provide some additional diagnostic information. In particular, there are some hints that the shape of the V wave may depend on the behavior of cochlea. The reaction of this organ to the stimulus seems to be reflected in the shape of the wave. The explanation of relation between the shape of the V wave and the behavior of an organ of hearing is still an open medical problem. The problem is both important and difficult.

Since syntactic methods are particulary useful in reflecting qualitative knowledge we decided to use these methods as a support for medical investigation [4, 5]. The achieved results prove that the ABR chart can be successfully modelled with the use of syntactic methods.

[5]The most important advantage of the ABR test is that it does not require any activity of a patient. Therefore the test could be performed on neonates, and it is also widely utilized in judical expertise. Nevertheless, the test is subjective because the results have to be analyzed by a human.

[6]In the hearing test, the stimulus of 100 dB, 90 dB, etc. is applied to a particular ear and the ABR is recorded. If the stimulus is strong enough, the V wave can be identified in the ABR chart. It is assumed that the presence of the V wave in the ABR chart means that a patient can hear sounds of a particular volume. The lowest volume level of such a stimulus is recognized as a level of hearing. Additional characterization of the hearing can be concluded from the timing (latency) of the V wave.

2 Basic definitions

2.1 GDPLL(k) grammars

Let us introduce two basic definitions corresponding to GDPLL(k) grammars [6, 10].

Definition 1. A *generalized dynamically programmed context-free grammar* is a six-tuple:

$$G = (V, \Sigma, O, P, S, M),$$

where: V is a finite, nonempty alphabet; $\Sigma \subset V$ is a finite, nonempty set of terminal symbols (let $N = V \setminus \Sigma$); O is a set of basic operations on the values stored in the memory (assignment, addition, subtraction, multiplication); $S \in N$ is the starting symbol; M is the memory; P is a finite set of productions of the form:

$$p_i = (\mu_i, L_i, R_i, A_i)$$

in which $\mu_i : M \longrightarrow \{TRUE, FALSE\}$ is the predicate of applicability of the production p_i defined with the use of operations ($\in O$) performed over M; $L_i \in N$ and $R_i \in V^*$ are left- and right-hand sides of p_i respectively; A_i is the sequence of operations ($\in O$) over M, which should be performed if the production is to be applied. □

A derivation for generalized dynamically programmed grammars is defined in the following way. Before application of a production p_i we test whether L_i occurs in a sentential form derived. Then we check the predicate of applicability of the production. The predicate is defined as an expression based on variables stored in the memory. If the predicate is true, we replace L_i with R_i and then we perform the sequence of operations over the memory. The execution of the operations changes the contents of the memory (memory variables). It is done with the help of arithmetical and assignment instructions.

Definition 2. Let $G = (V, \Sigma, O, P, S, M)$ be a dynamically programmed context-free grammar. The grammar G is called a GDPLL(k) grammar, if the following two conditions are fulfilled.

1. Stearns's condition of LL(k) grammars. (The top-down left-hand side derivation is deterministic if it is allowed to look at k input symbols to the right of the current position of the input head in the string).
2. There exists a a certain number ξ such that after the application of ξ productions in a left-hand side derivation we get at the "left-hand side" of a sentence at least one new terminal symbol.

The conditions have been more formally defined in [6]. □

2.2 ACLL(k) grammars

In this section we shall introduce the concept of ACLL(k) grammar.

Definition 3. A *Attribute Controlled Context Free grammar* is a five-tuple:

$$G = (V, \Sigma, \Delta, P, (S, x_0))$$

where: V is a finite, nonempty alphabet; $\Sigma \subset V$ is a finite, nonempty set of terminal symbols (let $N = V \setminus \Sigma$); Δ is a finite, nonempty alphabet of attributes; $P \subset 2^{\Delta^*} \times N \times V^* \times \Delta^{*\Delta^*}$ – is a finite, nonempty set of productions, such as if $(\mu, \cdot, \cdot, f) \in P$ then functions μ and f can be calculated in $O(1)$; $(S, x_0) \in N \times \Delta^*$ – is a starting symbol. □

A derivation for Attribute Controlled Context Free grammar is defined in the following way. The first sentential form is always (S, x_0). Before application of a production $(\mu_i, L_i, R_i, f_i) \in P$ to sentential form (γ, x) derived from (S, x_0), we test whether L_i occurs in γ (or $L_i = S$ when no productions were applied). Then we check the predicate of applicability of the production by testing if $\mu_i(x) = 1$ (or $\mu_i(x_0) = 1$ if no productions were applied). If the following conditions are met we replace the first occurrence of L_i in γ by R_i, and the function f_i is applied to x (or to x_0 if no productions were applied). The resulting sentential form is the $(wR_i\beta, f(x))$ (or $(R_i, f_i(x_0))$ in case of application of the first production).

Definition 4. Let $G = (V, \Sigma, \Delta, P, (S, x_0))$ be a Attribute Controlled Context Free grammar. The grammar G is called a ACLL(k) grammar, if the following four conditions are fulfilled.

1. Stearns's condition of LL(k) grammars. (The top-down left-hand side derivation is deterministic if it is allowed to look at k input symbols to the right of the current position of the input head in the string).
2. There exists a a certain number ξ such that after the application of ξ productions in a left-hand side derivation we get at the "left-hand side" of a sentence at least one new terminal symbol.
3. After application of any production we get no more than one new terminal symbol.
4. The derivation process can be conducted infinitely i.e. there is always at least one production that can be applied to a sentential form. □

The concept of ACLL(k) grammars and more formal definition above will be the subject of further publications.

3 Concept of the system

The system that is being developed consists of four main modules: Preprocessor (throughout this paper referred to as PR), Expert System (referred to as ES), V Wave Identification module (referred to as VWI) and V Wave Recognition / inference module (VWR). The main role of the PR is to analyze analog input to the system and convert it to symbolic representation [5]. The ES is a rule based expert system which gathers and process knowledge delivered by VWI and VWR. In this paper we shall limit ourselves to the presentation of the concepts of VWI and VWR.

The VWI have to main inputs: the input from PR, i.e. ABR chart transformed to the syntactic form, and the approximation of V wave parameters for the particular input. The task of the identification of the fragment of the

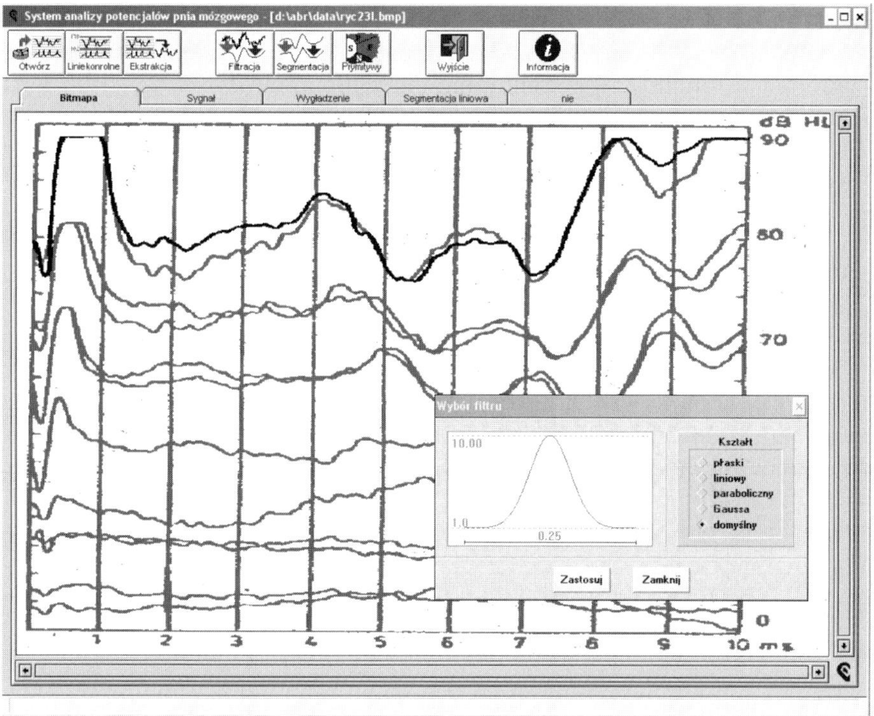

Fig. 1. The analysis of the ABR chart.

ABR chart where the V wave is present corresponds to the simulation of the expertise of a phisician[7]. Such an identification can rely both on numerical data about latency of the V wave and the analysis of the shape itself. One of the problems that has to be addressed at this stage is a determination whether a particular chart is a proper input for further analysis. The task could be solved by the use of ACLL(k) grammars. There are several reasons why ACLL(k) grammars have been chosen as the modelling tool for VWI. The predicates and actions in such grammars can be very flexibly defined and thus the expertise about ABR chart can be modelled properly and assure appropriate generating power. Additionally, the grammars assume an infinite derivation, which is needed if we consider the form of the ABR chart. Finally, the important practical reason behind this choice is that there is an efficient parsing scheme[8] for ACLL(k) grammars.

[7]Let us notice that the task of the identification of the V wave is related to the problem of automated analysis of ABR described in the introduction to this paper. The main difficulty is caused by the fact that the ABR chart for the stimulus that is close to the level of hearing is usually very distorted. The scope of our research is restricted to the most evident cases therefore such distorted patterns are not considered. That is why the goal of our research is not equivalent to automatic assessment of the level of hearing threshold.

[8]The computational complexity of the parser is $O(n)$.

There are two main operation modes for VWR: recognition and grammatical inference. In both modes the input to the system is the ABR chart converted to symbolic representation and the parameters provided by VWI describing the position, latency and amplitude of the V wave in this particular chart[9]. In parsing mode VWR analyzes the fragment of the input that was identified as V wave and checks if it matches any known grammar. If the parsing fails or the result is ambiguous[10] then the grammatical inference mode might be utilized. The task of the analysis of the V wave, which is the goal of VWR, is very difficult as there is no theory that could explain the particular shape of V wave. Actually, our task is to construct a system that could help to build such a theory. Since GDPLL(k) grammars generate a large subclass of context-sensitive languages and the languages can be parsed in $O(n)$, the GDPLL(k) grammars are a proper tool for the problem. The development of grammatical inference algorithm for this class of grammars is another strong reason behind the choice [11].

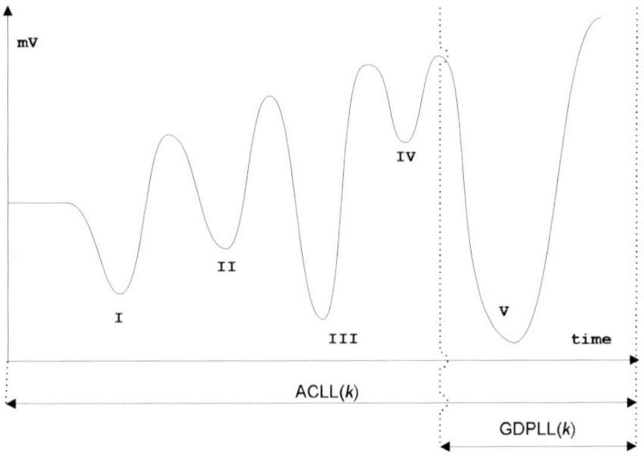

Fig. 2. The scope of ACLL(k) grammars (VWI module) and GDPLL(k) grammars (VWR module). The former is used to determine the position of the V wave, while the latter recognizes the V wave or performs the grammatical inference.

4 Modelling of the V wave of ABR based on GDPLL(k) and ACLL(k) grammars

Let us outline the way both subtasks described in the previous section can be solved with the use of GDPLL(k) and ACLL(k) grammars. Firstly, let us describe the concept of symbolic representation of the ABR chart. It is based on the following set of primitives: $\{s, p, f, n\}$. The definition of the primitives is given in the caption to the Figure 3.

---

[9]The symbolic representation of the ABR chart is the input string while the parameters passed by VWI initialize the memory.

[10]The result ambiguous if there are at least two different syndromes that can be matched with the particular input.

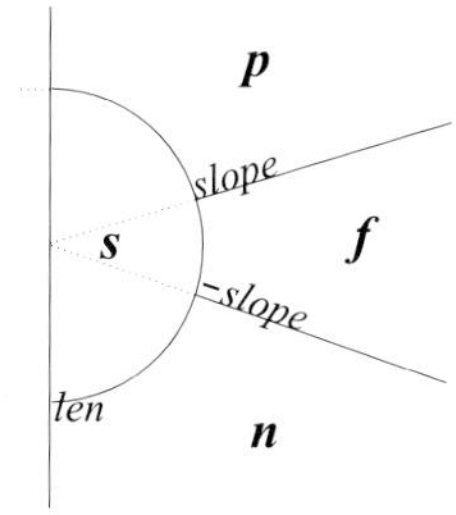

Fig. 3. The primitives are characterized by the following conditions:

s: $l \leq len$,
p: $(l > len) \wedge (a > slope)$,
f: $(l > len) \wedge (|a| < slope)$,
n: $(l > len) \wedge (a < -slope)$,

where length is denoted as l and slope angle as a and len and $slope$ are border parameters for l and a (correspondingly).

The exemplary ACLL(k) grammar described below is a simplified version of the actual grammars developed in the project[11]. It detects the if the `V` wave is present and if its parameters like starting point ($start$), amplitude ($ampl$), latency (lat) and ending point (end) match the limits provided by ES (i.e. $start_-$, $start_+$, $ampl_-$, $ampl_+$, lat_-, lat_+, end_-, end_+). As the result of the use of the grammar, the assessment of the parameters above is performed. The starting symbol is defined as (S, $start_-$, $start_+$, $ampl_-$, $ampl_+$, lat_-, lat_+, end_-, end_+):

$start < start_-$	$S \longrightarrow pS \vert fS \vert sS \vert nS$	$start++$
$start_- \geq start \geq start_-$	$S \longrightarrow pS \vert fS \vert sS$	$start++$
$start_- \geq start \geq start_-$	$S \longrightarrow nN$	$ampl++; lat++$
$ampl < ampl_- \vee lat < lat_-$	$N \longrightarrow nN$	$ampl++; lat++$
$ampl < ampl_- \vee lat < lat_-$	$N \longrightarrow fN$	$lat++$
$ampl < ampl_- \vee lat < lat_-$	$N \longrightarrow sN$	
$ampl_+ \geq ampl \geq ampl_- \wedge$ $\wedge lat_+ \geq lat \geq lat_-$	$N \longrightarrow pP$	$end := lat$
$end < end_-$	$P \longrightarrow pP \vert fP \vert sP$	$end++$
$end_+ \geq end \geq end_-$	$P \longrightarrow pP \vert fP \vert sP$	$end++$
$end_+ \geq end \geq end_-$	$P \longrightarrow nX$	
	$X \longrightarrow pX \vert fX \vert sX \vert nX$	

Let us also present a meta-scheme of a GDPLL(k) grammar, responsible for the analysis of the shape of the V wave. Auxiliary memory variables neg and pos are used to store information needed to determine the type of the shape.

(hill)	$H \longrightarrow PN \vert PSN$	$hill\_type := pos - neg$
(dale)	$D \longrightarrow NP \vert NSP$	$dale\_type := neg - pos$
(positive slope)	$P \longrightarrow p \vert Pp \vert PSp$	$pos++$
(negative slope)	$N \longrightarrow n \vert Nn \vert NSn$	$neg++$
(flat)	$F \longrightarrow f \vert Ff \vert FSf$	
(short)	$S \longrightarrow s$	

[11] For instance in the real world application parameters like relative latency and amplitude to other waves are taken into a consideration (if they are present).

5 Conclusions

The construction of the syntactic pattern recognition-based expert System for Auditory Brainstem Response Analysis (SABRA) presented in this paper summarizes the current state of the project. The overall concept of the system is outlined as well as the exemplary results. The system is based on ACLL(k) and GDPLL(k) grammars. The most rudimentary definitions in both cases have been formulated. The scope of use of both formalisms is the subject of more detailed analysis throughout the paper, as well as the cooperation between ACLL(k) and GDPLL(k) grammars. Although conceptual side of the project is merely complete, there are many open practical issues that have to be addressed in the future developments in the project.

References

1. Boston JR (1989) Automated interpreter of brainstem auditory evoked potentials: a prototype system, IEEE Trans. Biomed. Equip. IEEE Trans Biomed Eng. May, 36 (5), 528–32
2. Bruha I, Madhavan GP (1988) Use of attributed grammars for pattern recognition of evoked potentials, IEEE Trans. Syst., Man, Cybern., vol. 18, no. 6, 1046–1049
3. Dobie RA, Wilson MJ (1993) Objective response detection in the frequency domain, Electroencephalogr Clin Neurophysiol., 88 (6), 516–24
4. Flasiński M (1995) The Programmed Grammars and Automata as Tools for a Construction of Analytic Expert Systems. Archives of Control Sciences, no. 40, 5–35
5. Flasiński M, Reroń E, Jurek J, Wójtowicz P, Atłasiewicz K (2004) Mathematical Linguistics Model for Medical Diagnostics of Organ of Hearing in Neonates, Lecture Notes in Computer Science, no. 3019, 746–753
6. Flasiński M, Jurek J (1999) Dynamically Programmed Automata for Quasi Context Sensitive Languages as a Tool for Inference Support in Pattern Recognition-Based Real-Time Control Expert Systems. Pattern Recognition, Vol. 32 no. 4, 671–690
7. Fu KS (1982) Syntactic Pattern Recognition and Applications, Prentice Hall
8. Izworski A, Tadeusiewicz R (2003) System for Intelligent Processing and Recognition of Auditory Brainstem Response (ABR) Signals, Lecture Notes in Computer Science, no. 2690, 482–489
9. Jewett DL, Williston JS (1971) Auditory-evoked far fields averaged from the scalp of humans, Brain, 94, 681–696
10. Jurek J (2004) Recent developments of the syntactic pattern recognition model based on quasi-context sensitive languages, accepted for publication in Pattern Recognition Letters
11. Jurek J (2004) Towards Grammatical Inferencing of GDPLL(k) Grammars for Applications in Syntactic Pattern Recognition-Based Expert Systems, Lecture Notes in Computer Science, no. 3070, 604–609
12. Reroń E (1990) Badania kliniczne i elektrofizjologiczne narzadu słuchu u noworodków, Rozprawa Habilitacyjna, Akademia Medyczna im. M. Kopernika w Krakowie

Active Contour Technique in Post-segmentation Edge Smoothing Applied to Hand Radiograph Regions of Interest

Arkadiusz Gertych[1], Ewa Pietka[2], and H.K Huang[1]

[1] University of Southern California, Department of Radiology, Image Processing and Informatics Laboratory, Los Angeles, USA, gertych@usc.edu
[2] Silesian University of Technology, Department of Biomedical Engineering, Gliwice, Poland epietka@polsl.pl

Summary. In the current study two various segmentation methods have been implemented sequentially in computer-aided approach to the assessment of skeletal maturity, where correct location of borders of anatomical structures is a crucial step. The first segmentation stage, based on the Gibbs random fields technique, correctly segments out a bony structure and roughly outlines the edges of cartilage while the second one using the active contour strategy, smoothes them and prepares for the feature extraction stage. A synthetic region of interest has been designed to test and adjust weights of the snake energy functional. These weights have afterwards been applied to a real image data. In comparison to our previous works we observe a significant improvement of boundary location mainly when cartilage is included in the epiphyses.

1 Introduction

Image segmentation is a procedure often performed in computer-aided diagnosis related to radiological images. Usually the procedure initializes the image analysis or is preceded by a noise reduction function. It is performed over the entire image, or only a selected region of interest is subjected to it. In the current study two various segmentation methods have been implemented sequentially in computer-aided assessment of skeletal maturity – a procedure frequently performed in pediatric radiology [1, 2]. A comparison of a hand radiograph is compared with radiological pattern. The best match yields an assessment of the bone age. The computerized approach relies on a detailed analysis of a hand radiograph regions of interest (Fig. 1). The distal regions are of particular interest. In order to i mprove our computerized approach to the bone age assessment, an additional segmentation stage has been employed. The Gibbs random fields segmentation has been followed by a procedure based on the active contours technique.

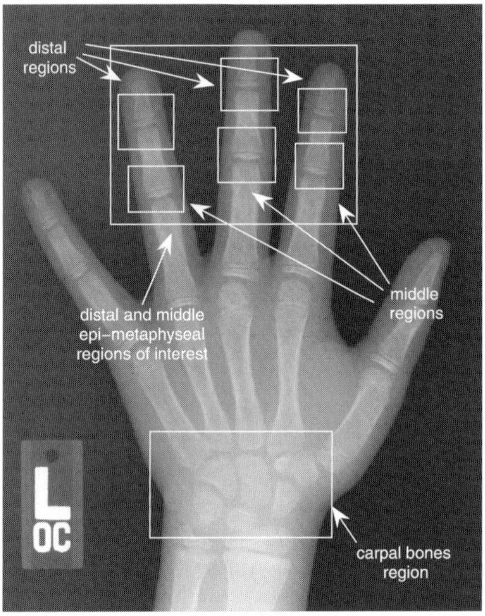

Fig. 1. Typical left hand radiograph with superimposed regions of interest

Both segmentation procedures are performed over the automatically selected [3] epiphyseal regions of interest (Fig. 2). The filtration stage reduces the noise and nonuniformity of bony and soft tissue structures caused by a presence of natural texture and scattered radiation [4].

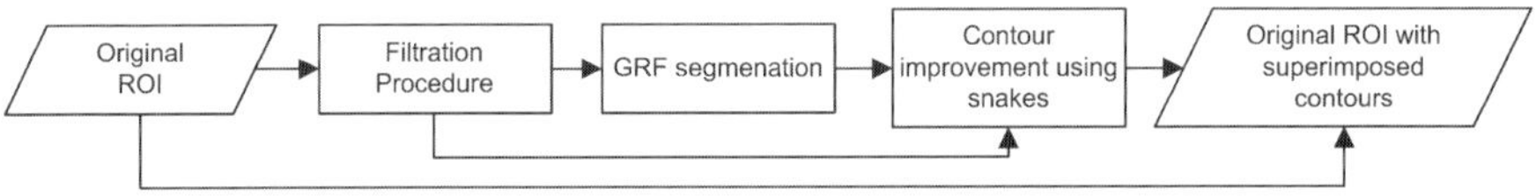

Fig. 2. Typical left hand radiograph with superimposed regions of interest

At the first step of segmentation an adapted Gibbs random fields (GRF) techniques separates the image pixels into 2 or 3 clusters (bony and soft tissues and cartilage, if any). Using the adaptive multigrid hierarchical implementation of segmentation algorithm [5], first the bony structure and soft tissue are segmented out, then, if necessary, a 3-rd region type is distinguished. The procedure yields a gray level image in which the bony structure appears as white objects and the cartilage as gray ones. These two areas are combined and turned to white whereas other areas (background and soft tissue) are turned to black. The contours of extracted regions of interest are continuous and often reflect location of well-outlined image details. Yet, in some cases, when the local contrast is of considerably low value, then the contour location does not meet expected results. This contour serves as an initial edge for the second segmentation procedure that employs the active contour model. Design of the active contour model, adjustment, and implementation in the overall project have been presented in this study.

2 Active contour approach in edge smoothing

Deformable or active contour models called "snakes" as introduced by Kass et. al [6] is are a special case of multidimensional deformable model theory developed by Terzopoulos [7]. Snakes are often used to determine the boundary of objects in images, based on an assumption that borders are piecewise continuous or smooth. The location and movement of contour pixels depend on the values of snake energy functional which is a linear combination of two terms: the image energy which pulls the snake toward the boundary and the internal energy which ensures smooth edge. The total energy is written as [6]

$$E^*_{snake} = \int_0^1 E_{snake}\big(\mathbf{v(s)}\big)ds = \int_0^1 E_{int}\big(\mathbf{v(s)}\big) + E_{image}\big(\mathbf{v(s)}\big)ds \tag{1}$$

where: E_{int} – represents the internal energy, E_{image} – external (image) energy.

The position of the snake in the image intensity $I(x, y)$, where $(x, y) \in R^2$ is the image plane is represented by location of the pixel curvature by $\mathbf{v(s)} = \big(x(s), y(s)\big)$. Both x and y are coordinate functions and $s \in [0, 1]$ is the parametric domain.

The internal energy is defined as:

$$E_{int} = \frac{1}{2}\left(\alpha(s)|\frac{\partial}{\partial s}\mathbf{v(s)}|^2 + \beta(s)|\frac{\partial^2}{\partial s^2}\mathbf{v(s)}|^2\right) \tag{2}$$

where: $\alpha(s)$, $\beta(s)$ are weighting parameters that control the tension and rigidity, respectively.

The external energy term consist of the following elements:

$$E_{image} = w_{line}E_{line} + w_{edge}E_{edge} + w_{term}E_{term} \tag{3}$$

where: $E_{line} = I(x, y)$ is the line energy, $E_{edge} = -|\nabla I(x, y)|^2$ is the edge energy, $E_{term} = (C_{yy}C_x^2 - 2C_{xy}C_xC_y + C_{xx}C_y^2)(C_x^2 + C_y^2)^{-3/2}$ is the terminal energy functional, and w_{line}, w_{edge}, w_{term} are weighting parameters. The energy coefficient E_{term} is an image subjected to a Gaussian filter [8], i.e.

$$C(x, y) = G_\sigma(x, y) * I(x, y) \tag{4}$$

Internal (elasticity) forces (Equ. 2) of the snake are responsible for keeping pixels of the curve close and preventing the contour from bending too much. They force the contour to be smooth and continuous. Weights $\alpha(s)$, $\beta(s)$ control the abilities of snake to bend. If $\beta(s)$ is locally set to zero then the second-order discontinuity appears and the snake forms a local corner. The image forces (Equ. 3) move the snake to a location of image features such as edges, lines or terminations of subjective contours.

In many approaches snakes require a proper initialization of the contour. An incorrect initialization may direct the contour toward an unwanted direction, or may tend to shrink it to a single point when only image forces are not strong enough. The target contour is found by minimizing the expression:

$$E^*_{snake} = \sum_{i=0}^{N-1} E_{snake}(\mathbf{v_i}) \tag{5}$$

where: N is the number of contour pixels. $v_N = v_0$ means that the contour is represented by a closed curve.

In the current study a minimization the energy functional has been performed by the Williams and Shah (greedy) algorithm [9].

Energy minimalization algorithm

The input image contains initial contour $p_1, \ldots, p_n$ (N is the number of contour pixels) located in the vicinity of the desired target contour.

Step 1 Search MxM neighborhood of each pixels p_i ($i = 1, \ldots, N$) location that minimizes the energy functional. Move the pixel p_i toward this location.

Step 2 Estimate the curvature of the snake at each point p_i and look for local maxima that exceeds the user-defined threshold. In that case turn β_i to 0.

Step 3 Update the value of contour energy.

Repeat steps 1-3 until the number of iteration is exceeded or the energy E^*_{snake} does not change significantly.

A normalization of every E^*_{snake} component is required for a proper minimization of the overall snake energy. For E_{int} all components at location p_i are divided by a largest value found in their neighborhood of size MxM. Similarly, for E_{line}, E_{edge}, E_{term} components of E_{image}, their local values are normalized using the expression: $[E_{\#} - \min(E_{\#\ M\times M})][\max(E_{\#\ M\times M}) - \min(E_{\#\ M\times M})]^{-1}$, where $E_{\#}$ is the current E_{image} component, and $E_{\#\ M\times M}$ is the appropriate energy value in the neighborhood of size $M \times M$, respectively. Typically M is set to 3 or 5.

Before minimizing the E^*_{snake}, the input image is subjected to a filtration procedure in order to blur the existing edges and enhance a gradient field. In this case we take advantage of the preprocessing step performed before the

GRF segmentation (Fig. 2). In order to calculate the E_{edge} energy component a Sobel gradient operator with a 7×7 mask is implemented.

In order to implement the active contour technique to the bone age assessment process flow, the already located epiphysis is subjected to a detailed analysis. The upper edge does not require any correction. Yet, the presence of cartilage requires the edge correction of the lower epiphyseal edge. Thus, before an active contour is initialized, the lower epiphyseal region is extracted and expanded outwards by a dilation procedure. This protects the active contour from being attracted toward stronger edges of the bony structure. Then, the snake pixels are evenly placed around the expanded contour.

3 Experiment and results

Two types of image data have been subjected to the procedure. First, a synthetic image analysis has shown the sensitivity to noisy. Synthetic image (Fig. 4) serves as a model of the epipyseal region of interest. Three gray level areas of 800, 500 and 200 pixel value simulate the anatomical structures of bones, cartilage and soft tissue, respectively. To imitate the influence of modulation transfer function the image is smoothed with a 5×5 mask Gaussian filter. In order to test the influence of various noise levels a noise standard deviation ranged from 20 to 70 has been added to the synthetic image. It reflects the range of normally distributed random noise in hand radiographs [10]. Blurred and noisy images are shown on Fig 3.

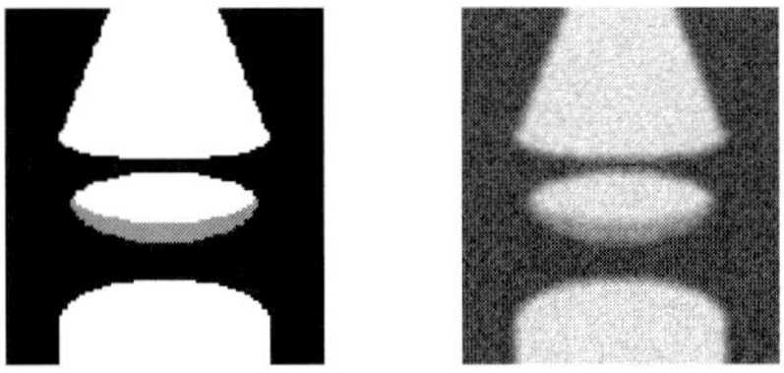

(a) (b)

Fig. 3. Synthetic ROI image model: **a)** the noise-free image model, **b)** blurred and noisy image model

The experiment has started by corrupting the model image with noise whose standard deviation σ has been set to 30. Beyond this level the GRF segmentation starts yielding inadequate results. The coefficients of the energy functional have been adjusted experimentally. The following parameters have been chosen: $\alpha = 0.3$, $\beta = 0.6$, $w_{edge} = 0.18$, $w_{term} = 1$, $w_{line} = 0.05$.

Then, synthetic images with different noise level have been subjected to the GRF segmentation only and active contours technique. For each standard deviation value, the segmentation procedure has been performed 10 times. Table 1 includes results of the location of boundaries of the synthetic image for both GRF segmentation and GRF segmentation followed by active contours technique. Each value shows the number of cases with correctly located boundaries around the epiphyseal model.

Table 1. Evaluation of boundary location by GRF and active contours algorithms

Method of contour location	Standard deviation of noise added to synthetic image 10	20	30	40	50	60	70
GRF	10	7	5	3	2	0	0
GRF + snake	10	10	10	10	10	10	9

One can note, that when the image noise exceeds a certain level, the GRF segmentation procedure fails in a correct outline of the cartilage tissue (Fig. 4). Two superimposed edges extracted by the GRF (black line) and combined method (white line) are shown in Fig. 4a. A profile located in a non-blurred image (Fig. 4b) reflects the quantitative analysis of the difference in performing the epiphyseal segmentation by both methods. The gap between lines pr3 and pr4 indicates the improvement of the segmentation achieved by implementing the additional stage (i.e. the active contour technique).

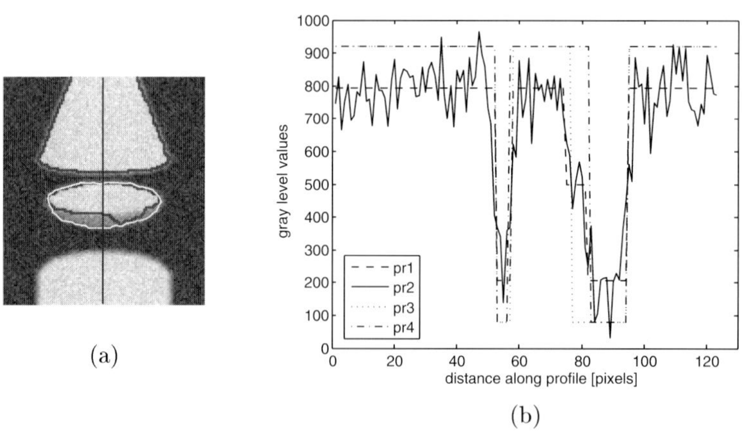

Fig. 4. The results of borders location for synthetic image: **a)** synthetic ROI with edges segmented out by different techniques, **b)** distance along profile in *a)*: pr1 – location of edges in non-blurred noise free image model, pr2 – gray level values along profile in noisy image, pr3 – edges detected by GRF segmentation, pr4 – improved edges detection by snake technique

As clinical data, 150 distal ROIs, reflecting different stage of development, have been selected from the hand radiographs database of 540 images. They serve as a testing set for the combined methodology. Four examples are shown in Fig 5. Gray lines reflect the boundary found by the GRF algorithm. Contour improved by snake technique are shown by light lines. An observable improvement in the contour location has been noticed in 62.3% of processed images. Due to the lack of cartilage, in 34.5% of cases no difference in contour location has been found. In the remaining 3.2% of cases some parts of contours have been attracted by the lower part of metaphyses whose edge has happed to be stronger (Fig. 5d).

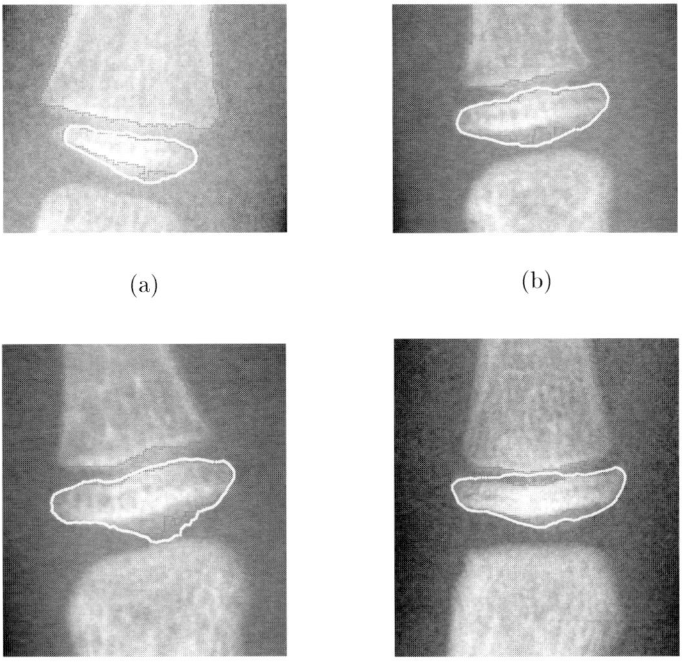

(a) (b)

(c) (d)

Fig. 5. a),b),c), d) Regions of interest in different stage of development after GRF segmentation (gray lines), and snake smoothing (white lines).

4 Conclusion

Results proof an advantage of implementing the snake method as a post-segmentation contour improvement technology. The GRF segmentation serves as a reasonable solution of the initial contour location, and has overcome the main drawbacks that prevents from implementing snake in many applications. This also implies that the entire routine remains objective and automated. Significant improvement in contour location and smoothing is observed in regions with cartilage developed at the bottom of the epiphysis. At this stage a simplest form of the snake has been tested. Its main disadvantage at a current stage of development is the subjective estimation of the set of parameters $\alpha(s)$, $\beta(s)$, w_{edge}, w_{term}, w_{line}. Future studies will address two areas: segmentation and extraction of carpal bones and adjustment of snake parameters as variables along the snake curvature.

References

1. Greulich WW, Pyle SI (1971) Radiographic Atlas of Skeletal Development of Hand Wrist. Stanford University Press, Stanford CA
2. Tanner JM, Whitehouse RH (1975) Assessment of Skeletal Maturity and Prediction of Adult Hight (TW2 Method). Academic Press, London
3. Pietka E, Gertych A, Pospiech S, Huang HK, Cao F (2001) Computer assisted bone age assessment: Image pre-processing and ROI extraction. IEEE Trans. on Medical Imaging 20:715–729
4. Pietka E, Pospiech S, Gertych A, Cao F, Huang HJ, Gilsanz V (2001) Computer Automated Approach to the extraction of epiphyseal regions in hand radiographs. Journal of Digital Imaging 14:165–172
5. Gertych A, Pietka E (2003) An automated segmentation and features extraction from hand radiographs. Archives of Theoretical and Applied Informatics 15 (workbook 3):315–326
6. Kass M, Witkin A, Terzopoulos D (1988) Snakes: Active Contour Models. International Journal of Computer Vision 321–331
7. Terzopoulos D, Fleisher K (1988) Deformable models. Visual Computer 4:306–331
8. Gonzales RC, Woods RE (1993) Digital Image Processing. Addison-Wesley, Reading MA
9. Williams DJ, Shah M (1992) A Fast Algorithm for Active Contours and Curvature Estimation. CVGIP: Image Understanding 55(1):14–26
10. Pietka E, Pospiech-Kurkowska S, Gertych A, Cao F (2003) Integration of Computer assisted bone age assessment with clinical PACS. Comp. Med. Img. Graph, 27(2-3):217–228

Classification of Medical Images in the Domain of Melanoid Skin Lesions

Zdzislaw S. Hippe[1], Jerzy Grzymala-Busse[1,2], Piotr Blajdo[1], Maksymilian Knap[1], Teresa Mroczek[1], Wieslaw Paja[1], and Mariusz Wrzesien[1]

[1] Department of Expert Systems and Artificial Intelligence, University of Information Technology and Management, 2 Sucharskiego St., 35-225 Rzeszow, Poland,
zhippe,pblajdo,mknap,tmroczek,wpaja,mwrzesien@wsiz.rzeszow.pl
[2] Department of Electrical Engineering and Computer Science, University of Kansas, Lawrence KS 66045, USA,
jerzy@ku.edu

Summary. In this paper we discuss computer-aided diagnosing and classification of melanoid skin lesions. The main goal of our research was to elaborate and to promote via Internet a new skin lesion diagnostic computer system. Its functionality and structure is described briefly in this report. In the current version of the system, five learning models are implemented to simultaneously supply five independent, partial results. Then, a special evaluation and voting algorithm is applied to select the correct class (concept) of the diagnosed skin lesion.

1 Research background

Available literature [1, 2, 3] shows that more than half of all identified cancer cases are different types of skin tumors. Recently in the United States about 800,000 cases of BCC (Basal Cell Carcinoma) tumors, 200,000 cases of SCC (Squamous Cell Carcinoma) tumors, and about 48,000 cases of malignant melanoma were diagnosed. Global data shows that the most dangerous type of tumor is malignant melanoma. It caused most fatal cases: 7,700 (4,800 men and 2,900 women) in comparison to 1,900 fatal cases caused by other types of tumors. Additionally, it should be emphasized that of all types of cancers, malignant melanoma cases increased the most between 1973 and 1994 (about 120% increase). Despite the comprehensive cancer research especially in the US progress has been limited in determining who is mostly at risk. Only recently a link was identified between the gene responsible for the Apaf-1 protein and skin cancer. Due to this progress, it is possible to use an immune therapy and more efficient and secure chemical therapy [1].
Recently, a decline in the cancer rate has been observed in Australia, Scotland

and Ireland [4]. Some reasons for this phenomenon can be attributed to: (i) self–diagnosis education and practices in Western Europe and United States; (ii) easy access to information about malignant melanoma symptoms including health risks based on atypical pigmented lessions, eye color, hair color and amount of exposure to ultraviolet radiation, and (iii) access to various survival prediction methods [1].

European research in the field has been focused on methods of classifying tumor types, identifying selected symptoms such as description of pigment lesions in preterminal skin cancer or in skin cancer cases demanding surgical intervention [2, 3].

Our current research in the classification of medical images is aimed at the development and promotion via Internet of a new skin lesions diagnosing computer program system, called by us the ***Internet Melanoma Diagnosing*** and ***Learning System, IMDLS***. It is assumed that present research and development efforts should be devoted to extending the functionality of our Internet–based system for diagnosing four categories of skin lesions: ***Benign_nevus***, ***Blue_nevus***, ***Suspicious_nevus***, and ***Melanoma_malignant*** [5]. Until now, our system has supported three methods (learning models) of diagnosis: (i) ***classic ABCD rule*** (based on ***TDS*** parameter) [6, 7], (ii) ***optimized ABCD rule***, (based on our own ***New_TDS*** parameter [8, 9]), and (iii) a ***decision tree*** (based on the ***ID3/C4.5*** algorithm) [10]. Recently, two other learning models have been implemented. The first model, called by us ***genetic dichotomization***, is based on a linear learning machine with genetic searching of the most important attributes [11]. The second model is based on an application of a new classifier from the family of ***belief networks*** [12]. Hence, the system being investigated now utilizes five different learning models (classifiers) to identify partially skin lesions. From these five partial results, the system predicts the final result, using a ***special evaluation and voting algorithm***.

2 Structure and operation of the system

Our diagnosis support system employs user interface in the form of a website (Fig. 1), to get the access to its three main working layers (Fig. 2). The first layer, for self-diagnosis, should be used by persons without medical background. This layer consists of two modules: the first one *(Module 1)* allows to determine in a very simple and clear way all symptoms required for correct classification of a given skin lesion. Thus, using this module, user can be easily acquainted with the knowledge required for correct assignment of all symptoms (**A**symmetry, **B**order, six types of **C**olors, five types of **D**iverse structures), related to a given lesion. The second module *(Module 2)* plays a role of an advanced calculator for non-invasive diagnosing of melanoid lesions. Input to this module creates a vector, conveying logical values of 13 previously pointed out descriptive attributes. These values, inputted by the user,

are processed to calculate the 14-th attribute, the ***TDS*** (in a fact, also the ***New_ TDS***). Then, five different algorithms described briefly in Section 3, are applied for development of five partial learning models (say, five partial classifiers). The classification process based on these models is also described in Section 3.
The second layer, dedicated to dermatology specialists, uses only Module 2. The third layer is planned for the next stage of our research. This layer will be based on automatic analysis and recognition of melanocytic images. The initial results, gained along this line, are subject of another paper presented at this conference [13].

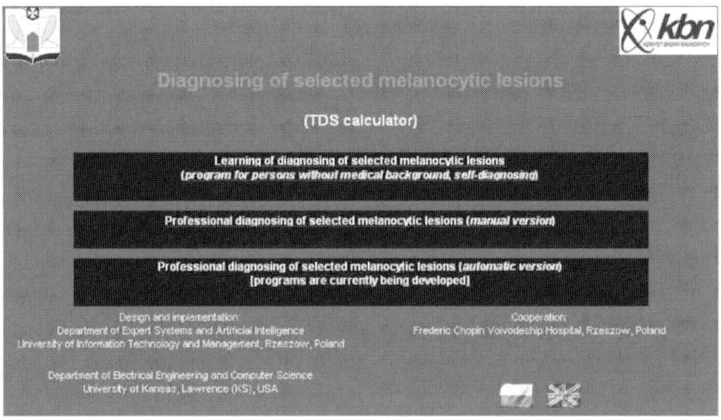

Fig. 1. SystemŠs interface

3 Recognition algorithms

3.1 Learning model based on a classic and optimized *ABCD* rule

Logical values of symptoms, inferred in the first or second layer, are processed using two different algorithms, for calculation of the ***TDS*** parameter, and for calculation of the ***New_ TDS*** parameter. It is worth to say, that both algorithms are based on a constructive induction [14], a very important methodology is machine learning. Then, the enlarged solution space (13+1 dimensions) is searched using the classic ***ABCD*** algorithm (see Equation 1),

$$\mathbf{TDS} = 1,3*\mathbf{A}symmetry + 0,1*\mathbf{B}order + 0,5*\sum \mathbf{C}olors + 0,5*\sum \mathbf{D}iversity \tag{1}$$

and simultaneously, using the optimized formula, for calculation the ***New_ TDS*** (see Equation 2)

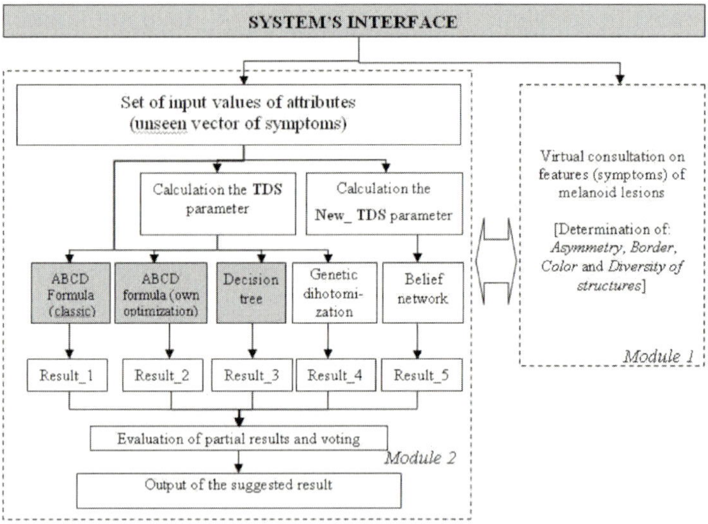

Fig. 2. SystemŠs structure

$$\begin{aligned}\textbf{New\_TDS} &= 0,8 * Asymmetry + 0,11 * Border + 0,5 * C\_White + \\ &+ 0,8 * C\_Blue + 0,5 * C\_DarkBrown + 0,6 * C\_LightBrown + \\ &+ 0,5 * C\_Black + 0,5 * C\_Red + 0,5 * Pigment\_Networks + \\ &+ 0,5 * Pigment\_Dots + 0,6 * Pigment\_Globules + \\ &+ 0,6 * Branched\_Streaks + 0,6 * Structureless\_Areas \end{aligned} \tag{2}$$

It was found, in numerous experiments, that the learning model based on the standard ***TDS*** parameter, classifies unseen objects with an error rate 9–11%, whereas learning model, based on optimized ***New_ TDS*** parameter, classifies the same set of unseen objects with an error rate 5%.

3.2 Learning model in form of *decision tree*

Our recent experiments pointed out that the learning model, based on co called certain decisions tree (see Fig. 3), classifies unseen melanoid skin lesions with an error rate on the on level of about 1,5%.

3.3 Learning model based on the *genetic dichotomization*

This learning model contains **n(n-1)/2** vectors (where generally, **n** is the number of identified concepts, in our case **n** = 4), trained outside the ***IMDLS***, capable to classify correctly four classes of melanoid lesions. These vectors,

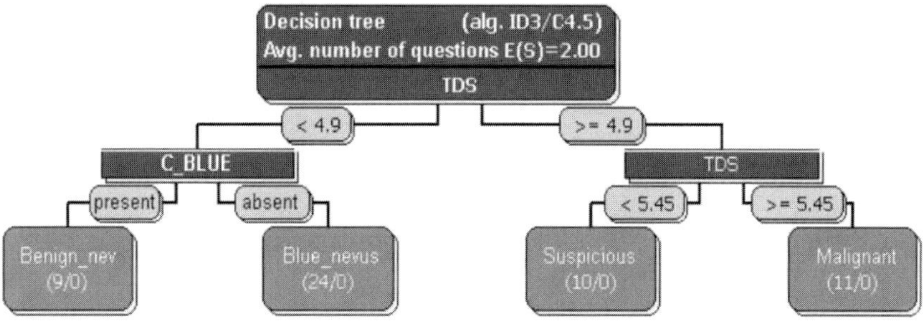

Fig. 3. Learning model in form of *decision tree*

Table 1. Recognition of on unseen case (here ***Melanoma malignant***) by the *genetic dichotomization* model

Vector	Capable to recognize	Class assigned to unseen case (Melanoma_malignant)	Final decision
1	Bening_nev or Blue_nev	Bening_nev or Blue_nev	
2	Bening_nev or Malignant	Malignant	
3	Bening_nev or Suspicious	Bening_nev or Suspicious	Melanoma
4	Blue_nev or Malignant	Malignant	Malignant
5	Blue_nev or Suspicious	Blue_nev or Suspicious	
6	Malignant or Suspicious	Malignant	

initially trained for searching dichotomous solution space, underwent to mutation and crossing in order to extend their recognition capability and efficiency.

Recognition process of unseen cases is executed automatically (see Table 1): the ***IMDLS*** program assigns to unseen case a category, pointed out by the maximal number of vectors.

3.4 Learning model in form of *belief network*

This learning model contains a belief network (see Fig. 4), trained like the genetic dichotomization algorithm – outside the ***IMDLS***. The belief network used showed an error rate roughly 4,5%. It was found, that from all 14 descriptive attributes (symptoms), the most important ones, having direct impact on the decision (diagnosis), were: classic ***TDS*** parameter (***TDS*** on Fig. 4), asymmetry *(ASYMMETRY)*, color blue *(C_BLUE)*, and a structure of the lesion, called pigment network *(D_PIGM_NETW)*. Recognition process of unseen cases is executed automatically: the ***IMDLS*** program assigns to unseen case a category, which displays the highest value of marginal likelihood [12].

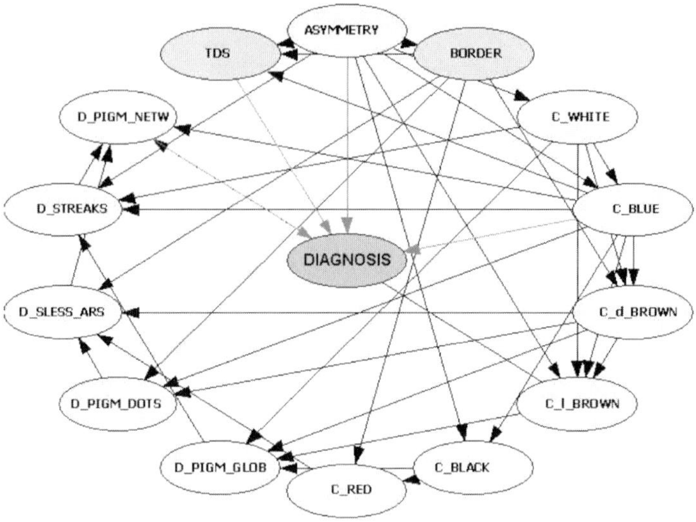

Fig. 4. Learning model in form of *belief network*

3.5 Evaluation and voting algorithm

The skin lesions diagnosing system supplies (five partial results) generated by the five learning models: classic ABCD rule, optimized ABCD rule, decision tree, genetic dichotomization and belief network. Each partial result is associated with its own weight parameter, dependent on error rate characterized for the learning model used (see Table 2). These weight parameters are computed from one general formula:

$$W_x = (1 - ErrorRateoftheModel) \quad (3)$$

Table 2. Weight parameters for learning models

No	Learning model	Weight parameter
1	ABCD formula (classic)	$W_x = 1 - 0,11 = 0,89$
2	ABCD formula (own optimization)	$W_x = 1 - 0,05 = 0,95$
3	Decision tree	$W_x = 1 - 0,015 = 0,985$
4	Genetic dichotomy process	$W_x = 1 - 0,06 = 0,94$
5	Belief network	$W_x = 1 - 0,04 = 0,96$

The final result is prepared depending on sum of weight parameters for suggested diagnosis. It should be stressed, that all learning models developed in our group (i.e. models 2, 3, 4 and 5) seems to be more accurate than the model 1 (world-wide accepted model of Braun-Falco and Stolz [7], called here classical ***TDS*** model). Operation of the evaluation and voting algorithm is explained in Table 3.

Table 3. Calculating of the weight parameters

No	Diagnosis	Bening nevus	Blue nevus	Suspicious nevus	Melanoma malignant
1	ABCD formula (classic)	0,89	0	0	0
2	ABCD formula (own optimization)	0,95	0	0	0
3	Decision tree	0	0,985	0	0
4	Genetic dichotomy process	0,94	0	0	0
5	Belief network	0	0,96	0	0
	Weight parameters	**2,78**	**1,945**	**0**	**0**

Here, ***Benign_nevus*** was identified three times, whereas ***Benign_nevus*** only twice. In addition, weight parameters (last row in Table 3) vote in ratio $2,78/1,946$ for ***Benign_nevus***.

4 Conclusions

In our research we followed the newest trend in diagnosing of skin lesions, namely, the turn to non–invasive identification methods. The ***IMDLS*** software, developed and extensively tested throughout our experiments, can be treated as a reliable and efficient tool that supports non–invasive classification of melanoid spots on the skin. Additionally, the ***IMDLS*** program displays some teaching functions, important for primary physicians. The ***IMDLS*** tool is now available on the website: http://www.wsiz.rzeszow.pl/ksesi.

References

1. http://dermoncology.med.nyu.edu.
2. Kreusch J., Rassner G.: *Standardisierte auflichtmikroskopische Unterscheidung melanozytischer und nichtmelanozytischer Pigmentmerkmale*, Hautartzt 42, 77-81(1991).
3. Kenet R.O., Kang B.J.,Fitzpatrick T.B., Sober A.J., Barnhill R.L.: *Clinical diagnosis of pigmented lesions using digital epiluminescence microscopy*, Arch. Dermatol. 129, 157-158 (1993).
4. Kirn T.F.: *Reasons Unclear for Worlwide Decline in Melanoma*, Skin & Allergy News 31 (5), 41-42 (2000).
5. Grzymala-Busse J.W., Hippe Z.S., Knap M., Paja W.: *Infoscience Technology: An Impact Of Internet Accessible Melanoid Data On Health Issues*, Proceedings of the 19th International CODATA Conference, The Information Society: New Horizons for Science, 7-10 November 2004, Berlin, Germany.

6. Braun-Falco O., Stolz W., Bilek P., Merkle T., Landthaler M.: *Das Dermatoskop. Eine Vereinfachung der Auflichtmikroskopie von pigmentierten Hautveranderungen*, Hautartzt 40, 131-136 (1990).
7. Stolz W., Harms H., Aus H.M., Abmayr W., Braun-Falco O.: *Macroscopic diagnosis of melanocytic lesions using color and texture image analysis*, J. Invest. Dermatol. 95, 491-497 (1990).
8. Alvarez A., Bajcar S., Brown F.M., GrzymalaĆa-Busse J.W., Hippe Z.S.: *Optimization of the ABCD Formula Used for Melanoma Diagnosis*, In: Klopotek M.A., Wierzchon S.T., K. Trojanowski (Eds.), Advances In Computing (Intelligent Information Systems and Web Mining), Physica-Verlag, Heidelberg 2003, pp. 233-240.
9. Bajcar S., Grzymala-Busse W.J., Hippe Z.S.: *Data Mining Analysis of the ABCD Formula Used for Diagnosis of Melanoma, International Workshop on Concurrency Specification and Programming*, Czarna (Poland) 25-27.09.2003.
10. Grzymala-Busse J.W., Hippe Z.S., Knap M., Mroczek T.: *A New Algorithm for Generation of Decision Trees*, TASK Quarterly 8 (2004, No 2) 243-247.
11. Hippe Z.S., Wrzesien M.: *Some Problems of Uncertainty of Data after the Transfer from Multi–category to Dichotomous Problem Space*, In: T. BurczyÅĎski, W. Cholewa, W. Moczulski (Eds.), Methods of Artificial Intelligence, Silesian University of Technology Edit. Office, Gliwice 2002, pp. 185-189.
12. Hippe Z.S., Mroczek T.: *Melanoma classification and prediction using belief networks*, In: Kurzynski M., Puchala E., Wozniak M. (Eds.), Computer Recognition Systems KOSYR 2003, Univ. of Technology Edit. Office, Wroclaw 2003, pp. 337-342.
13. Kwasnicka H., Paradowski M.: *Spread Histogram – A Method for Calculating Spatial Relations Between Objects*, Proceedings of 4th International Conference on Computer Recognition Systems CORES 2005, Wroclaw, Poland (in printing).
14. Michalski R. S., Bratko I., Kubat M.: *Machine Learning and Data Mining*, Methods and Applications, J. Wiley & Sons, London 1998, pp. 79-80, 83, 104.

Baseline and Acceleration Episodes - Clinically Significant Nonstationarities in FHR Signal: Part I. Coefficients of Inconsistency.

Janusz Jezewski, Janusz Wrobel, and Tomasz Kupka

Institute of Medical Technology and Equipment, 118 Roosevelt Str., 41-800 Zabrze, Poland,jezewski@itam.zabrze.pl

Summary. Fetal heart rate (FHR) is the main source of information on the fetal state. FHR is characterized by two components: the basal fetal heart rate (baseline) and the variability of FHR – mainly accelerations (A) and decelerations (D). A correctly determined FHR baseline is a precondition for correct identification of the acceleration and deceleration patterns. Even a small inaccuracy in the FHR baseline curve may significantly distort the detection of A/D, which may subsequently lead to false interpretation of clinical symptoms. It is necessary to develop a method and criteria, which would allow evaluation of the baseline algorithms. The algorithms could be evaluated either by means of direct comparison of baselines or by a comparison of A/D patterns. The paper presents a method which can be used to determine matched A/D patterns recognized using two various baselines in the meaning of their time location. The method uses the coefficients of inconsistency between these patterns.

1 Introduction

Analysis of fetal heart activity is the basic diagnostic tool in present-day perinatal medicine. There are two basic methods to obtain the fetal heart rate signal: direct electrocardiography – based on the heart's electrical activity, and the Doppler ultrasound method – based on the mechanical activity. The fetal heart rate signal (FHR) is determined as the inverse of the duration of the consecutive cardiac intervals expressed in milliseconds. Instantaneous FHR value is extrapolated into a one–minute interval and expressed in beats per minute (bpm), which meets the needs of the medical staff.

Fetal heart rate is characterized by two main components: the basal fetal heart rate (baseline) and the variability of FHR. The waveform representing "a kind of a mean" fetal heart rate over time is referred to as the baseline. FHR variability is associated mainly with short–lasting accelerations or decelerations (A/D) of fetal heart rate. The baseline and the A/D events are the main nonstationary features of the FHR signal. Classical signal processing algorithms require that the nonstationarities should be removed from the signal in the first step whereas the next step is usually the analysis of the signal in time or frequency domain. However, such approach is not very useful in everyday clinical practice. During routine fetal monitoring the FHR signal is usually analysed as one-hour recordings, and the nonstationarities are recognized to be the most clinically important features of FHR signal determined and interpreted both in classical visual and in computer-aided analysis.

A correctly determined FHR baseline is a precondition for correct identification of the acceleration and deceleration patterns in the fetal heart rate waveform. In 1986 the following definition of baseline was raised [5]: "Baseline is the mean level of the FHR when this is stable, accelerations and decelerations being absent. It is determined over a time period of 5 or 10 minutes and expressed in bpm". Acceleration is defined as a transient increase in heart rate of 15 bpm or more and lasting 15 sec or more. Deceleration is a transient episode of heart rate slowing bellow the baseline level of more than 15 bpm and lasting 10 sec or more. This baseline definition is only sufficient for visual analysis and it is completely useless for automated signal analysis. The computer cannot reject these patterns before they are detected so as to estimate the baseline in relation to which the A/D patterns are determined. Moreover, baseline estimation is the most difficult, and at the sometime, the most important in these segments of FHR where accelerations and decelerations occur. In the case of automated signal analysis, the definitions of A/D imply that a strict sequence must be kept, whereby the baseline estimation is the first stage of processing. Figure 1 presents a segment of FHR signal with its interpretation done by a computer-aided fetal monitoring system. The detailed parameters, including duration of accelerations, amplitude and area, are determined and displayed in a graphical form.

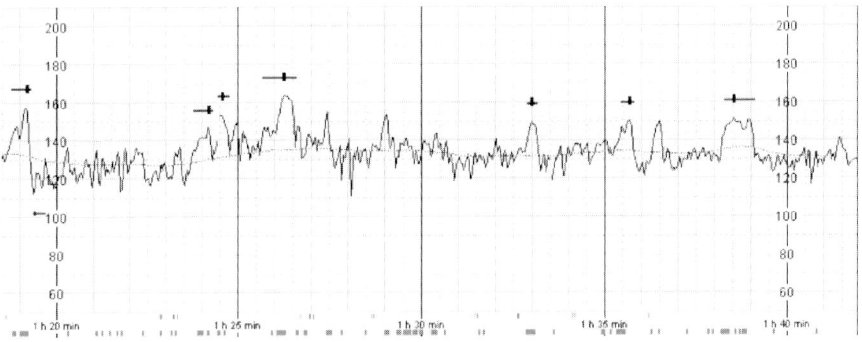

Fig. 1. Segment of FHR waveform and its automated interpretation: estimated baseline (dotted line) and recognized accelerations (graphical markers). Horizontal lines above A correspond to the pattern duration, and the vertical section represents the maximum peak.

There is a common opinion that algorithm for the FHR baseline estimation is responsible for the efficiency of the entire computer-aided fetal monitoring system [1], [4]. Even a small inaccuracy in the FHR baseline curve may significantly distort the detection of A/D, which may subsequently lead to false interpretation of clinical symptoms. Therefore, it is necessary to develop a method and criteria, which would allow evaluation of the baseline algorithms. The algorithms could be evaluated either by means of direct comparison of baselines or by a comparison of A/D patterns, i.e. using the effects of further interpretation of these baselines [3]. The latter approach shows local inconsistencies between the baselines in the places where clinically significant patterns occur. The paper presents a method which can be used to determine matched A/D signal patterns recognized using two various baselines in the meaning of their time location. The method uses the coefficients for inconsistency measurement between these patterns.

2 Baseline inconsistency

Three quantitative coefficients have been defined in order to evaluate the inconsistency between two baselines estimated for a given FHR trace. The following symbols have been used:

$BL = \{BL_1, BL_2, ..., BL_N\}$

$BL' = \{BL'_1, BL'_2, ..., BL'_N\}$

BL and BL'– two baselines estimated for the same FHR trace

N – number of baseline samples

The distance coefficient between the two baselines is a mean value of the absolute differences between the corresponding baseline samples; it is described by the following formula:

$$BLD = \frac{1}{N} \sum_{i=1}^{N} \left| BL_i - BL_i^{'} \right| \quad [bpm] \tag{1}$$

This coefficient ensures that small differences between the lines are accumulated when the two baselines cross each other frequently. Its weakness caused by averaging is that it masks the large but short lasting differences between the baselines.

The mean square distance coefficient for two baselines is determined by the formula:

$$BLS = \sqrt{\frac{1}{N} \sum_{i=1}^{N} (BL_i - BL_i^{'})^2 \quad [bpm]} \tag{2}$$

The coefficient highlights the significant, though short–lasting differences between the baselines. However, a high BLS value does not indicate whether the difference concerns clinically significant or insignificant segments.

The percentage coefficient of definite consistency of two baselines defines the percentage of duration of a given difference (greater than a set ΔBL value) in relation to the entire duration of the FHR record. It is defined as follows:

$$BLP = \frac{1}{N}\sum_{i=1}^{N} K_i \cdot 100\% \tag{3}$$

where:

$$\mathrm{K_i} = \begin{cases} 0 \text{ if } \left|\mathrm{BL_i} - \mathrm{BL_i'}\right| \leq \Delta\mathrm{BL} \\ 1 \text{ if } \left|\mathrm{BL_i} - \mathrm{BL_i'}\right| > \Delta\mathrm{BL} \end{cases} \tag{4}$$

The coefficient is a function of the set difference ΔBL, therefore it is written as follows:

$$\Delta BL = 1 \ or \ 5 \ bpm, BLP_{\Delta BL=1} = BLP1, BLP_{\Delta BL=5} = BLP5 \tag{5}$$

Both the above coefficients are absolute, they cannot be used to determine the degree of inconsistency between the lines, as there is no reference for establishing any limit values for these coefficients.

3 Acceleration inconsistency

The coefficients, that are used to evaluate the inconsistency of A/D determined for a given FHR signal using two baselines, have been defined on the assumption that they should reflect the three main components characterizing the inconsistency of the patterns found:

- difference in the number of patterns,
- difference in the location of the patterns for both lines,
- difference in the area of matched patterns.

The proposed sequence of components corresponds to the sequence in which they are determined. The coefficients were divided into two groups: direct and cumulative, accordingly to how a given component influences on the value of the coefficients. All the coefficients proposed are normalized into a range of 0 - 1, where 1 means full inconsistency whereas 0 means ideal consistency.

A direct coefficient is the measure of inconsistency of one isolated component and all the other components do not affect its value. When defining the cumulative inconsistency coefficient it has been recognized that it should evaluate a given inconsistency component, while the impact of the previous components should be preserved. Considering that acceleration and deceleration patterns differ only in direction of FHR changes, the same methodology can be applied to measure their inconsistency. So, the further description is limited to accelerations only.

The definitions of the acceleration inconsistency use the following symbols:

$A = \{A_1, A_2, ..., A_S\}$ – accelerations determined using the baseline BL
$A' = \{A'_1, A'_2, ..., A'_{S'}\}$ – accelerations determined using the baseline BL'
S– number of accelerations determined using BL
S'– number of accelerations determined using BL'
$P = \{P_1, P_2, ..., P_S\}$ – areas of A_i accelerations
$P' = \{P'_1, P'_2, ..., P'_{S'}\}$ – areas of A'_i accelerations.

In the case of inconsistency coefficients regarding the location and area components it is necessary to determine the criteria for matched location of the accelerations and to develop an algorithm for the detection of such accelerations [2]. Two accelerations have been assumed to be matched, if their common part is at least 5 sec. By introducing a condition for the common part instead of a condition for the location of peaks, we ensure correct interpretation, when FHR waveform temporary gets closer to the baseline within significant accelerations. It may happen that a large acceleration recognized using one baseline will simultaneously be divided into smaller accelerations when the second baseline is examined. Then, in accordance with the above assumption, all the partial accelerations are considered to be matched with the single longer acceleration, although their peaks do not overlap.

These criteria are then used to determine the Q matrix representing the location matching of two sets A and A'. The matrix element in k-th row and m-th column is defined as follows:

$$\mathrm{q_{km}} = \begin{cases} 1 \text{ if acceleration } \mathrm{A_k} \text{ has a location matching } \mathrm{A'_m} \\ 0 \text{ if acceleration } \mathrm{A_k} \text{ has a location not matching } \mathrm{A'_m} \end{cases} \tag{6}$$

Subsequently, the number C of accelerations recognized using the baseline BL, whose locations match the accelerations detected with BL' is calculated. This corresponds to the sum of non-zero rows in the matrix $Q(C \le S)$. A similar procedure is applied to calculate the number C' of the accelerations recognized from the line BL', where the locations match the accelerations detected with BL. It corresponds to the sum of the non-zero lines in the matrix Q $(C' \le S')$.

$$C = \sum_{i=1}^{S} W_i \qquad C' = \sum_{j=1}^{S'} W'_j \tag{7}$$

where:

$$\mathrm{W_i} = \begin{cases} 0 \text{ for } \sum\limits_{j=1}^{S'} q_{ij} = 0 \\ 1 \text{ for } \sum\limits_{j=1}^{S'} q_{ij} \neq 0 \end{cases} \qquad \mathrm{W_j'} = \begin{cases} 0 \text{ for } \sum\limits_{i=1}^{S} q_{ij} = 0 \\ 1 \text{ for } \sum\limits_{i=1}^{S} q_{ij} \neq 0 \end{cases} \tag{8}$$

The determination of the Q matrix and of the number of matching accelerations is illustrated in the following example. Figure 2 shows the accelerations recognized using two baselines, BL and BL'. The accelerations are represented by rectangles of the length proportional to the duration of acceleration. The accelerations constitute two sets:

$A = \{A_1, A_2, A_3, A_4\}$ and $A' = \{A_1', A_2', A_3'\}$. The number of accelerations detected on the basis of the line BL is $S = 4$, and on the basis of line BL' is $S' = 3$.

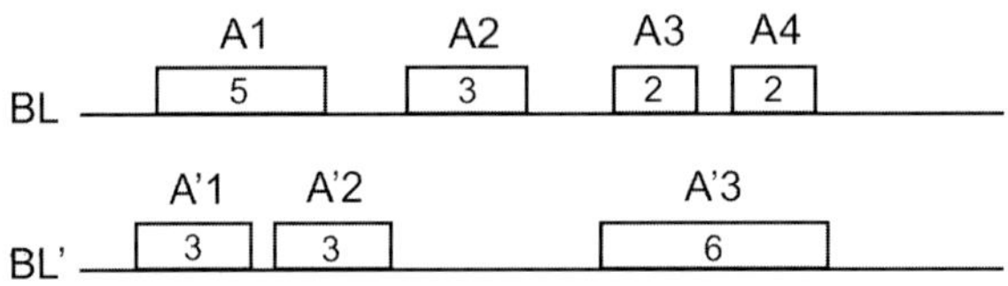

Fig. 2. Two sets of accelerations recognized using two baselines: BL and BL'

Based on the common parts of the accelerations we develop the agreement matrix:

$$[Q]_{4,3} = \begin{Bmatrix} 1\,1\,0 \\ 0\,0\,0 \\ 0\,0\,1 \\ 0\,0\,1 \end{Bmatrix} \tag{9}$$

This matrix is used to determine the numbers of the accelerations whose locations are matching for both lines: $C = 3$, and $C' = 3$. The definitions of the acceleration inconsistency coefficients (direct - D and cumulative - C) for the particular components are given in the Table 1.

The formulas:

$$PC_i' = \sum_{j=1}^{S'} q_{ij} \cdot P_j' \qquad PC_j = \sum_{i=1}^{S} q_{ij} \cdot P_i \tag{10}$$

occurring in the definitions correspond to the sum of the areas of partial accelerations recognized using BL' (BL) which locally match the acceleration of an area P_i (P_j').

Comparison of a set of inconsistency coefficients characterizing the individual inconsistency components is an obstacle to an efficient evaluation of the real range of the existing differences. Therefore, for the purpose of comparing the accelerations recognized using two baselines BL and BL', we have introduced two resultant inconsistency coefficients. The cumulative resultant coefficient AIR_C corresponds to the area inconsistency coefficient AIA_C. This coefficient incorporates all three components of inconsistency: number, location and area.

Table 1. Definitions of the acceleration inconsistency coefficients.

Component	Definition
Number	$AIN_C = \frac{\lvert S-S' \rvert}{S+S'}$
	$AIN_D = \frac{\lvert S-S' \rvert}{\max(S,S')}$
Location	$AIL_C = 1 - \frac{2 \cdot \min(C,C')}{S+S'}$
	$AIL_D = 1 - \frac{\min(C,C')}{\min(S,S')}$
Area	$AIA_C = 1 - \frac{1}{S+S'} \cdot \left[\sum_{i=1}^{S} \frac{\min(P_i,PC'_i)}{\max(P_i,PC'_i)} + \sum_{j=1}^{S'} \frac{\min(P'_j,PC_j)}{\max(P'_j,PC_j)} \right]$
	If min(C, C') = C then : $AIA_D = 1 - \frac{1}{C} \sum_{i=1}^{S} \frac{\min(P_i,PC'_i)}{\max(P_i,PC'_i)}$ IF min(C, C') = C' then : $AIA_D = 1 - \frac{1}{C} \sum_{j=1}^{S} \frac{\min(P'_j,PC'_j)}{\max(P'_j,PC'_j)}$ IF C = C' then : $AIA_D = 1 - \frac{1}{2C} \cdot \left[\sum_{i=1}^{S} \frac{\min(P_i,PC'_i)}{\max(P_i,PC'_i)} + \sum_{j=1}^{S'} \frac{\min(P'_j,PC_j)}{\max(P'_j,PC_j)} \right]$

If $S + S' = 0$ then AIN_C, AIN_D, AIL_C, AIL_D and AIA_C are equal to 0
If $S \cdot S' = 0$ then AIL_D is undefined
If $C \cdot C' = 0$ then AIA_D is undefined

The direct acceleration inconsistency coefficient AIR_D was defined on the basis of the direct inconsistency coefficients relating to the individual components:

$$AIR_D = 1 - \sqrt[3]{(1 - AIN_D) \cdot (1 - AIL_D) \cdot (1 - AIA_D)} \tag{11}$$

This is an "inconsistency oriented" coefficient, which means that if there is full inconsistency for one of the components, the resultant value points to full inconsistency. In other words, if one of the direct coefficients equals 1, the resultant inconsistency coefficient will also equal 1.

4 Conclusions

The nonstationarity of the fetal heart rate signal is subjected to analysis. The first and most important stage in the interpretation of the FHR signal is the estimation of the baseline. The method presented in the paper enables a comparison of two baselines provided by two different estimation algorithms. A direct comparison of the lines seems to be irrelevant to the evaluation of the algorithm. The results do not indicate whether the differences concern significant trace segments, with clinically important, i.e. accelerations/decelerations.

Therefore, we have suggested the method to evaluate the baseline algorithms that compares the A/D detected on the basis of the baselines being evaluated. A quantitative evaluation is possible using the proposed coefficients of patterns inconsistency. However such approach, although relevant from theoretical point of view, needs the experimental proof. The second part of this paper presents the methods applied to carry out these experiments. And the final results provide evidence for the fact that the evaluation of baseline estimation algorithms can be accomplished only by comparing A/D patterns.

Acknowledgement

Scientific study financed from the State Committee for Scientific Research resources in 2004-2006 years as a research project no. 3 T11E 017 26.

References

1. Dawes GS, Moulden M, Redman CWG (1990) Criteria for the design of fetal heart rate analysis systems. Int. J. Bio. Med. Comput. 25: 287–294
2. Jezewski J, Horoba K, Wrobel J, Gacek A (2002) Inconsistency in evaluation of clinical patterns in fetal heart rate waveforms. Journal of Medical Informatics and Technologies 4: 27–35
3. Jezewski J, Wrobel J, Horoba K (1996) Cardiotocographic image processing for extraction of FHR baseline drawn by clinical expert. Med. Bio. Eng. Comput. 35: 99–100
4. Mantel R, Ververs I, Colenbrander GJ, van Geijn HP (1994) Automated antepartum baseline FHR determination and detection of accelerations and decelerations. In: A critical appraisal of fetal surveillance, van Geijn HP and Copray FJK editors, Elsevier Science B.V. 333–348
5. Rooth G (1987) Guidelines for the use of fetal monitoring. Int. J. Gynaecol. Obstet. 25: 159–167

Baseline and Acceleration Episodes - Clinically Significant Nonstationarities in FHR Signal: Part II. Indirect Comparison.

Janusz Jezewski, Krzysztof Horoba, and Adam Matonia

Institute of Medical Technology and Equipment, 118 Roosevelt Str., 41-800 Zabrze, Poland, jezewski@itam.zabrze.pl

Summary. The analysis of fetal heart rate (FHR) trace is not based on evaluation of the baseline itself, but on the episodes which have been detected using this baseline - the acceleration (A) and deceleration (D) patterns. This implies the following approach: for a given trace the baseline is estimated automatically according to the algorithm selected, and once again manually by the clinical expert. For each baseline the A/D patterns are identified automatically by the fixed recognition algorithm. In the next step, the baseline interference procedure is used, which ensures constant difference between the primary baseline and the baseline with the interference superimposed. This difference corresponds to the average expertŠs inconsistency. The goal was to check the correlation between the inconsistencies of baselines and A/D episodes. The obtained results gave a proof that two baselines (from two different automated estimation methods) must be compared indirectly, solely on the basis of the A/D recognized when using them.

1 Introduction

There are many algorithms being used for automated estimation of the baseline. Since the baseline has a crucial influence on the quality of fetal heart rate (FHR) analysis, there is a need to evaluate the algorithms and indicate which is the most effective. However, because of lack of standards [1] defining the algorithm to produce the reference baseline, the only solution is to involve experienced clinical experts. Under certain conditions the expert could provide a baseline, which might be considered as the most closely approximating the ideal baseline [2].

It is important to recognize that the analysis of FHR trace is not based on evaluation of the baseline itself, but rather on the evaluation of the episodes which have been detected using this baseline [4]. The basis for assessment of fetal condition is the presence or absence of acceleration (A) and deceleration (D) patterns in a limited time period. During visual analysis, the clinical expert identifies these episodic patterns, using a hypothetical baseline, which is drawn only in his imagination. The baseline algorithms can also be evaluated by comparing the automatically identified A/D patterns with patterns identified manually by an expert. In other words, the comparison uses the results of the analysis of the baselines. However, as our earlier work [3] has demonstrated, experts do not conform with the standards on the detection of acceleration/decelerations; besides, inter and intraobserver agreement is low [1],[2].

In the light of the above, we may say that the evaluation of the impact of the estimation baseline algorithm on the A/D parameters should be accomplished as follows: baseline for the same record is determined twice: at first, automatically, according to the algorithm selected; and for the second time manually by the expert. For each baseline the A/D patterns are identified automatically by a fixed recognition algorithm. However, the inconvenience of this method is that the baseline drawn by the expert has to be acquired by the computer for further analysis. This implies the use of a scanner, and furthermore a method has to be developed to process baseline waveform from a graphic into a digital form. As a digital representation of the expert's baseline exists along the automatically estimated baseline, the following question may arise: should we not simply confine ourselves to comparing the two lines directly? If we know that the A/D patterns are in a way a derivative of the baseline, and if the interpretation of the threshold criteria applied in recognition procedure for these patterns is fixed, then any inconsistency of the lines should directly point to a corresponding inconsistency of the A/D. However direct evaluation of inconsistency of the baselines does not provide any information on where the differences are located, i.e. whether they occur in such segments of FHR signal that may be significant in the later analysis, i.e. for A/D patterns detection.

This paper presents a certain methodology, the purpose of which was to give a proof that a direct comparison of baselines is irrelevant in terms of usability of a given estimation algorithm in the recognition of the clinically significant nonstationary patterns, i.e. accelerations and decelerations episodes. The applied inconsistency coefficients describing differences between the A/D events are described in the first part of this paper.

2 Methodology

In the first step, we decided to evaluate whether it was possible to compare directly various baselines evaluated for a given FHR waveform. The goal was to compare the various automatically determined baselines and relate them to the baselines that were drawn by the experts. Since we had digital representations of various baselines we could directly compare different baselines. We proposed two baseline inconsistency coefficients calculated as a mean absolute error (BLD) and as mean squared error (BLS). We also defined $BLP5$ coefficient, which represents a percentage of differences between corresponding baselines samples that are greater than 5 bpm. They imply a certain hierarchy of inconsistency of a given baseline in relation to other baselines. However, it is not certain whether they assure the same hierarchy in relation to the inconsistency of the A/D detected using these baselines. In order to check that, we have designed and performed a dedicated research procedure. The main purpose was to evaluate the inconsistency of two baselines provided by a pair of experts with the inconsistency of the acceleration/deceleration patterns, determined automatically for the both lines.

The procedure was performed on 41 records collected between 31st and 41st week of gestation as part of routine clinical examinations, using a computer-aided fetal monitoring system. The length of the records varied from 31 to 60 min. Each of the five experts plotted their baselines on a FHR recording obtained as a printout from the system archive. The FHR recording was scanned and dedicated software converted the baseline from graphical into digital form. The output result was the file containing baseline samples averaged over 2.5 sec intervals, in the range from 50 to 210 bpm and with the accuracy of 0.25 bpm. For every expert's baseline entered to the system as the input parameter, the accelerations and decelerations were detected automatically for a given FHR record, and their detailed parameters were calculated. The baselines of each experts' pair were used to calculate the inconsistency coefficients of these baselines. The sets of detailed parameters corresponding to these lines were the input parameters for the procedure calculating the A/D inconsistency coefficients.

In the second part, for particular traces one of the expert's baselines that had been found as the most consistent in the previous studies was used as the input parameter to the baseline interference generator procedure. Each of the 16 different interferences was so modelled that the inconsistency between the expert's baseline and the baseline with the superimposed interference was constant. The generator output file contained baseline samples with superimposed interference. For both lines the system determined the detailed acceleration/deceleration parameters, which were the basis for inconsistency coefficients calculation. Baseline inconsistency coefficients were simultaneously calculated for both baselines. The purpose of this part of experiment was to evaluate whether the constant baseline inconsistency coefficient values imply the constant values of inconsistency of A/D recognized automatically on the basis of these baselines.

The main assumption underlying the selection of the baseline interfering function was that the difference between the original baseline and the baseline with the interference superimposed was constant and corresponded to the average inconsistency between the baselines from experts. The averaged values of expert's baseline inconsistency coefficients obtained in the first stage were as follows: BLD = 2.6 bpm, BLS = 3.6 bpm (BLS/BLD = 1.38) and $BLP5$ = 12 %. The BLD and BLS coefficients for the two lines BL and BL', where line BL' represents the line BL with a interference I added, can be expressed exclusively using the interference I:

$$BLD = \frac{1}{N}\sum_{i=1}^{N}|I_i| \qquad BLS = \sqrt{\frac{1}{N}\sum_{i=1}^{N}{I_i}^2} \tag{1}$$

Assuming that the interference values for the individual samples of the baseline are determined using the function $sin^{\mathrm{k}}(i)$, a value of the exponent k has to preserve the ratio $BLS/BLD \approx 1.38$. The amplitude was selected to ensure that the conditions for $BLD = 2.6$ bpm and $BLS = 3.6$ bpm were met. The function $6.1 \times sin^3(i)$ gave $BLD = 2.25$ bpm and $BLS = 3.41$ bpm, while $6.9 \times sin^4(i)$ gave 2.59 and 3.61, respectively. It seemed that the power of 4 gave values of BLD and BLS more closer to experts' results. However, the $BLP5$ should have been taken into account, which for k=3 was equal to 17 %, whereas for k=4 it reached 21 % (Table 1).

Table 1. Establishing the exponent k and amplitude A for the function $A \times sin^{\mathrm{k}}(i)$.

k	Amplitude of 1 bpm			Amplitude selection with respect to BLS, BLO and to $BLP5$ in ()			
	BLS/BLO	BLS [bpm]	BLO [bpm]	A [bpm]	BLS [bpm]	BLO [bpm]	$BLP5$ [bpm]
3	1.33	0.56	0.42	6.1 (5.8*)	3.4 (3.2*)	2.6 (2.5*)	17 (12*)
4	1.40	0.52	0.37	6.9 (5.9)	3.6 (3.1)	2.6 (2.2)	21 (12)

*selected parameters

Since that index was very sensitive to amplitude, a correction of amplitude was required. In order to obtain a $BLP5 = 12$ % it was necessary to assume the amplitude of 5.8 bpm for $k = 3$, and for $k = 4$ the amplitude of 5.9 bpm. At the end, it was checked once again how the amplitude adjustment affected the BLD and BLS values. Finally, k=3 and the amplitude of 5.8 bpm were established, providing the optimal matching of all the three coefficients for those settings. The general formula for particular interferences I_M is as follows:

$$I_M = A \cdot \sin^4(\varpi \cdot t + \varphi) \tag{2}$$

where:
M – interference marker, $1 \leq M \leq 16$
A – amplitude, $A \in \{3; 5.8\}$ $[bpm]$
φ – phase, $\varphi \in \{0; 1/2\pi; \pi; 3/2\pi\}$ $[rad]$
ϖ – period, $\varpi \in \{10; 20\}$ [min]

3 Results

The relationship between the baseline inconsistency coefficient BLS and the joined A/D resultant direct inconsistency coefficient A/DIR_D is represented in a form of histogram (Fig. 1). The A/DIR_D was calculated on a base of the acceleration and deceleration resultant coefficients as follows:

$$A/DIR_D = \frac{1}{2}(AIR_D + \frac{1}{2}DIR_D) \cdot 100\% \tag{3}$$

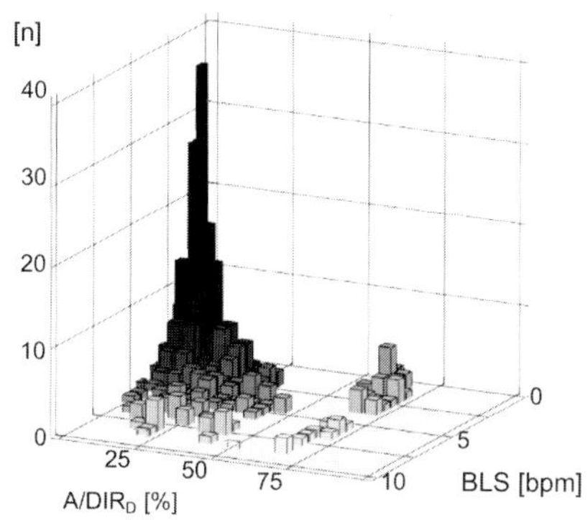

Fig. 1. Histogram of the joined A/D resultant direct inconsistency coefficient A/DIR_D and the baseline inconsistency coefficient BLS. The class width for A/DIR_D is 5 % and 0.5 bpm for BLS.

The histogram clearly shows that the mapping between the baseline consistency coefficient BLS and A/D consistency coefficient is not one-to-one. Values of BLS from the range of 1.5 to 2 bpm correspond to values of A/DIR_D from range of 0 to 20 %, but also to much higher values from the range 65 to 75 %. On the other hand, high values of A/DIR_D (70 to 75 %) are related with BLS values varying from 1.5 to 5 bpm. A similar observation was made for the BLO and $BLP5$ coefficients.

The changes of the acceleration/deceleration coefficient A/DIR_D as a function of particular traces and generated interferences are represented in Figure 2.

For a baseline inconsistency expressed by the value of $BLS = 3.2$ bpm, and corresponding to average inconsistency between the baselines drawn by experts, the influence of the changes of period and phase shift of the interfering function on the A/D parameters is obvious, although it looks different for different traces.

Change of baseline shape observed for the trace 8/2 (trace number from our database) caused large changes of A/D parameters that has been confirmed by values of the A/DIR_D varying from 9 to 72 % (Fig. 3). On the other

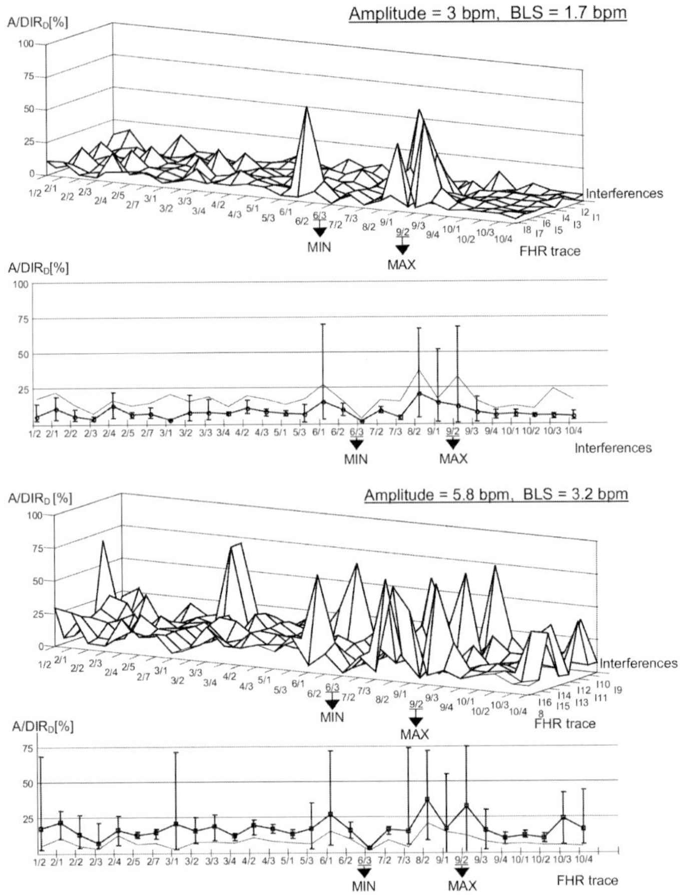

Fig. 2. Changes of the resultant A/DIR_D for the assumed baseline interferences in relation to the individual traces. The 3D diagram shows the influence of the changes of period and phase shift of the interference superimposed on baseline. The 2D diagram shows the A/DIR_D values (minimum, maximum and mean) for the whole set of interference functions. The additional thin line represents the diagram of mean A/DIR_D value for the alternative amplitude. The MIN and MAX indicate the traces that are the least sensitive and the most sensitive to baseline interferences.

hand, the trace 6/3 was the least sensitive to interferences, and the A/DIR_D varied in a range of 2-4 %. For interferences of 3 bpm amplitude (for BLS = 1.7 bpm) similar tendencies were noted, however range of the A/DIR_D variation was narrower. The FHR trace 8/2, as very sensitive, has been shown for interferences of a lower amplitude, whereas the trace 6/3 of low sensitivity, for the higher amplitude. That lets to emphasize various levels of sensitivity to baseline interference.

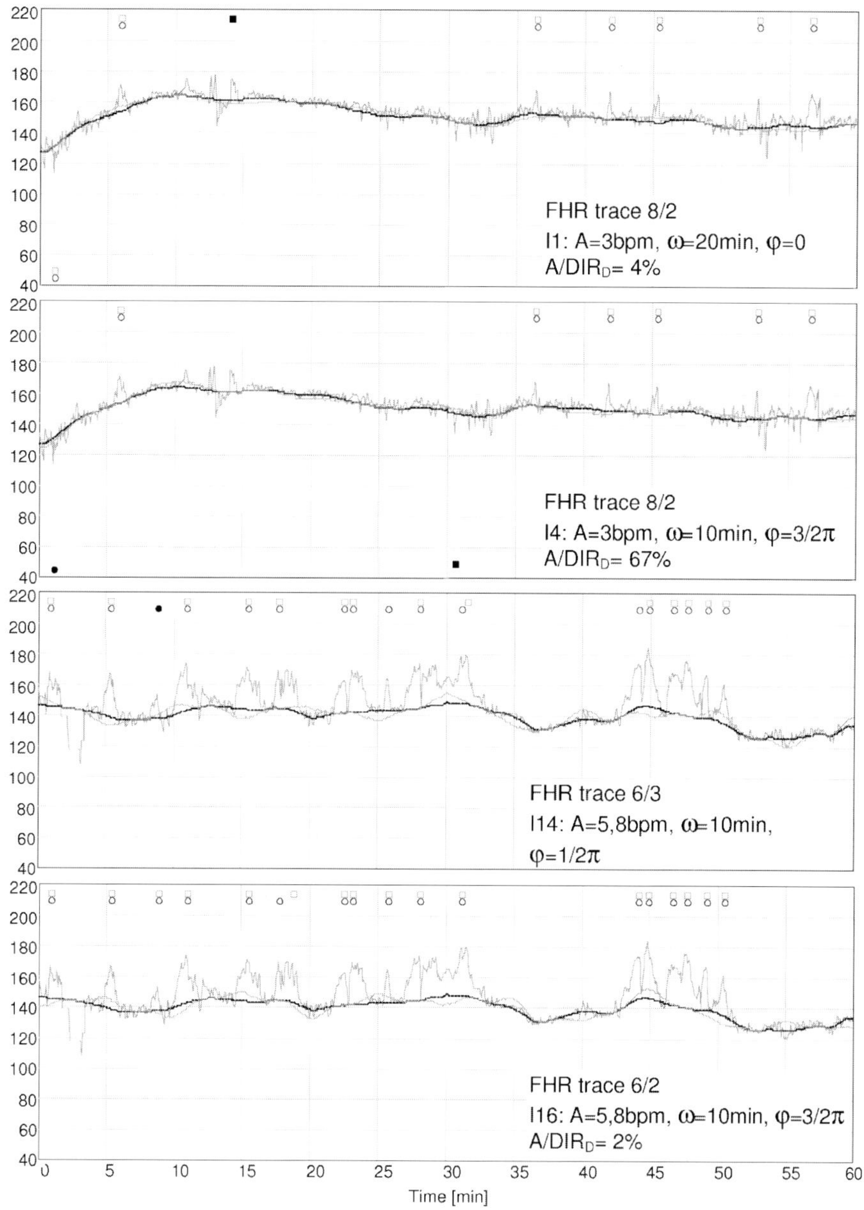

Fig. 3. FHR traces with various sensitivity to the baseline interference. Trace marked 8/2 as the most sensitive has been shown for interferences $I1$, $I4$ (low amplitude). Trace 6/3 characterized by the lower sensitivity has been shown for interferences $I14$, $I16$ (high amplitude). Marking: – expertŠs baseline BL, – baseline with superimposed interference BL, ○ episodes of accelerations decelerations recognized on a basis of BL, □ accelerations/ decelerations recognized on a basis of BL, •, ■ episodes inconsistent as regards to localization.

4 Conclusions

The relation between the inconsistency of baselines and inconsistency of acceleration/deceleration patterns is not a one-to-one mapping (Fig. 1). The baseline inconsistency may be low and it may be accompanied by a high inconsistency in the recognized patterns. This depends on whether the local line inconsistency concerns the segments with A/D events.

The results indicate that an identical inconsistency of two different baselines may demonstrate itself in a great differentiation of the A/D parameters identified on the basis of these lines. This means that direct baseline inconsistency evaluation is not a relevant measure of the usefulness of these lines for the automated acceleration/deceleration detection procedure. The main reason is that we do not have such baseline inconsistency coefficients, which would ensure measurement of local inconsistency, particularly for the clinically significant fragments of FHR signal, i.e. within the A/D events.

In conclusion it may be said that two baselines, a result of two different automated estimation methods, must be compared, or more accurately speaking, must be evaluated for their usage to fetal monitoring system, solely on the basis of the A/D recognized when using these baselines. The "gold standard "should be provided by the baseline from clinical experts, which already at the process of comparison is represented exclusively by the A/D patterns interpreted automatically using their baselines. Averaging of interpretations of the group of experts should rely on averaging the results of inconsistency between a given baseline estimation method and any particular expert.

Acknowledgement

Scientific study financed from the State Committee for Scientific Research resources in 2004-2006 years as a research project no. 3 T11E 017 26.

References

1. Arduini D, Rizzo G, Giannini F, Garzetti GG, Romanini C (1993) Computerized analysis of fetal heart rate: II. Comparison with the interpretation of experts. J. Matern. Fetal Invest. 3: 165–168
2. Gagnon R, Campbell MK, Hunse C (1993) A comparison between visual and computer analysis of antepartum fetal heart rate tracings. Am. J. Obstet. Gynecol. 168: 842–847
3. Jezewski J, Wrobel J, Horoba K (2002) Fetal heart rate variability: clinical experts versus computerized system interpretation. In: Proc. of 24th Int. Conf. IEEE EMBS, 1617–1618
4. Rooth G (1987) Guidelines for the use of fetal monitoring. Int. J. Gynaecol. Obstet. 25: 159–167

Bias Field Correction for MRI Images

Jaber Juntu[1], Jan Sijbers[2], Dirk Van Dyck[2] and Jan Gielen[3]

[1] Vision Lab, University of Antwerp, Groenenborgerlaan 171, B-2020, Antwerpen, Belgium. jaber.juntu@ua.ac.be
[2] Vision Lab, University of Antwerp, Groenenborgerlaan 171, B-2020, Antwerpen, Belgium. {jan.sijbers,dirk.vandyck}@ua.ac.be
[3] UZ, Antwerpen, 2650 Edegem, Belgium. jan.gielen@uza.be

Summary. Bias field signal is a low-frequency and very smooth signal that corrupts MRI images specially those produced by old MRI (*Magnetic Resonance Imaging*) machines. Image processing algorithms such as segmentation, texture analysis or classification that use the graylevel values of image pixels will not produce satisfactory results. A pre-processing step is needed to correct for the bias field signal before submitting corrupted MRI images to such algorithms or the algorithms should be modified. In this report we discuss two approaches to deal with bias field corruption. The first approach can be used as a preprocessing step where the corrupted MRI image is restored by dividing it by an estimated bias field signal using a surface fitting approach. The second approach shows how to modify the fuzzy c-means algorithm so that it can be used to segment an MRI image corrupted by a bias field signal.

1 Introduction

A bias field is a low frequency smooth undesirable signal that corrupts MRI images because of the inhomogeneities in the magnetic fields of the MRI machine. Bias field blurs images and thus reduces the high frequency contents of the image such as edges and contours and changes the intensity values of image pixels so that the same tissue has different graylevel distribution across the image. A low level variation will not have great impact on clinical diagnosis. However it degrades the performance of image processing algorithms such as segmentation and classification or any algorithm that is based on the assumption of spatial invariance of the processed image. A pre-processing step is needed to correct for the effect of bias field before doing segmentation or classification. The aim of the paper is to show how to restore the corrupted image and how to modify a classical fuzzy c-means clustering algorithm for MRI images corrupted by bias field signal.

2 Methods to Remove the Bias Field Signal

Different methods have been used for bias field correction such as using pre-scanned images, high-pass filtering and homomorphic filtering [1]. In this report, we discuss two techniques, namely, estimating bias fields by parametric surface fitting, and an iterative removal of bias field based on modifying the objective function of the fuzzy *c*-means algorithm. Experimental results show the effectiveness of these techniques.

2.1 The Parametric Surface Fitting Approach

A common practice for images that are unevenly illuminated is to divide the corrupted image by a background image represents an estimate of the variation in the illumination across the image. The same can be done for MRI images corrupted by bias field signal. The background image is normally estimated from the corrupted image by low pass filtering operation. Since it is very difficult to design an optimal low-pass filter that has sharp cut-off frequency and at the same time has no ripples in the pass-band and stop-band regions, the background image estimated this way has some noise introduced such as ripples in the image and ringing around the edges. To improve the quality of the background image, a two-dimensional surface equation is fitted to data points selected from the background image and then the fitted equation is used to generate the bias field signal. The bias field signal obtained this way is much smoother than the background image obtained using a low-pass filtering operation alone. Fitting the surface also averages out the noise introduced by the low-pass filtering. The form of the fitted surface equation and the criteria used for the fitting process have to be decided based on certain objectives.

In [2, 3] the *PA*rametric *BI* as field *C*orrection (PABIC) method is proposed. The surface equation to be fitted is chosen as the superposition of Legendre polynomials combined with a probabilistic parametric model for the tissue class statistics. The valley function, a nonlinear objective function, is used to find the parameters of the model. The procedure has no mechanism for controlling the overfitting and has large number of parameters to set a priori.

In this section we propose to fit much more simple surfaces than the Legendre polynomials to a background image and use the simple least-squares as an objective function. The steps of the algorithm are:

1. Extract a background image from the corrupted MRI image, for example, by smoothing the image with a Gaussian filter of a large bandwidth (about 2/3 the size of the MRI image) to filter out all the image details that correspond to high-frequency components.
2. Select few data points from the background image and save their coordinates and graylevel values into a matrix $D = (x_i, y_i, g_i), i = 1, 2, ...n$. It is recommended not to select points from the regions where there is no MRI signal since this regions has no bias field signal.

3. Select a parametric equation for the fitted surface . It is better to fit simple surfaces such as low order polynomial surfaces since they are very smooth and their parameters are very easy to estimate.
4. Estimate the parameters of the surface that best fits the data in matrix D by the method of nonlinear least-squares.
5. Use the fitted equation to generate an image of the bias field signal.
6. Divide the corrupted `MRI` image by the estimated bias field image in step 5.

Even though different surfaces can reasonably fit the data very well and it is not possible to tell which surface is most likely represents the actual bias field signal, however, in practice the bias signal estimated by fitting a smooth 2-dimensional polynomial surface to a background image can be used effectively to restore the corrupted `MRI` image.

How to Do the Surface Fitting?

Fitting of the surface is done by means of the Levenberg-Marquardt algorithm for nonlinear least squares fitting of a function $f(x, y; a_1, ..., a_m)$ of known form to n data points $\{(x_1, y_1, g_1), ..., (x_n, y_n, g_n)\}$. For example, a polynomial surface of degree three can be fitted which has the following equation:$f(x, y; \mathbf{a}) = a_1x^3 + a_2y^3 + a_3$, where $\mathbf{a} = \{a_1, a_2, a_3\}$ is the parameter vector that define the surface.

If we substitute the data points in the nonlinear function we get an overdetermined set of equations, i.e.,

$$\left\{ \begin{array}{l} g_1 = f(x_1, y_1; a_1, a_2, ..., a_m) \\ . \\ . \\ g_n = f(x_n, y_n; a_1, a_2 ..., a_m) \end{array} \right\}$$

These equations can be solved to obtains the unknown parameter vector $(a_1, a_2 ..., a_m)$ by minimizing the sum of the squares of the differences between the data and the fitted function

$$Q(\mathbf{a}) = \frac{1}{2} \sum_{i=1}^{n} (g_i - f(x_i, y_i; a_1, a_2, ..., a_m))^2 \tag{1}$$

Let $r_i(\mathbf{a}) = (g_i - f(x_i, y_i; a_1, a_2, ..., a_m))$, which is the residual vector of point i, then equation(1) can be written as:

$$Q(\mathbf{a}) = \frac{1}{2} \sum_{i=1}^{n} (r_i(\mathbf{a}))^2 \tag{2}$$

According to the Levenberg-Marquardt algorithm, Eq. 2 can be solved iteratively to find the values of the parameters vector ($\mathbf{a}$) starting from an initial estimate of the parameter vector ($\mathbf{a_0}$) using:

$$\mathbf{a}_{i+1} = \mathbf{a}_i - (H + \lambda\ diag[H])^{-1}\ \nabla Q(\mathbf{a}_i) \tag{3}$$

where H and $\nabla Q(\mathbf{a}_i)$ are Hessian matrix and the gradient of Eq. 2 both evaluated at $\mathbf{a}_i$, $diag[H]$ is the diagonal elements of the Hessian matrix. At each iteration, the algorithm tests the value of the residual error $Q(\mathbf{a})$ and adjusts λ accordingly (see [4] for more details). The result of applying surface fitting to a checker image corrupted by a bias field signal is shown in Fig. 1. The fitted surface is $z = a + bx^n + cy^m$. The estimated parameters are $a = -3.6765, b = 12.2149, c = 0.6581, n = 1, m = 0.5$. The fitted equation is used to generate the bias field image shown in Fig. 1b. Dividing the image in Fig. 1a by the generated bias field image in Fig. 1b results in the checker image in Fig. 1c. The improvement in SNR is 17.365dB.

Fig. 1. (a) A checker image corrupted by a bias field signal. (b) Surface representing the bias field signal as estimated by the Levenberg-Marquardt algorithm. (c)The results of dividing image a by image b.

The result of applying surface fitting technique to a real biased MRI image is shown in Fig. 2. The fitted equation is $z = a + bx + cx^2 + dy + ey^2 + fy^3$. The estimated parameters are $a = 9.2, b = 0.19, c = -0.023, d = 0.24, e = -0.061, f = 3.5$. The improvement in SNR is 16.7306dB.

The method works well even for images that are highly corrupted by a bias field signal like the image in Fig. 3. The improvement in SNR for this case is 19.01dB.

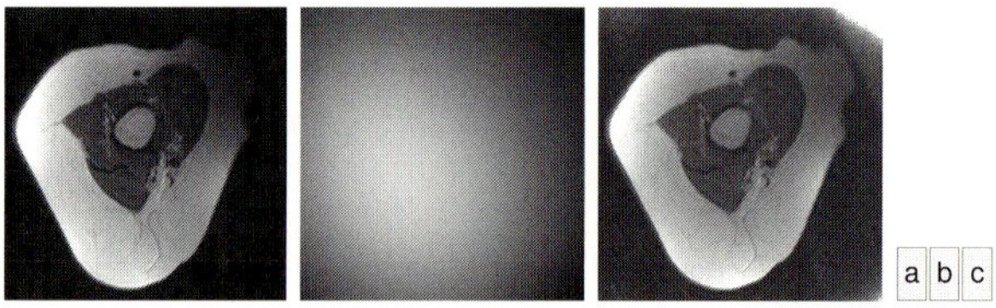

Fig. 2. A bias field correction of MRI image. (a) The original biased image. (b) The estimated bias field by the Levenberg-Marquardt algorithm.(c) The corrected image.

This technique works well for most biased MRI images; however, we have to specify beforehand the parametric form of the fitted surface. For example, we can fit: Quadratic surfaces, Polynomial surfaces, 2D Splines, etc. As a rule, it

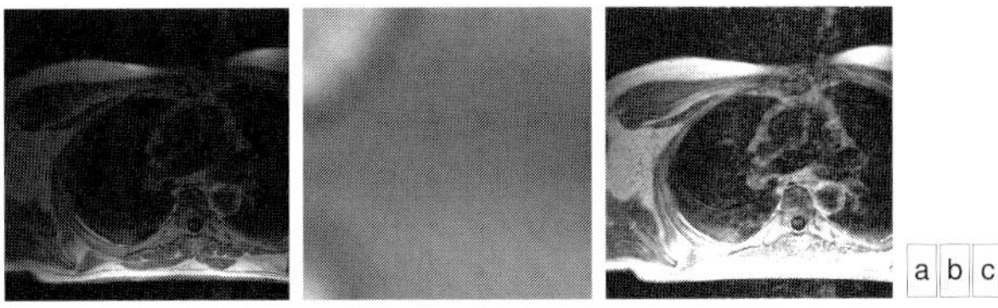

Fig. 3. (a) A severely biased MRI image. (b) The estimated bias field by the Levenberg-Marquardt algorithm. (c) The corrected MRI image.

is better to fit low order polynomial surfaces because they take less time to calculate and approximate well the underlying bias field signal. Instead of fitting a pre-defined surface, it is also possible to fit a general non-parametric surface using neural networks, radial basis networks or support vector machines to the data points.

2.2 Segmentation of MRI Images by the Fuzzy *c*-means Algorithm

Some image processing algorithms can be modified so that they can work with MRI images corrupted by bias field signal. For example, Wells [5] combined a classification algorithm and bias filed correction in an iterative algorithm. In each iteration he estimates the bias field for each pixel then uses Bayes' theorem to guess the class of each pixel. Another similar approach was reported by Ripley [6].

The fuzzy *c*-means algorithm is an iterative algorithm proposed by Dunn [7] and modified by Bezdek [8] for clustering. It has also been used for segmentation. Here, we present a simple modification of the fuzzy *c*-means algorithm to segment field biased MRI images.

To segment an MRI into c segments using fuzzy *c*-means algorithm we minimize the following cost function

$$J_m\left(\mathcal{X}, U, v\right) = \sum_{i=1}^{c} \sum_{k=1}^{N} \left(u_{ik}\right)^m d^2(v_i, x_k)\,. \tag{4}$$

where the data points $\mathcal{X} = \{x_1, x_2, ..., x_N\}$ represent the MRI pixels stacked as a vector by lexicographic operation, $U = \{u_1, u_2, ..., u_c\}$ are the membership values taking values in the interval [0,1]. $d(x, y)^2$ is a distance function which is mostly considered as the Euclidean distance.

Two constraints are usually used when optimizing Eq. 4, namely,

$$\begin{cases} \sum\limits_{k=1}^{N} u_{ik} > 0; \quad \forall i \in \{1, \ldots, c\} & \text{(a)} \\ \sum\limits_{i=1}^{c} u_{ik} = 1; \quad \forall k \in \{1, \ldots, N\}\,. & \text{(b)} \end{cases} \tag{5}$$

The first constraint ensures that no cluster is empty. The second constraint forces the summation of the membership values for every single data point to be equal to one.

Substituting back the Euclidean distance function in Eq. 4,

$$J_m(\mathcal{X}, U, v) = \sum_{i=1}^{c} \sum_{k=1}^{N} (u_{ik})^m \|x_k - v_i\|^2 . \quad (6)$$

The mathematical model for a biased `MRI` image is

$$y_k = x_k B_k . \quad (7)$$

In the last equation y_k, x_k, B_k are the corrupted image, the original image and the bias field signal stacked columnwise, respectively. If we substitute back Eq. 7 in Eq. 6 we obtain

$$J_m(Y, U, B, v) = \sum_{i=1}^{c} \sum_{k=1}^{N} (u_{ik})^m \left\| \frac{y_k}{B_k} - v_i \right\|^2 . \quad (8)$$

To optimize Eq. 8, use the Lagrange multipliers method and the constraints in Eq. 5

$$J_m(Y, U, B, v) = \sum_{i=1}^{c} \sum_{k=1}^{N} (u_{ik})^m \left\| \frac{y_k}{B_k} - v_i \right\|^2 + \lambda \left(1 - \sum_{i=1}^{c} u_{ik} \right) . \quad (9)$$

Equation(9) is solved by taking the derivative of $J_m(Y, U, B, v)$ with respect to U, B_k, v_i, λ and setting the result equal to zero

$$\frac{\delta J_m(.)}{\delta \lambda} = 0, \quad \frac{\delta J_m(.)}{\delta U_{ik}} = 0, \quad \frac{\delta J_m(.)}{\delta v_i} = 0 \text{ and } \frac{\delta J_m(.)}{\delta B_k} = 0.$$

After carrying out all the necessary calculation, eliminating λ by substituting it back, we arrive at the following three equations:

$$U_{ik} = \frac{1}{\sum_{j=1}^{c} \left(\left\| \frac{y_k}{B_k} - v_i \right\|^2 \Big/ \left\| \frac{y_k}{B_k} - v_j \right\|^2 \right)^{\frac{1}{m-1}}} , \quad (10)$$

$$v_i = \left(\sum_{k=1}^{N} (u_{ik})^m \frac{y_k}{B_k} \right) \Big/ \sum_{k=1}^{N} (u_{ik})^m , \quad (11)$$

$$B_k = \left(\sum_{k=1}^{c} (u_{ik})^m y_k \right) \Big/ \sum_{k=1}^{c} (u_{ik})^m v_i , \quad (12)$$

These three equations(10-12) are used iteratively to calculate the membership matrix (U), the prototype of each cluster (v_i) and the bias field signal (B_k), respectively. The following algorithm shows how to apply the fuzzy c-means algorithm to a biased `MRI` image:

1. Set the fuzziness index to a constant (empirically $m = 1.5$), the number of clusters c and the size of the spatial constraint window used to average the membership matrix.
2. Estimate an initial bias field signal B_0 from the corrupted `MRI` image by smoothing it with a low-pass Gaussian filter.
3. Calculate the membership matrix U_{ik} using Eq. 10.
4. For each pixel take the average of the membership matrix around a window of size 3×3 or 5×5. Normalize the membership matrix so for each image pixel the summation of the membership values corresponding to different clusters equals to one as in Eq. 5b.
5. Calculate the prototype centers v_i using Eq. 11 and estimate the bias field B_k using equation (12).
6. Test if the difference between the norm of the previous calculated membership matrix and the norm of the current membership matrix $\|U_l\| - \|U_{l+1}\|$ is less than a pre-specified error or if the algorithm exceeds certain number of iterations then exit, otherwise iterate again.

At each iteration the algorithm estimates a new approximation of the bias field signal and uses it to update the prototype centers and the membership matrix. The modified algorithm is sensitive to the number of iterations. If we iterate for long time, the algorithm is not guaranteed to converge. To overcome the convergence problem, a spatial constraint are imposed which forces neighbor pixels to be classified to the same class. This can be seen in step 4 of the algorithm where the membership matrix U_{ik} is averaged before calculating the prototype centers v_i .

The results of applying the classical c-means algorithm and the modified algorithm to the biased checker image and the biased `MRI` image are shown in Fig. 4 and Fig. 5, respectively.

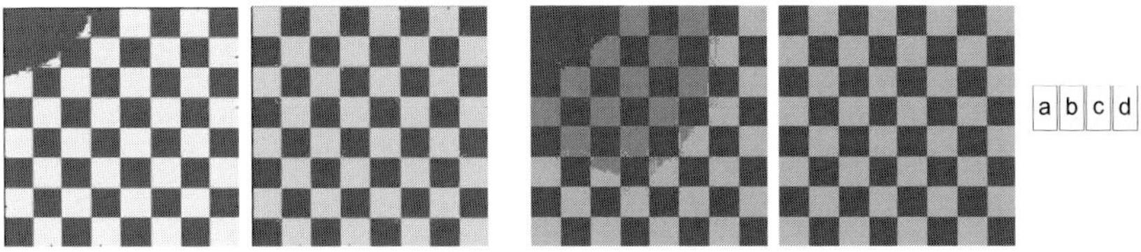

Fig. 4. The results of clustering using the classical and the modified fuzzy c-means algorithms. (a,b) Two-class clustering using the classical fuzzy c-means algorithm and the modified one, respectively. (c,d) Three-class clustering using the classical and the modified fuzzy c-means algorithm, respectively.

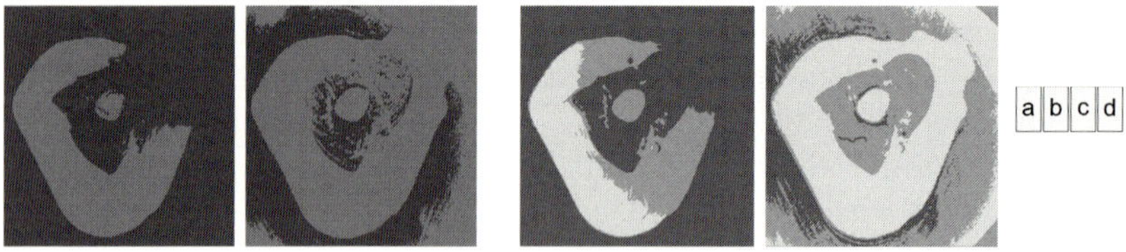

Fig. 5. The results of applying the classical fuzzy c-means algorithm and the modified fuzzy c-means algorithm on a field biased MRI image. The first set (a) and (b) are the result of two classes clustering using the classical and the modified c-means algorithms, respectively. The second set (c) and (d) are the result of three classes clustering using the classical and the modified c-means algorithms respectively. Notice how the modified algorithm better detected the boundaries in both cases.

3 Conclusion

The bias field signal signal changes the intensity values of the MRI image pixels and it should be reduced before doing segmentation or classification. Removal of the bias field is an ill-possed problem and some implicit or explicit assumptions should be made to obtain an approximate solution. For the surface fitting approach the assumption made here is the smoothness of the bias field signal. The least squares fits the smoothest surface to a given data points by averaging out the residual errors. For the modified fuzzy c-means algorithm, the objective is the minimization of the distances between the cluster centers and the data points. Including the bias field in the objective function of the fuzzy c-means algorithm disturbs the convergence of the algorithm. The solution adopted here is averaging the membership values of the neighborhood pixels which forces them to be clustered to the same class.

4 Acknowledgements

Thank to the *Universitaire ziekenhuis (UZA)*, department of *Magnetic Resonance Imaging* for providing the MRI images used in this research.

References

1. B.H. Brinkmann et al. Optimized homomorphic unsharp masking for MR grayscale inhomogeneity correction. *IEEE Trans. Med. Imag.*, (17):161–171, 1998.
2. Gabor Szekely et al. Parametric estimate of intensity inhomogeneities applied to MRI. *IEEE Trans. on Medical Imaging*, 19(3):153–165, March 2000.
3. C. Brechbuhler , G. Gerig and G. Szekely. Compensation of spatial inhomogeneity in MRI based on a parametric bias estimate. In *Proceedings of the Fourth International Conference on Visualization in Biomedical Computing*, Hamburg, Germany, 1996.

4. William H. Press et al. *Numerical Recipes in C++.* Cambridge University Press, second edition, 2002.
5. W.M. Wells III et al. Statistical intensity correction and segmentation of MRI data. In *proceedings of the SPIE: Visulaization in Biomedical Computing 1994*, October 1994.
6. Brian D. Ripley and Jonahthan Marchini. Statistical considerations in magnetic resonance imaging of brain function. Technical report, SCIA99 conference, Kangerlussuaq,Greenland, June 1999.
7. J.C. Dunn. A fuzzy relative of the ISODATA process and its use in detecting compact well-separated clusters. *J. cybern.*, 3(3):32–57, 1974.
8. James C. Bezdek et al. *Pattern Recognition with Fuzzy Objective Function Algorithms.* New York: Plenum Press., 1981.

SEM Image Analysis for Roughness Assessment of Implant Materials

Wlodzimierz Klonowski, Elzbieta Olejarczyk, and Robert Stepien

[1] Institute of Biocybernetics and Biomedical Engineering, Polish Academy of Sciences, 4 Trojdena Str, 02-109, Warsaw, Poland wklon@ibib.waw.pl
[2] GBAF-SENSATION, Medical Research Center, Polish Academy of Sciences, Warsaw, Poland gbaf@cmdik.pan.pl

Summary. We propose a new very simple method to determine roughness of a surface of an implant material from its scanning electron microscopy (SEM) image. For this purpose we have combined a preprocessing method that has been used in histopathology with fractal method used in nonlinear time series analysis. In the pre-processing step the image is transformed into 1-D signals ('landscapes') that are subsequently analyzed. Our method draws from multiple disciplines and may find multidisciplinary applications.

1 Introduction

Surface roughness plays an important role in cell adhesion to the surface. Quality of materials used for orthopedic prostheses and dental implants does depend on their surface properties. That is why treatment leading to modification of surface topographical properties shows potential for improving osteointegration of orthopedic prostheses and dental implants [1]. There is still a need for relatively simple methods of assessment of surface properties, based on analysis of experimental data such as microscopic images. The problem is that we all become more and more specialized in very narrow disciplines and we often do not know that the methods we want to apply in our research have been used for a long time in other disciplines. When we learn about it we are often amazed like Molier's Mr. Jourdain (*Le Bourgeois Gentilhomme* II.iv) who says: 'Good heaven! For more than forty years I have been speaking prose without knowing it'. What we do need is a multidisciplinary approach to the problems, and our method draws from multiple disciplines and may have multidisciplinary applications [2, 3, 4].

For example, epithelial roughness and texture play a central role in histopatological diagnosis of malignancy. T.Mattfeldt pre-processed microscopic 2-D images of tumor cells' epithelium into 1-D 'signals' (so called *landscapes*) and then embedded these signals in a phase space using 'time-delay' method; he found that *correlation dimension* differs considerably between benign and malignant mammary gland tumors [5]. Here we propose to use a similar simple method for pre-processing of the surface's 2-D image to construct 1-D landscapes, but in the second step we use *Higuchi's fractal dimension* method [6] for analysis of obtained landscapes (cf. [2]).

2 Methods

So called *shape-from-shading* method may be used for assessing the properties of a 3-D surface from its 2-D image (cf. [7]). Fractal dimension is a good predictor of people's perception of surface roughness ([8], [9]). We have developed a new method of inferring fractal dimension of a 3-D surface from a 2-D greyscale image of that surface without 3-D shape reconstruction – the image data are pre-processed to produce 1-D landscapes, which are further analyzed using methods we have been developing for signal analysis. This way dimensionality of the problem and so its computational complexity is drastically reduced. It is mistaken to suspect analogies of our method to classical methods of surface roughness analysis that make use of a mechanical sensor driven against the surface producing a 1-D scan called *roughness profile.*

A digitized image can be viewed as a surface for which $x-$ and $y-$ coordinates represent position and the $z-$ coordinate represents gray level (intensity). The fractal nature of this putative, statistically self-affine surface can then be characterized both in the spatial domain with fractal dimension, and in the frequency domain with spectral exponent β.

Fractal dimension is invariant with respect to linear scale transformations and it is simply related to power spectrum exponent β. If a fractal Brownian surface embedded in 3-D space has fractal dimension D_s and the power spectrum proportional to $f^{-\beta}$ its 2-D image shows power spectrum proportional to $f^{2-\beta}$, where $\beta/2 = (3 - D_s)$. So one can use power spectrum of the image to assess fractal dimension of the surface. β is also simply related to Hurst exponent [6, 7]. However, our method is much simpler than spectral or Hurst methods.

A digitized image is a pattern stored as a rectangular data matrix. Grayscale images are matrices where the matrix elements can take on values from 0 to $g_{max} = (2^b\text{-}1)$, where b denotes the number of bits (for b=8 g_{max} =255). The rendering on a video screen is a presentation of the values from black (0) to white (2^b-1). Most color images are overlays of three monochrome images. Stepping through a gray value image length of N pixels and height of M pixels *row by row* we calculate the sum of the gray values in each row, G_m (m=*1,...,M*), and normalize the numbers by using the largest of those values, G, so producing the 'horizontal' landscape

$$NGS_m = \frac{G_m}{G} \in [0,1] \tag{1}$$

Similarly, stepping through the same image *column by column* (n=*1,...,N*) we produce another, 'vertical' landscape. If necessary, other landscapes may be constructed using similar counting technique, stepping through the same picture in different directions, e.g. in diagonal directions or in some rectangular frames.

The resulting NGS series serve as input for the subsequent signal analysis using Higuchi's fractal dimension method (cf. [6]).

Higuchi's fractal dimension, D_f, is calculated directly from the time series, without embedding the data in a phase space like it is in the case of e.g. correlation dimension [10]. D_f is, in fact, fractal dimension of the *curve* representing the signal (landscape) under consideration, and so it is always between 1 and 2, since a simple curve has dimension equal 1 and a plane has dimension equal 2; the fractional part of D_f is a measure of the signal *complexity*. D_f should not be confused with fractal dimension of an attractor in the system's phase space [10]. D_f remains unchanged when the scale of the signal's amplitude is changed [10]; so normalization by factor G in (1) is only for convenience of graphical representation. Also if the length of a landscape becomes longer that a certain minimum of 200-250 further lengthening of the landscape practically does not change its D_f. Fractal methods become increasingly more important in image and signal analysis [11-15].

3 Results

We apply the method to SEM images from [1] for assessment of surface roughness of titanium-coated implant materials. The authors of [1] give several SEM images with different magnifications. For better clarity we have chosen only two magnifications – figures where a segment marked as '60 μm' has the length of ca. 4.2 cm, that is of magnification about 700x, and those where a segment marked as '10 μm' has the length of ca. 3.5 cm, that is of magnification about 3500x [1]. Since for further comparisons only a relative magnification is important, we denote lower magnification simply as '1x' and higher magnification as '5x'.

We pre-processed chosen images, constructing horizontal and vertical landscapes (cf. Fig. 1, Fig. 2). The landscapes were subsequently analysed by computing Higuchi's fractal dimension using 128 points window, moved in each step 1 point to the right (cf. [6], [10]); finally we calculated corresponding mean values - horizontal and vertical fractal dimensions, D_h and D_v.

Fractal dimension of landscapes obtained from surface images does change with the surface properties (cf. Fig. 3). The smoother is a surface, i.e. the smaller is its unevenness at any particular scale, the greater is fractal dimension of any landscape obtained from an image of the surface at given magnification. If a surface shows anisotropic roughness properties (texture) then fractal dimensions of its horizontal and vertical landscapes differ from one another.

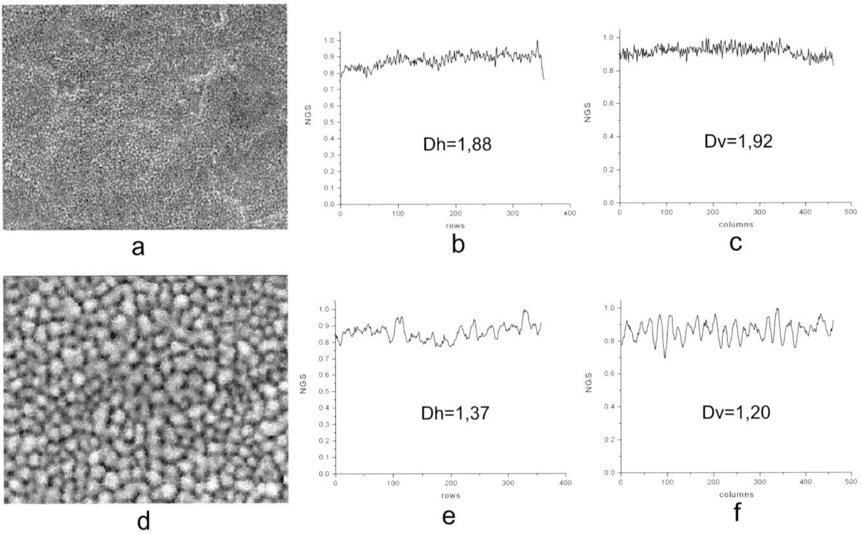

Fig. 1. Examples of SEM images and landscapes corresponding to them of 'naked' BSP treated surfac. e in two magnifications: **a.** magnification '1x' (Fig. 3 –a. from [1]): **b.** horizontal landscape (*row by row*); **c.** vertical landscape (*column by column*); **d.**magnification '5x' (Fig. 3 –b. from [1]): **e.** horizontal landscape; **f.** vertical landscape

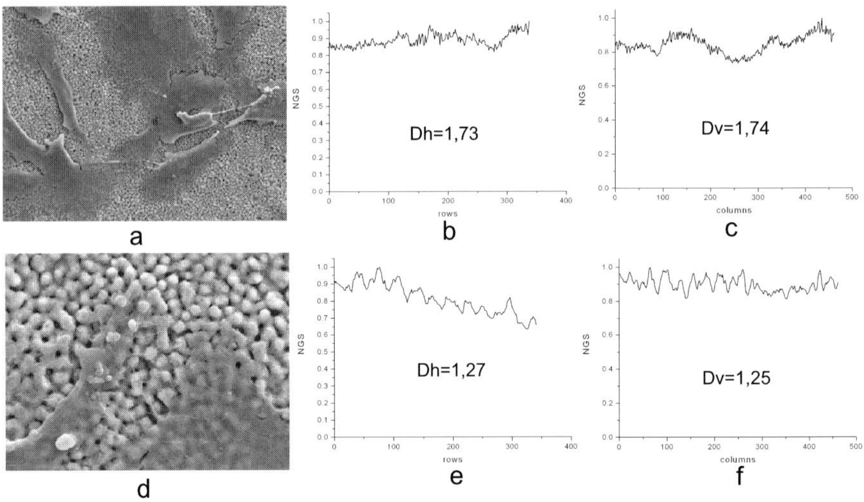

Fig. 2. Examples of SEM images and landscapes corresponding to them of BSP treated surface with 2 days MG63 cell culture on it in two magnifications: **a.** magnification '1x' (Fig. 8 –c. from [1]);**b.** horizontal landscape (*row by row*); **c.**vertical landscape (*column by column*); **d.** magnification '5x' (Fig. 9 –c. from [1]); **e.** horizontal landscape; **f.** vertical landscape.

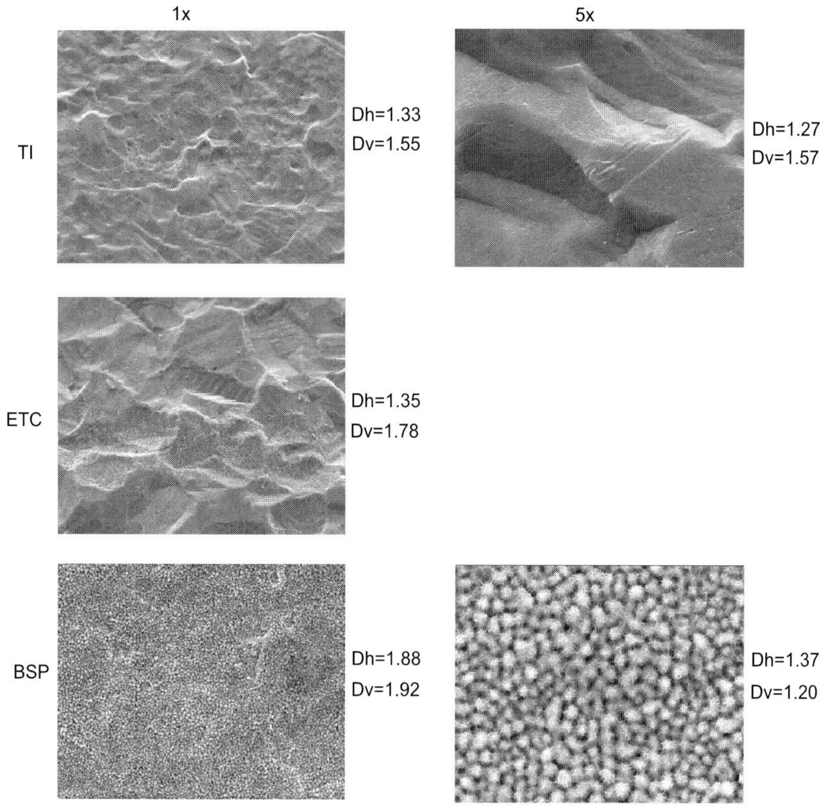

Fig. 3. SEM images (from [1]) and Higuchi's fractal dimension of surfaces of three titanium-coated implant materials at two different magnifications. There is no image in [1] for ETC with magnification '5x'.

4 Discussion

Fractal dimension of a surface is invariant with respect to scaling of the data. So, the normalization in (1) is convenient for presentation of the landscapes but it is not really necessary since it does not change the value of Higuchi's fractal dimension; thus the time necessary for calculations may still be reduced.

Fractal dimension of a surface depends on the corresponding dominant surface-forming process at any particular scale. That is why a surface may need *multifractal* description. The aim of measuring fractal dimensions is not only to add new structural parameters to already existing ones, possibly describing new structural characteristics; more important is a possibility to get insight into the development of complex structures and the processes that contribute to surface structure forming at particular scales.

Surface treated with BSP a novel bioactive titanium (obtained by BioSpark^{TM}) - shows good cell adhesion and significantly increases cell proliferation rate [1]. We find that such a surface has evident multifractal properties, with high D_f at lower magnification (so showing that in this scale the surface is quite smooth) and much smaller D_f at higher magnification (cf. Fig. 3). When a cell culture grows on such a surface D_f diminishes in comparison with 'naked' surface (cf. Fig. 4).

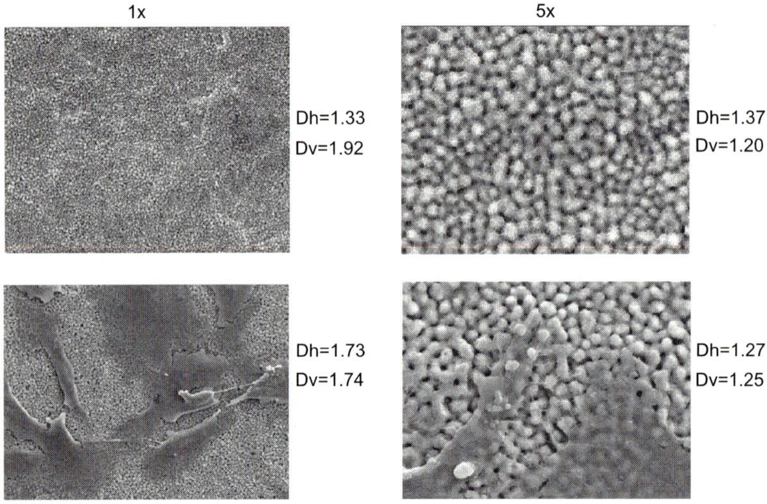

Fig. 4. SEM images (from [1]) and Higuchi's fractal dimension of a novel bioactive titanium-coated surface – 'naked' and with a MG63 two-day cell culture on it at two different magnifications.

Surfaces of two commercially available materials - no treated (TI) and chemically etched (ETC) titanium - are more rough and show worse cell adhesion [1]. Such surfaces in comparison with BSP have smaller D_f at lower magnification; they are also more anisotropic. Comparison of D_f at two magnifications shows that TI surfaces are rather monofractal.

5 Conclusions

The proposed method may be used for quality assessment of materials for orthopaedic prostheses and dental implants.The method is much simpler than other methods that make use of fractal properties for surface roughness assessment. It draws from multiple disciplines and may find multidisciplinary applications.

Acknowledgements

We thank the authors of [1] for kind permission to use the figures from their paper for our image analysis.

References

1. Giordano C, Sandrini E, Del Curto B, Signorelli E., Rondelli G, Di Silvio L (2004)Titanium for osteointegration: Comparison between a novel biomimetic treatment and commercially exploited surfaces. Journal of Applied Biomaterials & Biomechanics 2:35–44
2. Klonowski W, Olejarczyk E, Stepien R (2004) Fractal and Symbolic Methods for Nanomaterials Science and Nanosensors. E-MRS 2004 Fall Meeting Warsaw 6-10 September 2004 to be published
3. Klonowski W, Olejarczyk E, Stepien R (2003) New Methods of Nonlinear and Symbolic Dynamics in Sleep EEG-Signal Analysis. In: Feng DD, Carson ER (eds) Modelling and Control in Biomedical System. IFAC Publications, Elsevier, Oxford 241–244
4. Klonowski W (2004) W Signal and Image Analysis Using Chaos Theory and Fractal Geometry. Machine Graphics & Vision; 9: 403–431; also http://hrabia.ibib.waw.pl/~lbaf/PDF_Doc/gkpo2000.pdf
5. Mattfeldt T (1997) Spatial Pattern Analysis using Chaos Theory: A Nonlinear Deterministic Approach to the Histological Texture of Tumours. In: Losa GA, Merlini D, Nonnenmacher TF, Weibel ER (eds) Fractals in Biology and Medicine, Vol. II, Birkhäuser, Basel, Boston, Berlin 50–72
6. Higuchi T (1988) Approach to an irregular time series on the basis of the fractal theory. PhysicaD 31:277–283
7. Pentland A (1988) The transform method for shape-from-shading. MIT Media Lab Vision Sciences Tech Report 106
8. Pentland A (1984) Fractal-Based Description of Natural Scenes. IEEE Trans Patt Anal Mach Intel 6:661–674
9. Kube P, Pentland A (1988) On the Imaging of Fractal Surfaces. IEEE Trans Patt Anal Mach Intel 10:704–707
10. Klonowski W (2002) Chaotic dynamics applied to signal complexity in phase space and in time domain. Chaos, Solitons and Fractals 14:1379–1387
11. Ha SW, Gisep A, Mayer J, Wintermantel E, Gruner H, Wieland M (1997) Topographical characterization and microstructural interface analysis of vacuum-plasma-sprayed titanium and hydroxyapatite coatings on carbon fibre-reinforced poly(etheretherketone). Journal of Materials Science: Materials in Medicine 8:891–896
12. Kondev J, Henley CL, Salinas DG (2000) Nonlinear measures for characterizing rough surface morphologies. Physical Review E vol. 61, 1:105–125
13. Lung CW, Jiang J, Tian EK, Zhang CH (1999) Relation between fractal dimension and roughness index for fractal surfaces. Physical Review E vol. 60, 5:5121–5125
14. Tang YY, Tao Y, Lam ECM (2002) New method for feature extraction base on fractal behavior. Pattern Recognition 35:1071–1081

15. Chappard D, Degasne I, Hure G, Legrand E, Audran M, Basle MF (2003) Image analysis measurements of roughness by texture and fractal analysis correlate with contact profilometry. Biomaterials 24:1399–1407

Output Flow Estimation of Pneumatically Controlled Ventricular Assist Device with the help of Artificial Neural Network

Dariusz Komorowski[1,2] and Ewaryst Tkacz[1,3]

[1] Institute of Electronics, Division of Microelectronics and Biotechnology, Silesian University of Technology, Gliwice, Poland dkomorowski@polsl.pl, etkacz@polsl.pl
[2] Artificial Heart Laboratory, Institute of the Heart Prostheses, Zabrze, Poland
[3] Medical University of Silesia, Faculty of Pharmacy and Laboratory Medicine, Department of Bionics, Sosnowiec, Poland etkacz@slam.katowice.pl

Summary. The paper presents a novel approach to the problem of reliable estimation of the output flow going out from pneumatically controlled ventricular assist device (VAD). Among many possibilities, the application of artificial neural network (ANN) has been decided leading to the promising results. The basic difficulty however is to perform the suitable sufficiently exact measurement on the pneumatic side of the assisting system, which allows to avoid the application of the e.g. ultrasound measuring transducers on the hydraulic side, and which makes possible an implementation of the automatic control algorithm in the future for the whole measurement process. It is important however to underline that from the physical properties point of view these two mentioned sides i.e. pneumatic and hydraulic are completely different. Therefore, due to several nonlinearities, application of ANN gives an acceptable solution.

1 Introduction

Usually, pneumatically driven circulatory assistance systems equipped with VAD at its output require a possibility of output blood flow estimation based upon the control of pneumatic signals. On the other hand such an estimation can be easily performed with the help of suitable sensors. However, installation of such sensors e.g.hallotrons, inside VAD require additional electrical connections as well as changes in both VAD's construction and technological process. This might cause a certain complications from both technological and medical points of view, especially taking into account a possibility of infection. Eventual application of the wireless measurement transmission may not be allowed due to the presence of some other medical equipment surrounding patient's bed [1].

2 Methodology

Physical environments on the both sides of ventricular assist device membrane are totally different and strongly nonlinear. Therefore classical "black box" based identification methods applied to sort out the problem of output blood flow estimation fail. Below described method deals with an idea of artificial neural network (ANN) application to overcome the difficulties connected with strongly nonlinear signal dependencies on the both sides of VAD membrane. We have applied a classical perceptron with two hidden layers and different activation functions with control pneumatic signals i.e. pressure and flow to estimate a blood flow in the outlet cannula. A system for blood flow estimation needs training providing that both, wide range of pneumatic control input signals and previously measured expected level of hydraulic flow are delivered to the ANN input and output layers respectively. After the training is finished the structure of ANN giving best results is stored for further use in the "on-line" output blood flow estimation [2].

To validate the suggested method both suitable simulations and animals measurements (in the second stage) have been performed. First the lab stand including both VAD and pneumatically driven control unit has been designed with intension to gather data characterizing behavior of the system under application of the different loads at the output (figure(1)). Having the data of input and output pressures and flows the second stage of ANN structure choice and its training has been performed.

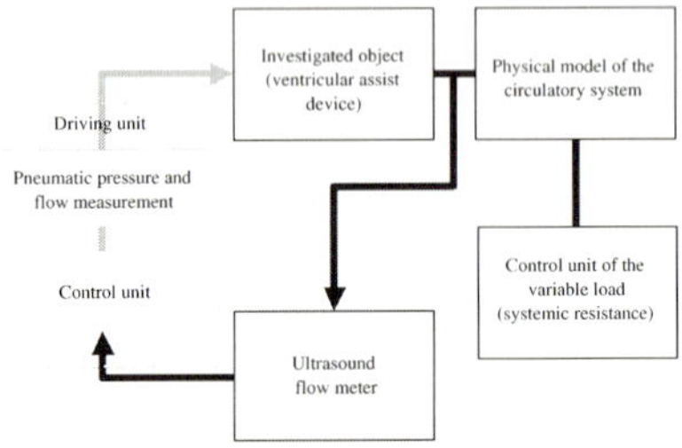

Fig. 1. Laboratory stand for basic measurements.

Figure(2) presents a model of the VAD, where: $u(t)$ is an input signal, P_p is a pneumatic driving pressure, F_p is a pneumatic driving flow (air), F_h is a hydraulic flow (under modeling) and $y_m(t)$ is a model response.

Generally the problem of identification can be presented in this case in the following way: we observe both the input data $u(t)$ and output data $y(t)$, which usually can be modeled in the dynamic system by the following equations:

$$\bar{u}(t) = [u(1), u(2), ..., u(t)]^T, \tag{1}$$

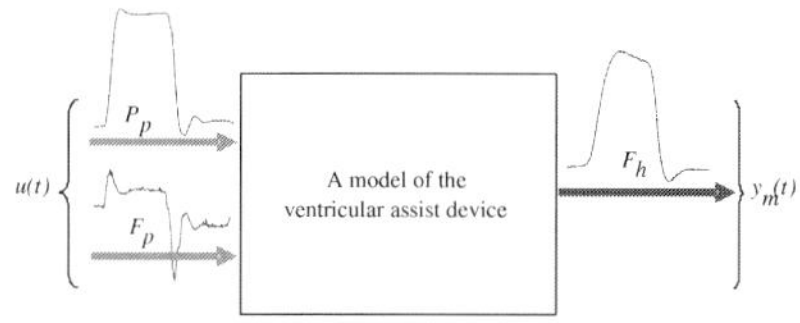

Fig. 2. A model of the ventricular assist device.

$$\bar{y}(t) = [y(1), y(2), ..., y(t)]^T \tag{2}$$

and then we search for the relationship between previous values of signals $[\bar{u}(t-1), \bar{y}(t-1)]$, and current output $y(t)$ [6][7]. Such a relationship can be expressed with a help of the equation:

$$y(t) = g\left(\bar{u}(t-1), \bar{y}(t-1)\right) + v(t). \tag{3}$$

The auxiliary variable $v(t)$, which has been placed in the last equation(3) means that the signal $y(t)$ is not only a function of preceding state of both inputs and outputs. The assumed goal is the minimization of $v(t)$, and obtaining in this way the function $g(\bar{u}(t-1), \bar{y}(t-1))$, which become a good predictor of the signal $y(t)$. Equation(3) can be considered as very good general description of any dynamic model. Let us consider a problem of finding the function g starting from equation(3). The first possible solution is acceptance of that function form the certain family of functions, later parametrization of that family using the parameter vector $\bar{\theta}$ with finite dimension. It can be presented in the following form:

$$g\left(\bar{u}(t-1), \bar{y}(t-1), \bar{\theta}\right). \tag{4}$$

The new task is to find the suitable structure as well as proper choice of vector parameters in such a way that defined error should reach its minimum value given by equation(5) [3]:

$$\sum_{t-1}^{N} (y(t) - g(\bar{u}(t-1), \bar{y}(t-1), \bar{\theta}))^2. \tag{5}$$

There are several methods of mentioned vector description, which would minimize the error expressed by equation(5), but the most frequent approach applied in the identification systems is the following: a function g can be presented as a function of two parameters: a vector of known parameters $\bar{\varphi}(t)$ and consequently a vector of unknown parameters $\bar{\theta}$. This can be expressed with the help of the following formula(6):

$$g(\bar{u}(t-1), \bar{y}(t-1), \bar{\theta}) = g(\varphi(t), \bar{\theta}), \tag{6}$$

where:

$$\bar{\varphi}(t) = \bar{\varphi}(\bar{u}(t-1), \bar{y}(t-1)). \tag{7}$$

Vector $\bar{\varphi}(t)$ is often called a regression vector and respectively its components are called regressors. The structure of the nonlinear model, in which all the regressors are included into the general nonlinear "black-box" type relation can be described by the following formula(8):

$$\widehat{y}(t) = g(\bar{\theta}, \bar{\varphi}(t)). \tag{8}$$

The structure of this model is known in the literature as a NARX model and its diagram is shown on figure(3) [7]. The generality of this structure allows for the description of an arbitrary nonlinear system. Unfortunately sufficiently good approximation of the system as well as its excellent robustness into noise presence require the application of the great number of samples of both preceding inputs $\bar{u}(t)$ and outputs $\bar{y}(t)$, i.e. a big dimension of the regressor $\bar{\varphi}(t)$. The other disadvantage,which occurs in this case is the lack of the interferences model. As a consequence the dynamics of the interferences is modeled together with the dynamics of the object. This is shown on figure(3).

Fig. 3. Block diagram presenting the structure of NARX model.

The application of Artificial Neural Network (ANN) can be considered as a one of the possible solution, allowing the application of the nonlinear relation g. Identification structure of dynamic model is presented on figure(4) [5], where: $y(t)$ is an object response, $\bar{u} = [u(t), u(t-1), ..., u(t-n_b)]$ and $\bar{y} = [y(t), y(t-1), ..., y(t-n_a)]$ are vectors of delayed samples of input signal and object response respectively, where n_b and n_a are values of delay, $e(t)$ is an error signal.

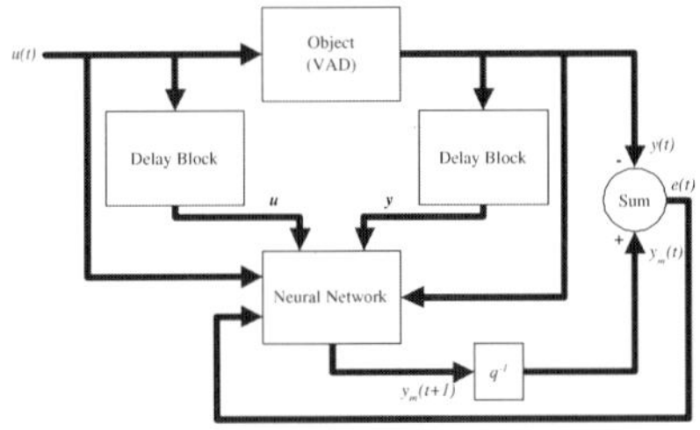

Fig. 4. Block diagram presenting the structure of elaborated method.

Concerning ANN structure, at this stage of investigation, the simplest back propagation structure with two hidden layers has been taken into account with number of inputs equal to the signal samples length i.e. 16 samples for input pneumatic pressure, 16 samples for input pneumatic flow and finally 16 samples for feedback. Both sampling frequency and amplitude resolution have been establish at 50 Hz and 12 bits respectively.

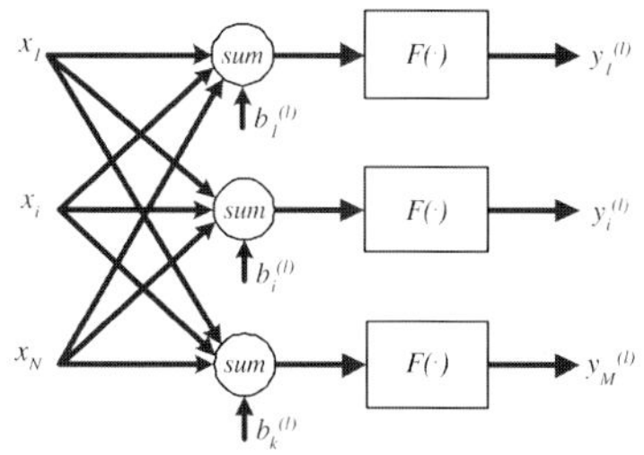

Fig. 5. Basic structure of the simplest perceptron like ANN.

The output signals have been calculated using the equation(9):

$$y_k^{(l)} = F(\sum_i w_{ik}^{(l)} s_i + b_k^{(l)}), \tag{9}$$

whereas the applied activation functions: linear, sigmoidal and tangsoidal are very known functions.

The VADS's model has been validated using the FIT coefficient, given by equation(10).

$$FIT = 100 \left(1 - \frac{\left(\sum_{i=1}^{N} [y_m(i) - y_o(i)]^2 \right)^{1/2}}{\left(\sum_{i=1}^{N} [y_o(i) - m_{yo}]^2 \right)^{1/2}} \right), \tag{10}$$

where y_o is a signal's sample of object response(VAD), y_m is a signal's sample of model response, m_{yo} is average value signal's samples of model response and N is number of samples.

The crucial point of the object identification process applying a model based upon the neural network refers to the proper preparation of both learning signals sets and verifying signals sets. The first attempts of the Ventricular Assist Device (VAD) model elaboration allow to conclude that it is relatively difficult to build up the VAD model, which would work properly in the whole range of VAD control parameters. The parameter which has significant influence into the model behavior is a Systolic Driving Pressure (SDP). Problem, which occurs in connection with SDP influence has been sorted out through the partitioning of the whole range of pneumatic driving pressure into proper subranges and following that construction of the individual VAD model for each subrange of SDP. There have been four subranges of SDP accepted for VAD models.

3 Result

The results obtained with help of elaborated method show that the reliable blood flow estimation in the outlet cannula is possible and does not require installation of any additional sensors for both direct and indirect measurements.

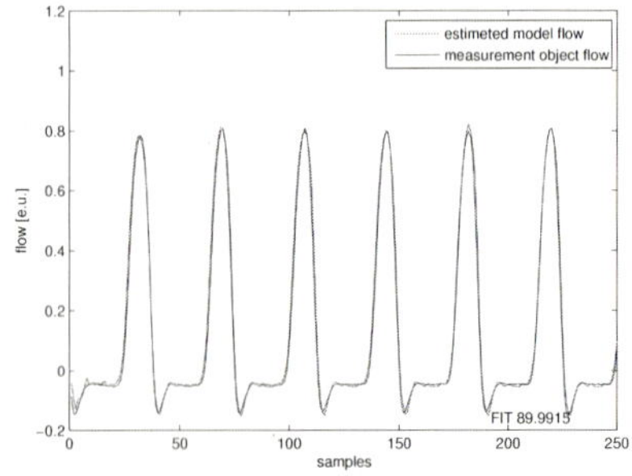

Fig. 6. Comparison between estimated hydraulic flow obtained from the model and measured flow obtained from the object.

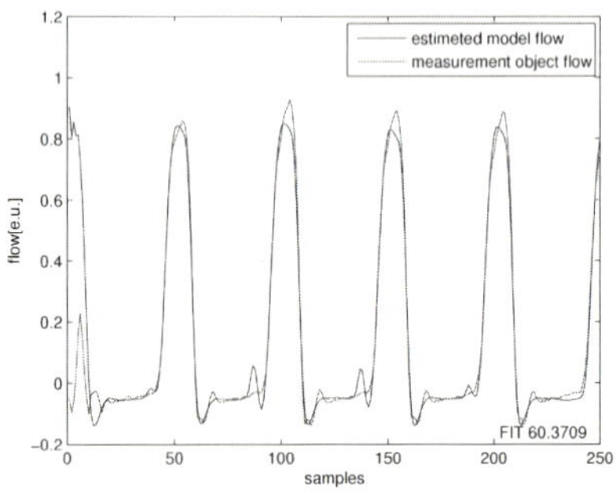

Fig. 7. Comparison between estimated hydraulic flow obtained from the model and measured flow obtained from the object.

One of the goal of this work refers to the elaboration of the method allowing reliable estimation of the VAD output blood flow based upon the control pneumatic signals. Below there are some results presented on figure(9), which have been obtained for different structures of ANN's applied in the VAD model in particular subrange of SDP parameter. These results are presented as a relative errors of blood flow measurements expressed in percents according to the equation(11).

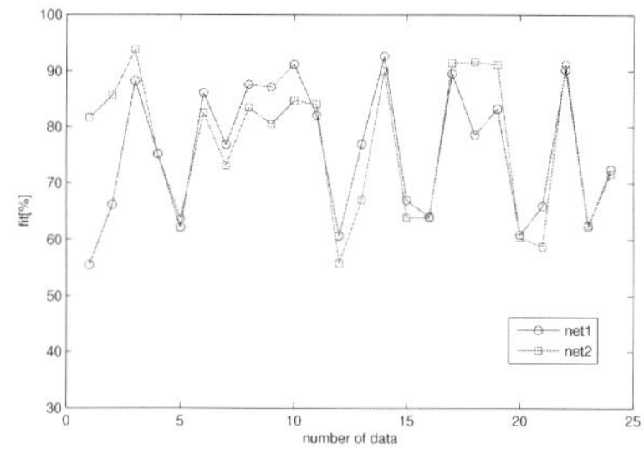

Fig. 8. An example fit for two different structures of neural network.

$$BFM = \frac{VFO - VFM}{VFM} 100\%, \tag{11}$$

where VFO is a value of object's flow output, VFM is a value of estimated model's flow output. All of these values have calculated through the integration of the flow curves within 60 seconds intervals.

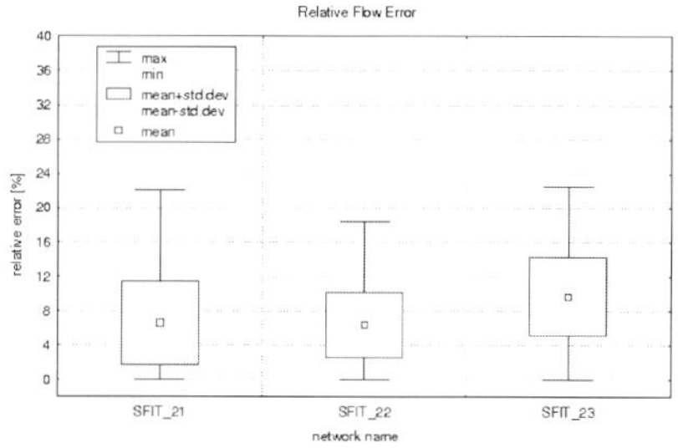

Fig. 9. Relative flow error for diffrent neural network.

4 Discussion

The presented approach concerning estimation of the output hydraulic flow going out from VAD allows obtaining important physiological information required by doctors during the circulatory system assistance without additional risk to the patient. As they are presented on figures (6) and (7) the level of agreement between measured value and estimated one is perfectly acceptable when taken into account the errors caused by most ultrasound transducers. The FIT for two different structures of neural network (VAD's model) for data obtained from animal experiment is showed on figure(8).

5 Conclusion

Current research is concentrated on implementation of the elaborated method with the help of state of the art DSP's, available on the market and afterwards if that implementation is successful the whole procedure will be installed in the control unit responsible for driving VAD. However before that, additional experimental animal research will be necessary especially considering further application in the clinical conditions.

References

1. Czak M, Komorowski D, Kustosz R (1998) An automatic control of the driving unit for pneumatic cardiac assist system. 25th Congress ESAO, November, Bologna
2. Komorowski D, Tkacz E, Kustosz R (2002) An Application of the Neural Network for Output Flow Estimation in the Pneumatically Driven Polish Ventricular Assist Device (POLVAD). 29th ESAO Congress European Society for Artificial Organs, Viena, Austria, The International Journal of artificial Organs, vol. 20, no. 10.
3. Ljung J (1987) System Identification: Theory for User. Prentice Hall, Englewood Cliffs, NY.
4. Ljung L, Soderstrom T (1983) Theory and Practice of Recursive Identification. MIT Press, Cambridge, Massachusetts.
5. Osowski S(1996) Sieci neuronowe w ujêciu algorytmicznym. WNT Warszawa
6. Sjoberg J (1993) Regularization issues in neural network models of dynamical systems. Linkoping studies in science and technology. thesis no.386, liu tek lic 1993:08, isbn 91-7871-072-3, issn 0280-7971, Department of Electrical Engineering, Linkoping University, Sweden.
7. Sjoberg J (1995) Non-Linear System Identification with Neural Networks'. PhD thesis Department of Electrical Engineering, Linkoping University, Sweden
8. Sontag E (1993) Neural networks for control. In H. Trentelman, and J. Willems, editors, Essays on Control: Perspectives in the Theory and its Applications, volume 14 of Progress in Systems and Control Theory, pages 339-380.

The Cell Structures Segmentation

Robert Koprowski[1] and Zygmunt Wrobel[1]

University of Silesia, Institute of Computer Science, 41-200 Sosnowiec, Bedzinska 39, Poland koprow@us.edu.pl, wrobel@us.edu.pl

Summary. In this article we have presented an attempt to segmentation of cell structures images acquired while histological slides microscopic observation. The described algorithm of segmentation is also applicable in other matters, where the image segmentation is an important part.

1 Introduction

Reproducibility of cell structures microscopic-based images measurements is an important issue of biological slides analysis. For the effective tissue slides estimation and documentation the computed images analysis techniques are applied. The first step should thus be the automatic or semiautomatic picture segmentation necessary for morphometric analysis of particular cells. Microscope settings like light intensity or magnitude influence the picture quality.

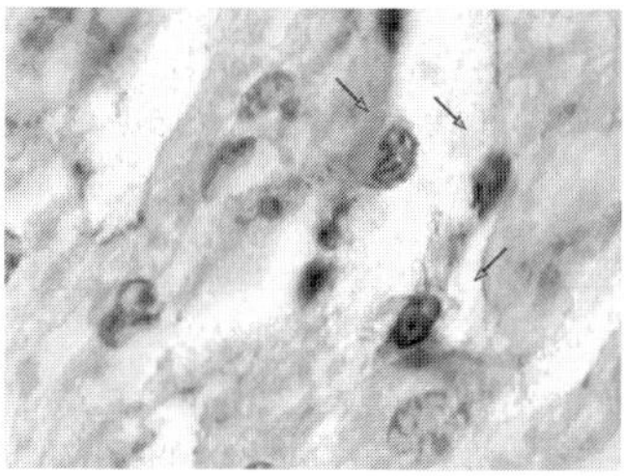

Fig. 1. An example of a microscopic- based cells image

Hence, it is impossible to set constant thresholds, even for the same operator and the same slide, e.g. for the digitalization process. Provided biological experiment aimed to delimitation the proliferation in the rat's liver regenerating cells. The slides made of examined tissue after the proper staining (H/E), can be viewed at immersion magnitude of NIKON E 600 microscope (the lens: Plan Fluor 100x/130 Oil, the eyepiece: 10x/22) and resulted image was transferred into the PC (the processor: 3.06 GHz) using CCD Panasonic Colour camera.

2 The automatic image segmentation

The colour input image L_{RGB} (m,n,o) where (m - row, n - column, o - R, G or B component) at $M \times N$ resolution is inserted into the Matlab package workspace with Image Processing tool [7]. The proceeding process - described in the articles Bibliography [2], [4], [7]- towards cell segmentation is questionable at many points referring both to selection threshold and the results. Thus the modification of the presented and its expanding to image recognition parts based on decision trees seems to be advisable

3 The employment of decision trees in the segmentation process

An image segmentation process is a vitally important point of the discussed issue. The value of measured quantities (such as the object saturation level or the object area) entirely depends on the segmentation process results. On the other hand, in case of a larger number of images, the capability to automation of the segmentation process would be a deciding factor in computation acceleration. Among many segmentation methods such as: the area extending method, the watershed method or the patterns analysis method, the decision tree based method has been selected. The decision tree was used in the form of classification knowledge description for segmenting the objects (the classes space) in the L_{HSV} image. The choice of the HSV colour space has been determined here by the image contents that includes both objects of different colours and object of the same colour but different saturation levels. The proceeding process, which is presented below, can be also applied to other colour spaces (e.g. like RGB or L*a*b). The decision nodes in the mentioned case are described by features, which are the saturation levels of HSV components (generally) x, y, z. The edges of a tree qualify the possible values for the feature (the saturation level). The tree leafs are the values of the classification feature in which in that case belongs a cluster (object index). Thus the segmentation (classification) is performed by the reviewing the tree from the root up to leafs by the edges featured with features values. Decision trees in this particular application (i.e. image segmentation) have important limitations, such as:

- a risk of excessive complexity of the tree (an excessive fit problem),
- there is no easy way to update in case of a different images set.

Excessive fit for input data causes small classification error but too large real error. Such a tree, due to its complex structure, usually reflects random relations in the learning data set (e.g. image noise). Due to their characteristic structure which makes them suitable to represent any hypothesis, decision trees are particularly exposed to dangers of that kind

The solution to this problem is trees truncation (i.e. truncation of their excessive fit), which could be simply explained as replacement of the original tree with its sub-tree. The advantages of the decision trees application for the discussed purpose include:

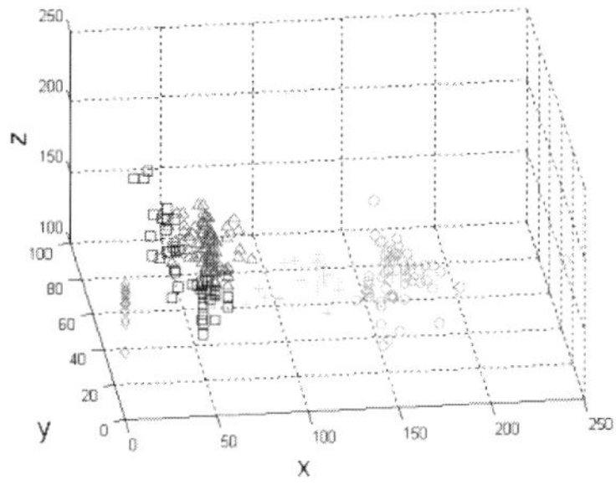

Fig. 2. The arranging of pixels to 5 clusters (marked with colours respectively)

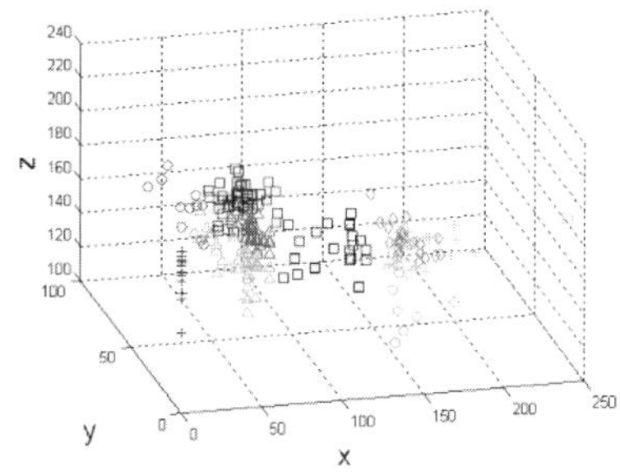

Fig. 3. The arranging of pixels to 10 clusters (marked with colours respectively)

- a capability for representing any hypothesis - a capability for performing image segmentation with any required precision,
- efficiency of the classification process - a capability for logic employment,
- efficiency in case of large learning sets - a capability for the classification trees employment for a segmentation process of any computation and logical complexity,
- a clear representation - assuming that the tree is not too complex and does not combine too many features.

The construction of the classification tree starts by L_{HSV} pixels clustering featured in the features space (L_{HSV}(m,n,1) the x coordinate, L_{HSV}(m,n,2) the y coordinate and L_{HSV}(m,n,3) the z coordinate) into 5 and 10 clusters (indexes) - see Fig. 2 and 3. The tests were performed for the following metrics: Minkowski, road, city and Chebyshev. In respect of strong influence of orders difference of features vector individual components, the normalization process was used. Finally the inner averages algorithm with Euclidean metric was used. The mentioned charts in the features space are shown below.

The object division created that way enables proceeding with decision tree creation. The decision tree generated that way is shown in the Fig. 4 and 5.

Based on this, the range of x, y, z values variability can be expanded against the entire features space containing the values range $x \in (0, 240), x \in (0, 100), x \in (100, 220)$, what was shown in Fig. 6, 7, 8 and 9.

According to mentioned disadvantages of excessively complex decision trees the MDL (Minimum Description Length) rule was used to the truncation purpose. The minimal cost rule allows to delimitation of optimal tree nodes quantity, what was shown on the chart in Fig. 11.

Based on this, in discussed case the optimal decision nodes quantity has been set to 9. The structure of truncated modified decision tree is shown in Fig. 12.

The result images, obtained on basis above, (each cluster reflects one index) are shown below (Fig. 13 - 17).

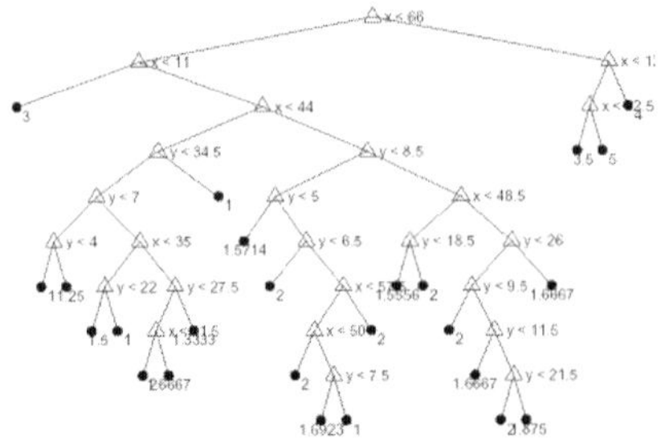

Fig. 4. The decision trees obtained for 5 clusters in the two features space (the H and S component)

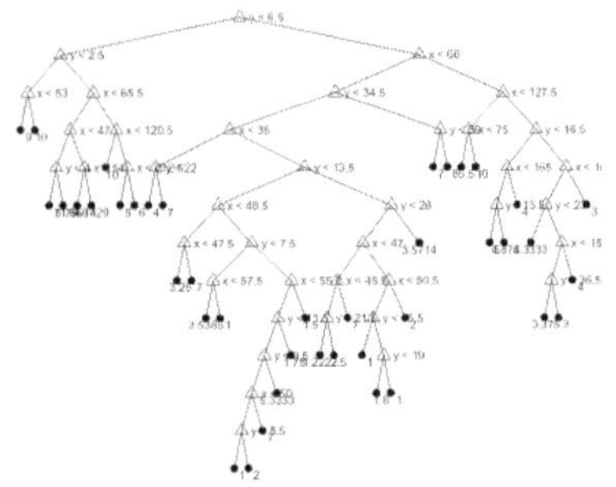

Fig. 5. The decision trees obtained for 10 clusters in the two features space (the H and S component)

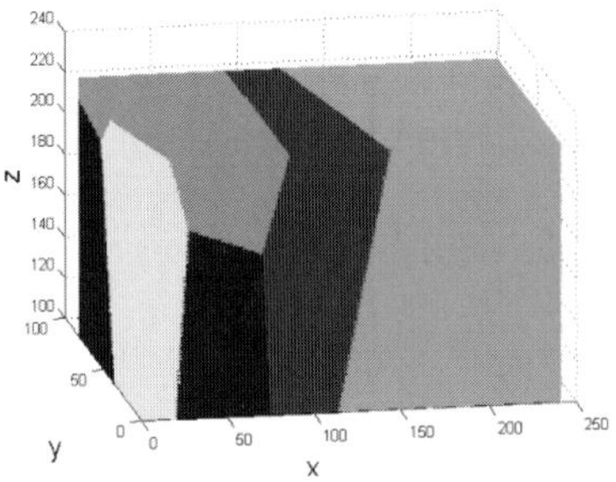

Fig. 6. The example of area boundaries in the features space for 5 clusters

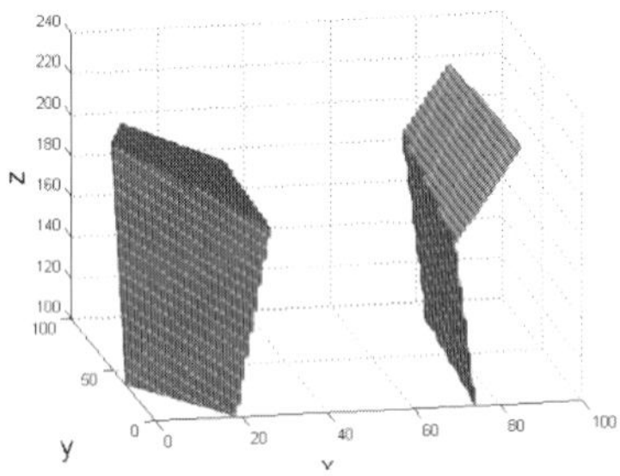

Fig. 7. The arranging of pixels to 5 clusters in the an example features space range (marked with colours respectively)

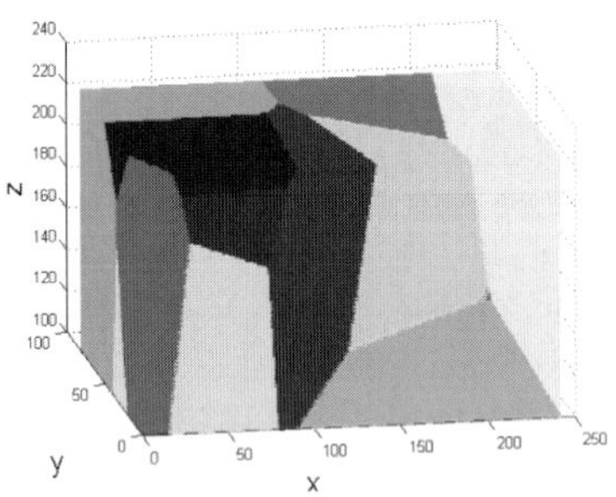

Fig. 8. The arranging of pixels to 10 clusters in the an example features space range (marked with colours respectively)

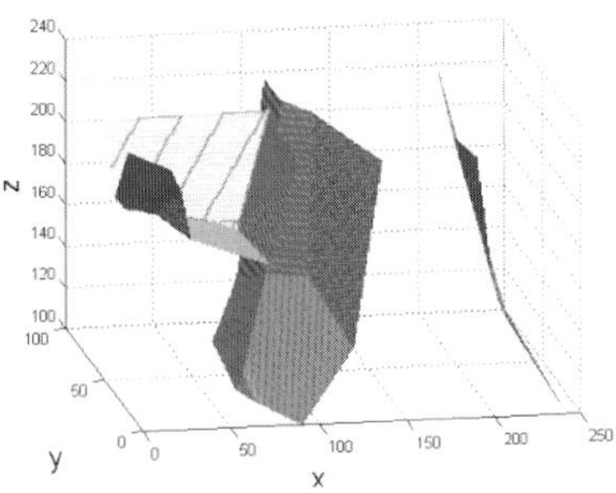

Fig. 9. The example of area boundaries in the features space for 10 clusters

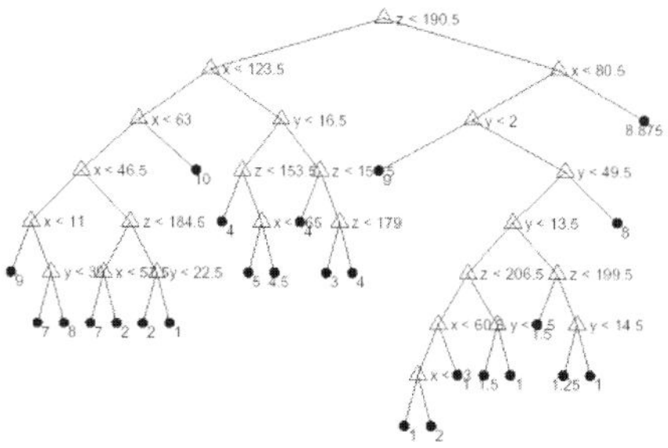

Fig. 10. The decision trees obtained for 10 clusters in respect of x, y and z variables

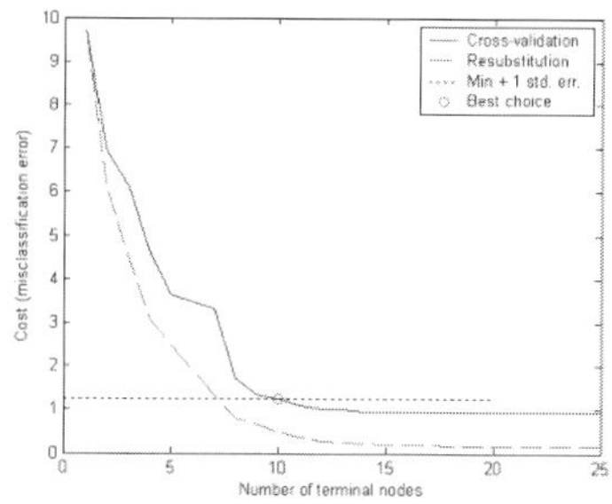

Fig. 11. The cost change chart as function of decision nodes quantity

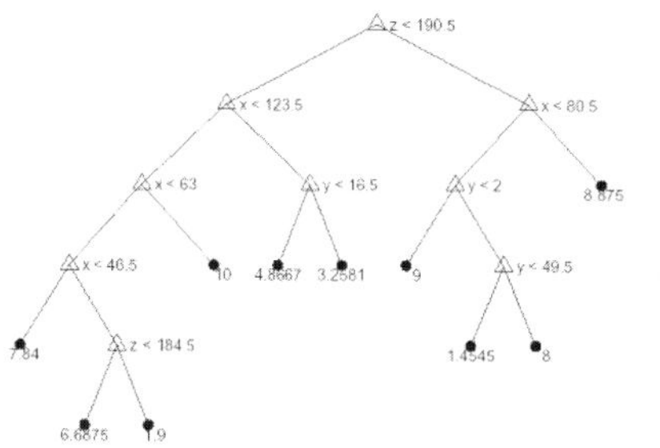

Fig. 12. The decision tree from Fig. 10 - truncated

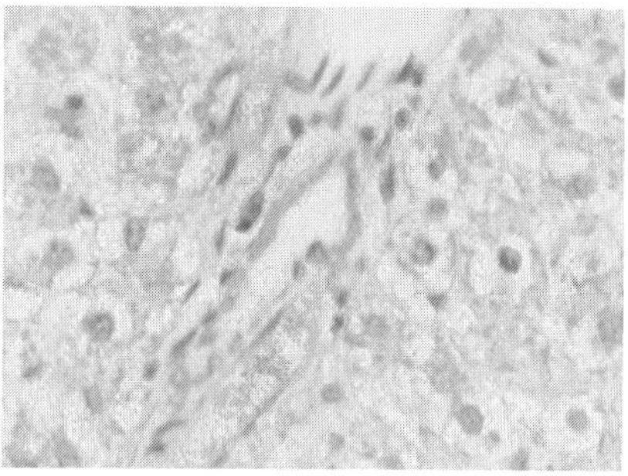

Fig. 13. The input image

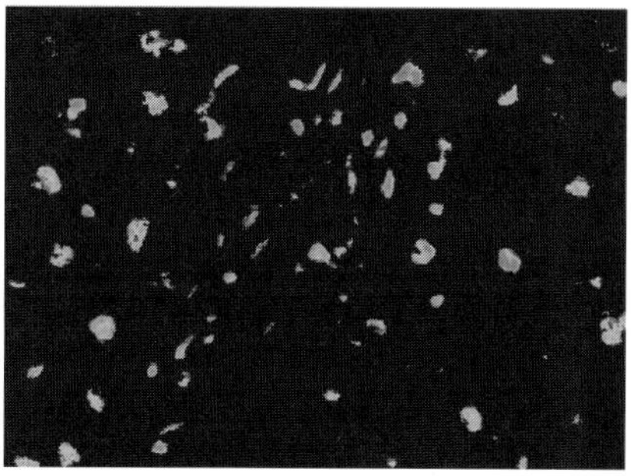

Fig. 14. The image obtained from the second cluster

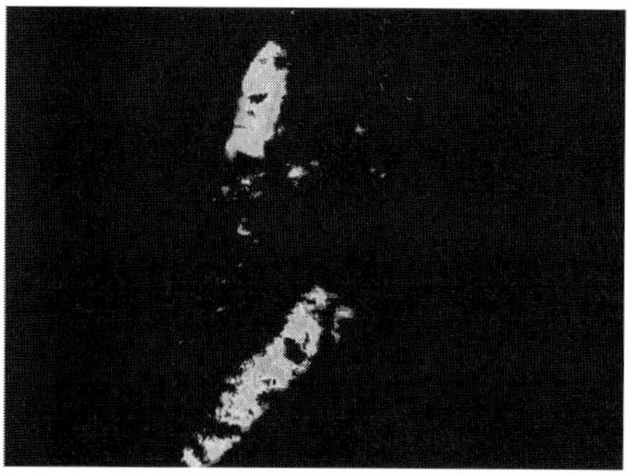

Fig. 15. The image obtained from the eight cluster

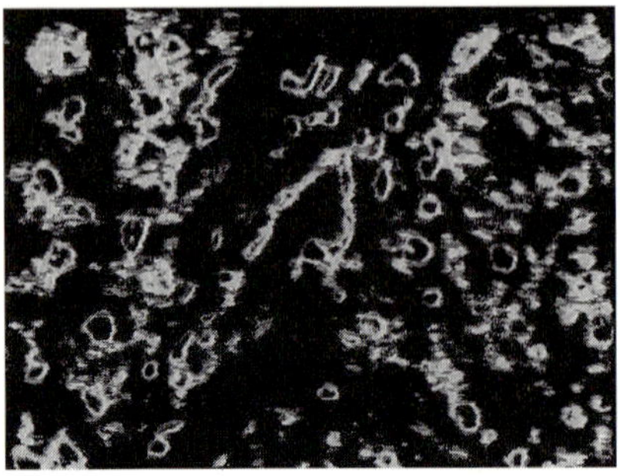

Fig. 16. The image obtained from the first cluster

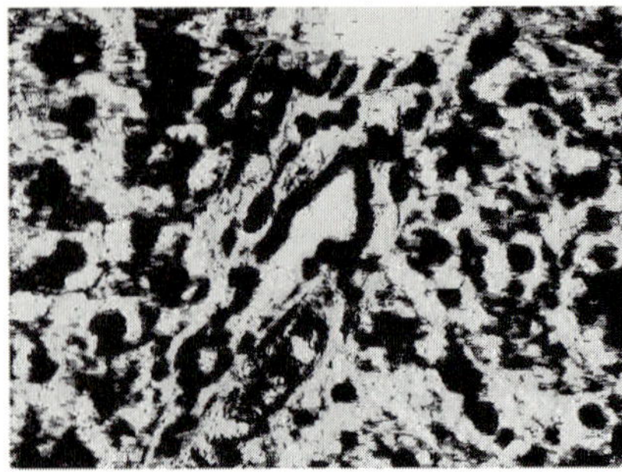

Fig. 17. The image obtained from the fifth cluster

In order to automate next algorithm phases the thresholding has been modified using the Nobuyuki Otsu formula. Presuming that grey scales or generally saturation levels i have the value i=0...k...g and searched threshold level is marked as k_c, we can write down:

$$\sigma_B^2(k) = \frac{[\mu_T - \omega(k)]^2}{\omega(k)[1 - \omega(k)]} \tag{1}$$

where:

$$\omega(k) = \sum_{i=0}^{k} \sum_{m=1}^{M} \sum_{n=1}^{N} \frac{p(i/(m,n))}{M \cdot N} \tag{2}$$

$$p(i/(m,n)) = \begin{cases} 1 & dla \quad L_{GRAY}(m,n) = i \\ 0 & dla \quad L_{GRAY}(m,n) \neq i \end{cases} \tag{3}$$

- the zero order moment

$$\mu(k) = \sum_{i=0}^{k} \sum_{m=1}^{M} \sum_{n=1}^{N} \frac{i \cdot p(i/(m,n))}{M \cdot N} \tag{4}$$

- the first order moment

$$\mu_T = \sum_{i=0}^{g} \sum_{m=1}^{M} \sum_{n=1}^{N} \frac{i \cdot p(i/(m,n))}{M \cdot N} \tag{5}$$

- the maximal existing value for chosen k_v

$$\sigma_B^2(k_v) = \max_{0<k<g} \sigma_B^2(k) \tag{6}$$

If k_v is a vector of F elements, in case of finding more then one maximal value, then:

$$k_c = \frac{1}{F} \sum_{w=1}^{F} k_v(w) \tag{7}$$

The results are the objects, which are the cells with the background. So it seems reasonable to perform the binary-splitting process properly to provided relation for delimited areas, which contain also the cell area, according to relation (1) for obtained level k'_c [5], [7]. The binary image L_{BIN} can be written as:

$$L_{BIN} = L_{GRAY}(m,n) > k'_c \tag{8}$$

The closing operation using the structure element $H(j,k)$ of $J \times K$ size is applied to the result image i.e.:

$$L'_{BIN}(m,n) = \begin{cases} L_{BIN}(m,n) & for \quad (k_w+1)\cdot k'_c > s_r \\ \max_{m_i,n_i \in H}(L_{BIN}(m_i,n_i)) & for \quad (k_w+1)\cdot k'_c \leq s_r \end{cases} \tag{9}$$

where:

$$s_r = \sum_{m_i=1}^{J} \sum_{n_i=1}^{K} \frac{L_{GRAY}(m_i,n_i)}{J \cdot K} \tag{10}$$

and:

$$k_w = [0, 0.1, 0.2, ..., 0.5] \tag{11}$$

The conditional erosion has been defined analogous way, i.e.:

$$L'_{BIN}(m,n) = \begin{cases} L_{BIN}(m,n) & for \quad (1-k_w)\cdot k'_c > s_r \\ \min_{m_i,n_i \in H}(L_{BIN}(m_i,n_i)) & for \quad (1-k_w)\cdot k'_c \leq s_r \end{cases} \tag{12}$$

The (9) and (12) relations were used for the closing operation using the H structural element with the k_w coefficient arbitrarily set at 0.2 level [1], [6], [2], [4]. The selectivity of the algorithm was increased by decreasing the H mask size for following iterations up to the 5×5 level $(J \times K)$. The results are visible at zoom (see Fig. 18 and Fig. 19).

The indexation process (L''_{BIN}) has been performed on the L''_{BIN} image, where pixels were included into an object if they are in the eight neighbour 'distance' from the others. Next the following morph-metric parameters were calculated for each of delimited objects (indexes) [1], [3]: area, the edge length using Crofton formula (according to Cauchy rule), shape parameter (Blair-Bliss). In selected instances of analysed cells the problem of connected objects occurs, which has been eliminated by the additional performance of the opening operation on objects which the square area stays in standard level while the diameters ratio is over threshold.

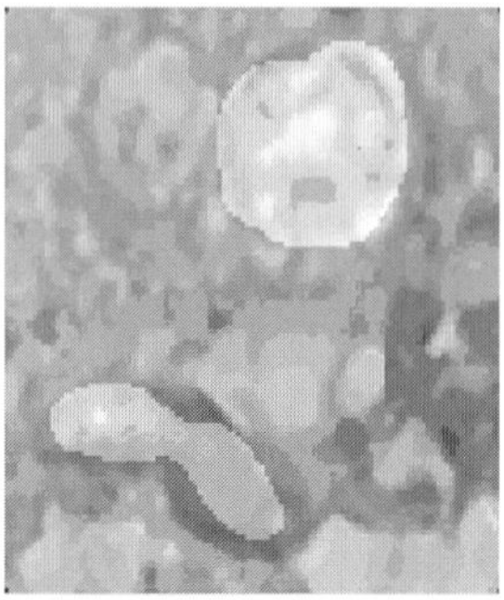

Fig. 18. The L_{BIN} binary image superimposed on a part of the L_{RGB} image

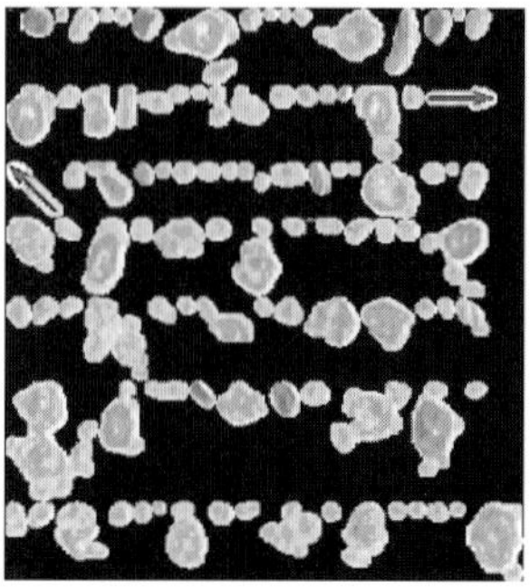

Fig. 19. The L''_{BIN} binary image superimposed on a part of the L''_{RGB} image

4 Conclusion

Fully automatic cell structures analysis gives new information like shape of particular objects. With employment of the presented decision trees algorithm, biological diagnostic support goes here fully automatically. Additionally created tools, which are a program option, allow on-line data import into the Excel spreadsheet. In addition, the described proceeding process of image processing - especially the image segmentation - can be extended on images of different contents. Independently from the fact that the way to a full expert system supporting diagnostic is still long, it seems that presented method is the beginning of new automatic morphometric factors measurements research direction.

References

1. Bajcsy R. (1973), Computer Description of Textured Surfaces, Proc, Int. Conf. Artificial Intell., Stanford, Calif, pp. 572-579.
2. Davis L.S. (1992), Hierarchical Generalized Hough Transforms and Line-Segment Based Generalized Hough Transforms, Pattern Recog., vol. 15, no. 4, pp. 277-285.
3. Forczek M. (2000), Colour Space Models in Colour Images Segmentation Algorithms, ZNPS, no. 38, pp. 171-195, Silesian Technical University Publisher, Gliwice (in Polish).
4. Fu, K.S., Mui J.K. (1981), A Survey of Image Segmentation, Pattern Recog., vol. 13, no. 1, pp. 3-16.
5. Gonzalez R.C. (1992), Computer Vision, Yearbook of Science and Technology, McGraw-Hill, New York, pp. 128-132.
6. Pavlidis T. (1972) Segmentation of Pictures and Maps Through Functional Approximation, Comp. Graph. Image Proc., vol. 1, pp. 360-372.
7. Wrobel Z., Koprowski R. (2004) The Practice of Image Processing in Matlab Application, Academic Publisher Exit, Warsaw (in Polish)

Analysis of Stem Cell Clonal Growth

Anna Korzynska[1], Marcin Jurga[2], Krystyna Domanska-Janik[2], Wojciech Strojny[1], and Darek Wloskowicz[1]

[1] Institute of Biocybernetics and Biomedical Engineering Polish Academy of Sciences Anna.Korzynska@ibib.waw.pl
[2] Medical Research Center Polish Academy of Sciences

Summary. The proper balance between a symmetric and asymmetric cell division is crucial for the neural stem cell maintenance both *in vitro* and *in vivo*. These conditions are provided by specific regions of the brain called neural stem cell niches and *in vitro* occur in neurospheres or adherent clones. A method and a tool for cell culture growth monitoring applied in the investigation of the clonally growth of HUCB-NSC (Human Umbilical Cord Blood derived Neural Stem Cells) line, as an *in vitro* model of the neural stem cell niche, is proposed.

1 Introduction

The goal of the investigation is to design tools to detect and describe two phenomena: the pattern of neural stem cells (NSC) division [1] and the pattern of their movement [2]. These tools will be used in the next step of the investigation to trace cell-cell interactions during the clone formation. Knowledge of these interactions and of all clone-forming cells history will be used on both the biological and mathematical levels. On the biological level it will allow to check the hypothesis of influence of cell-cell interactions on later fate of neural stem cell. On the mathematical level it will allow to propose a model of the neural stem cell culture development.

This knowledge is available only if the process of the cell culture growth is observed and measured many times. In order to obtain the measurement data, a method of extracting information from the images of the neural stem cell culture acquired several times in the course of the culture growth is proposed. The proposed method is based on the image processing and analysis methods that treat the images of the culture as a monitoring window, through which the immanent features of the observed phenomenon are extracted.

The observation of the cell culture requires a specimen, with the cell culture placed in the microscope and when the images of cells are acquired via the digital camera and are delivered into the computer memory. On the one hand, with this point in mind, the number and duration of these image acquisition sessions should be reduced to as short and as rare as possible. On the other hand, it is necessary to know which cell in the previous image corresponds to the cell in the current image (to minimize the errors in matching); the time between two consecutive image acquisition session should be adjusted to the quickest process under observation. This observation is related to various time increments due to the heterogeneity of the cell division process with respect to time. The cells divide more frequently in the first phase of cell culture growths than later on. A part of the sequence of cells culture development is presented in Fig.1.

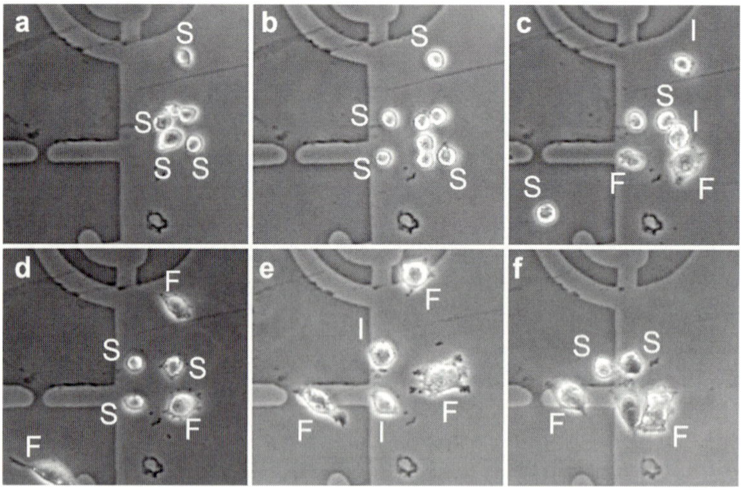

Fig. 1. An example of the sequence of images to analyse. Images were captured in the first (a), second (b), third (c), fourth (d), eighth (e) and tenth (f) day of cell culture grows. There are tree types of cells morphology in NSC culture: S - spherical, small neural stem cells F - flatten, large neural stem cell I - intermediate state between small and large cells

Several features should be extracted from the sequence of the images to describe the cell culture growth. Among them the number of cells of each image with their positions and types are extracted. It allows to follow several types of events that take place in the cell culture, e.g. the cell division, cell shape changes and the cell position changes.

The last event, the cell position changes, are important from the point of view of the cell movement detection and measurement. To achieve the required precision in the measurement of the distance of movement, registration of multisession images should be done. This problem is described later on in this paper.

The cell division process makes it difficult to determine of the number of cells in image (see Fig.1 marked with ellipse). The criteria of discrimination between two cases: - a cell is a single unit, but in an advanced division state, or - there are two separate cells very closely located just after division, are not obvious. This problem will be automatized in next step of development of proposed method.

There are cells of different morphology in Fig.1. They were classified into three categories: -small and compact (rounded, probably spherical) stem cells (see Fig.1 marked by S), -large, flatten cells with complicated, long outline (see Fig.1 marked by F), and a medium size cells, which seems to be the intermediate state between small and large cells (see Fig.1 marked by I). This paper concerns only the problems of cell segmentation and movement detection.

1.1 Registration of Cell Multisession Images

Registration is a process that allows to align one image with another, if the images show the same objects, and the images are acquired from different angles, distances or perspectives [3]. In our case, the cell culture images are acquired in multisession conditions. It is well known that even after a very careful localization of the microscopic slide on the microscopic stage, the image planes are very often translated and rotated one to the other and the lighting is distributed variously (see Fig.2).

It was decided to apply the registration with the external marker, because the objects under observation move and change their shape. The anchor points, localized on the crosses of the localization grid and/or the defects on microscopic slide (easy observed in Fig.2. as dark dots and a scratch) are chosen by the operator. The precision the procedure is visually examined by operator using the difference image (see Fig.3).

The images are enhanced and adjusted to the human visual system before the registration process. Taking only the dark part of full colour image, information about the cells not useful at this stage, is reduced and the grid crosses and defects on the microscopic slide are emphasized. The results of the process of registration are checked on the saturation channel Q of the YIQ colour mode because the colour mode transformation reduces the non-homogeneous light distribution in the image plane and lighting variation throught the sequence (see Fig.2c).

1.2 Cell Movement Detection

An important part of this investigation is the cell movement detection [4]. If all the images in the sequence are registered, and the cells in each image are segmented, two step matching process is done based on the analysis of the difference between two consecutive registered frames. All cells, that partly share the space in difference images are matched as the same cell in the consecutive observation. If any objects fail to share the space with one of objects in the next frame, the next step of the matching procedure is done. The nearest

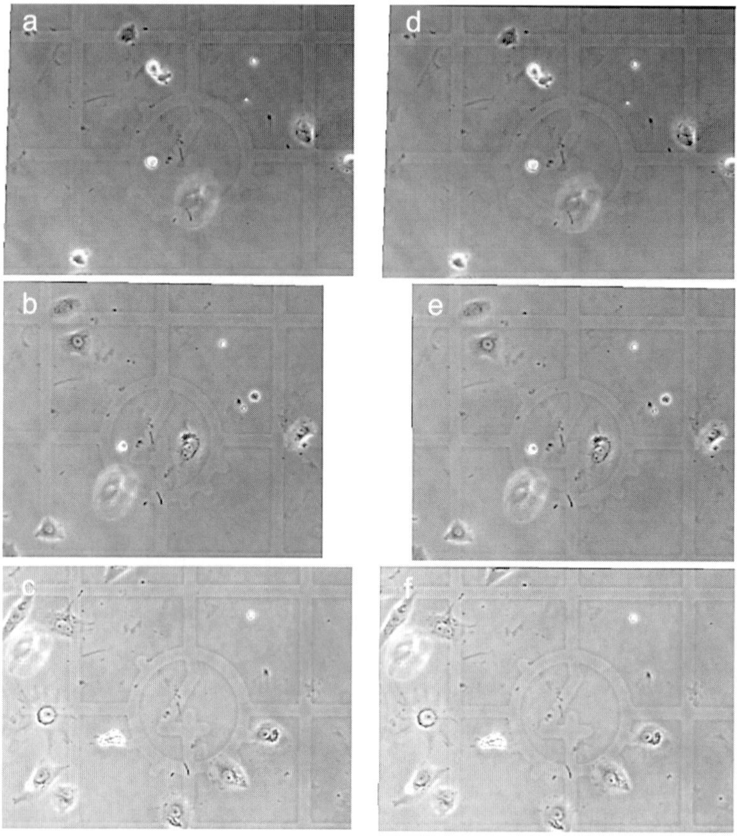

Fig. 2. Sequences of three source images (left) and the result of their registration to the coordinates of the first image (right)

neighbour located at a distance not longer than the threshold and not matched in first step of the procedure is matched to this cell. If there is no such a cell, the monitoring of the movement in respect of this cell is terminated.

This can be observed in Fig.3, where the top cell shared space in two consecutive positions (a and b), but bottom cell is located closely but separately in analysed frames what is well visible on difference image (c).

Based on the results of the matching procedure, the two types of distances between the two cell positions are calculated. It is assumed, according to the investigation on neutrophils [5], that the cell position is fixed in the centre of gravity of the cell binary mask, calculated as a result of the segmentation procedure.

Cell movement from t_0 to t_n is calculated with two types of measurement [4]: as trajectory length D

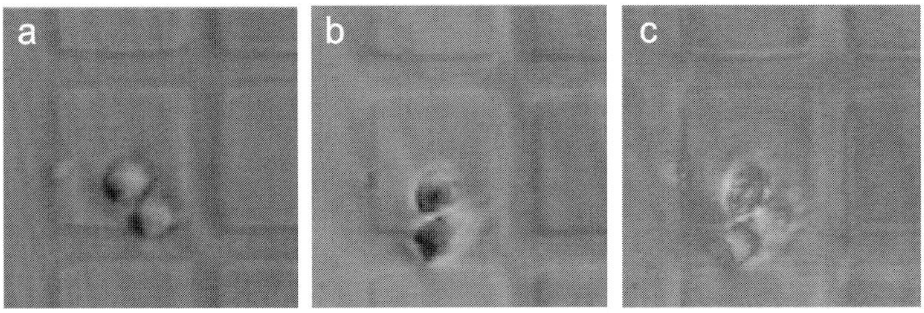

Fig. 3. Matching and distance of the cell movement measurement a - frame n; b - frame n+1; c - absolute difference between frame n and n+1

$$D_{0,n} = \sum_{i=0}^{n-1} d_{i,i+1} \tag{1}$$

or as a direct, the shortest distance d

$$d_{o,n} = \sqrt{(x_0 - x_n)^2 + (y_0 - y_n)^2} \tag{2}$$

Distances D and d for three consecutive cell positions are shown in Fig.4. The trajectory is marked by a thick line and the direct distance as a thin line (see Fig.4).

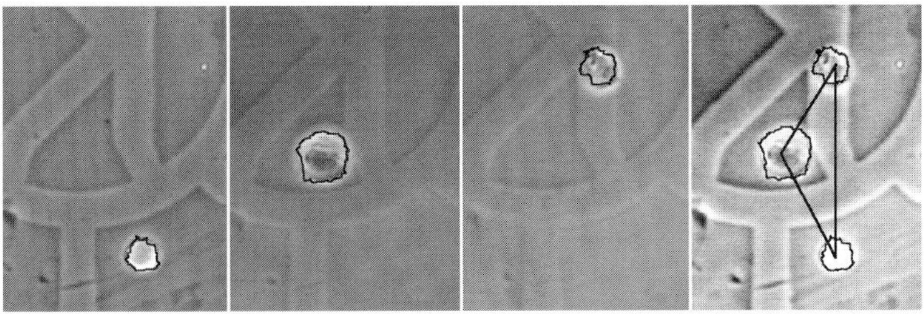

Fig. 4. Trajectory and direct distance of the stem cell movement

Both types of the distance can be used to describe movement patterns.

1.3 Cells Segmentation Method

The cell movement detection and measurement are based on the segmentation results. A hybrid method of the cell segmentation was used. This method is based on the texture and Prewitt gradient analysis on grey scale images [5]. The Y-channel of YIQ colour mode of NSC images become source image for segmentation procedure. The results of the segmentation are shown in Fig.5.

The proposed segmentation procedure segments individual cells placed separately in the image plane (see Fig.5 b, c and d) and cells located in clusters

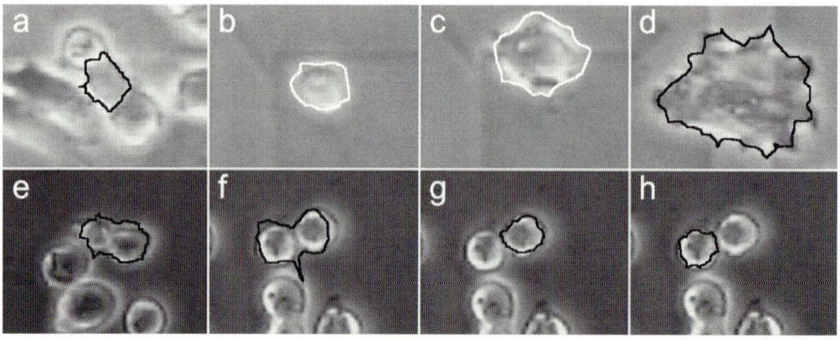

Fig. 5. Results of various types of NSC segmentation

(see a). Some results of the segmentation are good or even very good. However the procedure fails in 30% of the total number of the tested cells. It happens in the following cases: - when the cells are in early stages of the cell separation after the cell division (see f) and - when the cell is located too close to the other one with a higher gradient (sharper edge) then the analysed cell (see a thin peninsula in f) or - when a cell is located too closely to the border of the image. The last problem is the immanent feature of segmentation procedure and it causes the necessity of elimination of all the cells placed near the image border from the measurements. The other problems will be the subject of the future development of the segmentation method.

The number of cells in the culture is one of the most important parameter which describes the growth of the neural cell culture. This number is extrapolated based on the knowledge from the previous image in the sequence or the operator initialisation information in the first image in the sequence. There are several cases of cell division in our experimental material (lower row of Fig.5 and first two images in Fig.1) on which the segmentation procedure fails. On the actual stage of implementation of the segmentation procedure, the solution of the cell division problem is dependent on the operator's intervention. Operator moves a starting point of the segmentation procedure to the centre of one daughter cell and starts the first starting point of the segmentation procedure in the second daughter cell (see results in last two images in Fig.5).

2 Material and Results

The pre-study experimental material for future investigation was collected to verify the proposed method of the data analysis and the prepared tool.

The introductory experiments were performed on the line of Neural Stem Cells derived from Human Umbilical Cord Blood (HUCB-NSC) [6, 7]. The cells maintain high rate of proliferation, ability to form clones and to differentiate toward neuronal lineages for over two years *in vitro* [1].

The collected experimental material contains 43 sequences of 4-8 images of chosen plains from five microscopic slides. This material documents the behaviour of about 200 cells for from 4 to 10 days. The images were made in various time increments from 16 hours to two days, mostly in 24 hours (one day) time lapse.

According to assamption only low density plaines on mocriscopic plain has chosen to the computer analysys. The mean number of cells in the observed plane (698,88 μm x 524,16 μm = 0,366 mm^2) is 9, but the range is from 1 to 83 what causes that about 1480 cells positions were segmented and analysed. Some cells, which occupy a place near the image border in any step of observation, have been excluded from the collected data and only 31 sequences ware used to to analyse cell shape. In 39% cases cells were small (an area to 1300 pixels) and rounded, 24% cells were large (an area over 12000 pixels) and flat and 37% cells were somewhere between these two groups.

There are 15 cases of morphology changes found in collected data. In all these cases only changes from small and rounded to intermediate state (9 cases) and to large and flat (in the rest of cases).

There are cells division cases recorded in 13% of all sequences. In all of these cases this phenomenon is reocognised in two consecutive frames.

Under assumed treshold neibihood distance (mean value of diamiters of all small, rounded cells) cell matchng procedure detect movement only in four sequences. This movement is analysed as centre of gravity of cells' segmented area positions changes (see Tab. 1).

Table 1. Cell position in source image (x,y) and registered image (x',y')

	Sequence 1		Sequence 2		Sequence 3		Sequence 4	
Frame	(x,y)	(x',y')	(x,y)	(x',y')	(x,y)	(x',y')	(x,y)	(x',y')
1	1107,1138		706,794		1574,1123		631,587	
2	1001,954	1010,940	646,766	655,752	1334,1081	1327,1101	667,547	673,559
3	1082,802	1052,812	505,625	501,631	1139,991	1131,986	731,540	742,535

Using calibration rate 1 pixel it is 0,91μm it was found that the mean and standard deviation of all sequences for one day step is 131,04$\pm$89 μm, two days step is 148,33$\pm$62 μm.

The preliminary study of the experimental material shows that procedure needs any improvments in the collecting data and the data analysis procedure. The experimental material should be collected more frequently. This would allow us to collect more detailed information about a particular cell behaviour, its changes of morphology, its movement and division. The tool to analyse images is semi-automatic and requires a careful operator's supervision. It seems that full automatization of this process is impossible but the operator involvement in the process should be reduced. This reduction can be achieved by:

- development of semi-automatic registration procedure, based on the cross-correlation or entropy measurement

- changes in the segmentation procedure to avoid cases in which the highest gradient in the next cell causes extra area segmentation
- detection of an early state of the cell proliferation to extrapolate the cell division

3 Conclusions

The proposed method of the neural stem cell clone observation appears to be adequate and sufficient to notice all important events in the cell culture, including the cell proliferation, cell movement and cell morphology changes. The proposed tool collects information about each cell and sequence separately, and it is possible to extract the history of each monitored cell from these data using statistical software. The proposed method is time consuming and should be future developed to reduce the operator's engagement.

References

1. Buzanska L, Machaj EK, Zablocka B (2002) Human Cord Blood - Derived Cells Attain Neuronal And Glial Features In Vitro. Journal of Cell Sciences 115: 2131-2138
2. Chon J.H, Vizena A.D, Rock B.M, Chaikof E.L (1997) Characterization of Single-Cell Migration Using a Computer-Aided Fluorescence Time-Lapse Videomicroscopy System. Analytical Biochemistry 252:246–254
3. Castellanos N.P, Angel P.L.D, Medina V (2004) Nonrigid medical image registration technique as a composition of local warpings. Pattern Recognition 37:2141–2154
4. Korzynska A (2001) Computer Aided Neutrophil Granulocytes Movement and Shape Assessment. Ph.D. thesis, Institute of Biocybernetics and Biomedical Engineering Polish Academy of Sciences, Warsaw (in Polish)
5. Korzynska A, Hoppe A, Strojny W, Wertheim D (2004) Investigation Of A Combined Texture And Contour Method For Segmentation Of Light Microscopy Cell Images. Proceedings of the Second IASTED International Conference on Biomedical Engineering ISBN: 0-88986-379-2, pp 234-239
6. Buzanska L, Jurga M, Stachowiak EK, Stachowiak MK, Domanska-Janik K (2005) Growth Characteristics, Differentiation And Gene Expression Analyses Of A Human Umbilical Cord Blood-Derived Neural Stem Cell Line. Stem Cell (submitted)
7. Jurga M, Markiewicz I, Sarnowska A, Habich A, Kozlowska H, Lukomska B, Buzanska L, Domanska-Janik K (2005) Neurogenic Potential Of Human Umbilical Cord Blood - Neural Stem Cells Depends On Previous Long-Term Culture Conditions. Journal of Neuroscience Research (submitted)

Feature Extraction Optimization in Neural Classifier of Heart Rate Variability Signals

Pawel Kostka[1,2] and Ewaryst Tkacz[1,2]

[1] Institute of Electronics, Division of Microelectronics and Biotechnology, Silesian University of Technology, Gliwice, Poland pkostka@polsl.pl, etkacz@polsl.pl
[2] Medical University of Silesia, Faculty of Pharmacy and Laboratory Medicine, Department of Bionics, Sosnowiec, Poland etkacz@slam.katowice.pl

Summary. In this paper neural classifier system preliminary feature extraction and selection process using time-frequency representation of heart rate variability (HRV) signal is presented. The crucial point of described method is hybrid multi-domain feature set creation, combining different type parameters as well as feature selection based on the measure of class separability property, computed for each extracted feature. Regarding specific properties of non-stationary HRV signal, wavelet transform was chosen as time-frequency representation tool. Presented results are connected both with optimal feature extraction and selection of HRV signals from patient with coronary artery disease as well as classifier performance verification.

1 Introduction

In many classification problems dealing with real world data, high-dimensional classifier input vectors can be involved. Treating original signal samples obtained as a result of data acquisition process (e.g. ECG, EMG, EEG biomedical signals) as a primary N-element feature vector $x_l \in X \subseteq \Re^M$, able to describe whole complex objects, for a given narrow classification task it almost always consists of redundant components. On the other hand most pattern recognition algorithms work much better on non-redundant data, what allows to estimate the output class probability distributions with higher liability. So direct application of original values of sampled signals as a classifier input very often can not be carried out with satisfactory results. Additionally, assuming $y \in Y = \{y_1, y_2, ..., y_k\}$ as an output classifier vector with K - class labels in K-dimensional output space Y, the difference between input (N) and output (K) space size ($N >> K$) is very often unacceptable so original signal space is highly redundant with respect to classifier response space. Common way used to improve the classifier performance is the reduction of too high input feature vector size in intermediate feature extraction and selection stage. The basic goal is to reveal only discriminate features for given task and discard remain, reducing also the classifier complexity. So complex classification system should be created taking into consideration both feature extraction and selection step as well as final classifier stage with the same importance [1]. Proposed feature extraction tools almost always must depend on the specificity of classification task to be sensitive to features, which will be able to distinguish between health and pathology cases. Classifier presented in this paper was designed for the problem of coronary artery disease detection based on heart rate variability (HRV) signal analysis. This signal reflects interaction between cardiovascular and autonomous nervous system (ANS), which controls the hemodynamics and the heart work [2]. Presented in literature experimental results of synchronize HRV recordings before, during and after percutaneous transluminal coronary angioplasty [3] allowed to state, that when a coronary artery is blocked, the control is usually affected due to blood flow restrictions and to pressure changes induced mechanically in the affected area of the artery [4] [5]. Such experiments confirm the modulation of HRV signal by ANS mainly in frequency ranges: low-frequency band (LF,0-0.07 Hz), mid-frequency band (MF,0.07-0.15 Hz), and high-frequency band (HF, 0.15-0.45 Hz) [6]. Taking into consideration well known facts, that HRV features are included both in time and frequency domain the crucial point of feature extraction part proposed in this work is creation of hybrid - multitype feature vector combining time (T), frequency (F) and time-frequency (T-F) signal representation parameters. Assuming, that important HRV based features are characterized by local information in the duals domains of time and frequency and treating HRV as non-stationary signal from its nature, wavelet transform was chosen as T-F signal representation tool [7] [8]. In next feature selection stage the most representative feature set is created based on feature ranking competition algorithm described in section 2.4. Neural network structure based on multilayer perceptron (MLP) fulfils the role of nonlinear classifier of extracted signal representation parameters (section 2.5) in proposed method of screening examinations of coronary artery disease [9] [10].

2 Methods

2.1 General Structure

Proposed classifier for screening examinations of patient with coronary artery disease, based on their heart rate variability (HRV) signals consists of following stages (fig.1):

- HRV signal preprocessing i.e. continuous representation of HRV (Derivative Cubic Spline Interpolation method) HRV uniformly resampling (f_s = 5 [Hz]);
- New feature vector extraction, using hybrid, multi-type features from three groups i.e. Time domain features: statistical parameters of original HRV

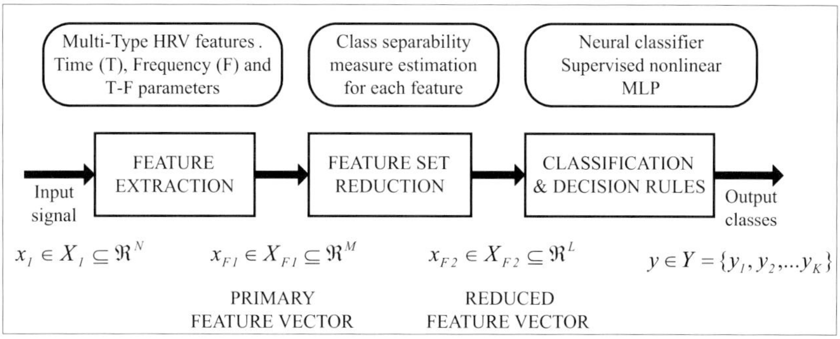

Fig. 1. General structure of feature extraction and classification stages of proposed method.

signal and its breath related component Respiratory Sinus Arrhythmia (RSA) and NRSA reflecting influence of remain factors (autonomous nervous system) on HR modulation frequency domain features: spectral parameters computed for analyzed HRV signal. Time-frequency features: T-F HRV representation, based on wavelet transform, which is suitable for non-stationary signals

- Classification - Supervisory learnt, nonlinear multilayer perceptron.
- Decision rules, which assign the neural network outputs to pathological or physiological groups based on elaborated norms.

2.2 Feature extraction

A generalized feature extraction method can be expressed as a map f: $X1 \rightarrow X_{F1}$, such that $X_{F1} \in \Re^M$ is the M-dimensional feature space, where $M << N$.

As presented in fig.2 three feature sets based on HRV signal were computed.

- The RSA and NRSA components were obtained from HRV and then set of statistical parameters (mean, std, range) characterizing these signals were calculated to create the $F1_A$ feature vector.
- The energy of LF and HF component of HRV spectrum as well as their ratio were included in $F1_B$ feature vector: $F1_B = [HRV_{LF-EN}, HRV_{HF-EN}, HRV_{LF-HF-R}]$
- The most complex feature set F1C was created based on time-frequency (T-F) HRV analysis

The specificity of HRV signal, which as it is well known has important features included both in time and frequency domain conditioned the area of

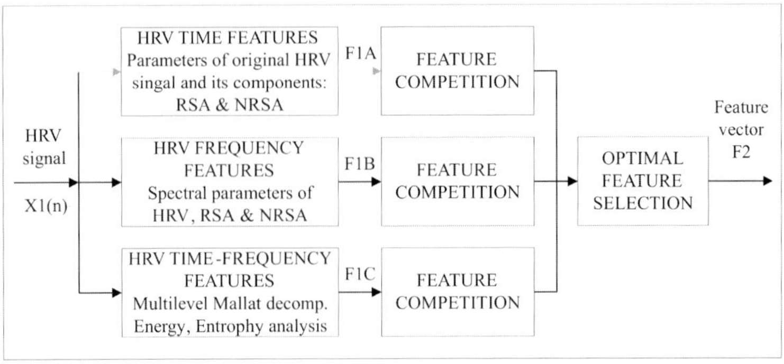

Fig. 2. Structure of hybrid multi-type HRV feature extraction and selection algorithm

appropriate feature extraction methods searching to the field T-F signal representation. Considering the possibility of using several T-F methods including: Short Term Fourier Transform, Wigner Distribution [11] [12] and their modification - Smoothed Wigner Distribution as well as Choi Williams Distribution their main limitation considering our classification task is that, they require the analyzed signals to be full or quasi - stationary. In case of HRV signal, which is non-stationary from its nature, mentioned above methods could not be able to reveal all important features for further classification. That's why we decided to choose as feature extraction tool verified in many applications wavelet transform, which is suitable to deal with non-stationary signals.

2.3 Wavelet transform as a feature extractor

As a feature extraction tool the wavelet transform based on multilevel Mallat signal decomposition [13] was used. Taking into consideration specific features of the HRV signal, especially that its significant frequency components are included in the range: $f_{HRV} < 0; 0.5 > [Hz]$, the grid of discrete wavelet scale - a values was created, corresponding to Mallat signal decomposition levels (for sampling frequency $f_S = 5[Hz]$, six levels corresponding to scale values: $a^i, i = 3..8$ were taken into consideration). Multilevel Mallat decomposition on every level corresponds to two-channel filtering using low and high pass filters to extract the detail and approximation signal component respectively (1),(1):

$$c_j^G = \frac{1}{2^j}\langle f(x), \Phi(\frac{x-k}{2^j})\rangle \tag{1}$$

$$c_j^H = \frac{1}{2^j}\langle f(x), \Psi(\frac{x-k}{2^j})\rangle \tag{2}$$

where: c_j - wavelet coefficient on $j-th$ decomposition level and $k-th$ translation, $\Phi(x)$ - scaling function, $\Psi(x)$ - wavelet function. As a next step, to create the new features vector, for every signal component obtained on each decomposition level a set of parameters was computed. For each subspace wavelet coefficients were squared and normalized to obtain the energy probability distribution (3):

$$p_i = \frac{c_i^2}{\sum_{k=1}^{n} c_k^2} \tag{3}$$

For each wavelet scale, the sorted series may be considered as an inverse empirical cumulative energy distribution function (ECDF). Base on this parameters the Shannon entropy (4) of energy distribution p_i (3) were calculated as a measure of energy unpredictability in each wavelet decomposition subspace.

$$E = \sum_i p_i log_2(p_i) \tag{4}$$

This procedure allowed to reveal a group of new features based on energy and entropy measures. The whole set of new feature vectors $\overline{Fl_{C1}}..\overline{Fl_{C5}}$, created as a result of multilevel Mallat signal decomposition, which is put to the input of classifier structure includes the following groups of parameters as a series for every of $i-th$ decomposition level:

- Mean values of wavelet coefficients in each subband (frequency distribution information) -$\overline{Fl_{C1}}$
- Standard deviations of wavelet coefficients (level of change of frequency distribution information) -$\overline{Fl_{C2}}$
- Energy of $i-th$ component (3) -$\overline{Fl_{C3}}$
- Shannon entropy of wavelet component (distribution of the amount of information included in every subband) -$\overline{Fl_{C4}}$
- Shannon entropy E of energy distribution p_i (4) -$\overline{Fl_{C5}}$

2.4 Feature selection

Feature set may be considered near to optimum if it minimizes chosen error based criterion function. There are two approaches to feature selection problem:

- Feature subset selection
- Feature projection, which tries to find optimal original feature combination (projection) into smaller set of new features. Principle component analysis (PCA) [14] or projection pursuit [15] are often used feature projection methods.

In presented HRV classifier structure, dimension reduction method based on selecting the best feature subset according to assumed criteria were used.

Because the evaluation of optimal cost function using probability of misclassification is too complex [16], in presented approach simpler criterion based on class separability (CS) were applied. Considering a two class problem with an original M-dimensional feature set space X_{F1} a feature selection algorithm used in presented work is following:

- Create two feature matrices M^p_{F1}, M^q_{F1}, representing two classes: p and q consisting of patterns (vectors) $x^{(p,m)}_{F1}$, $x^{(q,m)}_{F1}$ from learning data set (5):

$$M^p_{F1} = [x^{(p,1)}_{F1}, x^{(p,2)}_{F1}, ..., x^{(p,P)}_{F1}] \tag{5}$$

for class p, and

$$M^q_{F1} = [x^{(q,1)}_{F1}, x^{(q,2)}_{F1}, ..., x^{(q,P)}_{F1}] \tag{6}$$

for class q (6), where (7)

$$x^{(p,m)}_{F1} = [x^{(p,m)}_{F1-1}, x^{(p,m)}_{F1-2}, ..., x^{(p,m)}_{F1-M}]^T \tag{7}$$

is a mth pattern in class p.
- Assuming, that we are trying to evaluate the "discriminant power" of each single feature separately (not e.g. feature combination), according to class separability criteria the ability to discriminate of ith feature is represented by i^{th} row in feature matrices M^p_{F1} or M^q_{F1} (depending on class type).
- Define $DM(p_i, q_i)$ as a discriminate measure for the ith feature, which expresses what is the value of separability weight of the given feature in classification process. Different types of $DM(p_i, q_i)$ can be considered and several of them have further been tested [17]:
  1. Fisher's class separability index (8):

$$DM(p_i, q_i) = \frac{(mean(p_i) - mean(q_i))^2}{var(p_i) + var(q_i)} \tag{8}$$

     where $mean()$ and $var()$ are computed across i^{th} matrix row
  2. Relative entropy (9):

$$DM(p_i, q_i) = p_i log\frac{p_i}{q_i} \tag{9}$$

  3. Symmetric relative entropy (10):

$$DM(p_i, q_i) = p_i log\frac{p_i}{q_i} + q_i log\frac{q_i}{p_i} \tag{10}$$

  4. Euclidean distance (11):

$$DM(p_i, q_i) = \|p_i - q_i\| \tag{11}$$

- Sort obtained in previous paragraph $DM(p_i, q_i)$ values to create a feature rank as a results of feature competition.
- Choose the most discriminant L features to create a new feature vector X_{F2}

2.5 Classifier stage

Two layer well-known and verified supervised learnt, feedforwad perceptron structure with sigmoidal and linear activation functions fulfil the role of non-linear classifier of new L-element feature set.

3 Results

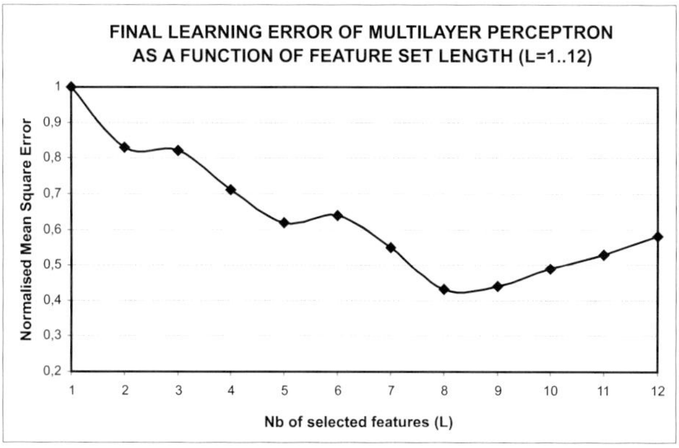

Fig. 3. Neural network classifier training error as a function of number of chosen features in new feature vector.

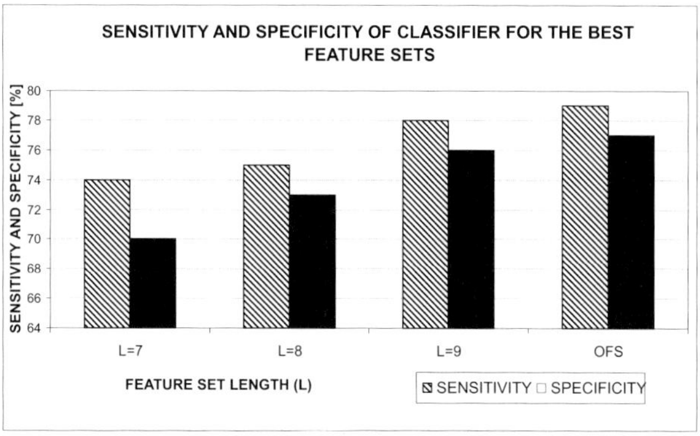

Fig. 4. Neural network based classifier performance for four the best subsets of new features, selected from HRV based parameters.

Proposed structures were tested using the set of clinically characterized heart rate variability (HRV) signals of 62 patients, as cases with a coronary artery disease of different level. Additionally similar control group of healthy patients was analyzed. Whole database was divided into learning and verifying set. Classification task was defined as the trial of two group (healthy and pathology cases) distinguish, based on new feature subsets obtained in feature extraction and selection stages of whole procedure. First group of results is connected with the search for optimal feature subset for given classification task. To find the most discriminant parameters obtained from input HRV signal analysis, for every feature included in time (T) domain HRV feature group - F_{1A}, frequency (F) domain feature set - F_{1B} and T-F HRV representation parameters: F1C1-F1C5 (see section 2.2 for detail feature description) a discriminate measure $DM(p_i, q_i)$ was computed. It can be expressed as the computed feature separability indicator distribution among all HRV based parameters group used in described classifier system. Fig.3 presents influence of the most discriminative features number included in new feature vector on final learning phase error of 2-layer perceptron classifier structure. Finally whole presented classifier system were verified using test set of HRV data from patient with coronary artery disease. Common used sensitivity and specificity classifier performance measures obtained for different new feature vector length (L) are presented in fig.4. Apart from feature subset included the L features with maximum class separability properties $DM(p_i, q_i)$, additional subset consisting of seven features - one, the best representative from each of HRV parameters group was created and this vector seemed to be optimal feature set (OFS) for HRV signal classification.

4 Discussion and Conclusion

New heart rate variability (HRV) signal representation belonging to the reduced space, obtained as a result of feature selection process from hybrid multi-domain: time (T), frequency (F) and T-F domain HRV parameters was presented. Evaluation of all extracted HRV features based on the measure of its class separability property showed, that the most discriminant features for given classification task are the parameters in F_{1C3}, F_{1C4} and F_{1C5} feature vectors. These feature sets include energy, entropy and Shannon entropy E of energy distribution parameters respectively of Mallat HRV signal decomposition components. The most significant features from these vectors are assigned to $d3$ and $d4$ level of discrete wavelet analysis (frequency subbands: (0.3125; 0.6250) and 0.1563; 0.3125 [Hz]). It corresponds to rather high frequency (HF) components of HRV PSD function. Results of optimal feature selection process based on feature discriminity measure ($DM(p_i, q_i)$) presented in fig.4 showed that the smallest learning error of neural network classifier was reached for 8 or 9 features with maximum class separability properties. Results of optimal feature selection based on learning phase error optimization were not until the end confirmed in final whole system verification step using training set of data. Assuming the measure of classifier sensitivity and specificity as system performance indicators as presented in fig.4 the best results were obtained for new feature vector created by taking the best one feature from each feature vectors: - F_{1A}, F_{1B} and F_{1C1}-F_{1C5} .

Described method of feature selection was limited to the case, that every analysed feature was taken as single feature (not e.g. several feature combination) what could affect the process of discriminaty measure $(DM(p_i, q_i))$ computation. That's why to create optimal feature vector simply taking the feature with maximal value of $(DM(p_i, q_i))$is not enough. It was shown in fig.4 where the best classification results was obtained for the feature combination consisting of the best feature representant from each group: F_{1A}, F_{1B} and F_{1C1}-F_{1C5} but not for the simply combination of feature with the highest value of class separability parameter $(DM(p_i, q_i))$, computed for every feature in feature selection stage (section 2.2). To conclude,obtained results showed, that classification procedure gave satisfactory results, considering presented classification algorithm as a contribution to coronary artery disease detection on preliminary screen examination stage. Before pattern classifier can be properly designed and effectively used, it is necessary to consider the feature extraction and data reduction problems. Feature extraction should consists in choosing those features, which are most effective for preserving the class separability. Presented classification procedure gave satisfactory results, considering described classification algorithm as a contribution to coronary artery disease detection on preliminary screen examination stage.

References

1. Duda, R.O. Hart, P.E., Pattern classification and scene analysis, John Wiley and Sons, New York, 1973.
2. Marciano, F. Bonaduce, D. Petretta, M. Valva, G. Migaux,M.L., Spectral behavior of heart rate variability in acute ischemic episodes," in Computers in Cardiology. Los Alamitos, CA: IEEE Comput. Soc. Press, 1995, pp.111-114.
3. Clariá F. Vallverdú M. Caminal P. The effects of coronary occlusion location on the RR signal. IEEE Engineering in Medicine and Biology. July/August 2002.pp.59-64.
4. Thakor N.V. Sherman D.L. Biomedical problems in time-frequency-scale analysis - new challenges, Proceedings of the IEEE-SP pp. 536-539, 1994.
5. Saul,J.P. Heart Rate Variability and Sudden Death: What's the Connection? Amsterdam, The Netherlands: IOS Press, 1995.
6. Láng, E., Caminal, P., Horváth, G., Jané, R., Vallverdú, M., Slezsák, I., Bayés de Luna, A., Spectral analysis of heart period variance (HPV)-A tool to stratify risk following myocardial infarction," J. Med. Eng. Techonol., vol. 22, no. 6, pp. 248-256, 1998.
7. Akay M., Time-Frequency and wavelet analysis, IEEE EMB Magazine 14(2), 1995.

8. Akay, Y.M., Akay, M. Welkowitz, W., Kostis, J., Noninvasive detection of coronary artery disease, IEEE Engineering in medicine and biology, 11/12, pp. 761-764, 1994.
9. Tkacz E. Kostka P., An Application of Wavelet Neural Systems for Classification Patients with Coronary Artery Disease Based on HRV Analysis, World Congress on Medical Physics and Biomedical Engineering - Chicago 2000.
10. Kostka P. Tkacz E., The comparison of the wavelet-neural systems and radial neural networks in classifier structure, 16th International Eurasip Conference Biosignal 2002, pp.46-48,June 2002, Brno,Czech Republic.
11. Hlawatsch, F., Boudreaux-Bartels, G.F., Linear and quadratic time-frequency signal representations, IEEE Signal Processing Mag., vol. 9, pp. 21-67, July 1992.
12. Cohen, L., Time-Frequency Analysis, Englewood Cliffs, N.J.: Prentice Hall Signal Processing Series, 1995.
13. Mallat S. (1989): A theory for multi-resolution signal decomposition: the wavelet representation, IEEE Transaction on Pattern Analysis, 7(11):674-693, 1989.
14. Karhunen, Joutsensalo, J. Generalizations of principle component analysis, optimization problems, and neural networks, Neural Networks, vol. 8, No. 4, pp. 549-562, 1995.
15. Friedman, W.J. Tukey, J.W. A projection pursuit algorithm for exploratory data analysis, IEEE Transactions on Computer, Vol. 23, pp. 881-889, 1974.
16. Fukunaga K. Introduction to statistical pattern recognition. 2nd edition. Academic Press, San Diego, CA,1990
17. Kullback, S. Leibler, R.A. On information and sufficiency, Annals of Mathematics and Statistics, 22, pp. 79-86, 1951.

Clustering DNA Microarray Data

Henryk Maciejewski[1] and Anna Jasinska[2]

[1] Institute of Engineering Cybernetics, Wroclaw University of Technology, ul. Wybrzeze Wyspianskiego 27, 50-370 Wroclaw, Poland, `Henryk.Maciejewski@pwr.wroc.pl`

[2] Laboratory of Cancer Genetics, Institute of Bioorganic Chemistry, ul. Noskowskiego 12/14, Poznan, Poland, `aniaj@rose.man.poznan.pl`

1 Introduction

This paper is devoted to cluster analysis of DNA microarray data, and specifically to clustering gene expression vectors. We show that settings of the clustering algorithm applied can have a deep effect on clusters obtained. This creates a challenge to interpret results of clustering. In this work we concentrate on hierarchical clustering algorithms, for which case we propose an approach to decide on usefulness of results of clustering obtained under a particular setting of the clustering engine. This approach is based on a measure that allows to compare clustering results by confronting them with biological knowledge. The measure can be taken into account by a biologist trying to decide on biological relevance of clusters obtained.

A number of different methods of clustering DNA have been reported in literature, e.g., hierarchical clustering [3], self-organizing maps (SOM) [11], or support vector machines [2]. A broad overview of methods available can be found in [7] and overview of methods in the context of DNA microarray experiments can be found in [9] and [10]. Although it has been recognized that different methods can have an effect on results of clustering, see e.g., [8], little effort has been reported in literature to tackle the 'different algorithm - different results' challenge. Here we aim to propose means to help biologists evaluate correctness of clustering results by confronting results with biological knowledge. Hence the degree of adherence to expected biological contents of clusters can be used to align different clustering results by biological relevance.

2 Clustering Gene Expression Data

2.1 Input Data

A DNA microarray (chip) allows to measure expression levels of up to several thousand genes in a single experiment involving one sample of genetic material. Typically, a microarray study consists of a number of samples, where each comes from a different patient or is representative to a different stage of a disease, etc. Data for analysis produced from such a study can be represented as a matrix with columns corresponding to subsequent samples (i.e. chips) and rows, roughly, corresponding to subsequent genes tested on a DNA chip concerned. More precisely, rows represent probesets, where a probeset is an area on a DNA chip designed to measure expression level of a particular gene. One gene may be represented by more than one probeset (redundant gene representations allowed).

The gene expression data used here for the purpose of illustration has been gathered in a study involving seven human genome U133A GeneChips from Affymetrix[1]. Thus data for analysis forms a matrix of seven columns and about 22 thousand rows (approx. the number of probesets on a U133A chip). Three columns correspond to patients suffering from fragile X syndrome (related to mutation of FMR1 gene, i.e. three FMR1 mutants in the experiment), three other columns represent samples with no mutation and one column corresponds to a premutant (no full mutation yet observed). Although this information is not directly used in the numerical studies performed, it is given here to highlight organization of a DNA study.

2.2 Task Formulation

Based on data outlined before, the task discussed here is to find groups of genes (probesets) that exhibit similar expression profile across a series of samples tested. In case of the FMR1 data, the task can be formulated as clustering rows of expression matrix (each row is a vector of dimensionality of 7). Sample rows of the matrix are shown in Table 1.

Table 1. Sample data from FMR1 study

probeset	sample1	sample2	sample3	sample4	sample5	sample6	sample7
1007_s_at	687.2	709.7	604.5	809.4	512	700.4	804.7
1053_at	870.6	479.1	613.4	636.1	667.8	728.7	691.3
117_at	194.6	326.8	183.7	151	910.2	438.9	264.2
121_at	1710.8	2217.9	1514.1	1748.5	1810.6	1522.1	2003.9

2.3 Hierarchical Clustering

Clustering process usually requires that raw data should be preprocessed which usually involves:

[1]This work was supported by the State Committee for Scientific Research, grant no PBZ/KBN/040/P04/2001.

- Normalization of data to ensure that vectors (probesets) cluster by expression profile rather than signal level. In the numerical study described here, rows of expression matrix were normalized to the length of 1 (every row divided by its length).
- Linear transformation of input variables (in our case - columns of expression matrix); the algorithm given by [1] was used for data preprocessing prior to clustering. The purpose of this stage is to make elongated or elliptical clusters more spherical, which improves performance of clustering algorithms. This is often required, since, as shown by [4], many clustering algorithms perform better with spherical clusters than with elongated clusters. Our prelimiary studies have shown that separation of clusters seems better with this step performed. As a result, seven canonical variables were obtained to be used on input to clustering algorithm.

The most often used version of hierarchical clustering is based on agglomerative method, which involves following steps:

- Step 1: Each row of (transformed) expression matrix is treated as a one element initial cluster.
- Step 2: Join two nearest clusters into one cluster, which replaces the two clusters joined.
- Repeat Step 2 until one cluster remains.

Distance between clusters can be defined in a number of ways, e.g.,

- Average linkage: distance between clusters is defined as the average distance between pairs of points, each in one cluster.
- Complete linkage: distance between clusters is the maximum distance between pairs of points, each in one cluster.
- Ward's minimum variance method: distance between clusters is ANOVA sum of squares between clusters added up over all the variables.

This list is far from complete. Clustering algorithms based on different definition of intercluster distance tend to produce clusters showing different characteristics in terms of cluster shape, number of members or dispersion. For instance, complete linkage is biased to producing compact clusters, usually similar in size, Ward's method also gives clusters of approximately same size, whereas average linkage gives clusters of similar variance. However, since it is virtually impossible to judge in advance what the real characteristics of groupings in data are, there are no clues available as to which clustering algorithm might produce results which best resemble real clusters in data.

2.4 Results

A convenient way to present results of hierarchical clustering is by using a dendrogram, which illustrates subsequent joins of clusters. Part of a dendrogram for the sample FMR1 study is shown in Fig. 1, this result has been obtained using Ward's method. This results shows a 23 element cluster built around a probeset `212133_at` (which represents an FMR1 related gene NIPA2). Interestingly, a cluster of similar size obtained around `212133_at` using complete linkage method includes only about 30 % probesets in common with Ward's method results; the same is observed when comparing results of Ward's vs. average linkage method. Probesets clustered together are shown in Table 2.

Table 2. Intersection of clusters built with different methods (Ward, Average linkage (AL), Complete linkage (CL))

Ward-AL	Ward-CL
202172_at 206507_at	202172_at 204580_at
212133_at 215228_at	212133_at 215228_at
218515_at 218974_at	217353_at 218974_at
220202_s_at 222146_s_at	220446_s_at 222146_s_at

This numerical example illustrates the profound influence of the clustering algorithm on clusters obtained. It is also noteworthy that some genes cluster together irrespective of the algorithm adopted, which probably indicates real clusters in data. Probably this indicates real groupings in data rather than numerical artifacts. This however requires further investigation.

3 Trace Clusters Approach

Here we introduce a procedure aimed to help biologists judge on biological relevance of clustering results obtained under different settings of clustering algorithms. This can be done by employing biological perspective/knowledge in decision process rather than trying to base the decision solely on algorithm specific cluster quality criteria (such as cubic clustering criterion), which turn out hardly useful for microarray data analysis.

The main assumption we make is that biological knowledge is available about at least two genes or probesets that are expected to exhibit similar expression profile across the series of samples (chips) tested. We denote this set of genes RG. The main idea of our approach is then to observe the hierarchical clustering process 'bottom-up' starting with seeds in RG, favoring results of clustering algorithm that tends to join members of RG earlier than other algorithms.

Let us introduce following notation: C_i is a cluster at level i ($i = 1$ denotes an initial one element cluster, i.e. leaf of dendrogram, see Fig. 1), B_i is a 'brother' cluster joined by the algorithm to form a 'parent' cluster P_i, of size $|P_i|$ (i.e. number of member points). Then the algorithm can be given as:

- Step 1: Initialize: select $C_1 \in RG, i = 1$, found=FALSE.

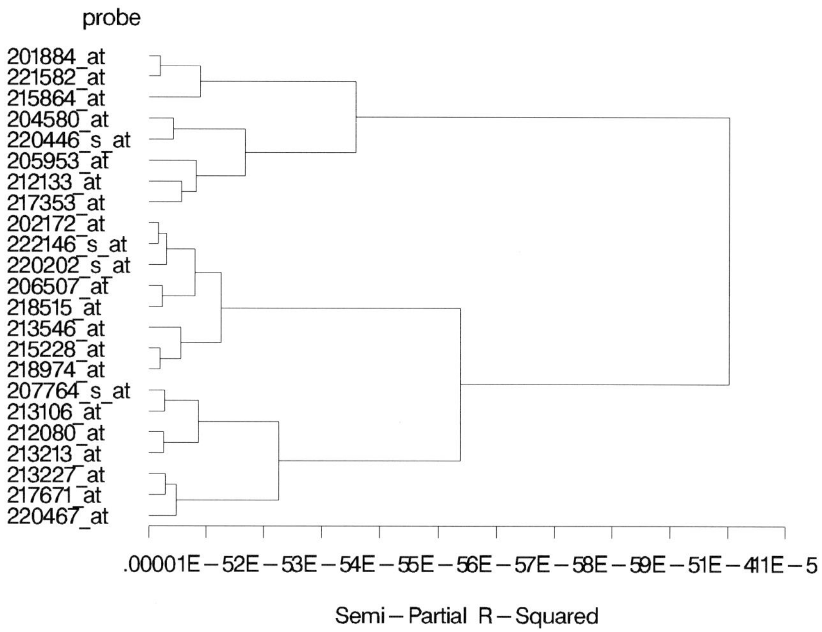

Fig. 1. Sample dendrogram illustrating a hierarchy of clusters

- Repeat while not found
- Step 2: Find B_i. Output B_i and output $level = i$.
- Step 3: Find P_i. If $RG \subset P_i$ then found=TRUE and output $|P_i|$ and output $level = i$. Else $C_{i+1} = P_i, i = i + 1$

The procedure gives on output the level at which group of believed-to-be-related genes first cluster together and the size of this first common cluster. This number can be used as a measure of performance of a clustering algorithm in terms of its ability to group related genes together.

This can be illustrated by the means of FMR1 data analysis. If we assume that the set RG included two FMR1 related genes CYFIP2 and NIPA2, then we receive for the corresponding probesets (`212133_at` and `220999_s_at`):

- Ward method: $level = 7$; $|P_7| = 74$
- Complete or average linkage: $level = 11$; $|P_1 1| = 1400$

Ward method apparently joined points in RG earlier than other methods, which may indicate that Ward results are more trustworthy in this case. Obviously, such firm conclusion require more investigation by a biologist, who starts with defining relevant RG sets.

Interestingly, Ward method also tends to produce the tightest clusters around its seed probeset (`212133_at`), which can be illustrated by average Euclidean distance (*avgdist*) computed for cluster members closest to

212133_at (in the example below first clusters P_i are taken exceeding 20 in size):

- Ward method: $avgdist = 0.15, N = 23$
- Complete linkage: $avgdist = 0.19, N = 46$
- Average linkage: $avgdist = 0.17, N = 23$

Another output of the procedure is a series of clusters B_i, which after being merged with known gene annotation information are to be analyzed by a biologist to make final decision of validity from biological perspective of clustering algorithm decisions. Sample of this output is given in Table 3.

Table 3. Subsequent clusters joined starting with a selected seed

probe	level	Gene	GeneTitle
212133_at	1	NIPA2	non imprinted in Prader-Willi/Angelman syndrome 2
217353_at	2	HNRPA1	heterogeneous nuclear rib nucleoprotein A1
205953_at	3	LRIG2	leucine-rich repeats and immunoglobulin-like domains 2
204580_at	4	MMP12	matrix metalloproteinase 12 (macrophage elastase)
220446_s_at	4	CHST4	carbohydrate (N-acetylglucosamine 6-O) sulfotransferase 4

4 Conclusions and Further Work

Proper interpretation of results of clustering of gene expression data from DNA microarray tests is one of major challenges in experiment data analysis. Interpretation problems arise due to the fact that different algorithms tend to produce different results, while some clusters appear to be invariant of an algorithm applied. A procedure described in this work can be a good starting point for a decision making process to evaluate biological relevance of clustering results obtained. In our view, any other similar approach aiming to discover biologically relevant clusters will have to include biological information. It would be probably beneficial if relevant biological knowledge could be incorporated on the input side of clustering algorithm rather than at the results post processing / interpretation stage, as described in this work. Making clustering algorithms make clustering decision biased towards biologically relevant groupings, thus forming 'supervised clustering' approach may be a motivation for further research in this area.

References

1. Art D, Gnanadesikan R, Kettenring R (1982) Data-based Metrics for Cluster Analysis. Utilitas Mathematica 21A:75-99
2. Brown M, et al. (2000) Proc. Natl. Acad. Sci. USA 97:262-267
3. Eisen M, et al. (1998) Proc. Natl. Acad. Sci. USA 95:14863-14868
4. Everitt B (1980) Cluster Analysis, Second Edition. Heineman Educational Books Ltd., London
5. Ewens W, Grant G (2001) Statistical Methods in Bioinformatics. Springer, Berlin Heidelberg New York
6. Faller D, et al. (2003) Journal of Computational Biology 10:751-762
7. Hastie T, Tibshirani R, Friedman J (2002) The Elements of Statistical Learning. Data Mining, Inference and Prediction. Springer, Berlin Heidelberg New York
8. Hoffmann R, Seidl T, Dugas M (2002) Profound effect of normalization on detection of differently expressed genes in oligonucleotide microarray data analysis. Genome Biology
9. Quackenbush J (2001) Nature Reviews Genetics 2:418-427
10. Shannon W, Culverhouse R, Duncann J (2003) Pharmacogenomics 4:41-51
11. Tamayo P, et al. (1999) Proc. Natl. Acad. Sci. USA 96:2907-2912

Cytomorphometry of Fine Needle Biopsy Material from the Breast Cancer

Andrzej Marciniak[1], Andrzej Obuchowicz[1], Roman Monczak[2], and Mariusz Kołodziński[2]

[1] Institute of Control and Computation Engineering, University of Zielona Góra
A.Marciniak;A.Obuchowicz@issi.uz.zgora.pl

[2] Department of Pathomorphology, Zielona Góra Hospital

Summary. A computer system has been developed for evaluating the morphometrical feature extraction. The features are derived directly from a digital scan of breast fine needle biopsy slides. First the background elimination by thresholding hue component is applied, then the actual segmentation is done with region growing technique. The quality of feature space is measured with classifier based on non-parametric density estimation. The automatic system of malignancy classification was applied on a set of medical images with promising results. The comparison of human accuracy in the cytological diagnosis of breast cancer with the accuracy of digital image analysis combined with computer-based classification is presented.

1 Introduction

Breast cancer is the most common cancer in women in USA, Europe and Australia. About 12 percent of all women at age 70 years are having or have already had the breast cancer. The presence of a breast lump is an alert sign, but it does not always indicate a malignant cancer. The mortality rate is estimated at about 50 percent and detection of disease in its early stage is crucial for keeping the chance of recovery. The prognosis is strongly dependent on the disease development before any treatment is applied. It is good if the tumor location is limited only to breast area and there are no metastasis to other parts of the body by way of blood or lymphatic vessel. In fact, breast cancer is an insidious disease. Even small tumor (e.g. with diameter of 0.5 cm) can be the site for distant transfer of cancerous cells whilst about 95 percent of tumors with diameter up to 1 cm is limited to breast area only [7].

Most breast cancers are detected as a lump on the breast, by so called triple-test which includes self examination (palpation), mammography or ultrasonography and *fine needle biopsy* (FNB). Any observed anomaly within the structure of breast needs a microscopic verification. Performing FNB under ultrasonographic control enables us to collect material even from tubercles

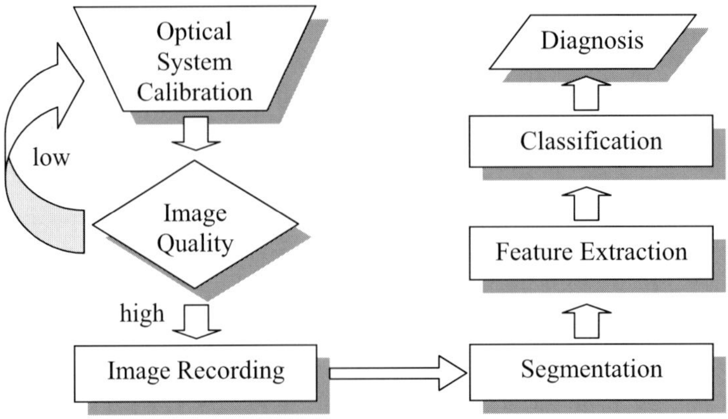

Fig. 1. General scheme of the diagnosis system

of below 1cm in diameter. FNB is a cost-effective, non-traumatic and minimally invasive diagnostic test that obtains information required to evaluate malignancy. However, the diagnostic efficiency of FNB is still under debate [5]. Its success depends on the experience of cytologist and objective possibilities of distinguishing malignant from benign cells in tumor. In general, the cytological diagnostic criteria are quoted commonly and well known, but they are rather imprecise and dependent on arbitral and subjective evaluation. In short, some pathologists diagnose better than others. In order to make the decision independent on arbitrament factor, the *morphometric analysis* can be applied.

Objective analysis of microscopic images of cells has been a goal of human pathology and cytology since middle of the 19th century. Early work in this area consisted of simple manual measurements of cell and nuclear size. Along with the development of advanced vision systems and computer science, the quantitative cytopathology has become established as a useful method for detection of diseases, infections as well as many other disorders. Historical development of this research can be found in [2].

In this paper, we present a new automatic attempt for recognition of malignancy of breast tumor. Our approach can be divided into several well-defined stages, presented in Fig.1.

2 Materials and Methods

2.1 Origin and Acquisition of the Images

Testing existing and new developed algorithms requires to have databases at disposal, on which tests and benchmarks can be realized, especially in the domain of image analysis where in many problems a domain knowledge

Table 1. Comparison of datasets

	WDBC	Zielona Góra
Number of samples (malignant)	569 (357)	50 (25)
Number of images	569	500
Format of files	GIF	BMP
Resolution	512×480	704×578
Number of colors	256	256^3
Dye	h+e	h+e

need to be taken into account. Probably, the most commonly known data set of FNB images is Wisconsin Database of Breast Cancer (WDBC) [8] which can be obtained from repository of machine learning database University of California. WDBC contains both raw images and visually extracted features, but the quality of images is rather poor and it does not fit for automatic feature assessment. In our study we decided to design a new data set that could be applied for completely automatic process of image analysis [3]. The comparison of parameters of both image sets is given in Table 1.

Morphometric examinations of cell nuclei were carried out on the cytological material obtained by FNB. Biopsy without aspiration was performed under the control of ultrasonograph with a needle with a diameter of 0,5 mm. Smears from the material were fixed in spray fixative (Cellfix of Shandon company) and dyed with hematoxylin and eosin (h+e). The time between preparation of smears and their preserving in fixative never exceeded three seconds. The smears were derived from 25 FNB of benign and 25 of malignant lesions collected from 50 patients of out-patient clinic ONKOMED in Zielona Góra. All cancers were histologically confirmed and all patients with benign disease were either biopsied or followed for a year.

The image for digital analysis was generated by SONY CCD IRIS color video camera mounted atop an AXIOPHOT microscope. The slides were projected into the camera with 10 and 160× objective and a 2,5× ocular. One image was generated for enlargement 100× and nine for enlargement 400×.

2.2 Segmentation of the Nuclei

Most of the criteria of malignancy are seen in the nuclei of the cells. Therefore, it is essential to isolate the nuclei from the rest of image. That can be done by *region growing* where each pixel in the image initially constitutes a separate region. In this study, we combined the classical *region growing* technique with *thresholding*. The modification of approach that has already been investigated in [1] is performed. The image is transformed from Red-Green-Blue color space into Hue-Saturation-Value (HSV) representation space. The only component that is taken into account in our segmentation algorithm is Value component (in other words intensity or brightness component).

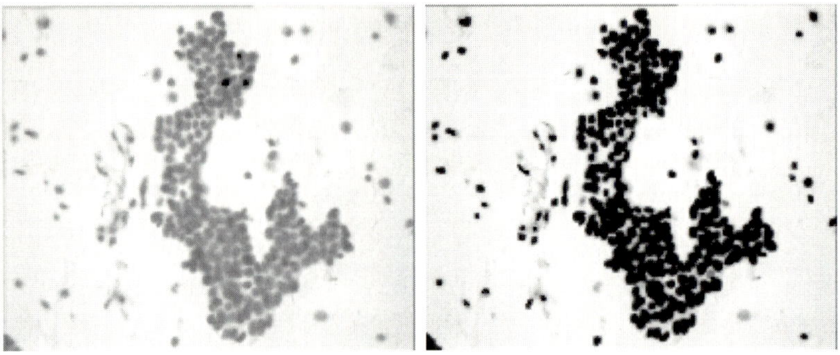

Fig. 2. Results of the region growing: (**a**) $T = 100$, (**b**) $T = 160$

The segments are maintained by active entities, called agents, that autonomously try to optimize the local criteria of region *homogeneity* (uniformity). In the initial state each pixel constitutes an agent. Agents are merged with their neighbors if proper conditions are met. A homogeneity criterion is formulated as

$$\overline{Value}(A_i), \overline{Value}(A_j) < T, T = 1, 2, \ldots, H_1 \tag{1}$$

$$\min\{\sharp(A_i), \sharp(A_j)\} < H_2 \tag{2}$$

$$|\overline{Value}(A_i) - \overline{Value}(A_j)| < H_3 \tag{3}$$

where $\overline{Value}(A_i)$ denotes the mean value of the intensity for agent A_i, $\sharp(A_i)$ is the number of pixels in i-th agent and $H_n, n = 1, 2, 3$; are arbitrarily given thresholds. For each value of T, all pairs of adjacent agents fulfilling the above conditions are merged.

Due to great number of agents in the initial state, a pre-initial stage of segmentation which could allow to eliminate all pixels belonging to the background is desired. We used Otsu's method which chooses the threshold to minimize the intraclass variance of the black and white pixels [4]. Certainly, only background pixels are excluded from being initial agents. Since the use of hematoxylin and eosin is giving a distinct contrast in color between objects in the image (nuclei, cytoplasm etc.) and its background, thresholding on the Hue component only can be performed. In addition, all adjacent initial agents that have exactly the same intensity value are merged. Applying the pre-segmentation stage leads to decreasing the number of initial agents from over 400 000 to 20 000 on average.

2.3 Features

In contrast to normal and benign cells, which are typically uniform in appearance, malignant cells are characterized by irregular morphology that is reflected in several parameters. Morphometric measurements characterizing the shape and size have been mainly used for feature extraction. The extracted

Table 2. Classification results for individual features (% correctness on testing set using *leave-one-out*)

No.	Feature	Statistics				
		Minimum	Maximum	Mean	Median	STD Deviation
1	Area	48	82	44	58	62
2	Circularity	56	66	60	52	36
3	Perimeter	80	72	48	62	50
4	Malinowska's Coeff.	46	54	66	48	56
5	Blair-Bliss' Coeff.	76	54	60	64	40
6	Danielsson's Coeff.	48	46	58	56	62
7	Haralick's Coeff.	56	62	58	46	48
8	MZ Coeff.	64	44	72	56	66
9	Lp1 Coeff.	44	62	66	66	66
10	Elipticity	64	50	56	72	38
11	Compactness	48	38	54	54	64
12	Length of longer axis	66	48	38	54	42
13	Length of shorter axis	54	56	48	46	48
14	Eccentricity	52	64	80	48	46

features are: size, circularity, perimeter, compactness, lengths of axis of ellipse circumscribing the nuclei, ellipticity and eccentricity of ellipse circumscribing the nuclei and the following shape coefficients: Malinowska's, Blair-Bliss', Danielsson's, Haralick's, Lp1, Mz. The details about these features can be found in [8, 6]. Five basic statistics of the values for the individual cell nuclei were calculated for each feature. This gives in total 70 features to be subject to classification.

2.4 Classification

A *k-nearest neighbor* classifier was used to test the effectiveness of the feature set in diagnosing new samples. The prospective accuracy of the resulting classifier was tested using the *leave-one-out* validation technique. In this approach, if N samples are available, N partitions are formed by leaving one single pattern for testing, and using the remaining $N-1$ to build the classifier. The N performance results obtained this way is then averaged and gives an accurate and unbiased estimate of the method's prospective accuracy. It is a measure of generalization ability of classifier (generalization to unseen samples).

Since the number of samples is relatively small, using a *k-nearest neighbor* with *leave-one-out* is computationally tractable and allows for accurate estimation of *Bayes error*. The discriminative power of individual features was estimated with *3-nearest neighbor* classifier and results in form of recognition rates are presented in Table 2. In order to reduce the dimensionality of feature space, *sequential forward selection* was applied and comparison of the whole set and its best subset is presented in Table 3. Ignoring redundant and

Table 3. Classification results achieved for the whole set and after feature selection

Distance measure	Subset S_d	% recognition for S_d	%recognition for whole set
Manhattan	Median of 1 STD of 3 Mean of 13	94	64
Tchebyshev's	Median of 1 STD of 3 Min.of 11	92	58
Euclidean	Median of 1 STD of 3 Min.of 11	92	62
Mahalanobis'	Median of 1 STD of 3 Min.of 11	94	54

irrelevant features leads to great improvement in recognition rates. One can observe that shape factors hardly discriminate malignancy and the statistical analysis shows that they have small range independently from the class. This property can be used for improving the segmentation process, i.e. objects with far different shapes can be rejected as not being the nuclei.

3 Conclusions

The first objective of the described work was to develop an automatic classification system for diagnosis the breast cancer. The results achieved in the experiments seems to be very promising. So far inspections of the segmented nuclei showed big differences in size between benign and malignant cases. Shape factors do not have good discriminative properties, however they can be used within the preprocessing stage segmentation. Hence, there are three challenges for the near future. First, the information obtained from shape factors will be applied in the process of nuclei detection. This could reduce the computational effort of region growing and thresholding approach. As a second challenge, the recognition rate in malignancy classification has to be improved. More features will be added and further research on the classifier will be performed. Finally, the data set of medical cases; i.e. set of cytologic images; has to be increased.

References

1. Jelonek J, Krawiec K, Slowinski R, Szymas J (1999) Intelligent decision support in pathomorphology. Polish Journal of Pathology 50:115–118

2. Koss LG (1987) Automated cytology and histology A historical perspective. Anal. Quantum Cytology 9:369–374
3. Marciniak A, Monczak R, Kołodziński M, Pr etki P, Obuchowicz A (2004) A benchmark for breast cancer diagnosis using fine needle biopsy. Proc. Artificial Intelligence in Biomedical Engineering, Kraków, Poland, CD-ROM (in Polish)
4. Otsu N(1979) A threshold selection method from gray-Level histograms. IEEE Transactions on Systems, Man, and Cybernetics, 9(1):62–66
5. Szylberg T, Sygut J, Kulig A (1997) Diagnostic value of fine needle aspiration biopsy for breast fibroadenoma diagnosis. Polish Journal of Pathology 48:79–86
6. Tadeusiewicz R (1992) Vision systems of industrial robots. WNT, Warszawa (in Polish)
7. Underwood JCE (1987) Introduction to biopsy interpretation and surgical pathology. Springer-Verlag, London
8. Wolberg WH, Street WN, Mangasarian OL (1993) Breast cytology diagnosis via digital image analysis. Analytical and Quantitative Cytology and Histology, 15:396–404

Feature Ranking for Protein Classification

Faouzi Mhamdi,[1] Ricco Rakotomalala[2] and Mourad Elloumi[3]

[1] URPAH, Faculty of Sciences of Tunis,Tunisia faouzi.mhamdi@ensi.rnu.tn
[2] ERIC, University of Lyon 2, Lyon, France rakotoma@univ-lyon2.fr
[3] URPAH, Faculty of Sciences of Tunis,Tunisia mourad.elloumi@fsegt.rnu.tn

Summary. In this paper, a knowledge discovery framework is used for protein classification. The processing is achieved in three steps: feature extraction, feature ranking and feature selection. Inspirited from text mining results for the first step, we use n-grams descriptors; descriptors are ranked from chi-2 statistical indices in the second step; and in the final step, the subset of descriptors is selected which will minimize the prediction error rate using a k-nearest neighbor classifier. Experiments show that this framework gives good results: the dimensionality reduction is effective and increases the classifier performances.

1 Introduction - The Protein classification process

In recent years, the area of machine learning techniques has been extended to include unstructured datasets such as text processing, combinatorial chemistry and biological data. The knowledge discovery framework [1], especially the text mining framework [2], gives a good guideline for this kind of analysis. Compared to a traditional approach, two supplementary steps are essential: extracting the feature from the original description in order to build an attribute-value table which is useful for data mining techniques, and selecting the best features from among them. Owing to the fact that the number of potential features is very high, the computational efficiency is of primary importance in this last step.

In this study, the text mining framework is used to solve a protein classification problem in their primary structures (Figure 1). The analogy with text classification is relevant in this case, indeed, the original description of the datasets are very similar. It is well-known that a protein sequence consists of a sequence of different characters called amino acids. There are 20 possible amino acids.

An example of a file describing a few proteins is shown in Figure 2. However, unlike the text classification, there is no *natural* separation in the character sequences, it is not possible to extract *words* to which we can easily

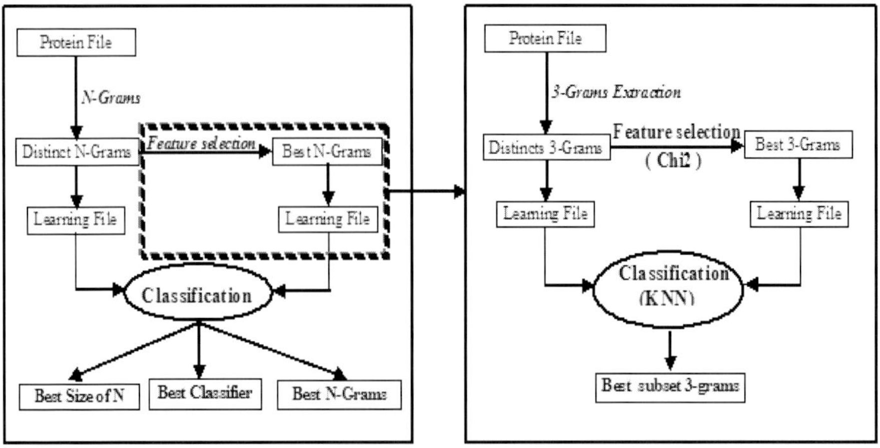

Fig. 1. a- General Protein Classification Framework, b- Subset Feature Selection.

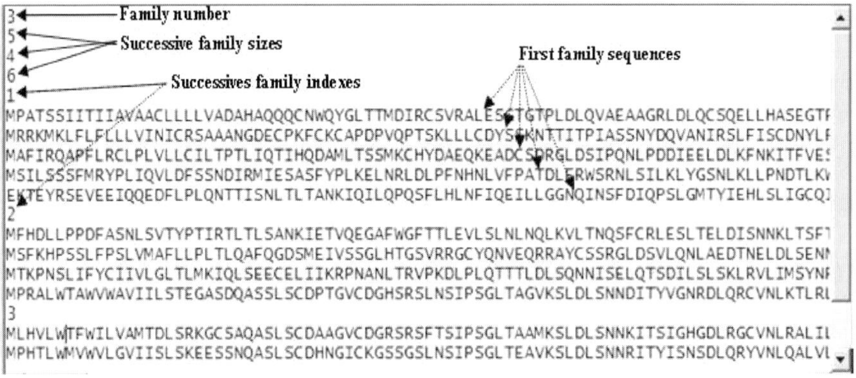

Fig. 2. Protein family file.

attach semantic properties. Therefore the n-grams have been used, namely a sequence of n characters, providing extraction techniques in order to produce descriptors. Table 1 demonstrates the parallels between the text classification and the protein classification properties.

In the following section, the choices for each basic step of the protein discrimination will be presented, particularly concerning feature extraction and feature selection. Section three will present experimental results from various families of proteins. The fourth and final section will serve as the conclusion.

Table 1. Equivalence between text and protein classification.

General Notion	Text classification	Protein classification
Alphabet	26 characters (Language)	20 amino acids
Descriptors	word, n-gram	n-gram
Individual	Text	Sequence
Class	Category	Family
Learning set	Corpus	Family set

2 Extract and select features from protein description

2.1 3-grams feature extraction

Initially, it is necessary to choose the correct features from original data description. Text mining results [2] show that character sequences (n-grams) give good results. This approach can be transposed in the protein classification field. The primary task was to define the best length n in the extracted sequences. In a previous work [3], various values of n were tested: if it were too low then the captured information was of too poor a quality. If it were too high, features were too specific and disturbed, and, most importantly, fast computing became impossible. For instance, if $n = 4$ was set, the theoretical number of features would be $20^4 = 160,000$. Eventually, a moderate size ($n = 3$) seemed to be a good compromise. It allows good information capture and avoids noise contained in the data. Computation time is reasonable.

The next step is the construction of the attribute-value table from the original unstructured dataset. Taking one example (a row in Figure 3), several kinds of values can be attributed to a feature (a column in Figure 3). Then from text mining domain, the occurrence of features can be counted, as their frequencies, or simply their presences/absences. This last data representation can appear rather rough, but several studies in the text mining domain show their effectiveness. Moreover, it can be used whatever the data mining learning algorithm : those which handle only continuous descriptors (presence/absence are coded 1/0) such as the neural network [5], linear and discriminant analysis [5]; and those which handle discrete descriptors (presence/absence are regarded as Boolean values true/false) such as the naïve bayes classifier [4].

The use of a 1/0 description was selected for this work. From the original data representation (Figure 2), an attribute value table can be constructed: a row represents a classified protein (an example), and the column, a 3-grams descriptor; at the intersection, we set "1" if the 3-gram appears in the protein (Figure 3).

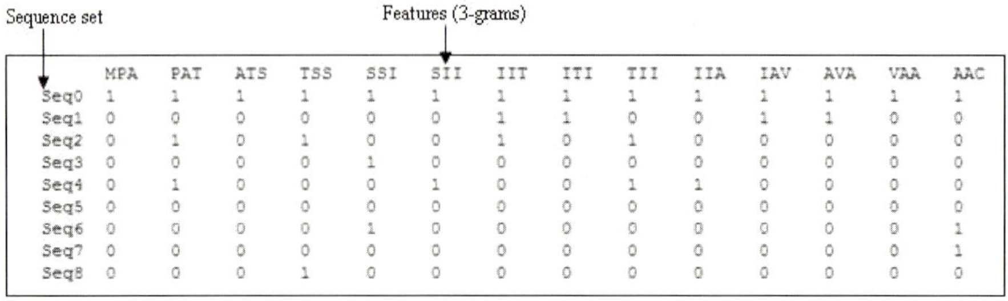

	MPA	PAT	ATS	TSS	SSI	SII	IIT	ITI	TII	IIA	IAV	AVA	VAA	AAC
Seq0	1	1	1	1	1	1	1	1	1	1	1	1	1	1
Seq1	0	0	0	0	0	0	1	1	0	0	1	1	0	0
Seq2	0	1	0	1	0	0	1	0	1	0	0	0	0	0
Seq3	0	0	0	0	1	0	0	0	0	0	0	0	0	0
Seq4	0	1	0	0	0	1	0	0	1	1	0	0	0	0
Seq5	0	0	0	0	0	0	0	0	0	0	0	0	0	0
Seq6	0	0	0	0	1	0	0	0	0	0	0	0	0	1
Seq7	0	0	0	0	0	0	0	0	0	0	0	0	0	1
Seq8	0	0	0	1	0	0	0	0	0	0	0	0	0	0

Fig. 3. Learning boolean file.

2.2 Feature ranking from chi-2 criterion

Owing to the fact that the number of descriptors is potentially high, a subset has to be selected from among them. In our 3-grams representation, maximum number of features is $20^3 = 8000$, experiments shows that this value is very close (Table 2). There are several reasons for this dimensionality reduction: (1) machine learning algorithms work badly when the dataset is too sparse; selecting a subset of relevant features often improves the classifier performances; (2) the complexity of the learning algorithms always depends on the number of input features; the elimination of the useless attributes allows to a considerable improvement in computing time; (3) a reduced number of features provides a better understanding of the classifier.

Table 2. Error rates with all 3-grams.

Families	3-grams number	Error rates (1-NN)
F_1_2	7145	0.057471
F_1_3	6998	0.053191
F_1_4	6945	0.041322
F_1_5	6828	0.046296
F_2_3	7143	0.108911
F_2_4	7107	0.062500
F_2_5	7011	0.095652
F_3_4	6860	0.177778
F_3_5	6740	0.262295
F_4_5	6688	0.161074

The majority of learning algorithms suffer from an overabundance of input features. Surprisingly, some learning methods take advantage of feature selection, such as a decision tree [5], for which the selection process is embedded in the learning strategy. There are two families of attribute selection [6]:

- Wrapper methods [8]: The wrapper method explicitly uses learning algorithms in the feature selection process. Through using resampling techniques, such as cross-validation[4] or leave-one out[4] for the error rate estimation, it attempts to detect the subset of features which will optimally minimise the error rate. In fact, because the learning process is repeatedly called, the wrapper method is too slow and so computationally impracticable in the context of the protein classification.
- Filter methods [8]: The filter method select the right subset of features before the learning process. They are particularly computationally advantageous and well adapted in this context. Being independent of the characteristics of the classifier, there is no guarantee that the selected feature subset will be powerful whatever the method used thereafter.

Feature ranking is an intermediate way [7]. Feature selection is carried out in two steps:

- In the first step, a chi-2 statistical criterion is used to sort features according to their importance. This statistic measures the dependence between the class attribute (protein family) and each feature.
- In the second step, candidate features are introduced one by one into the order which was defined by the chi-2 criterion and error rate is evaluated using a leave-one out procedure. There can be two kinds of stopping rules: setting, a priori, the maximum number of features to introduce, or stopping the process as soon as the error rate no longer decreases. Another strategy, which has been used in this study, would be to select the subset which minimizes the error rate.

This method combines the advantages of the two approaches described above. Indeed, the computing time is reasonable, and the performances of the selected subset is in relation to the characteristics of the learning algorithm.

Experimental results described in the next section show that this framework is powerful for protein classification. The number of selected features was dramatically reduced, and simultaneously, the performance of the classifier, using a K-nearest neighbor algorithm [4], was improved.

3 Experiments and results

To evaluate this framework, five protein families have been extracted at random from the data bank SCOP [9], the aim being to discriminate them two at a time. Firstly, the learning algorithm was applied ("K=1"-nearest neighbor was set) to all extracted 3-grams. The number of features and the estimated error rate are listed in table 2. Secondly, feature ranking and selection framework was applied. For each problem, the subset of features which minimize the estimated error rate was determined(Table 3)[4]. Results show that the number of selected features is very low compared to the initial number of 3-grams. In several cases, this dimensionality reduction improves the classifier performances.

Table 3. Error rates after feature selection.

Families	Feature subset sizes	Error rates after feature selection
F_1_2	11	0
F_1_3	12	0
F_1_4	5	0
F_1_5	2	0
F_2_3	30	0.019802
F_2_4	62	0.007813
F_2_5	83	0.026087
F_3_4	13	0.022222
F_3_5	16	0.040984
F_4_5	9	0.026846

Detailed results for the discrimination between families 4 and 5 can be seen(figure 4).

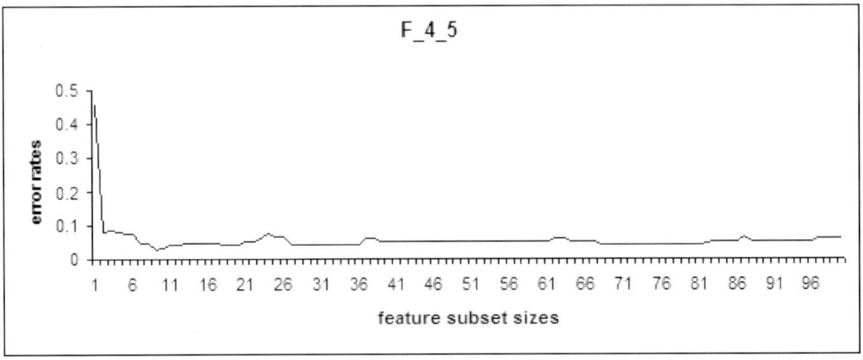

Fig. 4. Evolution of prediction error rates relatively to the selected feature numbers.

It can be noted that the first attributes, sorted according to chi-2 criterion, greatly reduce the error rate. The first nine features allow attainment of the

[4]When estimated error rate with leave-one-out procedure is zero, it is probably an artifact related to these datasets, In fact, we can say that the real error rate is very small.

minimum error rate. Thereafter, the error rate remains initially quite stable as some attributes are added, then it is degraded when the number of attributes becomes too high. This behavior confirms the results obtained in the other feature ranking studies [7]. Chi-2 criterion, and more generally, information based ranking, brings the most interesting features to light.

4 Conclusion

In this paper, the knowledge discovery framework was used, specifically the text mining framework, for a protein classification problem. Experimental results seem encouraging, feature reduction from feature ranking gives a good compromise between computational efficiency and classification performances. The number of selected features is very low compared to the initial number of features. In some cases, classifier performances are improved.

However, some questions remain open. Future studies will attempt to evaluate various stopping rules to improve computing time, particularly in those suggested above. Nothing guarantees that the error rate will not deteriorate if the simplicity bias is too restrictive. Another important problem is the feature redundancy. Because the features are independently evaluated, some selected features which are independently relevant can be redundant i.e. contribute to the same information in the learning classifier.

References

1. Fayyad UM, Shapiro G, Smyth P (1996) From data mining to knowledge discovery : An overview, Advances in Knowledge Discovery and Data Mining. AAAI Press and the MIT Press, Chapter 1 : 1-34
2. Sebastiani F (2002) Machine learning in automated text categorisation. In ACM Surveys, 34(1): 1-47
3. Mhamdi F, Elloumi M, Rakotomalala R (2004) Textmining, features selection and datamining for proteins classification. In IEEE/ICTTA'04, Damascus, Syria
4. Hastie T, Tibshirani R, Friedman J (2001) The Elements of Statical Learning:Datamining, Inference, and Prediction, Springer-Verlag
5. Lefébure R, Venturi G, (2001) Data mining : Gestion de la relation client personnalisation de sites web, Eyrolles
6. Molina LC, Belanche L, Nebot A (2002) Feature Selection Algorithms: A Survey and Experimental Evaluation, In ICDM'02, Maebashi City, Japan
7. Duch W, Wieczorek T, Biesiada J, Blachnik M (2004) Comparison of feature ranking methods based on information entropy Proc. of International Joint Conference on Neural Networks (IJCNN), Budapest, IEEE Press : 1415-1420
8. Isabelle G, André E (2003) An introduction to variable and feature selection. Journal of Machine Learning Research 3: 1157-1182
9. Murzin GA, Brenner ES, Hubbard T, Chothia C (1995) SCOP: a structural classification of proteins database for the investigation of sequences and structures. J. Mol. Bio.. 247 : 536-540

Recognition of the Medical Structures in Computerized Chest Radiograph Processing

Piotr Michalec[1], Wojciech Tarnawski[2], and Zygmunt Mazur[1]

[1] Institute of Applied Informatics, Wroclaw University of Technology, Skwer Idaszewskiego 1, 50-372 Wroclaw, Poland
zygmunt.mazur@pwr.wroc.pl, piotr.michalec@pwr.wroc.pl

[2] Chair of Computer Systems and Networks, Wroclaw University of Technology, Janiszewskiego 11/17, 50-372 Wroclaw, Poland
wojciech.tarnawski@pwr.wroc.pl

Summary. The study presents automatic identification of lungs area and rib borders detection in digital chest radiograph. There are used several segmentation methods, region labeling and region convex hull construction algorithm in lungs identification. Their final shape is derived from lungs identification algorithm, which uses machine learning. Ribs' borders detection is based on Canny's edge detection algorithm. Image of chest radiograph processed by Canny's algorithm is used as entry point to the ribs' border detection. Then, there are presented successive stages of borders detection for a sample rib. There is also described a method of gathering new patterns of ribs' borders.

1 Introduction

Chest radiograph examination makes up 20-40% of total number radiograph examinations. Chest radiographs remain the initial and often the most valuable means for investigating the chest. They are an essential part of the physical examination, which precedes many operations, and provide arguments to solve many medical diagnostic problems. Analyses of effectiveness, in chest radiograph examinations [9, 11], has been done so far, show that it has significant influence on diagnosis formulation, medical treatment and patient's health condition.

Appearing digital form of chest radiographs has opened a possibility of using computer image processing methods to support their analysis. In this study we present lungs selection and rib borders identification in standard digital chest radiograph pictures. It is obligatory step in analysis of chest radiography. This subject is widely described in [10, 11, 12].

2 Segmentation

Segmentation is conversion of monochrome picture to binary picture. Binary picture should include all essential information about number, positions and shapes of objects. Fundamental reason of monochrome pixels classification is that pixels with similar levels in nearby area usually belong to the same object. The most popular method used to conversion between monochrome and binary pictures is thresholding of picture's histogram (in fig.1 there is sample chest radiograph and its gray level histogram).

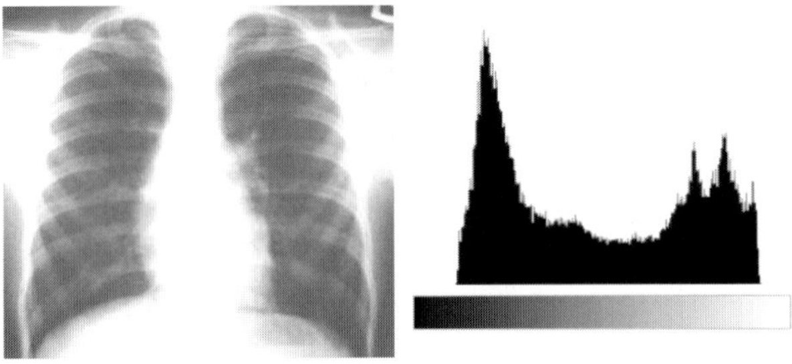

a b

Fig. 1. Sample chest radiograph picture (a) and its histogram(b)

Threshold selection can be formulated as a decision making problem where the optimal decision rule for a two class problem is required. Often the threshold is obvious, because it is in the lowest point between two peaks of histogram. If histogram has two peaks, it is good to choose that kind of threshold.

Automatic threshold selection methods are described by low computation complexity. However, not all of them fit to chest radiograph segmentation. The threshold computed by the Minimum Error Thresholding method [16] is too high and information content in picture is completely lost. Remaining methods (the Iterative Selection Method [15], the Method of Grey Level Histogram [14], the Using Entropy Method [17], the Fuzzy Sets Method [18]) give good results. Taking into consideration further analysis of chest radiograph and making use of segmentation to identify lungs area, the best choice is the Fuzzy Sets method. The threshold computed by this method has the lowest value, but it does not lose information about lungs area and its outline is exact. Segmentation by the Moving Averages method [19] does not meet our requirements. The lungs area is represented by lines in every second row; it results from specific construction of data stream used in this method. Fig.2 shows comparison of automatic threshold selection methods.

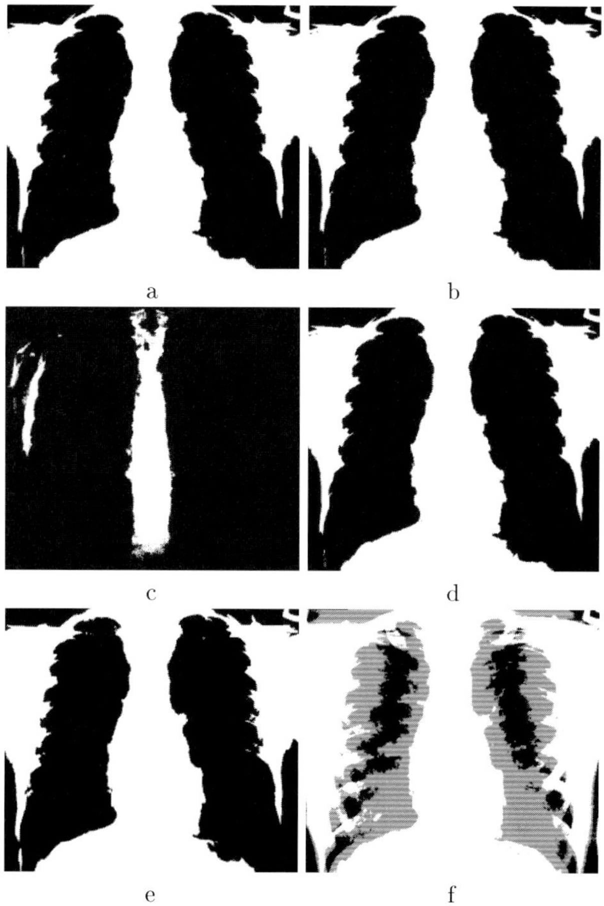

Fig. 2. Comparison of automatic threshold selection methods: (a) Iterative Selection T=129, (b) Otsu T=131, (c) Minimum Error Thresholding T=215, (d) Using Entropy T=131, (e) Fuzzy Sets T=121, (f) Moving Averages

3 Lungs identification

Applying gray level criterion, by segmentation, and knowing that soft tissues of lungs area have low levels we get image, which has objects connected with lungs area. There is one disadvantage - other areas with similar gray levels also create objects, which are not lungs, or they deform lungs area. Fig.3a shows chest radiograph after segmentation, threshold is determined by the Fuzzy Sets Method, for details see Sect.2.

In fig.3a we can see, that lungs are included in two largest areas. However, it is necessary to apply region labeling algorithm [4] to give them properties of different objects (fig.3b). According to [4] the segmented image R consists of m disjoint regions R_i, thus image R consists of objects and a background:

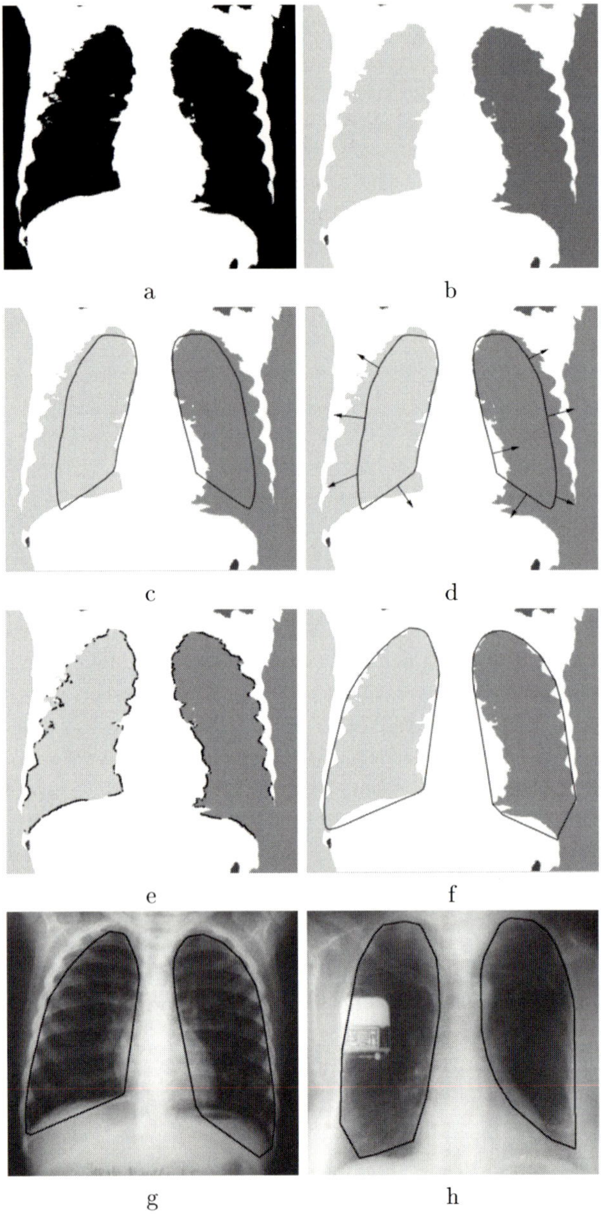

Fig. 3. Process of lungs identification: First step: (a) segmentation of chest radiograph, (b) regions labeling. Second step - introduction flexible pattern into image: (c) initial pattern adaptation, (d) rules of finding lungs' edges by pattern points (e) localized lungs' edges, (f) lungs' convex hull. Final result of lungs identification (g) and example of lungs identification for patient with artificial pacemaker (h).

$$\bigcup_{i=1, i \neq b}^{m} R_i = R_b^C \tag{1}$$

where R^C is the set complement, R_b is considered background, and other regions are considered objects.
Two largest areas, where lungs are included, do not model only lungs, but there are also included both side blackouts. Considering this, we apply machine learning [7] to mark off lungs area in labeled image. Very important is introduction of initial knowledge [7, 8], based on flexible lungs' pattern. Now we can insert flexible lungs' pattern to labeled image and using its properties lungs' edges can be found.
Pattern's operation consist in preliminary adapting to lungs' areas by moving whole pattern so that lung's area point, which is nearest to image centre, will be anchor for flexible pattern of left or right lung (fig.3c). Then every point of flexible pattern is moving in direction of lung's area border and perpendicular to pattern's line (fig.3d). In this way pattern's points finds lung's area border (fig.3e). Because of segmentation, bright parts of ribs superimpose lungs' areas. To get satisfied accuracy of lung's outline it is necessary to create convex hull [13] using pattern's points, which are on lungs edges (fig.3f).
After these transformations, every point of flexible pattern is modified a bit so that it stores the direction of found lung's edge. This operation causes that every process of identification of lungs' areas has influence on pattern's shape. After several operations of lungs identification the flexible pattern become more universal. Fig.3g shows final result of lungs identification.

4 Rib border identification

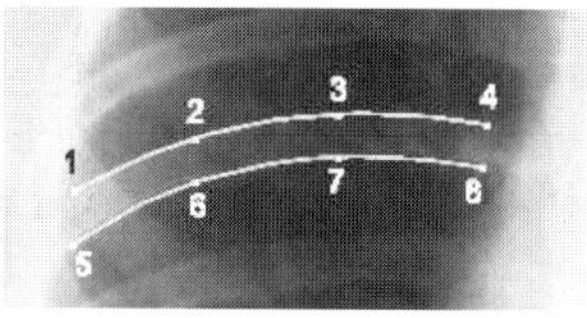

Fig. 4. Rib pattern

In rib borders identification it is assumed, that one border of rib can be described by four points (whole pattern has eight: four for upper and four for bottom border), which are connected with curves interpolated by Bezier polynomials (fig.4). Bezier curve [6] is defined by several points, which are included to this curve (base points) and by several control points. Bezier curve does not contain control points, but they have influence on curve's

shape. In our rib pattern, control points are calculated from base points. It allows precisely control curve's shape by using only a few points.
For better understanding and clearly presentation of rib's border identification concept, process will be presented for one sample rib.
First step of rib's border identification is finding edge pixels in image with Canny's edge detector (fig.5). By edge detection objects can be located, and their basic properties like borders and shape can be measured.

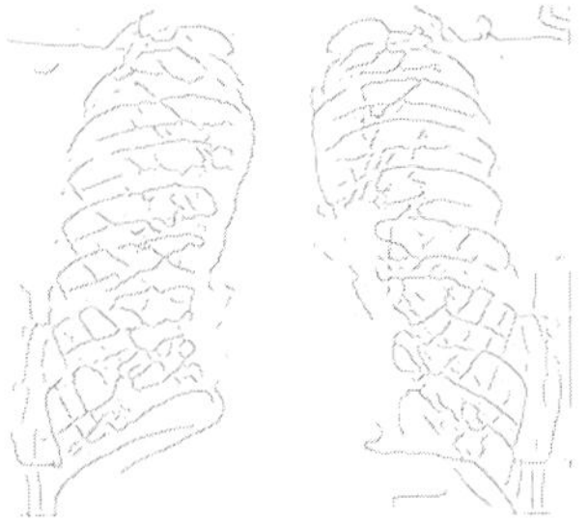

Fig. 5. Applying Canny's edge detector to chest radiograph picture.

Our algorithm has initial knowledge in form of rib's patterns database for each rib separately. Using this patterns in next step, each of patterns is matched to detected edges, by locating edges in base points neighborhood. Every point of one pattern locates edge independently of the others, and it is moved to match pattern to located edge (fig.6a). Possible moves are counted, and values of this moves are summed for every pattern. Base points, which do not find edges in their neighborhood, are assigned value equal to double allowed moving. It is necessary, so that pattern with such base points has not been marked as a good one. Patterns, which best match the edge (are closest to the edge) will gain the smallest values of $Moves$.

$$Moves = \sum_{i=1}^{n} d_i + \sum_{j=n+1}^{8} 2C \tag{2}$$

where: d_i - distance from i^{th} base point of pattern to nearest edge, C - maximum allowed moving distance, and n - number of located edges by base points of rib's pattern.
The disadvantage of this solution is possibility that pattern will match the area between ribs and gain the smallest value of $Moves$. To minimize danger of occurring such situation as candidates for rib are chosen two patterns with the smallest values of $Moves$ (fig.6b). Than, brightness of areas marked by chosen patterns is checked (fig.6c). Pattern, which marks brighten area, define the rib (fig.6d).

Still, there is possible situation, when both candidate patterns will mark the area between ribs, therefore there is introduced another condition for a

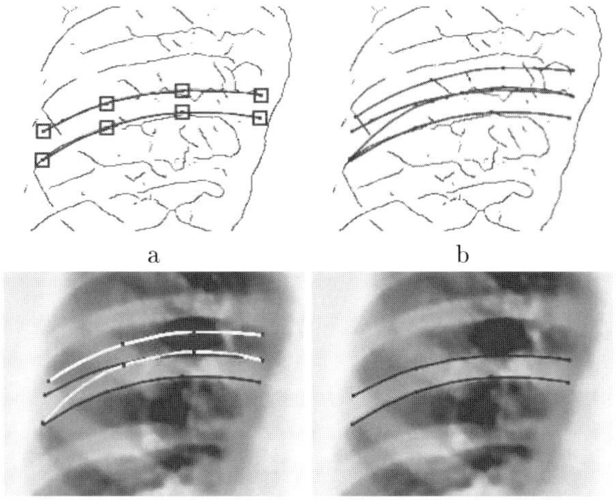

a b c d

Fig. 6. Rib's borders location process: a - location of edges by base points of rib's pattern in local neighborhood of each point; b - two patterns, which best matched the edge are candidates for rib; c - checking the brightness of areas marked by candidate patterns; d - finally chosen rib's borders

brightness of areas marked by candidate patterns, if brightness of both areas is lower than assumed threshold, both patterns are moved vertically down, according to:

$$\forall i \in [1,8]: \; bp_i.y = \begin{cases} bp_i.y + (bp_i.y - bp_{(i+4)}.y), & for\ i \leq 4 \\ bp_i.y + (bp_{(i-4)}.y - bp_i.y), & for\ i > 4 \end{cases} \tag{3}$$

where $bp_i.y$ is vertical coordinate of ith base point.
After that operation,the brightness of areas is checked again, and pattern which defines brighten area finally is chosen as the rib. The lack of pattern matching after last step reduces effectiveness of our algorithm. However, conducted experiments showed, that pattern will match the rib's border and next processing won't be necessary.
After rib's border identification, adequately rib patterns database is updated. Rib's pattern, which finally matches rib borders, and his sum of moving values exceeded prior defined threshold (it means that moving was significant) is qualified as a new pattern (taking into consideration all moves of his base points) and is added to rib patterns database. Number of patterns stored in database can be controlled by value of threshold, which qualify pattern as a new pattern. The higher value of that threshold, the fewer new patterns exceed the threshold and database size will be smaller. Additionally, to prevent out-of-control growing of database, there can be introduced supervision of system's user during adding new pattern to database, or there can be assumed maximum size of database.

5 Summary

Chest radiographs are initial, the most valuable and in many cases the only source of information to analysis ribcage. They are essential part of physical examination, which precede many medical treatments and gives data to solve many diagnostic problems. Physician's diagnosis can be supported by identification characteristic areas occurred on chest radiograph. First of all, lungs are these areas. In this study we used several segmentation methods, region labelling and region convex hull construction algorithm mark lungs' areas. Final shape of lungs is derived from lungs identification algorithm, which uses machine learning.
Next, ribs are very important because they provide reference system in chest. In that case in computer processing of rib chest radiograph essential is to automatically identify rib borders. In this study we adapted methods of image processing and image analysis, with some methods of machine learning to detect rib borders.

References

1. Jahne B (1997) Digital Image Processing. Springer-Verlag, Berlin New York
2. Klette R, Zamperoni P (1996) Handbook of Image Processing Operators. John Wiley & Sons, New York
3. Parker James (1996) Algorithms for Image Processing and Computer Vision. John Wiley & Sons, New York
4. Ostrowski M (1992) Image Information. WNT, Warszawa (in Polish)
5. Mokrzycki W (1992) Encyclopedia of Image Processing. Academy Publishing RM,Warszawa (in Polish)
6. Bezier P (1972) Numerical Control: Mathematics and Applications. John Wiley & Sons, New York
7. Cichosz P (2000) Machine Learning Systems. WNT, Warszawa (in Polish)
8. Mulawka J (1996) Expert Systems. WNT, Warszawa (in Polish)
9. Zgliczynski L (1989) Radiology. State Workshop of Medicine Publishers, Warszawa (in Polish)
10. Daniel B (1988) Atlas of Human's Radiological Anatomy. State Workshop of Medicine Publishers, Warszawa (in Polish)
11. Smajkiewicz L (1999) Polish Radiology Reviews 52:178–183 (in Polish)
12. Yezzi A, Kumar A, Olver P (1997) IEEE Trans on Medical Imaging 16:199–209
13. Melkman A (1987) Information Processing Letters 25:11–12
14. Otsu N (1979) IEEE Trans on Systems, Man, and Cybernetics 9:62–66
15. Ridler T, Calvard S (1978) IEEE Trans on Systems, Man, and Cybernetics 8:629–632
16. Kittler J,Illingworth J (1986) Pattern Recognition 19:41–47
17. Kapur JN, Sahoo PK, Wong AKC (1985) Computer Vision Graphics Image Processing 29:273–285
18. Huang LK, Wang MJ (1995) Pattern Recognition 28:41–51
19. Wellner P (1993) Communications of the ACM 36:87–96

Correlation-based Method for Automatic Mitotic Cell Detection in Phase Contrast Microscopy

Lukasz Miroslaw[12], Artur Chorazyczewski[3], Frank Buchholz[4], and Ralf Kittler[5]

[1] Institute of Engineering Cybernetics, Wroclaw University of Technology
lmir@diablo.ict.pwr.wroc.pl
[2] Max Planck Institute of Molecular Cell Biology and Genetics Dresden
miroslaw@mpi-cbg.de
[3] Institute of Engineering Cybernetics, Wroclaw University of Technology
achorazy@ieee.org
[4] Max Planck Institute of Molecular Cell Biology and Genetics Dresden
buchholz@mpi-cbg.de
[5] Max Planck Institute of Molecular Cell Biology and Genetics Dresden
kittler@mpi-cbg.de

Summary. A simple and fast method is presented which detects mitotic cells from two cell lines imaged in two phase-contrast microscopy techniques. Such detection is a first step to more sophisticated image processing tasks like determination of mitotic index or mitotic cell tracking in time-lapse movies. Detection algorithm is based on template matching approach that provides a list of candidates. The list is then pruned by validation algorithm that takes into account *a priori* information about mitotic cells. The method has been implemented as plugin for ImageJ and has been tested for several different data sets.

1 Introduction

Phase contrast microscopy is a common technique of imaging living cells over long period of time [2, 17]. Invented by Frits Zernike [18, 19, 20] in 1930 the method has been widely used in medicine [3, 5, 11, 12] and biology [1]. The phase contrast microscope enables visualizing components in a cell, tissue or bacteria, which are very difficult to be seen in an ordinary light microscope. This technique proves its usefulness in a cancer research where by using gene silencing method (siRNA) it is possible to identify the putative genes in biological processes by screening whole genomes [6]. The interpretation of images offered by this method relies on detection of abnormal cell types or behaviors. One of important factors used to describe the cell phenotype is the mitotic index defined as a quotient of the numbers of mitotic and normal cells present in the image at given time. The first step to set its value is to detect mitotic cells. There exist applications which allow a user to select cells either manually [16, 4] or in semi-automated, computer-assisted mode [9]. In the latter case, image processing algorithms are used to select candidate target objects. Afterwards, the user prunes the list of candidates. Both methods could be used for small number of images. However, when datasets contain hundreds or thousands of images, as in the case of large scale screening, manual or semi-automated selection of cells is hardly ever possible. Therefore, automated cell selection/detection algorithm processing the image is required.

Many solutions of this task have already been proposed originated in image processing methods like segmentation, pattern recognition, texture analysis, etc. Many of them have already been employed in the cryo-electron microscopy where automatic particle detection algorithms have been investigated for a long time. Several approaches with different degree of success were recently reviewed by Nicholson and Glaeser [13]. They compared methods that use template matching based on cross-correlation, edge detection, methods based on intensity comparisons, texture-based methods, and neural networks. They concluded, that none of the methods used alone performs good enough to decrease significantly high false positives ratio. Thus, it is presumed that combination of different approaches could provide better results. Template matching proved to be an efficient method of finding candidate particles. Therefore, we chose this approach to detect mitotic cells of various cell lines imaged by different microscopy techniques.

The later part of introduction presents a short review on template matching method. In Section 2 the proposed method is described. In Section 3 the method was tested on real images. Final remarks are collected in Section 4.

Template matching

In template matching approach the template image, which is presumed to be similar to the target object, is shifted with respect to the original image by the vector (x', y'). In the shifted position the scalar product of the two images is computed and placed in cross-correlation map at the position (x', y'). The vector (x', y') includes all possible positions on the sampling grid [13]. The object is detected if the ratio of similarity between both images $c(x', y')$ exceeds a given threshold. When performed in the spatial domain, this operation is described by the formula

$$c(x', y') = \sum_{x,y} f(x, y) g(x + x', y + y') \tag{1}$$

where $f(x, y)$ is the image and $g(x, y)$ is the template.
There are several disadvantages of using this operation: the range of c(x',y') depends on the size of the template, the correlation is not invariant to changes

in image intensity such as uneven illumination during imaging. The *correlation coefficient* overcomes those drawbacks by normalizing the image and the template. However, it also leads to increased computational cost.

Taking advantage of the correlation theorem the correlation can be also computed faster [8] in the frequency domain:

$$c(x', y') = F^{-1}\{F\{f(x, y)\}F^*\{g(x, y)\}\} \tag{2}$$

where F indicates the Fourier transform operation, F^{-1} its inverse, and $F^*\{g(x, y)\}$ the complex conjugate. Implementation of the FFT algorithm (the butterfly algorithm) requires the template and the image to be extended to the same size to a common power of two [15].

Important drawback of cross-correlation technique is dependence on rotation and scaling of the template. This imposes the use of multiple templates with different scale and rotation. The choice of number of templates is usually a difficult task because the trade off between the expected results and computational time is hard to determine. Moreover, due to spatial variations in image intensities and different peak hight, additional post-processing steps, like searching for correlation peaks have to be taken.

Peaks searches

The output of the correlation is higher at positions where the image matches the template. Those intensity peaks indicate possible positions of target objects. However, not only high but sometimes low peaks can correspond to target objects. Therefore, matched filters allow only to achieve reasonably efficiency in detection of candidate objects and further validation of detected peaks is needed [13]. There are many techniques of searching for peaks. Ludtke et al. [9] as well as Nicholson [14] combine different correlation-maps, each from a different template, by selecting a maximal value at each pixel location from the set of cross-correlation map. Since some particles can have multiple peaks really close to each other the cross-correlation image map is then low-pass filtered [9]. It overcomes the problem of choosing the same particle many times. Slightly different algorithm is presented in [14]. Candidate peaks are detected by a suitable threshold and then pruned on the basis of peaks adjacency. Peaks within user-selected distance (usually slightly larger than the size of the target object) are eliminated in favor of the highest peak. This leads to removal of contaminations and aggregates larger than true objects.

Another family of algorithms pruning the candidate objects are performed on the original image, provided the list of candidate peaks is present indicating the positions of candidate target objects. Kivioja et al. [7] propose a ring filter which calculates at every pixel position the average intensity inside the round particle, and intensity in a ring surrounding it. The filter values are formed by the difference between the average intensities inside the circle and those at the ring. Authors proved the method was successful in detection of spherical virus particles. In the same task Boier Martin et al. [10] described the cross-point

method which is performed directly on original image. The image is scanned twice in the same manner. The first scan is done starting from the top of the image, the second from the bottom. The output binary image is created according to difference in intensities of pixels pairs located at $r+1$ distance in horizontal and vertical direction. If in both cases the difference between pixels is larger than a given threshold the output pixel in this position is marked as hit. The final image is an average sum of two binary images.
Both of the methods are suitable mainly for the cryo-electron microscopy as they depend strongly on the characteristic of the original image and the particle itself.

2 Materials and methods

In this paper we have focused on the detection of mitotic cells. When imaged in the phase contrast microscopy (positive or negative) they express very regular, circular shape and are distinctive from other cell types. This *a priori* knowledge can be used to design the detection method based on template matching.

Materials

Two subclones of HeLa cell line, Kyoto and TDS, were imaged using phase contrast microscopy technique. They were imaged every 10 minutes in 96-well plates (38 wells filled with cells) resulting in 4 films for each experiment. Cells were imaged using CCD-camera attached to Axiovert 200 Microscope with 20X objective with negative (Kyoto) or positive (TDS) phase plates.

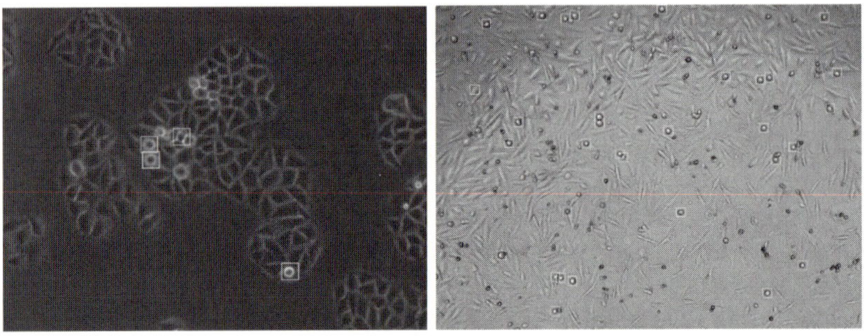

Fig. 1. Sample images with marked detected mitotic cells for Kyoto (left) and TDS (right) cell lines which were imaged by negative (left) and positive (right) phase contrast microscope.

From all 360 images per film available, four series of 31 8-bit images of the size 1392×1040 (TDS) or 736×570 (Kyoto) were selected randomly. The aim of the detection algorithm was to process a high number of movies. Thus, we have focused on using the cross-correlation algorithm (Eq. (1)) and limited number of templates.

The templates were created either from 3D model or from test data. In the latter case, the templates were created by cropping mitotic cells from test images. The 3D model of Kyoto mitotic cells was a black circle with white (Kyoto) boundary. Similarly, the model of TDS mitotic cells was a white circle with black boundary. Due to higher brightness of images the background was gray (intensity 128).

The experimental estimated radii of target cells varied from 20 to 32 pixels and the cell membrane was about 2 pixels thick. In both cases, radii of templates were chosen to cover the whole range by equi-length segments ($r = \{20, 24, 28, 32\}$).

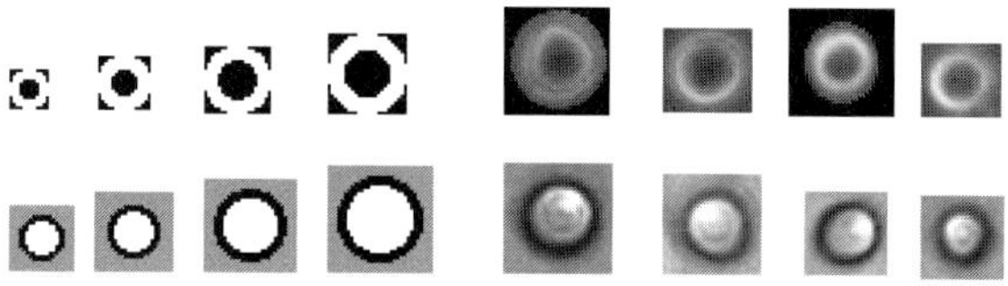

Fig. 2. Templates used for cross-correlation. Left side: artificial templates, right side: templates from test data, top row: Kyoto cell line, bottom row: TDS cell line

Methods

As a pre-processing step images were smooth by the (3×3) median filter to suppress local fluctuations in pixel intensities. The detection algorithm could be described with following steps:

Step 1: Correlation between the image $f(x, y)$ and each of the templates $g_k(x, y)$, $k = 1, \ldots, n$ according to Eq. (2).

Step 2: The choice of the maximum correlation value at each pixel position:

$$I_{max}(x', y') = \max_{i=1}^{n} c_i(x', y'). \tag{3}$$

This step ensured high and narrow peaks in cross-correlation map $c(x', y')$ when the target object matched any of the templates.

Step 3: Peaks detection and validation (Peak searches)

The highest peaks are detected by using a suitable threshold [14] determined by visual inspection of cross-correlation results. Because the peaks have different heights, this operation generates many false positives. Therefore, there is a need to validate detected objects.

Step 4: Pruning the list of candidates.
Presuming that peaks coordinates indicate positions of candidate target objects, the validation is performed on the original image $f(x, y)$ similar to cross-point method developed in [10]. False positive objects are identified by the method of the *modified local gradient* which is described below. Using the mean intensity of the image (which, we observed, is close to the intensity of the cell-inside) the gradient is calculated along four directions (up, down, left, right) originated at the object center. If a prescribed gradient is detected in all the directions at a certain distance from the center, the object is classified as a mitotic cell. Otherwise, the object is removed from the list of target objects. An object is also removed if the gradient is found to close to the center of the cell or if peaks are to close to each other (closer than the radius of the cell). The parameters of this step are the following: the difference in gradients between the inside of a cell and its boundary, the range where the gradient is to be observed. The method of validation peaks takes advantage of apriori knowledge. It was observed that mitotic cells, in both phase contrast techniques examined, have rather homogeneous intensity inside which changes dramatically at the cell membrane.

The resulting list of objects is assumed to be the list of detected mitotic cells. The output image is depicted in Fig. 1.

3 Results

Tests of the detection method were performed for many images taken in various conditions. A few sets of templates to ensure reliable evaluation were examined. The results are presented in Fig. 3 by means of TPF and FPF [6] for both imaging techniques. $TPF = TP/(TP + FN)$ and $FPF = 1 - FP/(FP + TN)$, where TP, TN, FN and FP mean true positive, true negative, false negative, and false positive respectively. These results were obtained using real images segmented by an expert for a reference classification. The figure shows that, in average, a correct classification can be derived for about 0.908 for positive phase contrast (TDS cell line) and 0.823 for negative phase contrast (Kyoto line). The algorithm was implemented in Java using Image Processing Library provided by ImageJ [16]. The whole process of detection for one image and one set of four templates took couple of seconds, when performed on Power Mac G5 with two 2 GHz processors and 2 GB RAM and up to 22 seconds when performed on PC with Intel Pentium 4 3.06 GHz processor and 2GB RAM.

[6] True and False Positive Factor, also known in literature as sensitivity and specificity

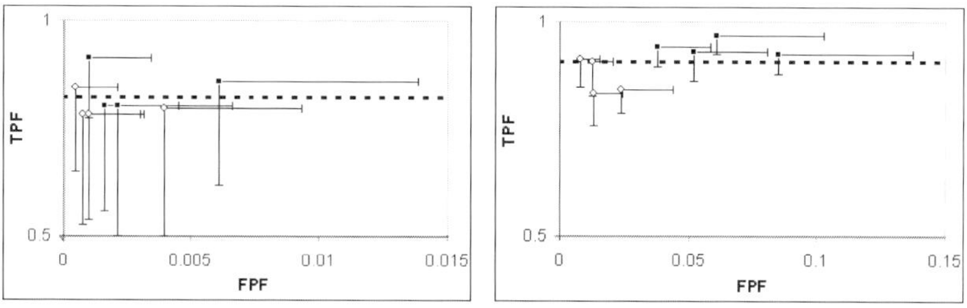

Fig. 3. TPF vs. FPF measured for negative (left) and positive phase contrast (right). Each element (quadrants for artificial templates, circles for real templates) represent mean value and standard deviation for one film. Dotted line represents the mean value of all results.

4 Discussion

This paper presents a simple and fast method aiming at detection of mitotic cells from two cell lines imaged in two phase-contrast microscopy techniques. The method has proven its reliability in finding mitotic cells. The main features of the algorithm are: simplicity, fast computation and a wide spectrum of possible applications. We expect the program can be also used for other types of cell lines as long as target objects have distinguishable morphology. In the case of mitotic cells, their round regular shape and characteristic intensity allowed to limit the number of templates to only a few.

We observed that four templates with an evenly spread radii range were sufficient for the reliable detection. It is also noticeable that artificial templates perform quite well comparing with templates derived from test data. It allows to use them for a preliminary detection despite different lighting conditions. The tests also revealed the importance of the validation algorithm which significantly decreases the high number of false positives obtained after first three steps of the algorithm (correlation and thresholding). On the other hand, a simple validation algorithm does not cover all cases when small fluctuations in intensity of the cell are present or when the center of the peak is not placed in the center of the cell. Therefore a more sophisticated approach is needed. Promising alternatives are presented in [7, 10], shortly described in Sec. 1. Another optimization may consist in using local threshold (for example calculated around each peak) instead of global one, which does not take into account uneven illumination present in the image.
It is also possible to set parameters such as threshold and gradient automatically, according to some a priori knowledge about the microscopy technique or features of the object itself. The only parameters to be set would be the size of a target object (minimal and maximal cell radius).

Designing a system that automatically selects cells is a very difficult task due to problems with modeling visual processes in recognition of faulty cells [7]. It is obvious, that the automatic detection can dramatically reduce human effort compared with manual selection and it may support more sophisticated tasks like tracking of trajectories of moving target cells.

References

1. Bennett A.ÊH. (1951), Phase Microscopy: Principles and applications, Wiley, New York
2. Bradbury S., Evennett P. (1996), Phase Contrast and Modulation Contrast. In: Contrast Techniques in Light Microscopy, BIOS Scientific Publishers, Ltd., Oxford
3. Chanwimaluang T., Fan G. (2003), An efficient blood vessel detection algorithm for retinal images using entropy thresholding. IEEE International Symposium on Circuits and Systems
4. Frank J., Radermacher M., at el. (1996), Spider and web: Processing and visualization of images in 3d electron microscopy and related fields. Journal of Structural Biology, 116:190–199
5. Glab G., Florczak K., Jaronski J., Licznerski T. (2001), Cyto-gynaecological diagnoses in phase contrast microscopy (in polish). Blackhorse Publishing
6. Kittler R., Buchholz F. (2003), RNA interference: gene silencing in the fast line. Seminars in Cancer Biology, 13:259–265
7. Kivioja T., Ravantti J., Verkhovski A., Ukkonen E., Bamford D. (2000), Local average intenstiy-based method for identifying spherical particles in electron micrographs Journal of Structural Biology, 131:126–134
8. Lewis P. (1995), Fast normalized cross-correlation. Vision Interface
9. Ludtke J., Baldwin P., Chiu W. (1999), EMAN: Semiautomated software for high-resolution single-particle reconstruction. Journal of Structural Biology, 128:82–97
10. Boier Marti I.M., Martinescu D. C., Lynch R.E., Baker T.S. (1997-2005), Identification of spherical virus particles in digitized images of entire electron micrographs. Journal of Structural Biology, 120:146–157
11. Miniello G. (1998), Colposcopy and Phase Contrast Microscopy. CIC Edizioni Internationale
12. Miniello G. (2001), Vaginal Fungal Infections by Phase Contrast Microscopy. CIC Edizioni Internationale
13. Nicholson W.V., Glaeser R.M. (2001), Review: Automatic particle detection in electron microscopy. Journal of Structural Biology, 133:90–101
14. Nicholson W., Malladi R. (2004), Correlation-based methods of automatic particle detection in electron microscopy images with smotothing by anisotropic diffusion. Journal of Microscopy, 213:119–128
15. Press W.H., Teukolsky S.A., Vetterling W.T., Flannery B.P. (2004), Numerical Recipes in C. Press Syndicate of the University of Cambridge
16. Rasband W. S. (1997-2005), ImageJ
17. Sanderson J. (2002), Phase contrast microscopy. Encyclopedia of Life Sciences
18. Zernike F. (1935), Phase contrast method in microscopic observations (in german). Z. tech. Phys., 16:454
19. Zernike F. (1942), Phase contrast, a new method for microscopic observation of transparent objects. Physica, 9:686
20. Zernike F. (1955), How I discovered phase contrast. Science, 121(3141):345–9

Watershed Extraction of the Exact Shape of Microcalcifications in Mammograms

Mariusz Nieniewski

Institute of Fundamental Technological Research, 00049 Warsaw, Poland
mnieniew@ippt.gov.pl

Summary. The presented method of extraction of the *exact* shape of the microcalcifications (MCs) is based on the watershed segmentation (WS) and region merging. Assuming that the locations of the MCs are known, we use two kinds of markers for the WS: the internal marker indicating the interior of the MC, and the external marker separating the MCs. Carrying out the WS in the area between these two markers we obtain a limited number of regions. These regions are merged into a mask of an MC by maximization of the average contrast between the mask and its surroundings. The obtained shapes of the masks agree with human intuition and can be used for the classification of MCs as to their malignancy.

1 Introduction

Detection and extraction of MCs from mammograms is described in many papers [1]. However, for the purposes of classification of malignancy of the detected MCs it is imperative to obtain their exact shape. The methods described in the literature may not give the exact shape since the detection of MCs is influenced by the size and shape of the structuring elements or by tuning parameters used in morphological or other detectors. In this paper, the assumption is made that a map of MCs has been obtained using the morphological detector [2], [3]. However, any other detector can be used as well. Since our aim is to obtain the precise shape of MCs, the WS seems the most appropriate approach [4]. In [5] the use of markers is proposed for controlling the WS process. Suppose we want to extract a dark object shown against lighter background. (For this purpose we would have to complement the mammogram since MCs are local brighter spots.) The marker is generated by assuming 0 brightness in the areas, where we want to have the minimum in the segmented image. All other pixels are set to the maximum 255. Subsequently we superpose the marker on the segmented image, which is usually the gradient of the original image. The markers allow us to indicate multiple objects as well as the background. If there is a sufficient contrast between the object to be extracted and the background, then there may be a single watershed contour indicating the boundary between the object and the background. However, the MCs usually have a very poor contrast and several watershed regions are obtained instead of a single one representing the MC, hence the problem of finding which regions should be included in the mask [6].

2 Sequence of Operations for Watershed Segmentation of a Mammogram

The WS of the mammographic image is conducted by means of two markers: the internal and external markers. The internal marker is obtained by interaction of regional maxima of the original image with a map of MCs obtained by an MCs detector. The external marker is obtained by a modification of the watershed lines of the original image (and not of its gradient, as is usually the case). The described approach is based on the assumption that a typical MC contains exactly one brightness maximum and some of its neighborhood and nothing more. Carrying out the WS in the area between the markers we obtain a small number of regions (comp. Fig. 1 below). Some of these regions are merged into a single mask of the MC based on the maximization of the average contrast between the MC and its surroundings.

Fig. 1(a) shows a window of size 512×512 pixels taken from the mammogram A_1108_1.LEFT_CC in the DDSM database [7]. Fig. 1(b) displays the same window after histogram stretching for better visibility.

Fig. 1(c) presents a map of MCs obtained by a morphological detector [2]. Fig. 1(d) shows the regional minima which were obtained for image in Fig. 1(a), complemented and filtered by means of the opening-closing with a structuring element of size 3×3. Obviously Fig. 1(d) contains masks of MCs as well as many other artifacts and should be cleaned. The intersection of images (c) and (d) is shown in Fig. 1(e). Fig. 1(e) includes only those parts of regional minima which overlap masks of MCs. Some of the regional minima are removed completely and others in part. Fig. 1(e) is an approximation of the internal marker, but it is not its final version yet.

Fig. 1(f) shows the complement of the morphological gradient of the filtered image. The filtered image is the same as that used for obtaining (d). The gradient was calculated using a structuring element of size 3×3. The obtained values of the gradient were multiplied by 10. Then the resulting image was complemented and the value of 20 was subtracted from the brightness in each pixel for better visibility. The watershed lines for the gradient image (f) are shown in Fig. 1(g). Heavy oversegmentation in (g) is due to the relatively weak filtering of the original image.

Fig. 1(h) presents the watershed lines of the filtered original image. The filtered image is the same as was used for obtaining (d) and (f). Fig. 1(h) will serve as the external marker.

Pixels of the internal and external markers should not be mutual neighbors since otherwise the watershed would be incorrect. Dilating the external marker by the structuring element of size 3×3 we obtain a larger area in which no pixel of the internal marker should be encountered (Fig. 1(i)). Fig. 1(j) shows the internal marker obtained as a logical difference of the approximate internal marker (e) and the dilated external marker (i). The complete marker (Fig. 1(k)) is generated by logical summation of the internal marker (j) and the external marker (h). The internal and external markers in Fig. 1(k) are separated.

Fig. 1(l) depicts areas which possibly contain masks of MCs. This image is obtained by reconstruction by dilation with the internal marker (j) used as a marker and the complemented image (h) as a mask. The final marker (Fig. 1(m)), used for the WS, is obtained as a logical intersection of (l) and the complement of the internal marker (j).

The modified gradient (Fig. 1(n)) of the filtered image is obtained by intersecting the gradient (f) with the final marker (m). (The image in Fig. 1(n) has been manipulated similarly to (f) for better visibility.) Fig. 1(n) shows the gradient in those areas which may contain MCs. The watershed lines for the modified gradient of Fig. 1(n) are depicted in Fig. 1(o). Fig. 1(p) shows the complement of intersection of (l) with the complement of (o). It can be observed in (p) that a mask of a single MC may consists of one or more watershed regions. Fig. 1(q) shows the restriction of the original image to the areas possibly containing the MCs, and Fig. 1(r) illustrates the superposition of watershed lines (p) on (q).

3 Region Merging for Generation of the Mask of MCs

The proposed method of region merging is adapted from solar images [8] to mammograms. The principle of the proposed approach is as follows. The algorithm starts with a bright region chosen by the user and supposedly belonging to an MC. The current mask of the MC is obtained by iteratively adding successive neighboring regions with the maximum average brightness. In each iteration the average brightness of the i regions belonging to the mask is calculated

$$\text{ave\_brightness}(mask) = \frac{\sum_{t=1}^{t=i} [\text{ave}_t * \text{area}_t]}{\sum_{t=1}^{t=i} \text{area}_t}, \quad (1)$$

and similarly the average brightness of the $(k-i)$ regions which are 8-neighbors (or 4-neighbors) of the mask is also calculated

$$\text{ave\_brightness}(boundary) = \frac{\sum_{t=i+1}^{t=k} [\text{ave}_t * \text{area}_t]}{\sum_{t=i+1}^{t=k} \text{area}_t}. \quad (2)$$

Equation 2 represents the average brightness of the regions surrounding the mask. The average contrast of the brightness of the current mask with respect to the background is defined as

$$\text{contrast} = \text{ave\_brightness}(\textit{mask}) - \text{ave\_brightness}(\textit{boundary}). \quad (3)$$

This contrast is calculated for successive iterations, and curves such as in Fig. 4 below are obtained. The maximum of the contrast in Fig. 4(a) or (b) determines the number of iterations corresponding to the required mask of the MC.

Two examples of region merging are shown in Figs. 2 and 3. The operations carried out for Fig. 2 are as follows. We select a region that may contain an MC by means of a reconstruction. For this purpose we choose an arbitrary pixel belonging to the required region (Fig. 1(l)) and set the value in this pixel to 255, whereas all other pixels contain 0. A binary image obtained in this way serves as a marker for the reconstruction. At the same Fig. 1(l) is used as a mask. The result of the reconstruction is shown in Fig. 2(a) in a window of size 80 × 80 pixels. Fig. 2(b) depicts the watershed lines in black, superposed on the image in Fig. 2(a). The labelling of the regions in Fig. 2(b) is carried out for the complement of Fig. 2b and the complement of appropriate window taken from Fig. 1(a) [8]. As a result of labelling, 29 regions are obtained with average brightness in the range from 66.48 to 95.42. In order to generate a mask of an MC we choose any pixel belonging to the internal marker for the starting region of the mask. The easiest way is to choose such a pixel from Fig. 1(m) and find the region corresponding to this pixel.

One example of the results of calculating the average contrast as a function of iterations is shown in Fig. 4(a). The maximum contrast in this case is obtained after three iterations and is equal to 14.94. Including subsequent regions results in a decrease of the average contrast, until after 18 iterations the contrast starts growing again. The masks corresponding to 0, 1, 2, and 3 regions are shown in Figs. 2(c)–(f). The second example of the results of calculating the average contrast as a function of iterations is shown in Fig. 4(b), and the relevant region and mask are illustrated by Figs. 3(a) and (b). Fig. 3(c) depicts the window taken from Fig. 1(b). Comparison of Figs. 3(b) and (c) confirms that the obtained shapes of the masks of MCs and shapes visible in Fig. 3(c) are similar. Strictly speaking the comparison should be made with the original image in Fig. 1(a), but in this latter figure visibility is poor.

4 Conclusions

A method for extracting the *exact* shape of MCs is presented. This shape is independent of structuring elements or tuning parameters used in the MCs detector. Because MCs typically have diameter less then 500 μm and the resolution of mammograms is, for example, 42.5 μm in the DDSM database, it follows that a diameter of an MC is on the order of several pixels. These data illustrate the fact that only a rough approximation of the contour of an MC is possible and a difference of one pixel in a dimension of an MC may significantly change the MC statistics and influence the classification of MCs. The proposed approach allows us to obtain a shape which remains in good agreement with our visual expectations and may improve the results of MCs classification. An example of MCs classification is given in [9].

References

1. Cheng H D, Cai X, Chen X, Hu L, Lou X (2003) Pattern Recognition 36:2967–2991
2. Nieniewski M (1999) Machine, Graphics and Vision 8:427–448
3. Ustymowicz M, Nieniewski M (2004) Morphological Method of Microcalcifications Detection in Mammograms. In: Proceedings of the International Conference on Computer Vision and Graphics 2004, Kluwer
4. Soille P (1998) Morphological Image Analysis. Springer, Berlin
5. Beucher S (1990) Segmentation d'Images et Morphologie Mathématique. PhD Thesis, École Nationale Supérieure de Mines de Paris, Paris
6. Nieniewski M (2005) Segmentation of Digital Images. AOW EXIT, Warsaw (in Polish)
7. DDSM, http://figment.csee.usf.edu/Mammography/DDSM
8. Nieniewski M (2004) IEEE Trans Systems, Man, a. Cybernetics, B, 34:796–801
9. Ustymowicz M (2005) Automatic Detection and Classification of the Microcalcification Clusters in Mammographic Images. PhD Thesis, Institute of Fundamental Technological Research, Warsaw (in Polish)

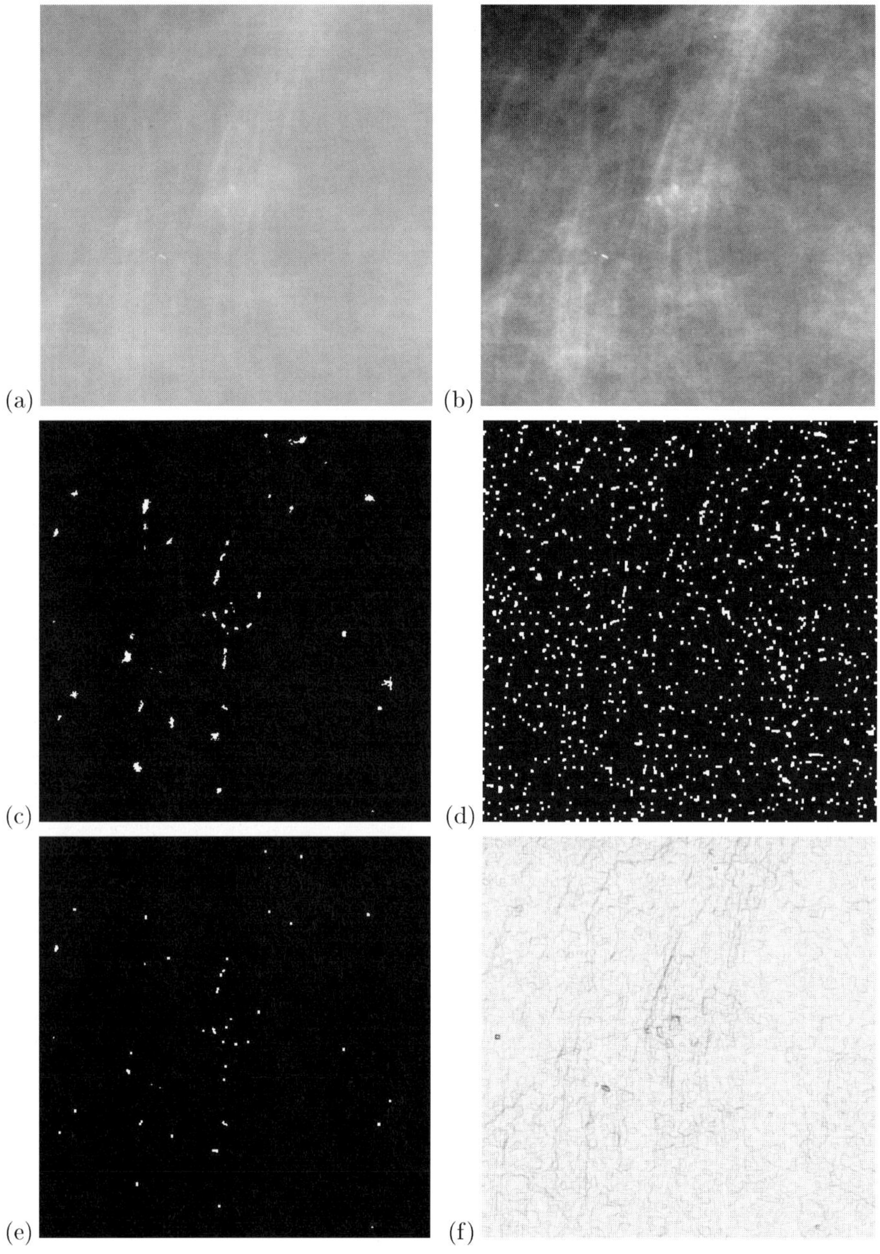

Fig. 1. WS – part I. (a) Original image. (b) Image (a) after histogram stretching. (c) Map of MCs obtained for (a). (d) Regional minima for filtered and complemented image (a). (e) Logical intersection of (c) and (d). (f) Complement of the gradient of the filtered original image.

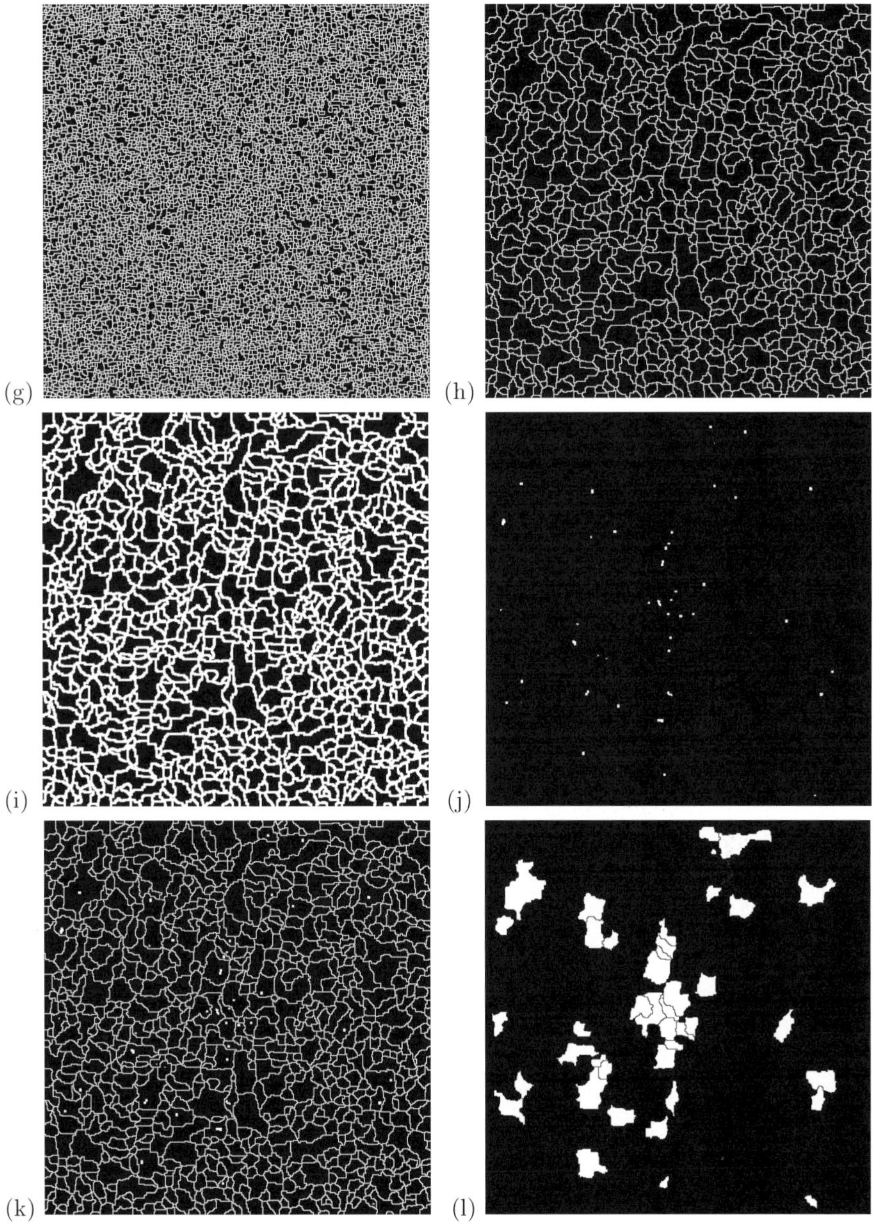

Fig. 1. WS – part II. (g) Watershed lines for the gradient of the filtered image from (a). (h) External marker. (i) Dilation of the external marker. (j) Internal marker. (k) Complete marker. (l) Regions of possible MCs.

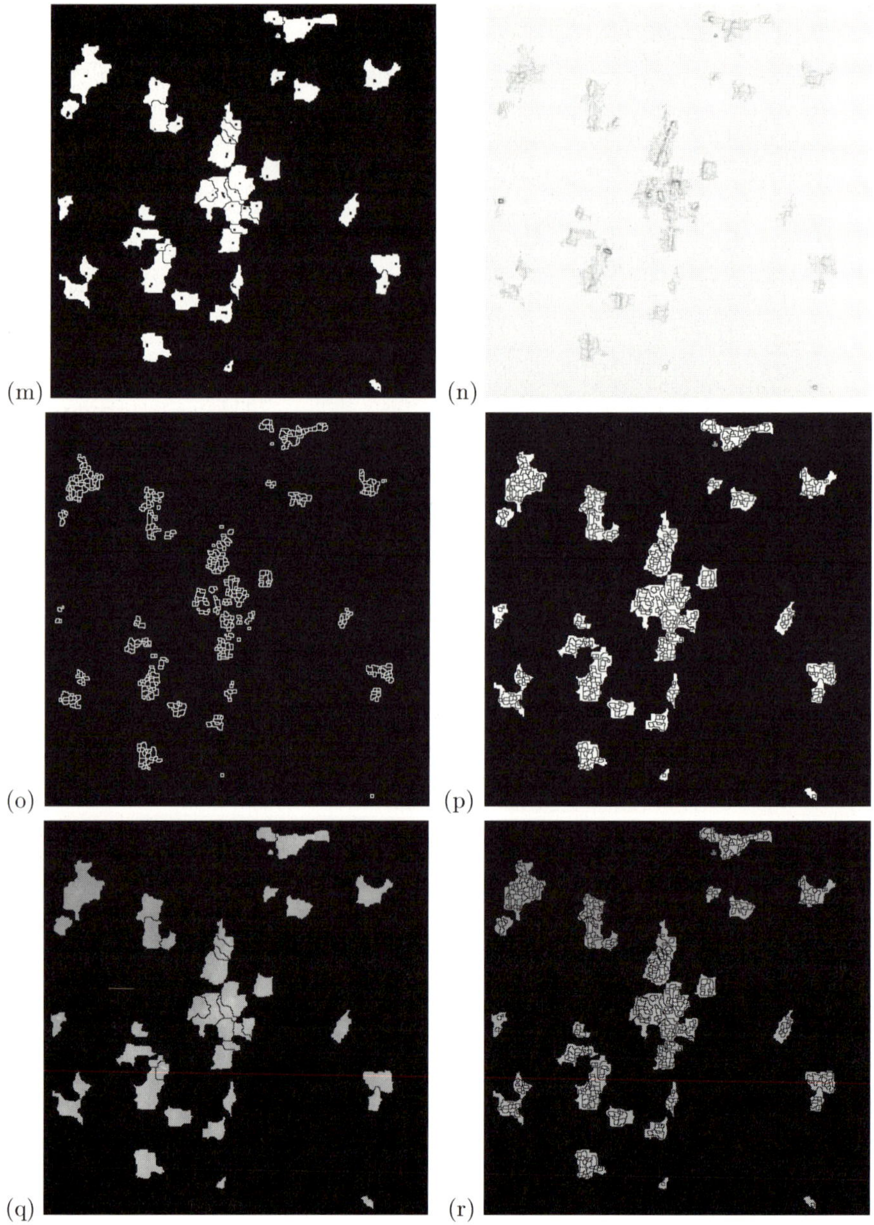

Fig. 1. WS – part III. (m) Final marker. (n) Complement of the image of the modified gradient. (o) Watershed lines of the modified gradient. (p) Watershed lines superposed on image (l). (q) Restriction of the original image. (r) Watershed lines superposed on (q).

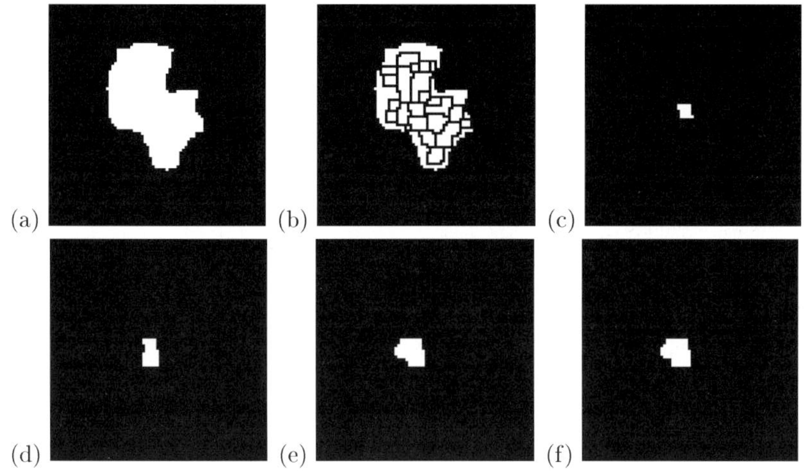

Fig. 2. Region merging. (a) Selected region from Fig. 1(l). (b) Respective region from Fig. 1(o). (c)–(f) Mask after 0, 1, 2, and 3 iterations, respectively.

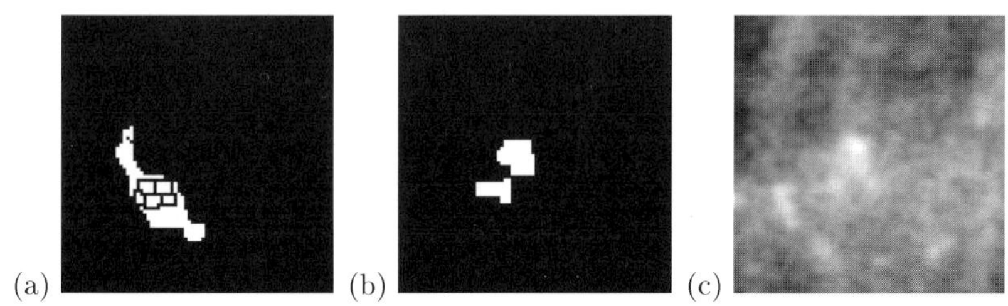

Fig. 3. Region merging. (a) Selected region from Fig. 1(o). (f) Logical sum of the mask for image (a), obtained after 2 iterations, and of the mask from Fig. 2(f). (c) Window taken from Fig. 1(b), after histogram stretching.

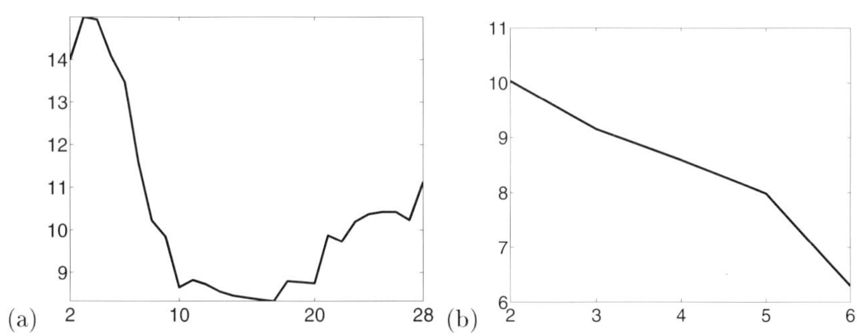

Fig. 4. Average contrast as a function of iterations. (a) For example in Fig. 2. (b) For example in Fig. 3(a).

Computer Recognition of Biological Objects' Internal Structure Using Ultrasonic Projection

Krzysztof J. Opielinski[1] and Tadeusz Gudra[2]

[1] Institute of Telecommunications and Acoustics, Wroclaw University of Technology, Wybrzeze Wyspianskiego 27, 50-370 Wroclaw, Poland
krzysztof.opielinski@pwr.wroc.pl

[2] Institute of Telecommunications and Acoustics, Wroclaw University of Technology, Wybrzeze Wyspianskiego 27, 50-370 Wroclaw, Poland
tadeusz.gudra@pwr.wroc.pl

Summary. In this paper, the possibilities of using ultrasonic projection for computer recognition of biological structures were analyzed. The computer simulation of the distribution of ultrasonic wave propagation velocity local values inside two three-dimensional objects dipped in the water was done, and next for a few projection planes the Radon transforms were calculated in the form of the mean value distribution in parallel geometry. The results of simulation were visualized in the form of the grey scale images and in the form of the pseudo-3D charts with lighting. The real measurements of three-dimensional biological objects were carried out on the elaborated research setup. The obtained projection images of the distribution of three different acoustic parameters in the object structure: propagation velocity, amplitude and mid frequency shift of ultrasonic wave were compared with the optical scan in the same plane of object cross-section. The conclusion is that computer-assisted ultrasonic projection enables the correct recognition of biological structures. On the basis of a few projection images of the examined structure obtained from many directions it is possible to attempt to computer-reconstruct in 3D heterogeneity boundaries inside the structure, which can be used e.g. in diagnosing the early stages of women's breast cancer.

1 Introduction

Ultrasonic imaging technique plays a more and more important role in medical diagnostics. In most applications echographic methods are used (ultrasonography, ultrasonic microscopy). By means of such measurements an image presenting the changes of the reflection coefficient within the examined structure is constructed. This paper presents an attempt to use information included in ultrasonic pulses running through the object in order to generate images showing a projection of the examined structure [1] in the form of the distribution of the mean values of the measured acoustic parameter, for one or several planes perpendicular to the direction of ultrasonic waves (just as in X-ray radiography). Fig.1 presents an example of computer recognition used in dentistry for imaging a tooth's inner structure using X-ray projection.

An advantage of the transmission method is the level of ultrasonic received pulses, which is twice as high compared to the echo method. Moreover, a few acoustic parameters can be imaged concurrently, indicated by a computer on the basis of information included directly in ultrasonic pulses running through the structure, allowing us to generate a few different projection images at a time, each of them visualizing a different feature of the structure. Such a complex characteristic can be of paramount importance e.g. in detecting and diagnosing cancerous changes in tissues.

a) b) c)

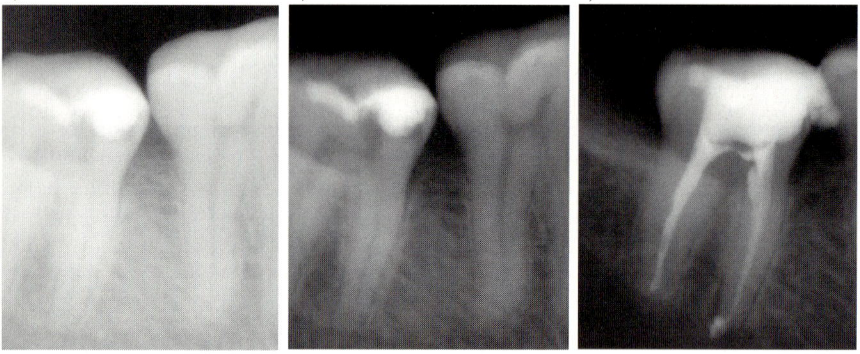

Fig. 1. Using computer recognition for X-ray diagnostics in dentistry: a) the X-ray image of a tooth without image processing - the canals are barely visible, b) the X-ray image of a tooth with image processing, c) the X-ray image of a tooth with image processing after filling the canals

Because of the possibility to obtain the images in pseudo-real time (e.g. with a constant delay deriving from the necessity to buffer and process data), a device using this method can be named an ultrasonic transmission camera (UTC) [2]. In the majority of known experimental elaborations, the projection parameter is the signal amplitude, and not its runtime, the latter seeming more attractive because of measurement simplicity and accuracy. In this paper, using computer simulation and real measurements, we analyzed the possibilities of using ultrasonic projection of three different acoustic parameters for computer recognition of biological structures, which can be used e.g. in diagnosing the early stages of women's breast cancer.

2 Projection

Using a definition from basic stereometry, a general statement can be made that projection is a transformation allowing us to image objects in a system with a lesser number of degrees of freedom (e.g. three-dimensional shapes on a plane). There are two classes of projection algorithms: parallel projection and central (divergent) projection. In the case of function value projection, every single projection can be defined as the following integral [3]:

$$\mathrm{F}_L = \int_L f(s)ds \tag{1}$$

where f is a function defined on the plane s, and L denotes every straight belonging to the plane. The straight L on the plane can be defined by means of two numbers: $p \in R$ and $\psi \in (0, 2\pi]$. Then:

$$\mathrm{L}_{p,\psi} = \{(x, y) : x \cos\psi + y \sin\psi = p\} \tag{2}$$

Fig.2 shows the schemes od two ways of projecting a two-dimensional function.

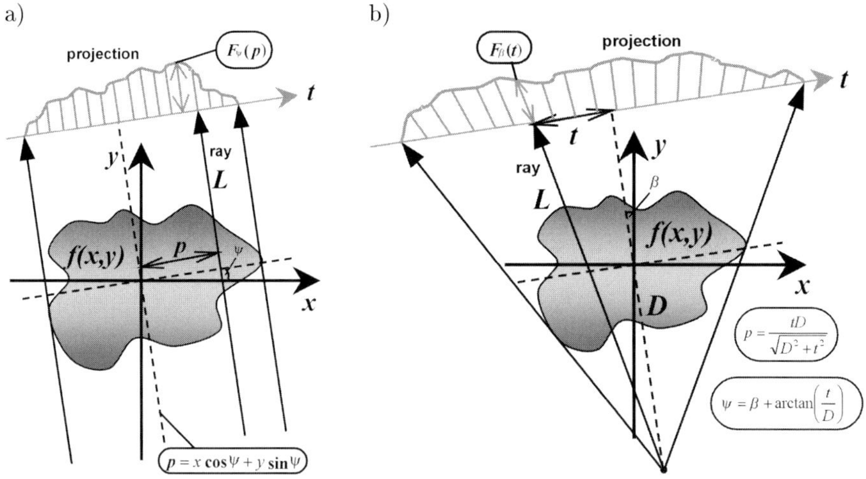

Fig. 2. The ways of projecting a two-dimensional function: a) parallel-ray projection, b) divergent projection

For the function f defined on the plane, the set of its projections along an infinite number of straights $L_{p,\psi}$, is known as the Radon transform [3]:

$$\mathrm{R}\{f(p, \psi)\} = \int_{L_{p,\psi}} f(x, y)ds \tag{3}$$

The Radon transform is used in many methods of medical diagnostics, e.g. in X-ray diagnostics where the examined biological structure is exposed to X-rays, and the differences in radiation intensity caused by differences of absorbtion capacities are registered on a photographic film, on a luminescent screen or are computer-imaged (Fig.1). A disadvantage of this method is its destructive effect on biological structures, as well as the phenomena of dissipation, diffraction, interference, polarization, refraction and reflection of X-rays, causing distortions of the projected image. Like in the case of the X-ray visualizing of biological structures, the projection method using ultrasonic waves can be used. The phenomena bound up with the propagation of ultrasonic waves in biological structures cause only small distortions of the projected image provided that the acoustic impedance local values in the examined structure vary considerably [4]. The greatest advantage of ultrasound is, however, their harmlessness, thanks to which it is possible to project the examined structure from many different directions, enabling us to reconstruct in three dimensions the boundaries of heterogeneity inside the structure [5]. This paper analyzes projection images only in parallel geometry, for only one projection plane.

3 Computer simulation

A computer simulation of the distribution of ultrasonic wave propagation velocity local values inside a few 3D objects dipped in the water, and next for a few projection planes the Radon transforms were calculated in the form of the mean value distribution in parallel geometry (with the fixed distancne of 20 mm between the surface of the transmitter and the receiver) [1]. In the space surrounding each of the objects the assumed ultrasonic wave propagation velocity was 1485 m/s, corresponding to the ultrasound velocity in the water at the temeperature of ca 21°C. In the case of ultrasonic projection, the use of water as the coupling medium is necessary because of matching with the examined biological structure acoustic impedance. Fig.3 presents two 3D objects used for the simulation. The object shown in Fig.3a is a homogenous sphere with a diameter of 14 mm, including 7 homogenous bubbles, each 0.5 mm in diamater, placed along the sphere diameter at the intervals of 2 mm between the centres of the bubbles except for the two extreme bubbles - invervals of 1.85 mm. The object shown in Fig.3b is a homogeneous ellipsoid with the large semiaxis of 7 mm and the small semiaxis of 1 mm, including 7 homogenous bubbles of 0.5 mm in radius each, placed in the same way as in the case of the sphere. The local values of ultrasonic wave propagation velocity are 1500 m/s for every point of the inside and the boundary of the sphere and the ellipsoid, and for the small bubbles these values are 1499 m/s.

Fig.4 shows the ultrasonic projection images of the objects from Fig.3, obtained on the basis of computer-simulated measurements of ultrasonic wave

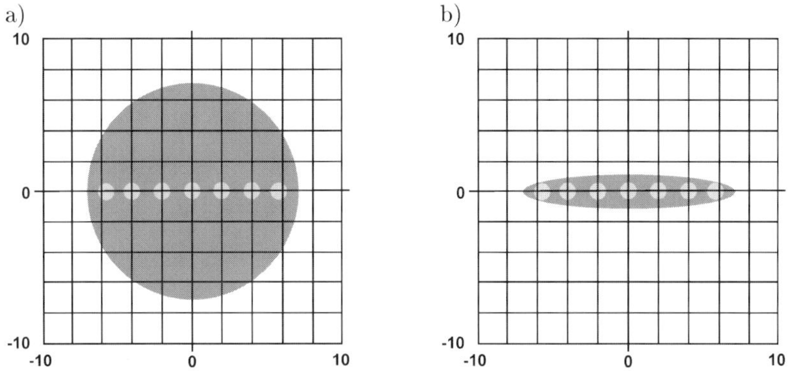

Fig. 3. The structure of the 3D objects used for the simulation: a) the sphere, b) the ellipsoid

propagation velocity mean values (projections). These values were visualized linearly in the grey scale, from black to white.

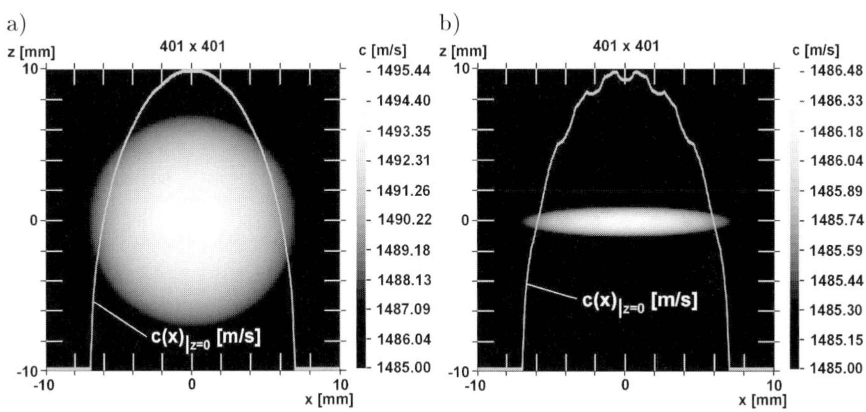

Fig. 4. The imaging of the ultrasonic projection of the sphere (a) and the ellipsoid (b), obtained on the basis of computer-simulated measurements of ultrasonic wave projection velocity mean values

What can be clearly seen in the images is the shape of the objects and their edges in the projection. The small spherical structures inside the ellipsoid (Fig.4b) are barely visible on the background of its projection (dark, round spots), and the inside of the sphere seems homogenous (Fig.4a). In order to improve the contrast of heterogeneous structures' visualization, Fig.5 shows the same ultrasonic projection images of the sphere and the ellipsoid in pseudo-3D with lighting, corresponding to the mean values of ultrasonic wave projection velocity on the Z axis in the Cartesian XYZ system of coordinates.

The projections of spherical heterogeneities are visible here in the shape of characteristic, oval concavities in the object structure.

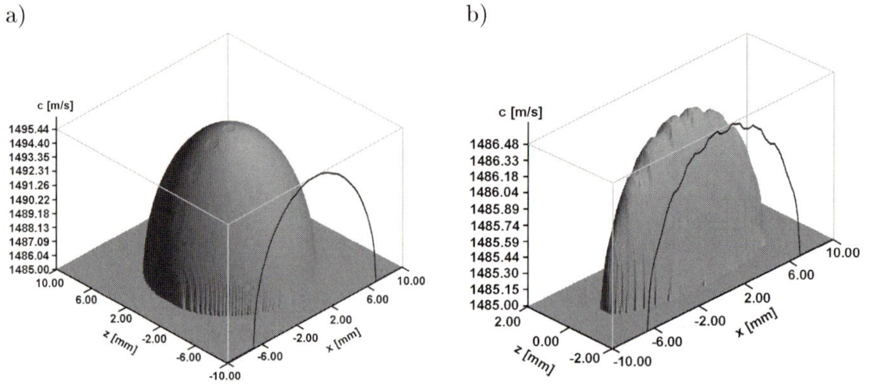

Fig. 5. The pseudo-3D imaging of ultrasonic projection: a) the sphere, b) the ellipsoid

4 Measurement set-up

To examine biological structures by means of ultrasonic projection, the computer-assisted measurement setup was elaborated. To scan objects dipped in the water two ultrasonic probes of 6 mm in diameter were used, functioning as the transmitter and the receiver of ultrasonic waves and working with the frequency of 5 MHz. The probes fixed on the axis facing each other are shifted with meandering movement with a definite step in the object's chosen plane. The probes' movement is software-controlled through the RS232 bus, using the XYZ shift mechanism. The transmitting probe is powered with a burst-type sinusoidal signal, and the pulses are received by the receiving probe by a digital oscilloscope, transmitted via RS232 to the computer and recorded on the hard disk. The appropriate parameters of particular pulses were visualized in the rainbow or gray scale as well as in pseudo-3D by means of special software enabling advanced image processing.

5 Measurement results

On the elaborated research setup a few measurements of three-dimensional biological objects in different projection planes were carried out. Fig.6 shows the obtained projection images and the optical image of one of the examined object structures - a hard-boiled hen's egg devoid of the eggshell.

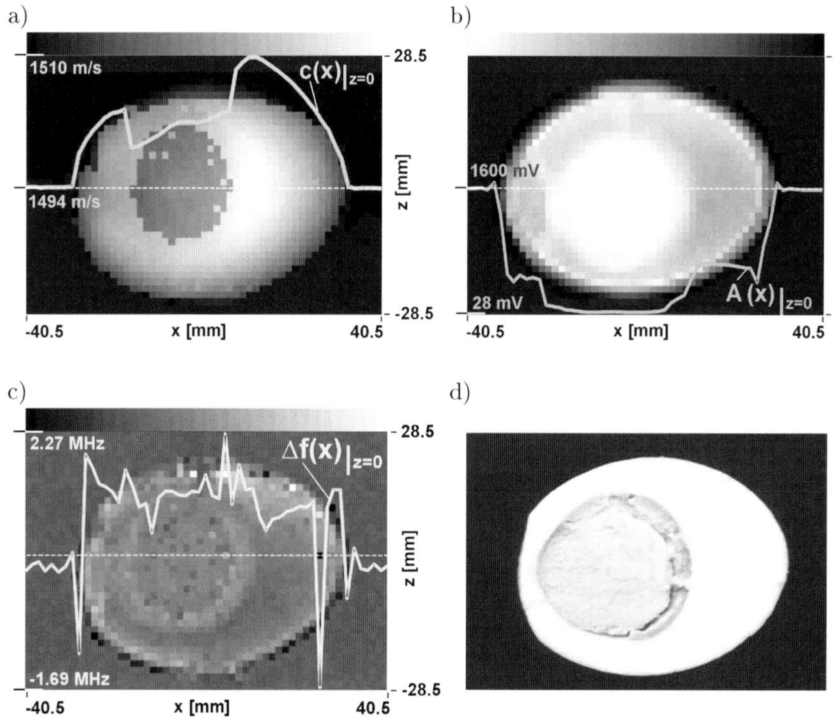

Fig. 6. Ultrasonic projection images of a hard-boiled hen's egg without eggshell obtained from the following measurements: a) ultrasonic wave propagation velocity mean values, b) ultrasonic wave amplitude mean values, c) the mean values of the ultrasonic wave mid frequency shift, d) optical scan of egg cross-section structure

On the basis of a set of registered received pulses in one scanning plane with the step of 1.5 mm x 1.5 mm what was obtained were images showing the projection of the distribution of three different acoustic parameters in the object structure: propagation velocity, amplitude and mid frequency shift of ultrasonic wave. The solid line in the images shows the distributions of particular parameters along the marked broken line.

6 Conclusions

The interpretation of projection images requires spatial imagination and a basic knowledge of spatial geometry. Particular images show the projections of the local values' distributions of the structure's measured acoustic parameter in the plane perpendicular to the scanning surface and are a mirror reflection for projection planes at the front and at the back of the object. Projection imaging is not entirely a quantitative imaging; nevertheless on the basis of particular pixels' values the differentiation of the examined structure's parameters can be defined. Comparing the images shown in Fig.6a,b,c and Fig.6d it can be definitely said that computer-assisted ultrasonic projection enables the correct recognition of biological structures (the yolk is clearly visible in the egg's structure). The advantage of this method is the possibility to obtain a few different images from one measurement setup, each of them characterizing slightly different features of the object structure. These images can be appropriately processed and correlated by special software, allowing us to diagnose structures invisible in single images. On the basis of a few projection images of the examined structure obtained from many directions it is possible to attempt to computer-reconstruct in 3D the heterogeneity boundaries inside the structure. What is planned within the framework of future research is the construction of a special measurement setup for visualizing biological structures using the ultrasonic projection method, enabling concurrent measurements of a few acoustic parameters in real time (multi-parameter ultrasonic transmission camera).

References

1. Opielinski K J, Gudra T (2004) Biological Structure Imaging by Means of Ultrasonic Projection. In: Structures - Waves - Human Health 13 (2). Polish Acoustical Society Division Krakow. Krakow, Poland.
2. Ermert H, Keitmann O, Oppelt R, Granz B, Pesavento A, Vester M, Tillig B, Sander V (2000) A New Concept For A Real-Time Ultrasound Transmission Camera. In: IEEE Ultrasonics Symp. Proc., San Juan, Puerto Rico.
3. Radon J (1917) Uber die Bestimmung von Funktionen durch ihre Integralwerte langs gewisser Mannigfaltigkeiten. In: Berichte Sachsische Akademie der Wissenschaften, Math.-Phys. Kl., 69, Leipzig.
4. Opielinski K J, Gudra T (2000) Ultrasonics 38: 424-429.
5. Opielinski K J, Gudra T (2004) Ultrasonics 42: 705-711.

Machine Learning Methods for Dialysis Therapy Decision Problem – Comparative Study

Wojciech Penar[1] and Michal Wozniak[2]

[1] Institute of Engineering Cybernetics, Wroclaw University of Technology
wojciech.penar@pwr.wroc.pl

[2] Chair of Computer Systems and Networks, Wroclaw University of Technology
michal.wozniak@pwr.wroc.pl

1 Introduction

First successful attempts to utilize computer techniques in medical diagnostics were undertaken in seventies of the last centaury, when first expert systems were set up. These systems answer questions using the rules were entered by experts. The main disadvantage of the systems was that experts had to prepare rules and enter them into the system.

Along with computer technique development, the artificial intelligence methods were developed. One of its main branches was machine learning. Assumption of machine learning method consists in self-modifying program to better results give (in next execution on the same data). There are two groups of machine learning algorithms, unsupervised learning and supervised one, also called exemplar based learning. The natural course of events was to employ the latter one to construct rules for expert systems from medical data.

This paper describes our research with regard to application of two machine learning algorithms (based on sequential covering idea and decision tree induction) for the chronic renal failure therapy problem.

The paper is organized as follows. In section 2 algorithms we used are presented, the next section describes medical problem and then its mathematical model is presented. In section 5 results of experimental investigations are presented. Last section concludes the paper.

2 Introduction to machine learning

In our research we have used two inductive learning algorithms. The first is the CN2 algorithm, written by P. Clark and T. Niblett [7], which is well known algorithm based on sequential covering concept. The second one is the decision tree induction algorithm C4.5 written by J. R. Quinlan [4], which was developed as improvement of ID3 algorithm [2]. We used R. Boswell's implementation of the CN2 algorithm [8] and the J48 classifier from WEKA toolset [6] which is implementation of C4.5 algorithm.

3 Medical problem

In our research we have focused on the problem of introducing a dialysis therapy for the patients with chronic renal failure. This treatment has to be started just before the end stage renal disease occurs, since it is onerous for the patients and expensive.

The therapy depends mainly on the results of biochemical tests, but special attention is paid to immeasurable factors such as patient's general condition or coexisting diseases. Patients with diabetes are treated with special attention because the procedure of introducing dialysis for them is quite different. It is important to note that there was no test which could unequivocally defines starting time of therapy - the evaluation depended only on the expert's experiences.

4 Mathematical model

The data contains 13 attributes and each record is labeled by one of two categories – OK and LATE – if the therapy has started on time or not. Each case was evaluated by an expert, according to medical records describing what happened with the patient after the therapy has started.

Formally the problem was to qualify a case to one of two classes (OK and LATE) according to 13 attributes.

In table 1 all attributes, their value ranges and normal values are presented.

5 Experimental investigations

In order to investigate and compare algorithms two criteria were used. The first was classifying accuracy defined as a number of correctly classified by obtained recognisers exemplars from test dataset. The other was essential value of classifier, which was designated by an expert. To compute classifying accuracy of the classifier, the cross validation method was used. In case of C4.5 we have used WEKA's built-in mechanism, in case of CN2 – we had to prepare our own program for computer experiments. The size of validation set is fixed on 10% of number of available elements.

The data for this research has been provided by Medical University of Wroclaw and it included information on 185 patients who have started a dialysis therapy within last 20 years.

Four datasets were used in each stage of computer experiments. First dataset contents all 13 attributes and the classification label and it is called

Table 1. The list of attributes used in experiments

Attribute	Values	Normal values
Age	[23; 94]	
Gender	{ K, M }	
Diagnosis	{ NC, non-NC }	
Protein	[36,4; 87,9]	[60,0; 83,0]
Albumin	[15,81; 53,51]	[35; 50]
Urea	[10,9; 75,0]	[2,8; 7,6]
Creatinine	[3,5; 22,8]	[0,40; 1,40]
Cholesterol	[1,8; 10,8]	[3,1; 5,2]
Calcium	[1,31; 3,70]	[2,20; 2,55]
Phosphorus	[2,1; 12,8]	[2,7; 4,5]
Ferrum	[6,7; 225,5]	[50; 170]
TIBC	[50,6; 453,4]	
Hematocrite	[16,1; 44,9]	[33; 35]

dataset "0". Then attribute "age" was removed, because it is considered as not differentiating for renal disease – dataset "1". Next two datasets contents attributes which are considered as most important in renal disease evaluation. Dataset "2" contains six attributes: "gender", "diagnosis", "creatinine", "urea", "calcium" and "phosphorus" , dataset "3" employs only four attributes: "gender", "diagnosis", "creatinine" and "urea". All datasets and used attributes are presented in table 2.

Table 2. Attributes used in datasets

Dataset	Attributes used
"0"	age, gender, diagnosis, protein, albumin, urea, creatinine, cholesterol, calcium, phosphorus, ferrum, TIBC, hematocrite
"1"	gender, diagnosis, protein, albumin, urea, creatinine, cholesterol, calcium, phosphorus, ferrum, TIBC, hematocrite
"2"	gender, diagnosis, urea, creatinine, calcium, phosphorus
"3"	gender, diagnosis, urea, creatinine

Experiments were run in two stages. In first stage original data were used. Results of this part of experiment were presented to the expert and rules were checked against consistency with medical knowledge. Expert mentioned some inconsistencies in obtained results and classification of all the cases were re-evaluated. These data were used in second stage of experiments. Results of second series of experiments were presented to the expert again and, as in first stage of investigations, they were checked against inconsistencies.

The classification accuracies obtained in first stage of investigations are shown on figure 1. The accuracy of default rule (all exemplars recognized as belonging to the most frequent class) is 63.3%. By using the CN2 algorithm, the classification accuracy is increased to about 78% (with datasets 0, 1 and 2) or - in the worst case - to 70.7%. The use of the C4.5 algorithm gives classification accuracy between 72.5 and 75.1%.

In case of CN2 algorithm, best results were achieved for dataset "2". In case of dataset "3" classifier is highly overfitted and consists of many specific rules. Reduction of attributes number causes induction of many highly specific rules that uses only one attribute and covers only few exemplars (usually two or one). Previously these exemplars were covered by other rules which employ removed attributes. In that case reduction of attributes number cause overfitting of the classifier and classifying accuracy falls.

In case of C4.5 algorithm reduction of attributes number gives usually better results. As expected, classification accuracy for dataset without attribute "age" (dataset "1") is better then for dataset with all attributes (dataset "0"). In case of dataset "3" classifying accuracy falls due to lack of attributes that could provide better differencing tests – the tree is overfitted. Pruning causes, that classifying accuracy for classifier induced from dataset "3" rises. In case of dataset "1" moderate pruning gives better classifying accuracy, but heavy pruning causes that classifying accuracy is as in case of unpruned tree (for dataset 1) or worse (dataset 2). It indicates that heavy pruning causes loss of important information.

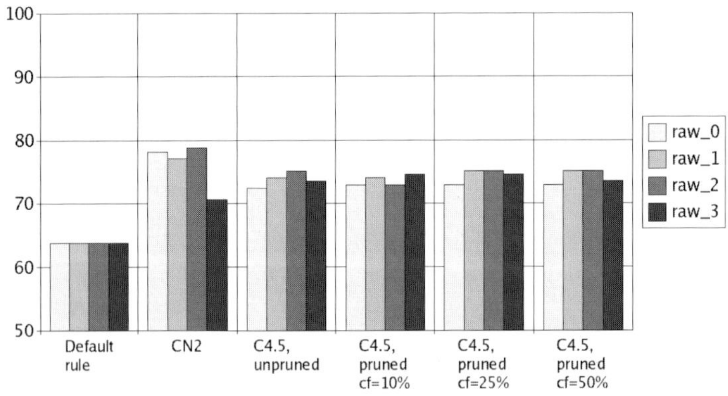

Fig. 1. Classification accuracy for classifiers induced from original data

After first series of tests, results were presented to an expert. Expert pointed at some inconsistencies in obtained rules. All of them were caused by errors in data – usually wrong classification. This indicates usefulness of machine learning methods for erroneous data detection. Highly specific, overfitted rules are induced usually for few (usually one or two) exemplars and indicate unusual cases.

Data that caused such anomalies was found in dataset and was corrected (reclassified). This data is used to produce four "revised" datasets, with the same attributes as in four original datasets. In second stage of investigations tests were repeated for corrected datasets.

On figure 2 the accuracy of classification for classifiers induced from revised datasets is presented. The default rule's accuracy is 61.0%, which is lower then in case of original datasets. In the CN2 algorithm testes, best results are given for datasets 0 and 1 with classifying accuracy of 88.4%.

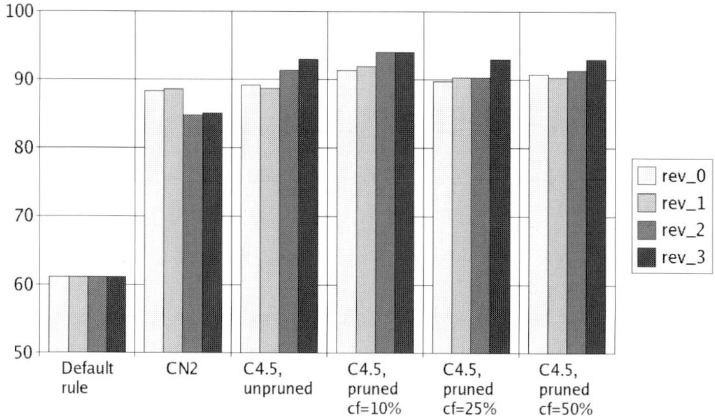

Fig. 2. Classification accuracy for classifiers induced from revised data

The best result was achieved for dataset "1" - with all attributes but "age". When datasets with smaller number of attributes were used, the classifying accuracy fell and many highly specific rules were induced. Some rules covered only one or two exemplars and tested only one attribute within very short range of values. As in case of original datasets, when there were no attributes which could be used to create good rule, that covers more exemplars, many small and highly specific rules were induced. The use of the C4.5 algorithm gives better classifying accuracy than the CN2 algorithm. The best results were achieved for datasets with smaller number of attributes. For each dataset, the best result is produced by the most pruned classifier. The classifying accuracy of the best classifier is 94.1%. Heavy pruning gives the best results for all datasets, because some specific rules were removed.

In expert's opinion rules provided by C4.5 are better than rules provided by CN2. The main advantage of these rules is that they contain more attributes and position of the test in the decision tree corresponds with its importance. The best decision tree induced from revised data is presented on figure 3.

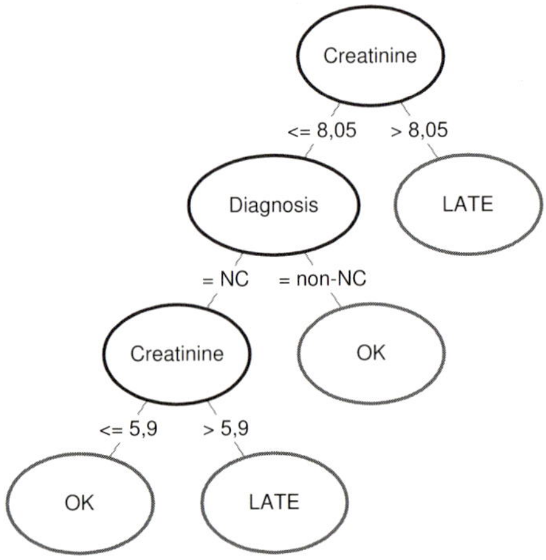

Fig. 3. The best decision tree induced from revised data

6 Conclusions

The main goal of our research was attempt to answer the question, if machine learning algorithms could be applied in computer aided medical diagnostics. Executed tests proven, that data mining methods based on machine learning can be used in medical diagnostics, but it can not substitute an expert, especially in case of rare diseases.

Classifiers induced from revised datasets have better classification accuracy. It is indicating that quality of training data has significant influence for induced classifier accuracy.

In expert opinion of expert, the most of obtained decision rules are consistent with medical knowledge. All cases of incorrect classification were caused by insufficient mathematical model. The evaluation of exact mathematical model in medicine is very difficult.

In this paper results of experiments on two popular machine learning algorithms were presented. The appliance of other classification methods, like Bayesian classifiers, neural networks and fuzzy sets, is subject of future research.

References

1. Januszewicz Wlodzimierz, Kokot F. (red.), Internal Medicine, Medical Publishers PZWL, Warsaw 2002 (in polish)
2. Mitchell Tom, Machine Learning, McGraw Hill, 1997
3. Orlowski Tadeusz (red.), Kidney Diseases, Medical Publishers PZWL, Warsaw 1997 (in polish)
4. Quinlan J. Ross, C4.5: Programs For Machine Learning, Morgan Kaufmann Publishers, San Mateo, California, 1993

5. Rutkowski Boleslaw, Czekalski S., Guidelines for Kidney Diseases Diagnosis and Treatment, Medical Publishers Makmed, Gdansk 2001 (in polish)
6. Witten Ian H., Eibe F., Data Mining: Practical Machine Learning Tools and Techniques with Java Implementations, Morgan Kaufmann Publishers, 2000
7. Clark Peter, Niblett Tim, The CN2 Induction Algorithm, Machine Learning Journal, 3(4), pp261-283, 1989
8. Clark Peter, Boswell R., Rule Induction Witch CN2: Some Recent Improvements, Machine Learning - Proceedings of the Fifth European Conference (EWSL-31), pp151-163, 1991

Removing Artefacts from Microscopic Images of Cytological Smears. A Shape-Based Approach.

Dariusz Pietka, Annamonika Dulewicz, Pawel Jaszczak

Image Processing Systems Laboratory
Biomedical Information Processing Methods Department
Institute of Biocybernetics and Biomedical Engineering, PAS
darek@ibib.waw.pl

Summary. Most of reports on computer supported cytological investigations focus on searching for objective, quantitative descriptors enabling an automated system to distinguish between "normal" and "pathological" objects, usually cells or their organelles. A great number of sophisticated tools have been developed and reported. However, few reports may be found concerning the problem of detecting artefacts in cytological smears and reducing their influence on the overall system performance. On the other hand, the problem is crucial for the whole system setup and if not properly solved may spoil any attempts to implement the system in practice. The paper addresses this neglected problem trying to point out some general rules and procedures that should be followed to reject artefacts from automatic cytological analysis.

1. Objective of the work

Most projects related to clinical implementation of computerised image processing in cytology face the common problem of distinguishing between artefacts and objects of interest which should be measured and analysed. Let us imagine automated detection of early cancer cells in a microscopic smear. Since most algorithms of cancer cell identification rely on some kind of abnormality detection, artefacts left in a sample would generate too many undesired, false-positive alarms, making such a system impractical. Thus, efficient detection and removal of artefacts from later analysis is crucial for the entire system setup and its overall performance [1]. Regardless of the microscopic enlargement used, considerable number of artefacts will always be present in collected images (Fig.1). In the next section we shall explain in more detail what is understood by the term "artefact" in our work. **Generally, these are undesired objects or phenomena influencing the appearance of a smear and obstructing or even preventing proper analysis of important factors of the cytological sample**. As such, artefacts or their influence on the experiment should be avoided. Objective of this work was to find and to point out some general rules and methods allowing efficient rejection of artefacts in the process of automatic analysis of cytological material. This is a trial to make a step towards breaking through one of the main obstacles in the practical implementation of computer-aided cytological screening. To make our deliberations more useful we illustrate them with the clinical material of Feulgen-stained epithelial cells from urinary bladder obtained by means of *bladder washing* technique in Medical University of Nijmegen, Holland. The material is used in our collective investigations on non-invasive, computer-aided detection of bladder cancer.

2. Artefacts in cytological smears

Let us define, for the purpose of this work, the objective of cytological investigations as measuring different morphological parameters of isolated nuclei found in a Feulgen-stained smear. The results of the measuring stage are usually used for successive statistical analysis and discrimination, but this is out of the scope of the paper.

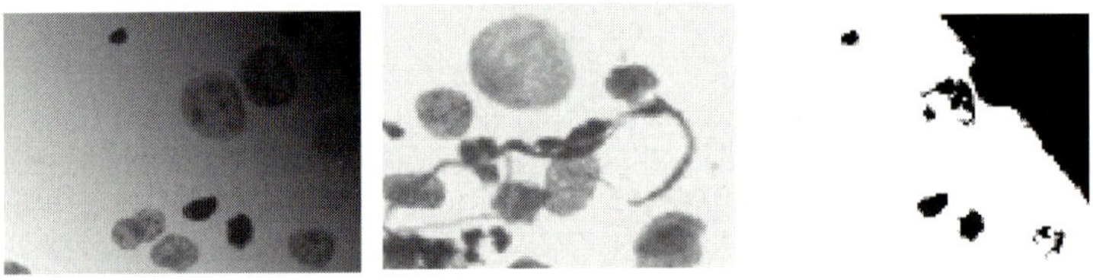

Fig. 1. Artefacts caused by different factors: non-uniform illumination of the scene, non-interesting objects appearing in a smear and an inadequate thresholding applied to an image.

(I) To get reliable measuring results it is expected that the visual appearance of nuclei are not affected by changeable physical factors of a system but only reflect their important biomedical features. To fulfil this requirement appropriate image correction and normalization algorithms should be applied. These are not covered in this paper but were the subject of our earlier works [5].
(II) It is necessary for correctness of the results that measured objects are isolated nuclei and only nuclei, not granulocytes or other biological or artificial objects encountered in a sample. Our method of shape-based discrimination between these interesting and non-interesting objects will be actually one of the main topics of this paper.
(III) As we are going to utilise shape features it is obvious that the objects should be properly extracted from the background. Otherwise, image processing algorithms used for objects extraction may themselves become a

source of artefacts. An efficient adaptive thresholding procedure built in the course of our investigations will be presented as it is crucial for preserving shapes of extracted objects.

To get an impression of the presence of artefacts in a smear a test was performed. 60 randomly chosen images from three smears of different persons were visually inspected by an expert who outlined each nucleus manually. Isolated nuclei size distribution was used to set up lower and upper limits for possible sizes of a single nucleus. Then, a rule for a human expert was established to classify artefacts in images. The order of artefacts identification steps is important as it simulates the way this algorithm is going to be implemented in a computer system. The script for checking off artefacts in a smear image is as follows:
1) mark all objects on the edge of an image as artefacts, then for the rest of objects
2) mark all objects smaller than nucleus lower size limit as artefacts, then for the rest of objects
3) mark all objects larger than nucleus upper size limit as artefacts, then for the rest of objects
4) mark all overlapping objects as artefacts, then for the rest of objects
5) mark each object that is not nucleus as artefact

By applying this artefact identification rule to each object or aggregate of objects, the isolated nuclei are found as those objects which are left unmarked. The overall results of this initial, interactively conducted experiment are very interesting. They are presented in the table (Fig.2). Let us summarize:
- 57% from all of 472 objects encountered in this particular material proved to be artefacts from the point of view of our study (not isolated nuclei),
- most of them (~41 %), three upper classes in the table, may be detected easily by means of simple and fast computer algorithms,
- the actual problem are overlapping objects (~16%) of overall area not exceeding acceptable sizes of a single nucleus.

Conclusions are straightforward. When implementing an automated cytological screening system, most efforts should be directed to detect overlapping objects and eliminate them from successive morphological and statistical analysis.

3. Adaptive thresholding of a smear image

As was stated earlier, to utilise shape features for discrimination it is a necessary condition that the objects were properly extracted from the background. Many of our images (dark objects on bright background) are characterised by a simple, bi-modal histograms. It is relatively easy to design an algorithm for

Total number of objects 472 (100%)	
Isolated nuclei	**Artefacts**
	Objects on the image edge thus not completely visible 146 (~ 31 %)
	Smaller than nucleus e.g. granulocytes 42 (~ 9 %)
203 (~ 43 %)	Larger than nucleus e.g. nuclear aggregates 4 (~ 0.8 %)
	Overlapping objects 75 (~ 16 %)
	Miscellaneous inclusions of the hard-to-define origin 2 (~ 0.4 %)

Fig. 2. Statistical summary of objects found in the smear.

finding proper global threshold for such cases. Unfortunately, real-life smears are not always so easy to analyze and global thresholding may lead to critical processing errors (see Fig.2.3). On the contrary, adaptive thresholding selects an individual threshold for each pixel based on the range of intensity values in its local neighborhood, allowing for thresholding of an image whose histogram doesn't contain distinctive peaks.

A general definition of a dynamic threshold $\mathbf{t_{xy}}$ that we are going to use can be written in as follows:

$$\mathbf{t_{xy}} = \mathrm{T}\ [\ \mathrm{x},\ \mathrm{y},\ \mathrm{f(x,y)},\ \mathrm{p(x,y)}\ ]$$

where $f(x,y)$ is the light intensity of point (x,y) in the original image, and $p(x,y)$ is some local property of this point. Several adaptive thresholding algorithms have been tested and the best one, adopted from Intel's Picture Processing Library (*Open Source licence*) [2], finally chosen. Let $f(x,y)$ be the input image. For every pixel (x,y) the mean $\mathbf{m_{xy}}$ and a measure of intensity variations in its neighborhood $\mathbf{v_{xy}}$ are calculated as follows:

where $\mathbf{p}$ is the half-size of pixel neighborhood. Local threshold for pixel (x,y) is computed as follows:
where $\mathbf{v_{min}}$ is some application specific minimum variance value.

To find optimal parameters for the algorithm an experiment was performed. Its idea is illustrated in Fig.3. An original and artificially degraded

$$m_{xy} = 1/(2p+1)^2 \sum_{s=-p}^{p} \sum_{t=-p}^{p} f(x+s, y+t)$$

$$v_{xy} = 1/(2p+1)^2 \sum_{s=-p}^{p} \sum_{t=-p}^{p} | f(x+s, y+t) - m_{xy} |$$

$$t_{xy} = m_{xy} + v_{xy} \quad \text{for } v_{xy} > v_{min}$$

and

$$t_{xy} = t_{xy-1} \quad \text{for } v_{xy} \leq v_{min}$$

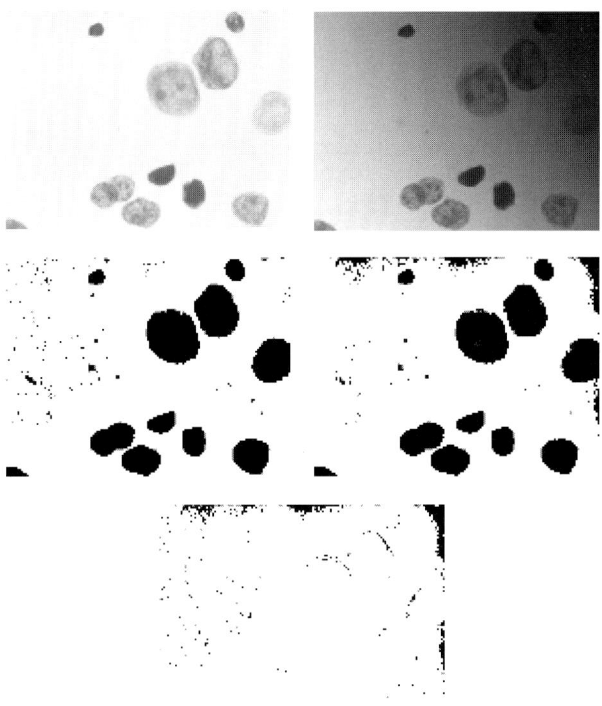

Fig. 3. Input images, thresholded binary images and their difference.

images were thresholded and subtracted to evaluate differences between them, especially differences between extracted regions of objects. Applying different values of parameters $\mathbf{p}$ and $\mathbf{v_{min}}$ for 30 randomly chosen images the best pair was found: $\mathbf{p=13}$, $\mathbf{v_{min}=4}$. The range of measured differences between extracted areas of the same objects not exceeded 4% of their sizes. Although indirect, it is a rather strong confirmation that our adaptive segmentation algorithm is strongly independent of background variations and preserves the shape of objects. The results of object extraction by means of adaptive, local thresholding give us a good base for successive shape-based analysis.

4. Shape-based artefacts detection

In this section we put a short description of the methods used to cope with those 16% of artefacts which can't be detected by means of simple methods based on object's position (on the edge) and size. As we have demonstrated experimentally, most of those "hard cases" are overlapping objects. After the im-

Fig. 4. Processing steps of the image with overlapping objects.

age processing steps and adaptive thresholding procedure we get the extracted contours in Fig.4 (object on the edge was also automatically removed).
An initial attempts using simple scalar shape descriptors (e.g. eccentricity, elongatedness, rectangularity, compactness [6]), although supported by multivariate discriminant analysis, were not promising, so abandoned. Nevertheless, visual inspection of objects in many thresholded images suggested applying of some advanced shape descriptors. Experience of the laboratory staff in two-dimensional spectral analysis directed us to *Elliptic Fourier Descriptors (EFD)* for shape-based discrimination between objects. Basically, our method does not make an *a priori* choice of the relevant features; it rather tries to automatically associate an importance degree. For that purpose the *Principal Component Analysis (PCA)* is used. The Fourier descriptors were defined in such a way, that they remain translation, rotation and scaling invariant [3,4]. They identify a shape, independent of its position, orientation or size. After Fourier transformation of chain-coded contours we get the feature vectors consisting of Fourier coefficients (20 harmonics are used). After that, the

PCA was applied to capture most of the essential relations from the data, creating new factors *(Principal Components)*. Five of them were supplied for final discriminant analysis in order to obtain satisfactory shapes separation results.

5. Experiments and results

The biological material used in this work consisted of urinary bladder epithelial cells. It was obtained by means of *bladder washing*. Only nuclei were visible in a smear. Image acquisition and processing were performed in a computer system equipped with a frame-grabber, CCD camera and moving stage optical microscope. For proper shape description of objects relatively large optical magnification had to be used (60x). Although results of this stage of the work were not intended to be implemented in a real-life screening system, the same material, consisting of 60 images (472 objects), as described earlier in section 2, was used to verify overall effects of the work. It may be very informative for final conclusions to compare an expert screening data in the table from section 2 (Fig.2) with the results of fully automated processing, equipped with advanced thresholding and shape-based object identification (Fig.5). What is really interesting and worth discussion it is ability of our shape-based analysis to identify artefacts in the form of overlapping objects. Assuming that the real number of overlaps in examined material was 75 (expert evaluation) the total *sensitivity* of advanced processing tools that were applied in the work may be reported to be high and equals to 92%. In other words, only 8% of overlaps were not identified as artefacts and were left as isolated nuclei for successive diagnostic steps. Examples of missed occlusions are shown in Fig.6. They are usually two or three clustered nuclei or granulocytes forming quite regular, nuclear shapes. Since we have concentrated on detecting and removing artefacts to reduce false-positive signals, *sensitivity* of the method was the most important parameter. Nevertheless, a question about nuclei misclassified as artefacts must not be left without a short discussion. Pathology, to be diagnosed, must be evident at some level. Therefore, it does not seem critical if some insignificant amount of nuclei are classified as artefacts and removed from analysis. However, to confirm that misclassification actually concerns inessential number of nuclei, *specificity* of the method has to be evaluated. 53, from the total number of 203 isolated nuclei were chosen randomly and went through the shape-based artefacts identification procedure. Three of them were identified as artefacts (Fig.7). An evaluation of specificity yields some 94%, which may be accepted as it implies insignificant number of misclassifications.

Total number of objects 472 (100%)	
Isolated nuclei	**Artefacts**
203 (~ 43 %) 209 (~ 44 %)	Objects on the image edge thus not completely visible 146 (~ 31 %) 146 (~ 31 %)
	Smaller than nucleus e.g. granulocytes 42 (~ 9 %) 42 (~ 9 %)
	Larger than nucleus e.g. nuclear aggregates 4 (~ 0.8 %) 4 (~ 0.8 %)
	Overlapping objects 75 (~ 16 %) 69 (~ 15 %)
	Miscellaneous inclusions of the hard-to-define origin 2 (~ 0.4 %) 2 (~ 0.4 %)

Fig. 5. Overall results of the automatic, computer artefacts identification (larger numbers) and comparison with the interactive, human-expert results (smaller numbers).

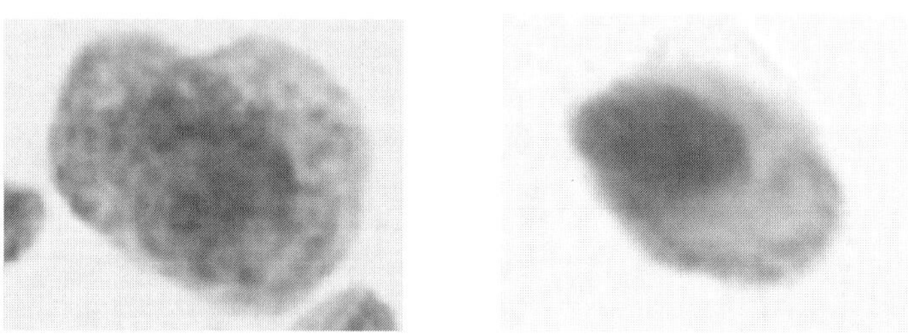

Fig. 6. Occlusions of two cytological objects forming regular, nuclear shapes.

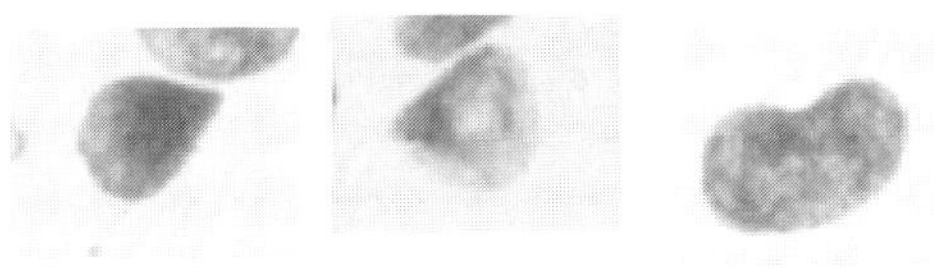

Fig. 7. Nuclei misclassified as artefacts by the shape-based analysis.

6. Conclusions

It was demonstrated that it is possible to efficiently detect artefacts in cytological smears by means of advanced shape analysis of properly extracted objects. Well known and powerful tools of *Fourier Shape Descriptors* and *Principal Component Analysis* have shown to be very useful in solving the problem. Sensitivity of our artefacts detection method yields some 92% and it seems we have reached, or are very close to, limits of shape-based approach to artefacts detection. The cases likely to be misclassified are usually clustered nuclei or granulocytes forming quite regular shapes. It is easy for the human visual system to recognize most of such occlusions, but it is extremely difficult to translate physiological algorithms into machine procedures. Fortunately, "difficult" does not imply "impossible". Detection of those "hardest cases" may be a challenge and direction for future work.

References

1. I. Al and J.S. Ploem, 1979, ŞDetection of suspicious cells and rejection of artefacts in cervical cytology using the Leyden Television Analysis SystemŤ, Journal of Histochemistry and Cytochemistry, Volume 27, Issue 1, pp. 629-634
2. INTEL Corporation, 2004, ŞOpen Source Computer VisionLibrary - OpenCVŤ, http://www.intel.com/research/mrl/research/opencv/
3. Iwata, H. and Y. Ukai, 2004, ŞSHAPE: A computer program package for quantitative evaluation of biological shapes based on elliptic Fourier descriptorsŤ, Journal of Heredity, in press
4. Wallace, T. P. and Wintz, P. A., 1980, An Efficient Three-Dimensional Aircraft Recognition Algorithm Using Normalised Fourier DescriptorsŤ, Computer Graphics and Image Processing, Vol. 13, 99-106
5. Dulewicz A., Pietka D., Jaszczak P., Nechay A., Sawicki W., Ko?mi?ska E., Borkowski A., 1998, Computer Analysis of Epithelium Cell Nuclei of Urinary Bladder for Cancer DetectionŤ, VIII Mediterranean Conference on Medical and Biological Engineering & Computing. MEDICON Ś98, Proceedings of the conference, Limassol, 14-17, June
6. M.Sonka, V.Hlavac, R.Boyle, 1998, ŞImage Processing, Analysis and Machine VisionŤ, PWS Brooks & Cole Publishing

Automatic Recognition of the Arterial Input Function in MRI Studies

Jacek Ruminski and Bartosz Karczewski

Department of Biomedical Engineering, Gdansk University of Technology, Narutowicza 11/12, 80-952 Gdansk, Poland {jwr,bart}@biomed.eti.pg.gda.pl

Summary. Quantitative perfusion imaging using Dynamic Susceptibility Contrast (DSC) MRI method requires to measure the Arterial Input Function (AIF) and deconvolve it from the measured tissue signal. We present a method for automatic recognition of the global AIF based on multistage algorithm. The method is validated using real world (clinically measured) DSC-MRI image series. Only 5% of all automatically generated AIFs (one series) were rejected by the expert. The method can be easily extended to produce a set of local AIFs and can be used as fully automatic or as an intelligent assistant tool for a neuroradiologist.[1]

1 Introduction

Parametric imaging become more and more popular. This includes DSC-MRI [1], ASL MRI [2], dynamic PET/SPECT [3], dynamic active thermography [4], etc. Parametric images represent values of reconstructed parameters for the assumed tissue/activity model. This extends the structural imaging towards the functional one. Qualitative parametric imaging could be extremely useful technique, however quantitative imaging could be even much more powerful, especially using the same modality as used for the structural imaging. This is a reason why DSC-MRI is an active area of research in the quantitative cerebral perfusion.

In the DSC-MRI brain studies, after injection of a bolus of the contrast agent (Gd-DTPA), a series of images is measured. This time-sequence data presents local voxel activity of the contrast (blood) flow and it's distribution. It is assumed, that measured MRI signal values are proportional to the contrast concentration. Contrast concentration as a function of time is measured for brain supported arteries. This function can be estimated as the arterial input function (AIF). Assuming ideal conditions this function should be an ideal impulse function, so measuring the output function (impulse response) one can

[1]This work was partly supported by the grant of Polish State Committee for Scientific Research (2003-2006) 4 T11E 042 25.

specify properties of the object under study, including mass flow, mass volume, and mean transfer time. Since AIF is not an ideal impulse function (dispersion and delay) and because in DSC-MRI measurements are done from volume of interest (VOI), deconvolution should be used to calculate VOI impulse response [5]

$$C_t(t) = \frac{\varrho}{Kh} \int_0^t C_a(\tau)(FR(t-\tau))d\tau, \tag{1}$$

where:
$C_a(t)$- contrast concentration in the artery (e.g., Middle Cerebral Artery) - Arterial Input Function AIF,
$C_t(t)$- contrast concentration in the tissue,
$\frac{\varrho}{Kh}$ - scaling factor (quantitative description) mean tissue density of a brain, ϱ=1.04 g/mol; Kh - hematocrit ratio (large to small arteries) Kh=(1-Hd)/(1-Hm); Hd=0.45; Hm=0.25;
$FR(t)$ - scaled impulse response (residue function) inside VOI,
$R(t)$ - represents fractional tissue concentration:

$$R(t) = 1 - H(t) = 1 - \int_0^t h(\tau)d\tau, \tag{2}$$

where:
$h(t)$ - a transport function - an impulse response (an ideal instantaneous unit bolus injection).
Distribution of transit times through the voxel; depends on the vascular structure and flow. The model is based on tracer kinetics for nondiffusable tracers - contrast material remains intravascular. Scaled impulse response could be calculated using Fourier transform (FFT) or matrix algebra (with matrix decomposition SVD to eliminate singularities). Since $R(t=0)$ should be equal to 1, then $FR(t) = F = CBF$(Cerebral Blood Flow).
Cerebral blood volume (proportional to the normalized total amount of tracer) can be calculated as

$$CBV = \frac{\int_0^\infty C_t(\tau)d\tau}{\frac{\varrho}{Kh}\int_0^\infty C_a(\tau)d\tau}. \tag{3}$$

Based on the central volume theorem, Mean Transit Time - MTT - (average time required for any given particle of tracer to pass through the tissue after an ideal bolus injection) can be estimated as

$$MTT = \frac{CBV}{CBF}. \tag{4}$$

Three types of quantitative parametric images (CBF, CBV, MTT), synthesized under strictly controlled procedure, offer additional and important information for brain studies.

2 Method

2.1 The AIF limitations

One of the most important task in DSC MRI perfusion imaging is appropriate extraction of the AIF. Based on the AIF the required signal ($FR(t)$) is deconvolved and used for quantitative maps synthesis. Theoretically the AIF describes concentration of the contrast agent in the feeding vessel to the VOI. Practically it could be localized far away from VOI (carotid artery, middle cerebral artery). The path between measured AIF source and true AIF localization is unknown. It can introduce the AIF delay and dispersion [6]

$$C_a^{true}(t) = C_a(t) \otimes h(t), \tag{5}$$

where:
h(t) - vascular transport function, e.g.:

$$h(t) = \frac{1}{t_D} exp(\frac{-t}{t_D}), \tag{6}$$

where:
t_D- dispersion constant.
The AIF delay and dispersion may be also described by absolute parameters: D - dispersion described by the Gauss distribution (standard deviation) in reference to the ideal impulse; t_d - delay time, equal to MTT (mean transit time between measured and real AIF location) for no dispersion; else

$$t_d = MTT \frac{D_{max} - D}{D_{max}}. \tag{7}$$

Delay and dispersion introduce problems in perfusion quantification - significant underestimation of CBF and overestimation of MTT. Correction of delay error may be done using bolus arrival time information. Delays of 1 to 2 seconds can introduce an approximately 40% underestimation of CBF and 60% overestimation of MTT [6]. Another problem with the AIF determination is reproducibility and dependence on the radiologist experience. The automatic determination of the AIF can offer an important improvement.

2.2 Preprocessing

The first step in the automatic AIF extraction is the preprocessing of the measured image time series. Excluding image enhancing procedures this step applies knowledge based decisions about possible AIF sources:
1. AIF candidate can not be localized outside the imaging object,
2. AIF candidate can not be localized in tissues with very limited vasculature

like bones, fluids, fat, etc.
Image segmentation is used in the masking of the possible AIF pixels. First, we calculate the difference image

$$I_D = I_0 - avg_{t_{min\mu}>t>t_0}(I_t), \tag{8}$$

where:
I_0 - an initial image at t=0;
$avg_{t_{min\mu}>t>t_0}(I_t)$- an average image calculated for images measured after Bolus Arrival Time (t_0) and before time calculated for the image with the minimum mean value ($t_{min\mu}$) in the image sequence (i.e., image with the highest concentration of the contrast agent in vessels). Practically we used up to 7 images counting down from $t_{min\mu}$ (for the sampling period - TR=1430 ms). The I_D image was thresholded to produce the binary mask. The threshold was set as a sum of the I_D mean value and standard deviation.

$$T = \mu_{I_D} + \delta_{I_D}. \tag{9}$$

In Fig. 1 illustration of the mask generation is presented.

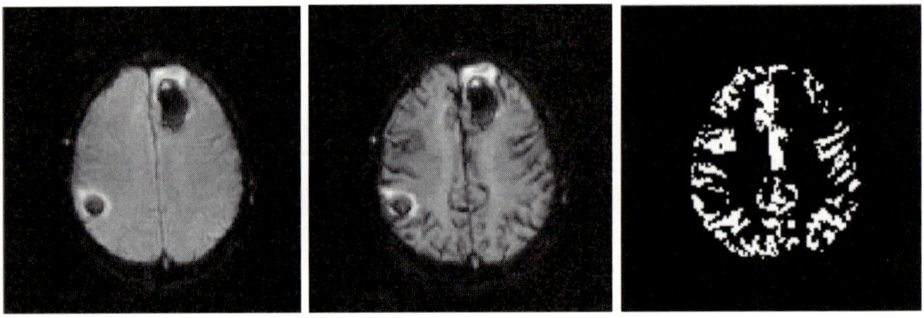

a) b) c)

Fig. 1. The AIF mask procedure illustration: a) - I_0 , b) I_t at $t_{min\mu}$, c) the final mask (in white - AIF candidates).

2.3 AIF description

The next step of the AIF extraction is description of the AIF signal. Descriptors formulation is based on the following set of conditions:
1. the BAT of the real AIF should be shortest (first contrast agent arrives to supplying arteries),
2. the AIF peak hight of the real AIF should be highest (an impulse estimate),
3. the AIF peak area of the real AIF should be smallest (an impulse estimate),

4. the AIF function should well correlate with the AIF model (Gamma - variate function [7]).
Different descriptors were evaluated. Based on high correlation with the expert decisions the chosen set of descriptors is $\{BAT, TTP, E\}$, where: TTP - Time To Peak (since Bolus Arrival Time, in seconds), E - the peak energy of the fitted AIF model. The peak energy is calculated as a sum of squared, normalized samples height. Normalization is performed using total peak area of the fitted AIF model (to reduce calculation time the peak area is evaluated starting with a sample for $t = BAT$).

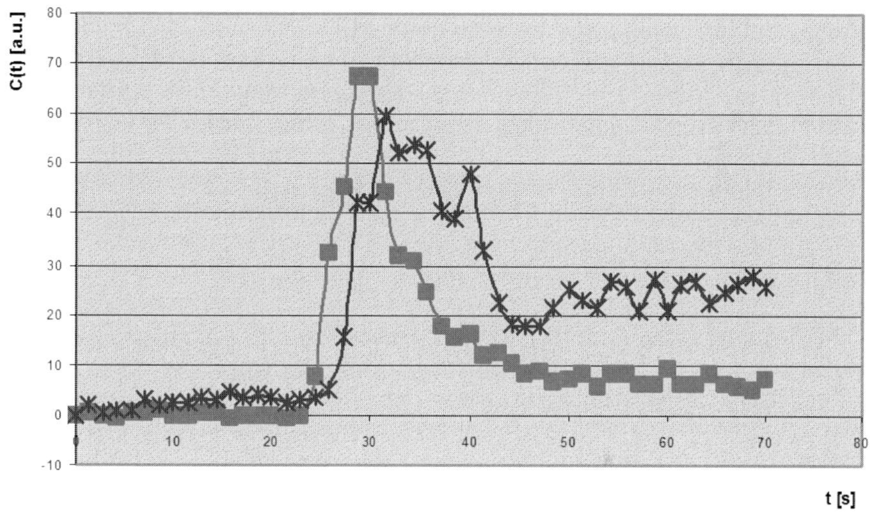

Fig. 2. Illustration of different concentration curves, candidates for the AIF.

Descriptors were calculated for real world data (i.e., clinically measured: 1.5T MRI SE-EPI with: 12 slices, 50 samples, TR=1.25-1.61s; TE=32-53ms; slice thickness 5-10 mm). Each image series was described by descriptor sets for AIF candidates and by the set of points - manually extracted AIFs. In our experiments we used 60 series (about 3000 images).

2.4 Knowledge extraction and decision rules

Manually extracted AIFs (10 series) were used as the reference for the two-class classification (AIFs, other signals). Using Mathematica software package (Wolfram Research) we discovered the mean cluster values for concentration curves descriptors for manually extracted AIFs $\{22.88, 5.72, 0.124\}$ and other $\{23.595, 8.10, 0.0956\}$. These means however cannot be used as universal reference, since MRI sequence parameters and procedures varies (e.g. TR, TE).

Additionally every patient is different, so absolute reference values are not useful. The discovered knowledge about means and distances of descriptor sets (feature vector) for manually extracted AIFs can be used to formulate metrics to weight AIF candidates. Taking into account that the AIF is searched for a one slice; based on the mined knowledge that the distribution of BAT of the manually extracted AIFs was within 3 seconds, we conclude that the first criterion of the automatic AIF detection is the elimination of all candidates with $BAT > (min(BAT)+3s)$. The second criterion is based on the proposed similarity measure. First each AIF candidate is labelled. Then two vectors are constructed: the firs one for Groups Of sorted E values (GOE); the second for Groups Of sorted TTP values (GOT). Unique identifier is assign for each group, so AIF labels point to those groups (in GOE and GOT). Identifiers for the GOE vectors are ordered numbers equal to $Ie = i * (S - 1)$, where $i = 1..N$, N - number of GOE, S - neighborhood range (in reported studies $S = 5$). The neighborhood range describes the influence of TTP on similarity scoring.

The complete algorithm, based on constructed descriptors, is following:
1. eliminate all candidates with $BAT > (min(BAT) + 3s)$
2. construct $GOE(S)$ and GOT
3. take the next group from GOE, take its identifier Ie
   for each label Li in the group Ie
   find location $It(It = 1M, M$ No of $GOT)$ of corresponding label Li in GOT
      $if(Ie - ((S-1)/2) \leq It \leq Ie, s[Li] = Ie - It$
      else $if(It > Ie), s[Li] = Ie + It$
      else $s[Li] = Ie - ((S-1)/2)$
   more groups in GOE - go to 3
4. sort s;
5. take K-nearest neighbors from s (or just a one AIF).

In the proposed algorithm the peak energy is dominating descriptor, because in the knowledge extraction step (i.e., described earlier clustering in relation to manually extracted AIFs treated as a golden standard) this descriptor was a leading one (higher influence/weight):

$$\frac{\delta_E}{\mu_E} < \frac{\delta_{TTP}}{\mu_{TTP}}, \tag{10}$$

where: μ - mean value, δ - standard deviation.
In reported studies the term from (11) was about 3 times lower than for TTP (e.g., $0.0514985 < 0.138846$).

3 Results and conclusion

The proposed method was applied for 20 series of images measured in the same conditions as series used in knowledge discovery part. After some post processing resulted AIFs were presented to be validated by the expert. Only in 1 case, the first proposed AIF, was rejected by the expert (which gives 5 % error - False Positives and Negative). Algorithms were implemented in Java (Sun JDK 1.5) and tested on Pentium IV PC (2.66GHz, 1GB RAM, Windows XP). Total time consumed by the implemented method to proceed was lower that 1s, fully acceptable. Further studies require to test the proposed method with data acquired using different type of the equipment (we tested SE-EPI for Siemens, and GE 1.5T MRI scanners). However, the method does not depend on sequence specific parameters (only on physiological/pathological). Another problem is the role of the global AIF: high sensitivity to dispersion (and delay, but it could be corrected using special algorithms). It could be interesting to create a set of regularly distributed local AIFs. The proposed method can be easily extended to do that, e.g. by successive elimination of s entries based on labels coordinates. The method is a part of the created software module for dynamic MRI data processing.

References

1. Calamante F, Gadian DG, Connelly A (2002) Quantification of Perfusion Using Bolus Tracking Magnetic Resonance Imaging in Stroke. Assumptions, Limitations, and Potential Implications for Clinical Use. Stroke 33:1146-1151
2. Wang J, Alsop DC, Li L, Listerud J, Gonzalez-At JB, Schnall MD, Detre JA (2002) Comparison of Quantitative Perfusion Imaging Using Arterial Spin Labeling at 1.5 and 4.0 Tesla. Magnetic Resonance in Medicine 48:242-254
3. Cai W, Feng DD, Fulton R (2000) Content based retrieval of dynamic PET func-tional images. IEEE Transactions on Information Technology in Biomedicine 4 (2)152-158
4. Ruminski J, Kaczmarek M, Nowakowski A (2001) Medical Active Thermography - A New Image Reconstruction Method. Lecture Notes in Computer Science LNCS2124 Springer, 274-181
5. Ostergaard L, Weisskoff RM, Chesler DA, Gyldensted C, Rosen BR (1996) High resolution measurement of cerebral blood flow using intravascular tracer bolus passages: I. Mathematical approach and statistical analysis, II. Experi-mental comparison and preliminary results. Magn. Reson. Med. 36 715-36
6. Calamante F, Gadian DG, Connelly A (2000) Delay and dispersion effects in dynamic susceptibility contrast MRI: simulations using singular value decomposition. Magn. Reson. Med. 44 466-73
7. Ruminski J, Bobek-Billewicz B (2004) Parametric imaging in Dynamic Susceptibility Contrast MRI - phantom and in vivo studies, Proc. of the 26th Int. Conference of the IEEE EMBS, CD-ROM edition, San Francisco

Mean Shift Segmentation, Genetic Algorithms and Support Vector Machines for Identification of Glaucoma in Fundus Eye Images

Katarzyna Stapor[1] and Adrian Brueckner[2]

[1] Institute of Computer Science, Silesian University of Technology, Gliwice, Poland delta@ivp.iinf.polsl.gliwice.pl
[2] Institute of Mathematics, Silesian University, Katowice, Poland

Summary. In this paper the new method for the automatic segmentation and classification of fundus eye images taken from classical fundus camera into normal and glaucomatous ones is proposed. The presented method consists of the following three stages: segmentation, feature selection, and classification. The mean sensitivity of the proposed method is 93%, while the mean specificity is 97%.

1 Introduction

Glaucoma is a group of ocular diseases characterized by the proceeding optic nerve neuropathy which leads to the rising diminution in vision field, ending with blindness. The optic disk structure (i.e. the exit of the optic nerve from the eye known as "blind spot" is comprised of a yellowish cup surrounded by a neuroretinal pink rim [6] (e.g. see Fig. 1a)). Glaucomatous changes in the retina appearance embrace various changes in the cup, as the result of nerve fibers damages. The spectrum of the cup damages ranges from the highly localized enlargements of the cup at the superior, or more commonly, inferior poles of the optic disk to the concentric ones. Searching for glaucoma damages during the routine examination is not an easy task and gives uncertain results even with the experienced ophthalmologist. The new methods of retina analysis based on a scanning-laser-tomography are expensive, accessible only in the specialized ophthalmic centers and do not result in a reliable diagnosis. That is why we have developed the new, objective and cheaper method that enables automatic classification of digital fundus eye images (FEI) taken from classical fundus-camera into normal and glaucomatous ones [10]. The method proposed in [10] relies on the fact that shape of the cup and its numerical characteristics correlate with glaucoma progress. In this paper, we present the improved version of the mentioned method, by using the new segmentation and classification algorithms that enable for better classification performance. The new method is composed of the following three main stages:

1. Mean shift segmentation of the cup region.
2. Selection of the cup features using genetic algorithms.
3. Classification of FEI using the support vector machine (SVM) classifier.

In the existing approaches to automatic segmentation of FEI for supporting glaucoma examinations [2, 7, 8, 9], researchers focused on the detection of the optic disk and its characteristics. The automatic extraction of the cup region from FEI was not the area of interest. Also, the automatic classification of FEI acquired from the fundus cameras into the normal and the glaucomatous ones has received no attention. We plan to integrate the proposed method into the classical fundus camera software to be used as a tool supporting glaucoma diagnosis in the routine examinations by ophthalmologists.

2 Methodology

2.1 Mean shift segmentation

The proposed cup segmentation is based on clustering by density estimation and mode seeking, i.e. clusters are identified by searching for regions of high density in feature space. Each mode is associated with a cluster center and each pattern is assigned to a cluster with a nearest center. We use a non-parametric estimator of density gradient, the mean shift with the associated iterative procedure of mode seeking originally proposed in [1]. The algorithm utilizes the multivariate kernel density estimate:

$$\hat{f}(x) = \frac{1}{Nh^d} \sum_{j=1}^{N} K(\frac{x - x_j}{h}) \tag{1}$$

with the Epanechnikov kernel which yields the minimum mean integrated square error:

$$K(y) = \begin{cases} \frac{1}{2} c_d^{-1} (d+2)(1 - yy^T) & \text{if } yy^T \leq 1 \\ 0 & \text{if } yy^T > 1 \end{cases} \tag{2}$$

The $\{x_j | j = 1, \ldots, N\}$ is an arbitrary set of N points in d-dimensional Euclidean feature space, h is the bandwidth of a kernel and c_d is a volume of a unit hypersphere. The estimate of the density gradient is defined as the gradient of the kernel density estimate:

$$\hat{\nabla} f(x) \equiv \nabla \hat{f}(x) = \frac{1}{Nh^d} \sum_{j=1}^{N} \nabla K(\frac{x - x_j}{h}) = \hat{f}(x) \frac{d+2}{h^2} M_h(x) \tag{3}$$

where:

$$M_h(x) = \frac{1}{|S_h(x)|} \sum_{x_i \in S_h(x)} (x_i - x) \tag{4}$$

is the sample mean shift in a neighborhood $S_h(x) = \{y : (y-x)^T(y-x) \leq h^2\}$ of a point x. Since mean shift vector always points toward the direction of the maximum increase in the density, it can define a path leading to a local density maximum (a mode). The mean shift procedure for each data point x_j is defined as:

$$x_j^{i+1} = x_j^i + \frac{h^2}{d+2} M_h(x_j^i) \tag{5}$$

We use the following stopping criterion:

$$max_{j=1,\ldots,N} \parallel x_j^{i+1} - x_j^i \parallel < \epsilon \tag{6}$$

The mean shift segmentation of data point set $\{x_j | j = 1, \ldots, N\}$ is defined in the following way:

1. For each $j = 1, \ldots, N$ run the mean shift procedure for x_j and store its convergence point in c_j
2. Identify clusters G_r for $r = 1, \ldots, c$ of convergence points by linking together all c_j which are closer than 0.2 from each other in the feature space
3. For each $j = 1, \ldots, N$ assign label r if $c_j \in G_r$

2.2 Feature selection using genetic algorithms

In our approach, 30 geometric features are computed on the extracted cup region. These are: different moment invariants, circularity coefficients, as well as some shape coefficients, like for example Haralick's, Feret's ones [11].

Genetic algorithms [3] are then used to select the most significant features characterizing the shape of cup region. A given feature subset is represented as a binary string (a chromosome) of length l, with a zero or one in position i, denoting the absence or presence of feature i in the set (l is the total number of available features). The initial population is generated in the following way: the number of 1's for each chromosome is generated randomly, then, the 1's are randomly scattered in the chromosome. Each chromosome is evaluated to determine its "fitness", which determines how likely the chromosome is to survive and breed into next generation. We propose the following fitness function:

$$Fitness = 10^4 accuracy + 0.4 zeros \tag{7}$$

where *accuracy* is the accuracy rate that the given subset of features achieves (i.e. the performance of a classifier on a given subset of features), *zeros* is the number of zeros in the chromosome. Reproduction is based on a random choice according to a fraction with repetitions method [3]. New chromosomes are created from old chromosomes by the process of crossover and mutation [3].

As a classifier we use SVM with Gaussian kernel (described in the next subsection). The accuracy of the SVM classifier on a given subset of features (i.e. chromosome) required for the calculation of the fitness function is measured as a generalization error G_e, calculated using the k-fold cross-validation method (k=10):

$$G_e = \frac{(TP + TN)}{(TP + TN + FP + FN)} \tag{8}$$

where TP — true-positive, FN — false-negative, TN — true-negative, FP — false-positive. The parameters we use in all experiments are as follows: 1) the length of each chromosome: 30, 2) the population size: 120, 3) the maximum number of generations: 500, 4) the cross-over rate: 0.6, 5) the mutation rate: 0.005. The best chromosome (i.e. the best feature subset) is the one which is the most frequent among the chromosomes in the last generation.

2.3 Support vector machines

Having a training set $S = (x_i, y_i, 1 \leq i \ \leq N)$ composed of the examples $x_i \in R^n$, each belonging to a class labeled by $y_i \in \{1, -1\}$, the goal of the SVM classifier [12] is to find the optimal separating hyperplane (OSH) — i.e the one which maximizes the separation margin which is a distance between the hyperplane and the closest data point. In the case when the data points are not linearly separable, a non-linear transformation $\phi(x)$ is used to map the data vector x into a higher dimensional space using a kernel function:

$$K(x_i, x_j) = \phi(x_i)\phi(x_j)) \tag{9}$$

In our experiment, a nonlinear SVM with a Gaussian radial basis kernel:

$$K(x, z) = \exp(-\gamma \cdot |x - z|^2) \tag{10}$$

where γ is a constant, was used. The problem of finding the OSH in general is equivalent to the maximization of the function:

$$W(\alpha) = \sum_{i=1}^{N} \alpha_i - \frac{1}{2} \sum_{i=1}^{N} \sum_{j=1}^{N} \alpha_i \alpha_j y_i y_j K(x_i, x_j) \tag{11}$$

subject to the constraints:

$$\sum_{i=1}^{N} y_i \alpha_i = 0, \ 0 \leq \alpha_i \leq C \tag{12}$$

where α_i are the N nonnegative Lagrange multipliers, C is a regularization parameter. Finally, the decision function for classifying a new data point x can be written as follows:

$$f(x) = \mathrm{sgn}(\sum_{i=1}^{Ns} y_i \alpha_i K(x_i, x) + b) \tag{13}$$

where Ns is the number of support vectors, α_i, b are constants, all determined through the numerical optimization during learning.

3 Experiments

3.1 Segmentation of the cup region

The data set used for this research consists of 100 digital fundus eye images of patients with glaucoma and 100 images of normal patients. These images are part of the data set acquired from the Department of Ophthalmology, Friedrich–Alexander–University Erlangen-Nuremberg, Prof. Dr George Michelson. To produce a "gold standard" segmentation, an ophthalmologist marked manually the boundary of the cup in each of the images. The 3-dimensional feature space (L, a, b) was used for clustering, i.e. each image pixel was described by three components of Lab color model. All features were normalized using z-score normalization [5]. To decrease the computational time, the cup segmentation described in section 2.1 was performed in a window, automatically computed based on the cup localization procedure described in [10]. We used the following values of the parameters: $h = 0.15, \epsilon = 0.01$. The cup in the segmented image was chosen as the region having the smallest value of a. Fig. 1b) presents the segmented image from the FEI shown in Fig. 1a) with the contour of the cup region imposed on it.

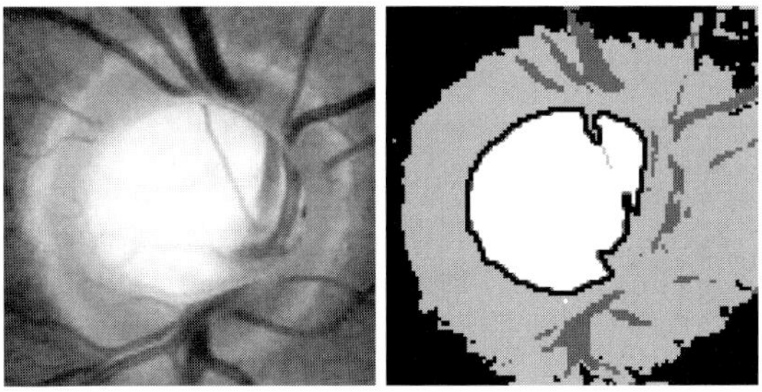

a) b)

Fig. 1. a) The automatically selected window from input FEI with the cup in the central part b) the segmentation result with the contour of the cup region imposed

3.2 Model selection and testing

The set of 200 segmented cup regions was divided into two disjoint subsets: 1) the training set: 150 images, 2) the testing set: 50 images. In each of those sets there were equal numbers of glaucomatous and normal cups. The training set was used for model selection: the suboptimal feature vector calculation based on genetic algorithms, setting SVM classifier parameters (performed by 10-fold cross-validation method) and final OSH learning. The feature selection described in subsection 2.2 was performed for different combinations of the classifier parameters C and γ. For each such combination we noted down the best subset of features with the corresponding value of the generalization error G_e. As the final subset of features we took the one with the smallest value of G_e:

$$v_o = (\phi_2, I_3, R_F) \tag{14}$$

where:

$$\phi_2 = (\eta_{20} + \eta_{02})^2 + 4\eta_{11}^2 \tag{15}$$

is Hu invariant moment, in which η_{20}, η_{02}, η_{11} are normalized central moments [5],

$$I_3 = \mu_{20}(\mu_{21}\mu_{03} - \mu_{12}^2) - \mu_{11}(\mu_{30}\mu_{03} - \mu_{21}\mu_{12}) + \mu_{02}(\mu_{30}\mu_{12} - \mu_{21}^2) \tag{16}$$

is compound, invariant moment,

$$R_F = \frac{L_h}{L_V} \tag{17}$$

is Feret coefficient, where:
L_h — the maximal diameter in the horizontal direction
L_V — the maximal diameter in the vertical direction.

The selected feature vector v_o corresponds to the combination of the classifier parameters: $C = 100, \gamma = 2.5$. Finally, the classifier was trained on the set composed of feature vectors v_o computed on the training set.

Classifier performance was tested on the feature vectors v_o calculated on the testing set. The following results were obtained: the mean sensitivity which is the percent of the correctly classified glaucomatous cases:

$$sensitivity = \frac{TP}{TP + FP} = 93\% \tag{18}$$

and the mean specificity which is the percent of the correctly classified normal cases:

$$specificity = \frac{TN}{TN + FN} = 97\% \tag{19}$$

4 Conclusions

In this work we demonstrated the improved method for the detection of the glaucomatous changes from 2D digital FEI taken from classical fundus camera. The proposed method enables automatic classification of digital FEI into normal and glaucomatous ones. The obtained classification results are encouraging. It is expected that the new method, after clinical tests, would support glaucoma diagnosis based on digital FEI obtained from fundus camera.

The Authors would like to thank prof. George Michelson from Department of Ophthalmology, Friedrich–Alexander–University Erlangen–Nuremberg for the possibility to access a database with fundus eye images, as well as for medical consultation.

References

1. Fukunaga K, Hostetler LD (1975) The estimation of the gradient of a density function with applications in pattern recognition. IEEE Trans. On Information Theory, 21(1), 32–40
2. Goh KG, Hsu W, Lee M, Wang H (2000) ADRIS: an automatic diabetic retinal image screening system. In: Cios KJ (ed) Medical Data Mining and Knowledge Discovery. Springer–Verlag, New York, 181–210
3. Goldberg D (1989) Genetic algorithms in search optimization and machine learning. Addison Wesley
4. Gonzalez RC, Woods R.E (2002) Digital image processing. Prentice-Hall, New Jersey
5. Jain AK, Dubes RC. (1988) Algorithms for clustering data. Prentice-Hall, Englewood Cliffs N.J.
6. Kanski J, McAllister JA, Salmon JF, Tarrant TR (1996) Glaucoma: a color manual of diagnosis and treatment. Butterworth–Heinemann Medical
7. Morris DT, Donnison C (1999) Identifying the neuroretinal rim boundary using dynamic contours. Image and Vision Computing. 17(3–4): 169–174
8. Pinz A, Prantl M, Datlinger P (1998) Mapping the human retina. IEEE Trans. Medical Imaging, 17(4): 210–215
9. Sinthanayothin C, Boyce J, Williamson CT (1999) Automated localization of the optic disk, fovea, and retinal blood vessels from digital colour fundus images. British Journal of Ophthalmology. 38(1): 902–910
10. Stapor K, Switonski A (2004) Automatic analysis of fundus eye images using mathematical morphology and neural networks for supporting glaucoma diagnosis. Machine Graphics & Vision, 13(1/2): 65–79
11. Trier O, Jain A, Taxt T (1996) Feature extraction methods for character recognition — a survey. Pattern Recognition. 29(4): 641–662
12. Vapnik V (1995) The nature of statistical learning theory. Springer Verlag, New York

Part V

SPEECH AND WORD RECOGNITION

Automatic Recognition and Verification of Voice Commands in Natural Language Given by the Operator of the Technological Device Using Artificial Neural Networks

Wojciech Kacalak and Maciej Majewski

Department of Mechanical Engineering, Technical University of Koszalin
Raclawicka 15-17, 75-620 Koszalin, Poland, {wk5, mmaj}@tu.koszalin.pl

1 Intelligent Two-Way Communication by Voice

Speech is the natural mode of communication for humans. It is a singularly efficient way for humans to express ideas and desires. Therefore, it is not surprising that we have always wanted to communicate with and command various technical devices by voice. Voice control is particularly appealing when the humanŠs hands or eyes are otherwise occupied [6].

According to the new conception, the intelligent layer of two-way voice communication of the technological device with the operator presented in Fig. 1, is equipped with the following intelligent mechanisms: operator identification, recognition of words and complex commands, command syntax analysis, command result analysis, command safety assessment, technological process supervision, and also operator reaction assessment [2,4,5].

If the operator is identified and authorized by the intelligent voice communication layer, a produced command in continuous speech is recognized by the speech recognition module and processed to the text format. Then the recognised text is analysed with the syntax analysis subsystem. The processed command is sent to the word and command recognition modules using artificial neural networks to recognise the command, which next is sent to the effect analysis subsystem for analysing the status corresponding to the hypothetical command execution, consecutively assessing the command correctness, estimating the process state and the technical safety, and also possibly signalling the possible error caused by the operator. Then the command is sent to the safety assessment subsystem for assessing the grade of affiliation of the command to the correct command category and making corrections. Next the command execution subsystem signalises commands accepted for executing, assessing reactions of the operator, defining new parameters of the process and run directives. The subsystem for voice communication produces voice commands to the operator [1].

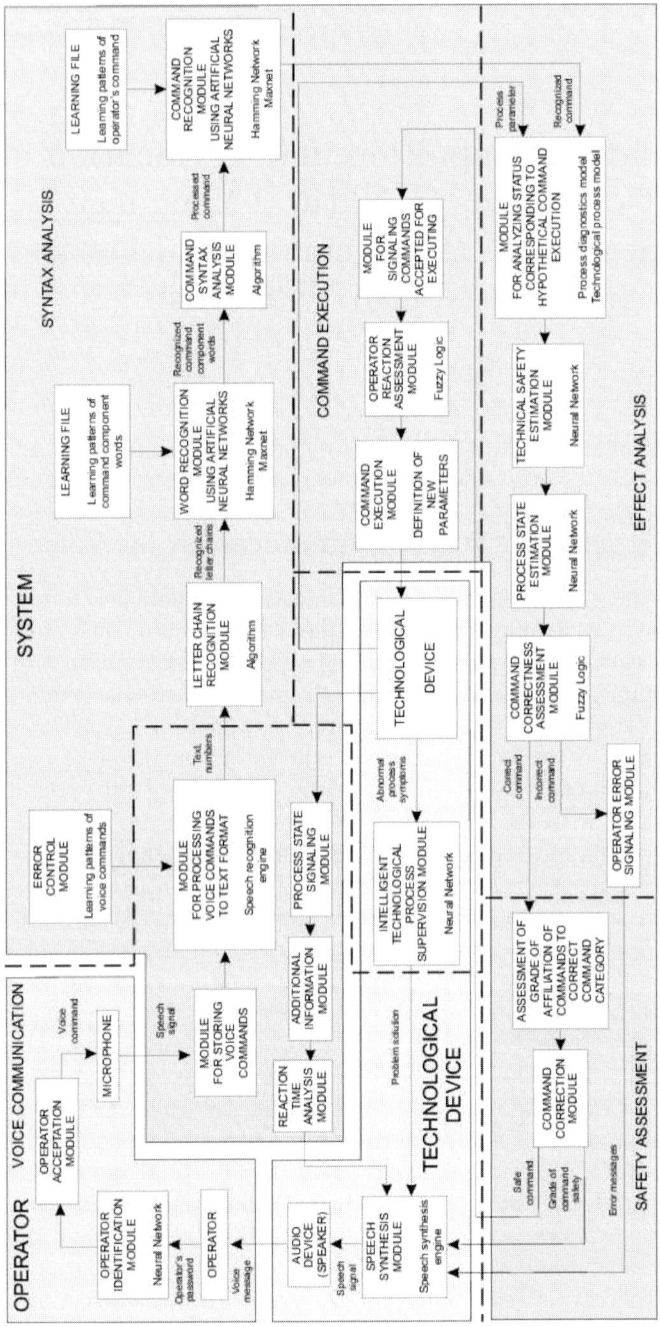

Fig. 1. Scheme of the intelligent layer of two-way voice communication of the technological device with the operator

2 Command Recognition and Safety Estimation

In the automatic command recognition system as shown in Fig. 2, the speech signal is processed to text and numeric values with the module for processing voice commands to text format using the speech recognition engine. The separated words of the text are the input signals of the neural network for recognizing words. The network has a training file containing word patterns. As the work result, the network recognizes words as the operator's command components, which are represented by its neurons. The recognized words are sent to the algorithm for coding words. Next the coded words are transferred to the command syntax analysis module. It is equipped with the algorithm for analyzing and indexing words. The module indexes words properly and then they are sent to the algorithm for coding commands. The commands are coded as vectors and they are input signals of the command recognition module using artificial neural network. The module uses the 3-layer Hamming neural network [7] either to recognize the operator's command or to produce the information that the command is not recognized. The neural network is equipped with a training file containing patterns of possible operator's commands.

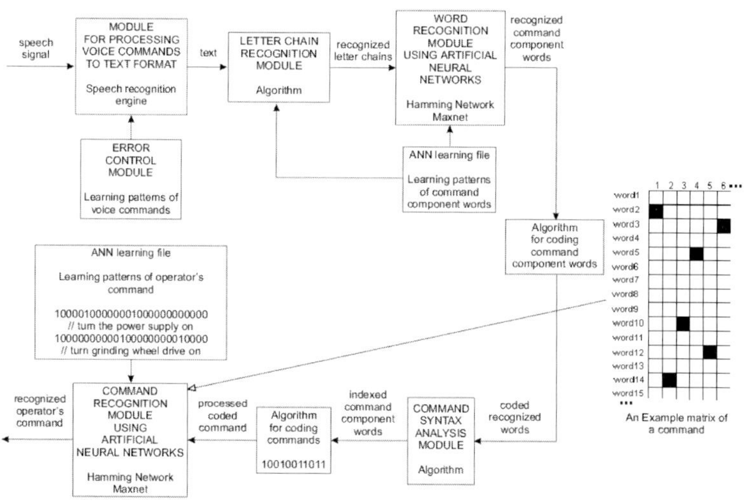

Fig. 2. Scheme of the automatic command recognition system

The recognised command given by the operator is processed and sent from the command syntax subsystem to the verification subsystems of effects and safety. The effect analysis module, shown in Fig. 3a, makes analysis of the recognised command. The technical safety of the technological device is checked by analysing the state of execution of the commands required to have been done as well as the commands to execute in next decisions. The process parameters to be modified by executing the command are checked and the allowable changes of the parameter values are determined. The analysis of the parameter values is based on the technological process features. The values of the parameter changes are the input signals of the neural network of the process state assessment system. The neurons of the neural network represent solutions to the diagnostics problem. The neural network also makes an estimation of the grade of safety of the recognised command. The system for checking the state of the automatic device for grinding of small ceramic elements that is shown in Fig. 3c, before executing next commands is presented in Fig. 3d. The technological safety assessment system, shown in Fig. 3b, is based on a neural network which is trained with the model of work of the technological device. New values of the process parameters are the input signals of the neural network. As the work result of the system, voice messages from the technological device to the operator about the possibility of executing of the command are produced [3].

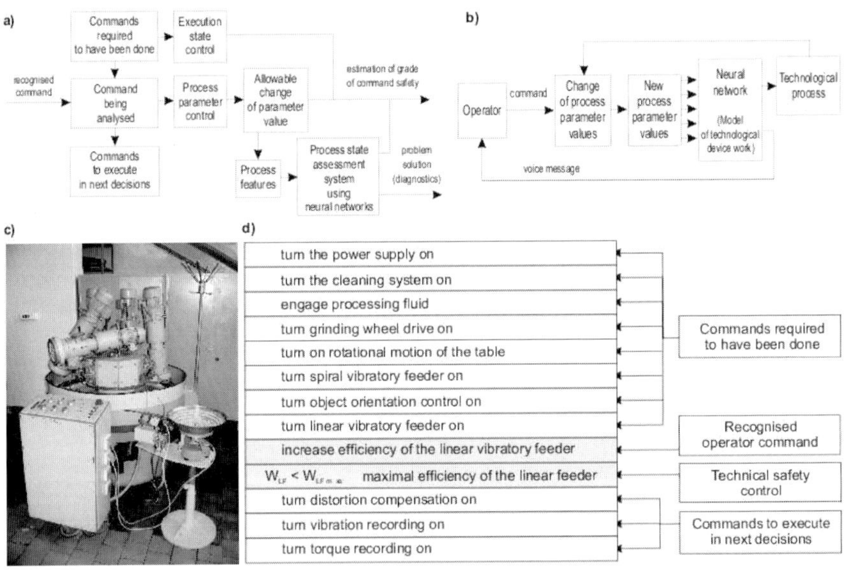

Fig. 3. Scheme of the command effect analysis and safety assessment system

There was an algorithm created for assessing the technological safety of commands. In Fig. 4, the lines present dependence of the force on the grinding process parameters for particular grinding wheels. Basing on the specified criteria, there is the grinding force limit determined for each grinding wheel. Basing on the grinding force limit, there is the table speed limit assigned. According to the operator's command, if the increase of the speed makes a speed of the table smaller than the smallest speed determined from the force limit for all the grinding wheels, then the command is safe to be executed.

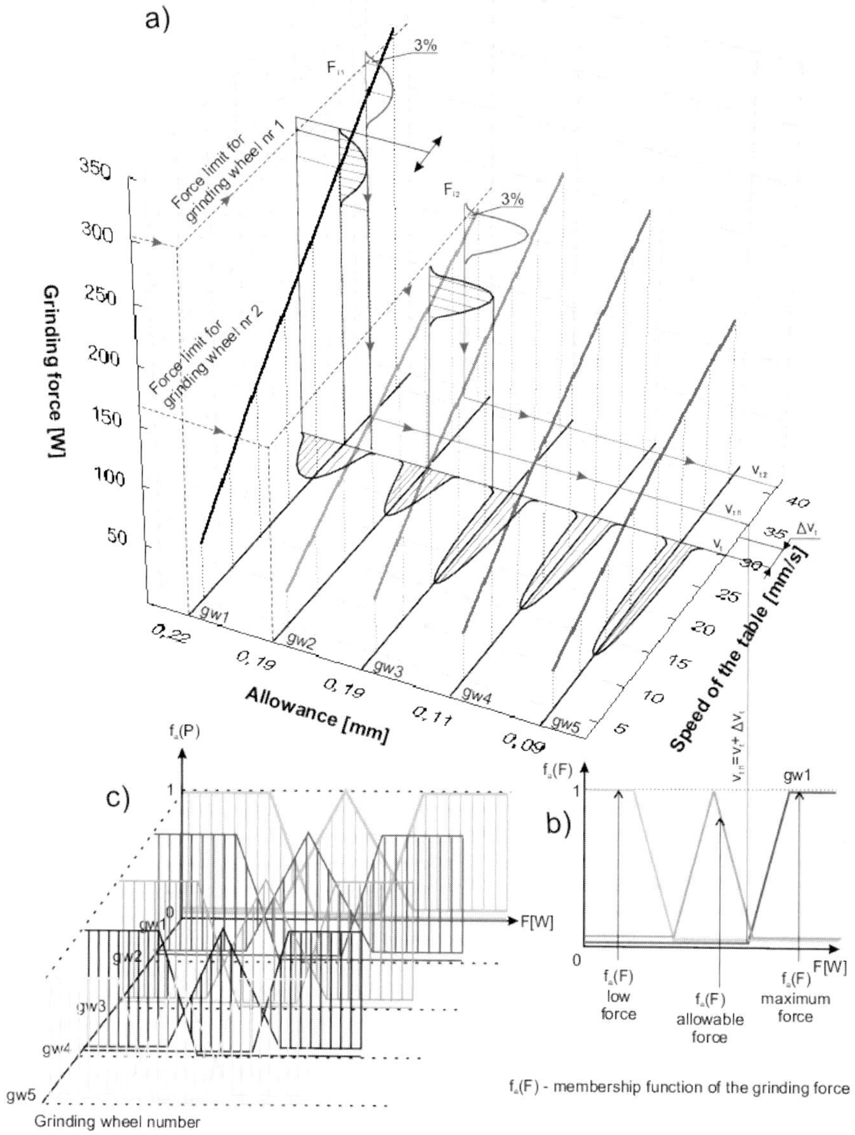

Fig. 4. Algorithm for assessing the technological safety of commands based on the real technological process

3 Research Results of Automatic Command Recognition

For the evaluation of research results of the automatic speech recognition, it has to be defined how to calculate the command recognition rate. The calculation is done after performing each case of recognition event. The recognition rate is calculated from the formula for the total number of errors and the error rate. The total number of errors is the sum of the insertion errors and the out-of-context errors. The error rate equals to the total number of errors divided by the total number of commands in a case.

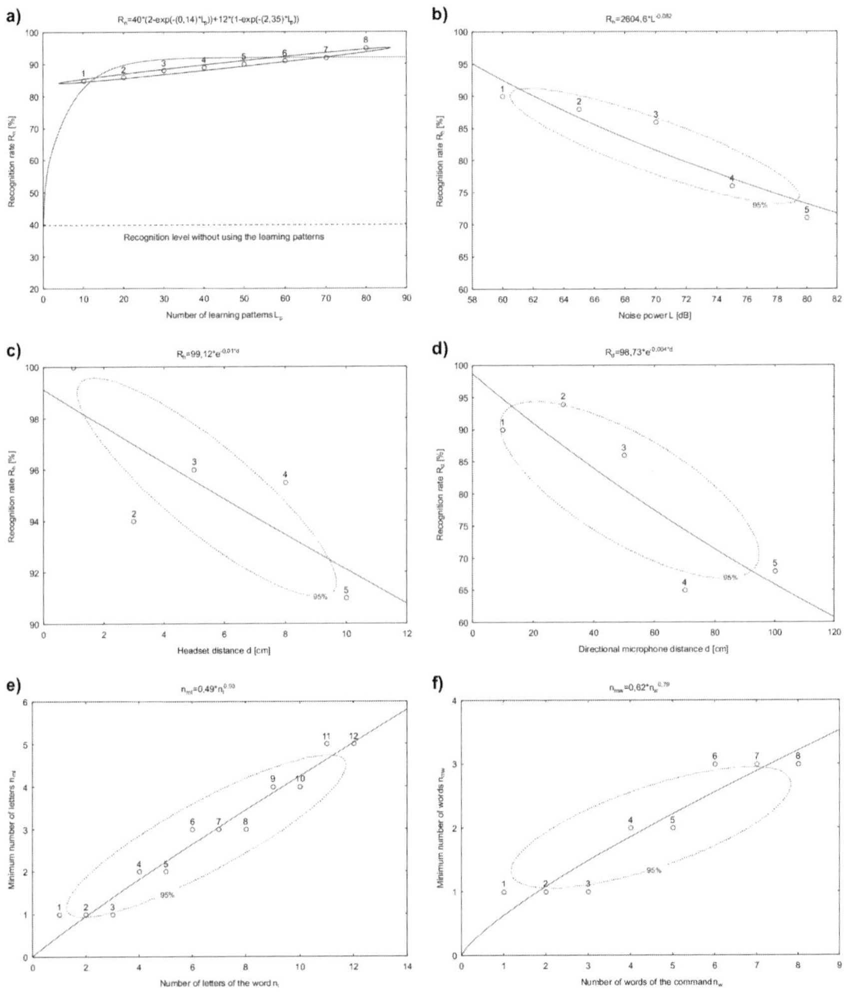

Fig. 5. Speech and command recognition rate

As shown in Fig. 5a, the speech recognition module recognizes 85-90% of the operator's words correctly. As more training of the neural networks is done, accuracy rises to around 95%.

For the research on command recognition at different noise power, the microphone used by the operator is the headset. As shown in Fig. 5b, the recognition performance is sensitive to background noise. The recognition rate is about 86% at 70 dB and 71% at 80 dB. Therefore, background noise must be limited while giving the commands.

For the research on command recognition at different microphone distances, the microphone used by the operator is the headset. As shown in Fig. 5c, the recognition rate decreases when the headset distance increases. The recognition rate has been dropped for 9% after the headset distance is changed from 1 to 10 cm. Also for the research on command recognition at different microphone distances, the microphone used by the operator is the directional microphone. As shown in Fig. 5d, the recognition rate after 50 cm decreases reaching rate about 65%.

As shown in Fig. 5e, the ability of the neural network to recognise the word depends on the number of letters. The neural network requires the minimal number of letters of the word being recognized as its input signals. As shown in Fig. 5f, the ability of the neural network to recognise the command depends on the number of command component words. Depending on the number of component words of the command, the neural network requires the minimal number of words of the given command as its input signals.

The command recognition module using Hamming Maxnet neural networks is capable of recognizing different commands of the same meaning in natural language. The ability of the 3-layer neural network to learn to recognise commands depends on the number of learning patterns of possible operator commands. The specified number of the patterns enables the network to learn and work efficiently and quickly. Based on the research, it could be said that the fewer patterns the neural network is trained with, the faster it works and learns.

4 Conclusions and Perspectives

In the future, voice messages in natural language will undoubtedly be the most important way of communication between humans and machines. Great progress is made in many fields of science, where communication between the technological devices and the operator is an important task, e.g. motorization, road traffic, etc. The condition of the effectiveness of the presented intelligent two-way voice communication system between the technological device and the operator is to equip it with mechanisms of command verification and correctness. In the automated processes of production, the condition for safe communication between the operator and the technological device is analyzing the state of the technological device and the process before the command is given and using artificial intelligence for assessment of the technological effects and safety of the command. In operations of the automated technological processes, many process states and various commands from the operator to the technological device can be distinguished. A large number of combined technological systems characterize the realization of that process. In complex technological processes, if many parameters are controlled, the operator is not able to analyze a sufficient number of signals and react by manual operations on control buttons. The research aiming at developing an intelligent layer of two-way voice communication is very difficult, but the prognosis of the technology development and its first use shows a great significance in efficiency of supervision and production humanization.

Acknowledgements: The support of the Polish State Committee for Scientific Research (KBN grant 5 T07D 036 24) is highly appreciated.

References

1. Kacalak, W., Majewski, M.: Automatic recognition and safety estimation of voice commands in natural language given by the operator of the technical device using artificial neural networks, Proceedings of the ANNIE 2004 Conference, Artificial Neural Networks in Engineering ANNIE 2004, Vol. 14: Smart Engineering Systems Design, 7-10 November 2004, St. Louis, ASME Press, New York 2004, 831-836.
2. Kacalak, W., Majewski, M.: Intelligent Layer of Two-Way Voice Communication of the Technological Device with the Operator, ICAISC2004 7-th International Conference on Artificial Intelligence and Soft Computing , Zakopane 7-11 June 2004, Lectures Notes in Artificial Intelligence 3070, Subseries of Lecture Notes in Computer Science, Springer-Verlag 2004, 610-615.
3. Kacalak, W., Majewski, M.: Selected problems of effect analysis and safety assessment of commands given by the operator of the technological device using artificial neural networks, International Industrial Simulation Conference ISC2004, 7-9 June 2004, Malaga, Spain, Eurosis Ghent 2004, 35-39.
4. Kacalak, W., Majewski, M.: Intelligent two-sided voice communication system between the machining system and the operator, Proceedings of the ANNIE 2003 Conference, Artificial Neural Networks in Engineering ANNIE 2003, Vol. 13: Smart Engineering System Design, St. Louis 1-4 November 2003, ASME Press, New York 2003, 969-974.
5. Kacalak, W., Majewski, M.: Supervising of technological process using two-sided voice communication between the machining system and operator, Modern Trends in Manufacturing CAMT2003, Wroclaw 20-21 February 2003, Wroclaw 2003, 175-182.
6. O'Shaughnessy, D.: Speech Communications: Human and Machine, IEEE Press, New York 2000.
7. Principe, J. C., Euliano, N. R.: Lefebvre W. C., Neural and Adaptive Systems: Fundamentals through Simulations, John Wiley and Sons, Inc., New York 2000.

Recognition of Isolated Words of the Polish Sign Language

Tomasz Kapuscinski[1] and Marian Wysocki[2]

[1] Rzeszow University of Technology, Computer and Control Engineering Chair
tomekkap@prz-rzeszow.pl
[2] Rzeszow University of Technology, Computer and Control Engineering Chair
mwysocki@prz-rzeszow.pl

Summary. The paper considers recognition of isolated words of the Polish Sign Language using a canonical stereo system that observes the signer from a frontal view. Recognition is based on human skin detection and Hidden Markov Models. Several feature vectors taking into account information about the hand shape and 3D position of the hand with respect to the face are examined. To improve the recognition rate the classifiers are combined by voting or by fuzzy integral. We focus on 101 words that can be used at the doctor's and at the post office.

Introduction

Sign language is the natural language of the deaf people. It is a visual language, different from the spoken language, but serving the same function. It is not a universal language. Regionally different languages have been evolved. The Polish Sign Language (PSL) is an adaptation of the sign language used in our country to the rules of the native language [4].

Inability to use spoken language considerably complicates life of the deaf people. The aim of sign language recognition is to provide an efficient and accurate mechanism to transcribe sign language into text or speech so that communication between deaf and hearing society could be more convenient. Furthermore, sign language recognition generally serves as a good basis for the development of gestural human-machine interfaces.

Automatic gesture recognition has recently acquired much attention. Good overviews about such systems can be found in [8]. There exist two main approaches to collecting data for the classification process: instrumented glove-based data collection and video-based data collection. The video-based approach leads to a more natural interface, although some video-based recognition systems require the signer to wear coloured cotton gloves.

One of the first video-based systems was presented by Tamura and Kawasaki [12]. Recognition was performed using a hierarchical analysis of parameters of a gesture such as: position of the hand at the beginning and at the end of the gesture, the direction of the hand motion and the hand shape classified to one of two defined types. The system recognized twenty one-handed words of the Japanese Sign Language. Approximately 45% of them were recognized correctly. Similar hierarchical approach was presented by Charyaphan and Marble, who considered interpretation of the American Sign Language (ASL) [1]. The classification was based on the sequence of three tests: the stop hand position, simple shape of the trajectory, and the shape of the hand at stop position. The proposed technique successfully classified a test sample of 31 ASL signs. Grobel and Assam [3] recognized isolated signs of German Sign Language (GSL) collected from video recordings of a signer wearing coloured cotton gloves. 91.3% accuracy out of a 262 sign vocabulary was reported. The authors used Hidden Markov Models (HMM). Starner et al. [10] presented a video-based system for real-time recognition of ASL sentences. They employed a single video camera as part of two different setups. The first system observed the signer from a frontal view (desk mounted camera), while the second system used a cap mounted camera for image recording. The vocabulary included 40 signs. The system was tested on a corpus of 94 sentences for the desk-based system and 100 sentences for the cap-based system. Recognition accuracy ranged between 74.5% and 97.8% depending on the camera position and the grammar used.

This paper considers recognition of 101 words of PSL using a stereovision-based system. Our earlier article [5] refers to a smaller word set and a single camera system. To our knowledge it is the first approach related to PSL recognition. The reverse issue, i.e. translation of written (spoken) sentences into PSL using graphical animation was considered in [11]. We use a canonical colour camera system observing the signer from a frontal view. The signer is not required to wear any colour gloves but he/she should wear long-sleeved clothes. The clothes and the background should be of different colours from the skin. Recognition is based on human skin detection and HMM. Several feature vectors taking into account information on the hand shape and 3D position of the hand with respect to the face are examined. To improve the recognition rate the classifiers are combined by voting or by fuzzy integral.

1 Characteristics of PSL

In PSL, similarly to other sign languages, a sign is the equivalent of the word. Every sign can be analysed by specifying at least three components: (i) the place of the body against which the sign is made, (ii) the shape of a hand or hands, (iii) the movement of a hand or hands. Although in practical sign language communication some additional features (such as lip shape, etc.) are often used, we do not consider them in this paper.

We focus on 101 words that can be used at the doctor's and at the post office. Fig. 1. presents starting and final phases of the gestures denoting *temperature* and *nurse* [4].

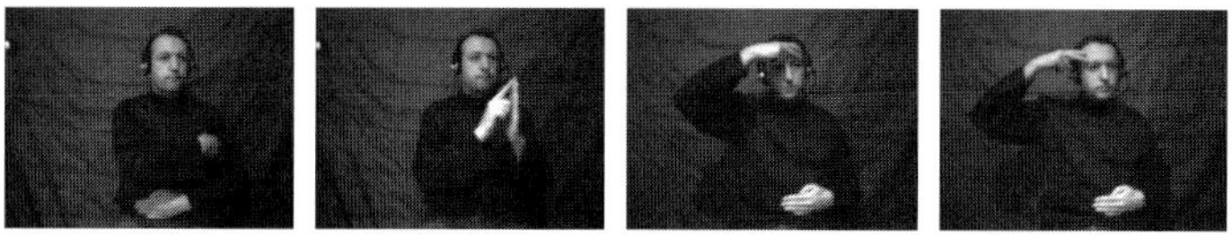

Fig. 1. Sample gestures of PSL

The considered 101 gestures are either static or dynamic. Most of them are two-handed. For one-handed signs the so-called dominant hand performs the sign, whereas for two-handed signs the second hand, the non-dominant hand is also used. The non-dominant hand is often still, but in some gestures both hands move. The hands often touch each other or appear against the background of the face. The motion can be single or repeated.

2 Construction of the Feature Vectors

The following problems are important in our recognition task: (i) determinig and tracking the position of the hands and the face of a signer, (ii) feature selection and determining.

To identify regions of the hands and the face we used a histogram-based chrominance model of human skin in the normalized RGB space. We assumed that the signer was facing the front of the camera and was not changing his/her distance and orientation with respect to the camera. In order to ensure correct segmentation there were some restrictions for the clothing of the signer and the background, particularly, other people should not appear in the background.

Areas of objects toned in skin colour, their centres of gravity and ranges of motion were analysed to recognize the right hand, the left hand and the face. Comparison of neighbouring frames helped to notice whether the hands (the hand and the face) touched or partially covered each other.

The following twelve features were computed (see also fig. 2): (1) the length l_r of the line segment connecting gravity centres of the right hand and the face, (2) the orientation φ_r of the aforementioned line, (3) the area S_r of the right hand, (4) compactness γ_r of the right hand, (5) eccentricity ε_r [13] of the right hand, (6) difference z_r between average depth of the face and the right hand, (7) – (12) corresponding parameters for the left hand. The parameters l and φ characterise the position of the hands in the picture, the S, γ and ε represent the hand shape. We notice that for a circle the parameters γ and ε are equal to 1 and 0, respectively. For an ellipse they are functions of its aspect ratio. The parameters z contain the information about the depth. In our system the so called sparse disparity map was computed by a correlation method [6]. Before, the monochromatic pictures were rectified, i.e. transformed to the form that would be obtained in the ideal canonical stereo system, and the LOG filtering was performed. We used 15 feature vectors, denoted here with the numbers 1 – 15. All vectors contain the elements l and φ, vectors 2 – 15 include the information about the hand shape (see tab. 1), vectors 9 – 15 are constructed as the vectors 2 – 8, respectively, with additional information about the depth.

Table 1. Information about the hand shape

feature vector	2, 9	3, 10	4, 11	5, 12	6, 13	7, 14	8, 15
elements used	S	γ	ε	S, γ	S, ε	γ, ε	S, γ, ε

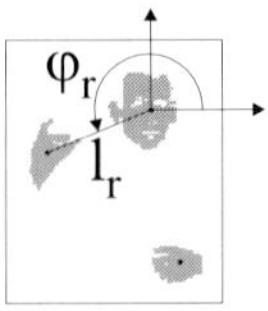

Fig. 2. Elements of the feature vector describing position of the hand

3 Hidden Markov Models

A Hidden Markov Model is a statistical model used to characterize the statistical properties of a signal [9]. An HMM consists of two stochastic processes: one is an unobservable Markov chain with a finite number of states, an initial state probability distribution and a state transition probability matrix; the other is a set of probability density functions associated with observations generated by each state.

Human hand gestures are spatiotemporal entities. The performance of the gesture is usually not perfect. The same gesture changes in time and space even if one person performs it twice. Human performance involves two distinct stochastic processes: human mental states and resultant actions. The mental state is immeasurable, the action is measurable. Therefore many researches use HMM to recognize hand gestures [3, 8, 10].

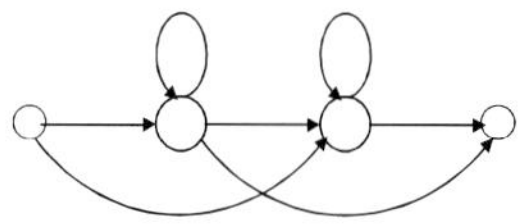

Fig. 3. A HMM with two emitting states and the non-emitting start and end state

4 Experiments

In our experiments the HTK software [14], originally prepared for speech recognition with HMM, was used. The models had the form shown in fig. 3. Two states in the model generate observation (the emitting states) and two additional start and end states do not generate any observation (the non-emitting states) The start and end state facilitate construction of composite models e.g. models for recognition of sentences on the basis of models of isolated words [5, 14]. Continuous output probability distributions were assumed to be mixtures of two Gaussians.

Colour stereo head of the Videre Design was used in our system. Recognition was performed in real time. We carried out experiments using a vocabulary of 101 words and three data sets A, B and B'. Each set consisted of 20 realizations of each word, registered as sequences of images with the resolution of 320*240 pixels and the frequency of 25 frames/s. Signs in the sets A and B were performed in good lighting conditions by signers SA and SB, respectively. The set B' was prepared in worse lighting conditions by the signer SB. The SB is a PSL teacher, the SA has learnt PSL for the purpose of this research. Each set was divided into two separate subsets, i.e. a training set and a test set, both with 10 realizations of each word. We will use the suffix tr for training and te for testing further on in this paper. Thus, for instance, Atr denotes the training set obtained from the data set A and Ate the test set obtained from the data set A. Table 2 shows the recognition rates on test sets for each of 15 feature vectors defined in section 2.
The columns A, B, B' refer to the recognition results on the Ate, Bte, $B'te$, respectively, with HMMs trained on the corresponding training sets. The column A/B is for Bte and the models trained on Atr. B/A has similar interpretation. Finally AB/A and AB/B are for testing of models trained on a composition of Atr and Btr. We also considered the cross-validation on the sets A and B. Each of them was divided into four independent subsets of the equal size. Then, three subsets were used to train the HMM, and the remaining one to test them. This process was repeated for each of four possible choices and the test recognition rates were averaged for all four results and shown in the columns CVA and CVB.

Information about the motion of the hand towards the direction parallel to the optical axis of the camera is partially included in the area of the hand seen in successive images (compare fig. 4a and 4d). But when detection of the skin is imprecise (as the result of worse lighting conditions) the area of binary

Table 2. Recognition accuracy on test sets [%]

feature vector	CVA	A	CVB	B	B'	A/B	AB/B	B/A	AB/A
1	90.1	83.0	90.8	84.0	73.0	81.8	83.9	82.7	83.5
2	91.6	85.0	92.2	85.3	74.6	82.9	84.9	85.0	85.2
3	91.8	86.0	92.4	87.0	76.0	85.2	86.6	85.0	86.5
4	92.0	84.0	92.5	85.2	74.6	83.3	84.7	84.9	85.0
5	92.3	87.0	93.1	87.8	76.0	85.5	87.0	86.0	86.9
6	93.0	87.0	93.6	88.1	77.2	86.0	87.7	86.0	87.3
7	93.1	85.0	93.5	85.7	75.3	82.8	85.2	83,3	85.7
8	93.1	88.0	93.8	88.5	77.4	85.5	88.1	87.0	87.6
9	92.8	87.0	93.5	88.5	86.7	84.8	87.8	85.4	87.2
10	93.4	85.0	93.8	86.8	85.7	84.6	86.6	84.8	86.2
11	92.2	87.0	92.8	88.3	85.9	85.1	87.7	86.0	87.6
12	93.1	88.0	93.7	88.4	86.9	84.9	87.8	86.7	87.5
13	93.4	88.0	93.9	88.5	87.8	86.2	88.2	87.0	87.9
14	93.3	87.0	93.9	88.2	86.9	86.3	88.1	86.3	87.7
15	93.5	87.0	94.1	88.0	87.0	85.3	87.3	86.0	86.7

object representing the hand does not reflect that kind of motion correctly (see fig. 4b, 4e). Then, additional information about the depth, based on disparity maps obtained using stereovision, can improve recognition significantly. This is evident from fig. 4c and 4f. Similar situations brought about the fact that feature vectors 1 – 8 turned out significantly worse in recognition on the set B'.

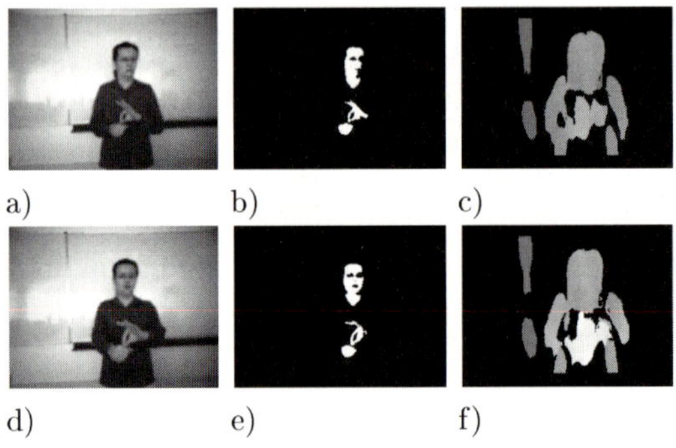

Fig. 4. The gesture *referral* registered in worse lighting conditions a), b), c) – in turn: monochromatic image, binary image, and disparity map, at the beginning of the gesture, c), d), e) – similar images at the end phase

Gestures were performed in the ways that were characteristics for the signers' individual manners. We used models trained on one person's gestures for recognition of signs presented by other signer. As one should expect the results turned out worse than those obtained with the models trained on gestures of the same person whose gestures were tested. Combining the training sets improved the results.

We examined recognition rates depending on feature vectors used in construction of classifiers. Global results presented in tab. 2 show rather slight differences. A prior closer look at the recognition of particular words let us state that different feature vectors may be best for different words. Comparison of the columns (c) and (e) in tab. 3 shows it evidently.

Table 3. Recognition rates of various classifiers on the test sets [%]

test set	indiv. worst	indiv. best	class. 1	glob. best	voting	50% +1	Borda	averaging	fuzzy integral
(a)	(b)	(c)	(d)	(e)	(f)	(g)	(h)	(i)	(j)
Ate	67.0	96.0	83.0	88.0	92.5	89.5	89.7	90.7	92.1
Bte	67.0	96.9	83.2	88.5	92.5	89.1	89.2	90.5	92.2
$B'te$	50.4	97.9	73.0	87.8	91.5	87.0	63.5	63.2	91.2

(b): each word was recognized using a feature vector that turned out worst for that word, (c): each word was recognized using a feature vector that turned out best for that word, (d): classifier used the feature vector 1, (e): classifier used the feature vector that turned out best for the whole test set, (f) - (j): classifiers combined, in turn, by: majority vote rule, absolute majority rule, the Borda count, averaging, the fuzzy integral with the degree of importance of each classifier equal 0.75.

Furthermore, looking at the columns (d), (b) we can note that the simplest feature vector 1 does not have to be the worst for each word. Taking this into account we also examined a few fusions of classifiers based on different feature vectors. Combination of classifiers by: majority vote rule, absolute majority rule, the Borda count, averaging, and the fuzzy integral [2] are presented in columns (f) – (j). The last two methods required normalization of outputs of the HMMs. The majority rule and the fuzzy integral turned out best in the experiments.

5 Conclusions and Future Work

In this paper an HMM-based PSL recognition system was introduced. We considered vocabulary of 101 words used in typical situations at the doctor's and at the post office. Each sign was modelled by a single HMM. The proposed feature vectors were composed of features taking into account information about the hand shape and 3D position of the hand with respect to the face, determined on the basis of sequences obtained in a stereovision system. Recognition with classifiers based on different feature vectors as well as with fusions of those classifiers were discussed. We used the data set of 6060 sequences prepared by two signers in various lighting conditions.

Information about the hand posture taken into consideration in this paper was quite rough and it may, therefore, be insufficient in some situations, e.g. when spelling a name with the finger alphabet. Some results related to recognition of the Polish finger alphabet were reported in [7]. This research will be continued in the future and it is planned to consider simple sentences recognition as well. Preliminary experiments were presented in [5].

Acknowledgement

The work was supported by the State Committee for Scientific Research under grant 4T11C 006 25.

References

1. Charayaphan C, Marble A E (1992) Image Processing System for Interpreting Motion in American Sign Language. J. Biomed. Eng, 14:419–425
2. Cho S B, Kim J H (1995) Multiple Network Fusion Using Fuzzy Logic. IEEE Trans. On Neural Networks, 6, 2:497–501
3. Grobel K, Assam M (1997) Isolated Sign Language Recognition Using Hiden Markov Models. Proc. of the IEEE Int. Conf. on SMC, Orlando:162–167
4. Hendzel J K (1997) Dictionary of the Polish Sign Language. OFFER Press, Olsztyn (in Polish)
5. Kapuscinski T, Wysocki M (2003) Vision-Based Recognition of Polish Sign Language. Proceedings of the Symposium on Methods of Artificial Intelligence AI-METH, Gliwice, Poland:145–148
6. Konolige K (1997) Small Vision System: Hardware and Implementation. Proc. of the Int. Symp. on Robotic Research, Hayama, Japan:111-116
7. Marnik J, Wysocki M (2004) Hand Posture Recognition Using Mathematical Morphology. Archives of Theoretical and Applied Informatics, 16, 4:279-293
8. Pavlovic V I, Sharma R, Huang T S (1997) Visual Interpretation of Hand Gestrures for Human-Computer Interaction: A Review. IEEE Trans PAMI, 19, 7:677–693
9. Rabiner L R (1989) A Tutorial on Hidden Markov Models and Selected Applications in Speech Recognition. Proc. of the IEEE, 77, 2:257–286
10. Starner T, Weaver J, Pentland A (1998) Real-time American Sign Language Recognition Using Desk and Wearable Computer Based Video. IEEE Trans. PAMI, 20, 12:1371–1375
11. Suszczanska N, Szmal P, Francik J (2002) Translation Polish Text into Sign Language in the TGT System. Proc. of the 20th IASTED International Multiconference Applied Informatics, Insbruck:282–287
12. Tamura S, Kawasaki S (1988) Recognition of Sign Language Motion Images. Pattern Recognition, 21, 4:343–353
13. Theodoridis S, Koutroumbas K (1999) Pattern Recognition. Acad. Press, New York
14. Young S et al. (2000) The HTK Book. Microsoft Corporation

Simple Measure of Typewriter Prints Quality

Jacek Lebiedź

Gdańsk University of Technology, Faculty of Electronics, Telecommunications and Informatics, ul. G. Narutowicza 11/12, 80–952 Gdańsk, Poland
jacekl@eti.pg.gda.pl

Summary. The paper presents author's method of shape evaluation adapted to quality estimation of characters printed by typewriter. The shown method is based on statistical analysis of the Maximal Square Map (MSM) described in detail in the paper. This method has been elaborated for quality evaluation of computer aided information retrieval from archival machine typed paper documents.

1 Introduction

The far-reaching goal of the international project Memorial (5FP EU Grant IST–2001–33441–MEMORIAL) [5] is to enable creation of virtual archives based on documents existing in libraries, archives, museums, and public record offices. The current goal of the Memorial project is a computer aided information retrieval from machine typed paper documents from former Nazi concentration camp museums realized in cooperation with State Museum Stutthof in Sztutowo near Gdańsk. To achieve these goals the estimation of processing quality in recognition of machine typed paper documents is needed. Loss of quality can appear at different stages of information retrieval [3]. The presented method permits evaluating a quality of scanned characters after preprocessing (background elimination) and segmentation (text separation). According to systematization in [3] it belongs to the category GM2 of goodness measures.

New measure of typewriter prints quality results in remark that letters and numerals originated as handwritten marks. Therefore their shapes have the form of lines that have constant width corresponding with thickness of the pen. Traditional typewriter characters have also the form of lines. Any deviation from this form means loss of letter quality. Hence dispersion or variance of hypothetical pen width (fiber width) may serve as a measure of quality of restored characters. A pen path length (fiber length) may be used as an additional criterion.

Calculations of the pen (fiber) parameters may seem to be complex (e.g. because of expected need of constructing a skeleton). Fortunately some simple transformation (map) described below permits evaluating of them.

It is proper to add that in [2] I discussed the well-known classical shape measures [1, 4] and I showed that they are insufficient for this application.

2 Maximal Square Map

Consider a map, where each pixel is assigned a value that is the diameter (or the radius) of maximal disk containing the pixel and belonging to the figure entirely. This map counts for every pixel a size of the wider pen that can draw the given pixel. Because of calculation complexity it would be better to change disks for squares. The results should be similar.

A map, where every pixel is assigned a value that is the side of maximal square containing the pixel and belonging to the figure entirely, is called the Maximal Square Map (MSM). It was introduced by the author in [2]. Figure 1 presents an example of the MSM applied to some restored letters.

	2	2	2											3	3	3							
	2	2	2											3	3	3							
		2	2	2										3	3	3							
		2	2	2		2	2	2	2	1					2	2			2	2	2	2	
		2	2			2	2	2	2					1	1		2	2	2	2	2	2	
			3	3	3	3	2		1						2	2	2	2					
			3	3	3	3									2	2	2	2					
		1	3	3	3	3									1			2	2				
			1			1	1								1			2	2				
			1				2	2						2	2	2			2	2			
	2	2	2	2	2		2	2	2	2			1	2	2	2	1		2	2	2	2	
	2	2	2	2	2		2	2	2	2				1							2	2	

Fig. 1. The MSM calculated for two restored letters **k**

The MSM can be obtained in a similar way like the Euclidean Distance Map (EDM). The MSM procedure implemented by the author uses three passes. The first pass calculates for each pixel minimal its distance (i.e. number of pixels) from the background for all directions between the left and the top (Fig. 2). Figure 3 shows a way of calculation for the first step of the three-pass MSM algorithm. The second pass calculates horizontal segments of the maximal squares stretching pixel values from the first step to the left (Fig. 4).

The last pass calculates vertical segments of the maximal squares stretching pixel values from the second step to the top and as a consequence it finds out whole maximal squares (Fig. 5).

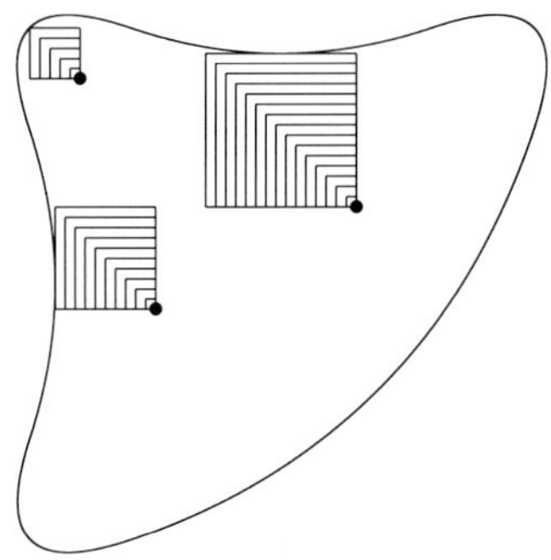

Fig. 2. Geometrical interpretation of minimal pixel distance (i.e. number of pixels) from the background for all directions between the left and the top (in three pixels)

a)

```
// foreach(x, y)
//   pixel(x, y) = ((x, y) ∈ shape)?1 : 0;
foreach(x, y) in ascending order
  if(pixel(x, y) ≠ 0)
    pixel(x, y) = min(pixel(x-1, y-1),
                      pixel(x-1, y), pixel(x, y-1)) + 1;
```

b) $y \downarrow$

		1	1	1			
		1	2	2			
		1	2	3			
	1	1	2	3	1	1	
	1	2	2	3	2	2	
	1	2	3	3			
	1	2	3	4	1		

Fig. 3. The first pass of the three-pass MSM algorithm (**a**) and an exemplary result of its executing (**b**)

Author hopes to reduce presented algorithm to two passes (like the EDM algorithm [4]).

3 Dispersion of the MSM as a criterion

For a shape like typewriter print that has a form of a line with constant width the MSM should produce almost the same values for most pixels. Dispersion (or variance) of the MSM should be close to zero for shapes like that. Therefore

a)

foreach(x, y) in descending order
  if(pixel$(x, y) \neq 0$)
    pixel$(x, y) = \max_{k \in \mathrm{N} \cup \{0\}}$(pixel$(x{+}k, y)$:
      $k <$ pixel$(x{+}k, y)$);

b) $y \downarrow$

		1	1	1			
		2	2	2			
		3	3	3			
	1	3	3	3	1	1	
	2	3	3	3	2	2	
	3	3	3	3			
	4	4	4	4	1		

Fig. 4. The second pass of the three-pass MSM algorithm (**a**) and an exemplary result of its executing (**b**)

a)

foreach(x, y) in descending order
  if(pixel$(x, y) \neq 0$)
    pixel$(x, y) = \max_{k \in \mathrm{N} \cup \{0\}}$(pixel$(x, y{+}k)$:
      $k <$ pixel$(x, y{+}k)$);

b) $y \downarrow$

		3	3	3			
		3	3	3			
		3	3	3			
	4	4	4	4	2	2	
	4	4	4	4	2	2	
	4	4	4	4			
	4	4	4	4	1		

Fig. 5. The third pass of the three-pass MSM algorithm (**a**) and an exemplary result of its executing (**b**)

it can be used as a criterion of typewriter restored prints quality. The ratio of the dispersion to the mean value of the MSM give us better criterion because it is independent of a line width. Line length, calculated as the ratio of the number of pixels (figure area) to the squared mean value of the MSM (area of averaged square), can be treated as an additional criterion.

Tests have shown that the criterion based on the MSM works in practice. The experiments were performed with archival machine typed documents scanned with the resolution 100 dpi, 200 dpi, and 300 dpi. Because the experiment results were analogous, there are enclosed only results for the smallest resolution where distortion is the most noticeable. Figures 6–8 show three characters sets apiece: manual designed model characters (upper rows) and real typewriter restored prints subjectively arranged in two visual qualities – rather good (middle rows) and rather poor (lower rows). Tables 1–3 present described parameters of the letters and numerals from Figs. 6–8. All model characters have dispersion equal to zero. They are perfect in the sense of proposed measure. Generally speaking, this measure agrees also with subjective character classification in respect of quality. The ratio of the dispersion to the mean value of the MSM for typewriter restored prints with better quality is mostly less than this ratio for characters with poorer quality. The average ratio for all characters from sets of rather good quality is 0.194 (for upper case: 0.205, for lower case: 0.199, for numerals: 0.159), whereas the average ratio for all signs of rather poor quality is greater and it is equal to 0.268 (for upper case: 0.281, for lower case: 0.260, for numerals: 0.256). Note also that some more complicated letters (like M or W) were restored with relative poorer quality and they have relative greater ratio of the dispersion to the mean value of the MSM.

Fig. 6. Upper case: manual designed model letters (upper row) and real typewriter restored prints subjectively arranged in two visual qualities – rather good (middle row) and rather poor (lower row)

Table 1. Parameters of the upper case letters from Fig. 6
(n – number of pixels, m – mean value of the MSM, σ – dispersion of the MSM)

letters	model quality					good quality					poor quality				
	n	m	σ	$\frac{\sigma}{m}$	$\frac{n}{m^2}$	n	m	σ	$\frac{\sigma}{m}$	$\frac{n}{m^2}$	n	m	σ	$\frac{\sigma}{m}$	$\frac{n}{m^2}$
A	56	2	0	**0**	14	54	2.093	0.482	**0.230**	12.332	62	2.387	0.656	**0.275**	10.881
B	69	2	0	**0**	17.25	60	1.833	0.373	**0.203**	17.851	59	1.898	0.630	**0.332**	16.373
C	52	2	0	**0**	13	53	1.943	0.231	**0.119**	14.033	34	1.471	0.499	**0.339**	15.722
D	64	2	0	**0**	16	62	1.968	0.177	**0.090**	16.012	47	1.723	0.447	**0.260**	15.824
E	74	2	0	**0**	18.5	72	2.042	0.455	**0.223**	17.273	58	1.690	0.463	**0.274**	20.316
F	58	2	0	**0**	14.5	58	1.741	0.438	**0.251**	19.127	59	1.712	0.453	**0.265**	20.133
G	60	2	0	**0**	15	58	2.069	0.583	**0.282**	13.549	56	2.054	0.610	**0.297**	13.279
H	72	2	0	**0**	18	59	1.644	0.479	**0.291**	21.828	65	1.677	0.468	**0.279**	23.115
I	40	2	0	**0**	10	46	1.957	0.204	**0.104**	12.017	37	1.811	0.392	**0.216**	11.284
J	52	2	0	**0**	13	48	1.958	0.200	**0.102**	12.516	49	1.959	0.198	**0.101**	12.766
K	74	2	0	**0**	18.5	62	1.758	0.428	**0.244**	20.060	64	1.875	0.625	**0.333**	18.204
L	50	2	0	**0**	12.5	48	1.854	0.353	**0.190**	13.962	42	1.714	0.452	**0.264**	14.292
M	86	2	0	**0**	21.5	86	2.000	0.591	**0.295**	21.500	79	2.266	0.853	**0.376**	15.388
N	74	2	0	**0**	18.5	79	2.316	0.586	**0.253**	14.722	71	1.944	0.669	**0.344**	18.794
O	56	2	0	**0**	14	57	1.982	0.131	**0.066**	14.503	56	2.143	0.610	**0.285**	12.196
P	62	2	0	**0**	15.5	60	1.850	0.357	**0.193**	17.531	54	1.741	0.438	**0.252**	17.821
R	72	2	0	**0**	18	60	1.750	0.433	**0.247**	19.592	58	1.914	0.624	**0.326**	15.836
S	56	2	0	**0**	14	61	2.049	0.493	**0.241**	14.527	53	1.679	0.467	**0.278**	18.795
T	56	2	0	**0**	14	54	1.833	0.373	**0.203**	16.066	48	1.667	0.471	**0.283**	17.280
U	60	2	0	**0**	15	55	1.945	0.227	**0.117**	14.532	61	2.098	0.432	**0.206**	13.854
W	74	2	0	**0**	18.5	60	1.900	0.700	**0.368**	16.620	64	2.063	0.704	**0.341**	15.045
Z	60	2	0	**0**	15	56	1.857	0.350	**0.188**	16.237	61	2.279	0.576	**0.253**	11.748

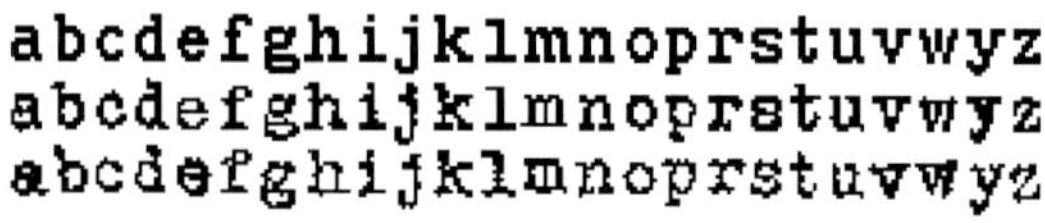

Fig. 7. Lower case: manual designed model letters (upper row) and real typewriter restored prints subjectively arranged in two visual qualities – rather good (middle row) and rather poor (lower row)

Table 2. Parameters of the lower case letters from Fig. 7
(n – number of pixels, m – mean value of the MSM, σ – dispersion of the MSM)

letters	model quality					good quality					poor quality				
	n	m	σ	$\frac{\sigma}{m}$	$\frac{n}{m^2}$	n	m	σ	$\frac{\sigma}{m}$	$\frac{n}{m^2}$	n	m	σ	$\frac{\sigma}{m}$	$\frac{n}{m^2}$
a	48	2	0	**0**	12	48	1.833	0.373	**0.203**	14.281	54	2.407	0.562	**0.233**	9.317
b	52	2	0	**0**	13	50	1.940	0.237	**0.122**	13.285	49	1.959	0.198	**0.101**	12.766
c	34	2	0	**0**	8.5	37	1.919	0.273	**0.142**	10.048	30	1.600	0.490	**0.306**	11.719
d	54	2	0	**0**	13.5	55	1.927	0.260	**0.135**	14.807	46	1.717	0.450	**0.262**	15.596
e	46	2	0	**0**	11.5	35	1.371	0.483	**0.352**	18.609	55	1.891	0.312	**0.165**	15.382
f	46	2	0	**0**	11.5	41	1.780	0.414	**0.232**	12.933	38	1.526	0.499	**0.327**	16.312
g	74	2	0	**0**	18.5	60	1.733	0.442	**0.255**	19.970	63	1.683	0.465	**0.277**	22.254
h	62	2	0	**0**	15.5	60	2.067	0.478	**0.231**	14.048	44	1.523	0.499	**0.328**	18.976
i	34	2	0	**0**	8.5	38	2.158	0.539	**0.250**	8.161	32	1.844	0.363	**0.197**	9.413
j	42	2	0	**0**	10.5	46	1.891	0.311	**0.165**	12.860	37	1.649	0.477	**0.290**	13.613
k	61	2	0	**0**	15.25	62	2.081	0.548	**0.263**	14.322	54	2.037	0.543	**0.267**	13.014
l	40	2	0	**0**	10	37	1.919	0.273	**0.142**	10.048	44	2.045	0.601	**0.294**	10.517
m	71	2	0	**0**	17.75	52	1.615	0.487	**0.301**	19.927	59	1.847	0.360	**0.195**	17.286
n	52	2	0	**0**	13	47	1.979	0.144	**0.073**	12.004	42	1.571	0.495	**0.315**	17.008
o	40	2	0	**0**	10	41	2.000	0.000	**0.000**	10.250	33	1.576	0.494	**0.314**	13.290
p	55	2	0	**0**	13.75	46	1.717	0.450	**0.262**	15.596	45	1.444	0.497	**0.344**	21.568
r	43	2	0	**0**	10.75	44	1.955	0.208	**0.107**	11.518	38	1.711	0.454	**0.265**	12.987
s	50	2	0	**0**	12.5	53	1.943	0.231	**0.119**	14.033	42	1.762	0.426	**0.242**	13.530
t	42	2	0	**0**	10.5	44	1.864	0.343	**0.184**	12.669	34	1.853	0.354	**0.191**	9.903
u	46	2	0	**0**	11.5	47	2.085	0.539	**0.259**	10.810	37	1.730	0.444	**0.257**	12.366
v	40	2	0	**0**	10	41	2.122	0.550	**0.259**	9.106	44	2.114	0.532	**0.251**	9.849
w	50	2	0	**0**	12.5	49	1.857	0.350	**0.188**	14.207	55	2.018	0.556	**0.275**	13.503
y	54	2	0	**0**	13.5	54	2.611	0.951	**0.364**	7.920	47	1.766	0.423	**0.240**	15.071
z	43	2	0	**0**	10.75	41	1.878	0.327	**0.174**	11.624	36	1.583	0.493	**0.311**	14.360

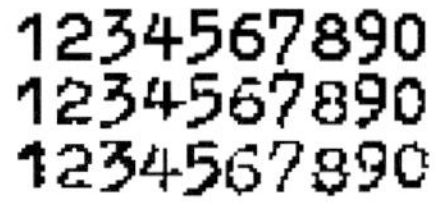

Fig. 8. Numerals: manual designed model digits (upper row) and real typewriter restored prints subjectively arranged in two visual qualities – rather good (middle row) and rather poor (lower row)

Table 3. Parameters of the numerals from Fig. 8
(n – number of pixels, m – mean value of the MSM, σ – dispersion of the MSM)

digits	model quality					good quality					poor quality				
	n	m	σ	$\frac{\sigma}{m}$	$\frac{n}{m^2}$	n	m	σ	$\frac{\sigma}{m}$	$\frac{n}{m^2}$	n	m	σ	$\frac{\sigma}{m}$	$\frac{n}{m^2}$
1	34	2	0	**0**	8.5	37	2.216	0.473	**0.213**	7.533	42	2.286	0.452	**0.198**	8.039
2	61	2	0	**0**	15.25	54	1.889	0.314	**0.166**	15.135	45	1.689	0.463	**0.274**	15.776
3	54	2	0	**0**	13.5	54	1.907	0.290	**0.152**	14.842	56	2.036	0.533	**0.262**	13.513
4	50	2	0	**0**	12.5	46	1.891	0.311	**0.165**	12.860	39	1.538	0.499	**0.324**	16.477
5	60	2	0	**0**	15	55	1.927	0.260	**0.135**	14.807	60	1.933	0.249	**0.129**	16.052
6	62	2	0	**0**	15.5	53	1.792	0.406	**0.226**	16.496	46	1.609	0.488	**0.303**	17.775
7	49	2	0	**0**	12.25	48	1.958	0.200	**0.102**	12.516	35	1.543	0.498	**0.323**	14.703
8	70	2	0	**0**	17.5	54	1.852	0.355	**0.192**	15.746	57	1.754	0.430	**0.245**	18.519
9	61	2	0	**0**	15.25	61	1.918	0.274	**0.143**	16.581	50	1.860	0.347	**0.187**	14.453
0	56	2	0	**0**	14	54	1.963	0.189	**0.096**	14.014	45	1.578	0.494	**0.313**	18.077

4 Conclusions

The efficient criterion for evaluation of typewriter prints quality is proposed. It is based on the new MSM transformation given in the paper in detailed description. Shown test results of the new criterion reveal its correctness.

References

1. Choraś R. S. (2003) Object Recognition Based on Shape, Texture and Color Information. Proceedings of the 3rd Conference on Computer Recognition Systems KOSYR'2003, Wrocław University of Technology, Wrocław 181–186
2. Lebiedź J. (2004) Shape Similarity to Alphanumeric Sign. Proceedings of the 2nd International Conference on Computer Vision and Graphics ICCVG'2004
3. Lebiedź J., Podgórski A., Szwoch M. (2003) Quality Evaluation of Computer Aided Information Retrieval from Machine Typed Papers Documents. Proceedings of the 3rd Conference on Computer Recognition Systems KOSYR'2003, Wrocław University of Technology, Wrocław 115–121
4. Russ J. C. (2002) The Image Processing Handbook. CRC Press, Boca Raton
5. Wiszniewski B. The Virtual Memorial Project. http://docmaster.eti.pg.gda.pl
6. Zhang D., Lu G. (2004) Review of Shape Representation and Description Techniques. Pattern Recognition 37 1–19

Conversion of Textual Information to Speech for Polish Language

Bozena Piorkowska, Janusz Rafalko and Edward Shpilewski

Institute of Computer Sciences, University of Bialystok, Sosnowa str. 64, 15-887 Bialystok, Poland. edszp@ii.uwb.edu.pl

Summary. An approach to solving the problem of the high-quality system Text-to-Speech (TTS) for Polish language synthesis is considered in this paper. Synthesis of phonemic speech characteristics is based on Polish language linguistic resources analysis and Allophones Natural Waves (ANW) method of speech signal concatenation.

1 Introduction

A creation of the computer synthesizer of speech is very important in various spheres of life [1]. The aim of the research is the making of Polish speech synthesizer on the basis of text.

The high-quality synthesis of speech from text TTS is based on the linguistic resources and acoustical voice database of the Polish language. To this date, certain experience in creating linguistic and acoustic resources has been gained for Polish. The research is to further develop the available linguistic resources (both vocabulary and grammar) and voice acoustical databases. The development of voices acoustical database is based on original methodology. Methods and algorithms for solving the problem of the high quality system Text-to-Speech (TTS) for Polish language synthesis is considered. Synthesis of phonemic speech characteristics is based on the Polish language linguistic resources analysis and Allophones Natural Waves (ANW) method of speech signal concatenation. These procedures consist of Polish speech allophones database, methods of their cutting-out from the text and the automatization of the cutting-out process. Next, on the basis of phonetic, orthographic and grammatical rules, a synthesizer changing written text into speech is built. Output signal speech should not differ from natural speech. The working out of voice "cloning" technology that is at the core of TTS synthesis provides basis for further high quality personalization of speech [2]. The problem of speech recognition and notation in the form of text is highly important. Conversion of the information Speech-to-Text (STT).

2 Textual processor

The first operation with the source text is removal of all signs that do not have influence on the pronunciation and accentuation of words or sentences. These are e.g.: paragraphs, empty lines, i.e. empty signs. Moreover, all the unnecessary punctuation marks and redundant signs are being removed. Furthermore, symbols, numbers and fractions are substituted by their verbal representations, and all the abbreviations are spelled out. The symbols are substituted according to a symbol chart and their verbal representation appended to the application.

Table 1. Symbol substitution chart for the Polish language

Symbol	Polish text	Symbol	Polish text	Symbol	Polish text
-	minus	@	małpa	±	plus minus
#	hasz	|	lub	÷	dzielone na
$	dolar	~	tylda	§	paragraf
%	procent	+	plus	©	copyright
&	oraz	<	mniejszy	®	registered
*	razy	=	równa się	‰	promil
/	łamane przez	>	większy	€	euro

Number substitution is an easy operation. However, it is worth remembering that the form of numerals such as thousands, millions etc. depends on the value of a number.

The process of spelling out the abbreviations is similar to the substitution of symbols.

The processed text is then divided into syntagmae and the word stress is marked.

The syntagm marking consists in the marking of the places where in the reconstructed speech there would appear pauses of appropriate length. The aim of this operation is to convey the aesthetics and the rhythm of the speech in the generated speech. Syntagmae are marked "strong" – longer pauses, and "weak" – shorter pauses. "Strong" syntagmae are placed instead of the punctuation marks (, ... ! ? () " " ; :) and instead of the conjunctions: *i, a, oraz, ani, zarazem, też, także, ni, albo, bądź, czy, lub, ale, zaś, jednak, natomiast, lecz, więc, toteż, zatem, dlatego, że, iż, żeby, co.* "Weak" syntagmae are marked whenever there appears a syntagmae that is too long (longer than 4 words).

The marking of word stress makes the synthesized speech sound naturally, free of the artificiality caused by the reading out of all words with the same colour and voice intensity.

The stress in Polish means distinguishing a certain section of a text by a stronger pronunciation, i.e. by strengthening the exhalation. A great majority of Polish words have stress on the second syllable from the end. This is the so-called paroxytonic stress. However, in a standard pronunciation there appear certain lexemes or groups of lexemes with a different stress. The exception is related to the structure of lexemes, etymology or pronunciation practice. There are several groups of words that are stressed differently.

1. Verbs with movable particles (in past tense forms and conditional) are stressed as if they did not have personal endings (past tense) and the morphemes of mode. As a result the stress falls:
   - e.g., [**gra**-li-śmy, wy-ko-**na**-li-ście], [na-ry-**so**-wał-bym, prze-**ko**-nał-byś, wy-**stą**-pił-by, na-u-**czy**-li-by];
   - e.g., [po-dzi-**wia**-li-byś-my, na-ma-lo-**wa**-li-byś-cie].

   These types of verbal forms may carry paroxytonic stress in colloquial speech.
2. The nouns which are adopted from Latin or through Latin, ending with *-ika* or *-yka*, in standard pronunciation have antepenultimate stress in nominative case and in these inflectional forms which have the same as number of syllables as nominative, e.g., nominative [**e**-ty-ka, ga-**lak**-ty-ka, po-**li**-ty-ka; accusative singular [**e**-ty-ce, ga-**lak**-ty-ce, po-**li**-ty-ce]. If declensional forms are longer than nominative, the stress is penultimate; it applies to the instrumental plural [e-ty-**ka**-mi, ga-la-kty-**ka**-mi, po-li-ty-**ka**-mi]. It is possible to stress the penultimate syllable of all the forms of these wards, including the nominative case.
3. Monosyllabic nouns, preceded with particle *arcy-*, *eks-*, *wice-*, carry stress on the last syllable (oxytonic), e.g., [arcy**mistrz**] (collaterant to [**arcy**mistrz]), [wice**król**], [eks**mąż**].
4. Compounds cardinal bisyllabic numerals and particles *-kroć, -sta, -set* carry the same stress as without the particles, i.e., on the antepenultimate syllable, e.g., [**czte**-ry-sta, **sie**-dem-kroć, **o**-siem-kroć, **sie**-dem-set, **o**-siem-set, **dzie**-więć-set].
5. Conjunctions joint with movable personal endings of verb *(-śmy, -ście)* as well as with morphemes of the conditional carry antepenultimate stress, e.g., [**a**-byś-my, **że**-byś-cie, je-**że**-li-by, **jeś**-li-by, po-**nie**-waż-by].

All borrowings with the status of quotations are pronounced according to the source language convention, e.g., French words *aperitif, emploi, foyer, tournee, vinaigrette* (a-pe-ri-**tif**, amp-**lła**, fła-**je**, tur-**ne**, wi-ne-**gret**]), they carry ulimate stress (oxytonic).

Prepositions, personal pronouns and conjunctions are a separate group of words where the stress has to be marked. Prepositions: *to, do, ku, na, nad, od, pod, przed, przy, u, za, bez, spod, to* are stressed in conjunction with the word following them (if nothing follows, they are stressed as independent words) e.g. *do-**ra**-na, nad-**rze**-ką, przy-sa-mo-**cho**-dzie.* Personal pronouns: *Ci, cię, go, ich, im, ją, je, jej, mi, mu, nas, was, się* are added to the preceding word that is stressed e.g. ***po**-każ-go, po-**da**-łeś-mi, za-pro-wa-**dzi**-my-was.* Conjunctions: *i, a, oraz, ani, zarazem, też, także, ni, albo, bądź, czy, lub, ale, zaś, jednak, natomiast, lecz, więc, toteż, zatem, dlatego, że, iż, żeby, co,* are stresses as independent words.

This is an example of the text with marked stress and syntagmae:

Da+wniej zapodstawo+we dziedzi+ny | zastoso+wań teo+rii maso+wej | obsłu+gi uważa+no telefo+nię, ‡ usłu+gi handlo+we ‡ czy+ usłu+gi sie+ci | słu+żby zdro+wia.

3 Properties of Polish phonetic system

What is phonetics? It is a study that deals with describing characteristic features of a set of sounds used in a language. The function of organs of speech and ear is the field of phonetics [3].

3.1 Phoneme

In every language we can distinguish larger parts we fully pronounce and their component part: moods, sentence, words, syllables, and finally sounds we can identify with the phoneme. Sound is the smallest linguistic element, capable of being distinguished from the structure word. The dependence of sound on a language is so important, that when people are getting familiar with a new language, they usually calculate and estimate its supply of sounds differently than its native speakers. Polish **cz** or **c** is considered by foreigners don't have this sounds in their languages as a cluster of **t** + **sz** (eng. sh) or **t** + **s**, when every Pole distinguishes perfectly between **czy** and **trzy**, or **Czech** and **trzech**.

Names vowel and consonant are functional names. It means that sounds of first category are syllabic and they can, and even have to form syllables. The sounds of the second category come in composition of syllables, but they do not form them. In reality we do not distinguish vowels and consonant by their function, but on the basis of their acoustic and articulatory characteristics. Acoustically, vowels are tones; consonants are murmurs (in which laryngeal tone may show up).

The tongue is the most restless organ of speech. It is the main regulator of the tone of vowels because while changing its position it changes the height of the basic tone and thus shortens or lengthens resonance space. Besides pure vowels that are mouth vowels, there are nasal vowels. Numerous tints of nasal vowels exist. Polish language has well-known **ę** and **ọ** (nasal o), but there also exist **ą** (nasal a), **ų**, **į**, **y̨**, although they are not marked in writing.

The significance of articulation of consonants is the creation in the mouth of a cavity articulatory obstruction which expiratory stream has to overcome. Consequently, the type of organs, their articulatory neighbourhood, the way its obstruction is formed as well as the gender of the dam created, all that influences the character of murmur and the consonant sound.

3.2 Polish phonemes

Polish alphabet consists of 32 letters of Latin alphabet, in this 23 are usual letters: **A, B, C, D, E, F, G, H, I, J, K, L, M, N, O, P, R, S, T, U, W, Y, Z**, and 9 letters additional marks: **ó, ś, ż, ź, ć, ń, ą, ł, ę**. Moreover in adopted words we can come across the letters: **Q, V, X**. Besides letters occurring individually in the speech sounds mark we also use 7 digraphs: **Cz, Dz, Dż, Dź, Rz, Sz, Ch**. We have 51 phonemes altogether (in brackets, notations used in our project can be found):

8 vowels: **U**(u), **O**(o), **A**(a), **E**(e), **Y**(y), **I**(i), **Ą**(ą), **Ę**(ę)

43 consonant: **p**(p), **b**(b), **f**(f), **w**(w), **m**(m), **ł** (ł), **t**(t), **d**(d), **s**(s), **z**(z), **c**(c), **dz**(Ď), **sz**(š), **ż**(ż), **cz**(č), **dż**(Đ), **r**(r), **n**(n), **l**(l), **k**(k), **g**(g), **h**(h), **p'**(P), **b'**(B), **f'**(F), **w'**(W), **m'**(M), **t'**(T), **d'**(D), **r'**(R), **l'**(L), **k'**(K), **g'**(G), **h'**(H), **s'**(S), **z'**(Z), **c'**(C), **ś**(ś), **ź**(ź), **ć**(ć), **dź**(đ), **ń**(ń), **j**(j)

The exchange of orthographical text into phonemic text is a very important stage. It is the kind of translation of written word to spoken. This is very important and necessary because clear differences step out among them. The fact that the pronunciation often differs from the writing can be easily seen in these examples. We write e.g. *prośba, szczaw, krzesło*, and we tell *prożba, ščaf, kšesło*. The after-effect of these differences is a set of phonetic rules, thanks to which orthographical text can be transcribed into phonetic text.

And here is the example of phonemic text:

da+wNej zapotstawo+we đeđi+ny | zastoso+wań teo+Rji maso+wej | opsłu+Gi uważa+no telefo+Nę, ‡ usłu+Gi handlo+we ‡ čy+ usłu+Gi śe+ći | słu+żby zdro+Wa.

4 Allophones natural waves database

Depending on neighbouring vowels and consonants, phonemes can have different sound. Therefore, one phoneme can have many variants of the so-called allophones. Theoretically, every combination of phonemes corresponds to a different allophone. However, in practice, it is possible to divide the phonemes into certain groups applying suitable criteria, such as the place of articulation, the way of articulation and, in the case of vowels, stress. The phoneme division is different for consonants and vowels. It was our aim to find as many allophones to make the synthesis work correctly. Using the division into groups we can make suitable tables to find these allophones and create the database. Thus, considering the word stress, we have got three tables for each vowel. We have stressed vowel, vowel before the stressed syllable, as well as vowel after the stressed syllable. In the case of consonants it is possible to simplify the procedure. We do not have the stressed consonants, which reduce the number of allophones, besides a very detailed division like the one in case of vowels is not necessary.

Table 2. Allophones of phoneme p

after \ before	# 0	c. voiceless 1	c. voiced 2	v. not stressed 3	v. stressed 4
# 0	——— p00	**p**rzepływ p01	**p**rogramowe p02	**p**opularnym p03	**p**awie p04
c. voiceless 1	p10	w**p**rzeszłość p11	zpost**p**rocesorem p12	s**p**ecjalistycznych p13	ws**p**omagania p14
c. voiced 2	hel**p** p20	Kom**p**tona p21	współ**p**racuje p22	kom**p**uterowego p23	kom**p**uter p24
vowel 3	galo**p** p30	skry**p**tami p31	o**p**rogramowania p32	po**p**ularnym p33	La**p**onia p34

In order to find these allophones the tables should be completed, not all combinations occur in Polish, e.g. we will always have voiceless consonant at the end of the word. Further, these words should be recorded and then suitable allophones should be cut out and recorded it under a suitable name. For this purpose indices are being created in order to allocate a given allophone to a suitable group. The whole base contains about 5000 allophones.

By taking into account the phonemes' neighbourhood in phonemic text, it is possible to extract allophonic text. Only allophonic text can be used in speech synthesizing. Every speech sound in this text is being assigned an allophone from the database of allophones. This speech sounds are then being connected with each other and reproduced.

And here is an example of an allophonic text:

d20 a102+ w10 ń21 e143 j12 # z20 a110 p22 o021 d12 s11 t21 a012 w22 o201+ w20 e101 # Đ20 e133 Đ22 i203+ n20 y102 # |10 # z20 a120 s12 t21 o022 s22 o202+ w20 a131 ń12 # t20 e122 o202+ R20 i103 # m20 a121 s22 o202+ w20 e141 j12 # |10 # o110 p12 s11 ł21 u205+ G20 i103 # u110 w22 a021 ż22 a202+ n20 o102 # t20 e122 l22 e012 f22 o201+ ń20 e103 # ‡10 # u120 s12 ł21 u205+ G20 i103 ...

5 Acoustical processing of allophonic text

After receiving allophonic text suitable sound files should be opened. Every allophone is recorded in a separate file in wav format. During the playbacking of the allophones, millisecond pauses between the allophones can be heard, which does not happen in normal speech. In order to avoid the problem the wave files (allophones) ought to be put together to make one unit and them to be replayed. Such connecting of allophones goes for every word. We would therefore say that the process of reproduction of speech is made word after word. As final effect suitable intonation is added.

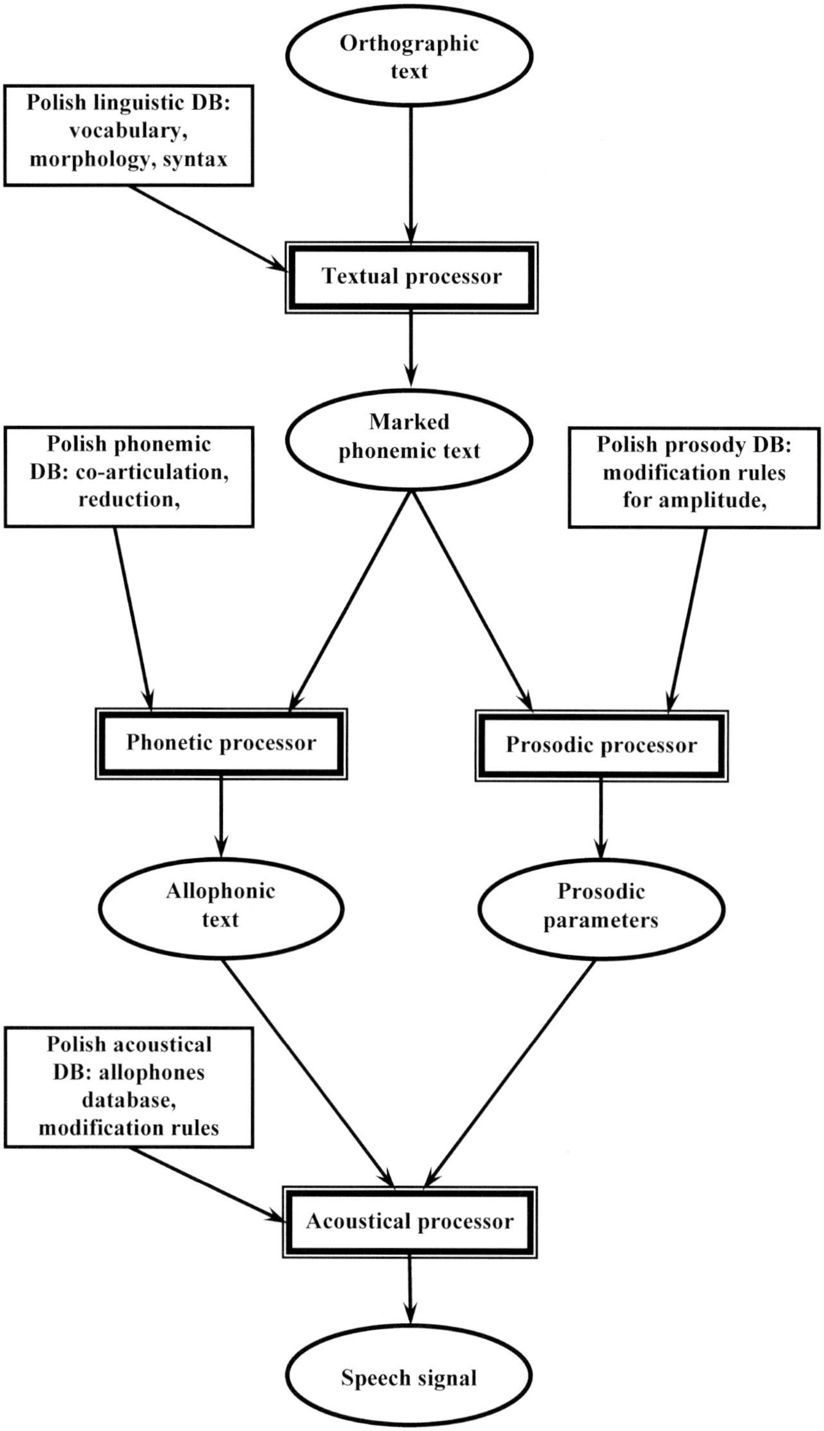

Fig. 1. General structure of the TTS-synthesizer

6 General structure of the TTS-synthesizer

The general scheme of textual synthesis is presented below, on Figure 1. The source text is being reorganised (clearing the text, changing the symbols and numbers into their verbal equivalents, marking the accents, writing full word forms and dividing into syntagmae. Textual processor realizes it. Next, according to phonetical rules the text is changed into phonemic one and finally transformed into allophonic. This stage realizes phonetic processor. At the same time the process of sign of suitable intonation is realized, together with choosing their parameters. It realizes prosodic processor. In this way we receive allophonic text together with prosodic parameters. Having the database of allophones is possible reproduction of text in form of signal speech [4].

On present stage of work over synthesizer, it is impossible to give general opinion about synthesizer and algorithm of his works yet. This fact results from the track of intonation is not realized yet. In present phase of realization the synthesized speech is devoid any intonation, wherethrough sounds rather artificially. The intonation track is in progress of realization obviously.

7 Conclusion

A computer synthesizer may be applied in various spheres of life:

1. It can be used in audio servers to provide information to the users in telephone banking, cultural and tourist information telephone services.
2. A speech synthesizer makes possible a round-the-clock telephone transmission of the required information by means of speech.
3. The expected results of the project can be applied in further research in applied linguistics, especially, in the study of phonetics and prosody of the Polish language, in expanding the theoretical framework for multilingual speech communication systems. The results of the project will be presented in the form of a book-length study, in a series of articles as well as in university lectures on the theory and applications of speech technologies.
4. The project has great relevance for economic and social fields. The obtained results will facilitate the development of new areas of business activities and services in Poland.

The extension of this work is project of application executing opposite process, which is speech recognizing and notation one in the form of text. Conversion of the speech information into text: Speech-to-Text (STT). Recognition and synthesis methods of the audio-visual patterns will be development.

Acknowledgement

This paper was supported by the EUROPEAN COMMISSION under grant INTAS Ref. number 04-77-7404. The author wish to express their thanks for the support.

References

1. Lobanov B., Karnevskaya H. "*MW Speech Synthesis from Text*" Proc. of the XII International Congress of Phonetic Sciences. Aix-en-Provense, France, 1991, pp. 406-409.
2. Boguslavsky I., Lobanov B. and Karnevskaya H. "*Generation of Intonation and Accentuation of SyntheticSpeech on the Base of Morpho-Syntactic Knowledge*", Proceedings of the International Workshop"Integration of Language and Speech", Moscow, 1996, pp. 11-28.
3. Dłuska Maria, "*Polish Phonetics*", PWN Warszawa – Kraków, 1981.
4. 4. Shpilewski E., Piurkowska B., Rafalko J., Lobanov B., Kiselov V., Tsirulnik L. "*Polish TTS in Multi-Voice Slavonic Languages Speech Synthesis System*", Proceedings of the 9th International Conference "Speech and Computer" – SPECOM'2004. Saint Petersburg, Russia, 2004, pp. 565-570.

Multilevel Recognition of Structured Handprinted Documents - Probabilistic Approach

Jerzy Sas[1] and Marek Kurzynski[2]

[1] Wroclaw University of Technology, Institute of Applied Informatics, Wyb. Wyspianskiego 27, 50-370 Wroclaw, Poland jerzy.sas@pwr.wroc.pl
[2] Wroclaw University of Technology, Faculty of Electronics, Chair of Systems and Computer Networks, Wyb. Wyspianskiego 27, 50-370 Wroclaw, Poland marek.kurzynski@pwr.wroc.pl

Summary. In the paper the multilevel probabilistic approach to handprinted form recognition is described. The form recognition is decomposed into three levels: character recognition, word recognition and form contents recognition. On the word and form contents level the probabilistic lexicons are available. The decision on the word level is performed using probabilistic properties of character classifier and the contents of probabilistic lexicon. The novel approach to combining these two sources of information about classes (words) probabilities is proposed, which is based on lexicons and accuracy assessment of local character classifiers. Some experimental results and examples of practical applications of recognition method are also briefly described.

1 Introduction

Handwritten text recognition is one of principal areas of interest of artificial intelligence. Despite four decades of intensive research, there are still no sufficiently reliable methods and techniques assuring acceptable error rate of handwritten text recognition. Handprinted form recognition problem is relatively simple in comparison to cursive script analysis. The need for automatic form recognition and processing appears in many practical areas of applications, e.g. mail sorting, banking operations, education, polling, medical information systems, to name only a few. Typical form being considered here has precisely defined structure. It consists of separated data fields, which in turn consist of character fields. Due to fixed geometric layout, the contents of data fields (let us call it word) can be easily extracted. For this reason the problem of text segmentation is easy to solve, although still not quite trivial. In our approach we assume that the whole form contents describes an object from the finite set of items and the ultimate aim of form recognition is selecting of relatively small subset of objects. Therefore, instead of using the classic pattern recognition approach consisting in indicating a single class, we will apply "soft" recognizer ([5]) which fetches the vector of soft labels of classes, i.e. values of classifying function.

Despite the accuracy of recently developed separate character classifiers exceed 99% for digits and 90% for letters, it is still not sufficient to reliably recognize longer texts. In order to improve the overall recognition quality compound recognition methods are applied. Two most widely used categories of compound methods consist in combining classifiers based on different recognition algorithms and different feature sets ([1]). Another approach divides the recognition process into levels in such a way, that the results of classification on lower level are used as features on the upper level ([7], [2]). Two-level approach is typical in handwriting recognition, in which the separate characters are recognized on the lower level and next on the upper level the words are recognized, usually with the use of lexicons. In some approaches the third level is introduced where some syntactic or semantic properties of the language are utilized in order to select most probable sequence of words fetched as alternatives by word level classifier.

In this paper, the method which uses both classifier combination and multilevel recognition is described. Probabilistic properties of lexicon and character classifier are typically used to build Hidden Markov Model(HMM) of the language ([4], [3], [10]). We propose another approach to the word recognition, in which probabilistic lexicon is treated as a special kind of classifier based on a word length, and next result of its activity is combined with soft outcomes of character classifier based on recognition of character image. Different methods of fusion of both classifiers lead to the several word classifiers which differ in concepts, activity and - as it results from experimental investigations - also in recognition quality. Soft outcomes of a word classifier can be used next as data for semantic level classifier, which recognize the object described by the whole form. Since relation between word classifier and semantic classifier with lexicon is exactly the same as relation between the character recognizer and word classifier with word lexicon, hence in this paper we pay our attention on the first and the second level of three-level form recognition systems because results presented here can be easy extended on the last level of recognition process. More details of using analogous approach to complete semantic level can be found in [9]. Application of the presented concept to automatic processing of laboratory test order forms in hospital information system is described in [8].

The contents of the work are as follows. Section 2 introduces necessary background and provides the problem statement. In section 3 the different concepts of fusion strategies of character-based and lexicon-based classifiers are discussed. The proposed algorithms were practically implemented in the problem of recognition of polish names and surnames and results of classification accuracy obtained on the real data are given in section 4.

2 Preliminaries and the Problem Formulation

Let us consider a paper form F designed to be filled by handprinted characters. The form consists of data fields. Each data field contains a sequence of characters of limited length coming from the alphabet $\mathcal{A}$. Data fields do not have to be filled completely - only the leading part of each field must be filled with characters. We assume that the actual length of filled part of data field can be faultlessly determined. The set $\mathcal{A}$ can be different for each field. Typically we deal with fields that can contain only digits, letters or both of them. For each data field there exists a probabilistic lexicon $\mathcal{L}$. Lexicon contains words that can appear in the data field and their probabilities:

$$\mathcal{L} = \{(W_1, p_1), (W_2, p_2), ..., (W_N, p_N)\}, \tag{1}$$

where W_j is the word consisting of characters from $\mathcal{A}$, p_j is its probability and N is the number of words in the lexicon.

The completely filled form describes an object from finite set $\mathcal{B}$. For instance in e-learning system the part of the test form containing name, surname and credit book number describe a student coming from finite students group. Similarly as for words, here we also know the appearance probabilities of elements from $\mathcal{B}$. For example, in our student group some students may prefer to attend more less scored test and in result their forms appear in the system more frequently. Other prefer to attend lower number of highly scored tests, so their forms appear less frequently.

Our aim is to recognize the object $b \in \mathcal{B}$ on the base of scanned image of a form F. The recognition process can be divided into three levels, naturally corresponding to the three-level form structure:

- character (alphabetical) level - where separate characters are recognized,
- word level - where the contents of data fields is recognized, based on the alphabetical level classification results, their probabilistic properties and probabilistic lexicon (1),
- semantic level - where the relations between fields of the form being processed are taken into account to further improve the recognition performance.

We assume next that on the alphabetical level a classifier Φ is given which recognize character $c \in \mathcal{A}$ on the base of its image x, i.e. $\Phi(x) = c$ and furthermore characters in sequence of data fields are recognized independently. Probabilistic property (quality) of Φ is described by conditional probabilities of character $w \in \mathcal{A}$ appearance

$$p_{wc} = P(w \mid \Phi(x) = c), \tag{2}$$

which for all $w, c \in \mathcal{A}$ form the confusion matrix $\mathcal{P}_\Phi$ of a rule Φ. In practice, probabilities (2) can be got from manually verifying results of form processing.

Any classifier can be used on character level. In further experiments we have applied near neighbor classifier using a vector of directional features [6].

Let the length $\mid W \mid$ of currently recognized word $W \in \mathcal{L}$ be equal to n. This fact defines the probabilistic sublexicon $\mathcal{L}_n$

$$\mathcal{L}_n = \{(W_k, q_k)_{k=1}^{N_n} : W_k \in \mathcal{L}, \mid W_k \mid = n\}, \tag{3}$$

i.e. the subset of $\mathcal{L}$ with modified probabilities of words:

$$q_k = P(W_k / \mid W_k \mid = n) = \frac{p_k}{\sum_{j:|W_j|=n} p_j}. \tag{4}$$

The sublexicons (3) can be considered as a soft classifier Ψ_L which maps feature space $\{\mid W_k \mid : W_k \in \mathcal{L}\}$ into the product $[0,1]^{N_n}$ or equivalently, for each word length n produces the vector of decisions support

$$s = (s_1, s_2, ..., s_{N_n}), \tag{5}$$

where for Ψ_L support s_k of decision W_k is equal q_k.

Let suppose next, that classifier Φ, applied n times on the character level, has recognized the sequence of characters (word) $C = (c_1, c_2, ..., c_n)$ on the base of character images $X = (x_1, x_2, ..., x_n)$, namely:

$$\Phi(X) = C. \tag{6}$$

Such an activity of classifier Φ will be further treated as an action of soft classifier Ψ_C, which - as previously - produces vector of decision support (5) for words $W_k = (w_1^{(k)}, w_2^{(k)}, ..., w_n^{(k)}) \in \mathcal{L}_n$, where now

$$s_k(\Psi_C) = P(W_k \mid \Phi(X) = C) = \prod_{j=1}^{n} P(w_j^{(k)} \mid \Psi(x_j) = c_j) = \prod_{j=1}^{n} p_{w_j^{(k)}, c_j}. \tag{7}$$

Now our purpose is to built soft classifier Ψ_W for word recognition as a fusion of activity of both lexicon-based Ψ_L and character-based classifier Ψ_C. In the next chapter a number of possible combination methods are discussed.

3 Combining Lexicon-based and Character-based Classifiers on the Word Level

3.1 Simple Classifier Selection

In the simplest approach the final classifier Ψ_W is equal either Ψ_C or Ψ_L. The choice depends on an evaluation criterion $Q(\Psi)$ of candidates. The following two methods seem to be intuitively substantiated:

- the maximum value of decision supports (5) produced by Ψ:

$$Q(\Psi) = \max_{k:W_k \in \mathcal{L}_n} s_k(\Psi), \tag{8}$$

where $s_k(\Psi_L)$ and $s_k(\Psi_C)$ are equal to (4) and (7), respectively
- the normalized entropy of the support vector s calculated by Ψ:

$$Q(\Psi) = 1 - \frac{\sum_{k:W_k \in \mathcal{L}_n} s_k(\Psi) log_2(s_k(\Psi))}{log_2 \frac{1}{N_n}}, \tag{9}$$

which is frequently used as a measure of discriminative power of a classifier ([3]).

The choice (combination) rule is obvious: *If* $Q(\Psi_C) > Q(\Psi_L)$ *then* $\Psi_W = \Psi_C$ *else* $\Psi_W = \Psi_L$.

3.2 Linear Interpolation of Classifiers

In this method the support vector of a word classifier $s(\Psi_W)$ is obtained as a linear interpolation of $s(\Psi_L)$ and $s(\Psi_C)$, namely:

$$S(\Psi_W) = \alpha s(\Psi_L) + (1 - \alpha) s(\Psi_C), \tag{10}$$

where coefficient α can be determined using evaluation criterion (8) or (9):

$$\alpha = \frac{Q(\Psi_L)}{Q(\Psi_L) + Q(\Psi_C)}. \tag{11}$$

3.3 Classifiers Interleaving

Let $\mathcal{N} = \{1, 2, ..., n\}$ be the set of numbers of character positions in a word $W \in \mathcal{L}_n$ and $\mathcal{I}$ denotes a subset of $\mathcal{N}$. In the fusion method with "interleave" first the algorithm Φ is applied for recognition of characters on positions $\mathcal{I}$ and next - using these results of classification - the lexicon $\mathcal{L}_n$ (or algorithm Ψ_L) is applied for recognition of a whole word W.

The main problem of proposed method consists in an appriopriate division of $\mathcal{N}$ into sets $\mathcal{I}$ and $\bar{\mathcal{I}}$ (complement of $\mathcal{I}$). Intuitively, subset $\mathcal{I}$ should contain these positions for which character recognition algorithm gives the most reliable results. In other words division of $\mathcal{N}$ should lead to the best result of classification accuracy of a whole word. Thus, subset $\mathcal{I}$ can be determined as a solution of an appropriate optimization problem.

Let $C^{\mathcal{I}} = \{c_i, i \in \mathcal{I}\}$ be the set of characters on positions $\mathcal{I}$ which have been recognized by classifier Φ, i.e. $\Phi(X^{\mathcal{I}}) = C^{\mathcal{I}}$. Hence for any set of characters $W^{\mathcal{I}} = \{w_i, i \in \mathcal{I}, w_i \in \mathcal{A}\}$ we have:

$$P(W^{\mathcal{I}} \mid \Phi(X^{\mathcal{I}}) = C^{\mathcal{I}}) = \prod_{i \in \mathcal{I}} P(w_i \mid \Phi(x_i) = c_i) = \prod_{i \in \mathcal{I}} p_{w_i c_i}. \tag{12}$$

The above formula determines conditional probability that on positions $\mathcal{I}$ of word to be recognized are characters $W^{\mathcal{I}}$ provided that rule Φ recognized characters $C^{\mathcal{I}}$. Since the whole word $W_k \in \mathcal{L}_n$ consists of characters $W_k^{\mathcal{I}}$ and $W_k^{\bar{\mathcal{I}}}$, i.e. characters on positions $\mathcal{I}$ and $\bar{\mathcal{I}}$, respectively, hence we have the following support vector (5) of combined rule Ψ_W:

$$s_k(\Psi_W) = P(W_k \mid C^{\mathcal{I}}) = P(W_k^{\mathcal{I}} \cap W_k^{\bar{\mathcal{I}}} \mid C^{\mathcal{I}}) = \frac{P(W_k^{\bar{\mathcal{I}}} \mid W_k^{\mathcal{I}} \cap C^{\mathcal{I}}) P(W_k^{\mathcal{I}} \cap C^{\mathcal{I}})}{P(C^{\mathcal{I}})}, \tag{13}$$

and after simple transformations we get:

$$s_k(\Psi_W) = P(W_k \mid C^{\mathcal{I}}) = P(W_k^{\mathcal{I}} \mid C^{\mathcal{I}})\, P(W_k \mid W_k^{\mathcal{I}}) \tag{14}$$

The first factor of (14) is given by (12), whereas the second one can be calculated as follows:

$$P(W_k \mid W_k^{\mathcal{I}}) = \frac{q_k}{\sum_{i: W_i contains W^{\mathcal{I}}} q_i}. \tag{15}$$

Since the support vector (14) of the rule Ψ_W strongly depends on the set $\mathcal{I}$ hence we can find such a set $\mathcal{I}^*$ which maximizes criterion $Q(s(\Psi_W))$, where Q can be defined as in (8) or (9). The number of solutions of above problem is equal to 2^n, thus - except the case of very short words - the exhaustive search is rather infeasible method. Therefore we suggest the following suboptimal method which was applied in the further experimental investigations.

Initial Conditions: $C = \{c_1, c_2, ..., c_n\}$, $\mathcal{I} = \oslash$
Step 1: Find in C character c^* for which $\max_w p_{wc^*} = \max_c (\max_w p_{wc})$,
$C \leftarrow (previous\ C) - c^*$, $\mathcal{I} \leftarrow (position\ number\ of\ c^*)$
Step 2: calculate $Q(\mathcal{I})$,
If $Q(\mathcal{I}) > Q(previous\ \mathcal{I})$ then go to Step 1, else STOP.

4 Experiments

In order to study the performance of the proposed word recognition concepts and evaluate their usefulness to the practical structured handprinted forms recognition, several computer experiments were made in which polish names and surnames were applied as recognized words. The recognizer performance was evaluated in three ways using:

- the number of test cases where actual word has highest score in the support vector,
- the number of cases where actual word is among 3 highestly scored words,
- the number of cases where actual word is among 5 highestly scored words.

At the character level the NN recognizer was applied based on gradient features set according to the procedure described in [6]. As the learning set for NN rule 354 character prototypes were selected. Using the testing set of 9200 character images probabilities of confusion matrix P_Φ were evaluated. The average accuracy of the NN character recognizer was equal to 86.3 %. The surname and name lexicons contained 884 and 9944 items, respectively, were created on the base of hospital information database system containing 36,290 patient records. In the experiments first the word from lexicon was randomly selected and next it underwent a recognition procedures according to the applied algorithms. In each experiment we calculated frequency of correct classification for investigated algorithms and for three adopted criteria. Outcomes are presented in Table 2 and 3 for names and surnames, respectively.

Table 1. Results of empirical tests - names recognition

Criterion	Single classifier	Classifier selection	Linear combination	Interleaving
1 of 1	78.9%	79.7%	76.1%	86.6%
1 of 3	85.1%	86.4%	90.2%	93.5%
1 of 5	89.0%	89.4%	91.3%	97.6%

Table 2. Results of empirical tests - surnames recognition

Criterion	Single classifier	Classifier selection	Linear combination	Interleaving
1 of 1	54.6%	55.6%	54.0%	66.7%
1 of 3	67.9%	69.2%	69.2%	74.0%
1 of 5	71.3%	71.7%	72.4%	78.7%

The simple combination methods (classifier selection, linear combination of classifiers) did not increase the word recognition accuracy significantly. It is probably caused by domination of character-based classifier over lexicon-based classifier, especially in case of names. Greater performance boost could be expected if the performances of both combined classifiers would be more balanced, e.g. in unconstrained script recognition, where character recognition quality is much lower due to unambiguous segmentation problems. Significant improvement has been observed in case of classifiers interleaving. The recognition quality increased on average by 22% in case of surnames and by almost 40% in case of names.

5 Conclusions

In this paper we have focused our attention on the two-level words recognition in structured handprinted documents via connection of results of character classifier on lower level and probabilistic lexicon treated as a special classifier on upper (word) level. Taking the probabilistic model of classification task, we have discussed different concepts of fusion of both classifiers which lead to the soft word classifier producing vector of support values for all words from the lexicon instead of a single hard decision.

Presented algorithms have been experimentally tested on the real data containing a set of polish names and surnames. Their results, especially comparision with recognition quality of separated character image recognition algorithm, demonstrate the effectiveness of the proposed word recognition concepts and yield some recommendation for a wide range of practical applications which deal with problem of structured handprinted text recognition.

Acknowledgement. This work was financed from the State Committee for Scientific Research (KBN) resources in 2005 - 2007 years as a research project No 3 T11E 005 28.

References

1. Kuncheva L.I. (2002) A Theoretical Study on Six Classifier Fusion Strategies, IEEE Trans. on Pattern Anal. and Machine Intelligence, Vol. 24, No 2 : 281-286
2. Chen W.T., Gader P., Shi H. (1999) Lexicon-Driven Handwritten Word Recognition Using Optimal Linear Combination of Order Statistics, IEEE Trans. on Pattern Anal. and Machine Intelligence, Vol 21, No 1 : 71-82
3. Grandidier F., Sabourin R. (2000) A New Strategy for Improving Features Set in a Discrete HMM-based Handwriting Recognition System, In: Schomaker L.R.B., Vuurpijl (eds) Proceedings of the Seventh International Workshop on Frontiers in Handwritting Recognition, Sept. 11-13 2000 Amsterdam : 113-122
4. Kim J. H., Kim K.K., Suen C. Y. (2000) An HMM-MLP Hybrid Model for Cursive Script Recognition. Pattern Analysis and Applications, No 3 : 312-324
5. Kuncheva L.I. (2001) Using Measures of Similarity and Inclusion for Multiple Classifier Fusion by Decision Templates, Fuzzy Sets and Systems, 122 (3) : 401-407
6. Liu C., Nakashima K., Sako H., Fujisawa H. (2003) Handwritten Digit Recognition: Benchmarking of State-of-the-Art Techniques, Pattern Recognition, Vol. 36 : 2271-2285
7. Lu Y., Gader P., Tan C. L. (2002) Combination of Multiple Classifiers Using Probabilistic Dictionary and its Application to Postcode Generation, Pattern Recognition, Vol 35 : 2823-2832
8. Sas J. (2004) Handwritten Laboratory Test Order Form Recognition Module For Distributed Clinic, Journ. of Medical Informatics & Technologies, Vol 8 : 59-68
9. Sas J. (2004) Three-Level, Lexicon-Based Handwritten Form Recognition Method, In: Klopotek M., Tchorzewski J. (eds) Proc, of VI Int. Conf on Artificial Intelligence AI-19'2004, Vol. 1 (23) : 113-124
10. Vinciarelli A., Bengio S., Bunke H. (2004) Offline Recognition of Unconstrained Handwritten Texts Using HMMs and Statistical Language Models, IEEE Trans. on Pattern Anal. and Machine Intelligence, Vol 26, No 6 : 709-720

Application of Statistic Properties of Letters Succession in Polish Language to Handprint Recognition

Jerzy Sas[1] and Marek Kurzynski[2]

[1] Wroclaw University of Technology, Institute of Applied Informatics, Wyb. Wyspianskiego 27, 50-370 Wroclaw, Poland jerzy.sas@pwr.wroc.pl

[2] Wroclaw University of Technology, Faculty of Electronics, Chair of Systems and Computer Networks, Wyb. Wyspianskiego 27, 50-370 Wroclaw, Poland marek.kurzynski@pwr.wroc.pl

Summary. In the paper the method of handprinted word recognition is described, which combines statistical lexical language model and character classifier properties in order to improve the recognition accuracy. The statistical lexical model determines the conditional probabilities of letters succession in the language. For some letters in polish language only very small subset of successors appears with significant conditional probability. If the confidence of predecessor recognition is assessed as high then the recognition of successor can be reliably supported by utilizing probabilistic lexical properties. In contrast to many other approaches, the method is not based on lexicons, so it can be used in these cases where the exhaustive lexicon is not available or its usage is inefficient, e.g. due to great number of elements.

1 Introduction

Despite significant progress in handwriting recognition in recent years, existing methods and techniques are still not satisfactory as far as natural language is being concerned. In some applications we deal with relatively easy case, where text is handprinted, i.e. the writer is forced to write separated characters, hence difficult problem of text segmentation does not appear. The recognition process consists then in recognizing separate characters and building words consisting of recognized letters.

Most methods of handwritten text recognition use lexicons in order to improve the recognition quality ([4], [6], [7], [8]). The lexicon defines the set of words that can appear in the text. In this way the set of allowed letter combinations is reduced drastically. Lexicons are very useful in such applications where the acceptable words count is moderate, as e.g. in personal data processing, mail address recognition, etc. In natural language recognition however the lexicon may contain tenths thousands of words, what leads to inefficiency in its usage. Firstly, matching in large vocabulary is time consuming. What is more important, there are many similar words in large lexicon, so discriminative abilities of the lexicon decrease. In some applications, lexicons must be assumed incomplete, or there are no lexicons at all. It such cases the word recognizer should utilize the results of character classification, known properties of the character classifier and general statistical properties of the language to elaborate most reliable rank of candidate words.

In the method described in this paper the approach utilizing statistical language model concept ([1], [9]) transposed to lexical level is applied. Statistical language model determines the conditional probabilities $p(w_i \mid (w_{i-n+1}, w_{i-n+2}, ..., w_{i-1})$ of word w_i appearance provided that the sequence of n words $(w_{i-n+1}, w_{i-n+2}, ..., w_{i-1})$ appeared just before it. The sequence of n successive words is called n-gram. Analogous concept can be applied to letters constituting words. By analyzing sufficiently large corpus of texts, the conditional probabilities of letters succession can be estimated. In experiments we investigated only bi-grams, i.e. the probabilities $p(c_i \mid c_{i-1})$ of letter c_i appearance, provided that preceding letter is c_{i-1} The statistical analysis of polish language corpus indicates, that for many letters only a few of successors appear with significant conditional probability. This observation can be used to improve the word recognition based merely on the results of character classification. The decision if to use lexical language model or just apply results of character classifier for given character is the crucial element of the recognition method. In the approach presented here it depends on comparison of the expected accuracy assessments for the character subject to decision.

The contents of the work are as follows. Section 2 provides necessary background and gives the problem statement. In section 3 we discuss different concepts of soft character classifiers leading to the concrete propositions of character classifier outcomes. In section 4 combined algorithm of word recognition is presented, which uses results of character recognition, quality evaluation of character classifier and statistical properties of letters succession in polish language. Section 5 describes results of experimental investigations of proposed algorithm and concludes the paper.

2 Problem Formulation

Let us consider two-level word recognition problem. On the lower (character) level separate letters are being classified. On the upper (word) level the results of letters classification are used to recognize the whole word. The word is handprinted, i.e it is written in block capital letters, enclosed in the separated regions (character fields) located in fixed positions on the printed form. Hence it is easy to extract isolated images of subsequent characters and the problem of text segmentation is almost trivial and not discussed here. We also assume that the text image analyzer can perfectly distinguish between empty and nonempty character fields, so the length m of the word is fixed before recognition begins. In each nonempty character field there can appear the image of the letter from the alphabet $\mathcal{A} = \{a_1, a_2, ..., a_M\}$. The separate character images X_i, $i = 1, ...m$ are inputs to the soft character classifier $\Phi(X_i)$, which recognizes individual characters independently. Classifier $\Phi(X_i)$ produces the vector of normalized support values (ranks) for letters in the alphabet, namely:

$$\Phi(X_i) = (s_1^{(i)}, s_2^{(i)}, ..., s_M^{(i)}), \quad s_1^{(i)} \geq 0, \sum_{j=1}^{M} s_j^{(i)} = 1 \tag{1}$$

The soft classifier (1) can be constructed in various ways. Some proposals will be given in the next section.

We do not know the set of all words that can appear, but probabilistic lexical model of the language is available (PLLM for short). The term PLLM means the set of probabilities of letters succession and precedence. The conditional probability of letters succession

$$\underline{p}_{cd} = p(c_i = c \mid c_{i-1} = d), \quad c, d \in \mathcal{A} \tag{2}$$

is the probability that on the i-th character position is the letter c provided that on the preceding position there is letter d. In the same way the probabilities of letters precedence can be defined:

$$\overline{p}_{cd} = p(c_{i-1} = c \mid c_i = d), \quad c, d \in \mathcal{A}. \tag{3}$$

PLLM consists of two $M \times M$ matrices $\underline{P}$ and $\overline{P}$ containing probabilities (2) and (3), respectively.

Now, our aim is to construct the soft classifier $\Psi(X_1, X_2, ..., X_m)$, which using results of character classifier Φ and PLLM matrices $\underline{P}$ and $\overline{P}$ recognizes the whole word. Thus, for given the actual word length m and the sequence of character field images $(X_1, X_2, ..., X_m)$, classifier Ψ produces support values s_i for fixed number N of the most likely words w_i, viz.

$$\Psi(X_1, X_2, ..., X_m) = ((w_1, s_1), ..., (w_N, s_N)). \tag{4}$$

Word recognition algorithm Ψ will be presented in section 4, but first we will discuss the construction methods of soft character classifier Φ.

3 Constructing Soft Classifier for the Character Level

We have different possibilities to determine the output vector of classifier (1) on character level. Generally, the nature of extracted features, classification criteria (discriminant functions of classifier) or classifier statistical properties can suggest some solutions. Let us consider some proposals of support vector of classifier Φ.

3.1 Support Vector Based on Confusion Matrix

For any "hard" character classifier $\Phi(X)$, confusion matrix P_Φ of the size $M \times M$ contains the following probabilities [6]:

$$p_{z,c} = P(z \mid \Phi(X) = c), \quad z, c \in \mathcal{A}, \tag{5}$$

which can be easily got from manually verifying results of character recognition. As the support factor for letter z and character image X we directly adopt the element of confusion matrix $p_{z,\Phi(X)}$.

The advantage of this method is that it can be applied for any "hard" classifier, provided that the sufficiently large set of empirical data has been collected. Disadvantage is that the method ignores individual properties of the character image X.

3.2 Support Vector for MLP Character Recognizer

As a character recognizer very often the multi-layer perceptron (MLP) is applied. We used this concept on character level in the further experimental investigations. For the polish alphabet containing 35 letters MLP has been constructed with 35 outputs. MLP was trained in "1 of M" manner, i.e. first neuron on and rest off to the character a_1, second neuron on and rest off to the character a_2 and so one. In this way, in the classification phase the value appearing on the ith position corresponding to the letter $a_i \in \mathcal{A}$ can be interpreted as the similarity measure between character image to be recognized and prototypes of a_i presented to MLP in the training phase. Support factors can be calculated by appropriately normalizing the MLP output values vector $(o_1, o_2, ..., o_M)$. First, the values o_i, $i = 1, 2, ..., M$ are clamped to $(0, 1)$ range:

$$\bar{o}_i = \begin{cases} 0 & \text{if } o_i < 0 \\ 1 & \text{if } o_i > 1 \\ o_i & \text{otherwise} \end{cases} \tag{6}$$

and next we put:

$$s_i = \frac{\bar{o}_i}{\sum_{j=1}^{M} \bar{o}_j}. \tag{7}$$

3.3 Support Vector for Dissimilarity-based Methods

In order to determine support factors for character classifiers with distance (dissimilarity) measure $d(X, X^{'})$ between character images X and $X^{'}$, we first calculate $o_i = d(X, X^*_{a_i})$, where $X^*_{a_i}$ denotes the closest (the most similar) prototype of letter a_i in the learning set. Next o_i is normalized $\bar{o}_i = \exp(-o_i)$ and finally we get s_i according to (7).

4 Soft Word Recognition Algorithm

The task of word recognition is equivalent to evaluating the support factors appearing in (4) for most likely words, using results of character classifications for all character fields and PLLM matrices. The idea of iterative method proposed here consists in extending string of recognized characters in successive iterations starting from the single character position for which character level recognizer provides the most reliable result. The support vector for subsequent character position is either direct result of character soft recognizer or it is derived using PLLM. The choice of method depends on the value of the normalized entropy of the support vector obtained after the next iteration of procedure is made. The algorithm starts with single character field for which normalized entropy of support vector (1) fetched by character classifier is minimal.

Let $I_k = (i_1, i_2, ..., i_k)$ denote the subsequence of character fields already processed in steps 1 to k of the iterative algorithm. $\widehat{S}_k$ is the set of subwords of the length k corresponding to subsequence I_k with their support factors:

$$\widehat{S}_k = ((w_1^k, s_1^k), ..., (w_l^k, s_l^k)), \tag{8}$$

i.e. $\widehat{S}_k$ is a result of word partial classification on the positions I_k, which can be assessed by normalised entropy of values $(s_1^k, ..., s_l^k)$

$$Q(s_1^k, ..., s_l^k) = \frac{\sum_{i \in \{1,l\}} s_i^k log_2(s_i^k)}{log_2 \frac{1}{l}} \tag{9}$$

In the $k + 1$ step the next character field position is appended to the set I_k. It can be done either on the left or on the right side giving $I_{k+1} = (i_1 - 1, i_1, i_2, ..., i_k)$ or $I_{k+1} = (i_1, i_2, ..., i_k, i_k + 1)$, correspondingly. Let us consider only the first case - the second one is analogous. Our aim is now to determine the set of subwords of the length $k + 1$ with their support factors (8) for the next, $(k + 1)$-th step. It can be done either using the results of soft classification $\Phi(X_{i_1 - 1})$ of the character on the new position or using PLLM. Which approach is applied depends on the value of entropy (9) for both methods.

4.1 Using Character Level Classification

Let

$$\Phi(X_{i_1-1}) = (s_1^{(i_1-1)}, s_2^{(i_1-1)}, ..., s_M^{(i_1-1)}). \tag{10}$$

The new set of subwords is created by appending (from left side) each letter $a_j \in \mathcal{A}$ to each word $w_i^k \in \widehat{S}_k$. Taking into account the independence of individual character classification, the support factor $s^{(k+1)}$ for the word $a_j w_i^k$ is equal

$$s^{(k+1)} = s_i^k * s_{a_j}^{(i_1-1)}, \tag{11}$$

where s_i^k is the support value for word w_i^k in the sequence (8) evaluated in the previous step and $s_{a_j}^{(i_1-1)}$ is the support value for letter a_j in (10). To avoid enormous growth of the sequence $\widehat{S}_{k+1}$ only l subwords with greatest $s^{(k+1)}$ remain for further processing.

4.2 Using PLLM

In this method set $\widehat{S}_{k+1}$ is created in similar way as previously, but now the support factors for $k+1$ length words are calculated using PLLM. Let in subword $w_i^k \in \widehat{S}_k$ the letter d be on the first position. The support value of the word $c_j w_i$ is calculated as:

$$s^{(k+1)} = s_i^k * \bar{p}_{a_j d}, \tag{12}$$

where $\bar{p}_{a_j d}$ is the value from letters precedence matrix (3). If the word w_i^k is extended on the right side (i.e. we consider suffix a_j to word w_i^k) then the letter d on the last position of w_i^k is important and the letters succession probability $\underline{p}_{a_j d}$ should be used in (12).

4.3 Soft Classification Algorithm for the Whole Word

We can now define the complete algorithm of soft word classification for the word of length m, as follows:

Calculate the normalized entropy (9) of support vector (1) for all character fields of the word; let r be the index of the character field with minimal entropy;
Create the set $\widehat{S}_1 = (((c_1), s_1^{(r)}), ..., ((c_M), s_M^{(r)}))$ as the set of single letter sequences (c_i) and their support values;
for k=2 to m
    Create temporary set $\widehat{S}_k^1$ using PLLM for words in set $\widehat{S}_{k-1}$ prefixed with all letters from the alphabet;
    Calculate Q_1 - entropy (9) for the set $\widehat{S}_k^1$;
    Create temporary set $\widehat{S}_k^2$ using character level classification for words in set $\widehat{S}_{k-1}$ prefixed with all letters from the alphabet;
    Calculate Q_2 - entropy (9) for the set $\widehat{S}_k^2$;
    Create temporary set $\widehat{S}_k^3$ using PLLM for words in set $\widehat{S}_{k-1}$ suffixed with all letters from the alphabet;
    Calculate Q_3 - entropy (9) for the set $\widehat{S}_k^3$;
    Create temporary set $\widehat{S}_k^4$ using character level classification for words in set $\widehat{S}_{k-1}$ suffixed with all letters from the alphabet;
    Calculate Q_4 - entropy (9) for the set $\widehat{S}_k^4$;
    Calculate $i = arg\min_{j\in\{1,2,3,4\}} Q_j$;

```
    Remove from the set Ŝ_k^i all elements except the l ones
    with the highest support values;
    Adopt the reduced set Ŝ_k^i as the set Ŝ_k;
end for
```

5 Experiments and Concluding Remarks

In order to evaluate the proposed method of words recognition computer experiments were made, in which the method has been compared to simple algorithm based just on character level recognition and to word recognizer based on probabilistic lexicon ([6]). In experiments as character classifier the MLP was applied which used directional features extracted according to procedure presented in [5]. The support vectors were evaluated according to idea described in section 3.2. MLP was trained using the set of 8,795 isolated letter images. PLLM was prepared using the corpus of texts in polish language consisting of about 3,200,000 characters. The correctness of character classifier was equal to 89.9%. In the experiments, polish surnames coming from the lexicon containing 9,944 elements were tested. The experiments were performed using simulated data according to the following scheme.

First, the word to be recognized were randomly selected from the lexicon. Then, for each character field of the selected word, appropriate letter image was randomly selected from the set of 2,000 letter images others than images used to train MLP. The word images obtained in this way were subject of recognition. Three word classifiers were compared:

- combined classifier described in section 4.3,
- the classifier using lexicon of recognized words,
- the classifier based merely on classification of isolated characters.

Results of experiments (frequency of correct classifications in percent) are depicted in the Table 1.

Table 1. Classifiers accuracy - surnames recognition

Criterion	Character-based	Combined	Lexicon-based
1 of 1	48.1%	79.7%	59.5%
1 of 3	57.0%	68.4%	71.4%
1 of 5	65.6%	73.0%	77.2%

On the one hand, the performance of combined classifier is noticeably better than performance of simple classifier based just on character recognition, on the other however, it is close to lexicon-based recognizer accuracy. This is probably due to relatively low lexicon-based recognition quality caused by large number of elements in the lexicon.

Experimental investigations confirm that the performance of proposed combined method is close to the performance of methods based on lexicon, in case of large lexicons, hence it can be treated as an alternative approach in cases where lexicons are large or are not available at all. Obviously, the results of experiments presented here have not an exhaustive character. More tests are necessary to fully assess the method. In particular it should be compared with approaches utilizing Hidden Markov Models. Nevertheless, obtained results suggest some perspectives for further investigations, both in theoretical and experimental aspects.

Acknowledgement. This work was financed from the State Committee for Scientific Research (KBN) resources in 2005 - 2007 years as a research project No 3 T11E 005 28.

References

1. Marti U.V., Bunke H. (2001) Using a Statistical Language Model to Improve the Performance of an HMM-based Cursive Handwriting Recognition System, International Journal of Pattern Recognition and Artificial Intelligence, Vol 15, No 1 : 65-90
2. Bote-Lorenzo M.L., Dimitriadis Y.A., Gomez-Sanchez E. (2003) Automatic Extraction of Human-recognizable Shape and Execution Prototypes of Handwritten Characters, Pattern Recognition, Vol 36 : 1605-1617
3. Hanmandlu M., Murali Mohan K.R. (2003) Unconstrained Handwritten Character Recognition Based on Fuzzy Logic, Pattern Recognition, Vol 36 : 603-623
4. Koerich A. L., Sabourin R., Suen C.Y. (2003) Large Vocabulary off-line Handwriting Recognition: A Survey, Pattern Anal. Aplic., No 6 : 97-121
5. Liu C., Nakashima K., Sako H., Fujisawa H. (2003) Handwritten Digit Recognition: Benchmarking of State-of-the-art Techniques, Pattern Recognition, Vol. 36 : 2271-2285
6. Sas J., Kurzynski M. (2005) Multilevel Recognition of Structured Handprinted Documents - Probabilistic Approach, Proc. Int. Conf. on Computer Recognition Systems CORES'05, Springer Verlag (in this Volume)
7. Sas J. (2004) Handwritten Laboratory Test Order Form Recognition Module For Distributed Clinic, Journ. of Medical Informatics & Technologies, Vol 8 : 59-68
8. Sas J. (2004) Three-Level, Lexicon-Based Handwritten Form Recognition Method, In: Klopotek M., Tchorzewski J. (eds) Proc, of VI Int. Conf on Artificial Intelligence AI-19'2004, Vol. 1 (23) : 113-124
9. Vinciarelli A, Bengio S, Bunke H (2004) Offline Recognition of Unconstrained Handwritten Texts Using HMMs and Statistical Language Models, IEEE Trans. on Pattern Anal. and Machine Intelligence, Vol 26, No 6 : 709-720

Speaker Recognition for VoIP Transmission Using Gaussian Mixture Models

Piotr Staroniewicz

Wroclaw University of Technology, Institute of Telecommunication and Acoustics, Wybrzeze Wyspianskiego 27, 50-370 Wroclaw, Poland
Piotr.Staroniewicz@pwr.wroc.pl

Summary. The paper presents the speaker recognition problem in the background of voice transmission via Internet. The Gaussian Mixture Models (GMM) classification, the voice feature extraction, the Internet speech transmission standards and the packet loss simulation methodology applied in the tested system were overviewed. Speaker identification scores obtained for the tested GMM based text-dependent system has revealed a minor significance of packet loss problem in this aspect.

1 Introduction

The Internet is evolving into a universal communication network and it is contemplated that it will carry all types of traffic, including voice, video and data. Among them, telephony, namely VoIP (Voice over IP) is an application of a great importance. The speaker verification problem was partly solved for transmission over traditional PSTN networks (Public Switched Telephone Network). It is also important to assess how specific conditions and distortions of Internet transmission (like packet delay and loss) can influence the speaker verification problem. Gaussian Mixture Models (GMMs) are dominant classifiers in nowadays text-independent speaker recognition [2, 6] and is used as a generic probabilistic model for multivariate densities. GMM-based systems have been applied to the annual NIST Speaker Recognition Evaluation (SRE) which have produced state-of-the-art performance [6]. The advantages of using a GMM are that it is computationally inexpensive and based on well-understood statistical model. What is the most important for text-independent tasks is that the GMM is insensitive to the temporal aspects of the speech, modeling only the underlying distribution of acoustic observation from a speaker [2].

2 GMM speaker identification system

The classical speaker recognition system consists of two main procedures: feature extraction and classification. MFCC (Mel Frequency Cepstral Coefficients) parameterization method was chosen. In the Fig.1 the applied feature extraction procedure is presented. The speech signal is first preemphasized

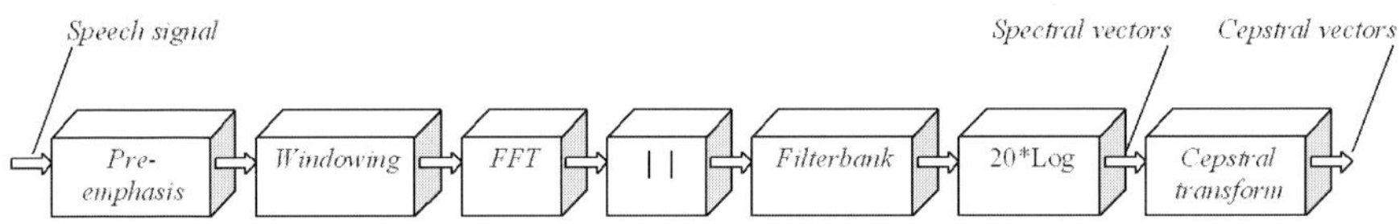

Fig. 1. Scheme of the filterbank-based cepstral parameterisation

to enhance the high frequencies of the spectrum. After windowing with the Hamming window the signals fast Fourier transform (FFT) is calculated. Finally the modulus of FFT is extracted and a power spectrum is obtained. To realize the smoothing and get the envelope of the spectrum in an auditory scale (similar to the frequency scale of human ear) we multiply the spectrum by a Mel scale filterbank. After obtaining spectral envelope in dB as a final step of parameterisation procedure, the cosine discrete transform is performed and yields cepstral coefficients. Such received parameters vectors are given to the classification procedure. The GMM belong to statistical methods of classification. For D-dimensional feature vector $\mathbf{x}$, the mixture likelihood density function (Fig.2) is defined as a weighted linear combination of M unimodal Gaussian densities $p_i(\mathbf{x})$:

$$p(\mathbf{x}|\lambda) = \sum_{i=1}^{M} w_i p_i(\mathbf{x}) \tag{1}$$

Each density is parameterized by a $D \times 1$ mean vector $\boldsymbol{\mu}$ and $D \times D$ covariance matrix Σ_i:

$$p_i(\mathbf{x}) = \frac{1}{(2\pi)^{D/2}|\Sigma_i|^{1/2}} \exp\left((-1/2)(\mathbf{x} - \boldsymbol{\mu}_i)'\Sigma_i^{-1}(\mathbf{x} - \boldsymbol{\mu}_i)\right) \tag{2}$$

The mixture weights w_i satisfy the constraint: $\sum_{i=1}^{M} w_i = 1$. Given a collection of training vectors, maximum likelihood model parameters are estimated using the iterative expectation-maximum (EM) algorithm [2]. The EM algorithm iteratively refines the GMM parameters to monotonically increase the likelihood of the estimated model λ. Under the assumption of independence feature vectors, the log-likelihood of model λ for a sequence of feature vectors $\mathbf{X} = \{\mathbf{x}_1, \ldots, \mathbf{x}_T\}$ is computed as follows:

$$\log p(\mathbf{X}|\lambda) = \frac{1}{T}\sum_t \log p(\mathbf{x}_t|\lambda) \tag{3}$$

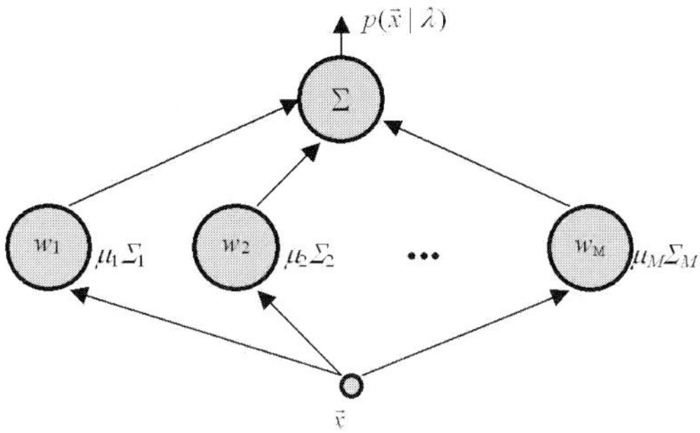

Fig. 2. Mth-order Gaussian model

3 Packet loss problem in VoIP

Figure 3 presents the VoIP transmission scheme [7]. The sender digitizes/encodes the media content and transmits it via the network as packets at regular intervals. The receiver gets the media packets and schedules an appropriate playout time in order to produce a smooth output media stream. The playout buffer delivers a continuous stream of packets to the depacketizer and eventually to the decoder which reconstructs the speech signal. Decoders often implement Packet Loss Concealment (PLC) that produces a replacement for a lost packet, similar to the original one, by filling in silence or noise, by interpolating or even by regenerating the packet from the surrounding ones. Error concealment works best for small loss rates and durations. Voice data can be compressed by a coder to take as little IP-network bandwidth as possible. VoIP voice compression standards are presented in Table 1 [8]. The G.711 standard describes the type of voice transmission used in ISDN. The voice sampled at 8kHz and 12 bits analog/digital conversion is converted to 8 bits according to a-law (Europe) or μ-low (North America) characteristics. The G.726 Adaptive Differential PCM (ADPCM) relies on the consecutively following values varying only slightly from each other, so it is better to send an absolute value first and then the deviations from this value. Additional compression is achieved by changing the bit coding according to the statistical evaluation. The difference in quality compared to uncompressed PCM is hardly noticeable. CELP (Codebook Excited Linear Predictive Coding) technology (G.728 LD-CELP, G.729/G.729A CS-ACELP) realizes a very efficient

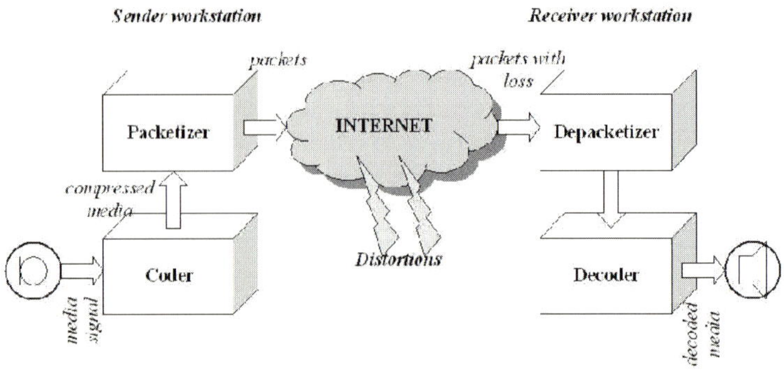

Fig. 3. VoIP voice transmission

use of the bandwidth designed on the basis of the mathematical model of the human speech system. The G.723/G.723.1 MP-MLQ were developed from the ITU standard H.323 which was developed in the field of video conferencing.

Table 1. Standards of voice compression in the VoIP

Code type	Transfer rate (kbps)	Processor load (MIPS)	Voice quality	Delay
G.711 PCM	64	-	Very good	Nominal
G.723 MP-MLQ	6.4/5.3	20	Good to poor	High
G.723.1 MP-MLQ	6.4/5.3	20	Good to poor	High
G.726 ADPCM	40/32/24/16	8	Good to poor	Very slight
G.728 LD-CELP	16	40	Good	Slight
G.729 CS-ACELP	8	30	Good	Slight
G.729A CS-ACELP	8	20	Satisfactory	Slight

The quality of VoIP transmission is primarily determined by packet loss and delay. If packet is lost, the quality degrades and on the other hand, if a packet delay is too high and misses the playout deadline, it leads to a late loss. If a packet has a large delay, next packet is also likely to do so. This burstiness effect can not be captured by simple metrics such as an average loss and delay. Therefore, the metrics that can characterize the packet loss and delay process have been established. Packet losses are not independent on the frame-by-frame basis, but appear in burst. Studies on the distribution of the packet loss in the Internet [4, 7] have concluded that this process could be approximated by a Markov models. Two states Markov model, also known as the Gilbert model is used most often to capture the temporal loss dependency (Fig.4). In Fig.4, p is the probability that the next packet is lost, provided the previous one has arrived, q is the opposite. $1 - q$ is the conditional loss probability.

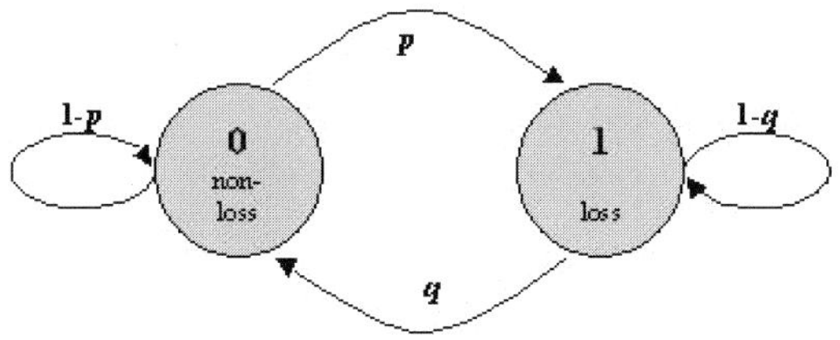

Fig. 4. Gilbert model

From the definition, we can compute π_0 and π_1, the state probability for state "0" and "1", which also represent the mean arrival and loss probabilities, respectively.

$$\pi_0 = \frac{q}{p+q}, \pi_1 = \frac{p}{p+q} \tag{4}$$

Let m_i, where $i = 1, 2, \ldots, n-1$ denote the number of loss burst having length i, where $n-1$ is the length of the loss burst. Let m_0 denote the number of delivered packets. Then p and q probabilities can be calculated as follows:

$$p = \frac{\sum_{i=1}^{n-1} m_i}{m_0}, q = 1 - \frac{\sum_{i=2}^{n-1} m_i \cdot (i-1)}{\sum_{i=1}^{n-1} m_i \cdot i} \tag{5}$$

4 Experiment description

The system was tested with the SV-POL database [5] which consists of speech samples of 22 speakers recorded at 16bit/48kHz in acoustically good conditions (a recording studio). The speech material included isolated digits and vowels, phonetically rich sentences and strings of digits. In all experiments the sentences (below noted as "S") and strings of digits (noted as "D") were used. For tests the original signals were down-sampled to 8kHz and transmitted via two types of encoders typical for VoIP transmission:

- G.711 with a-law (64 kbit/sec.)
- G.723 (5.3 kbit/sec.)

The process of packet loss was simulated with the two states Gilbert model (Fig.4), where state "0" represents the case when the packet is lost and state "1" when the packet is correctly transmitted. Probabilities p and q represent going from state "0" to "1" and from "1" to "0". Two conditions were simulated: bad network conditions ($p = 0.25$, $q = 0.4$) and average network conditions ($p = 0.1$, $q = 0.7$) [1]. The packet length was 30ms in both cases. In the front-end procedures of the voice recognition system experimentally selected feature extraction settings were used: pre-emphasis parameter 0.95, window length of 256 samples, overlap of 128 samples and finally the feature vector consisted of 12 MFCC parameters extracted with the bank of 26 mel-filters. The GMM classifier had 16 Gaussian densities. The number of iterations in the EM algorithm was experimentally set for 15.

5 Results and discussion

Table 2 presents speaker identifiaction scores for the two tested speech items ("S"-sentences and "D"-digit strings) and three network conditions: with packet loss, average and poor (as defined in 4).

Table 2. Speaker identification scores for G.711 and G.723 encoders for three network conditions: with no packet loss, average and poor.

Encoder	No loss "S"	No loss "D"	Average "S"	Average "D"	Poor "S"	Poor "D"
G.711	97.18%	98.30%	97.02%	97.72%	94.93%	96.81%
G.723	96.03%	92.64%	95.52%	87.40%	99.34%	85.30%

For both tested coding types (G.711 and G.723) packet loss does not affect the identification scores. For the low bit rate encoder G.723 (5.3kbit/sec.) there is the maximum fall of 11.51%. The scores of G.723 encoder are on average 4% lower than for G.711.

6 Conclusions

Results obtained with the tested text-independent system has shown a minor influence of the packet loss problem on speaker identification scores (this confirms the results of preliminary experiments presented in [1, 3]). Beside expanding the research to other aspects of speech recognition such as speaker verification and authentication the main topic of further experiments would probably be testing the influence of the packet loss on the text-dependent speaker recognition. Despite the fact that the packet loss problem does not affect the text-independent speaker recognition scores, it has probably a bigger impact on the text-dependent recognition which is similarly to the automatic speech recognition, more sensitive to time distortions (including packet loss) in speech signal.

References

1. Besacier L, Mayorga P, Bonastre J F, Fredouille C (2002) Methodology for evaluating speaker robustness over IP networks. Proc. of a COST 275 workshop The Advent of Biometrics on the Internet, Rome, Italy, 43-46
2. Bimbot F, Bonastre J F, Fredouille C, Gravier G, Magrin-Chagnolleau I, Meignier S, Merlin T, Ortega-Garcia J, Petrovska-Delacretaz D, Reynolds D A (2004) A Tutorial on Text-Independent Speaker Verification. EURASIP Journal on Applied Signal Processing 4, 430-451
3. Evans N, Mason J, Auckenthaler R, Stamper R (2002) Assesment of speaker verification degradation due to packet loss in context of wireless devices, Proc. of a COST 275 workshop The Advent of Biometrics on the Internet, Rome, Italy, 43-46
4. Jiang W, Schlzrinne H (2000) Modeling of Packet Loss and Delay and Their Effect on Real-Time Multimedia Service Quality, Proc. NOSSDAV, North Carolina, USA
5. Majewski W, Staroniewicz P, Sadowski J (2003) Speaker Verification via Internet vol.1, Institute of Telecommunication ans Acoustics Report I28/S-005/2003
6. Reynolds D A, Quatieri T F, Dunn R B (2000) Speaker Verification Using Adapted Gaussian Mixture Models. Digital Signal Processing 10, 19-41
7. Staroniewicz P (2004) Creation of Real Conditions VoIP Database for Speaker Recognition Purposes, Proc. of 2nd Cost275 Workshop in Vigo, Spain, 23-26
8. ITU-T Recommendation H.323, Packet-based multimedia systems

Semi-Automatic Segmentation of Speech: Manual Segmentation Strategy. Problem Space Analysis.

Marcin Szymanski and Stefan Grocholewski

Poznan University of Technology, Institute of Computing Science
ul. Piotrowo 3a, 60–965 Poznañ, Poland
mszymanski@cs.put.poznan.pl

Summary. The important element connected with today's speech recognition/synthesis systems is the speech database – the set of fully annotated wave-files. Since the manual segmentation of speech is a very time-consuming task, the automatic segmentation algorithms are needed. However, the manual segmentation still outperforms the automatic one and at the same time the quality of resulting synthetic voice highly depends on the accuracy of the phonetic segmentation. In this paper we concentrate on a semi-automatic approach, in which a human expert, unlike in the common approach, manually allocates the selected boundaries *prior* to the automatic segmentation of the rest of the corpus. In the paper we quest for the appropriate strategy for an expert. We check if locating some boundary classes influence the rest of the annotations. It is done for two different quality measures.

1 Introduction

In the process of constructing speech recognition and synthesis systems it is essential that the proper set of prerecorded utterances is available. Moreover, it should additionally contain precise information, such as the sequence of phoneme labels and subsequent unit durations. In case of recognition systems the accuracy of phoneme boundaries is not crucial, however the errors in segmentation seriously affects the quality of obtained *synthesis* system.

The manual segmentation of speech is a very labor-intensive process, moreover it should be performed by an expert (usu. in phonetics) and it is prone to inconsistencies. The simplest idea is to implement an algorithm which will do this task automatically. In this task (in contrast to the phoneme recognition problem), we assume that the phone sequence is known. Obviously, the obtained automatic boundary points will not be faultless.

The basic algorithmic solution of the segmentation is to run a HMM recognizer in *forced alignment* mode[6]. The segmentation can now be considered a special case of recognition, where the word– and model-net are the simple concatenation of units, corresponding to the imposed phonetic transciption of an utterance.

The important limitation of HMM's application to speech processing is its ignoring the probability densities of phoneme duration. Since the state transition probability in standard HMM is represented by one constant value, the state duration have an implicit geometric probability density, which most probably is inadequate as the duration model. For this reason, the observed phoneme duration should be modeled. In [7], we expanded[1] the *Token Passing* algorithm to consider more suitable pdf's (context-dependent triple gamma mixture was used in this work). Although this extension implicates a higher complexity, we think that calculation time is not crucial in the segmentation task.

In this work we propose and test the approach in which a human expert performs the manual segmentation of selected transition cases and the rest of boundaries are calculated automatically. Initially, we cluster the monophones into several classes, which implies about 100 different phoneme transition classes. Then we analyze the effect of "revealing" selected expert annotated boundaries from one (or more) of these classes on the segmentation accuracy measured on the whole corpus. For the needs of the task we did further (rather straightforward) extension to the segmentation algorithm to consider the predefined transition points.

It is expected that inserting annotations considering one class will influence the error measures of other classes, because shifting one boundary can "pull" neighbouring transitions (particularly since the introduction of the duration models).

The basic goal is to find a strategy of performing the manual segmentation so that maximum error reduction is obtained compared to the expert labor required. It is clear that there can be many "optimal" solutions on different levels of allowed labor or required accuracy.

The analysis is done for 2 error measures: number of gross errors and deviation/variance of error. The gross error occurs when an automatically located boundary passes beyond the adjacent manually labelled segments[2]. The standard deviation of error is a suitable measure since we want to minimize the spread of usual boundary displacements and not necessarilly the mean error (bias). For the calculation of error *reduction*, however, we use variance instead of deviation.

The rest of this paper is organized as follows: Section 2 describes the clustering of phonemes into classes that has to be done prior to the main task; in Section 3 we discuss some theoretical considerations of this problem; in Section 4 we present the experimental results. The paper is concluded in Section 5.

---

[1]Technically, in every state we maintain a separate list of N best paths (represented by tokens) for every different time of entering the current model (N is hypothesis list length, here N=1). See [4] for the equivalent approach.

2 Clustering of phonemes

Our phonetic system contains of 39 polish phonemes. Of course, n phoneme groups implicates n^2 transition groups (a bit less actually, since some cannot be found in the training set, e.g. $sil \rightarrow sil$). Having too many groups would cause the resulting matrix to be too large to analyze (since we want to check the impact of "revealing" some boundary types on the segmentation error). On the other hand, having too few groups might produce too large clusters of boundaries.

We start with a "scattered" partition: {sil sp}, {i y}, {e}, {a}, {ê ź} {o u} {j ş l}, {r}, {m n ñ N}, {w z § £}, {f h}, {s IJ sz}, {dz d§ d£}, {c æ cz}, {b d g}, {p t k}. We concentrated on the matrix of standard error deviation and calculated the horizontal and vertical correlation coefficients between every 2 clusters. We performed the clustering in a greedy manner – at each step two classes with the highest weighted sum of horizontal and vertical error correlation were merged, provided that the resulting cluster did not exceed 25% of the phoneme database. The stop condition was reaching less than 10 clusters or a maximum correlation below 0.6 (both were satisfied at the same time). Practically it means that if two clusters had a low correlation, they were problematical in *different* contexts. We do not want to create a class of boundaries with comparable numbers of both coarse and fine errors because then we run a higher risk of forcing an expert to manually provide boundaries which are quite well derived automatically. Obviously, this method is by no means optimal but testing different phoneme partitions with respect to manual segmentation strategy would form a too complex task.

Finally, we achieved a partition of 9 clusters presented in Table 1. The two vowel clusters still have a very high correlation coefficient (over 0.8), but they have not been merged because the resulting cluster would be too large (besides, a lookup at the gross error matrix showed that, among 4 inter-vowel transition classes, there had been gross errors only for $VU \rightarrow VU$). What is suprising here is that voiced plosives were included in one cluster with unvoiced fricatives and affricatives. It should be noted once again that the clustering is not based on acoustical similarity, but on the context-dependent segmentation error. After performing the greedy clustering, simple validations were made, in which (1) 'ê' was taken out of the VLD set (to be possibly included in VU); (2) 'b', 'd', 'g' were taken out to form a separate cluster. The partition from Table 1 has not been altered as a result of those validations.

3 Problem space complexity

First, some definitions must be introduced. We analyze a matrix of 80 ($9\times9-1$) phoneme transition (boundary) classes. One class (e.g. VU→CVX) will be denoted by one small italic letter (in this case "m", which is not to be confused with a phoneme 'm'). If we choose to "reveal" several transition classes at once

Table 1. Final phoneme clusters used in this paper

Symbol	Phonemes	Symbol	Phonemes
S	sil sp		
VU	i y e	VLD	o u a ź ê
CVX	j ş l	CVN	m n ñ N
CV	dz d§ d£ r w z § £	CLAF+VP	c æ cz s IJ sz b d g
CLP	p t k	CLF2	f h

we have a set denoted in curly brackets (e.g. "$\{abcdef\}$") and sometimes by capital letters ($\mathcal{F}$), whereas parentheses denote a *sequence* of actions (e.g. *(c, d, a, f, b, e)*). One set is referred to as an *object* and corresponds to one test performed in this work.

Each object has *two* associated parameters, the total size of included classes (as the percentage of all boundaries in the corpus) and an accompanying accuracy boost or error reduction (E.R.) – e.g. "$\{abd\}$" contitutes 1.4% of transitions, yielding a 4.3% reduction of error variance. We want to minimize the total size of the object and maximize the E.R. This means we deal with a multi-criteria problem (see [5]). Such problems are usually analysed by a dominance relation. We say that object $\mathcal{X}$ *strictly dominates* $\mathcal{Y}$ with respect to the criteria set C if $\mathcal{X}$ is not worse on any criterion in C and it is better than $\mathcal{Y}$ on at least one criterion. An object that is not dominated by any other object is called Pareto–optimal. An object that is not dominated by any other *known* object is called potentially Pareto–optimal (PPO).

Our objective could be to find as many PPO objects as possible. In fact, we are first of all interested in finding one *convex* path of PPO – that is the expert strategy. When we say *path*, we assume that each set can be "reached" from any other set that is the subset of the former smaller by one class. *Convexity* means that the normalized error reduction decreases as subsequent classes are added.

The ideal, globally optimal, solution would require a search in which we test all possible subsets of class matrix (up to the desired size), hence it would require performing $\binom{N}{L}$ tests, where N is the number of classes ($N = 80$) and L is the number of levels or a maximum size of a set (we assume $L \approx 20$). The result would be the set of *all* Pareto–optimal objects not larger than L.

We can also think of a sub-optimal strategy, where the extension would consider only those objects, which yield a maximum E.R. (absolute or normalised by a total number of boundaries) among all objects of the same size. That requires performing $L \times (2 \times N - L + 1) \div 2$ tests. However, that is still over 1000 tests.

The simplest method consists of two rules. *(Rule 1.)* Sorting all transition classes by a decreasing error measure and greedily extending the set of manually provided ("revealed") classes. *(Rule 2.)* If it is detected that extending any set $\mathcal{A} \cup \{x\}$ by a class $\{y\}$ gives higher E.R. compared to adding $\{x\}$ to $\mathcal{A}$

(normalized by a number of inserted boundaries), testing the object $\mathcal{A} \cup \{y\}$ is also performed; this way a convexity of a path is assured – $\{y\}$ will precede $\{x\}$ in the final sequence if adding $\{y\}$ to $\mathcal{A}$ gives higher E.R. than adding $\{x\}$ to $\mathcal{A} \cup \{y\}$.

Of course this method does not guarantee that the found objects will be Pareto–optimal in the global sense. We think, however, that it is a good heuristic, especially if we expect that the impact of inserting one class is additive and remains similar irrespective of the set that is expanded by the class.

Additional complexity comes from the necessity to analyze two quality criteria. In this work, a separate search "thread" will be developed for both of them.

4 Experimental results

We used a part of Polish Corpora [1] database. It consisted of a total of 5 hours of speech, inside 28 folders of 365 separate sentences each, coming from 24 different speakers. Hence, we deal with a speaker-independent segmentation.[2] The baseline HMM models were trained for the MFCC target rate of 10 milliseconds, however, due to the simple trick in the segmentation algorithm[7] the final precision of 5 ms was achieved, which corresponds the precision of expert annotations. The accuracy of this system can be described by the following parameters: 308 gross errors per one million boundaries, 16.6 ms of standard deviation of error, 72.9% of boundaries correctly located within 10 ms tolerance and 89.2% within 20 ms tolerance.

All tests were performed in 7–fold cross-validation, repeated 4 times.

4.1 General observations

Tables 2 and 3 present the baseline error for different transition classes. A few remarks can be made: (1) the most problematic transitions are from speech to silence (usually systematically calculated ca. 20 ms before the expert annotation);[3] (2) horizontally, unvoiced fricatives CLF2 were highly error-prone, esp. on transitions to silence and consonants; (3) vertically, the highest errors were associated with transitions *to* silence (S) as well as CVX (additionally, CVX→CVX contained many gross errors) and CLF2; (4) inter-vowel boundaries were also problematic, however the gross errors were observed only for VU→VU; (5) *all* of the 8 "homogeneous" classes, i.e. representing transition

[2]It may be noted that the database was specifically designed to contain as much different diphones as possible. As some sentences were not of daily occurence, this might have influenced the statistical models used in this work.

[3]This is *not* caused by the introduction of the prosody-insensitive duration models, since the basic HMM version suffers from the same phenomenon.

Table 2. Number of gross errors per one million boundaries

from \ to	S	VU	VLD	CVX	CVN	CLAF+VP	CV	CLF2	CLP	(all)
S	-	0.0	0.0	0.0	0.0	518.9	566.3	3355.7	0.0	380.0
VU	0.0	6476.7	0.0	1163.8	289.4	152.9	0.0	0.0	0.0	345.1
VLD	428.9	0.0	0.0	969.0	0.0	105.7	0.0	0.0	0.0	219.7
CVX	1430.6	1356.2	55.0	4166.7	0.0	0.0	0.0	0.0	0.0	622.3
CVN	863.6	302.8	0.0	0.0	0.0	0.0	0.0	0.0	0.0	157.7
CLAF+VP	0.0	251.3	59.0	0.0	0.0	0.0	892.1	0.0	0.0	147.2
CV	4629.6	634.5	0.0	0.0	0.0	275.9	1710.9	0.0	0.0	534.1
CLF2	0.0	0.0	0.0	0.0	0.0	2247.2	0.0	0.0	0.0	446.3
CLP	0.0	0.0	264.6	0.0	0.0	0.0	0.0	6060.6	0.0	230.5
(all)	593.7	564.7	62.8	691.4	90.1	212.7	328.7	892.6	0.0	317.5

Table 3. Mean variance of error

from \ to	S	VU	VLD	CVX	CVN	CLAF+VP	CV	CLF2	CLP	(all)
S	-	121.7	90.0	99.0	120.8	228.5	249.8	868.3	92.7	201.1
VU	601.4	1160.0	440.4	542.0	113.3	110.0	151.1	111.5	157.8	306.9
VLD	508.4	448.5	861.1	324.3	124.2	94.8	121.0	119.1	118.2	290.3
CVX	465.6	255.8	206.4	534.2	259.8	125.5	238.3	93.5	164.2	249.2
CVN	495.0	260.0	81.0	156.6	1579.7	131.3	307.3	245.3	87.7	296.3
CLAF+VP	174.5	74.5	64.8	102.9	175.3	655.6	283.3	258.9	106.4	159.0
CV	2254.3	160.2	117.6	264.1	229.8	163.4	601.4	290.8	192.2	321.6
CLF2	3373.4	95.6	80.2	45.4	87.6	378.4	214.8	432.4	180.3	737.8
CLP	880.9	86.9	67.3	70.3	185.7	194.0	202.6	296.5	397.5	225.6
(all)	856.4	187.7	134.3	310.1	192.7	190.3	233.5	301.0	145.4	285.0

from a phoneme cluster into the same cluster, were among 19 most missegmented classes.

4.2 Experiments

Table 4 shows the 21 most problematic classes. As a result of applying the greedy Rule 1 (Sect. 3) using the measure of error variance, we obtain a path of object which can be described as a sequence of classes: *(d, c, f, a, i, b, h, j, k, l, m, g, n, o, r, q, s, p, u, t, e)*. All 21 classes in total constitute 18.6% of the database and yield 57.2% of variance-of-error reduction. As it can be seen, this sequence does not form a convex path. Hence, we apply the Rule 2, yielding a sequence *(c, d, a, f, h, n, l, i, o, b, j, k, g, m, p, q, t, r, s, u, e)*. This means that the human expert should manually do all CV→S transitions first, then, if labor resources still allow it, annotate CLF2→S class, then VU→VU, then CVN→CVN and so on. It can be found that this sequence is not perfectly convex, as adding "*t*" to {*abcdfghijklmnopq*} gives lower accuracy boost than

Table 4. The most problematic transition classes

Transition	Gr.errors	Std.dev.of error	Class size (%)	Symbol
VU→VU	6476.68	34.06	0.173	*a*
S→CLF2	3355.71	29.47	0.533	*b*
CV→S	4629.6	47.48	0.290	*c*
CLF2→S	0	47.68	0.657	*d*
CLP→CLF2	6060.61	17.22	0.148	*e*
CVN → CVN	0	39.75	0.329	*f*
CVX → CVX	4166.67	23.11	0.376	*g*
VLD → VLD	0	29.35	0.374	*h*
CLP → S	0	29.68	0.962	*i*
CLAF+VP→CLAF+VP	0	25.61	1.059	*j*
CV → CV	1710.86	24.52	1.046	*k*
VU → S	0	24.52	1.394	*l*
VU → CVX	1163.78	23.28	2.114	*m*
VLD → S	428.92	22.55	3.650	*n*
CVN → S	863.56	22.25	1.036	*o*
CLF2 → CLF2	0	20.79	0.040	*p*
VLD → VU	0	21.18	0.199	*q*
CVX → S	1430.62	21.58	0.625	*r*
VU → VLD	0	20.99	0.533	*s*
VLD → CVX	968.99	18.01	2.769	*t*
CLP → CLP	0	19.94	0.300	*u*

inserting "*r*" into {*abcdfghijklmnopqt*}, but swaping "*t*" and "*r*" in the sequence does not give a convex path either.

Analogously, we got a sequence *(a,g,b,c,h,m,k,o,q,t,j,n)* [4] for the number of gross errors measure. It requires segmentation of 13.6% transitions and gives 75.4% E.R.

4.3 Characteristics of error reduction

As the method was already tested while it was still developed, there were several aberrations from the above rules and there were much more tests performed. They allowed us, however, to analyze the impact of "revealing" a particular class on the rest of the automatically segmented transitions. The complete results of those tests are not presented here, except for the Figure 1. Out of 192 tested objects, 33 were PPO for the gross error meausure and

[4]Class CLP→CLF2 ("*e*") has been left out of a sequence, because correcting the "gross error" boundaries for that class was *impossible* due to the HMM topology limitations. The Token Passing algorithm *penalyzed* tokens breaking the "expert boundary" instead of simply deleting them, so it did not return the "no token survived" answer. The situation of impossible expert trancription can be easily detected, but this issue was not dealt with in this paper.

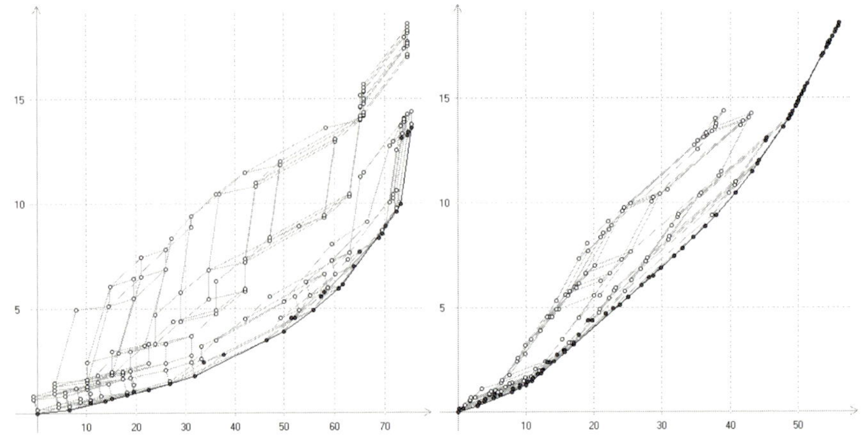

Fig. 1. Required percentage workload in function of (left) desired percentage reduction of gross errors and (right) in function of desired variance E.R. Solid points indicate PPO objects, expert strategies are represented by black lines.

82 were PPO for variance of error measure. Major observations concerning the tests include: (1) inserting one class (to the set of manually segmented) practically does not change the error calculated for the other problematic classes from Table 4, as far as the variance of error is concerned; in general, the influence on the other classes from the matrix is very small; (2) for the number of gross errors, there were examples of correcting boundaries that were not "revealed" (e.g. adding VU→VU implicated the E.R. for the VU→CVN class); (3) we observed very little effect of combining classes, i.e. the error reduction was similar (almost additive, with a few exceptions), no matter how large a class was expanded; this is demonstrated on Fig. 1 by numerous parallel connections.

5 Conclusions

We have confirmed that some classes of boundaries are more error-prone than other classes and should be the first to be manually annotated; they were mostly sentence-final and homogeneous transistions. We have proposed two manual segmentation strategies, depending on the minimized error measure. We have found that inserting one class has very little impact on the segmentation error among other classes.

Further work include, concerning the semi-automatic segmentation: testing other ideas for the "prior" segmentation (e.g. force every nth transition to be manually segmented), developing boundary confidence measure and proposing the interactive, possibly multi-pass, segmentation procedure; concerning the automatic alignment stage: introducing prosody-dependent and speaking-rate normalized phoneme duration models and applying the boundary-specific post-processing[3].

References

1. Grocholewski S. (1997), *CORPORA – Speech Database for Polish Diphones*, Proc. Eurospeech'97, pp. 1735-1738
2. Kvale K. (1993), *Segmentation and Labelling of Speech*, Ph.D. Thesis, Institutt for Teleteknikk, Trondheim
3. Matousek J., Tihelka D., Psutka J. (2003), *Automatic Segmentation for Czech Concatenative Speech Synthesis Using Statistical Approach with Boundary-Specific Correction*, Proc. Eurospeech 2003, pp. 301-304, Geneva
4. Ostendorf M., Digalakis V.V., Kimball O.A. (1996), *From HMM's to Segment Models: A Unified View of Stochastic Modeling for Speech Recognition*, IEEE Trans. on Speech and Audio Proc., Vol. 4, No. 5, September 1996
5. Steuer R.E. (1986), *Multiple Criteria Optimization Ũ Theory, Computation and Application*, Wiley, New York
6. Szymański M., Grocholewski S. (2003), *Automatic Speech Segmentation Based on Transcription*, RB–023/03, Poznan Univ. of Tech., Inst. of Computing Sc. (in Polish)
7. Szymański M., Grocholewski S. (2005), *Implementation of Speech Segmentation Algorithm with Statistical Duration Models. Tuning the Model Parameters*, RB–004/05, Poznan Univ. of Technology, Inst. of Computing Science (in Polish)

Part VI

FINGERPRINT AND FACE RECOGNITION

HMM and WT Fusion for Face Identification

Janusz Bobulski

Czestochowa University of Technology, Institute of Computernad Information Science, Dabrowskiego Street 73, 42-200 Czestochowa, Poland
januszb@icis.pcz.pl

Summary. This paper describes the original *FaMar* method of user's identification. The method bases on the fusion of Wavelet Transform (WT) and Hidden Markov Models (HMM), which is used for three parts of the face (eyes, nose, and mouth) separately.

1 Introduction

A problem of persons' identification is a leading issue of many research centres. The interest of this domain results from the potential possibilities applying the new approach to person's identification in systems that require access authorizations to resources [1, 2]. Research on the face recognition systems has lasted for over twenty years. However, there is still no 100% effective method, which could be used to access authorizations. In recent years, considerable progress has been made on the problem of face detection and face recognition [3] and efficient algorithms have been created especially for stable conditions such as: small variations in lighting, facial expression and pose. These methods can be roughly divided into two different groups: geometrical features matching and template matching. In the first case, some geometrical measures about distinctive facial features such as eyes, mouth, nose and chin are extracted [4]. In the second case, the face image, represented as a two-dimensional array of intensity values, is compared to a single or several templates representing a whole face. The earliest methods for template matching are correlation-based, thus computationally very expensive and require a great amount of storage, and for a few years, the Principal Components Analysis (PCA) method is successfully used in order to perform dimensionality reduction [5, 6]. Other popular methods are using Wavelet Transform (WT) [7] or Hidden Markov Models (HMM) [8]. Analysis of the existing solution revealed their defects, which caused their weak effectiveness. The disadvantages of these methods are as follow:

- In case of the new user's registration, process of learning and addition his facial image to a database require repeated learning of whole system.
- They work with whole face image.
- They are computationally very expensive.

The work concerns creation of the original *FaMar* method of user's identification on the basis of the frontal facial image, in which the fusion of the WT and HMM are used for three parts of face (eyes, nose and mouth); the decision is made on the basis of the sum maximalisation of likelihood of generating of models observation [9].

2 The proposed method

The proposed method is combination two mathematical tools, Wavelet Transform (WT) and Hidden Markov Model (HMM). Here, WT is used for features extraction, and HMM for identification. This system works in two modes, learning and testing. These modes are different from each other. The algorithm of this method consists of four main parts:

1. Pre-processing: normalization and face division on three parts.
2. Features extraction: WT of face image.
3. Training: generating and learning HMM for each part of the face.
   Testing: testing models from the database.
4. Training: saving to database the learned models of face.
   Testing: making a decision - maximum likelihood of model.

Pre-processing

The normalization consists in fixing the centres of eyes, and then respective scaling of the face so that the distance between them equals 60 pixels. The second part of this process is the division of the normalized face into three parts of: the area of eyes, nose and mouth (Fig. 1).

2.1 Features extraction

WT is used to features extraction. Using 2D WT (Fig. 2) [10], the face image is decomposed into four subimages via the high-pass and low-pass filtering. The image is decomposed along column direction into subimages to high-pass frequency band H and low-pass frequency band L. Assuming that the input image is a matrix of m x n pixels, the resulting subimages become $m/2$ x n matrices. At second step the images H and L are decomposed along row vector direction and respectively produce the high and low frequency band HH and HL for H, and LH and LL for L. The four output images become the matrices of $m/2$ x $n/2$ pixels. Low frequency subimage LL ($A1$) possesses

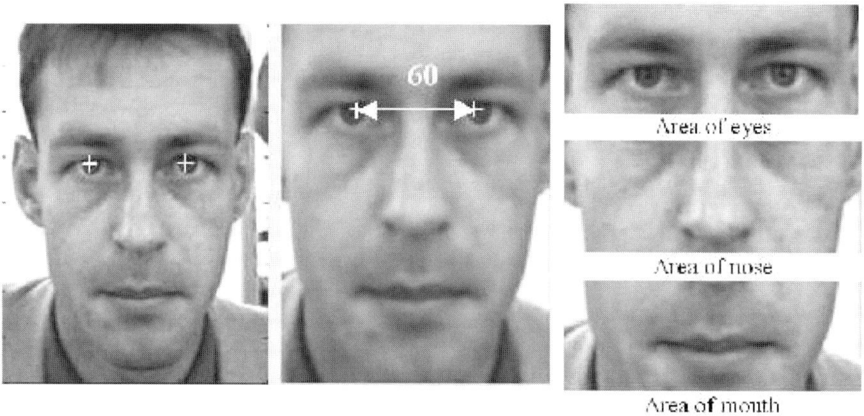

Fig. 1. Pre-processing of the face image

high energy, and is a smallest copy of original images (*A0*). The remaining subimages *LH*, *HL*, and *HH* respecticly extract the changing components in horizontal (*D11*), vertical (*D12*), and diagonal (*D13*) direction [7]. Wavelet Transform of second level (Fig. 3) is used to features extraction in propose technique. After first level wavelet decomposition, the output images become input images of second level decomposition. The results of two-level 2D WT are coded in this way, so that they can be applied in HMM (Fig. 5). One of the simplest methods of reduction and information coding is the calculating of standard deviation or mean value. Each part of the face is transformed separately by discrete wavelet transform (Fig. 4). The bank filters' selection is an important thing in this transformation. It guarantees a good recognition rate. More information about it can be found in [11].

2.2 Training

HMM is used to the identification process. A HMM is a double stochastic process with underlying stochastic process that is not observable (hidden), but can be observed through another set of stochastic processes that produce a sequence of observation. Let $O = \{O_1, \ldots, O_T\}$ be the sequence of observation of feature vectors, where T is the total number of feature vectors in the sequence. The statistical parameters of the model may be defined as follows [10].

- The number of states of the model, N
- The transition probabilities of the underlying Markov chain, $A = \{a_{ij}\} 1 \leq i, j \leq N$ where a_{ij} is the probability of transition from state i to state j subject to the constraint

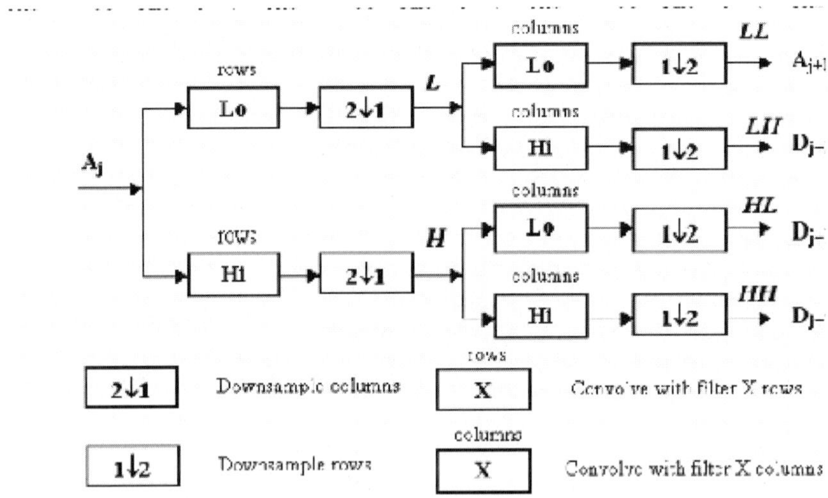

Fig. 2. One-level two-dimensional wavelet transform

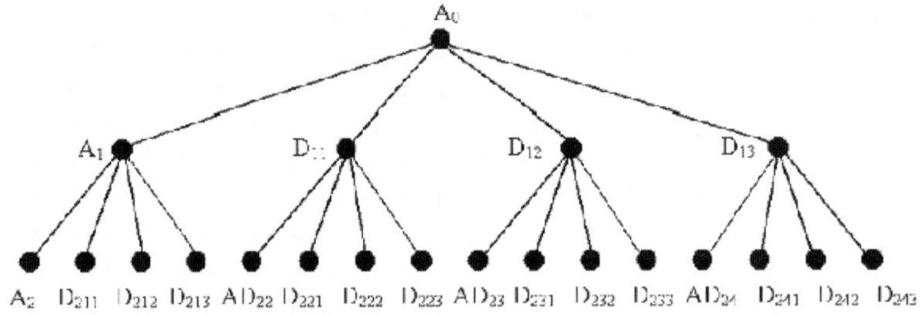

Fig. 3. The wavelet decomposition tree

- The observation probabilities, $B = \{b_j(O_T)\}, 1 \leq j \leq N, 1 \leq t \leq T$ which represents the probability of the t_{th} observation conditioned on the j_{th} state.
- The initial probability vector, $\Pi = \{\pi_i\}, 1 \leq i \leq N$.

Hence, the HMM requires three probability measures to be defined, A, B, π and the notation:

$$\lambda = (A, B, \pi) \tag{1}$$

is often used to indicate the set of parameters of the model.

In proposed method, one model is made for each part of the face. The parameters of the model are generated at random at the beginning. Then they are estimated with Baum-Welch algorithm, which is based on the forward-

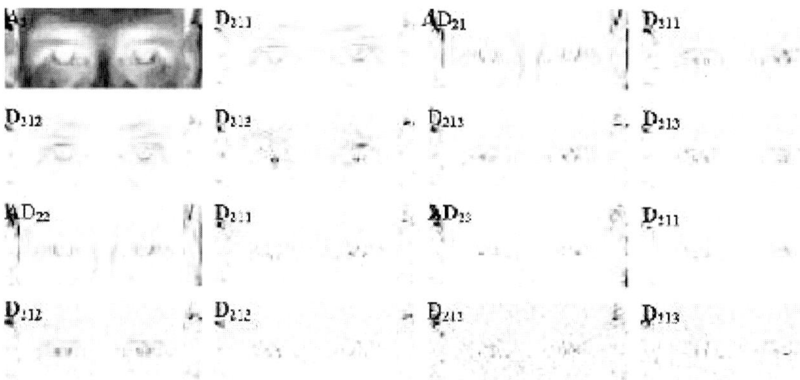

Fig. 4. Example of level 2 of the wavelet decomposition of image

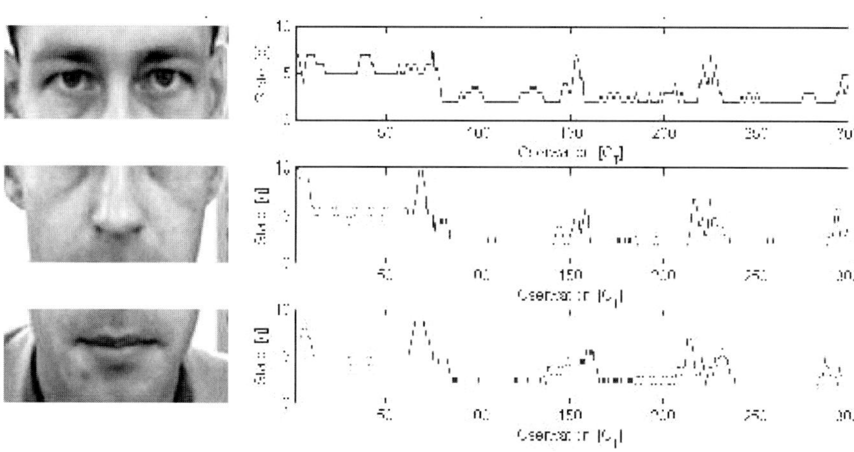

Fig. 5. Parts of face and correspond them sequences of observation

backward algorithm. The forward algorithm calculates the coefficient $a_t(i)$ (probability of observing the partial sequence $(o_1, \ldots, o_t)$ such that state q_t is i). The backward algorithm calculates the coefficient$b_t(i)$ (probability of observing the partial sequence $(o_{t+1}, \ldots, o_T)$ such that state q_t is i). The Baum-Welch algorithm, which computes the λ, can be described as follows [10]:

1. Let initial model be λ_0.
2. Compute new λ based on λ_0 and observation O
3. If $\log P(O \mid \lambda) - \log P(O \mid \lambda_0) < DELTA$ stop
4. Else set $\lambda_0 \rightarrow \lambda$ and goto step 2.

The parameters of new model λ, based on λ_0 and observation O, are estimated from equation of Baum-Welch algorithm [13], and then are recorded to the database.

2.3 Testing

The testing process consists of computing the probability of observation generating by the models saved in database and choosing this model for which the likelihood is maximum. In the proposed method, probabilities are calculated separately for each of the three models representing parts of the face, then they are added. The face, for which the sum of probability is maximum, is chosen as the correct face. The probability of generating sequences of observations is computed from the equations 2–4 [13].

$$P(O \mid \lambda) = \sum_{q} P(O \mid q, \lambda) P(q \mid \lambda) \tag{2}$$

$$P(O \mid q, \lambda) = \prod_{i=1}^{T} P(o_t \mid q_t, \lambda) = b_{q_1}(o_1) b_{q_2}(o_2) \ldots b_{q_T}(o_T) \tag{3}$$

$$P(q \mid \lambda) = \pi_{q_1} a_{q_1 q_2} a_{q_2 q_3} \cdots a_{q_{T-1} q_T} \tag{4}$$

3 Experimenting

The following constrains are imposed to guarantee the reliability of face identification: a frontal face view is located in the centre of an input image; eyes are open and mouth is closed; the face should not be covered with shadow, the rotate angle of an input image must be less than ten degree. The results of three experiments are presented here. The first one was carried out on the basis of the BioID face database in which there are 24 subjects [14]. The second and the third experiment were carried out using my own face database FaDab in which there are 150 subjects [15]. The first two experiments were carried out on the basis of the three parts of a face, whilst the third one only on the area of the eyes, which means that the face was represented by one model. The results of experiments are shown in Tab. 1. [16].

Table 1. The results of experiment

Face database	Number of face parts	Number of persons	Error rate [%]
BioID	3	24	12,51
FADAB	3	150	10,00
FADAB	1 (eyes)	150	8,00

4 Conclusion

The new method of face identification and the effective identification system were presented. On the basis of experimental research it was stated the area of eyes contains the most useful information for the persons' identification, and it could be successfully applied in specific methods of identification (e.g. detection). The method is characterized by following novelties:

1. The usage of the three areas of the face for identification and creating for each of them one independent HMM (which it is possible to use separately or together). This procedure gives possibility to short calculation request and permit obtaining a recognition rate as good as in modern method.
2. The transition from 2D pictures to 1D-WT of the facial areas. This procedure permits to obtain the recognition rate as good as in modern method and gives possibility to short calculation request also.
3. The fusion of WT and HMM (see p.1 and p.2) with using the assumption of maximalization of the likelihood's sum of generating of the observation.

References

1. Wu Ch.J., Huang J.S. (1990) Human face profile recognition by computer. Pattern recognition, vol. 23, No 3/4: 255–259.
2. Blaszczyk M. (1993) Automatyczne rozpoznawanie twarzy. Rozprawa doktorska, Politechnika Ňlźska, Gliwice.
3. Garcia C., Zikos G., Tziritas G. (2000) Wavelet pocket analysis for face recognition. Image and Vision Computing 18: 289–297.
4. R. Brunelli, T. Poggio (1993) Face Recognition: Features versus Templates. IEEE Transactions on Pattern Analysis and Machine Intelligence, 11(6): 1042–1052.
5. Kirby M., Sirovich L. (1990) Application of the Karhunen-Loeve Procedure and the Characterization of Human Faces. IEEE Transactions on Pattern Analysis and Machine Intelligence, 12(1): 103–108.
6. Belhumeur P., Hespanha J., Kriegman D., (1997) Eigenfaces vs. Fisherfaces: Recognition using class specific linear projection. IEEE Transactions on Pattern Analysis and Machine Intelligence, 19(7): 711–720.
7. Chien J.T., Wu Ch.Ch. (2002) Discriminant Waveletface and Nearest Feature Classifiers for Face Recognition. IEEE Transactions on Pattern Analysis and Machine Intelligence, 24(12): 1644–1649.
8. Samaria F., Young S. (1994) HMM-based Architecture for Face Identification, Image and Vision Computing, Vol. 12 No 8.
9. Bobulski J. (2003) The Face Recognition Technique Using Wavelet-face, Hidden Markov Model and Maximum Likelihood Principle, ACS'2003, Miedzyzdroje Poland.
10. Misiti M., Misiti Y., Oppenheim G., Poggi J.-M. (2000) Wavelet Toolbox User's Guide, MatLab 5.0, The MathWorks.
11. Kozuani A.Z. He f. Sammut K. (1997) Wavelet Packet Face Representation and Recognition. Proc. IEEE Conf. Systems, Man, and Cybernetics: 1614–1619.

12. Rabiner L. R. (1989) A tutorial on hidden Markov models and selected application in speech recognition. Proc. IEEE 77: 257–285.
13. Kanungo T. Hidden Markov Model Tutorial, http://www.cfar.umd.edu/ kanungo.
14. Face Database BioID, http://www.humanscan.com/bioid.htm.
15. Face Database FaDab, http://icis.pcz.pl/ januszb/baza.htm
16. Bobulski J. (2004) The Method of User's Identification Using the Fusion of the Wavelet Transformand Hidden Markov Models, PhD Thesis, Czestochowa University of Technology.

Restoration of Partially Occluded Shapes of Faces Using Neural Networks

Christina Draganova[1], Andreas Lanitis[2], and Chris Christodoulou[3]

[1] Dept. of Computing, Communication Technology and Mathematics, London Metropolitan University, 100 Minories, London EC3 1JY, UK
c.draganova@londonmet.ac.uk

[2] School of Computer Science and Engineering, Cyprus College, P.O. Box 22006, Nicosia, Cyprus alanitis@cycollege.ac.cy

[3] School of Computer Science and Information Systems, Birkbeck College, University of London, Malet Street, London WC1E 7HX, UK
chris@dcs.bbk.ac.uk

Summary. One of the major difficulties encountered in the development of face image processing algorithms, is the possible presence of occlusions that hide part of the face images to be processed. Typical examples of facial occlusions include sunglasses, beards, hats and scarves. In our work we address the problem of restoring the overall shape of faces given only the shape presentation of a small part of the face. For this purpose a novel technique which utilizes combination of Hopfield and Multi-Layer Perceptron (MLP) neural networks was used. According to the experimental results it is possible to recover with reasonable accuracy the overall shape of faces even in the case where a substantial part of the shape of a given face is not visible. The presented technique could form the basis for developing face image processing systems capable of dealing with occluded faces.

1 Introduction

There has been substantial research in the areas of automatic face recognition, face detection and face reconstruction in recent years [13]. One common problem for such applications is when a face image is occluded by other objects (e.g., sunglasses). This results in decreased performance and robustness of systems dealing with face recognition, detection or reconstruction tasks.

This paper addresses the occlusion problem in the case of reconstructing the shape of an occluded facial region. The motivation of our work comes from important applications relying on robust face recognition and reconstruction, free of restrictions such as lighting, expression, pose, size and occlusion. Such applications include among others, human-robot-interaction, human-computer-interaction, information security, CCTV access control, automated surveillance, suspect tracking and investigation (see [13] for a review).

In our experiments the shape of a face is represented with the co-ordinates of 68 landmarks characterizing the shape of the overall face and the shape of individual facial features (see Fig. 1). In our work we assume that the positions of the landmarks on the visible facial region are available.

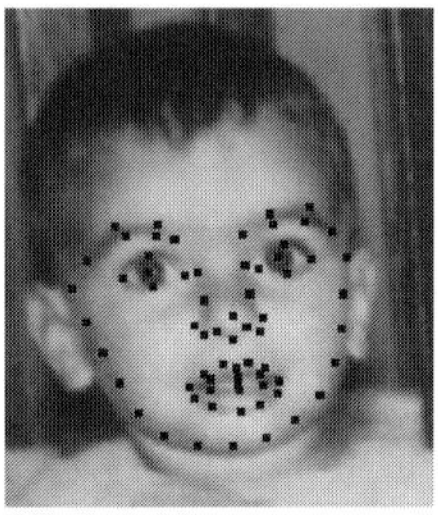

Fig. 1. Original contour face image defined by 68 landmarks

The aim of our work is to reconstruct the shape of an occluded facial region, given a shape representation of a visible region. We consider several different cases of occlusion, grouped in two settings, examples of which are given in Fig. 2 and Fig. 3.

The first setting consists of six sets of occluded face shapes corresponding to occlusion of different parts of a face. For example, in Case 1 a small part of the right lower part of a face is missing, in Case 2 the right part of a face is missing and in Case 3 the entire lower part of a face is missing, as shown in Fig. 2. Excluding all the points that were not excluded in Cases 1 – 3 and including all the points that were excluded in Cases 1 – 3, respectively, generates the Cases 4 – 6, as shown in Fig. 2. The second setting of occluded face shapes consists of five sets which are generated by randomly replacing the co-ordinates of 10%, 30%, 50%, 70% and 90% of the points in the original face shapes with random numbers in the range between the minimum and the maximum of the co-ordinates of the visible points. Typical examples of this occlusion cases are shown in Fig. 3.

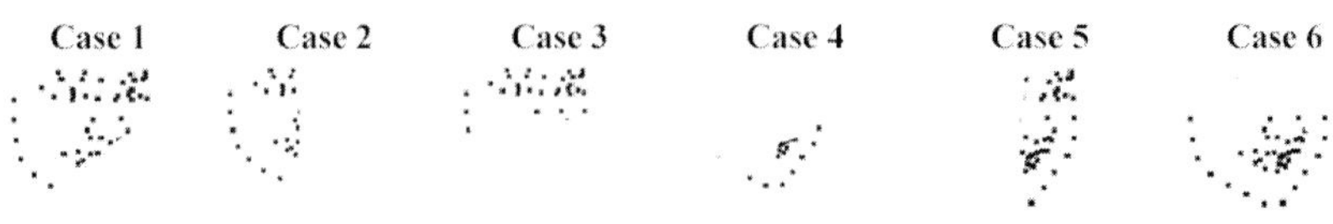

Fig. 2. Cases 1 – 6: Occlusion of different parts of the face

A model which combines Hopfield and Multi-Layer Perceptron (MLP) neural networks is considered. The experiments are run on a publicly available face image database, the FG-NET Aging Database [14].

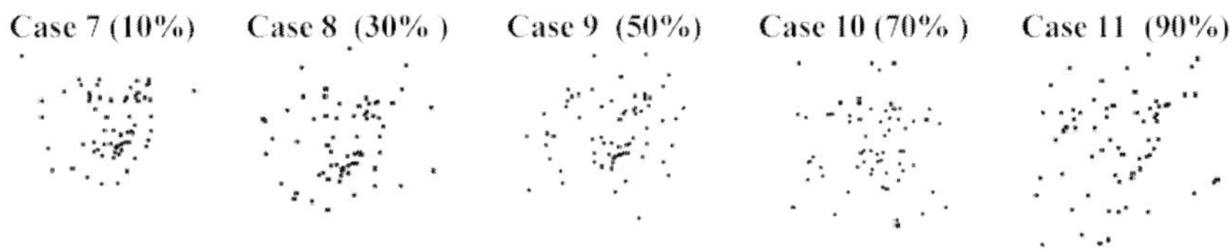

Fig. 3. Cases 7 – 11: Occluded face shapes obtained by replacing of an increasing number of points in the original face shapes with random numbers

The restoration of face shapes presents a demanding test scenario for our work because face shapes display significant variation arising by differences in the shapes between different individuals. On top of that, face shapes undergo within-individual variations caused by changes in the 3D orientation and expression of faces.

The remainder of the paper is organised as follows: in Sect. 2 we present a brief overview of the relevant literature; in Sect. 3 we describe the method used in our experiments; in Sect. 4 we describe the experimental set-up and present the results obtained, and finally in Sect. 5 we give our conclusions. A full comparative evaluation of the Hopfield-MLP technique with other techniques for restoration of partially occluded face shapes is presented elsewhere [2].

2 Literature review

A number of researchers have recently made contributions for resolving the problem of occlusion in face recognition, face detection, face identification and face reconstruction.

Kurita et al. [7], suggest recursive use of an auto-associative MLP network for reconstruction of occluded faces. Subsequently this approach is applied to face recognition and face detection. The idea of using an auto-associative neural network is based on the observation that auto-associative memory can recall a whole image from its partial image [6]. The suggested system is built with the aim of performing face recognition and the restoration of occluded face images and is limited only to the cases where the occluded face images belong to individuals whose images have already been seen during the training.

Martinez [8], [9] used a variation of the eigenface approach [12] in order to deal effectively with the problem of recognizing occluded face images. They divide the facial region into six local regions and use a PCA (Principal Component Analysis) based local model for each local part. During recognition, the contribution of each part is weighted by the distance of the corresponding PCA coefficients from the centroid of the distribution, so that the contribution of the occluded facial regions in the recognition process is minimized.

Park et al. [10], use recursive PCA for removing spectacles from face images. Given a face image of a subject wearing spectacles, they code and subsequently reconstruct the face. The difference between the reconstructed and original image is processed in order to enhance the occluded regions. The occluded regions detected in the difference image are replaced by the corresponding pixels from the mean image among the training set. This procedure is repeated until the resulting image converges.

Hwang and Lee [5] describe a method for restoring the appearance of occluded faces in images. Given an occluded face image and information about the location and size of the occlusion, they use least squares analysis for estimating the optimum weights required for decomposing the appearance of the non-occluded regions as a weighted sum of basis images. The same weights are used in conjunction with basis images of the occluded region for restoring the appearance of the occluded facial regions.

3 Method

Combining different neural network architectures is a common approach for improving generalization performance and efficiency of neural network models. We propose a combination of the Hopfield and MLP neural networks which to the best of our knowledge has not been previously employed elsewhere. The proposed approach includes two steps: initial face shape restoration using the Hopfield and tuning the obtained face shape using the MLP.

The Hopfield Neural Network [3], [4] is a binary artificial network which is used to store patterns in an associative or content-addressable way so that when the network is presented with noisy or partial information the full pattern can be recovered. In order to apply the Hopfield model to the problem of recovering occluded face shapes we first convert the facial co-ordinates into a binary form and represent each shape as a binary vector. The co-ordinates (x, y) corresponding to the facial shape landmarks are scaled so that each x and y is in the interval [0, 31] and subsequently converted to 5 bit representation using the natural binary encoding. In this way each face shape is represented as a binary vector of 680 bits (2 x 68 x 5). For addressing the capacity limitation of the Hopfield network we train the model only with 102 face representations (15% of 680 neuronodes) from the data set. The natural binary encoding is applied to each of the distorted sets of face shapes. The trained Hopfield model is presented consequently with the different sets of distorted encoded face representations, which correspond to the different occlusion cases (Case 1 – Case 11) described in the Introduction (see Fig. 2 and Fig 3). The recovered patterns are decoded back to present each face shape as 68 decimal (x, y) co-ordinates. The resulting shapes form the initial approximation of the restored occluded face shapes.

Next we use the MLPs with the backpropagation learning algorithm [11] to train eleven different models for each occlusion case corresponding to the case settings described in Sect. 1. Input vectors for each MLP model are the 136 dimensional vectors corresponding to the 68 (x, y) co-ordinates representing the respective occluded face shape in the specific occlusion case. The co-ordinates of the occluded points are replaced with random numbers in the range of the co-ordinates of the visible points. Output vectors contain the 136 elements corresponding to the 68 (x, y) co-ordinates representing the shapes of the original sample faces. Based on the training sets each type of network is evaluated in order to establish the optimal architecture and optimal parameters.

The co-ordinates of the initial face shapes restored with the Hopfield method are fed as inputs into one of the already trained MLP networks described above. The resulted output is the desired restored face shape.

4 Experimental Evaluation

In our experiments we use the FG-NET Aging Database [14], which is publicly available. The image database in question contains 1002 face images from 82 different individuals. On average there are 12 images available per subject. For each face image in the FG-NET Aging Database, a detailed shape annotation consisting of 68 landmarks (see Fig. 1) is also publicly available. In our experiments we use the shapes of 102 face images from 8 subjects in the database for training (due to the Hopfield network capacity limitation, see Sect. 3) and the remaining 900 face shapes of the remaining 74 subjects for testing. Prior to our experiments the shapes of all faces were normalized so that all face shapes had the same centre of gravity and approximately the same height. It is important to highlight that in our experiments we did not use face shapes of the same subject in either the train or test sets.

We use two error measures to evaluate the performance of the different methods. The first error measure is the mean Euclidean distance between the shape of the original faces and the corresponding recovered face shapes for the different occlusion cases, which we call the *overall error*. The second is the mean Euclidean distance between the original points from the occluded parts and the corresponding recovered points, which we call *restricted error*.

The training and testing time for the Hopfield model is in the order of seconds. The training time for the MLP method is in the order of minutes (1 – 2 min) and the testing time in the order of seconds. The experiments using the MLP method are performed with learning rate varying 0.1 and 0.2, momentum between 0.7 and 0.9, number of hidden units between 10 and 25, and number of iterations between 1000 to 1500. According to our experiments the optimal network architecture has one hidden layer with 15 hidden neuronodes, learning rate equal to 0.1 and momentum equal to 0.7.

We are mainly testing the Hopfield-MLP method but we also provide results for the each method Hopfield and MLP individually for benchmarking. Table 1 displays the testing set results of applying the Hopfield, MLP and Hopfield - MLP methods to the different occlusion cases 1 – 11 described in Sect. 1.

Method	Case number 1 mean	1 st. dev.	2 mean	2 st.dev.	3 mean	3 st. dev.	4 mean	4 st. dev.	5 mean	5 st. dev.	6 mean	6 st.dev.	Average of means Cases 1-6
Overall errors													
Hopfield	0.89	*2.33*	3.21	*1.68*	4.09	*1.77*	6.11	*1.9*	2.69	*1.72*	2.33	*1.93*	3.22
MLP	0.41	*1.3*	2.6	*1.93*	2.97	*1.96*	4.42	*2.4*	1.88	*1.45*	1.25	*1.09*	2.26
Hopfield – MLP	0.55	*2.14*	2.05	*1.67*	2.65	*1.97*	3.91	*2.22*	1.67	*1.74*	1.45	*2.01*	2.05
Restricted errors													
Hopfield	6.03	*0.34*	5.90	*0.91*	6.63	*1.09*	7.16	*1.62*	5.89	*0.78*	6.09	*0.73*	6.28
MLP	2.88	*0.19*	4.93	*1.02*	4.95	*1.17*	5.33	*1.99*	4.24	*0.64*	3.37	*0.41*	4.28
Hopfield – MLP	3.71	*0.31*	3.76	*0.91*	4.29	*1.21*	4.58	*1.89*	3.67	*0.79*	3.80	*0.77*	3.97

Method	Case number 7 mean	7 st. dev.	8 mean	8 st.dev.	9 mean	9 st. dev.	10 mean	10 st. dev.	11 mean	11 st. dev.	Cases 7-11
Overall errors											
Hopfield	0.48	*2.25*	1.34	*1.77*	2.19	*1.81*	3.19	*1.76*	3.59	*1.85*	2.16
MLP	0.22	*0.91*	0.77	*0.94*	1.17	*0.97*	2.23	*1.57*	2.77	*1.85*	1.43
Hopfield – MLP	0.31	*1.82*	0.87	*1.66*	1.4	*1.79*	2.02	*1.77*	2.32	*1.94*	1.38
Restricted errors											
Hopfield	5.47	*0.19*	5.06	*0.47*	5.52	*0.72*	5.86	*0.95*	6.10	*1.09*	5.60
MLP	2.52	*0.08*	2.99	*0.24*	3.04	*0.37*	4.22	*0.83*	4.84	*1.05*	3.52
Hopfield – MLP	3.03	*0.16*	3.29	*0.44*	3.54	*0.69*	3.71	*0.96*	3.94	*1.14*	3.50

Table 1. Mean and standard deviation of the Euclidean distances between recovered face shapes and original face shapes for testing set in Cases 1 - 11

Image representations of a typical recovered face shape from the testing set for some of the occlusion cases using the Hopfield, MLP and Hopfield-MLP methods are shown in Fig. 4.

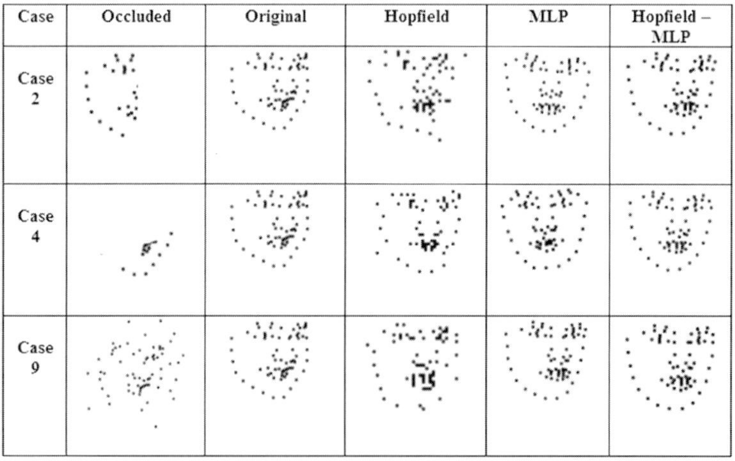

Fig. 4. Sample face shapes as seen visually: occluded, original and reconstructed by the different methods

The numerical results in Table 1 show, as expected, that in cases where only a small part of the face shape is missing (e.g. Cases 1 and 7) the *overall error* is small, and it gradually increases when larger parts of the face shape are missing (e.g. Cases 2, 5, 6, 9, 10 and 11). However, this increase of the *overall error* remains limited. This is also valid for the *restricted error* as we can see in Table 1.

The *overall* and *restricted errors* for the Hopfield method are slightly higher than the other two methods in all of the cases. It is well known that Hopfield networks have two main drawbacks: limited capacity (15% of the network units) and local energy minima occurrences referred as spurious attractors. Theoretically these limitations can be overcome by using the Boltzmann Machine neural network [1], which is a stochastic neural network with hidden neurons. In practice, however, the learning in the Boltzmann Machines is extremely slow. Due to this and given that we address one of the problems of the Hopfield network, namely the limited capacity (see Sect. 3), we decided not to employ Boltzmann Machines in this study. As suggested in Sect. 3, we use one of the trained MLP networks for the occlusion cases 1-11, and feed the results from the Hopfield network as inputs to the chosen MLP network. This results in considerable improvement for the *overall* and the *restricted errors* over the Hopfield method and over the MLP method in six cases, namely cases 2, 3, 4, 5, 10 and 11. The last columns in Table 1 shows that the average of the *overall errors* and the average of the *restricted errors* for the cases 1-6 and 7-11 are smallest for the combined Hopfield-MLP method. The visual results in Fig. 4 demonstrate that when the Hopfield-MLP method is employed the reconstructed shapes retain both the geometrical structure and the 3D orientation of the original face shape. The low standard deviations of the *restricted errors* obtained, indicate that there is uniform performance over the testing set for the considered methods.

5 Conclusions

We have presented an experimental evaluation of the problem for restoration of occluded face shapes where the performance of classical neural network methods, such as Hopfield, MLP, and a combination of these, were evaluated.

Numerical results based on the mean Euclidean distance between: (i) the original face shapes and the corresponding recovered face shapes and (ii) the original face points and the recovered occlusion points, as well as visual results were used to compare the performance of the different methods.

The combined Hopfield-MLP method gives better performance than the individual Hopfield and MLP methods in most of the occlusion cases and according to the visual results, it also produces the best reconstructions.

The results show that it is feasible to develop a system based on classical methods for the automatic prediction of the shape of occluded facial features.

With the suggested approach, an occluded face shape of an unseen individual can be restored even when a large part of the face shape is missing.

The problem of restoring occluded face shapes in such general forms of occlusion like the ones considered in our experiments has not been addressed up to now. This makes it impossible to compare directly the results from our approach to previously reported methods for restoration of occluded face shapes. Systems reported in [7], [8] and [9] for restoration of partly occluded face images are developed with the aim of face recognition and in general are limited to restoration of face images of individuals whose images were used in the learning process, which is not the case with ours.

The results show that it is possible to recover with reasonable accuracy the overall shape of faces even in the case where a large proportion of the shape of a given face is not visible. The feasibility of such predictions is based on the strong correlation between the appearances of individual facial features.

References

1. Ackley D, Hinton G, Sejnowski T (1985) Cognitive Science 9:147–169
2. Draganova C, Lanitis A, Christodoulou C (2005) submitted to IEEE Trans. on Systems, Man and Cybernetics Part B
3. Hopfield J (1982) Proc. Nat. Acad. Sci. USA 79:2554–2558
4. Hopfield J (1984) Proc. Nat. Acad. Sci. USA 81:3088–3092
5. Hwang B, Lee S (2002) Proc. of the 16th ICPR'02 2:366–369
6. Kohonen T (1989), Self-Organization and Associative Memory, 3rd Edition, Springer-Verlag, Berlin
7. Kurita T, Pic M, Takahashi T (2003) Proc. IEEE Conf. AVSBS, 53–58
8. Martinez A (2002) IEEE Trans. Pattern Anal. Machine Intell. 24:748–763
9. Martinez A(2000) Proc. of IEEE CVPR'2000 I:712–717
10. Park J, Oh Y, Ahn S, Lee S (2003) AVBPA, Springer-Verlag, 2688:369–376
11. Rumelhart D, Hinton G, Williams R (1986) Nature 323:533–536
12. Turk M, Pentland A (1991) Journal of Cognitive Neuroscience 3:1:71–86
13. Zhao W, Chellappa P, Phillips P, Rosenfeld A (2003) ACM Computing Surveys 35:4:399–458
14. FG-NET(2004) http://sting.cycollege.ac.cy/~alanitis/fgnetaging

Comparison of Minutiae Matching Techniques

Maciej Hrebień, Andrzej Marciniak, and Józef Korbicz

Institute of Control and Computation Engineering
University of Zielona Góra
ul. Podgórna 50
65-246 Zielona Góra
{a.marciniak,j.korbicz}@issi.uz.zgora.pl

Summary. This paper presents comparison of three minutiae matching techniques, i.e. Hough transform, global star method and orientation correlation. Short description of the pre-processing stage based on filtering, thinning and minutiae extraction is presented. The investigations are performed with a high quality fingerprint scanner.

1 Introduction

In recent decade security systems based on biometric technologies have played an important role in our community. Fingerprint identification is one of the most important biometric approaches. The uniqueness of a fingerprint is exclusively determined by the local ridge characteristics called minutiae. The automatic fingerprint matching depends on the comparison of these minutiae and their relationships.

This paper is focused on minutiae matching presented in Sect. 4. However, one can find here brief description of image pre-processing (Sect. 2) and minutiae extraction (Sect. 3) methods applied in this work. Preliminary results of minutiae matching are summarized in Sect. 5.

2 Image Enhancement

The very common technique of reduction the quantity of information received from fingerprint grayscale image is known as Gabor filtering [2]. The filter based on local ridge orientation and frequency estimations produces a near-binary output - the intensity histogram has U-shaped form [1].

The Gabor filter is defined by

$$g(x, y, f, \theta) = exp\Big\{ -\frac{1}{2}\big(\frac{x_\theta^2}{\delta_x^2} + \frac{y_\theta^2}{\delta_y^2}\big)\Big\} \cos(2\pi f x_\theta), \tag{1}$$

where $x_\theta = x\sin\theta + y\cos\theta$, $y_\theta = x\cos\theta - y\sin\theta$, and θ is local ridge orientation, i.e. the angle that fingerprint ridges form with the horizontal axis when crossing through an arbitrary small block.

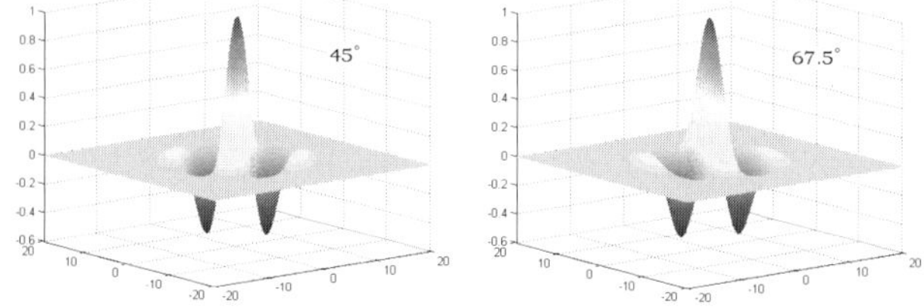

Fig. 1. Graphical representation of Gabor filter in two exemplary directions ($f = \frac{1}{10}$, $\delta_x = 4$, $\delta_y = 4$)

Because fingerprint ridges are not directed, 225° is the same as 45°. The ridge orientation for a specified block centered at position (i, j) can be estimated using (2), where $\delta_x(u, v)$ and $\delta_y(u, v)$ are gradients of pixel at position (u, v) (e.g. estimated with Sobel's mask). Ridge orientations can also be estimated by more direct mask method described in details in [3].

$$\theta(i,j) = \frac{1}{2}\tan^{-1}(\frac{V_y(i,j)}{V_x(i,j)}) \tag{2}$$

$$V_x(i,j) = \sum_{u=i-\frac{w}{2}}^{i+\frac{w}{2}} \sum_{v=j-\frac{w}{2}}^{j+\frac{w}{2}} 2\delta_x(u,v)\delta_y(u,v)$$

$$V_y(i,j) = \sum_{u=i-\frac{w}{2}}^{i+\frac{w}{2}} \sum_{v=j-\frac{w}{2}}^{j+\frac{w}{2}} (\delta_x^2(u,v) - \delta_y^2(u,v))$$

Local ridge frequency f can be estimated by counting the average number of pixels between two consecutive peaks of gray-levels along the direction normal to the local ridge orientation. The idea is based on $w \times l$ (where $w < l$) oriented windows centered at the center of each block and rotated with the angle θ. The frequency for each block is given by

$$f(i,j) = \frac{1}{T(i,j)} \tag{3}$$

where $T(i, j)$ is an average number of pixels between two consecutive peaks in the so-called *x-signature* obtained from

$$X_k = \frac{1}{w}\sum_{d=0}^{w-1} W(d,k), \qquad k = 0, 1, \ldots, l-1. \tag{4}$$

The three-dimensional graphical representation of Gabor filter is illustrated in Fig. 1 and the exemplary fingerprint image enhancement is shown in Fig. 2.

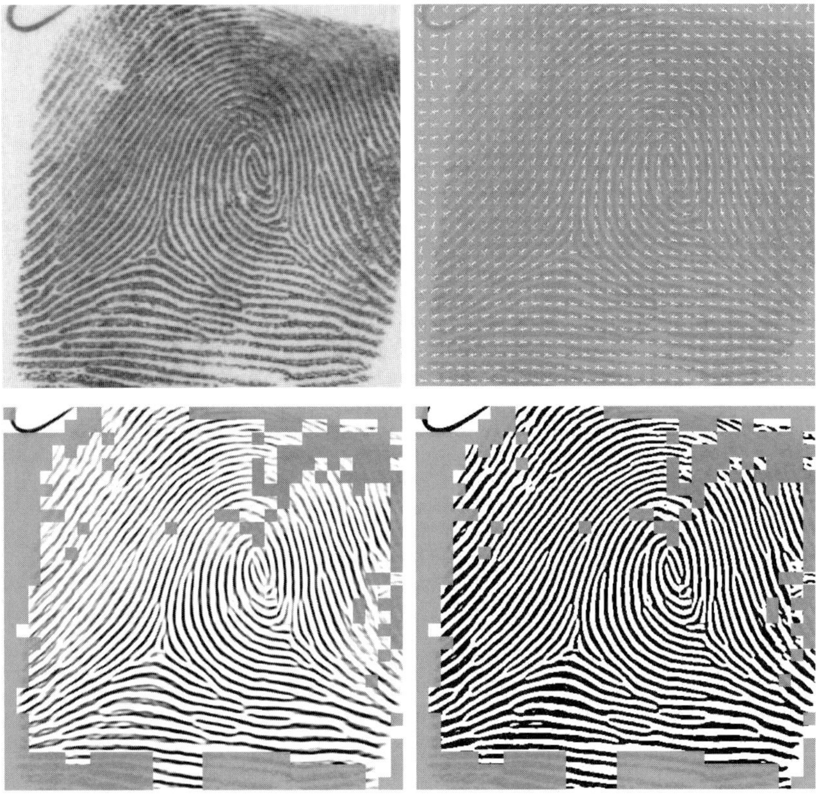

Fig. 2. An example of image enhancement and binarization based on Gabor filter

3 Minutiae detection

3.1 Image Thinning

Image thinning can be considered as a process of erosion. All pixels from the edges of an object (fingerprint ridge) are removed only if they do not affect coherence of the object as a whole and left untouched in the other case. The skeleton form of fingerprint is generated (see Fig. 3) if there are no more surplus pixels to remove. Thickness of ridges in the resulting image has to be equal to one pixel and the shape and run of original ridges should be preserved.

Fig. 3. An example of thinned form of a fingerprint image

3.2 Coordinates and types

To determine whether a pixel at position (i,j) in skeleton form of fingerprint is a minutiae point we have to deal with the rules illustrated in Fig. 4.

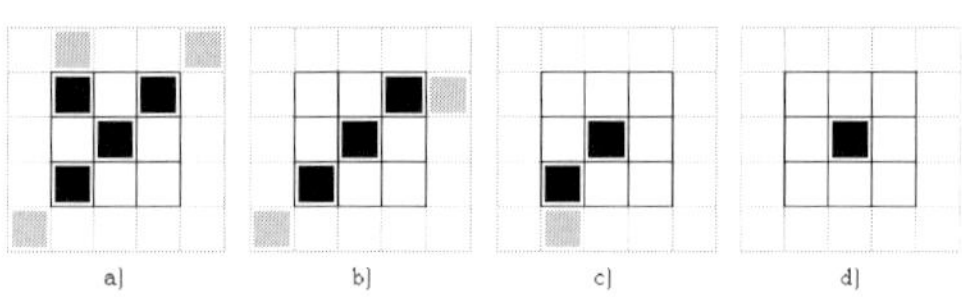

Fig. 4. An example of 3×3 masks used to define: (**a**) bifurcation, (**b**) non-minutiae point, (**c**) ending, (**d**) noise

3.3 Minutiae orientation

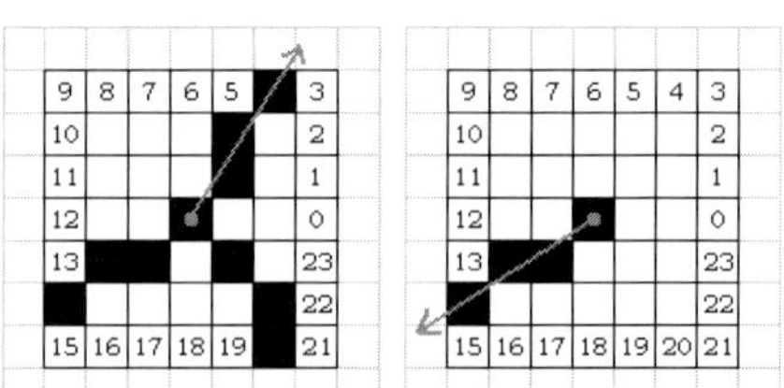

Fig. 5. Bifurcation (60°) and ending (210°) point orientation example

To define orientation of each minutiae we can use 7×7 mask technique with angles quantized to 15° and with the center placed in a minutiae point. The orientation of an ending point is equal to the point where a ridge is crossing through the mask. The orientation of a bifurcation point can be estimated with the same method but only the leading ridge is considered, that is the ridge with maximum sum of angles to other two ridges of the bifurcation as shown in Fig. 5.

4 Minutiae Matching

4.1 Hough Transform

Let $M_A = \{m_1^A, m_2^A, \ldots, m_m^A\}$ and $M_B = \{m_1^B, m_2^B, \ldots, m_n^B\}$ denote minutiae sets determined from images A and B. Each minutiae is described by image coordinates (x, y) and orientation angle $\theta \in [0 \ldots 2\pi]$, that is: $m_i^A = \{x_i^A, y_i^A, \theta_i^A\}_{i=1\ldots m}$ and $m_j^B = \{x_j^B, y_j^B, \theta_j^B\}_{j=1\ldots n}$.

Hough transform, which was adopted for fingerprint matching (see for instance [4]), is performed to find the best alignment of set M_A and M_B, including possible scale, rotation and displacement of image A versus B. The transformation space is discretized – each parameter of geometric transformation $(\Delta x, \Delta y, \theta, s)$ comes from a finite set of values. A four dimensional accumulator A is used to accumulate evidences of alignment between each considered pair of minutiae. The best parameters of geometric transform, that is $(\Delta x^+, \Delta y^+, \theta^+, s^+)$, are arguments of maximum value in A, and can be obtained with the procedure as below.

$\forall_i \forall_j \forall_k \forall_l \; A(i,j,k,l) \leftarrow 0$

$FOR\ \{x_i^A, y_i^A, \theta_i^A\} \in M_A, \quad i = 1 \ldots m$
$\quad FOR\ \{x_j^B, y_j^B, \theta_j^B\} \in M_B, \quad j = 1 \ldots n$
$\quad\quad FOR\ \theta \in \{\theta_1, \theta_2, \ldots, \theta_k\}, \quad k = 1 \ldots K$
$\quad\quad\quad IF\ \min(|\theta_i^A + \theta - \theta_j^B|, 360° - |\theta_i^A + \theta - \theta_j^B|) < \theta_0$
$\quad\quad\quad\quad FOR\ s \in \{s_1, s_2, \ldots, s_l\}, \quad l = 1 \ldots L$
$\quad\quad\quad\quad \{$

$$\begin{bmatrix} \Delta x \\ \Delta y \end{bmatrix} \leftarrow \begin{bmatrix} x_i^A \\ y_i^A \end{bmatrix} - s \begin{bmatrix} \cos\theta & \sin\theta \\ -\sin\theta & \cos\theta \end{bmatrix} \begin{bmatrix} x_j^B \\ y_j^B \end{bmatrix}$$

$\Delta x^\#, \Delta y^\#, \theta^\#, s^\# \leftarrow discretize(\Delta x, \Delta y, \theta, s)$
$A(\Delta x^\#, \Delta y^\#, \theta^\#, s^\#) \leftarrow A(\Delta x^\#, \Delta y^\#, \theta^\#, s^\#) + 1$
$\}$

$\Delta x^+, \Delta y^+, \theta^+, s^+ \leftarrow \arg\max(A)$

After performing transformation, minutiae are juxtaposed to calculate the matching score with respect to their orientation and type. An exemplary result of Hough transform for minutiae matching is shown in Fig. 6a.

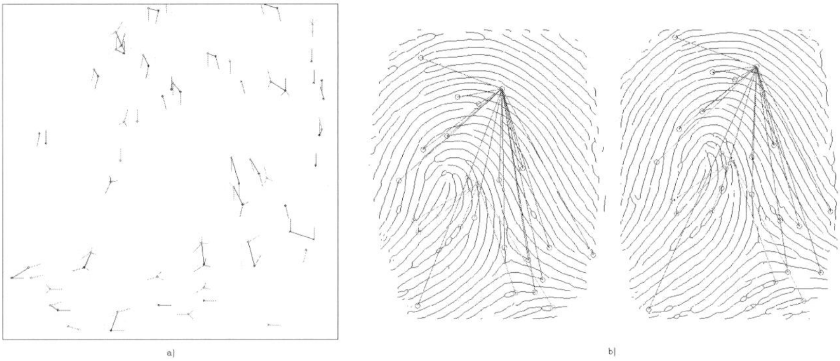

Fig. 6. Minutiae matching by: (**a**)Hough transform and (**b**) global star method

4.2 Global Star Method

The method is based on structural model of fingerprints. Distinguishing between types of minutiae (ending or bifurcation) and including possible scale, rotation and displacement of images a star can be created with the central point in one of minutiaes and arms directed to the other ones (as shown in Fig. 7). Assuming that $M_A = \{m_1^A, m_2^A, \ldots, m_m^A\}$ and $M_B = \{m_1^B, m_2^B, \ldots, m_n^B\}$ denote sets of minutiae of one type m stars for image A and n stars for image B can be created: $S^A = \{S_1^A, S_2^A, \ldots, S_m^A\}$, $S^B = \{S_1^B, S_2^B, \ldots, S_n^B\}$, where $S_i^A = \{m_1^A, m_2^A, \ldots, m_m^A\}_{i=1\ldots m}$ with center in m_i^A and $S_j^B = \{m_1^B, m_2^B, \ldots, m_n^B\}_{j=1\ldots n}$ with center in m_j^B. In opposition to local methods [5], voting technique for selection of the best aligned pair of stars (S_{wi}^A, S_{wj}^B) can be performed, including matching such features like between-minutiae angle K and ridge count D (as shown in Fig. 7):

$$
\begin{array}{l}
\forall_i \forall_j \; A(i,j) \leftarrow 0 \\
\\
FOR \; S_i^A \in S^A, \quad i = 1 \ldots m \\
\quad FOR \; S_j^B \in S^B, \quad j = 1 \ldots n \\
\quad\quad FOR \; m_k^A \in S_i^A - \{m_i^A\} \\
\quad\quad\quad assuming \; that: \; m_l^B \in S_j^B - \{m_j^B\} \\
\quad\quad\quad IF \; \exists_l (|D(m_j^B, m_l^B) - D(m_i^A, m_k^A)| < d_0 \; \& \\
\quad\quad\quad\quad\quad \& |K(m_j^B, m_l^B) - K(m_i^A, m_k^A)| < k_0) \\
\quad\quad\quad\quad A(i,j) \leftarrow A(i,j) + 1 \\
\\
S_{wi}^A \leftarrow S^A(\arg_i(\max(A))) \\
S_{wj}^B \leftarrow S^B(\arg_j(\max(A)))
\end{array}
$$

In the final matching decision also orientation of minutiae is taken into account. An illustration of star method is shown Fig. (6b).

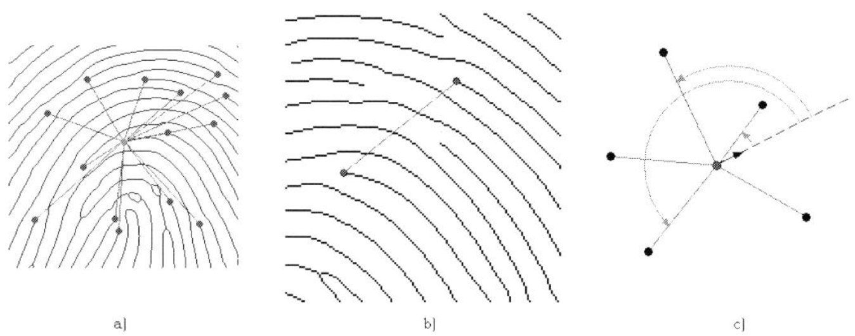

Fig. 7. General explanation of star method: (**a**) star created for endings of thinned fragment of fingerprint, (**b**) ridge counting (here equal 5), (**c**) determining relative angles between central minutiae and the remaining ones

4.3 Correlation

The correlation of fingerprints can not be applied directly for fingerprint matching because of many problems. Non-linear distortion, skin condition or finger pressure cause varying of image brightness and contrast of the same finger [2]. Moreover, including possible scale, rotation and displacement between images, an intuitive sum of squared differences is computationally very expensive.

To eliminate or at least reduce some above mentioned problems, a binary representation of fingerprint can be used. To speed up a process of preliminary alignment, a segmentation mask can be used with conjunction to the center of gravity of binary images. Also quantization of geometric transform features can be applied with considering scale and rotation only at the first stage (since displacement is the difference between centers of gravity).

After finding nearly best alignment of segmentation masks, looking for the best correlation is limited to much more reduced area. Including rotation, vertical/horizotal displacement, stretch and arbitrary selected granularity of these features, the best correlation can be found as shown in Fig. 8.

Fig. 8. Result of correlation between two impressions of the same finger – gray color underlines best alignment

Table 1. Summary of matching results

	match percentage	time relation
Hough Transform (HT)	88%	1× HT
Global Star Method	76%	$\sim$ 6× HT
Correlation	70%	$\sim$ 14× HT

Because fingerprint correlation does not tell anything about minutiae matching, a thinning process with minutiae detection should by applied on both binary images from the best correlation. Then two sets of minutiae can be compared to sum the matching score.

5 Experimental Results

The experiment were performed on a PC with a Digital Persona U.are.U 4000 fingerprint scanner. The database consist 20 fingerprint images with 5 diffrent impressions (plus one for registration phase). All images were enhanced with Gabor filter and matched using the algorithms described in Sect. 4. Matching rates in accordance with Polish regulations concerning fingerprint identification based on minutiae[6] and estimated time relations are summarized in Table 1.

6 Concluding Remarks

In this paper several methods for minutiae matching is reviewed. A complete fingerprint identification scheme is introduced. The results show quality differences and time relations between considered matching algorithms. Most of images from the error set of Hough transform and correlation technique were mismatched in opposition to global star method where nearly all unmatched images did not cross the given threshold (minutiae match sum) but were matched in the first stage of the algorithm. Considering the achieved results, there is still challenge to use global optimization techniques for setting the parameters of each method.

References

1. Hong L, Wan Y, Jain A (1998) Fingerprint Image Enhancement: Algorithm and Performance Evaluation. IEEE Trans. PAMI, 20(8):777–789
2. Maltoni D, Maio D, Jain A. and Prabhakar P (2003) Handbook of Fingerprint Recognition. Springer, New York
3. Stock R., Swonger C. (1969) Development and evaluation of a reader of fingerprint minutiae. Cornell Aeronautical Laboratory, Technical Report
4. Ratha N, Karu K, Chen S, Jain A (2004) A Real-time Matching System for Large Fingerprint Databases. IEEE Trans. PAMI, 18(8):799–813
5. Wahab A, Chin S, Tan E (1998) Novel approach to automated fingerprint recognition. IEE proc.-Visual Image Signal Processing, 145(3):160-166.
6. Grzeszyk C (1992)Dactyloscopy. PWN, Warsaw (in Polish)

Application of Active Shapes to the Structural Face Model

Andrzej Kasinski and Maciej Krol

Institute of Control and Information Engineering, Poznan University of Technology, str. Piotrowo 3A, Box 1, 60-965 Poznan, Poland
andrzej.kasinski@.put.poznan.pl,maciej.krol@interia.pl

Summary. In the article, the system for fitting the face-shape model to the image is described. Two model fitting approaches have been used: ACM and ASM. ACM has been applied in order to create a face-contours base of models. Starting with that base, an ASM model has been computed, which then has been used to obtain the face-shape descriptions from the provided images. ASM gives the explicit and structured shape description from the source image. Implementation of the ASM-based recognition system has been described and its performance evaluated. Some conclusions related to the choice of the local support are given, based on the experimental validation of the ASM method.

1 Face Modeling with Active Contours

The considered problem consists of finding contours on the image, given the vector $\mathbf{x}$ of n control points defined in image coordinates. Active contour method (ACM) [8] is based on fitting a selected curvilinear model to the appropriate samples of the image function under study. Usually considered contour shape piece-wise primitives are a line and a circle-arc. To generalize the ACM a model based on control points is introduced. It is assumed, that the preliminary vector of control points $\mathbf{x}_0$ on the image plane is available. The shape of the possible contour controlled by $\mathbf{x}_0$ is determined by A - a matrix, which is satisfying the equation $A\mathbf{x}_0 = 0$. It is usually a matrix having a special band-like structure (1) - as its non-zero elements are distributed over a predetermined number of its diagonals (to reflect the fact, that locations of particular control points are related only to the locations of their immediate neighbors in the vector). To each control point two shape-parameters are associated, namely α_i and β_i. For $\alpha_i = \beta_i = 0.5$ a local contour model passing through the i-th control point is a line-segment, while for $\alpha_i = \beta_i = \cos^{-1}(\frac{2\pi}{n})$ it is a circle-arc.

$$A = \begin{bmatrix} 1 & -\alpha_1 & -\beta_1 & 0 & 0 & \dots & \dots & \dots \\ -\alpha_2 & 1 & -\beta_2 & 0 & 0 & \dots & \dots & \dots \\ 0 & 0 & -\alpha_3 & 1 & -\beta_3 & 0 & \dots & \dots \\ \vdots & \vdots & \vdots & \vdots & \ddots & \vdots & \vdots & \vdots \\ 0 & \dots & \dots & \dots & 0 & -\alpha_{n-1} & 1 & \beta_{n-1} \\ 0 & \dots & \dots & \dots & 0 & -\beta_n & -\alpha_n & 1 \end{bmatrix} \tag{1}$$

The deformation of the fitted contour w.r.t. the starting contour is measured with the internal energy function E_{int} (2). The data fitting error is expressed as the external energy functional E_{ext} (3).

$$E_{int} = \mathbf{x}^T A^T A \, \mathbf{x} \tag{2}$$

$$E_{ext} = \varphi(I, \mathbf{x}) \tag{3}$$

Given the image I with apparent contours, one fit the contour represented by the vector of its control points $\mathbf{x}$ by minimizing the sum of those energies, as in (4).

$$\hat{\mathbf{x}} = \arg\min_{\mathbf{x}} \{E_{int}(\mathbf{x}) + E_{ext}(I, \mathbf{x})\} \tag{4}$$

2 Active Shape-Based Models

As in the human face description task shapes are case-dependent, it is reasonable to introduce some statistical cues to the modeling process. The availability of the appropriate statistics, estimated from the testing data-set (a population of parametric contours carefully extracted from the test photographs) lets to introduce such modeling concepts as Active Shape Model (ASM) or Active Appearance Model (AAM) [2]. Here we focus our attention to the ASM. ASM is a two-stage method. First, a Point Distribution Model (PDM) is produced, which is then used for the validation of the contour shape. Second, a Local Gray-Level Model (LGLM) is generated to best fit the contour points to the local image-context.

2.1 Point Distribution Model

It is assumed that the distribution of contour control points, extracted from the set of test images is Gaussian. After determination of its mean $\bar{\mathbf{x}}$ and covariance $C_{\mathbf{x}}$, one can calculate and order in decreasing mode the eigenvalues of the covariance matrix λ_i as well as their corresponding eigenvectors $\boldsymbol{\phi}_i$. Eigenvectors determine the directions of most efficient dislocation of control points, while fitting the contour. Selecting only first k eigenvectors (corresponding to the largest eigenvalues) and putting them as columns of the matrix P, one can get the approximate shape model (5).

$$\mathbf{x} \approx \bar{\mathbf{x}} + P\mathbf{b} \tag{5}$$

where b is a reduced (to k-dimensions) contour-vector satisfying (6).

$$\mathbf{b} = P^T(\mathbf{x} - \bar{\mathbf{x}}) \tag{6}$$

PDM minimizes the residual error $\mathbf{r} = \mathbf{x} - \bar{\mathbf{x}} - P\mathbf{b}$ and similarly to E_{int} (in ACM) is responsible for the spatial relationships between contour control-points. One can define here a new pseudoenergy E_{int} in the following way:

$$E_{int} = \mathbf{r}^T\mathbf{r} \tag{7}$$

2.2 Local Gray Level

Assuming that image function samples taken along the normals to the contour at the control points have 1-D Gaussian statistics, one can estimate the corresponding means $\bar{\boldsymbol{\delta}}_i$ and dispersions $S_{\boldsymbol{\delta}_i}$. To that goal, l closest to the contour pixels are sampled along normals on both sides of every control point $\boldsymbol{\chi}_i$. In that way one get local models of image-function distributions $N(\boldsymbol{\delta}_i, S_{\boldsymbol{\delta}_i})$.

The quality of the local fitting of the contour model is measured by the Mahalanobis metric (8).

$$E_{ext}(\boldsymbol{\chi}_i) = (\boldsymbol{\delta}_i - \bar{\boldsymbol{\delta}}_i)^T S_{\bar{\boldsymbol{\delta}}_i}^{-1}(\boldsymbol{\delta}_i - \bar{\boldsymbol{\delta}}_i) \tag{8}$$

3 Building the ASM

To apply the ASM, one has to know the preliminary location of contours on the image. Manual determination of it is cumbersome and not really exact enough. Automatic determination of initial location of preliminary contours without any a priori knowledge available is unfeasible. To overcome these difficulties a semi-automatic method is proposed, making reference to the previously extracted face-landmarks.

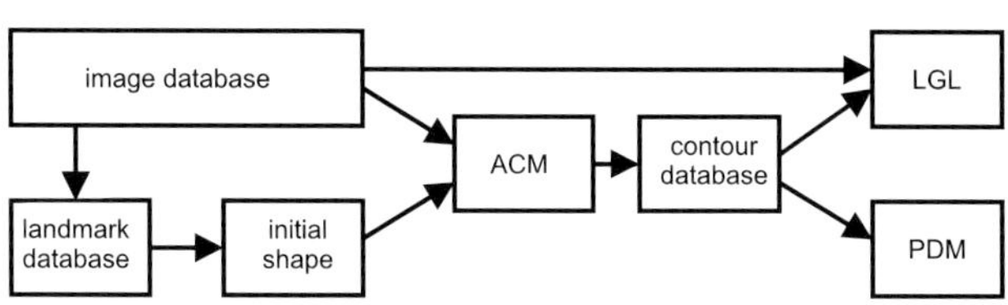

Fig. 1. Block diagram of the ASM

The following elements of the face have been described with contour models: eyes, nose, lips, eye-brows and face-outline. To that goal ACM (Sec. 1) has been used at the building stage of the ASM method. To obtain initial contours, which are apparently similar to typical shapes of above-mentioned elements some a priori knowledge has been applied. The appropriate contours have been interpolated with single-variable Hermit's splines $h(\bullet, t)$ [5]. Control points of those contours $\boldsymbol{\kappa}_i$ have been calculated as linear combinations of coordinates of landmarks $\boldsymbol{\lambda}_i$ (9), which has been manually determined from the test image-base.

$$\mathbf{k} = [\boldsymbol{\kappa}_1^T, \boldsymbol{\kappa}_2^T, \ldots, \boldsymbol{\kappa}_m^T]^T = W[\boldsymbol{\lambda}_1^T, \boldsymbol{\lambda}_2^T, \ldots, \boldsymbol{\lambda}_k^T]^T \tag{9}$$

Weighting matrix elements has been selected as the result of experimental search procedure. Contours, being the models of particular face elements are obtained as the concatenation of the appropriate Hermit's curves with control points $\boldsymbol{\kappa}_i$ (Fig. 2).

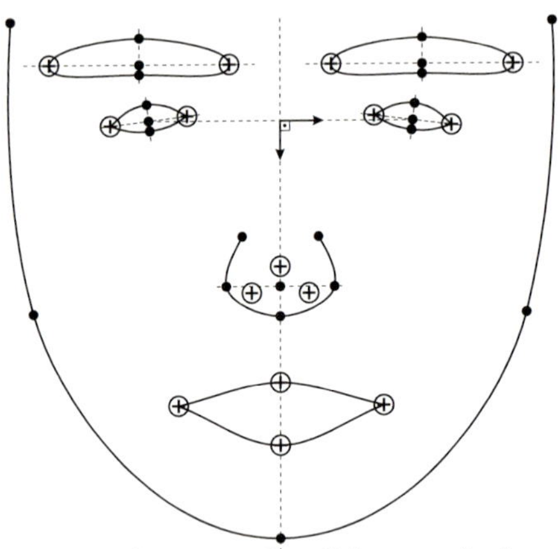

Fig. 2. Contours describing particular face elements, landmarks ($\oplus$) and curve control points ($\bullet$)

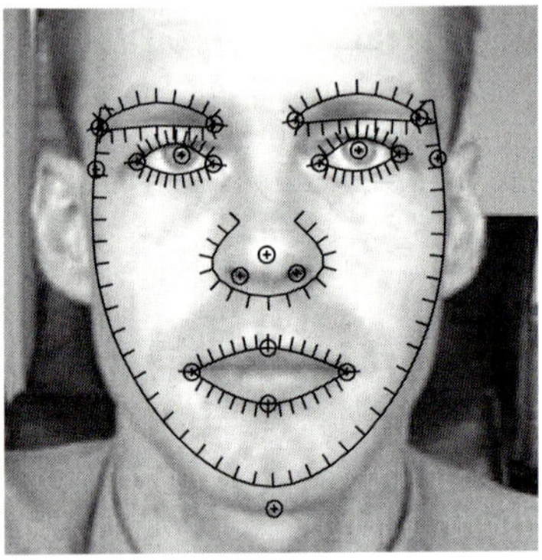

Fig. 3. An example of contours distribution on some face with marked landmarks ($\oplus$) and contour control points (short normals)

ACM and ASM models refer to the parametrically discretized image-locations of points, belonging to contours $k(t)$. Coordinates have been determined for some discrete set of t values. In the case considered in the paper, for 7 contours of interest 166 such points have been determined. After location of initial contour models on the image (Fig. 3), one has to fit particular contour models to the given image. It is done by minimizing (4) with use of a shape matrix A (1) evaluated for the initial contour under focus. Dynamic programming has been applied and the results of first two iteration steps are demonstrated in Fig. 4 a,b respectively.

For the given face image-base a base of contour models has been evaluated with ACM. With that method, a process of contours fitting could not be fully automatic. It was necessary to supervise the computation process and

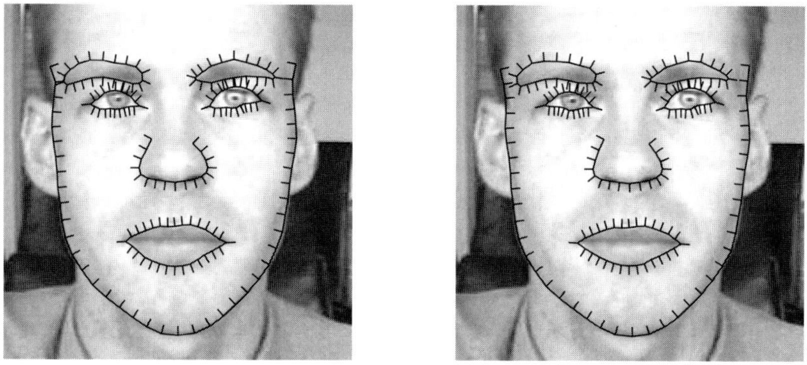

(a) Iteration 1. (b) Iteration 2.

Fig. 4. Iteration steps of ACM dynamic programming energy minimization

manually tune some optimization control parameters. The resulting set of contour models consisted of 183 face-shape descriptions. For those cases PDM and LGLM models has been estimated. In Fig. 5 an average shape and selected LGL profiles are presented.

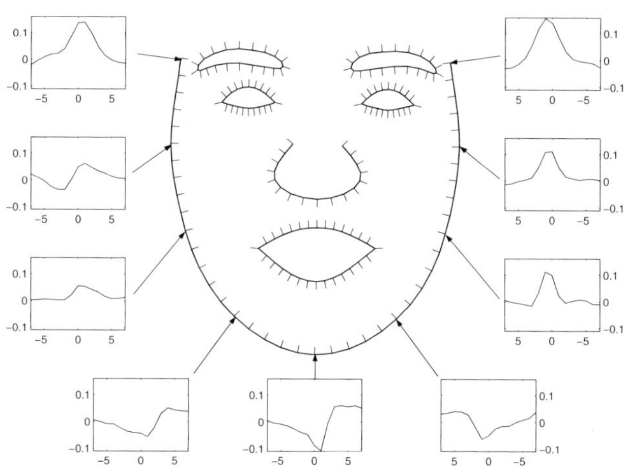

Fig. 5. Average shape model with selected LGL profiles for the face-outline model (15 profile samples)

4 The Performance of the ASM

The block diagram of the ASM procedure is given in Fig. 6. Preliminary location of the face on the image was based on eyes detection with simple bitmap-template matching. As the initial shape the average description, presented in Fig. 5 was used. It has been transferred to the location fixed by the coordinates of detected eyes. In the consecutive step LGLM-base location adjustments and PDM-based contour validations have been alternated until the iteration process converged. In average 5-10 iteration steps were enough.

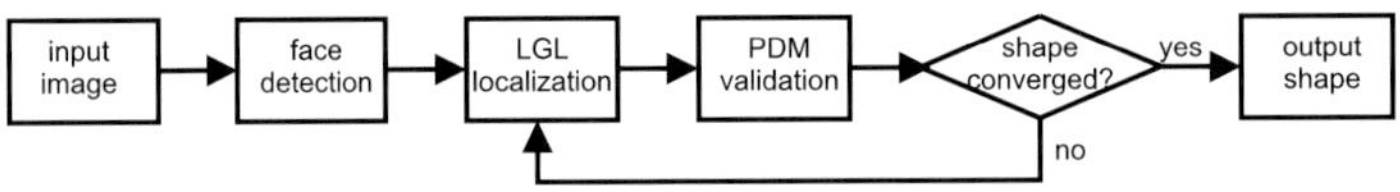

Fig. 6. The block diagram of the ASM procedure

It was interesting to investigate the impact of LGL profiles length on the quality of a particular face-features location. To that goal, for all images in the image-base, the LGLM-base localization process has been run and the obtained shapes have been compared with their correspondents in the contour-base. Location error was measured as the sum of squared distances between corresponding control points of the instance-contour and the appropriate model-contour from the contour-base. As goodness-of-fit measures three characteristics have been checked, namely Mahalanobis and Euclidean metrics and the steepest descent criterion. The results are given in Fig. 7. It is clear and not surprising, that the localization error almost uniformly decreases with the increasing profile-length. On the other hand, while comparing the results of Mahalanobis-metric-based local fitting with Euclidean-metric-based local fitting, one can notice that the last one case gives significantly worse results. From the graph in Fig. 7 one can conclude also that the use of profiles consisting of less than 5-samples in both cases is not efficient. For extremely short profiles gradient-based localization method performs better.

The overall classification performance of the ASM has been investigated in a simple experiment. Shapes extracted with ASM method have been compared with shapes in the contour-models base (manually extracted). If the difference was below a given threshold the shape obtained with ASM method was treated as True-Positive. The total classification efficiency of the ASM obtained with HumanScan image-base [4] was 98%. In Fig. 8 an example is given to illustrate the evolution of the active shape model as a function of a number of iterations. It is evident, that the method is insensitive w.r.t. the significant difference of the initial model scale and shape as well as w.r.t. the initial face location error. After 15 iterations the ASM fit is almost perfect.

Fitting the shape model to the particular image can be obtained, provided that fitted face elements are visible in the image. Nevertheless, with ASM it

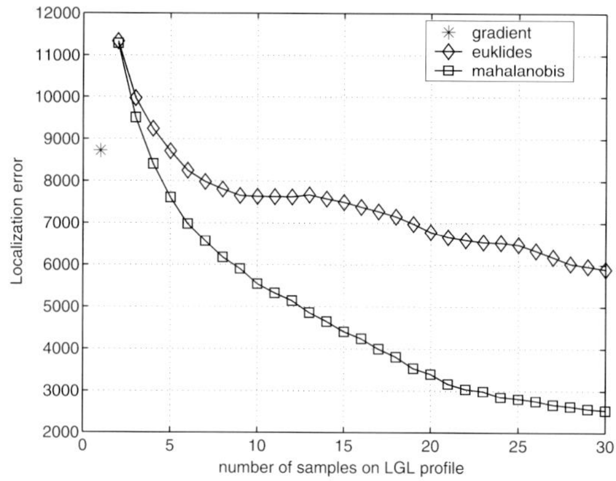

Fig. 7. Localization error as a function of a number of samples on LGL profile

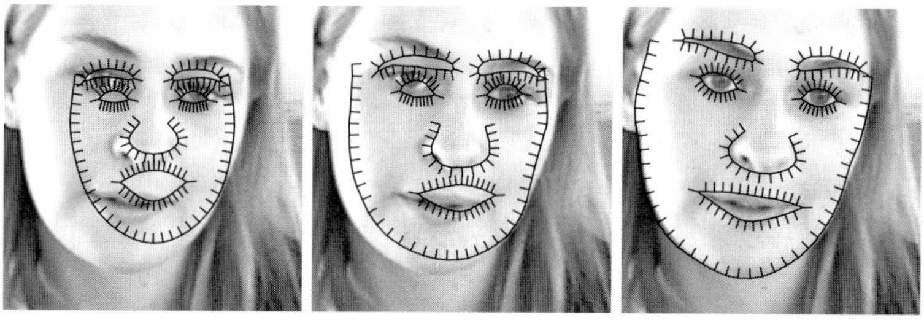

(a) initial shape (b) iteration 5. (c) iteration 15.

Fig. 8. Iteration steps of the ASM procedure

has been noticed, that the method is quite robust w.r.t. minor occlusions of the face in the image. It is due to the parallel nature of the fitting process. All model contours of the face features are fitted simultaneously but independently. However, large-area occlusions or strong contrast disturbances caused by glasses or hairs in important way disturb the process of correct fitting. Examples of difficult cases are provided in Fig. 9.

5 Summary

In the article, the system for fitting the face-shape models to the images is described. Two model fitting approaches have been used: ACM and ASM. ACM has been applied in order to create a face-contours base of models. Starting with that base, an ASM model has been computed, which then has been used to obtain the face-shape descriptions from the provided images. ASM gives

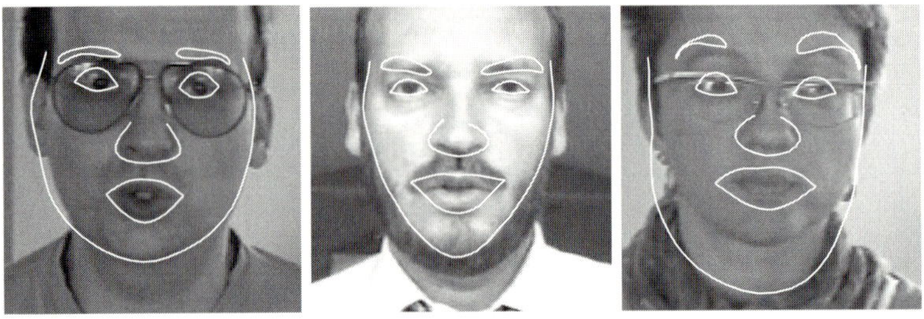

Fig. 9. Examples of incorrect ASM fits

the explicit and structured shape description from the source image. Such a description can be used for the study of the face mimics as well as to verify or identify the subjects [6]. Probabilistic models (local image samples distributions) are encoding the information enabling the identification of the subject (but also related to its pose while acquiring the image) in very implicit way. Projective transform of the image registration method introduces additional nonlinear distortions to the stochastic data. This is said to underline the potential difficulties with improving the performance of the ASM method. The Gaussian hypothesis used throughout the method is evidently a rough simplification. It is therefore desirable to study the real nature of local stochastic models in order to find the appropriate representation, which is sensitive and separable w.r.t. the recognition task goals.

References

1. A. Blake and M. Isard. *Active Contours.* Springer-Verlag, 1998.
2. T.F. Cootes and C.J. Taylor. Statistical models of appearance for computer vision. Technical report, Imaging Science and Biomedical Engineering, University of Manchester, October 2001.
3. D.A. Forsyth and J. Ponce. *Computer Vision: A Modern Approach.* Prentice Hall, August 2002.
4. HumanScan. www.bioid.com.
5. P. Kiciak. *Basis of curves and surfaces modelling.* WNT, 2000 (in Polish).
6. M. Krol. Face recognition with use of deformable contour models. Master's thesis, Poznan University of Technology, Institute of Control and Information Engineering, 2004 (in Polish).
7. K.F. Lai. *Deformable Contours: modeling, extraction, detection and classification.* PhD thesis, University of Wisconsin-Madison, 1994.
8. D. Terzopoulos M. Kass, A. Witkin. Snake: Active contour models. *International Conference on Computer Vision*, pages 259–268, June 1987.
9. D.G. Stork R. O. Duda, P. E. Hart. *Pattern Classification.* A Wiley-Interscience Publication, 2002.
10. A. Psarrou S. Gong, S. McKenna. *Dynamic Vision: From Images to Face Recognition.* Imperial College Press, 1999.

Comparison of Statistical Classifiers as Applied to the Face Recognition System Based on Active Shape Models

Maciej Krol and Andrzej Florek

Institute of Control and Information Engineering, Poznan University of Technology, str. Piotrowo 3A, Box 1, 60-965 Poznan, Poland
andrzej.florek@put.poznan.pl,maciej.krol@interia.pl

Summary. In this paper, a face recognition algorithm based on statistical model of Active Shape (ASM) is presented. A 31 degree-of-freedom shape model was used. The model was derived from a set of 183 faces shapes and named the learning set. Criteria of selection of face to model classifiers were evaluated. Classification was implemented in the shape space, in its Principal Component Analysis (PCA) and Multiple Discriminant Analysis (MDA) transformations. In the shape space the *Euclidean* and *Mahalanobis* metrics were used. *Euclidean* metric was used in PCA and MDA spaces as well. The results were based on experiments carried out on the set of 651 images of eight persons. Further proceedings in the case of ambiguous classification results were suggested.

1 Rationale, Motivation

Complex scenes are becoming a subject of growing interest in image recognition field. Analyzed images require more and more sophisticated recognition algorithms. Definition of features and criteria of object recognition in such cases like face and medical image analysis encounters difficulties. It is a complex task to create precise mathematical model of rules on which human beings are able to recognize and identify faces. Another problem is rised by the dimensionality of the feature vector describing face image model. One of the approaches for solving this problem is to introduce the Active Shape Model (ASM) [2]. Image analysis leads to evaluation of face description in form of contour parameters of face elements. In order to recognize face it is required to determine affiliation of its description to one of the classes in feature space.

2 Active Shape Model

This paper is focused on application of Active Shape Model to face recognition task. It is based on notion of shape which is defined as an ordered set of points describing face elements. ASM consists of two submodels. The first one, Point Distribution Model (PDM), is responsible for shape validation. The task of shape validation is to find a plausible transformation of a given shape in terms of shapes from a learning set. The second submodel, Local Gray Level (LGL), describes shape points neighbourhood on the image plane. The LGL submodel is used to locate descriptive features of the image function.

2.1 Dimension Reduction

The shape in ASM method is represented as an ordered set of control points placed on contours describing face elements and it is given by the following vector

$$\mathbf{x}_i = (x_1, y_1, x_2, y_2, \ldots, x_n, y_n)^T \tag{1}$$

where x_j and y_j are coordinates of shape control points, expressed in common coordinate frame for all m shapes in a given set. The shape vector $\mathbf{x}_i$ can be approximated by the reduced shape vector $\mathbf{b}_i$

$$\mathbf{b}_i = P^T(\mathbf{x}_i - \bar{\mathbf{x}}_\mathbf{i}) \tag{2}$$

being the result of the PCA method applied to covariance matrix $C_\mathbf{x}$ of the set of shapes $\mathbf{x}_i$ with mean $\bar{\mathbf{x}}_i$.

Matrix P contains k eigenvectors corresponding to the largest eigenvalues of $C_\mathbf{x}$. We select such k that vectors from matrix P cover sensible part of total variance of position for a given shape vectors set. It is possible to approximate $\mathbf{x}_i$ with the given reduced shape vector $\mathbf{b}_i$,

$$\mathbf{x}_i \approx \bar{\mathbf{x}}_i + P\mathbf{b}_i \tag{3}$$

2.2 Implementation

The ASM algorithm was implemented to the face model consisting of 7 contours, including 166 points. This implies 2n=322 dimensional shape space. After applying dimensionality reduction (2) shape space was reduced to 31 dimensions. This size gives 98% coverage of total variance of $C_\mathbf{x}$.

The implementation diagram is presented in figure 1. First step of algorithm is the face localization based on template matching. Next, a shape model is initialized by placing the mean face shape model at the position of localized face (Fig. 2). The ASM algorithm consists of two stages. The first stage, the shape model is deformed to fit the image function. The LGL model is used to determine the best fit function value. The second stage of the ASM algorithm is a PDM validation. At this stage, forward (2) and reverse (3) space transformation are consecutively applied. This is to reduce the distortions of fitted shape from the model. It gives an effect of attenuation of the small noise in local shape position. Two above mentioned stages are run one after another until the fixed-point is achieved.

The result of the algorithm is a shape model fitted to a face on the image (Fig. 3). The reduced shape vector $\mathbf{b}_i$ can be used to identify a given face.

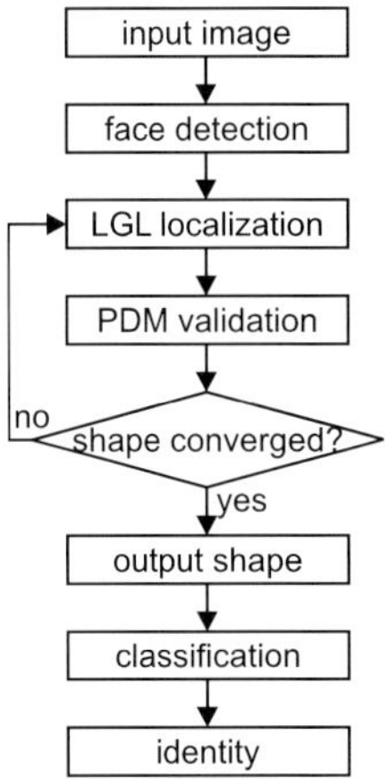

Fig. 1. Implementation diagram

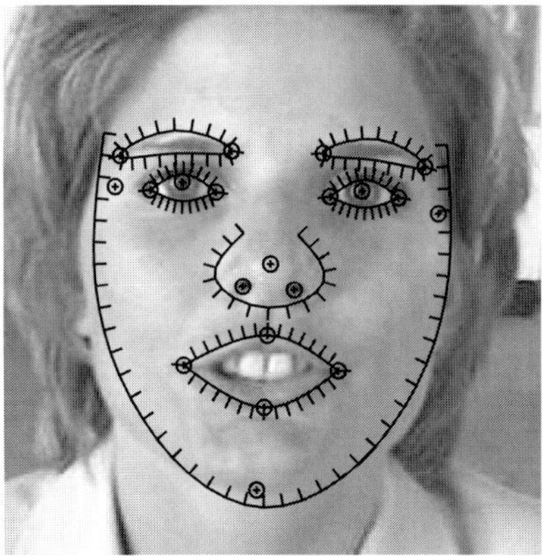

Fig. 2. Initial shape

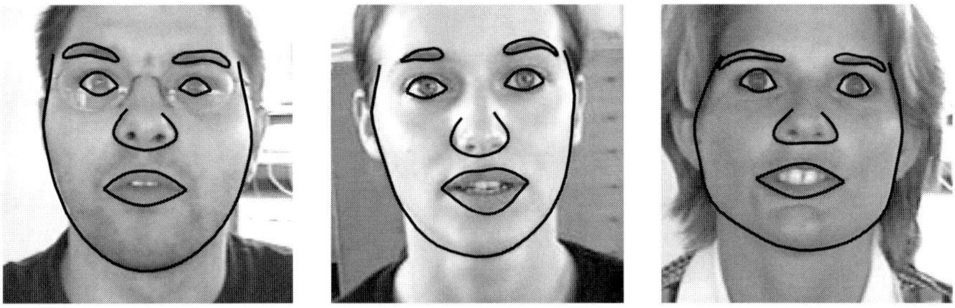

Fig. 3. Results of the ASM algorithm

3 Statistical Classifiers

In order to recognize a face on the image, an information about the face identity must be extracted. The shape model fitted to the face image describes some features of the personal identity. Face identification corresponds to a problem of shape vector classification. Classification in this case is a task of assigning a shape vector to the one of the c-classes representing face identity. To minimize the dimensionality of required feature space, reduced shape vector $\mathbf{b}_i$ is used.

3.1 Classifiers Definition

The shape vectors representing a given identify are distributed in clusters in the feature space. If assumption of *Gaussian* distribution of shape vectors $\mathbf{b}_i$ in those clusters is made, classification of an analyzed shape, represented by its reduced shape vector $\mathbf{b}$, can be based on the distance from the shape to some i-clusters. This kind of analysis is applied in Nearest Neighbourhood Classifier (NNC). The analyzed shape is assigned to the nearest class represented by the cluster and the average of the reduced shape vector $\bar{\mathbf{b}}_i$. In this paper four NNC with different feature spaces and metrics were used. The following metrics were considered:

- *Euclidean* distance
$$d_{Ei}^2(\mathbf{b}) = (\mathbf{b} - \bar{\mathbf{b}}_i)^T(\mathbf{b} - \bar{\mathbf{b}}_i) \quad (4)$$
- *Mahalanobis* distance
$$d_{Mi}^2(\mathbf{b}) = (\mathbf{b} - \bar{\mathbf{b}}_i)^T C_{\mathbf{b}}^{-1}(\mathbf{b} - \bar{\mathbf{b}}_i) \quad (5)$$
where $C_{\mathbf{b}}$ is a covariance matrix of a set of vectors $\mathbf{b}_i$
- distance in the MDA space
which is *Euclidean* distance $d_{Ei}^2(\mathbf{y})$ in $c-1$ dimensional *Fischer* discriminant space, where
$$\mathbf{y} = W^T\mathbf{b} \quad (6)$$
The MDA method determines such matrix $W \in \mathbb{R}^{2n \times c-1}$, that maximizes the between-class variance and at the same time minimizes the inner-class variance.
- distance in the PCA space
which is an *Euclidean* distance $d_{Ei}^2(\mathbf{z})$ expressed in the subspace of covariance matrix $C_{\mathbf{b}}$ eigenspace, where
$$\mathbf{z} = P^T(\mathbf{b} - \bar{\mathbf{b}}_i) \quad (7)$$
and $P \in \mathbb{R}^{2n \times k}$ consists of the eigenvectors corresponding to k largest eigenvalues of $C_{\mathbf{b}}$. This space was used in comparison with MDA.

Computation time for the face classification stage is insignificant as compared to the computation time for the ASM algorithm.

3.2 Classification Experiment

We distinguish two kinds of the shape sets: the learning set used in model derivation and the testing set used in the face identification experiment. Four classifiers, defined in 3.1, were used to the testing set of shapes. Number of correct and incorrect classifications was examined.

First two classifiers (*Euclidean* and *Mahalanobis* distances) make use of a 31 dimensional shape vector **b**. Classification was performed for 8 persons, hence for MDA and PCA classifiers space dimensions are equal to 8-1 = 7. Thus, a choice of directions reduction for a shape vector (reductions of degree-of freedom) could be investigated. For our case, the shape vector PCA reduced to 7 dimension covered 85% of a total variance of all **b** from the learning set (the measure being the ratio of a sum of 7 largest eigenvalues of $C_{\mathbf{b}_i}$ to the sum of all eigenvalues).

Robustness of the above mentioned classifiers was evaluated. To eliminate the influence of the preceding processing stages in the face recognition algorithm (such as the face detection and ASM procedures) the experiments with the use of shapes from the learning set were performed. In single experiments, face images from a given testing set were identified. In the first experiment we used a complete learning set as a testing set. Classifiers in this experiment were evaluated with respect to the complete learning set. In order to evaluate generalization capabilities of classifiers, two experiments were performed. The learning set was divided into two subsets of the same cardinality. With the use of the first one subset, classifiers were evaluated. Both of the subsets were then used in classification experiments as a testing sets. In table 1 results of experiments were presented, where the following notation has been used:

- TP - number of True Positive identification
- FP - number of False Positive identification
- R - identification ratio equal to quotient TP to the number of elements in the testing set

Table 1. Classification results for the learning set

	Euclidean			Mahalanobis			MDA			PCA		
	TP	FP	R	TP	FP	R	TP	FP	R	TP	FP	R
learning set	173	10	95%	183	0	100%	183	0	100%	149	34	81%
first subset	85	6	93%	91	0	100%	91	0	100%	69	22	76%
second subset	87	5	95%	92	0	100%	92	0	100%	80	12	86%

Identification rates for the *Mahalanobis* and MDA classifiers are equal to 100%, hence *a posteriori* we can justify our assumption of the *Gaussian* distribution of clusters in the shape space. This assumption for *Euclidean* and PCA classifiers turned out to be too restrictive.

3.3 Classification Results

The total efficiency of the whole recognition and identification system is a product of efficiencies of its sub-elements. The efficiency of the complete face recognition algorithm (Fig. 1) with the use of 651 face images of 8 persons was examined. Number of images for particular persons was different, dependent on the number of pictures in the image database. Two types of statistics for R value are presented (Table 2) . The first one has been calculated for the valid face position detection and the second one for all results of face detection.

Table 2. Classification results

Person	No. of Images	Valid face detections Eukc.	Mah.	MDA	PCA	No. of Images	All face detections Eukc.	Mah.	MDA	PCA
1	99	63%	82%	82%	42%	121	60%	74%	74%	40%
2	51	39%	63%	61%	35%	59	36%	54%	53%	32%
3	76	80%	93%	95%	40%	78	78%	90%	91%	38%
4	84	75%	92%	92%	70%	88	73%	89%	89%	68%
5	76	73%	84%	83%	37%	98	62%	71%	70%	33%
6	59	58%	75%	75%	42%	61	56%	72%	72%	41%
7	63	94%	98%	97%	89%	106	68%	67%	65%	68%
8	26	96%	100%	100%	81%	40	75%	80%	78%	65%
Total	534	71%	86%	85%	52%	651	64%	75%	74%	48%

3.4 Summary

In this paper an overview of the working face classification system based on the ASM algorithm was presented. It was found that the nearest neighbourhood classifiers based on the *Mahalanobis* metric in the reduced shape space and the *Euclidean* distance in MDA space yield good results. Effectiveness of the other two considered classifiers, based on the *Euclidean* distance in the reduced shape space and in its PCA transformation has been low. It is caused by the too high information content reduction.

It is advisable to examine influence of such factors like acquisition disturbances, choice of the ASM parameters, shape space choice and the initial shape (initial model of shape for starting the ASM procedures, when its the best fit face is calculated to the face under analysis). Thus we propose the initial shape selection system. The choice of the initial shape would be based on the results of classification given a set of classifiers. Voting scheme would be implemented. Lack of coherent decision amongst classifiers would cause the algorithm to restart with such initial shape that increases the probability of a correct decision settlement.

References

1. A. Blake and M. Isard. *Active Contours*. Springer-Verlag, 1998.

2. T.F. Cootes and C.J. Taylor. Statistical models of appearance for computer vision. Technical report, Imaging Science and Biomedical Engineering, University of Manchester, October 2001.
3. J.P. Marques de Sa. *Patter Recognition.* Springer-Verlag, 2001.
4. D.A. Forsyth and J. Ponce. *Computer Vision: A Modern Approach.* Prentice Hall, August 2002.
5. HumanScan. www.bioid.com.
6. M. Krol. Face recognition with use of deformable contour models. Master's thesis, Poznan University of Technology, Institute of Control and Information Engineering, 2004 (in Polish).
7. K.F. Lai. *Deformable Contours: modeling, extraction, detection and classification.* PhD thesis, University of Wisconsin-Madison, 1994.
8. D. Terzopoulos M. Kass, A. Witkin. Snake: Active contour models. *International Conference on Computer Vision*, pages 259–268, June 1987.
9. D.G. Stork R. O. Duda, P. E. Hart. *Pattern Classification.* A Wiley-Interscience Publication, 2002.
10. A. Psarrou S. Gong, S. McKenna. *Dynamic Vision: From Images to Face Recognition.* Imperial College Press, 1999.
11. A. Sulkowski. Methods of computer face recognition. Master's thesis, Poznan University of Technology, Institute of Control and Information Engineering, 2003 (in Polish).

Face Recognition Using DCT and LDA

Adam Nowosielski

Szczecin University of Technology, Zolnierska 49, 71-210 Szczecin, Poland
anowosielski@wi.ps.pl

Summary. In the article new method of face recognition is considered. It exploits two well-known approaches namely DCT and LDA. Using LDA on selected spectral components of the DCT better separation of classes can be achieved. Scaling problem of the face images was addressed and appropriate solution proposed. Experiments on the ORL [10] database of faces were carried out. Results were compared with the individual DCT approach and one of the most frequently used approach nowadays: PCA+LDA.

1 Introduction

Two most significant aims in face recognition are data dimensionality reduction and class separability. Reduction of the data dimensionality relies on coding information content in a face image by a small number of coefficients. Appropriate class separation means that in a final feature space, face images of every person are grouped and separated from each other.

One of the most frequently used approaches in face recognition takes advantages from transforms. Transforming an image to frequency domain changes the way the information is stored. Most information is retained in a relatively small subset of spectrum coefficients. This way, other coefficient (with small variance) can be discarded without great influence on the reconstruction process.

Among transforms Karhunen-Loeve transform (KLT) is an optimal one for a specific data set. This transform is very popular and used in such approaches like PCA and LDA [5]. The second group consists of transforms independent from the input data. The most popular here are: discrete cosine transform (DCT), discrete fourier transform (DFT) and wavelet transform. They differ in base functions (constant independent from the input set).

In this paper a combination of the DCT and LDA is being considered. This is an interesting approach which can be considered in two ways. First, as an improvement of the individual DCT approach (higher dimensionality reduction and much better class separability). Second, as a modification of the PCA+LDA approach. In the last case, advantage is taken from the base cosine functions (of the DCT) which approximate eigenvectors of the Karhunen-Loeve transform with high accuracy.

There is an abundance of work devoted to both DCT and LDA individually. Discrete cosine transform is used most frequently as a feature extraction method (only one approach in the system) [3, 7, 8] or together with other approaches: pseudo 2D Hidden Markov Models [2], neural network [9]. Additionally, two solutions are used: local DCT [2, 9, 11] (partitioning on the same size blocks, similarly as in the JPEG standard) and global DCT [3, 6, 7, 8] (the whole image transform). The LDA method is most frequently used in combination with the PCA method. PCA and LDA together form two-stage procedure of initial data reduction. Example solution can be found in [4, 5, 12].

The rest of the article is structured as follows. Section 2 describes classic dimensionality reduction methods. In Sect. 3 DCT+LDA approach is considered. Section 4 presents experiments and evaluations. The article ends with a summary.

2 Classic Methods of Dimensionality Reduction

PCA (Principal Component Analysis) and LDA (Linear Discriminant Analysis) both are classic methods of dimensionality reduction. The aim of the PCA is to represent the changes in face images by the smallest set of features possible [5]. LDA aims to increase the distance between the centers of face classes (persons) while in the same time decreasing the distance of the individual element (face image) from the center of the class it belongs. Schematic representation of possible variants of the above methods are presented in Fig. 1.

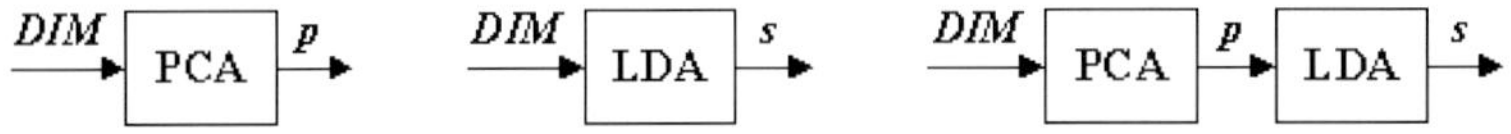

Fig. 1. Classic methods of data dimensionality reduction

When building a face recognition system with the use of the above approaches one has to secure fulfillment of the following conditions [5, 12]:

$$KL > (DIM + K) > p > s \; for \; DIM \leq MN \; and \; s \leq K - 1, \qquad (1)$$

where: K - number of classes, L - number of face images in each class (for system synthesis), M - number of rows in face image, N - number of columns in face image, p - dimension of a feature vector after PCA reduction, s - dimension of a feature vector after LDA reduction, DIM - input data dimensionality.

It occurs that it is hard to fulfill the above conditions. Usually, there is an input feature vector for an individual face image (matrix) which is created by concatenation of rows or columns. In such a situation we have: $DIM = MN > KL$. The initial conditions (1) are not fulfilled. In case of the ORL face database $DIM = 10304$ (112x92), and $KL = 400$ (40 persons, 10 face image for each person). Dealing with such a situation requires adjusting the dimensionality DIM so that conditions (1) are met [5]. This can be achieved by a change of the data structure by decomposition on "row-images" and "column-images". A simpler solution is to scale initial images. Both methods, however, have drawbacks. In the first case, one gets face images of very small resolution. These images lack details. In the second case, face image is treated as a set of independent rows or columns. The reader is referred here to [5] for details of the above procedures.

3 DCT and LDA

DCT+LDA approach is very similar to PCA+LDA approach. It is presented schematically in Fig. 2.

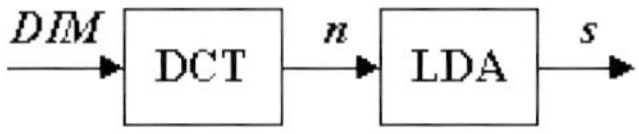

Fig. 2. Scheme of the DCT+LDA approach

Two dimensional discrete cosine transform C of the initial image X of dimensions MxN can be calculated using the following equations:

$$
\begin{aligned}
&C(p,q) = \alpha_p \alpha_q \sum_{m=0}^{M-1} \sum_{n=0}^{N-1} X(m,n) \cos \frac{(2m+1)\pi p}{2M} \cos \frac{(2n+1)\pi q}{2N} \\
&\alpha_p = \begin{cases} 1/\sqrt{M} \; if \; p = 0 \\ \sqrt{2/M} \; if \; 1 \le p \le M-1 \end{cases}; \; \alpha_q = \begin{cases} 1/\sqrt{N} \; if \; q = 0 \\ \sqrt{2/N} \; if \; 1 \le q \le N-1 \end{cases}
\end{aligned}
\tag{2}
$$

where: p, q - dimensions of the resulting spectrum with $p = 0..M-1$ and $q = 0..N-1$.

Low frequency components (components with small indexes) have the highest variance. They are selected from the left triangle of C ("triangle method"). This method guarantees the best choice of coefficients [7]. For example, 11 diagonals, i.e. $p = 0..10$, $q = 0..10$ and $q < p$, mean 66 features in a feature vector.

Using only these DCT features and Euclidean metric good results can be achieved e.g. [7, 8]. However, DCT feature vector can be treated as an intermediate feature vector which can be further processed and reduced using

LDA. In such an approach analysis of conditions (1) is required. The number of selected coefficients n has to fulfill the following condition: $n > s$ and $s \leq K - 1$.

After DCT features are calculated for a database face images the LDA procedure follows. Two matrices: a between-class and a within-class are created. Using these matrices the optimal projection matrix W is chosen as a matrix which maximizes the ratio of the determinant of the between-class scatter matrix of the projected data to the determinant of the within-class scatter matrix of the projected data [1]. This procedure is described in detail in [5].

In a real-life case the size of the face is not known a priori. Left side of Fig. 3 presents two face images of the same person. They only differ in scale. On the right hand side, there are two charts presenting values of the DCT components. The upper one depicts original values of both face images. Dots denote components of the bigger face image. The line connects corresponding values of the smaller image. The form of the two graphs is identical. Values differ by a scaling factor which in presented case equals 2. To avoid this problem face images are scaled to a fixed size. Example of such solutions can be found in [3, 6] where the distance between the eyes is used as a scaling factor. However, the distance between the eyes changes during left-right head rotations while the head size remains constant. What is more, additional operations for locating coordinates of eyes are required. It is therefore more stable and practicable to use width of the face image as a normalization factor. The DCT coefficients are divided by this width. Such solution is more precise because initial face image is not scaled (has more details). After normalization step both graphs from the bottom chart of Fig. 3 are identical.

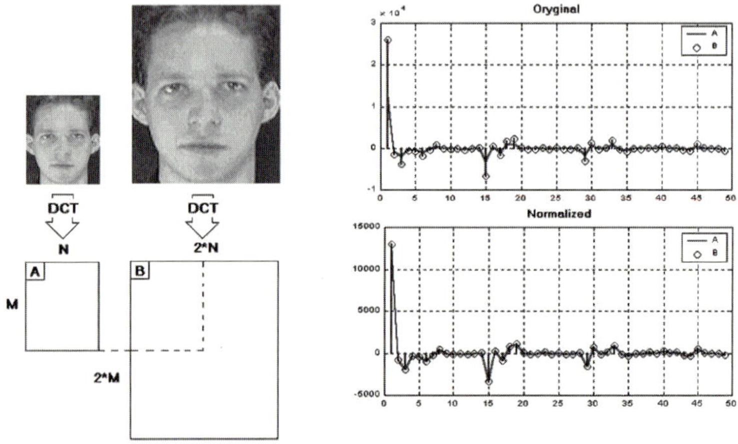

Fig. 3. Scaling problem of DCT components

4 Experiments

To test the considered solution the ORL Database of Faces [10] (10 pictures of 40 persons) was used. The test database was rearranged to be representative (to capture variation in the pose with the minimal number of face images). This was achieved by sorting procedure (like in [5]). Faces with the highest variance (within class) were assigned the smallest indexes.

In the first test 7 face images from each class were used to build face database. The remaining 3 face images from each class were used to test recognition accuracy. Classification of test images – their membership in a given class – was carried out with the minimal Euclidean distance criteria. The recognition results (RR%) were evaluated by the proportion of correctly recognized images to the total number of test images and presented as percentages.

Results of the simple DCT method are presented in Table 1. The coefficients from the discrete cosine transform were selected using the triangle method with different number of coefficients n (number of diagonals in brackets). After that, the DCT+LDA approach was tested for different number of DCT coefficients n and different dimensions of reduced feature space s (after LDA). Results are presented in Table 2.

Table 1. Recognition results of the DCT approach

n	28(7)	36(8)	45(9)	55(10)	66(11)	78(12)	91(13)	105(14)	120(15)
$RR\%$	99,17%	98,33%	99,17%	99,17%	98,33%	98,33%	98,33%	98,33%	97,50%

Table 2. Recognition results of the DCT+LDA approach

	n=45	n=55	n=66	n=78	n=91	n=105	n=120	n=136
s=9	98,33%	99,17%	99,17%	98,33%	99,17%	97,50%	97,50%	98,33%
s=10	98,33%	99,17%	99,17%	99,17%	98,33%	98,33%	97,50%	97,50%
s=11	99,17%	100,00%	100,00%	99,17%	99,17%	99,17%	97,50%	98,33%
s=12	99,17%	100,00%	100,00%	100,00%	100,00%	100,00%	98,33%	97,50%
s=13	99,17%	99,17%	100,00%	99,17%	99,17%	99,17%	98,33%	100,00%
s=14	99,17%	99,17%	100,00%	99,17%	99,17%	99,17%	98,33%	100,00%
s=15	99,17%	99,17%	100,00%	100,00%	99,17%	99,17%	100,00%	100,00%
s=16	100,00%	97,50%	100,00%	100,00%	100,00%	99,17%	100,00%	100,00%
s=17	100,00%	100,00%	99,17%	100,00%	100,00%	100,00%	100,00%	100,00%
s=18	100,00%	100,00%	100,00%	100,00%	100,00%	100,00%	100,00%	100,00%

It can be seen from presented experiments that using only 11 features for representing face image in a DCT+LDA approach equal or better results can be obtained in comparison with DCT approach alone. What is more, in DCT+LDA a perfect recognition was achieved for different parameters. LDA can improve recognition accuracy and secures better data clasterization as can be seen in Fig. 4.

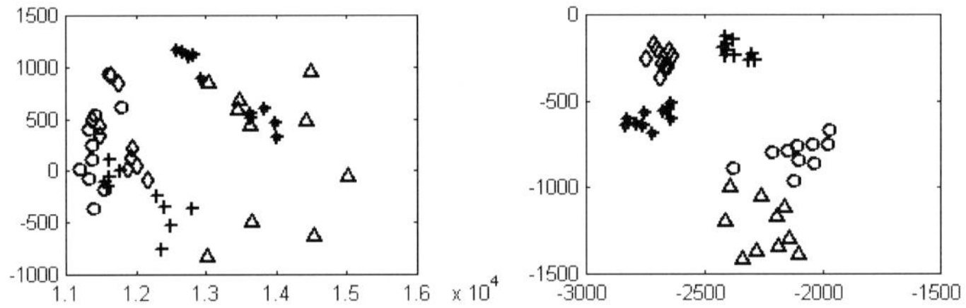

Fig. 4. Graphic representation of face images in DCT and DCT+LDA subspaces

Figure 4 graphically presents data clasterisation. The whole ORL database was considered but only first 5 classes are depicted (for clarity reasons). 66 spectral coefficients of the DCT were calculated (left image presents two first coefficients). Groups are hard to distinguish. However, considering other features it is possible to perform correct recognition (see Table 1). Using DCT coefficients LDA procedure was carried out (final dimension $s = 11$). Right image presents results. It is evident that face images of the same class are grouped in the DCT+LDA approach. Only two features are presented and there is still possibility for correct recognition.

DCT+LDA approach is finally compared to the most similar PCA+LDA approach. For the PCA+LDA conditions (1) has to be considered. In current test we have: 40x7 $> DIM + 40$. So $DIM < 240$ which means that using scaling variant (and preserving original proportions of the ORL database face images) we can process images of the size of 17x14 ($DIM = 238 < 240$). Results for different values of s and p parameters are presented in Table 3.

Results obtained from PCA+LDA approach are slightly better. 100% recognition was acquired for 10 coefficients while in DCT+LDA 11 coefficients were required. There is, however, not much difference in this result.

DCT+LDA approach was at last checked on its ability to reject not registered faces. Database of known individuals was built from the first 30 classes (7 images were used). The remaining 90 images (3 faces, 30 classes) were used to test recognition accuracy. The remaining 10 classes (100 images) were used to test the false acceptance rate. Results are presented in Table 4. In these tests 66 DCT and 29 LDA coefficients were used.

As can be seen from Table 4 very high recognition rates (RR=96.67%) can be achieved with zero false acceptance rate (FAR=0%). To further investigate these results all tests were repeated but every test image was scaled with the random scaling factor from 0.33 to 3.0. Similar results to that presented in

Table 3. Recognition results of the PCA+LDA approach

	p=45	p=50	p=55	p=60	p=65	p=70	p=75	p=80
s=9	98,33%	99,17%	97,50%	97,50%	99,17%	99,17%	98,33%	98,33%
s=10	99,17%	100,00%	98,33%	98,33%	99,17%	99,17%	99,17%	98,33%
s=11	99,17%	100,00%	99,17%	100,00%	99,17%	99,17%	100,00%	99,17%
s=12	98,33%	100,00%	100,00%	100,00%	100,00%	100,00%	99,17%	100,00%
s=13	99,17%	100,00%	100,00%	100,00%	100,00%	100,00%	99,17%	100,00%
s=14	99,17%	100,00%	99,17%	100,00%	100,00%	100,00%	99,17%	100,00%
s=15	97,50%	100,00%	100,00%	100,00%	100,00%	100,00%	100,00%	100,00%
s=16	97,50%	100,00%	100,00%	100,00%	100,00%	100,00%	100,00%	100,00%
s=17	97,50%	100,00%	100,00%	100,00%	100,00%	100,00%	100,00%	100,00%
s=18	97,50%	98,33%	100,00%	100,00%	100,00%	100,00%	100,00%	100,00%

Table 4. RR% and FAR% in the DCT+LDA approach

Threshold	6.8	7.0	7.2	7.4	7.6	7.8	8.0
RR%	92.22%	95.56%	**96.67%**	96.67%	97.78%	97.78%	97.78%
FAR%	0.00%	0.00%	**0.00%**	1.00%	2.00%	4.00%	6.00%

Table 4 were reported. Resulting DCT coefficients were scaled using the width of the face image.

5 Conclusions

In the article new face recognition method was examined. It exploits two well-known approaches namely DCT and LDA. This solution resembles the PCA+LDA approach and has almost as good results as this commonly used method. However, DCT+LDA approach has the following advantages: it does not require adjustment of the input data dimensionality (conditions (1) for PCA+LDA), it is less computationally expensive and easier to implement. In comparison with the individual DCT technique, better recognition accuracy, better clasterization and further dimensionality reduction were obtained.

Perfect recognition rates can be achieved using only 11 features for the database consisting of 40 classes of 7 face images each (40x3 test images, no threshold). This accounts for almost 1000 times decrease (937) in data dimensionality (112x92/11).

Scaling problem of the face images was also addressed and appropriate solution presented. Proposed DCT+LDA method is able to process images of different sizes without initial scaling. It has also high abilities for rejecting faces not registered in the database. In the test where face images of different sizes of known and unknown individuals were presented to recognize, 96.67% of correct recognition and 0% of false acceptance were registered.

Acknowledgement

Author of this paper would like to thank his promotor Prof. G. Kukharev for precious advices and great support during scientific research.

References

1. Belhumeur P N, Hespanha J P, Kriegman D J (1997) Eigenfaces vs. Fisherfaces: Recognition Using class Specific Linear Projection. IEEE Transactions on Pattern Analysis and Machine Intelligence 19(7): 711–720
2. Eickeler S, Müller S, Rigoll G (1999) High Performance Face Recognition Using Pseudo 2-D Hidden Markov Models. European Control Conference (ECC), Karlsruhe
3. Hafed Z, Levine M (2001) Face Recognition Using the Discrete Cosine Transform. International Journal of Computer Vision 43(3): 167–188
4. Kuchariew G, Forczmanski P (2003) Hierarchical Method of Reduction of Features Dimensionality for Image Recognition and Graphical Data Retrieval. Proceedings of the Sixth International Conference PRIP'2001: 57–71, Minsk
5. Kukharev G, Kuzminski A (2003) Biometric Techniques Part I: Face Recognition Methods. Szczein University of Technology Publisher, Szczecin (in Polish)
6. Kukharev G, Nowosielski A (2004) Visitor Identification – Elaborating Real Time Face Recognition System. WSCG'2004 Short Communications: 157–164, Plzen
7. Nowosielski A (2003) Feature Extraction from Face Images by Two Dimensional Discrete Cosine Transform. Szczecin University of Technology Publisher: 153–162, Szczecin (in Polish)
8. Nowosielski A, Miklasz M (2004) Face Recognition System in Client/Server Architecture. 11th International Workshop on Systems, Signals and Image Processing Proceedings IWSSIP'04: 359–362, Poznan
9. Pan Z, Adams R, Bolouri H (2001) Image Recognition Using Discrete Cosine Transforms as Dimensionality Reduction. IEEE EURASIP Workshop on Nonlinear Signal and Image Processing (NSIP01), Baltimore, Maryland
10. Samaria F, Harter A (1994) Parameterisation of a stochastic model for human face identification. 2nd IEEE Workshop on Applications of Computer Vision, Sarasota FL
11. Sanderson C, Paliwal K K (2003) Fast Features for Face Authentication Under Illumination Direction Changes. Pattern Recognition Letters 24(14): 2409–2419
12. Swets D L, Weng J (1999) Hierarchical discriminant analysis for image retrieval. IEEE Transaction on Pattern Analysis and Machine Intelligence (5): 386–401

The New Algorithm of Fingerprint Reference Point Location Based on Identification Masks

Piotr Porwik[1] and Krzysztof Wrobel[2]

[1] Institute of Informatics, Silesian University, 41-200 Sosnowiec, ul. Bedzinska 39
porwik@us.edu.pl
[2] Institute of Informatics, Silesian University, 41-200 Sosnowiec, ul. Bedzinska 39
kwrobel@zsk.tech.us.edu.pl

1 Introduction

Fingerprint identification is one of the most important biometric technologies especially in Fingerprint Identification System (AFIS). A fingerprint is the pattern of ridges and valleys on the surface on a fingertip. Each human has unique fingerprints. The uniqueness of a fingerprint allows to build systems which can recognize personal fingerprint feature automatically. Unfortunately it is a difficult task, because fingerprint images have very poor quality very often. In practice, due to variations in impression conditions, ridges configuration, skin condition, acquisition devices, etc. many restrictions have to considered. The ridge structures in poor-quality fingerprint images are not always well-defined, therefore they cannot be detected. Fortunately a fingerprint experts are often able to correct such corrupted images as long as the ridges and valleys structure is not corrupted completely. Each analyzed digital fingerprint image should has area where structure of finger will be visible. Generally, fingerprint images can be divided into three types: well-defined region, recoverable corrupted region and unrecoverable region [1,2,7]. In the first region ridges and valley are clearly differentiated, in the second region ridges and valleys are corrupted by a small amount of creases or smudges but are still visible. In the third area ridges and valleys are strong corrupted - such image is rejected. In our paper only two first areas are analyzed and each of them can be used to fingerprint recognition. In practice, fingerprint recognition and classification based on two techniques: the reference point location and the Gabor filtering or the ridge characteristics (called minutiaes). In this paper a new method of the reference point alignment has been presented. A new approach of reference point localization is based on so-called identification masks which have been composed on the basis of analysis of biometric characteristic of human finger. Construction of such masks has been presented.

2 Fingerprint enhancement

Fingerprint enhancement is used to recover the topology structure of ridges and valleys from the noisy image. An image enhancement allows to improve quality of digital fingerprint representation but this method should not results in any spurious ridge structures and configurations. This assumption is important because any artificial ridges deformation may change the base imprint. In fingerprint processing two types of images are accepted: binary or gray-level images. The binary image can be obtained of course from gray-level image. Unfortunately in binarization procedures some information about ridges can be lost therefore a gray-level images are preferred. The more advanced fingerprint enhancement procedure is described in [1]. In mentioned paper among other things orientation image, ridge frequency image and filtering have been explained. Such description for fingerprint minutiae's has been dedicated but can also be used for our experiments. A gray-level image I is defined as $N \times N$ matrix where $P(x, y)$ represents the intensity of pixel at the (x, y) point. In presented implementation, fingerprint image has dimension 256×256 and with 500 dpi resolution have been scanned. Such resolution is recommended by FBI. In the first stage a gray-level image to black-white format is exchanged, where the binarization threshold has been fixed on the level:

$$T = \frac{1}{N^2} \sum_{x=0}^{N-1} \sum_{y=0}^{N-1} P(x, y) \tag{1}$$

The constant 30 which occurs in equation (1) was fixed experimentally, and allows us to get the better binarization effect for scanned fingerprint images with dark background.

3 Direction image determination

In general, analyzed fingerprint is a well-defined orientation field, which has to compared with pattern. In most cases the reference point detection methods operate on the orientation field of fingerprint. The Poincare index analysis is well known method described in [5]. Unfortunately, this method is sensitive to noise of orientation field. Very efficient method proposed in [2] is based on multiresolution analysis, but this method is sensitive to the fingerprint rotation. In presented method, the orientation field, with some additional modifications, to reference point indication has been exploited. In our approach, base on orientation field, the special masks are proposed, which allow to detect a unique reference point consistently and accurately for all types of prints. To decide the ridge direction of each pixel in the image, the ridge direction of a given pixel $P(x, y)$ into directions i $(i = 0, 1, ..., 7)$ has been divided in a 9×9 window with the pixel in the center [3] (Fig. 1b). The gray values of pixels in eight directions (at positions marked by numbers 0,1,...,7) are added together to obtain sums $s_0, s_1, ..., s_7$. The sums s_i are equivalent to convolving the image with 9×9 masks, where each mask has value of 1 at positions where block shown in Fig. 1b has values i and 0 elsewhere. In next stage, for indices i min/max values are estimated:

$$i = \begin{cases} s_d = \underset{i=\{0,1,..,7\}}{\arg} \{\max(s_i)\} & \text{for} \quad p_{xy}^{bin} = 1 \\ s_l = \underset{i=\{0,1,..,7\}}{\arg} \{\min(s_i)\} & \text{for} \quad p_{xy}^{bin} = 0 \end{cases} \tag{2}$$

where: p_{xy}^{bin} is a pixel value at point (x, y), in fingerprint digital image. The direction at pixel is defined by means of s_d value if the central pixel is located on a ridge (black area), and by the s_l value if the central pixel is located in a valley. From equation (2), indices image can be prepared. Such image can also be called direction image. Unfortunately, obtained values treated as direction for each pixel are usually noisy, therefore they should be smoothed and averaged in a local neighborhood. In our application, as smoothing operation the mode function has been applied. The function mode computes the mode of the given data and the mode is defined as observation with the highest frequency. In other words, mode function can be treated as measure of central tendency. For example $d = mode(a, b, c, d, d, d, e, f, g)$. If there is more than one observation with the modal frequency, then choice is arbitrary. Results of mentioned operations are shown on Fig. 1.

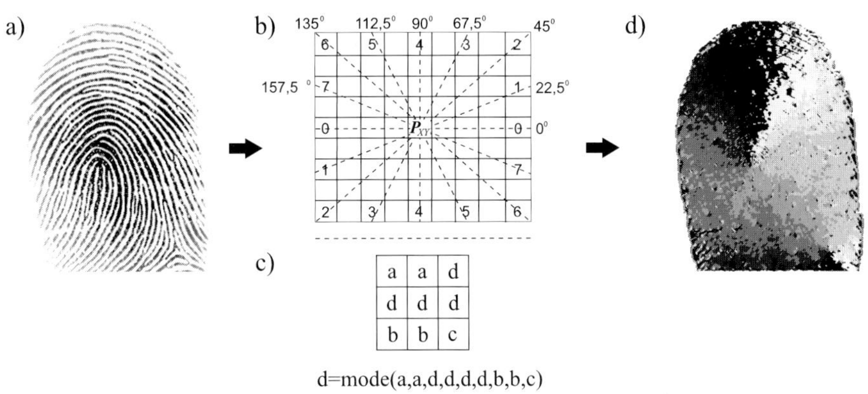

Fig. 1.
a) a given fingerprint, b) the 9×9 mask to compute the ridge directions, c) 3×3 mode filter mask, d) directions image.

4 The reference point determination by means of masks

For the first time, described bellow method in [7] has been presented. In present paper that method has been extended and new researches were carried out. From directions image, the orientation field has been prepared [2,4]. The orientation represents the local orientation of the ridges contained in the fingerprint. Directions image from Fig. 1d into 9×9 blocks (windows) has been divided. In each block, dominant direction (the most frequently occurring direction) like previously was calculated. From the gradient information the orientation angle is estimated. The 8 directions (see Fig. 1b) in each block is stored as table. This table can be presented as image (although it is not necessary), where each direction is performed as appropriate line. For example, for angle 0^0 horizontal line is performed, for angle 90^0 - vertical, and so one by step 22.5^0 for another direction lines. In next stage background is eliminated and only foreground is used. If in a given block 9×9 at least one black pixel can be found, then whole block create part of orientation field, otherwise such block is treated as background and is rejected. After background elimination

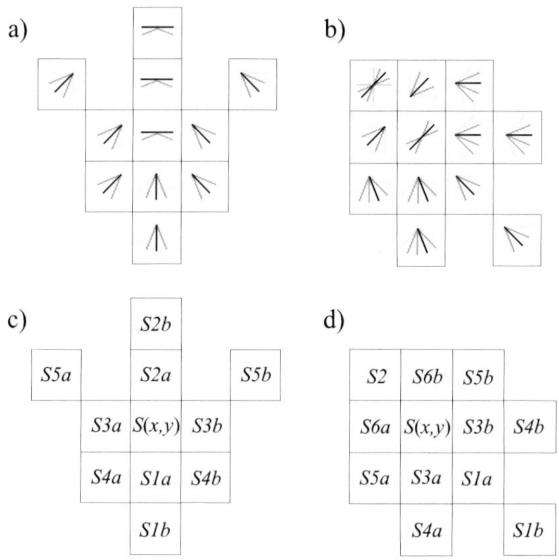

Fig. 2.
a,b) features of identification mask,
c,d) the identification masks for 0^0 and $+45^0$, respectively

the orientation field is smoothing with the aid of mode mask, similar as previously. In the last stage the reference point is determined. Because print of finger can be impressed in different manner, influence of rotation should be eliminated. In our method orientation field image is filtered by means of appropriates masks. That masks have special construction, what allows to collect fingerprints located in $0^0, \pm 45^0, \pm 90^0$. Proposed masks are called *identification masks.* Fig.2a,b present identification features of each mask. A special arrangement of short lines in each cell characterizes ridges in a human fingerprint. Such ridges distribution has been established experimentally. Such

approach is very efficient and it is equivalent to minutiae detection method [5], but unlike another methods [1,4,6] our solution gives faster reference point detection. By means of identification masks, which explore whole orientation field, fingerprint features represent by appropriate mask are sought. The masks for angles 0^0 and 45^0 present Fig. 2c,d. Because it is not known, how fingerprint was impressed, orientation field is filtered in turn by all masks. Finally, each identification mask more than one reference point can indicate. All identification masks into cells are split. The cell indicate direction at point x, y in orientation field. The 8 directions in orientation field are represented by means of numbers $0, 1, 2, 3, 4, 5, 6, 7$. Such numbers are substitution of angles $0^0, 22.5^0, 45^0, 67.5^0, 90^0, 112.5^0, 135^0, 157.5^0$, respectively (see Fig. 1b).
From orientation field (Fig. 2) the reference point for the angle 0^0 can be determined if the global condition is fulfilled:

$$\begin{aligned}
&if((S(x,y) = 0)or(S(x,y) = 1)or(S(x,y) = 7))and \\
&(((S1a = 2)or(S1a = 3)or(S1a = 4)or(S1a = 5)or(S1a = 6))and \\
&((S1b = 2)or(S1b = 3)or(S1b = 4)or(S1b = 5)or(S1b = 6)))and \\
&(((S2a = 0)or(S2a = 1)or(S2a = 7))and \\
&((S2b = 0)or(S2b = 1)or(S2b = 7)))and \\
&(((S3a = 0)or(S3a = 1)or(S3a = 2)or(S3a = 3)or(S3a = 4))and \\
&((S3b = 0)or(S3b = 4)or(S3b = 5)or(S3b = 6)or(S3b = 7)))and \\
&(((S4a = 1)or(S4a = 2)or(S4a = 3)or(S4a = 4))and \\
&((S4b = 4)or(S4b = 5)or(S4b = 6)or(S4b = 7)))and \\
&(((S5a = 0)or(S5a = 1)or(S5a = 2)or(S5a = 3)or(S5a = 4))and \\
&((S5b = 0)or(S5b = 4)or(S5b = 5)or(S5b = 6)or(S5b = 7)))
\end{aligned} \tag{3}$$

For remained masks, performed condition (3) should be modified by appropriate rotation of the masks. In the worst case, each identification mask can point different reference points, but it is well known that for fingerprint only one reference point can be indicated. From this reason, for all potential reference points, detected by means of identification masks (Fig. 2c), so-called influence coefficients have been estimated. The influence coefficients (*inco*) is calculated as follow:
For angle 0^0:

$$\begin{aligned}
&if(S(x,y) = S2a)\ then\ inco := inco + 1; \\
&if(S5a <> 4)or(S5b <> 4)\ then\ inco := inco + 1; \\
&if(|S1a - S(x,y)| = 4)or(|S1b - S(x,y)| = 4)\ then\ inco := inco + 1;
\end{aligned} \tag{4}$$

For angle $+45^0$:

$$\begin{aligned}
&if(S(x,y) = S2)\ then\ inco := inco + 1; \\
&if(S5a \leq 1)or((S5b \geq 4)and(S5b \leq 6))\ then\ inco := inco + 1; \\
&if(|S1a - S(x,y)| = 4)or(|S1b - S(x,y)| = 4)\ then\ inco := inco + 1;
\end{aligned} \tag{5}$$

Remained the influence coefficients are calculated similarly, and only appropriate values of angles should be changed: for $-45^0, \pm 90^0$, respectively. Additionally, the next principles are considered:

- if two (or more) references points is located in a local neighborhood, then value of the *inco* coefficient is increased of 4,
- for the lowest located reference point, its *inco* value is increase of 5,
- potential reference point which lies at a distance less than 8 pixels from edge background is rejected.

Finally, the point which has the largest *inco* coefficient, will be classified as reference point. Proposed detection of reference point together with the $2D$ Gabor filtering method [1,2] can be used to fingerprint matching.

5 Experimental results

In experiments fingerprint images from FVS database have been used [9]. Two type experiments has been conducted. The first experiment by means of graphic form has been performed. Fig. 3 presents results of such experiment, where various fingerprint images have been compared. Such images with two methods have been tested. The first method is described in this paper. The second method is full described in [8], where MatLab program is included. Such program based on method presented in [2]. Fig. 3 presents differences between reference point location. The reference points by signs $'+'$ or $'x'$ have been marked. Our reference point location by means of $'x'$ sign is indicated. The results of method [2,8] as $'+'$ is marked. The results of the second experiment presents Table 1. In this table reference point is defined by point (x, y), where x, y are pixels coordinates. Mentioned point by three manner has been determined. The column A by police expert has been stated. All values i Table 1 should be interpreted as pairs of numbers, which indicate appriopriate (x, y) coordinates. The values in the columns B and C by algorithm [2,8] and our method have been computed. The columns $D1, D2$, and the last column show differences between methods. Presented differences it is the Euclidean distance between appropriate x and y coordinates. For example distance in the column $D1$ is computed from formula $[(x_a - x_b)^2 + (y_a - y_b)^2]^{0.5}$. The values in the column $D2$ are calculated similarly. The last column shows difference between methods. From obtained results follow (all positive values) that in most of the cases our method gives the better results. In other words reference point location lies closer to expert's reference point.

6 Conlusions

In this paper, a new method to locate a unique reference point has been proposed. Since human experts may not be able to locate the pixel wise accurate reference point, we propose the new method which allow to determine such point. The our method base on so-called *identification masks*, which was designed on the basis of human finger print analysis. For proposed masks, the

Table 1. Results of fingerprint identification

Image	Expert's assesment (x_a, y_a) A	Alghoritm [8] (x_b, y_b) B	Our method (x_c, y_c) C	Distance $D1 = A - B$	Distance $D2 = A - C$	$D1 - D2$
19_7.bmp	125, 111	120, 136	130, 94	25, 50	17, 72	7, 78
37_3.bmp	153, 92	146, 106	157, 85	15, 65	8, 06	7, 59
37_5_2.bmp	143, 144	152, 167	139, 139	24, 70	6, 40	18, 30
37_7.bmp	116, 102	96, 151	121, 94	52, 92	9, 43	43, 49
1_1.bmp	168, 160	*Error*	167, 149	–	11, 05	–
11_1.bmp	115, 153	123, 153	112, 148	8, 00	5, 83	2, 17

influence coefficients have been stated. Proposed method with complete algorithm described in [2,8] has been compared. The described method with the aid of the *FVS* fingerprint database has been tested. Mentioned data collection includes varying quality fingerprint images.

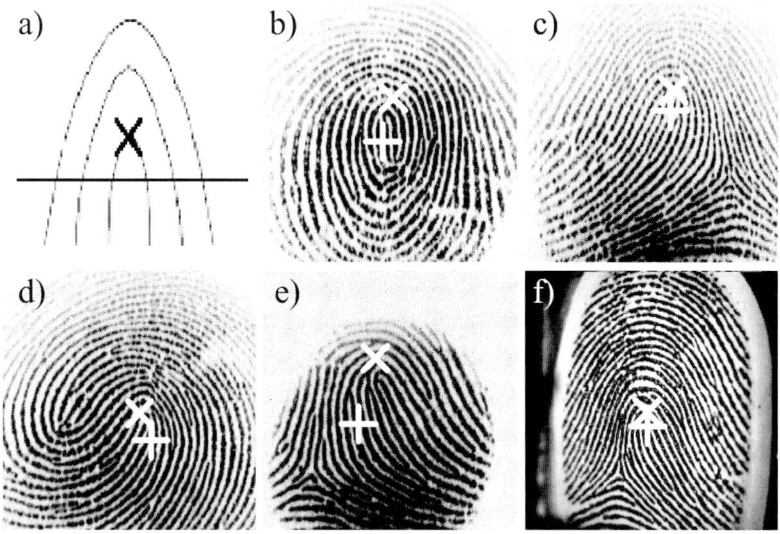

Fig. 3.
a) The perfect reference point location,
b)-f) reference point determination for different fingerprints quality from the *FVS* database.

References

1. L. Hong, Y. Wan, A.K. Jain (1998) Fingerprint image Enhancement: Algorithm and Performance Evaluation. IEEE Trans. On Pattern Analysis and Machine Intelligence, vol. 20, no.8, pp. 777-789.
2. A.K. Jain, S. Prabhakar, L. Hong, S. Pankanti (2000) Filterbank-Based Fingerprint Matching. IEEE Trans. on Image Processing, vol. 9, no. 5.
3. A. K. Jain, S. Prabhakar, L. Hong (1999) A multichannel approach to fingerprint classification. IEEE Trans. Pattern Analysis and Machine Intelligence, Vol. 9, no. 4 pp. 348-359.
4. X. Jiang, M. Liu, A.C., Kot (2004) Reference Point Detection for Fingerprint Recognition. Proc. of the 17*th* Int. Conf. On Pattern Recognition (ICPR'04) Cambridge, pp. 540-543, UK.
5. K. Karu, K. Jain (1996) Fingerprint Classification. Pattern recognition, vol.29, no. 3, pp. 389-404.
6. C.H. Park, S.K. Oh, D.M. Kwak, B.S. Kim, Y.C., Song, K.H. Park (2003) A New Reference Point Detection Algorithm Based on Orientation Pattern Labeling in Fingerprint Images. Proc. of 1st Iberian Conf. On Pattern Recognition and Image Analysis, Puerto de Andratx, pp.697-703, Spain.
7. P. Porwik, L. Wieclaw (2004) A new approach to reference point location in fingerprint recognition. IEICE International Journal Electronics Express Vol. 1, No. 18, pp. 575-581, Japan.
8. Luigi Rosa (2004) Fingerprint Recognition System, http://www.mathworks.it/matlabcentral/fileexchange/loadFile.do ?f4addg_fingerprint&objectId=4239&objectType=file
9. Shivang Patel (2002-2004) Fingerprint Verification System, http://fvs.sourceforge.net/download.html

Person Identification Using Fast Face Learning of Lifting Dyadic Wavelet Filters

Shigeru Takano, Koichi Niijima

Kyushu University, 6-1, Kasuga-koen, Kasuga, Fukuoka, JAPAN,
{takano,niijima}@i.kyushu-u.ac.jp

Summary. A person identification system based on fast face learning of lifting wavelet filters is proposed. The real power of our system lies in fast learning of lifting wavelet filters adaptive to facial parts such as eyes, nose and lips, in a set of training faces. In our system, free parameters in the lifting filter are learned fast by using Newton's method. The learned parameters are memorized in a database together with the training faces. The lifting filters with the learned parameters in the database are applied to each of video frames which contain faces of a person, and the faces are detected by measuring some kind of distance. A person whose face is detected in a maximum number of frames is identified as a target person. To realize fast face detection, the learned filters are applied only to the skin areas separated from background by using color segmentation. Simulation results show that our person identification algorithm is accurate and fast.

1 Introduction

Person identification by the recognition of faces is essential to intelligent human computer interaction. A great deal of methods for recognizing faces have been proposed: color segmentation, template matching, principal component analysis, support vector machine, Kohonen's self organizing map and their mixture [4]. The techniques combining color segmentation with template matching can localize the face position easily. However, it is difficult to identify a person from the localized face. On the other hand, face recognition systems based on PCA, SVM and SOM have enough ability to recognize a person face. These systems require a huge amount of computation to train the features of the face. Recently, Takano and Niijima [2] proposed an algorithm for extracting faces by the learning of lifting dyadic wavelet filters. Furthermore, we implemented a person identification system based on the algorithm [3]. However, a lot of time is needed in the learning process of lifting dyadic wavelet filters adaptive to faces, because the learning process is based on the steepest decent method for solving a minimization problem.

In this paper, we improve the learning algorithm by the use of Newton's method to solve the minimization problem fast. We also achieve fast face detection by applying the learned filters only to the skin areas separated from background exploiting color segmentation. Our algorithm involves four processes: color segmentation, learning, detection and identification. In color segmentation process, we extract skin color regions including faces from an input image. This preprocessing reduces the computational time for remaining three main processes. Learning process is to learn free parameters in the lifting filter so as to minimize the angle between a vector whose components are the lifting dyadic wavelet filters and a vector of facial lowpass components in each of multiresolution levels. This minimization problem is solved fast by using Newton's method. In detection process, we measure the angle between the learned filter and a detected skin color region from a video frame image at each resolution level, and find a face position by comparing the cosine of the angle with the maximum cosine value. Identification process is to apply our detection algorithm to each of video frames, and judge a person whose face is detected in a maximum number of frames to be a target person.

The outline of the paper is as follows. In Section 2, we describe a method for separating skin areas from background by the use of color-based segmentation method. Section 3 introduces lifting dyadic wavelet filters. In Section 4, we present a person identification system which consists of face learning, detection and identification processes. Section 5 involves experimental results. We close in Section 6 with concluding remarks and plans for future work.

2 Skin color segmentation

Color segmentation is simple but very effective for face localization in color images [1]. The color images are typically represented in RGB space. In this paper, we convert RGB components into YCrCb components. We find a region of the skin in an input image and choose the range of Cr and Cb values included in this area, i.e., $\mathrm{Cr}_{\mathrm{min}} < \mathrm{Cr} < \mathrm{Cr}_{\mathrm{max}}$ and $\mathrm{Cb}_{\mathrm{min}} < \mathrm{Cb} < \mathrm{Cb}_{\mathrm{max}}$. Based on the range of Cr and Cb values, we build a face color model. Skin color segmentation is done by checking whether or not the pixels of the input image fall within the face color model. The centroid of the skin area separated from background is regarded as the central point of a facial part in the input image. By trimming a region around the central point, we extract candidates of facial parts from the regions.

3 Lifting dyadic wavelet filters

In this section, we define lifting wavelet filters which serve as the foundation of our person identification system. In this paper, lifting dyadic wavelet filters mean dyadic wavelet filters constructed by adding a lifting term, which contains free parameters to initial dyadic wavelet filters. Let us denote lowpass and highpass analysis filters by $h^o[k]$ and $g^o[k]$, respectively, and lowpass and highpass synthesis filters by $\tilde{h}^o[k]$ and $\tilde{g}^o[k]$, respectively. We also denote the discrete Fourier transforms of the filters $h^o[k]$, $g^o[k]$, $\tilde{h}^o[k]$ and $\tilde{g}^o[k]$ by $\hat{h}^o(\omega)$, $\hat{g}^o(\omega)$, $\hat{\tilde{h}}^o(\omega)$ and $\hat{\tilde{g}}^o(\omega)$, respectively. These filters are called dyadic wavelet filters when they satisfy the following condition

$$\hat{\tilde{h}}^o(\omega)\hat{h^o}^*(\omega) + \hat{\tilde{g}}^o(\omega)\hat{g^o}^*(\omega) = 2, \qquad \omega \in [-\pi, \pi], \tag{1}$$

where the symbol * denotes complex conjugation. We call the condition (1) a reconstruction condition.

Let $c_0[i]$ be an original signal. Mallat's fast algorithm for computing the p-th lowpass and highpass components $c_p[i]$ and $d_p[i]$ of $c_o[i]$ is given by

$$c_p[i] = \sum_k h^o[k]c_{p-1}[i + 2^{p-1}k],\ p = 1, \cdots, P, \tag{2}$$

$$d_p[i] = \sum_k g^o[k]c_{p-1}[i + 2^{p-1}k],\ p = 1, \cdots, P. \tag{3}$$

Conversely, by virtue of the reconstruction condition (1), we can restore the lowpass components $c_{p-1}[i]$ from $c_p[i]$ and $d_p[i]$ exploiting

$$c_{p-1}[i] = \frac{1}{2}\sum_k \tilde{h}^o[k]c_p[i - 2^p k] + \frac{1}{2}\sum_k \tilde{g}^o[k]d_p[i - 2^p k]. \tag{4}$$

By iterating (4) p times, the original signal $c_0[i]$ can be reconstructed. This fact implies that $c_0[i]$ is equivalent to $\{c_p[i], d_p[i]\}, p = 1, \cdots, P$.

We construct a new set of filters $\{h[k],\ g[k],\ \tilde{h}[k],\ \tilde{g}[k]\}$ as follows:

$$\begin{aligned} h[k] &= h^o[k], \quad \tilde{h}[k] = \tilde{h}^o[k] + \sum_m s[-m]\tilde{g}^o[k-m], \\ g[k] &= g^o[k] - \sum_m s[m]h^o[k-m], \quad \tilde{g}[k] = \tilde{g}^o[k], \end{aligned} \tag{5}$$

where $s[m]$'s denote free parameters. We call (5) lifting dyadic wavelet filters and $s[m]$ lifting parameters. It can be proved that the Fourier transforms of the filters defined by (5) also satisfy the reconstruction condition (1). Therefore, the formulae (2), (3) and (4) hold for the new filters.

4 Person identification system

In this section, three processes, learning, detection and identification, are described. In Section 2, we presented the face localization method based on the color segmentation. This method may extract other skin areas such as hands as well as faces. Therefore, we need to extract only facial parts such as eyes, nose and lips, from the extracted skin area. After the detection of the facial parts, we have to identify persons. We carry out these two things by exploiting the learned lifting filters.

4.1 Learning process

Let $C_0[i,j]$ denote Y values in YCrCb components of a training facial part. Its lowpass components $C_p[i,j], p=1,\cdots,P$ are computed by applying repeatedly the initial lowpass filter $h^o[k]$ in horizontal and vertical directions to $C_0[i,j]$, respectively. Furthermore, applying $h[k]$ from (5) to $C_{p-1}[i,j]$ in vertical direction, we obtain

$$C_p^{row}[i,j] = \sum_k h[k]C_{p-1}[i+2^{p-1}k, j], \quad p=1,\cdots,P. \tag{6}$$

Next, we apply $g[k]$ from (5) to $C_p^{row}[i,j]$ in horizontal direction to get

$$D_p[i,j] = \sum_{l=N_1}^{N_2} g_{d,p}[l]C_p^{row}[i, j+2^{p-1}l], \quad p=1,\cdots,P. \tag{7}$$

Here $g_{d,p}[l]$ denotes $g[k]$ from (5), where $s[m]$ in $g[k]$ is denoted by $s_{d,p}[m]$. Similarly, we compute highpass components $E_p[i,j]$ in vertical direction as

$$E_p[i,j] = \sum_{l=N_1}^{N_2} g_{e,p}[l]C_p^{col}[i+2^{p-1}l, j], \quad p=1,\cdots,P. \tag{8}$$

Here $g_{e,p}[l]$ denotes $g[k]$ described in (5), where $s[m]$ in $g[k]$ is replaced by $s_{e,p}[m]$, and $C_p^{col}[i,j]$ is computed by

$$C_p^{col}[i,j] = \sum_k h[k]C_{p-1}[i, j+2^{p-1}k], \quad p=1,\cdots,P. \tag{9}$$

Now, we describe how to learn lifting filters $g_{d,p}[l]$ and $g_{e,p}[l]$, that is, free parameters $s_{d,p}[m]$ and $s_{e,p}[m]$ contained in $g_{d,p}[l]$ and $g_{e,p}[l]$. For convenience of expression in onward discussion, we define the following four vectors

$$\begin{aligned} g_{d,p} &= (g_{d,p}[N_1],\cdots,g_{d,p}[N_2]), \qquad g_{e,p} = (g_{e,p}[N_1],\cdots,g_{e,p}[N_2]), \\ C_{p,i,j}^{row} &= (C_p^{row}[i, j+2^{p-1}N_1],\cdots,C_p^{row}[i, j+2^{p-1}N_2]), \\ C_{p,i,j}^{col} &= (C_p^{col}[i+2^{p-1}N_1, j],\cdots,C_p^{col}[i+2^{p-1}N_2, j]). \end{aligned}$$

Using inner product symbol '$\cdot$', the lifted highpass components $D_p[i,j]$ and $E_p[i,j]$ in (7) and (8) are represented by the following forms

$$D_p[i,j] = g_{d,p} \cdot C^{row}_{p,i,j}, \qquad E_p[i,j] = g_{e,p} \cdot C^{col}_{p,i,j}.$$

Let $\theta_{d,p}$ and $\theta_{e,p}$ denote the angles between $g_{d,p}$ and $C^{row}_{p,i,j}$, and between $g_{e,p}$ and $C^{col}_{p,i,j}$, respectively. Then, the cosine for each of the angles $\theta_{d,p}$ and $\theta_{e,p}$ is defined as

$$\cos\theta_{d,p} = \frac{D_p[i,j]}{|g_{d,p}||C^{row}_{p,i,j}|}, \tag{10}$$

$$\cos\theta_{e,p} = \frac{E_p[i,j]}{|g_{e,p}||C^{col}_{p,i,j}|}, \tag{11}$$

where the symbol $|\cdot|$ denotes the Euclidean norm of the vectors.

We learn free parameters $s_{d,p}[m]$ and $s_{e,p}[m]$ so as to approximate $C^{row}_{p,i,j}/|C^{row}_{p,i,j}|$ by $g_{d,p}/|g_{d,p}|$, and $C^{col}_{p,i,j}/|C^{col}_{p,i,j}|$ by $g_{e,p}/|g_{e,p}|$. This implies that (10) and (11) tend to 1. We notice that the learned filter becomes a lowpass filter though the initial filter is a highpass filter. These approximations lead us to minimization problems of the functionals

$$J_{d,p} = \left(D_p[i,j] - |g_{d,p}||C^{row}_{p,i,j}|\right)^2, \tag{12}$$

$$J_{e,p} = \left(E_p[i,j] - |g_{e,p}||C^{col}_{p,i,j}|\right)^2, \tag{13}$$

where (i,j) represents the selected point of facial parts such as eyes, nose and lips. In our papers [2, 3], we solved these minimization problems using the steepest decent method. However, this method has slow convergence rate. In this paper, we employ Newton's method to seek stationary points of (12) and (13) fast. Newton's method has convergence rate of order two.

Differentiating $J_{d,p}$ and $J_{e,p}$ with respect to each of the free parameters, we obtain the following nonlinear systems of simultaneous equations

$$\frac{\partial J_{d,p}}{\partial s_{d,p}[m]} = 0, \;\; m = -m_0, \cdots, m_0, \tag{14}$$

$$\frac{\partial J_{e,p}}{\partial s_{e,p}[m]} = 0, \;\; m = -m_0, \cdots, m_0. \tag{15}$$

We solve (14) and (15) by Newton's method. These equations may have many solutions. So, we find a solution close to zero vector starting Newton iteration from the zero vector. Practically, to achieve accurate detection, 8 kinds of free parameters $s^{\nu}_{d,p}[m]$ and $s^{\nu}_{e,p}[m]$ are determined, where $-2\nu \le m \le 2\nu$ and $\nu = 1, \cdots, 8$ at the resolution level p.

Applying this learning algorithm to facial parts such as eyes, nose and lips of training faces for a variety of target persons, we learn free parameters in a lifting filter. These learned parameters are memorized in a server together with the training faces.

4.2 Detection process

We detect facial parts in a test image using the learned parameters $s^{\nu}_{d,p}[m]$ and $s^{\nu}_{e,p}[m]$ described in Section 4.1.

First, skin areas from a test image using the color-based segmentation method are extracted and Y values of the skin area are denoted again by $C_0[i,j]$. For $C_0[i,j]$, we compute lowpass components $C_p[i,j], p = 1, \cdots, P$ and highpass components $D^o_p[i,j]$ and $E^o_p[i,j]$ in horizontal and vertical directions, respectively, by using an initial filter. Combining the highpass components $D^o_p[i,j]$ and $E^o_p[i,j]$ with the learned parameters $s^{\nu}_{d,p}[m]$ and $s^{\nu}_{e,p}[m]$ in the databese, we compute new lowpass components $D^{\nu}_p[i,j]$ and $E^{\nu}_p[i,j]$ as

$$D^{\nu}_p[i,j] = D^o_p[i,j] - \sum_{m=-2\nu}^{2\nu} s^{\nu}_{d,p}[m] C_p[i, j + 2^{p-1}m], \qquad p = 1, \cdots, P,$$

$$E^{\nu}_p[i,j] = E^o_p[i,j] - \sum_{m=-2\nu}^{2\nu} s^{\nu}_{e,p}[m] C_p[i + 2^{p-1}m, j], \qquad p = 1, \cdots, P.$$

To extract the facial parts from $C_0[i,j]$, the following quantity is introduced:

$$R[i,j] = \sum_{p=1}^{P} \left(\sum_{\nu=1}^{8} (Q^{\nu}_{d,p}[i,j] - 1)^2 + \sum_{\nu=1}^{8} (Q^{\nu}_{e,p}[i,j] - 1)^2 \right). \tag{16}$$

Here $Q^{\nu}_{d,p}[i,j]$ and $Q^{\nu}_{e,p}[i,j]$ represent

$$Q^{\nu}_{d,p}[i,j] = \frac{D^{\nu}_p[i,j]}{|g_{d,p}||C^{\nu,row}_{p,i,j}|}, \qquad Q^{\nu}_{e,p}[i,j] = \frac{E^{\nu}_p[i,j]}{|g_{e,p}||C^{\nu,col}_{p,i,j}|},$$

respectively. If the quantity (16) is minimal at the point (i_0, j_0), then $C_0[i_0, j_0]$ provides a facial part. First, nose is found and next, eyes and lips around the nose are searched to reduce computational time. If the sum of $R[i,j]$ in (16) at the detected facial parts is less than a certain threshold, the current image is regarded as the image containing a face.

4.3 Identification process

For person identification, we prepare video frames including faces of a target person. Our identification process involves the following steps:

1. Extract skin areas from each of the video frames by exploiting color segmentation.
2. Compute p-th resolution level of lowpass and highpass components of the skin area by using an initial filter.
3. Detect facial parts following the detection process described in the previous section.

4. Detect faces from all the video frames by repeating Steps 1 through 3.
5. If the face of a person is detected in a maximum number of frames, as is detected in Step 4, then he/she is the target person.

5 Experimental results

In our experiments, we used the B-spline dyadic wavelet filters as intial filters. Training faces were extracted from twelve different person's images, each of which has depth of 24 bits (YCrCb color), and a size of 208×160 pixels. These images were captured from the mobile robot AIBO developed by SONY. We built a face color model from the captured images by the method described in Section 2. Using the face color model, we computed the centroid of the skin area in the images, and extracted the training faces, which has depth of 8 bits (grayscale), and a size of 128×128 pixels, as shown in Fig. 1. Applying

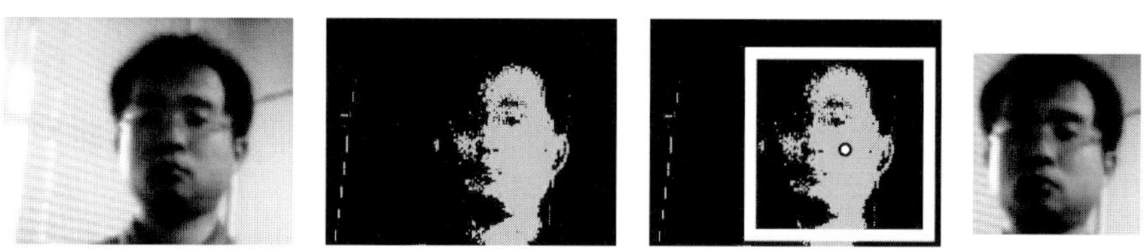

Fig. 1. Skin color segmentaion

the learning algorithm described in Section 4.1 to facial parts of the training faces, we learned free parameters $s^{\nu}_{d,p}[m]$ and $s^{\nu}_{e,p}[m]$ in a lifting filter.

For experiments of detection, five different facial images for each of twelve persons were selected from video frames. From the images, skin areas were extracted by performing the skin color segmentation. We applied the detection algorithm described in Section 4.2 to these skin areas. Figure 2 shows the detection results of facial parts. Firstly, we found a nose position and then,

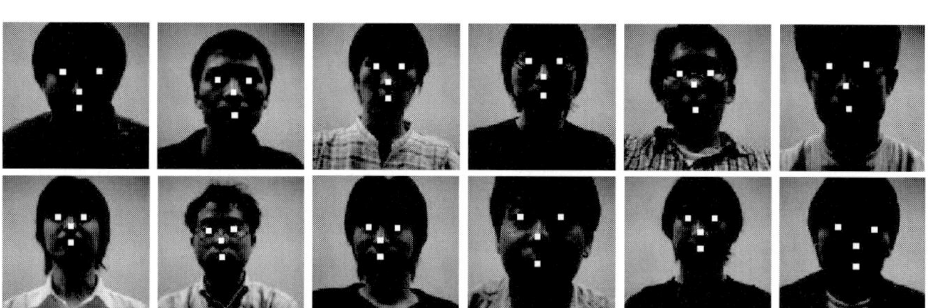

Fig. 2. Detection results of facial parts

searched a block area with 5×5 size around the nose for eyes and lips. Furthermore, we tried to search a small area at the lower resolution level around the facial parts detected at the higher resolution level to reduce the computational time for the detection algorithm.

We computed the sum of $R[i,j]$ in (16) at eyes, nose and lips extracted by applying the detection algorithm to each of the five different facial images for each of twelve persons, at the resolution levels $P = 1, 2, 3$. By checking the number of faces detected from the these images, we could identify all the persons at all the levels.

Simulation was done on a personal computer with Pentium M, 1.3GHz and 768 MB SDRAM. Computational time of our learning and identification algorithms are shown in Table 1. For comparison, we listed learning time by the steepest descent method.

Table 1. Computational time of learning and identification.

	P=1	P=2	P=3
Steepest descent method	695.0 sec	1388.9 sec	5667.9 sec
Newton's method	7.3 sec	14.9 sec	22.4 sec
Person identification	3.5 sec	5.9 sec	8.3 sec

6 Conclusion

We proposed a person identification method based on lifting dyadic wavelet filters. By doing skin segmentation in advance, we reduced the computational effort of the following process, and avoided the false detection of the facial parts. In the learning process for the free parameters, we used Newton's method to solve the minimization problem, which is much faster than the steepest descent method. Our detection process was performed efficiently as follows: (i) find nose first and then, search around the nose for eyes and lips, (ii) after the second resolution level, search a small region around the detected face at the higher resolution level for facial parts. Since the proposed detection method is fast and accurate, our person identification was carried out in realtime. In the future, we will implement the presented person identification system on AIBO robot, as it can identify persons by herself.

References

1. Chai D. and Ngan K.N. (1998), Locating Facial Region of a Head-and-shoulders Color Image, Proceedings of the Third International Conference Automatic Face and Gesture Recognition, pp. 124–129.
2. Takano S., Niijima K. and Abdukirim T. (2003), Fast Face Detection by Lifting Dyadic Wavelet Filters, Proceedings of the IEEE International Conference on Image Processing, pp. 893–896.
3. Takano S., Niijima K. and Kuzume K. (2004), Personal Identification by Multiresolution Analysis of Lifting Dyadic Wavelets, Proceedings of the 12th European Signal Processing Conference, CD-ROM.

4. Yang M.-H., Kriegman D. and Ahuja N. (2002), Detecting Faces in Images: A Survey, IEEE Transactions on Pattern Analysis and Machine Intelligence (PAMI), vol. 24, no. 1, pp. 34–58.

Part VII

VARIOUS APPLICATIONS

Dynamic Design with the Use of Intelligent Agents

Ewa Grabska, Grażyna Ślusarczyk and Paweł Grześ

Institute of Computer Science, Jagiellonian University, Nawojki 11, 30-072 Kraków, Poland, uigrabsk@cyf-kr.edu.pl, grazyna@ii.uj.edu.pl

Summary. In this paper ideas of the dynamic character of design are presented by using intelligent agents to conceptual design aided by computer. These ideas are illustrated by the application related to decorative art. Two cooperative curious agents are used to generate periodic patterns. Agents' curiosity controls evaluation of their actions, specification of new aims and planning future behavior. The interaction between agents and the environment strongly determines the course of designing.

1 Introduction

Recent frameworks for design focus on dynamic character of the context in which the designing takes place. In this paper ideas of the dynamic character of design are presented by using intelligent agents to conceptual design aided by computer. In our approach the interaction between agents and the environment strongly determines the course of designing. Each agent has a set of available actions which represent its ability to modify the environment.

By an intelligent agent we understand a system which is able both to create its own world representation on the basis of experience gained from interactions with the environment and to evaluate the creativity of solutions produced by itself. Our design agent is equipped with functions which enable it to remember the decisions taken in the previous states of the world and a function which determines the degree of interest taken in the present situation. Agent's curiosity controls evaluation of its actions, specification of new aims and planning future behavior.

The proposed approach is illustrated by the application related to decorative art. Two cooperating agents assist in the process of designing periodic patterns. The first agent generates new motifs and the second one creates patterns composed of these motifs. The evaluation of the obtained patterns performed by the second agent influences the types of motifs which are generated by the first agent.

2 A model of a curious agent

Creativity can be understood as an ability to produce something new (original, interesting, unexpected) and useful (valuable, proper, conformable). Creativity can be evaluated by a single person or by society (see the notions of P-creativity and H-creativity in [2]). In this paper we are interested in agents which can evaluate their solutions by themselves. Their ability is connected with situated creativity [6], where the solution is original when it is new in respect to the situation of agents. Our agent evaluates creativity of the obtained solution on the basis of the degree of interest taken in it and the curiosity raised by the present situation. The agent's interest depends on the possessed knowledge and the ability to use it, while its curiosity can be understood as a motivation to do further research. Curiosity makes the agent modify the solution and check if a new one satisfies the previously defined criteria.

An agent is a computer module situated in some environment and capable of autonomous action in this environment in order to satisfy design requirements. Its inside structure consists of processes connected with sensing, perception, evaluation, planning, action and influence on the environment.

The environment of a design agent is described as a set of possible states of the world $W = \{w_0, w_1, ...\}$. The effectoric capability of an agent is assumed to be represented by the set $A = \{a_0, a_1, ...\}$ of actions [7].

An agent can be characterized by:

- the way in which the agent senses the environment which is described by the function $\sigma : W \rightarrow S$, where $S = \{s_0, s_1, ...\}$ is a set of sensor states,
- the perception process which is the interpretation of data obtained from sensors and can be defined by the function $\pi : S \rightarrow P$, where $P = \{p_0, p_1, ...\}$ is a set of perceptual states,
- the evaluation and decision making process, during which a sequence of perception states is evaluated and new goals are specified, that can be described by the function $\chi : P^* \rightarrow C$, where $C = \{c_0, c_1, ...\}$ is a set of conceptual states,
- the action process which translates goals specified by a conceptual state to an action that should be taken and can be defined by the function $\alpha : C \rightarrow A$, where $A = \{a_0, a_1, ...\}$ is a set of actions.

To be able to learn an agent must have some type of memory and be equipped with functions which would enable it to remember the decisions taken and the previous states of the world. We assume that such an agent possesses two types of memory: a short-time memory and a long-term one. The short-term memory M_S, which is a subset of $P \times C \times A$, allows the agent to remember a few recent perceptual and conceptual states and actions taken. The content of M_S changes according to a function $\beta : P \times C \times A \times M_S \rightarrow M_S$. The long-time memory M_L stores agent's generalized experiences from the past.

In case of a curious agent the novelty of the current state and agent's interest in the current situation must be determined [5]. Then new goals of the agent are specified on the basis of the agent's interest in the current situation and past situations. The degree of novelty and unexpectedness of the current situation is computed by comparing the current conceptual state with agent's generalized experiences gathered in a long-time memory and can be described by a function $\nu : C \times M_L \rightarrow N$, where $N = [0, 1]$ is a set of situation novelty degrees. Computing the situation novelty the agent assumes that future situations will be similar to the past ones. Therefore, first it categorizes the current situation using the long-time memory and then it determines the categorization probability and computes the novelty degree as the reciprocal probability. It means that if there is a similar situation in the memory the novelty degree of the current one will be low.

The degree of interest in the current situation is computed on the basis of the novelty degree using a hedonistic function $\iota : N \rightarrow I$, where $I = [0, 1]$ is a set of interest degrees. The hedonistic function can be specified for example as a linear function or the Wundt's curve which expresses the fact that a given situation is interesting for the agent if it is very similar to and not very different from the previous ones.

Agent's curiosity is a conceptual process which controls the evaluation and planning process χ and can be described by a function $\xi : C \times M_L \times M_S \times I \rightarrow X$, where $X = [0, 1]$ is a set of curiosity degrees. The decisions taken by the agent depend on the degree of its curiosity.

A curious agent with a short-time and long-time memory can be characterized by:

- the environment sensing process $\sigma : W \rightarrow S$,
- the perception process $\pi : S \times M_S \rightarrow P$,
- the evaluation and decision making process $\chi : M_L \times X \times M_S \rightarrow C$,
- the action process $\alpha : C \rightarrow A$.

In our approach the short-time memory is a set of variables representing recent agent's states which have led to the current situation. The long-time memory has a form of a self-learning neural net called SOM (Self-Organizing Map)[4]. It is composed of a lattice of neurons, each of which represents a different category of input data. Each neuron has a vector of weights with a dimension equal to the number of input data. For each neuron a degree of similarity of its weights vector to the input data vector is computed. Learning process consists in decreasing weights difference between the best fitted neuron together with its neighbouring neurons and the input vector. SOM enables the agent to predict solutions which are located in the space regions which have not been searched yet.

The novelty of a solution is computed as a Euclidean distance between the vector of weights of the best-fitted neuron and the input vector. This distance is called a categorization error. Agent's interest in a current situation is modeled using a hedonistic function [1].

3 Creative design with curious agents

The proposed approach to curious agents is illustrated by the application related to decorative art. We consider computer generation and evaluation of periodic patterns by means of two agents. One of them generates new motifs and the second one creates patterns composed of these motifs. The evaluation of the obtained patterns performed by the second agent influences the types of motifs which are generated by the first agent.

The first agent has a buffer where the currently considered motifs are stored (at least two motifs should be present). The perception process of the agent consists in scaling the motifs stored in the buffer to the bit maps of low resolution. Bit maps of the shapes are transformed to input vectors of the agent's SOM. SOM classifies input vectors by assigning to them categories to which they belong and the categorization errors. The novelty degree of each input shape is computed on the basis of these two parameters. The interest degree taken in the input pattern is computed on the basis of the novelty degree and the interest assigned to this shape by the second agent. The first agent selects the shape to be learnt by SOM in respect to its interest degree. If this degree is high then the pattern with a low degree of novelty is chosen. Otherwise, the pattern with the high degree of novelty is chosen.

The agent creates a new motif on the basis of the two most interesting ones. It rotates these two shapes independently about their centers. Then the second shape is translated in respect to the first one. The parameters of rotations and translation are determined randomly. The sum of the transformed shapes gives a new motif which is added to the buffer. The agent's action consists not only of generating a new motif but also of transferring the three shapes which are most interesting from its point of view to the second agent.

Two initial motifs and the third one obtained using them are shown in Fig. 1, while SOM of the agent which generates motifs is presented in Fig. 2.

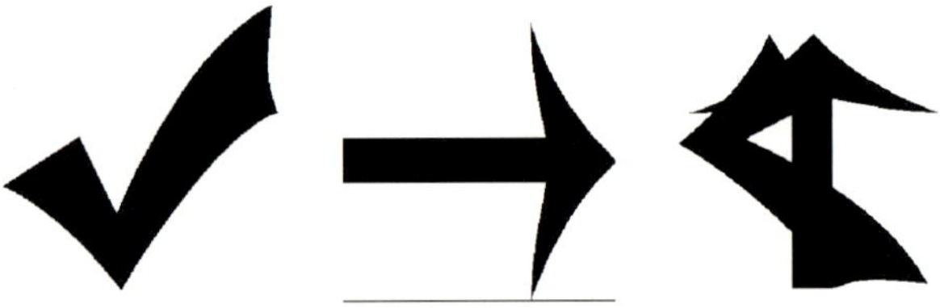

Fig. 1. An example of three different motifs

The second agent generates three rosettes of the motifs received from the first agent. Each rosette design is inscribed in a circle centered at a given point. To create a rosette with n-fold rotational symmetry a chosen motif is rotated n times about a fixed central point. An example of a 5-fold rotational symmetry rosette with its motif is presented in Fig. 3.

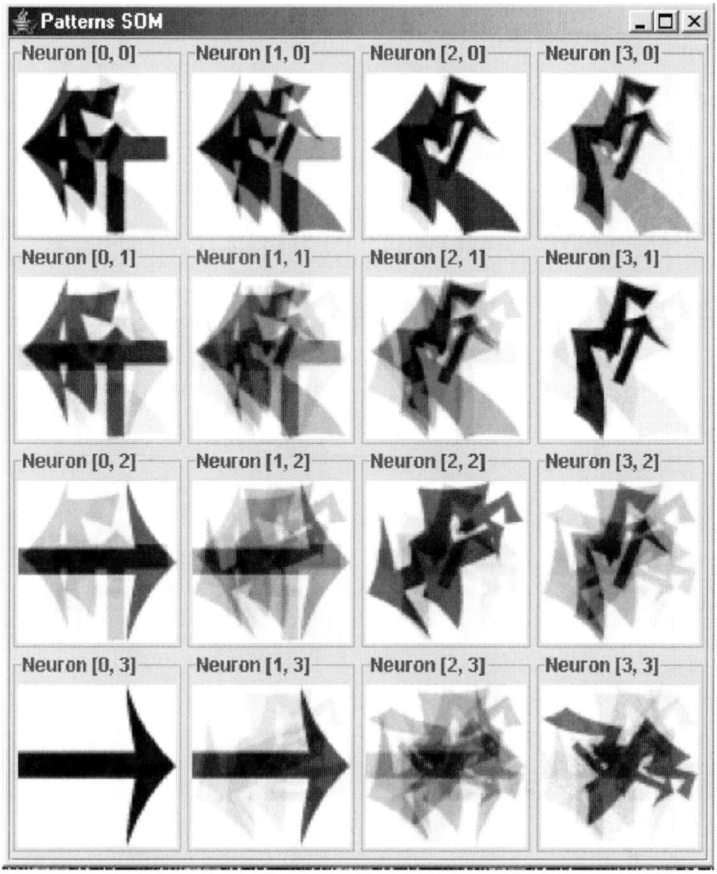

Fig. 2. SOM of the agent which generates motifs

In order to create a rosette of a given motif the agent specifies random values of the rotation angle for the motif and of the rotation centre. The perception process of the second agent consists in transforming the obtained rosettes to the bit maps of the low resolution. The novelty and the interest degrees of input patterns are computed on the basis of their classification performed by the agent's SOM. The agent chooses the rosette to be learnt by SOM in the analogous way as the first agent does. The action of the agent consists in transferring the degree of interest taken in rosettes to the first agent. SOM of the agent which generates rosettes is presented in Fig. 4.

The window of the application which enables the user to trace the behaviour of both agents is shown in Fig. 5. Three motifs which are currently most interesting are presented in panel 2. The framed pattern is the one which is chosen by the first agent to be learnt. This pattern belongs to the category represented by the best-fitted neuron (shown in panel 3) of the first agent's

Fig. 3. A 5-fold rotational symmetry rosette and its motif

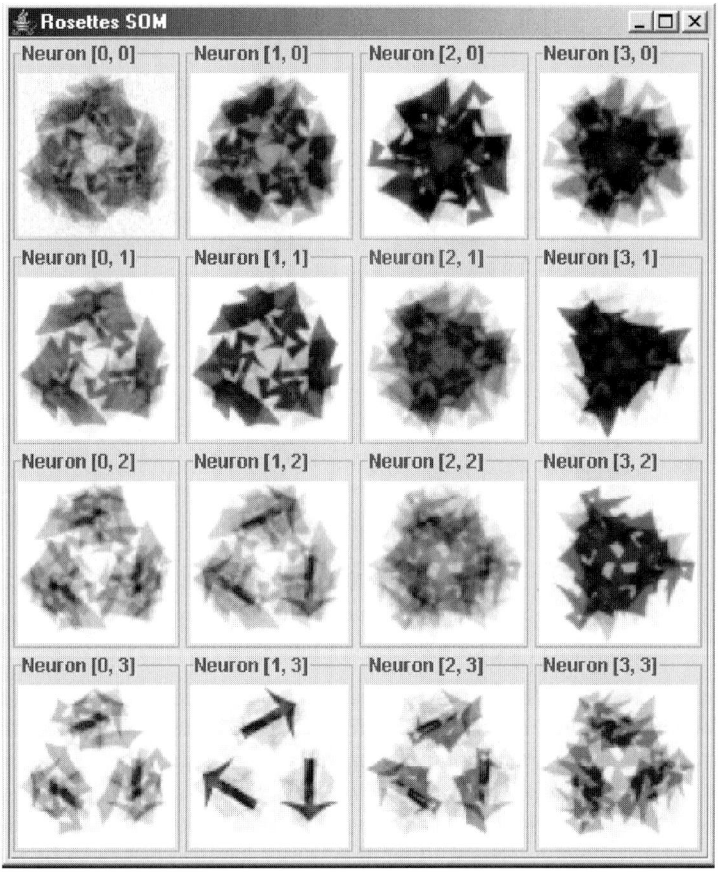

Fig. 4. SOM of the agent which generates rosettes

SOM. The rosettes which are generated of the motifs shown in panel 2 and currently considered by the second agent are presented in panel 4. The framed rosette is the one which is chosen by the second agent to be learnt. This rosette belongs to the category represented by the best-fitted neuron (shown in panel 5) of the second agent's SOM. Panel 6 shows the rosette which the second agent's SOM was learnt at the previous step. Panels 7 and 8 illustrate the degrees of patterns' novelty and interest taken in them in a few previous steps by the first and second agent, respectively.

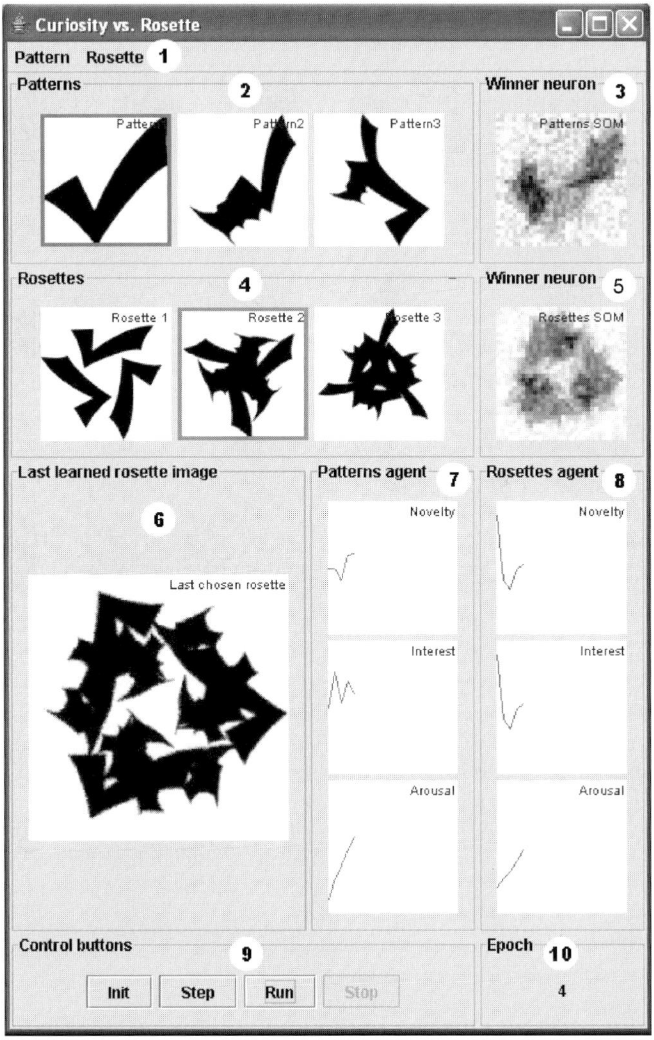

Fig. 5. The application window presenting the behaviour of both agents

4 Future Prospects

In our future work the process of generating new patterns by agents will be aided by animation. We performed experiments, where the designer evaluated rosettes animated by the system DARTAN (Decorative ART ANimation) [3]. In our new experiments in designing decorative patterns we intend to support the designer by three agents. The first agent will generate rosette motifs in the analogous way as it was described above. Each rosette initially generated by the second agent will be a starting point for animation. The animation of motifs rotation causes that the appearance of the rosette changes as new configurations composed of the copies of the same motif are formed. Our experiments with DARTAN system show that new interesting visual effects with great probability appear after every 18 degrees of motifs rotation. Therefore the second agent will evaluate the twenty configurations obtained for each of the three input rosettes during animation. The six most interesting rosette patterns of the sixty evaluated ones will be transferred to the third agent. This agent will be responsible for creating plane designs composed of the selected rosettes.

References

1. Berlyne DE (1971) Aesthetics and Psychobiology, Appleton-Century-Crofts, New York
2. Boden MA (1990) The Creative Mind: Myths and Mechanisms. Cardinal, London
3. Grabska E, Ślusarczyk G, Szłapak M (2004) Animation in Art. In: Gero JS (ed) Design, Design Computing and Cognition'04, Kluwer Academic Publishers, Netherlands
4. Kohonen T (1995): Self-Organizing Maps, Springer-Verlag, Berlin
5. Saunders R (2001) Curious Design Agents and Artificial Creativity. PhD Thesis, Faculty of Architecture, The University of Sydney
6. Suwa M, Gero JS, Purcell T (1999) Unexpected discoveries and S-invention of design requirements: A key to creative designs. In: Gero JS,Maher ML (eds.) Computational Models of Creative Design IV, Key Centre of Design Computing and Cognition, University of Sydney
7. Wooldridge MJ (1999)Intelligent Agents. In: Weiss G (ed.) Multiagent Systems: A Modern Approach to Distributed Artificial Intelligence, MIT Press, Cambridge, MA

Optical Music Recognition: the Case Study of Pattern Recognition

Wladyslaw Homenda

Faculty of Mathematics and Information Sci, Warsaw University of Technology, pl. Politechniki 1, 00-661 Warsaw, Poland,
homenda@mini.pw.edu.pl

Summary. The paper presents a pattern recognition study aimed on music notation recognition. The study is focused on practical aspect of optical music recognition; it presents a variety of methods applied in optical music recognition technology. The following logically separated stages of music notation recognition are distinguished: acquiring music notation structure, recognizing symbols of music notation, analyzing contextual information. The directions for OMR package development are drawn.

1 Introduction

Last two decades are witnesses of brisk development of computing technologies aimed on information and knowledge processing. A special interest of computing technologies development is focused on automation of paper-to-computer memory information transfer. Some paper-to-computer technologies have been successfully developed as, for instance, OCR technology. However, such areas as recognition of printed music, of handwritten text and handwritten music, of geographical maps are examples of fields opened for research and technology development.

In this paper we outline music notation recognition as a case of pattern recognition. The discussion is focused on the three most important aspects of OMR: document structure identification, music symbols' recognition and context knowledge mining. In Section 2 score structure identification is discussed: locating staves, systems, measures and blocks of texts. In Section 3 topics related to music symbols' recognition are discussed. Section 4 includes remarks about music knowledge processing and context information mining.

2 Acquiring Music Notation Structure

Music notation is a highly structured music knowledge container. The structure could be interpreted from two perspectives: logical and geometrical. Music

Fig. 1. Score - an example excerpt

notation describes a sequence of events along time axis, usually long sequence of events in long time interval. Conversely, music notation is printed on sheets of paper, so time axis has to be broken in order to fit page width. These two aspects form two perspectives of music knowledge structuring: logical and geometrical structures, c.f. [6] for detailed discussion.

2.1 Logical structuring

Parts and Staves Music notation describes a series of events that could be interpreted as notes and their attributes. These events are placed in three dimensional space. Time and frequency are the obvious parameters forming two dimensions: every note is played in a given time interval and has its own pitch. Music notation usually describes music played by an ensemble. In such a case notes and their attributes related to one instrument create a part of music and usually fill in one stave, sometimes two or three staves (piano, organ). Separation of music events with regard to instruments creates discrete points of a third dimension of music notation. In this interpretation the time axis is placed along staves while frequency/pitch axis is defined by note position on staff lines and by respective clef. Staves representing parts of music form points of the third dimension. The time axis is scaled in music units: beats and measures where measures are separated by barlines. The set of long staves, not broken by sheets size, form a logical system. Due to its nature logical system consists of staves representing all instruments.

2.2 Geometrical structuring

A piece of music is partitioned into rhythmic entities - measures. In music notation measures are separated by barlines. Besides measure separation barlines play additional roles denoting repetitions of some parts of music and grouping relative instruments as strings or winds. Repetitions are defined by usage of special barlines. Groups of parts are specified by sections of barlines connecting respective staves while groups separation is indicated by broken barlines between groups of staves, c.f. Figure 1.

Pages and systems Music notation, as printed on sheets of paper, is split into parts fitting page's width. The logical system, as described in Section 2.1, is split into parts defined by time intervals which - after necessary stretching - fit page width. This way we get sections of logical system creating systems in the meaning of geometrical structuring. In most cases geometrical systems do not break measures, though it may happen that a long measure is divided between two systems. Unlike logical system, geometrical systems often drop empty staves creating irregular systems, c.f. Figure 1.

Texts Music notation - besides music symbols - also includes texts and alphanumeric characters. The top part of the first page of music score is usually occupied by a kind of music specification: title of the piece, composer and poet names, arranger name, dates, etc. The bottom part of the first page is usually taken by copyright and publisher data. All this information is useless from musical point of view, but are useful for other purposes as, for instance, creating an index of compositions and composers. In contrast, music score often includes lyric as well as separated words or alphanumeric symbols which are important from musical perspective. As examples of such symbols we can take guitar chords, tempo and dynamic markings, fingering, instrument and part names, etc. Locating texts in music notation will give assured advances in further steps of alphanumeric symbols recognition.

2.3 Acquiring music notation structure

Staff lines, long horizontal equidistant lines, are the most characteristic elements of music notation. This feature makes that projections are the most popular method used for staff lines location, c.f. [3]. Horizontal projections computed on page width will distinguish staff lines with peaks of lines length. Localization of staff lines is easy for ideal notation. However, staff lines of music notation are usually distorted, so then more complex analysis of narrow, local horizontal projections must be done, c.f. [5, 6].

Finding systems is the next step of notation's structure analysis. Since staves of any geometrical system are joined with the beginning barline, finding this vertical line drawn at the beginning of system's staves results in locating system. Vertical projections are usual method applied in localization of barlines what leads to systems location. And again, for distorted notations analysis of local projections avoids all kinds of deformations.

Big systems often have their own internal structure. Staves are clustered collecting instruments of the same group. Such groups may be distinguished with brackets and braces placed at the front of the systems. Therefore, finding brackets and braces is an extension of system location task. Since internal system structuring is also defined by barlines, then both methods can cooperate in inner system's structure location.

3 Recognizing Symbols of Music Notation

Practical recognition process has two crucial stages: image segmentation and symbol classification. The first stage is - roughly speaking - aimed on two tasks: pattern localization and separation patterns from their environment. Localization of patterns most often results in finding bounding box of a given pattern, which is then passed to next steps of recognition process. Recognized patterns may touch each to other, may overlap each other or their bounding boxes may not be disjoint. Separation of such objects significantly simplifies classification and increases recognition rate, c.f. [10]. But objects' separation may be too expensive comparing with recognition rate improvement.

3.1 Segmentation

Since music notation is built around staves, staves' localization indicates placement of other symbols of music notation. Several categories of symbols can be distinguished from the perspective of their placement.

The first class includes symbols that are placed in a strictly determined position. Clefs are always placed at the beginning of stave in fixed vertical position. Key and time signatures appear just at the right sight of clefs while changes of signatures just follow barlines. Key signature always precede time signature if both appear in a measure. Key signature is a sequence of sharps or flats occupying strictly defined vertical position. As a result, finding location of symbols of this class should utilize rules of their placement in addition to image analysis. This kind of domain knowledge facilitates not only localization, but also classification of potential symbols emerging in an investigated area. Local projections and histograms as well as analysis of vertical and horizontal sections of black pixels are the basic methods utilized in such tasks, c.f. [3, 6]

The second class includes symbols with features that are easy to be detected from analyzed image area. All but whole notes have stems - vertical stick - that could be comparably easy filtered from local vertical projection. Similarly, regions that potentially include sharps, flats and naturals could also be found by vertical projection filtering. On the other hand, all sorts of symbols that may have horizontal sections as arcs (ties and slurs), dynamic hairpins, etc. could be located or partially located by utilizing horizontal projections. Horizontal projections are always affected by staff lines. However, having staves localized, it could be easy remove traces of staff lines from projections.

The third class includes symbols with location determined by other symbols. Among them are so called connectors (beams joining stems of beamed group, symbols of rhythmic groupings, arpeggios), but also accidentals, staccato and other articulation markings, etc. Such symbols usually have their own characteristic features, but also could be placed only according to other, previously found, symbols of music notation. And as above, local projections and histograms as well as vertical and analysis of horizontal sections of black pixels are the basic methods utilized in such tasks.

And, finally, there are symbols that - from recognition perspective, but not music perspective - are randomly placed. Such symbols as rests, change of clefs, dynamic markings, articulation and ornamentation symbols, etc. are examples of irregularly placed symbols. Finding placement of such symbols is often simplified by cleaning the image from already recognized symbols.

3.2 Classification

Compact music symbols Such symbols as clefs, notes, rests, accidentals, signatures, change of clefs have size comparable with stave height. Localization of pattern most often results in finding bounding box of a given pattern, which is then presented to classification module of recognition process. Compact music symbols can be either placed on staff lines or out of them. Therefore, classifiers must deal with random influence of staff lines. Symbols' classifiers cope with a part of original bitmap - bounding box of investigated symbol. Classification decision is made on the basis of direct investigation of bitmap or analysis of extracted features of classified symbol. So, withdrawing from staff lines influence is done either by removing staff lines from original image, by removing staff lines traces from extracted features or by selecting features insensitive to staff lines. Removing staff lines from original image is always run time consuming and may damage other symbols of the image. Since placement of staff lines is usually known, the last two methods overcome the former one.

There is a wide spectrum of features that can be used to characterize classified symbols. These features are extracted from original bitmap restricted by bounding box of analyzed symbol. Besides the simplest features, as bounding box width/height proportion, more complex features based on projections and moments are utilized, c.f. [7, 11] for detailed discussion on this topic.

As mentioned above, classification can be based directly on original bitmap or on a set of extracted features. Unfortunately, there is no universal classification method that could be successfully applied in music symbols classification. A variety of classifiers are utilized: neural networks, statistical classifiers, centroids and clustering, classification trees, c.f. [7, 8, 10].

Connectors Music symbols may create logical or musical groups. In many cases symbols of such groups are joined with so called connectors. For instance, noteheads of consecutive notes may be connected with ties creating one note of summed total duration, stems of a sequence of eight or shorter notes - with beams instead of having flags. Triplets and other rhythmic groupings outlined as horizontal bracket or arc (with a digit describing rhythmic grouping) are another examples of connectors. Such symbols are usually recognized by investigation of geometrical features of symbols that can potentially be connected. For instance, detection of beams is typically done by exploration of stems' endings opposite to noteheads. Detection of ties is based on finding horizontal arcs, usually flat in their middle parts, which connect consecutive noteheads of the same pitch. Investigation is done by utilization of simple methods: analyzing projections, finding horizontal or slightly sloped line sections, checking geometrical relations between connector and connected symbols.

Non-local symbols Such symbols like slurs, dynamic hairpins and octave modifiers are horizontal or sloped, solid or dashed lines and arcs that cannot be investigated with typical pattern recognizers. Their recognition is based on finding their placement and is usually done as late stage of recognition process. And again, as in case of connectors, non-local symbols are recognized by analysis of such features as local projections, runs of horizontal pixels, tracing their shape, etc. Having score structure identified and other symbols recognized and having position and shape's features of non-local symbols, an analysis of geometrical relations between all of them allows for classification of non-local symbols and - in further processing of acquired music information - describing musical function of them.

OCR Since symbols of music notation vary in size and shapes, some of them can be mistaken with letters. Finding text areas will significantly reduce misrecognitions between alphanumeric and music symbols. Lyric is a distinctive type of text in music notation. It is usually placed under respective staff line. Lyric words are split into syllables with hyphens and underscores extended between syllables. Lyric can appear in one row as well as in several rows. The number of rows may change between systems on a page as well as inside a system. As a result, text location module must cope with such irregularities in order to support recognition of alphanumeric characters.

Recognition of texts obviously applies an OCR technology. Subsequently, adaptation of an OCR package is the simplest solution of texts recognition in music notation. However, such a solution may not be acceptable due to cost of OCR purchasing. As a result, music recognition developer should consider design and implementation of his own OCR package having in mind that domain knowledge would be an important advantage in characters recognition. In [9] a simple technology aimed on text recognition in musical scores is presented. It is based on hierarchical classification method and seems to be adequate for music notation. Some types of music texts use special fonts what makes that it is necessary to apply specialized text recognizer instead of general OCR, c.f. [4].

4 Analysis of Context Information

Analysis of context information is a kind of syntactical pattern recognition. Music notation, despite that is very flexible, must satisfy some strict rules. Of course, due to its flexibility, music notation cannot be restricted by any global description as - for instance - a context free grammar. Nevertheless, there are local rules that could have simple description and are easily verified. Some of such rules were already presented in Section 3. Examples of more complex and very important rules are signatures and voices.

Signatures Key and time signatures put restrictions on the whole piece of music. Time signature defines rhythmic value of consecutive measures (with exceptions of possible upbeat and downbeat). Time signature constraint is strict and does not allow for exceptions. Consequently, music notation recognition - if correct - must satisfy this context constrain. Therefore, time signatures constraint is a tool of recognition's verification and is a possible correction tool. Similarly, key signature affects respective notes in all octaves and its control may be canceled by natural in the octave to the end of the measure of the natural. And as in time signature case, key signature constraint is a verification and - possibly - correction tool of notation recognition.

Voices A part is split to measures along time axis. Alternatively, a part may be divided to voice lines which play important role in music as, for instance, piano extract of orchestral music. The task of voice lines extraction is knowledge processing based on methods of syntactical pattern recognition rather then on optical pattern recognition.

5 Conclusions

Optical music recognition has been intensively developed for last two decades gaining promising results. However, practical realizations in this field are still far from perfection. The field of music notation recognition is still open for research and further improvements of OMR technology are still sought. This paper gives brief overview of OMR technology from the perspective of pattern recognition paradigm. Three important aspects of recognition process are distinguished: structure of music notation analysis music symbol recognition and context knowledge acquisition. The brief survey of optical music recognition methods is extended for a list of most suitable papers on this subject.

Acknowledgments This work is supported under State Committee for Scientific Research Grant no 3T11C00926, years 2004-2007.

References

1. Bainbridge D, and Bell T (2001) The challenge of optical music recognition, Computers and the Humanities 35:95-121
2. Barton L, W. G (2002) The NEUMES project: Digital Transcription of Medieval Chant Manuscriptis, In: Second International Conference on WEB Delivering of Music, Darmstadt, Germany, IEEE Computer Society Press
3. Fujinaga I (1988) Optical music recognition using projections, MSc thesis, McGill University, Montreal, Canada
4. Gezerlis V, Theodoridis S (2002) Optical character recognition of the Orthodox Hellenic Byzantine Music notation, Pattern Recog. 35: 895-914
5. Homenda W (1996) Automatic recognition of printed music and its conversion into playable music data, Control and Cybernetics, 25(2):353-367
6. Homenda W (2002) Granular Computing as an Abstraction of Data Aggregation - a View on Optical Music Recognition, Archives of Control Sciences, 12(4):433-455.
7. Homenda W, Luckner M (2004) Automatic Recogniton of Music Notation Using Neural Networks, In: Inter. Conf. On AI and Systems IEEE AIS'04, Divnomorskoye, Russia, Sept. 3-10, Proc. Physmathlit, Moscow:74-80.
8. Homenda W, Luckner M (2004) Automatic Recogniton of Music Notation Using Methods of Centroids and Classification Trees, Proc. of the Intern. Symposium ISCIIA'2004, Haikou, China, December 20-24.
9. Homenda W, Luckner M (2005) Hierarchical OCR System for Texts in Musical Scores, submitted, IFSA'2005 World Congress, Beijing, China, July 28-31.
10. Homenda W, Mossakowski K (2004) Music Symbol Recognition Neural Networks vs. Statistical Methods, In: Baets B. et al. (Eds.), Current Issues in Data and Knowledge Engineering, EXIT, Warszawa.
11. Luckner M (2003) Automatic Identification of Selected Symbols of Music Notatio, MSc thesis, Warsaw University of Technology, Warsaw, Poland.
12. Luth N (2002) Automatic identification of music notations, In: Second International Conference on WEB Delivering of Music, Darmsradt, Germany, IEEE Computer Society Press.
13. McPherson J R (2002) Introducing feedback into an optical music recognition system, In: Third Internat. Conf. on Music Information Retrieval, Paris, France.
14. Pinto J C, P. Vieira et al. (2003) A new graph-like classification method applied to ancient handwritten musical symbols, In: International Journal of Document Analysis and Recognition 6(1): 10-22.
15. Rossant F (2002) A global method for music symbol recognition in typeset music sheets, Pattern Recognition Letters 23:1129-41.

An Application for Tyre-Ground Contact Area Analysis

Klaudia Jankowska[1], Tomasz Krzyzynski[1], Andreas Domscheit[2]

[1] Technical University of Koszalin, ul.Raclawicka 15-17, 75-620 Koszalin, Poland
klaudia@tu.koszalin.pl, tkrzyz@tu.koszalin.pl

[2] Continental AG, Jaedekamp 30, 30419 Hannover, Germany
andreas.domscheit@conti.de

1 Introduction

To conduct analysis of tyre-ground contact area (called footprint) is essential since tyres are responsible for giving support for the vehicle and for transferring forces necessary to obtain required kinematic behavior of the vehicle. As track tyres are concerned footprint shape and pressure distribution within contact area are the foremost factors for wear and mileage performance, and have significant influence on braking behavior. They have also to be taken under consideration during tyre optimization for specific applications (load variations, inflation pressure, wheel position). On the other hand footprint shape analysis is performed during quality control of tyre production.

In daily work of tyre engineers many similar footprint images have to be compared, assessed and classified. As visual inspection of footprint images is time consuming, complex, unreliable and day-form-dependant an effort has been undertaken to automatise this process.

In this paper we present abilities of developed application equipped with Graphical User Interface. Reader interested in details of used methods should refer to our previous paper [3].

Programme is designed to work with original measurements data. Footprint images acquisition starts with pressing a tyre against an illuminated glass plate. Thin film is placed in between the tyre and the glass plate. Due to dispersion bright marks appear in the contact area of the tyre. Images are acquired in grayscale where bright levels depict higher pressure (fig.1).

Developed programme allows the user to make analysis of both tyre footprint shape and contact pressure distribution (fig.1). Before any analysis can be started image preprocessing is performed: background separation - net shape determination, gross shape determination and noise removal. Using calibration data preprocessed gray scale images are transformed to pressure scale.

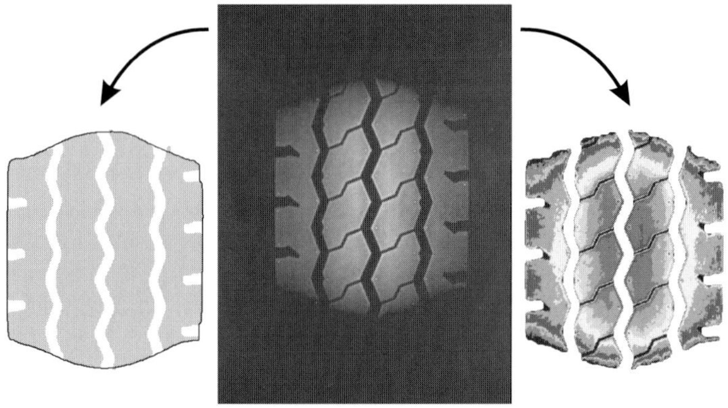

Fig. 1. Center - original image, left: calculated net contact shape (gray) with gross shape contour (black line), right: calculated contact pressure distribution.

2 Footprint shape analysis

Shape dimensions

For extracted footprint shapes their dimensions are calculated. From tyre properties point of view important are:

- footprint length, width and the ratio of the two,
- net and gross contact area and their ratio.

Scaling from pixels to metric values is done using pixel height and width obtained during calibration of the test stand.

Symmetry

Symmetry is one of the parameters describing the shape of the tyre footprint. For the ideal tyre we would expect footprint shape to show:

- symmetry of left and right part,
- symmetry of up and bottom part,
- skew symmetry.

In order to evaluate degree of shape symmetry appropriate parts are added. As a measure of symmetry number of overlapping pixels to all shape pixels ratio was used (fig. 2).

Shape conicity

In order to evaluate degree of footprint shape conicity angles between lines fitted to top and bottom footprint shape contours and horizontal line are calculated. Maximum value of this angles is used as a measure of the footprint shape conicity (fig. 3).

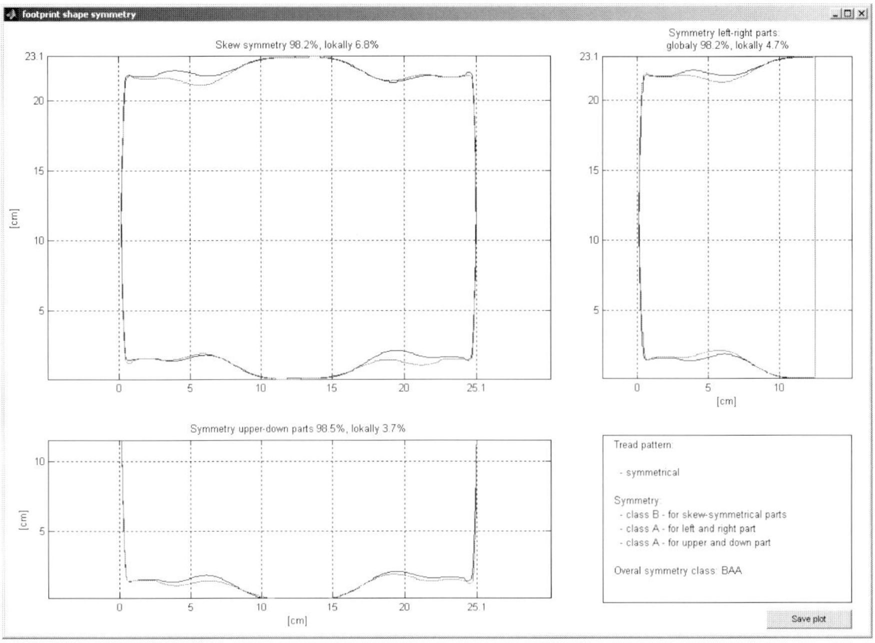

Fig. 2. Example of the tyre footprint shape symmetry assessment.

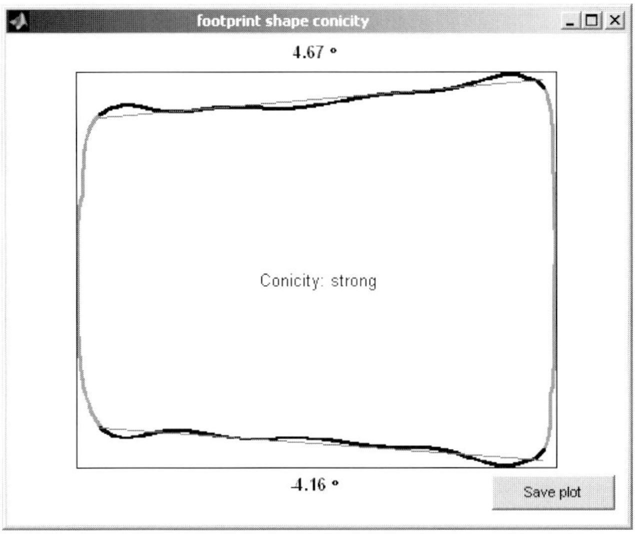

Fig. 3. Example of the tyre footprint shape conicity assessment.

Shape type

Gross footprint shapes are classified to one of the twelve groups (fig. 4). This classification is based on the work of A. Domscheit and F. J. Dopheide [2]. To describe existing types of footprint shape this classification takes into consideration two most important features:

- form of footprint shape: flat, round, depressed or waved, and
- appearance of footprint shape in shoulders: up, horizontal or down.

It considers only the edges of the shapes. Length to width ratio is unimportant for classification.

3 Contact pressure analysis

Contact pressure distribution

On the pressure distribution plot (fig. 5) one can see:

- contact area image in color scale corresponding to the pressure,
- graphs of average pressure along footprint width and length (their values can be saved as an `*.xls` file.),
- maximum pressure - spikes points are not taken into account,
- average pressure - all points belonging to the net contact area are taken into account,
- standard deviation of pressure values.

Contact pressure percentage and range

Percentage of the footprint area under specific pressure range can be displayed as a pie plot or a bar plot. If the pressure percentage plots are not detailed enough for specific pressure distribution evaluation one can use additional tool (fig. 6) to specify interesting pressure range. Footprint parts being under given pressure range are highlighted and the corresponding area is calculated and displayed in figure title.

Contact pressure profile

To have really close look into measurement data one can use tool shown on figure 7. This tool gives possibility to analyse:

- pressure along the line - horizontal or vertical line can be chosen, line can be moved to any footprint position,
- average pressure in the region - size and orientation of the region can be chosen; for specified region contained net area and corresponding load are displayed,
- pressure along segment line plotted by the user - one can plot single segment or multi-line.

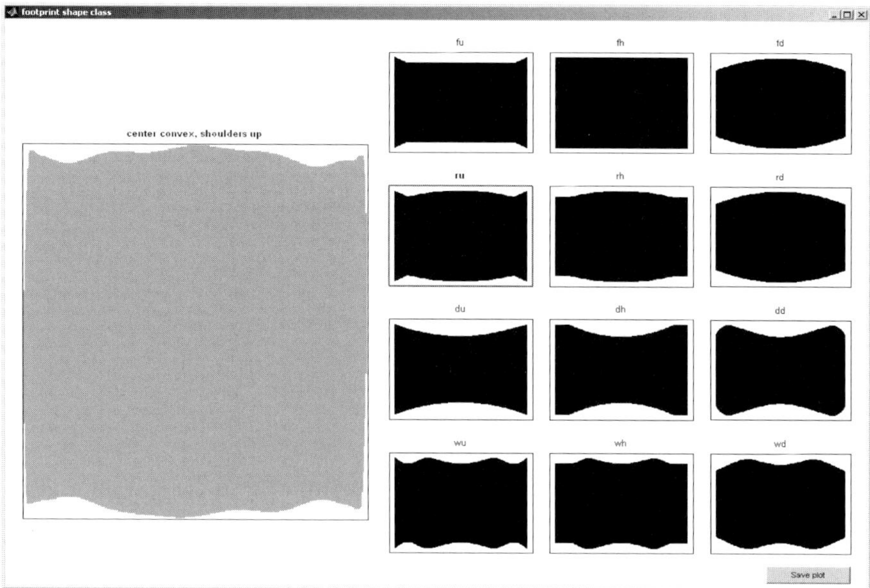

Fig. 4. Example of footprint shape classification.

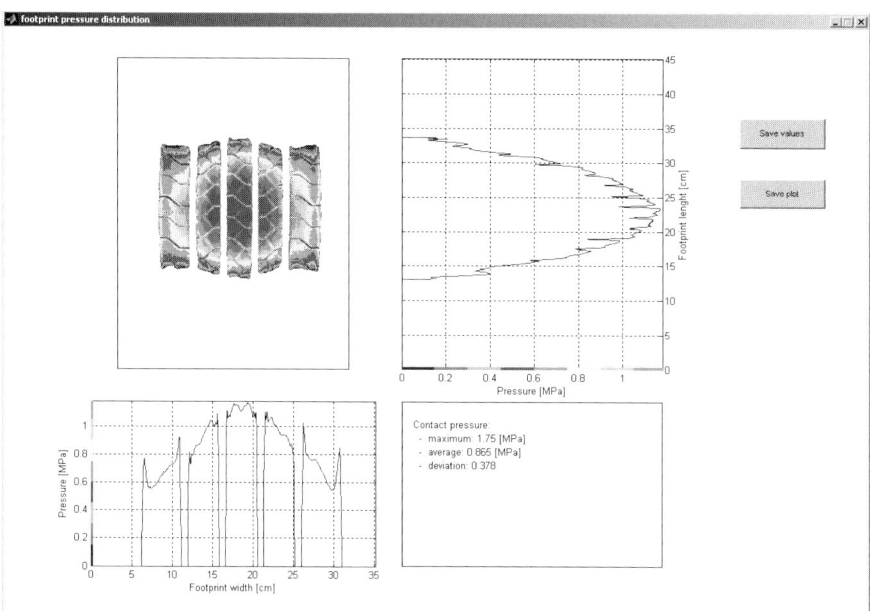

Fig. 5. Example of pressure distribution plot.

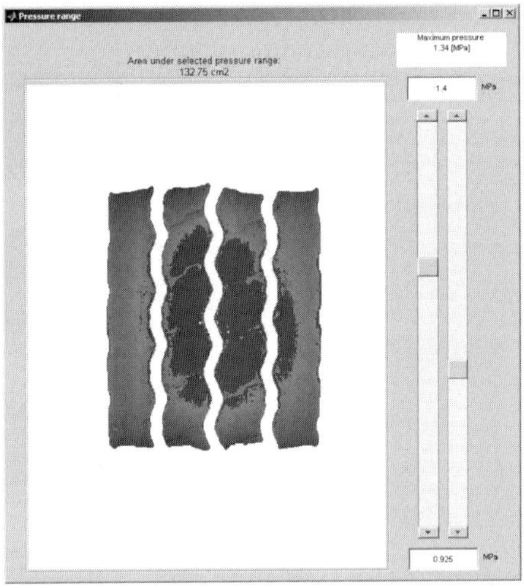

Fig. 6. Example of pressure range tool.

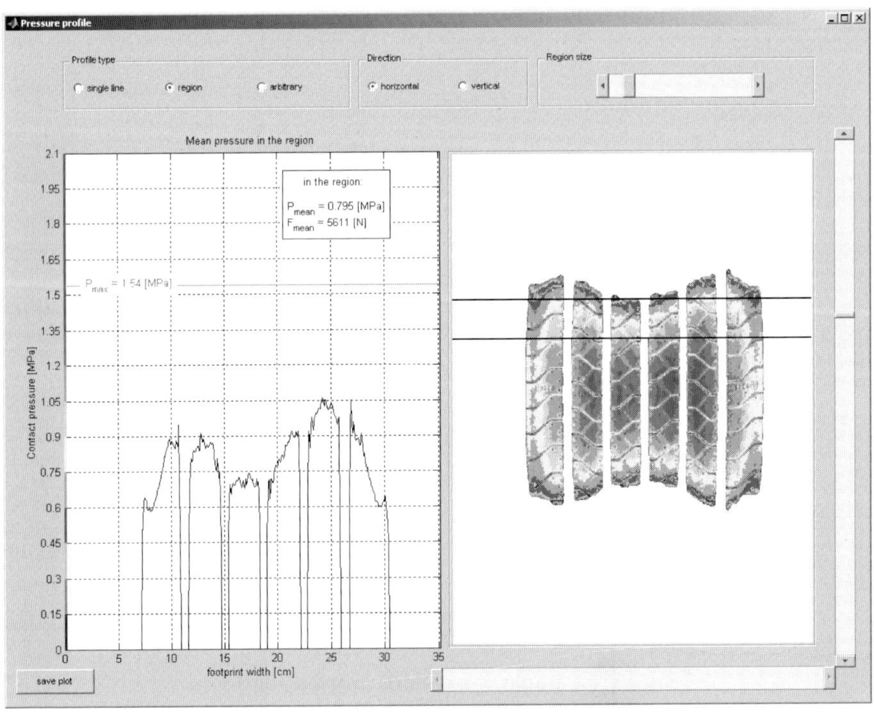

Fig. 7. Example of pressure profile tool.

4 Ribs analysis

Programme divides footprint shape into ribs. For each rib the following are calculated:

- area in centimeters and as a percentage of the net shape area,
- average pressure and pressure deviation,
- load as a percentage of whole load applied to a tyre.

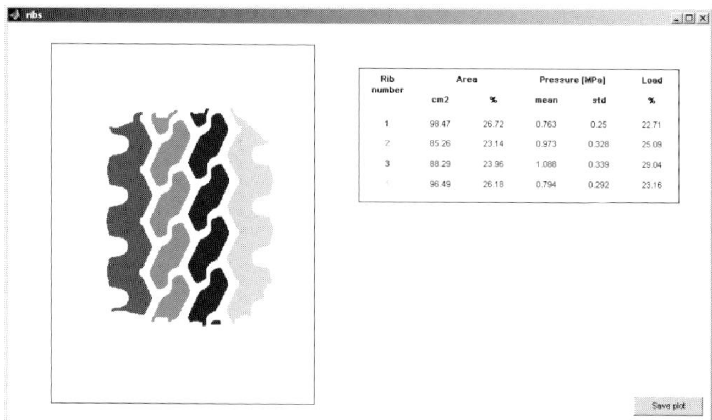

Fig. 8. Example of rib analysis plot for a tyre with block tread.

5 Comparison of two measurements

5.1 Images registration

In order to perform automatic footprint shape comparison images need to be registered as in particular image contact area appears in different position within an image. Correlation method based on FFT is used [4].

5.2 Comparison

Effective registration method gives possibility to indicate parts of footprint where change of shape and contact pressure occurred. This is valuable information additional to shape dimension and pressure values change. On the figure 9 one can see two footprints and difference between them, shown as:

- shapes difference plot - with colors indicating parts of contact area which undergo changes,
- pressure difference plot - colors indicate pressure difference,
- average pressure along footprint width and length,
- summary of measurable differences.

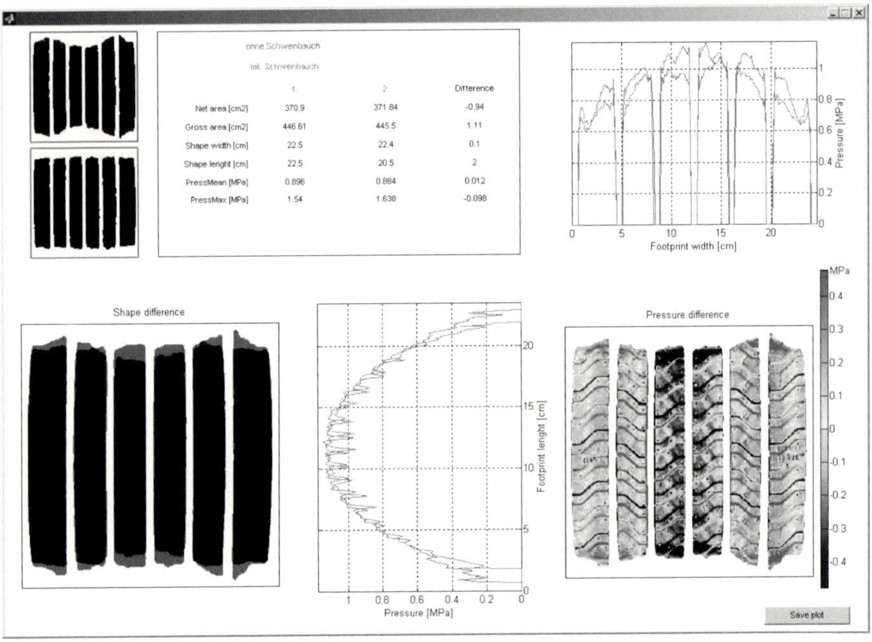

Fig. 9. Comparison of two RPDC measurements.

6 Summary

Most of the presented procedures were verified, the others are still under development. Representative set of 400 footprint measurements of tyres varying in construction, tread pattern, size, load, inflation pressure and degree of wear is used for verification.

Presented application gives practical benefit for tyre engineers who have to analyse many similar footprint images. It allows automatic assessment, classification and comparison of tyres on the basic of its footprint shape.

References

1. J. C. Dixon: Tires, suspension and handling. Warrendale, Pa. : Society of Automotive Engineers, second edition, 1996.
2. A. Domscheit, F. J. Dopheide: Target Footprint Design Guide EU/US, 2001.
3. K. Jankowska, T. Krzyzynski and A. Domscheit: Vision-based Analysis of the Tire Footprint Shape, International Conference on Computer Vision and Graphics, in print, 2004.
4. J. P. Lewis: Fast Template Matching, Vision Interface, pp. 120÷123, 1995.
5. J. R. Parker: Algorithms for image processing and computer vision. Wiley Computer Publishing, New York, 1997.

On the Use of Syntactic Pattern Recognition Methods, Neural Networks, and Fuzzy Systems for Short-Term Electrical Load Forecasting *

Janusz Jurek and Tomasz Peszek

Institute of Computer Science, Jagiellonian University
Nawojki 11, 30-072 Cracow, Poland

Summary. Several artificial intelligence methods of short-term electrical load forecasting are discussed in the paper. The model of a hybrid system based on syntactic pattern recognition, neural networks, and fuzzy techniques is introduced. The application of the model and the experimental results of short-term electrical load forecasting are presented.

1 Introduction

Short-term electrical load forecasting (STLF)[2] is very important for the safe and cost effective operation of the national power system. In particular, it is essential for energy suppliers and electrical distribution companies economics. The electricity load (or demand) is a function of many parameters like: time, type of a day (a weekday, Saturday, Sunday, or a holiday), weather (eg. temperature, humidity, and insolation), and random effects. This makes the prediction of the load very difficult. The research into construction of "intelligent" systems able to predict the load are being conducted in many institutes and companies all over the world. Although plenty of practical techniques and software tools have been already developed [10], there is still a need of constructing more accurate methods and systems.

About one year ago the research concerning a construction of a system supporting electrical load forecasting started in Institute of Computer Science, Jagiellonian University, Cracow, Poland. The research is conducted with cooperation with the Polish Power Grid Company, and the Cracow Power Distribution Company. The goal of our research is to provide useful tools for STLF on the basis of pattern recognition and artificial intelligence methods.

*This work was supported by the Polish State Committee for Scientific Research (KBN) under Grant No. 3 T11C 054 26.

[2]Such forecasts are usually defined as the prediction of the hourly or half-hourly electricity load from one to several days ahead.

In the paper we present a hybrid approach to the construction of the prediction system. The approach is based on the use of syntactic pattern recognition, neural networks, and fuzzy methods. Let us notice that neural networks and neuro-fuzzy systems are widely used for STLF [7, 6, 3], but there are practically no scientific reports on applications of syntactic pattern recognition methods for this purpose.

In section 2 we present the architecture of the predicting system. Section 3 contains the experimental results and some remarks on them. Conclusions are included in the final section.

2 Construction of the hybrid system for STLF

2.1 Overall architecture

In this section we present the overall architecture of the hybrid system for short-term electrical load forecast. The system consists of four main modules: neural network module, fuzzy reasoning module, syntactic pattern recognition module, and supervision module (see: Figure 1).

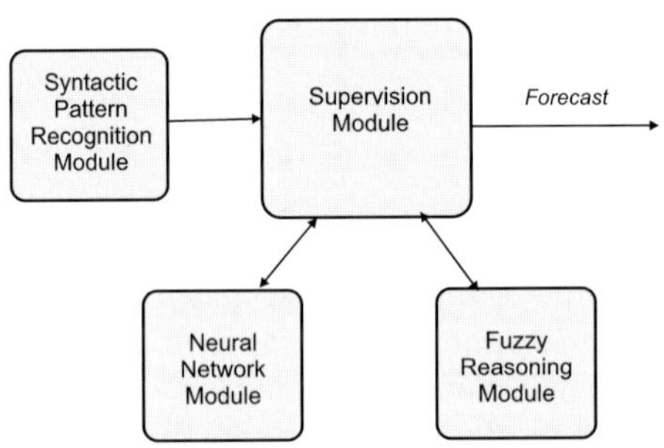

Fig. 1. The general scheme of the hybrid system for STLF.

Syntactic pattern recognition module is used to recognize *types* of the electricity load (see: section 2.2). This information is used by the supervision module. The module controls the work of the other components: neural network module (see: section 2.3) and fuzzy reasoning module (see: section 2.4). It executes the procedures for verification and synthesis of partial forecasts. In case of predictions exceeding the statistically probable range supervision module decides to reject the forecasts and generate them in an alternative way.

2.2 Syntactic pattern recognition module

One of the modules in the hybrid system for STLF is based on syntactic pattern recognition approach. We will use so-called GDPLL(k) grammars and parsers for the analysis of the *types* of electrical load functions. Let us introduce two basic definitions corresponding to GDPLL(k) grammars [1, 4].

Definition 1. A *generalized dynamically programmed context-free grammar* is a six-tuple $G = (V, \Sigma, O, P, S, M)$, where: V is a finite, nonempty alphabet; $\Sigma \subset V$ is a finite, nonempty set of terminal symbols (let $N = V \setminus \Sigma$); O is a set of basic operations on the values stored in the memory; $S \in N$ is the starting symbol; M is the memory; P is a finite set of productions of the form: $p_i = (\mu_i, L_i, R_i, A_i)$ in which $\mu_i : M \longrightarrow \{TRUE, FALSE\}$ is the predicate of applicability of the production p_i defined with the use of operations ($\in O$) performed over M; $L_i \in N$ and $R_i \in V^*$ are left- and right-hand sides of p_i respectively; A_i is the sequence of operations ($\in O$) over M, which should be performed if the production is to be applied. □

Definition 2. Let $G = (V, \Sigma, O, P, S, M)$ be a dynamically programmed context-free grammar. The grammar G is called a *GDPLL(k) grammar*, if the following two conditions are fulfilled.

1. Stearns's condition of LL(k) grammars. (The top-down left-hand side derivation is deterministic if it is allowed to look at k input symbols to the right of the current position of the input head in the string).
2. There exists a a certain number ξ such that after the application of ξ productions in a left-hand side derivation we get at the "left-hand side" of a sentence at least one new terminal symbol. □

The algorithm of the parser for GDPLL(k) grammars (GDPLL(k) parser) has been described in [1] and [4]. We will not present it in the paper, but let us notice that the algorithm reflects the way how the derivation in the grammars is performed (the algorithm uses top-down approach).

There are three main reasons why GDPLL(k) grammars and parsers have been chosen for the application in a hybrid system for STLF. First of all, the GDPLL(k) grammars are characterized by very good discriminative properties (they are able to generate a large class of context-sensitive languages) [1]. Secondly, it is possible to construct an efficient parser for GDPLL(k) grammars (of the linear computational complexity) [4]. Finally, there is a grammatical inference algorithm for GDPLL(k) grammars [5].

The application of syntactic pattern recognition methods to the analysis of the electrical load functions is done in an analogical way, as in case of ECG or EEG analysis [2]. We "translate" the functions to the symbolic form, and then we analyze the string of symbols obtained.

A GDPLL(k) parser is used to recognize *types* of the load charts by their beginning segments (see: Figure 2). In this way it provides an additional information which helps to prepare a better forecast (especially in case of some "atypical" days, like a day between holidays).

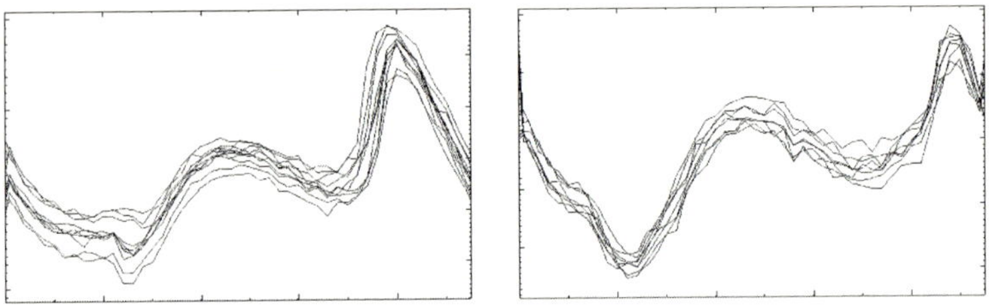

Fig. 2. Different types of electrical load charts.

2.3 Neural network module

Numerous prognostic tools based on neural networks have been already developed. There are several reasons why the neural technology is so popular in terms of application to STLF. The cost of reaching the outcome is low relatively to the volume of knowledge accumulated in the system. Secondly the neural networks possess great generalization abilities and do not require precise mathematical model of solution *a priori*. The neural technology is reliable as a result of distributed knowledge representation. Last of all some network types allow hardware implementation as a result of simple architecture.[3] The architecture and applications are widely described in [9, 8].

The neural network module consists of the one-layer Kohonen type network. Its main task is to match the presented partial pattern to the one of the already known patterns. The vectors presented to the network contain information on the history of electrical demand in the previous day, weather factors (air temperature), the particular type of weekday and the initial recognition hint derived from the syntactic recognition module. The lifetime of the network can be divided into two phases. In the learning phase the network learns to classify the presented, known patterns and accumulates information about them in the neurons. In case of occurrence of the unknown pattern new neuron is added. The real work phase follows the learning process and is based on the result of the preceding phase. The difference between phases on input data type is that we do not have the complete patterns. The initial forecast must be generated by means of statistical tools.

Although we deal with a neural network the problem of the forecast reasoning is not very difficult in this particular case. The simple algorithm explains the reasons for the particular predicted value. The neural network module turns out to be the only component of the system that actively acquires new facts. This knowledge serves as an information source not only to the network but also to the fuzzy reasoning module.

[3]For example perceptrons are such simply built networks.

2.4 Fuzzy reasoning module

The fuzzy systems are applied in many predicting solutions involving problems that are difficult to describe by a precise mathematical model. They allow to formulate the heuristic knowledge to the $IF-THEN$ rules built by use of language which is very similar to the common language. They work successfully in the partially noised environment and allow conclusion reasoning without support of computationally expensive algorithms. The parameters of the fuzzy system work are scalable for the particular application. However, there are some disadvantages of the fuzzy system that have to be mentioned. The cost of reaching the outcome is high relatively to the volume of knowledge contained in the system. The fuzzy systems can be not reliable in some situations. They must be secured in case pattern occurs that has not been recognized yet. Many examples of fuzzy systems applications not only to STLF are analyzed and discussed in [11].

The fuzzy reasoning module is the part of the system using the language of fuzzy sets, fuzzy rules and linguistic variables to describe and predict short term load demand. As far as its architecture is concerned it is a classical fuzzy rule processing engine operating on $IF - THEN$-type-rules. Applied defuzzyfication method is the *center of area* i.e.

$$\bar{x} = \frac{\sum_{i \in I} i\mu(i)}{\sum_{i \in I} \mu(i)},$$

where μ is the function of partity of the defuzzyficated fuzzy set and I is the domain. The predecessors of these rules consists of the natural language terms describing the factors influencing electrical demand in the short term i.e. the electrical load demand on the previous day, data on maximal and minimal air temperatures, weekday type as well as suggestion on probable electrical load demand received from the syntactic recognition module. The certain associations can effect in the new domain knowledge. However, the explicit knowledge representation results in significant growth on computational expenses on the side of fuzzy component.

2.5 Supervision module

Apart from classical concepts based on artificial intelligence effecting in many working prognostic solutions we should mention statistics as a domain offering a wide spectrum of tools that allow to verify the generated outcomes *ex ante*. Convincing example is the analysis of the growth of exactness as a result of use statistic tools to verification and aggregating of partial forecasts.

In course of experiments creation of an extra module turned out to be grounded and necessary. The supervising component of the system have several task to take care of. The first problem it has to deal with is aggregating of partial forecasts up to the final prediction. This turns out to be one of the crucial problems. Making use of elementary statistical tools it first determines the statistically probable range in which the generated forecast should be sought. To be more precise we take advantage of available history of electrical load demand and the special property of the function describing it. This function is namely cyclic i.e. it fluctuates within the weekly period. The fact implies that also the variable containing the change on electrical load demand between adjacent days is cyclic. Let us denote the difference $\Delta_k = E_k - E_{k-1}$, where E_k is the daily energy consumption in the day k. In this notation we can easily express the obvious dependance

$$E_t \approx E_{t-1} + \Delta_{t-7}. \tag{1}$$

By means of formula 1 we can estimate the interval which the expected electrical load demand should match. We will consider the value computed on base of formula 1 as the *initial forecast* of the system and denote it by $\widetilde{E}_t$. As δ we will denote the maximal, assumed in advance, expected distance between the real value of energy consumption and the first forecast. The forecasted value returned by one of the subsystems will be called *allowable* if it is contained in the interval $(\widetilde{E}_t - \delta, \widetilde{E}_t + \delta)$. Forecasts which occur not allowable will be rejected by the supervising module.

Another assignment to be conducted by the module is to stimulate components to learning when it seems reasoned. Last item to be mentioned is the security issue which is also maintained be the supervising module. The hybrid should be prepared for the situation when both components fail to generate a proper prediction. In this case it must be able to create prediction in an alternative way.

3 Experimental results

In the current paragraph we analyze the outcomes of the prediction system based on components described in the preceding sections i.e. syntactic pattern recognition module, fuzzy reasoning module, neural network module and the supervising module managing the whole hybrid.

Let us characterize the data used for the implementation and tests of the systems's prototype. The solution has been developed and tested on data received from the Cracow Power Distribution Company. The data contain the hourly electrical load demand in the period of the year 2002 and 2003. All weather data have been received from Institute of Geography, Jagiellonian University, Cracow, Poland and concern the maximal and minimal daily air temperatures denoted on the Jagiellonian University Climatic Station in Gaik-Brzezowa.

As it was already mentioned in the preceding paragraph a crucial problem when dealing with various forecast sources is the ability to combine the partial predictions in an efficient way to the final forecast. The importance of this aspect is easily realizable when considering different strategies for the synthesis of forecasts from different modules. In course of tests of the prototype system we analyzed following strategies:

1. *Strategy 1.* „Neural network". In this strategy we ignore the results of the fuzzy systems preferring the forecasts of the neural network.
2. *Strategy 2.* „Fuzzy system". In this strategy we ignore the results of the neural network preferring the forecasts of the fuzzy component.
3. *Strategy 3.* „Better yesterday, better today". The strategy prefers the component that turned out to guess better yesterday.
4. *Strategy 4.* „Weights sums" is based on the synthesis of the partial forecasts by use of the linear function. Its coefficients are determined by use of the least-squares method.
5. *Strategy 5.* „Strategy of allowable intervals". The approach was described in the section 2.5.

The best results are achieved by using the strategy number five used as default. There is a great difference between the default strategy and the other ones. This fact reveals the dominance of a hybrid systems based on neural, fuzzy and statistical components over homogenous models. The algorithm of partial forecast aggregation applied in the fifth strategy is more sophisticated than the ones used in strategies 3 and 4. It effects in the growth on efficiency and exactness of the system work.

Measure / Strategy	Avg(AE) (100 MWh)	Dev(AE) (100 MWh)	Max(AE) (100 MWh)	Avg(APE)	Dev(APE)	Max(APE)
Strategy 1.	12.7	12.3	147.4	7.44 %	7.11 %	66.24 %
Strategy 2.	12.71	15.17	230.2	7.31 %	7.88 %	100 %
Strategy 3.	11.68	9.90	53.17	6.91 %	6.35 %	37.55 %
Strategy 4.	11.85	11.17	97.60	7.03 %	6.97 %	55.57 %
Strategy 5.	6.03	6.33	38.99	3,46 %	3,82 %	25,71 %

Table 1. The exactness of hybrid forecast with different strategies. Error statistics. The common daily electrical load demand fluctuates between 14000 and 23000 MWh. Variable AE denotes the sum of absolute values of differences between real and forecasted demand value, APE contains the relative values of these differences.

Analyzing other measures of exactness such as the number of best answers[4] or convergency between time series of real and forecasted electrical load demand we can claim that the default strategy of the system scores the best results, too.

As far as the computational expenses are concerned we can easily realize the strong correlation between the volume of knowledge in the system and the number of operations to be conducted by the system.

[4]The best answer is the answer that is closest to the real electrical load demand.

4 Conclusions

In the paper we have presented the recent results of the research into the construction of a hybrid system for short-term electrical load forecasting (STLF). The hybrid system utilizes several methods of artificial intelligence: syntactic pattern recognition, neural networks, and fuzzy techniques.

Although the results of the experiments performed on the real data are very promising, there is still much to be done. First of all, the model has to be verified in practice. We plan to use the model in the Polish Power Grid Company and in the Cracow Power Distribution Company as a support tool for STLF. Then we are going to concentrate on the problem of "atypical" days (holidays, single days between holidays), because the forecast for such days is significantly worse. The development of the system will be a subject of further publications.

References

1. Flasiński M, Jurek J (1999) Dynamically Programmed Automata for Quasi Context Sensitive Languages as a Tool for Inference Support in Pattern Recognition-Based Real-Time Control Expert Systems. Pattern Recognition, 32 (4), 671–690
2. Fu KS (1982) Syntactic Pattern Recognition and Applications, Prentice Hall
3. Hippert SH, Pedriera CE, Souza RC (2001) Neural networks for short-term load forecasting: a review and evaluation, IEEE Trans. Power Systems, 16 (1), 44–55
4. Jurek J (2005) Recent developments of the syntactic pattern recognition model based on quasi-context sensitive languages, accepted for publication in Pattern Recognition Letters
5. Jurek J (2004) Towards Grammatical Inferencing of GDPLL(k) Grammars for Applications in Syntactic Pattern Recognition-Based Expert Systems, Lecture Notes in Computer Science, 3070, 604–609
6. Mastorocostas PA, Theocharis JB, Kiartzis SJ, Bakisrtzis AG (2000) A hybrid fuzzy modeling method for short-term load forecasting, Mathematics and Computers in Simulation, 51, 221–232
7. Papadakis SE (1998) A novel approach to short-term load forecasting using fuzzy neural network, IEEE Trans. Power Systems, 13 (2), 480–492
8. Tadeusiewicz R (1993), Sieci neuronowe, Akademicka Oficyna Wydawnicza, Warszawa.
9. Tadeusiewicz R, Flasiński M (1991) Rozpoznawanie Obrazów, Państwowe Wydawnictwo Naukowe PWN, Warszawa.
10. Zieliński J (1997) Survey of short-term electrical load forecasting methods, Mat. Konf. APE'97 Aktualne Problemy w Elektroenergetyce, Gdańsk, Jurata 11-13 czerwca 1997, tom IV, 121–129
11. Zieliński J (2000), Inteligentne systemy w zarzadzaniu. Teoria i praktyka, Wydawnictwo naukowe PWN, Warszawa.

A Real-Time Head Tracker Supporting Human Computer Interaction

Bogdan Kwolek

Rzeszów University of Technology, W. Pola 2, 35-959 Rzeszów, Poland
bkwolek@prz.rzeszow.pl

Summary. This paper describes a fast and completely automatic algorithm for human face tracking. The tracked face is represented by a weighted histogram. The current histogram is compared to histograms at the particles' positions. The weight of each particle is determined on the basis of Bhattacharyya distance and intensity gradient along the ellipse's boundary. The incorporation of information about the distance between the camera and the face undergoing tracking results in robust tracking even in presence of skin colored regions in the background. The initialization of the tracker is realized by means of face detection. The detection is carried out using Haar-like features, followed by the verification of face distance to the camera and face region size heuristics.

1 Introduction

Fulfilling the idea of machines that interact face to face with people forces us to think in new ways about computers that could be used in daily life. Within the past decade, significant advances in machine learning and perception open up the possibility of understanding human actions. To obtain a high level interpretation of human actions one must first detect humans. There are a variety of approaches to human detection, mainly focusing on face detection [18].

The visual tracking of objects of interests has become an elementary task in many applications, including surveillance, human-machine interfaces, smart environments, and many more. However, the majority of available algorithms assume that the camera is mounted at a fixed location. Most existing vision-based tracking algorithms give correct estimates of the state in a short span of time and often fail if there is a significant inter-frame change in object appearance. These methods generally fail to precisely track regions that share similar statistics with background regions. To improve the reliability of tracking in such circumstances we integrated in probabilistic manner the edge strength along the elliptical head boundary and color within the observation model of the particle filter. Particle filters provide a means to track the state of an object even if the dynamics and observations are non-linear/non-Gaussian [6][7].

The incorporation of information about the distance between the camera and the face undergoing tracking results in robust tracking on the basis of images acquired from a moving camera even in presence of skin colored regions in the background. In order to initialize the tracker, or reinitialize the system if the tracking fails, we adopt the fast and efficient face detecting method of Viola and Jones [17]. The face detector finds the location and size of each region containing the frontal face in an input image. Next, using the face location, the eigenfaces algorithm [16] is utilized to identify the robot user.

In tracking techniques [1][2][4], the current frame is searched for a region whose colors content best matches a reference color model. The searching starts from the final location in the previous frame and proceeds iteratively to find the minimum distance to the reference color histogram. Global color reference models and Bhattacharyya coefficient as a similarity measure between the color distribution of the model and target candidates have been used in a particle filter-based tracker [10]. A histogram representation of the region of interest has been extracted in a rectangular window. In work [3] an ellipse is used to approximate the head outline during 2D tracking on the basis of a particle filter. Darrell, at al. [5] combine stereo and color via an intensity pattern classification method to track people. The CMU face detector [12] has been used to distinguish the frontal face from other body parts. Over the years various strategies for face detection have been proposed in the literature [18]. The Viola-Jones system [17] was the first for real-time frontal face detection.

The remainder of the paper is organized as follows. In the next section we briefly outline particle filtering. In section 3 we present all ingredients of our tracker and demonstrate how color and contour cues can be integrated to improve the performance of the tracker. Then we describe the face detection algorithm. Section 4 reports results which were obtained in experiments with a moving camera. Finally, some conclusions follow in the last section.

2 Particle Filtering for Visual Tracking

For nonlinear models, multi-modal, non-Gaussian or any combination of these models the particle filter provides a Monte Carlo solution to the recursive filtering equation $p(\mathbf{x}_t \mid \mathbf{z}_{1:t}) \propto p(\mathbf{z}_t \mid \mathbf{x}_t) \int p(\mathbf{x}_t \mid \mathbf{x}_{t-1}) p(\mathbf{x}_{t-1} \mid \mathbf{z}_{1:t-1}) d\mathbf{x}_{t-1}$, where $\mathbf{x}_t$ and $\mathbf{z}_t$ denote the hidden state of the object of interest and the observation vector at discrete time t, respectively, whereas $\mathbf{z}_{1:t} = \{\mathbf{z}_1 ... \mathbf{z}_t\}$ denotes all the observations up to current time step. With this recursion we can calculate the posterior, given a dynamic model $p(\mathbf{x}_t \mid \mathbf{x}_{t-1})$ describing the state propagation and an observation model $p(\mathbf{z}_t \mid \mathbf{x}_t)$ describing the likelihood that a state $\mathbf{x}_t$ causes the measurement $\mathbf{z}_t$. Starting with a weighted particle set $S = \left\{(\mathbf{x}_{t-1}^{(n)}, \pi_{t-1}^{(n)}) \mid n = 1...N\right\}$ approximately distributed according to $p(\mathbf{x}_{t-1} \mid \mathbf{z}_{1:t-1})$ the particle filter operates through predicting new particles from a proposal distribution. To give a new particle representation

$S = \left\{(\mathbf{x}_t^{(n)}, \pi_t^{(n)}) \mid n = 1...N\right\}$ of the posterior density $p(\mathbf{x}_t \mid \mathbf{z}_{1:t})$ the weights of particles are set to $\pi_t^{(n)} \propto \pi_{t-1}^{(n)} p(\mathbf{z}_t | \mathbf{x}_t^{(n)}) p(\mathbf{x}_t^{(n)} | \mathbf{x}_{t-1}^{(n)}) / q(\mathbf{x}_t^{(n)} | \mathbf{x}_{t-1}^{(n)}, \mathbf{z}_t)$.

From time to time the particles should be resampled according to their weights to avoid degeneracy. The resampling selects with higher probability particles that have a high likelihood associated with them, while preserving the asymptotic approximation of the particle-based posterior representation. Without resampling the variance of the weight increases stochastically over time [6]. When the proposal distribution is chosen as the distribution conditioning the state at the previous time step, the importance function reduces to $q(\mathbf{x}_t^{(n)} \mid \mathbf{x}_{t-1}^{(n)}, \mathbf{z}_t) = p(\mathbf{x}_t^{(n)} \mid \mathbf{x}_{t-1}^{(n)})$ and in consequence the weighting equation takes the form $\pi_t^{(n)} \propto p(\mathbf{z}_t \mid \mathbf{x}_t^{(n)})$. This simplification leads to a variant of a well-known particle filter in computer vision, CONDENSATION [7].

3 State Space and Observation Model

The observation model integrates two different visual cues. We construct a likelihood model for each of the cues. The motion model will be presented as the first topic in this section. The observation model in which the multiple cue integration takes place will be discussed in detail later. The model adaptation over time will be presented afterwards. An outline of face detection algorithm ends this section.

3.1 State Space and Dynamics

The outline of the head is modeled in the 2D-image domain as a vertical ellipse that is allowed to translate and scale subject to a dynamical model. The object state is given by $\{x, \dot{x}, y, \dot{y}, s_y, \dot{s}_y\}$, where $\{x, y\}$ denotes the location of the ellipse center in the image, $\dot{x}$ and $\dot{y}$ are the velocities of the center, s_y is the length of the minor axis of the ellipse and $\dot{s}_y$ is the rate at which s_y varies.

Our objective is to track a face in a sequence of images acquired from a moving camera. To achieve robustness to large variations in the object pose, illumination, motion, etc. we use the first-order auto-regressive dynamic model $\mathbf{x}_t = A\mathbf{x}_{t-1} + w_t$, where A is a deterministic component describing a constant velocity movement and w_t denotes a multivariate Gaussian random variable.

3.2 Shape and Color Cues

As demonstrated in [1][3], the contour cues can be very useful to represent the appearance of the tracked objects with distinctive silhouette when a model of the shape can be learned off-line and then adapted over time. The shape of the head is one of the most easily recognizable human parts and can be quite well approximated by an ellipse. Therefore a parametric model of the ellipse with a fixed aspect ratio equal to 1.2 is utilized to verify the oval shape of head candidates. During tracking the oval shape of each head candidate is verified using the sum of intensity gradients along the ellipse's boundary.

When the contour information is poor or is temporary unavailable color information can be very useful alternative to extract the tracked object. Color information can be particularly helpful to support detection of faces in image sequences because color as a cue is computationally inexpensive [14], robust towards changes in orientation and scaling of an object being in movement. The discriminative ability of color is especially worth to emphasize if a considered object is partially occluded because edge-based methods can be ineffective.

A color histogram including spatial information can be extracted on the basis of a 2-dimensional kernel centered on the target [4]. The kernel weights the color of the pixel according to its distance from the kernel center. In order to assign smaller weights to the color of pixels that are further away from the center of the kernel a nonnegative and monotonic decreasing function $k : [0, \infty) \rightarrow R$ can be utilized [4]. The probability of particular histogram bin u at location $\mathbf{x} = \{x, y\}$ is determined by the following formula:

$$d_{\mathbf{x}}^{(u)} = C_r \sum_{j=1}^{L} k \left(\left\| \frac{\mathbf{x} - \mathbf{x}_j}{r} \right\|^2 \right) \delta \left[h(\mathbf{x}_j) - u \right] \tag{1}$$

where $\mathbf{x}_j$ are pixel locations, L is the number of pixels in the considered kernel, constant r is the radius of the kernel, δ is the Kronecker delta function, and the function $h : R^2 \rightarrow \{1...K\}$ associates the bin number. The normalization factor C_r ensures that $\sum_{u=1}^{K} d_{\mathbf{x}}^{(u)} = 1$. This normalization factor can be precalculated [4] for the utilized kernel and assumed values of r. The 2-dimensional kernels have been prepared off-line and then stored in lookup tables for the future use. The color representation of the target has been extracted by quantizing the ellipse's interior colors into K bins and extracting the weighted histogram. To make the histogram representation of the tracked head less sensitive to lighting conditions the V component obtained the 4-bin representation while the remaining components of the HSV color space have been represented by 8 bins [9].

To compare the histogram Q representing the tracked face to a histogram I obtained from the particle configuration we utilized the metric $\sqrt{1 - \rho(I, Q)}$, which is derived from Bhattacharyya coefficient $\rho(I, Q) = \sum_{u=1}^{K} \sqrt{I^{(u)} Q^{(u)}}$. The work [4] demonstrated that the utilized metric is invariant to the scale of the target and therefore is superior to other measures such as histogram intersection [14] or Kullback divergence. Using the Bhattacharyya coefficient we defined the color observation model as $p(\mathbf{z}^C \mid \mathbf{x}) = (\sqrt{2\pi}\sigma)^{-1} e^{-\frac{1-\rho}{2\sigma^2}}$. Thanks to such weighting we favor head candidates whose color distributions are similar to the distribution of the tracked head. The second ingredient of the observation model reflecting the edge strength along the elliptical head boundary has been weighted in a similar manner $p(\mathbf{z}^G \mid \mathbf{x}) = (\sqrt{2\pi}\sigma)^{-1} e^{-\frac{1-\phi_g}{2\sigma^2}}$, where ϕ_g denotes the normalized gradient along the ellipse's boundary.

3.3 Probabilistic Integration of Cues

The aim of probabilistic multi-cue integration is to enhance visual cues that are more reliable in the current context and to suppress less reliable cues. The correlation between location, edge and color of an object even if exist is rather weak. Assuming that the measurements are conditionally independent given the state we obtain the equation $p(\mathbf{z}_t \mid \mathbf{x}_t) = p(\mathbf{z}_t^G \mid \mathbf{x}_t) \cdot p(\mathbf{z}_t^C \mid \mathbf{x}_t)$ which allows us to accomplish the probabilistic integration of cues. To achieve this we calculate at each time t the L2 norm based distances $D_t^{(j)}$, between the individual cue's centroids and the centroid obtained by integrating the likelihood from utilized cues [15]. The reliability factors of the cues $\alpha_t^{(j)}$ are then calculated on the basis of the following leaking integrator $\xi\dot{\alpha}_t^{(j)} = \eta_t^{(j)} - \alpha_t^{(j)}$, where ξ denotes a factor that determines the adaptation rate and $\eta_t^{(i)} = 0.5*(\tanh(-aD_t^{(j)})+b)$. In the experiments we set $a = 0.3$ and $b = 3$. Using the reliability factors the observation likelihood has been determined as follows:

$$p(\mathbf{z}_t \,|\, \mathbf{x}_t) = [p(\mathbf{z}_t^G \,|\, \mathbf{x}_t)]^{\alpha_t^{(1)}} \cdot [p(\mathbf{z}_t^C \,|\, \mathbf{x}_t)]^{\alpha_t^{(2)}} \quad 0 \leq \alpha_t^{(j)} \leq 1 \tag{2}$$

3.4 Adaptation of the Color Model

The largest variations in object appearance occur when the object is moving. Varying illumination conditions can influence the distribution of colors in an image sequence. If the illumination is static but non-uniform, movement of the object can cause the captured color to change alike. Therefore, a tracker that uses a static color model is certain to fail in unconstrained imaging conditions. To deal with varying illumination conditions the histogram representing the tracked head has been updated over time. This makes possible to track not only a face profile which has been shot during initialization of the tracker but in addition different profiles of the face as well as the head can be tracked. Using only pixels from the ellipse's interior, a new color histogram is computed and combined with the previous model in the following manner $Q_t^{(u)} = (1-\gamma)Q_{t-1}^{(u)} + \gamma I_t^{(u)}$, where γ is an accommodation rate, I_t denotes the histogram of the interior of the ellipse calculated from the estimated state, $Q_{t-1}^{(u)}$ is the histogram of the target from the previous frame, whereas $u = 1...K$.

3.5 Depth Cue

The length of the minor axis of the considered ellipse has been determined on the basis of depth information. The length has been maintained by performing a local search to maximize the goodness of the observation match. Taking into account the length of the minor axis resulting from the depth information we considered smaller and larger projection scale of the ellipse about two pixels. Thanks to verification of face distance to the camera and face region size heuristics it is possible to discard many false positives that are generated through the face detection module.

3.6 Supporting the tracking through face detection

The face detection algorithm can be utilized to form a proposal distribution for the particle filter in order to direct the particles towards most probable locations of the objects of interest. The employed face finder is based on object detection algorithm described in work [17]. Using a training set of positive and negative images the Real AdaBoost [15] has been utilized both to select features and to train a robust classifier. A 18 layer cascaded classifier has been trained on images of size 20x20 pixels to detect frontal faces in gray images. The detector has been trained on 1500 frontal faces. All training images were manually aligned by eyes position. The aim of the detection algorithm is to find all faces and then to select the highest scoring candidate that is situated nearby a predicted location of the face. Next, taking the location and the size of the window containing the face we construct a Gaussian distribution $p(\mathbf{x}_t \,|\, \mathbf{x}_{t-1}, \mathbf{z}_t)$ in order to reflect the face position in the proposal distribution. The formula describing the proposal distribution has the following form:

$$q(\mathbf{x}_t \mid \mathbf{x}_{t-1}, \mathbf{z}_t) = \beta p(\mathbf{x}_t \mid \mathbf{x}_{t-1}, \mathbf{z}_t) + (1-\beta)p(\mathbf{x}_t \mid \mathbf{x}_{t-1}) \quad (3)$$

The parameter β is dynamically set to zero if no face has been found. In such a situation the particle filter takes the form of the CONDENSATION [7].

4 Experiments

4.1 The system

The experiments described in this section were carried out with a mobile robot Pioneer 2DX [11] equipped with commercial binocular MegaPixel Stereo Head. The dense stereo maps are extracted in that system thanks to small area correspondences between image pairs [8] and therefore poor results in regions of little texture are often provided. The depth map covering a face region is usually dense because a human face is rich in details and texture, see a depth subimage in Fig. 1. a). Thanks to such a property the stereovision provides a separate source of information and considerably supports the process of approximating the tracked head with an ellipse of proper size.

A typical laptop computer equipped with 2.5 GHz Pentium IV is utilized to run the software prepared in C/C++ and operating at images of size 320x240. During tracking, the control module keeps the user face within the camera field of view by coordinating the rotation of the robot with the location of the tracked face in the image plane. The linear velocity has been dependent on person's distance to the camera. In experiments consisting in person following a distance 1.3 m has been assumed as the reference value that the linear velocity controller should maintain. To eliminate needless robot rotations as well as forward and backward movements we have applied a simple logic providing necessary insensitivity zone. The PD controllers have been implemented in the Saphira-interpreted Colbert language [11].

4.2 Experiments on Real-World Situations

To test the prepared software we performed various experiments with the moving camera. After detection of possible faces, see Fig. 1. a), the system can identify known faces among the detected ones. In tracking scenarios the user moved about a laboratory, walked back and forth as well as around the mobile robot. The aim of such scenarios was to evaluate the quality of ellipse scaling in response of varying distance between the camera and the user, see Fig. 1. e),f). Our experimental findings show that thanks to stereovision the ellipse of proper size approximates the tracked head and in consequence, sudden changes of the minor axis length as well as ellipse's jumps are eliminated. The greatest variability is in horizontal motion, followed by vertical motion. Ellipse's size variability is more constrained and tends towards the size from the previous time step. By dealing with multiple cues the presented approach can track a head reliably in cases of temporal occlusions, see Fig. 1. b),c), and varying illumination conditions, see Fig. 1. e),f), even when person moves in front of skin-like colors of window-panes, see also Fig. 1. e). During a typical experiment with person following the user typically rounds the laboratory in 70 s and goes a distance about 35 m.

The tracker runs with 400 particles at frame rates of 12-13 Hz. The face detector can localize faces in images of size 320x240 in about 0.1 s. The full cascade consist of 820 weak classifiers. The first five stages of the cascade consists of 80 classifiers and the first ten stages is comprised of 250 classifiers. The recognition of single face takes about 0.01 s. These times allow the system to process about 6 frames per second when the information about detected faces is used to generate the proposal distribution for the particle filter.

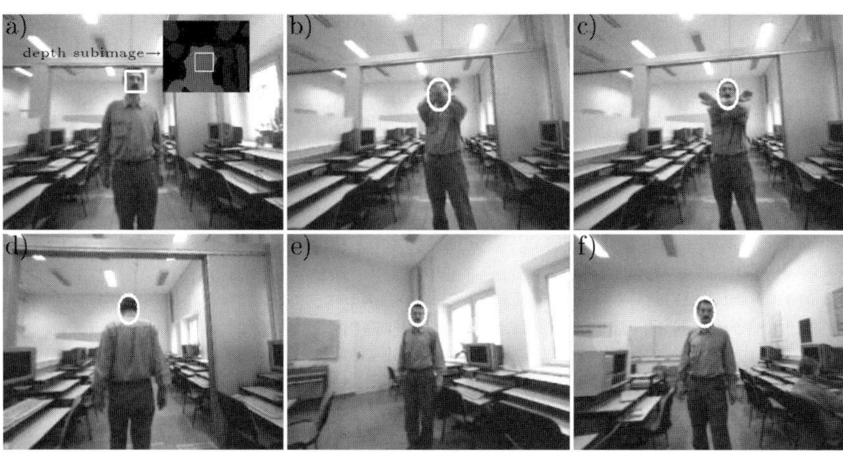

Fig. 1. Face detection and tracking, frames #9, #110, #111, #168, #317, #970

5 Conclusions

We have presented a vision module that robustly tracks and detects a human face. By employing shape, color, stereovision as well as elliptical shape features the proposed method can track a head in case of dynamic background. These features make it general enough to be useful for many human-machine as well as surveillance applications. Experimental results on tracking faces in long indoor video sequences demonstrate the robustness of the tracking system.

References

1. Birchfield S. (1998) Elliptical head tracking using intensity gradients and color histograms, The IEEE Conf. on Comp. Vision and Patt. Rec., 232–237
2. Bradski G. R. (1998) Computer vision face tracking as a component of a percepŋ tual user interface, In Workshop on Applications of Computer Vision, 214–219
3. Chen Y., Rui Y., Huang T. (2002) Mode–based multi–hypothesis head tracking using parametric contours, In Proc. IEEE Int. Conf. on Aut. Face Rec., 112–117
4. Comaniciu D., Ramesh V., Meer P. (2000) Real–time tracking of non–rigid objects using Mean Shift, The IEEE Conf. on Comp. Vision and Patt. Rec., 142–149
5. Darrell T., Gordon G., Harville M., Woodfill J. (1998) Integrated person trackŋ ing using stereo, color, and pattern detection, Proc. of the Conf. on Comp. Vision and Patt. Rec., 601–609
6. Doucet A., Godsill S., Andrieu Ch. (2000) On sequential Monte Carlo sampling methods for bayesian filtering, Statistics and Computing, 10:197–208
7. Isard M., Blake A. (1998) CONDENSATION ŋ conditional density propagation for visual tracking, Int. J. of Computer Vision, 29:5–28
8. Konolige K. (1997) Small Vision System: Hardware and implementation, Proc. of Int. Symp. on Robotics Research, 111–116
9. Kwolek B. (2004) Stereovision–based head tracking using color and ellipse fitŋ ting in a particle filter, 8th European Conf. on Computer Vision, 192–204
10. Perez P., Hue C., Vermaak J., Gangnet M. (2002) Color–based probabilistic tracking, European Conf. on Computer Vision, 661–675
11. Pioneer 2 mobile robots (2001) ActivMedia Robotics
12. Rowley H., Baluja S., Kanade T. (1996) Neural network–based face detection, Proc. of IEEE Conf. on Comp. Vision and Patt. Rec., 203–207
13. Schapire R., Singer Y. (1998) Improved boosting algorithms using confidenceŋ rated predictions, Proc. 11th Ann. Conf. Computational Learning Theory, 80ŋ91
14. Swain M. J., Ballard D. H. (1991) Color indexing, Int. J. of Computer Vision, 7:11–32.
15. Triesch J., Malsburg Ch. (2001) Democratic integration: Self–organized inteŋ gration of adaptive cues, Neural Computation, 13:2049–2074
16. Turk M. A., Pentland A. P. (1991) Face recognition using eigenfaces, In Proc. of Conf. on Comp. Vision and Patt. Rec., 586–591
17. Viola P., Jones M. (2001) Rapid object detection using a boosted cascade of simple features, The IEEE Conf. on Comp. Vision and Patt. Rec., 511–518
18. Yang M. H., Kriegman D., Ahuja N. (2002) Detecting faces in images: A survey, IEEE Trans. on Pattern Analysis and Machine Intelligence, 24:34–58

Wavelet Packets Features Extraction and Selection for Discriminating Plucked Sounds of Violins

Ewa Lukasik

Poznan University of Technology, Institute of Computing Science
ul. Piotrowo 3a, 60-965 Poznan, Poland
elukasik@cs.put.poznan.pl

Summary. Plucked sounds of musical instruments from chordophones group are examples of non-stationary sounds having both tonal and transient Plucked soun-character. The experiments presented in this paper had Plucked sounto answer to the question if the wavelet packet transform based strategy of features extraction and selection that proved useful in many other classification tasks will be also useful for distinguishing differences of sounds produced by master quality violins played pizzicato.

1 Introduction

In order to use pattern recognition methods effectively, the step of pre-processing is needed, in which data is processed before it is presented to any learning, discovering or visualizing algorithm. Many learning methods have been proposed for selecting, extracting, or constructing features. They usually improve the classification results, but still many algorithms perform poorly in domains with large number of irrelevant and/or redundant features. The need for additional methods to overcome the difficulties is constant. One of the strategies is using the domain-related cues for finding features being the most consistent with a specific application. This is also the case of various fields of signal processing, including musical signal processing.

The problems of classification in the domain of music have been recently broadly investigated in the field of the music information retrieval (MIR), e.g. in [8]. The range of topics includes problems of recognizing music using audio or semantic description, music representation and indexing, estimating similarity of music using perceptual and musicology criteria, auditory scene analysis, automatic musical phrases and genres recognition as well as classification of musical instrument sounds.

The effort of research teams in the field of automatic recognition of musical instruments brought to life the standardized (within MPEG-7 [6]) set of low-level descriptors of musical instruments timbre. Timbre constitutes the great challenge to measurement and specification due to the inherent multidimensional nature, perceived by humans by means of the interaction of static and dynamic properties of sound grouped into a complex set of auditory attributes. Sets of various parameters appeared successful in distinguishing isolated notes of different musical instruments, however relatively little work has been reported focusing on the perceptive mechanisms leading to the discrimination of voices of musical instruments from the same family.

The work presented in this paper is focused on the distinction of the elements of timbre of the set of master quality violins presented during the X International Violinmakers Competition in Poznan, Poland, in 2001. It is one of the series of reports, e.g. [12], [13] on searching for the method for representing and distinguishing the variety of sonorities produced by violins. This special interest in the analysis of violin voice is due to the richness of its acoustical phenomena and the great distinction the instrument acquired in the world of music. It also adds to the variety of techniques used including modal analysis, finite element methods, scanning electron microscopy, acoustic spectroscopy, holography and physical methods of chemical analysis [14].

The analysis is concentrated on plucked sounds (pizzicato). Being tonal, they also have character of transients.

Transient signals of short duration are well suited for analysis using wavelets. Wavelet analysis became very popular signal processing tool in many scientific fields in 90-ties, when, combined with multiresolution analysis methods, produced the fast wavelet decomposition algorithm. This algorithm allows for an efficient computation of wavelet coefficients using a cascade of digital filters. It consists of iterative decomposition of a signal into the coarse and detail approximation. The algorithm develops hierarchically in a tree-like mode. If the decomposition is full, i.e. if it concerns both low- and high frequency bands, the decomposition is called wavelet packets. Mother wavelets with the compact support are well suited for the pre-processing of short signals of transient character. The tree-like decomposition scheme also allows for the efficient methods of selection of significant coefficients (features) on which standard processing methods may be applied. Such methods, as e.g. best basis selection, gave very good results when applied for myoelectric signals [3], or unvoiced speech signals of transient character (stop consonants), that, being unvoiced, have similar to the plucked sounds time envelope [10].

Dealing with violin voices we apply unsupervised methods of machine learning to try to find clusters of similarly sounding instruments. Certainly most of information about instrument sounding abilities would be from the piece of music played by the good violinist, however analysis of individual sounds is more straightforward and enables concentration on fewer aspects of sounds. Our goal will be to try to find similarly responding instruments, so that the preferences of the jury members performing ranking of violins could be mirrored by a set of reliable features. Multidimensional scaling (MDS) [2] proved to be a useful method for performing the visualization of the original placement of multidimensional objects in reduced, 2-dimensional space, with the mutual distances between objects retained [12] and will be applied also in this project.

The paper is structured as follows: in the second section collection of plucked sounds used for analysis is characterized, in Section 3 methods of wavelet packet based features reduction are introduced, Section 4 describes the results of multidimensional scaling of plucked sounds represented by wavelet packets based feature set and Section 5 concludes the paper.

2 Characteristics of violin plucked sounds

Dealing with the instruments of the same type gives new constraints to the problem of musical instruments recognition. The differences in their timbre may be minute, hardly distinguished even for experienced human listeners.

The set of instruments analyzed in the paper comes from AMATI database [11] that contains digitized recordings of violins presented at the 10th Henryk Wieniawski International Violinmakers Competition held in Poznan, Poland, in 2001. A subset of 26 instruments ranked as the best medium and the worst by the musicians' jury has been chosen for experiments. However we have to remember, that competing violins generally are of a very good quality. Also, assessing the timbre of two instruments as equal does not mean, that their timbre is similar – there may be different perceptual features weighed up in the same way. What is more, contemporary violinmakers usually follow the similar Stradivarius model for instruments construction, use similar materials for their assembly causing the violins sound alike. All that makes the goal of finding distinct clusters of similarly sounding violins difficult.

Transient response to a pluck over the frequency range of radiated sound from a typical note on a violin shows clear spectral peaks up to at least 5-6 kHz. The evolution of individual partials over time shows a big irregularity revealing the decaying harmonics of the string (Fig. 1).

The decay is one of the perceptual features of plucked string tones, the pluck and the body response being the others. The change in the beginning of the tone, the attack, is very audible. Attacks of plucked violin sound may be considered as transients. From signal processing point of view a transient is localized over a very short time region of signal. A transient can be considered either as a deterministic signal or a stochastic and highly non-stationary signal. Recently Molla and Torresani [15] showed that using wavelets it is possible to determine how much of transientness the signal manifest. Starting point is the assumption that transient signals admit a sparse expansion in a wavelet basis and that tonals admit a sparse expansion in local cosine basis. Without entering into details let us assume the hybrid character of plucks of violins and try to apply the wavelet packet mechanism for features extraction. Sinusoidal analyses are very powerful and flexible but are also known to miss

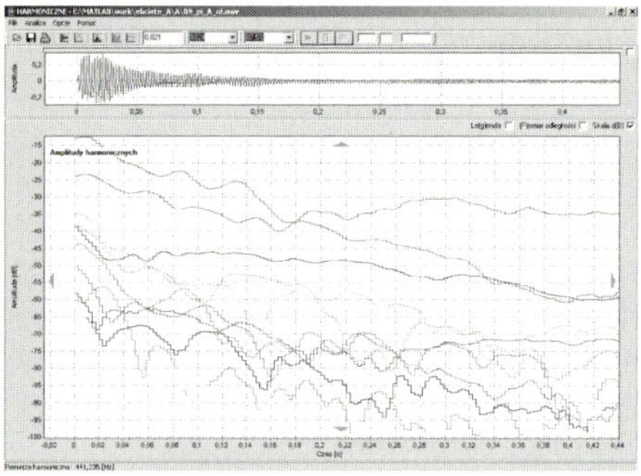

Fig. 1. Plucked sounds of an exemplary violin (in upper window) Plucked sounand evolution of the amplitude of the first ten partials Plucked sounin time (440ms)

certain features of the signals [5]. However, we will also combine harmonic analysis with wavelet packets analysis.

3 Features extraction and selection based on wavelet packets

Wavelet Packet Transform (WPT) [1] can be viewed as a generalized version of the wavelet transform providing level by level transformation of a signal from the time domain into the frequency domain. It is calculated using a recursion of filter-decimation operations leading to the decrease in time resolution and increase in frequency resolution. The frequency bins, unlike in wavelet transform, are of equal width, since the WPT divides not only the low, but also the high frequency subband. A full wavelet packet decomposition constitutes the initial feature set characterizing the signal. The full decomposition tree at level J is given by 2J possible decompositions with two orthogonal filters. Each node corresponds to the projection on a different function. Each basis function is orthogonal to the other bases.

The task of feature selection is to obtain the features that are essential for class separation. Three different approaches may be applied: Best Basis algorithm, Local Discriminant Base search algorithm and Singular Value Decomposition of wavelet packets coefficients matrix.

Coifman-Wickerhauser algorithm of Best Basis search [1]
Although the original method of best basis selection was motivated by signal compression application, it may be also used for feature selection. The method prunes the WPT binary tree by eliminating branches according to selected entropy cost function. The (non-normalized) Shannon entropy function is usually used

$$H(\mathbf{x}) = -\sum_i x_i^2 \log(x_i^2) \tag{1}$$

where $\mathbf{x}$ is the signal and x_i the coefficients of $\mathbf{x}$ in the orthonormal basis. The cost function is additive, so having calculated entropy function in each node of the decomposition tree the prunning of the tree is performed by searching the overal minimum entropy. If the sum of entropy of neighboring nodes is smaller than the entropy of the parent node, than these two nodes are removed. The operation is iteratively repeated on each level of decomposition. The resulting best basis tree is often represented with the branches length proportional to the entropy of the node.

Local Discriminant Basis search algorithm [17]
The basic idea of Local Basis discrimination can be described as best basis search algorithm over the calculated discriminant measure D between classes. It represents the measure of class separability. The input parameters to D are the time-frequency energy maps of each class calculated by accumulating the squares of the WPT coefficients for each entry in the binary packet tree and normalized by the total energy of the signal belonging to given class. Then the distance measure (cost function) has to be introduced in our case being or relative entropy (S), or Euclidean Distance (E). In our case it is relative entropy:

$$D_{s,q} = s_j \log(s_j/q_j) \tag{2}$$

or Euclidean distance:

$$D_{s,q} = |s_j - q_j| \tag{3}$$

where s_j and q_j are the features characterizing elements of two classes, $j = 1, \ldots, n$.

To compute the discrepancy between the distributions of the m classes of objects under consideration, one must sum up $\binom{m}{2}$ pairwise combinations of D.

Singular Value Decomposition
The method is based on the Singular Value Decomposition (SVD), of the (non normalized) Shannon entropy (or energy) matrices calculated from the WPT coefficients for m classes of objects, one for each class (columns represent entropies of signals within the class). Then for each class the singular vectors are calculated. Assuming a single (first) dominant singular value only one representative entropy vector is considered for each class: $\mathbf{u}_{c,1}$ Selection of features is performed on the basis of the biggest values of a separability vector $\Delta\mathbf{p}$ calculated as a sum of Euclidean distances between feature vectors of each class. Selection of features may be done on the threshold basis (e.g. choosing the coefficients corresponding to elements that fall above some percentage of the largest one).

4 Experiments and the results

First 11,6 ms of the attack has been analyzed for all 25 instruments plucked sounds used in experiment. The wavelet packet decomposition has been performed with Beylkin mother wavelet. Shift invariance has been assured by the thorough mark out of the starting point of the signal. J=5 levels of decomposition have been used, giving 63 frequency bins. Fig. 2 (left) shows, that the resulting decomposition is rather sparse, and not very different for various instruments. Since possible classes of signal had only to be discovered during the analysis, the natural choice for the method of features selection would be search for the best basis. Exemplary best basis trees obtained using [19] are shown in Fig. 2 (right).

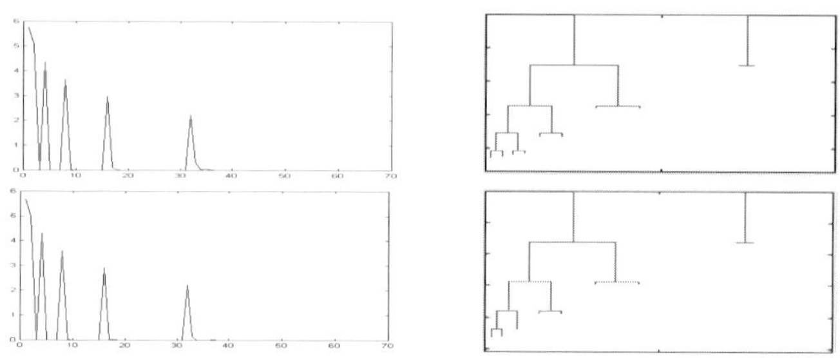

Fig. 2. Wavelet packets entropy (left)and best basis tree (right) Plucked sounof A-string plucked sounds of two exemplary violins

Fig. 3 displays the location of violin plucked sounds described by coefficients originated from the best basis decomposition in the reduced 2-D feature space. The distances between objects remain similar to those in the original multidimensional space thanks to the application of Multidimensional Scaling algorithm (MDS) performed using [4]. MDS minimizes so called stress function measuring how well the new configuration matches given dissimilarities; the lower is its value, the better is the match [16]. Starting point for the MDS algorithm was PCA performed on multidimensional features. Numbers in the figure denote identifiers of the instruments. The results are shown for open A-string. No distinct clusters can be observed, however most of the instruments ranked by the jury of musicians as the best are located close to each other. Second part of the Fig. 3 is devoted to the MDS result of reduced feature set equivalent to the one obtained by the SVD. The biggest values of entropy at the lowest level of wavelet packets decomposition have been taken and transformed by PCA to perform further the MDS procedure. Again some grouping of the best instruments may be observed.

The last series of experiments concerned comparison of wavelet packets and spectral harmonic features performance (Tristimulus 1, Tristimulus 2, Tristimulus 3 as defined in [7]). These tonal parameters roughly divide partials into

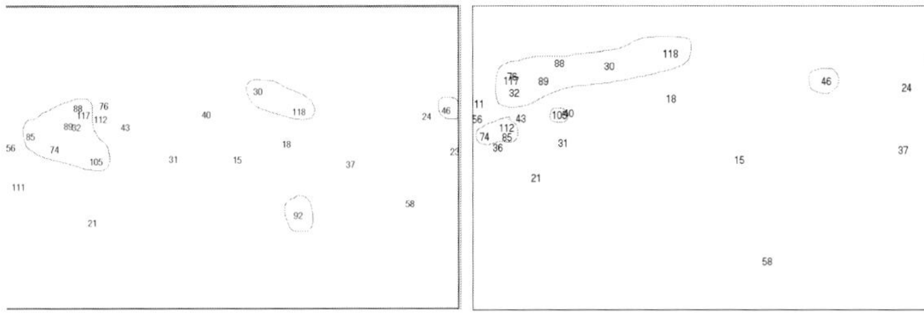

Fig. 3. MDS visualization of plucked open A-string sounds Plucked souncharacteristics. On the left - best basis selecion, on the Plucked sounright– biggest coefficients from the lowest level of Plucked soundecomposition. Objects encircled have got the highest Plucked sounmarks during violinmakers competition.

low, medium and high frequency components and may be very vaguely treated as harmonic spectral equivalents of wavelets. Fig. 4 presents results of the separate MDS analysis of spectral harmonic features (tristimuli and odd/even contents of sound) as well as harmonic features combined with wavelet packets best basis. Still, however no distinct borders between excellent and worse instruments may be observed.

Slightly different grouping of pizzicato sounds by different methods may be explained by the fact, that each of them emphasizes somewhat different characteristics of the signal. However it is interesting to note, that some distinct instruments (both – of good quality, as e.g. 30 and 118 or of worse quality, as e.g. 56 and 21) are marked out by each method similarly or as single outliers, or as neighbours. Merging wavelet and spectral harmonic techniques seems to improve the separability of possible object classes.

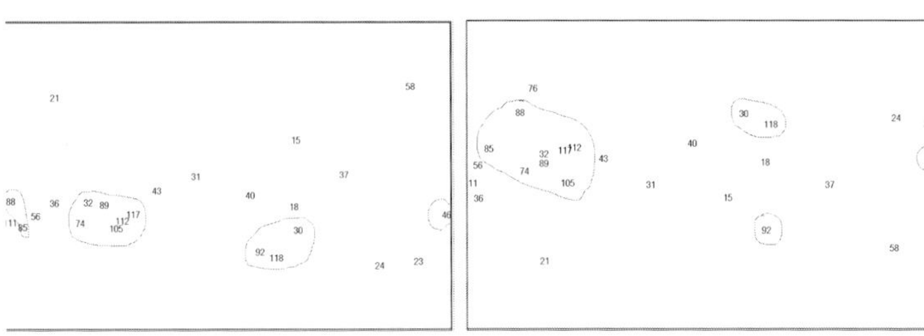

Fig. 4. Result of the MDS procedure – the map of Plucked sounviolins for a reduced set of harmonic spectral features Plucked sounand (to the right), these tonal features combined with Plucked sounbest basis set.

5 Conclusions

Wavelet Packet Transform is one of the alternate means of generating a set of features having the property of high level of *information packing* [18]. In the paper the method of feature selection for wavelet packets representation has been discussed for the distinction of specific violin sounds played pizzicato – examples of acoustic signals of non-stationary, transient, but also tonal character. The goal of this work was to give the contribution to the answer to the question of the usefulness of those sounds to the recognition of individual instruments from the same family and also the usefulness of wavelet packets to characterize those sounds. The answers to both questions are partly positive, as probably wavelet packets coefficients should be combined with other parameters and plucked sounds should be accompanied with other sounds in the analysis. Although machine learning methods of features selection are developing, still however acoustic cues seem to be necessary at early stages of sound analysis.

Acknowledgements This work was supported by the Scientific Research Committee (KBN) grant 4 T11 F 002 22.

References

1. Coifman R, Wickerhauser M (1992) Entropy-based algorithms for best basis selection. IEEE Trans. Information Theory, vol.38, No 2, pp. 713–718
2. Cox T.F, Cox M.A (1994) Multidimensional Scaling. Chapman and Hall, London
3. Englehart K (1998) Signal Representation for Classification of the Transient Myoelectric Signal. Ph.D. Thesis, University of New Brunswick, Fredericton, New Brunswick
4. GhostMiner Developer. http://www.fqspl.com.pl/ghostminer/
5. Gouyon F (1999) Detection and modeling of transient regions in musical signals. PhD Thesis. Stanford University
6. Coding of Moving Pictures and Audio: MPEG-7 overview. ISO/IEC JTC1/SC29/WG11 International Organization for Standardization http://www.chiariglione.org/mpeg/standards/mpeg-7/mpeg-7.htm
7. Kostek B (1999) Soft Computing in acoustics. Physica Verlag
8. Kostek B (2001) Musical Instrument Classification and Duet Analysis Employing Music Information Retrieval Techniques. Proc. IEEE Vol. 92, No 4, 712-729
9. Learned R.E, Willsky A.S (1995) A Wavelet Packet Approach to Transient Signal Classification. Applied and Computational Harmonic Analysis. Academic Press, Vol. 2, No 3, 265-278
10. Lukasik E (2000) Classification of Voiceless Plosives Using Wavelet Packet Based Approaches. In: Gabbouj M, Kuosmanen P (eds.) Signal Processing X, Theories and Applications, Proc. EUSIPCO 2000, 1933-1936
11. Lukasik E (2003) AMATI – Multimedia Database of Violin Sounds. In Roberto Bressin (ed.) Proc. Stockholm Music Acoustics Conference SMAC 03, 79-82

12. Lukasik E, Susmaga R (2003) Unsupervised machine learning methods in timbral violin charateristics visualization. In Roberto Bressin (ed.) Proc. Stockholm Music Acoustics Conference SMAC 03, 83-86
13. Lukasik E (2004) Searching for Similarities of Contemporary Concert Violin Voices Using Maximum Sound Level Curves as Descriptors. Proc. 11th International Workshop on Systems, Signals and Image Processing, 223-226
14. McLennan J, The Art, History and Science of Violin Making, http://www.phys.unsw.edu.au/music/publications/mclennan/arthistoryscience.pdf
15. Molla S, Torresani B (2004) Determining local transientness of audio signals. IEEE Signal Processing Letters, Vol. 11, No 7, 625–628
16. Naud A (2001), Neural and statistical methods for the visualization of multidimensional data. PhD Thesis, UMK Torun, http://www.phys.uni.torun.pl/publications/kmk/01phd-an.pdf
17. Saito N (1994) Local Feature Extraction and Its Applications Using a Library of Bases. PhD thesis, Yale University
18. Theodoridis S., Koutroumbas K (1999) Pattern Recognition. Academic Press, London
19. Wavelab 802, http://www-stat.stanford.edu/ wavelab/

Pattern Recognition and Fault Detection in MEMS

Karim Mohammadi and Reza Asgary

Electrical Engineering Dep. Iran University of Science and Technology
Mohammadi@iust.ac.ir, R_Asgary@iust.ac.ir

Summary. Micro Electro Mechanical Systems will soon usher in a new technological renaissance. Just as ICs brought the pocket calculator, PC, and video games, MEMS will provide a new set of products and markets. Learn about the state of the art, from inertial sensors to microfluidic devices [1]. Over the last few years, considerable effort has gone into the study of the failure mechanisms and reliability of MEMS. Although still very incomplete, our knowledge of the reliability issues relevant to MEMS is growing. One of the major problems in MEMS production is fault detection. After fault diagnosis, hardware or software methods can be used to overcome it. Most of MEMS have nonlinear and complex models. So it is difficult or impossible to detect the faults by traditional methods, which are model-based. In this paper different Neural Networks are used to classify and recognize faults. Different faults are recognized whilst considered as different patterns. We use different Neural Networks to classify different faults and fault free data. Two RF MEMS, which are RF Low pass filter and RF Inter digital capacitor are simulated by EM3DS, a MEMS software simulator. At last the results are compared.

1 Introduction

Diversity of application fields and properties of new materials generate new failure mechanisms in MEMS. Now if we take into account the lessons from the past in microelectronics, we note that failure analysis played a major rule not only in development time reduction but also in qualification and reliability evaluation. Most of the researches, which have been done in MEMS reliability, are about new material properties and fabrication technologies [2]. Only a few fault detection methods have been introduced for fault detection in MEMS. Some of these methods can be used only for special MEMS [3]. Additionally, most of them need a precise model of system [4, 5]. There are both electrical and nonelectrical parts in MEMS, so in most of them finding an accurate model is difficult or impossible. Using neural networks, there is no need to find accurate model. Neural networks are being learned by rich learning data set. Faulty and fault free patterns will be separated after finishing learning step. Faulty classes have different distances from each other and from fault free class.

2 Fault recognition methods

The work on fault diagnosis in the AI community initially focused on the expert system or knowledge-based approach where heuristics are applied to explicitly associate symptoms with fault hypothesis. The short coming of a pure expert system approach led to the development of model-based approaches based on qualitative models in the form of qualitative differential equations, signed digraphs, qualitative functional and structural models. Other approaches assume the availability of process history based data which are then used to develop neural network approaches [6-8]. Generally speaking, there are four types of neural networks:
-Back propagation Neural Network (BPNN).
-Probabilistic Neural Network (PNN).
-Self-Organizing Mapping (SOM).
-Radial Basis Function Neural Network (RBF).
There are some drawbacks to BPNN and SOM. The BPNN requires a large number of training patterns to let network learn the underlying mapping function. The second problem is that the accuracy of the training patterns should not be a measure of whether a model is good or not. BPNN has a low reliability with novel data. SOM is known as a topological mapping algorithm, in which patterns with similar characteristics cluster together automatically. Output nodes will thus be ordered by competitive learning. The learning rate and neighbor size of SOM have to be optimally selected by experience, and a SOM net needs a large time to converge. Two RF MEMS have been simulated by EM3DS software. Different kinds of faults consisting analog, digital and parametric faults have been considered. We used simulation data to make two pattern sets; the learning data set and testing data set. These data sets are used to learn and test BPNN, Gaussian RBF, multiquadric RBF, Probabilistic neural network and Heteroscedastic Probabilistic neural network. Two different training algorithms used in RBF network learning. First derivative based method is used and in other algorithm Extended Kalman filter is used to estimate optimum values of centers and variances of Gaussian Radial Basis Functions.

3 Radial Basis Functions

There have been a number of popular choices for the $g(.)$ function at the hidden layer of RBFs. The most common choice is a Gaussian function.Hidden layer should have the following properties:
1-The response at a hidden neuron is always positive.

2-The response at a hidden neuron becomes stronger as the input approaches the prototype.
3- The response at a hidden neuron becomes more sensitive to the input as the input approaches the prototype.

3.1 Derivative Based Optimization

The response of an RBF, with the hidden layer function, $g(.)$ can be written as follows:

$$\hat{y} = \begin{bmatrix} \omega_{10} & \omega_{11} & \cdots & \omega_{1c} \\ \omega_{20} & \omega_{21} & \cdots & \omega_{2c} \\ \vdots & \vdots & \vdots & \\ \omega_{n0} & \omega_{n1} & \cdots & \omega_{nc} \end{bmatrix} \begin{bmatrix} 1 \\ g(\|x - \nu_1\|^2) \\ \vdots \\ g(\|x - \nu_c\|^2) \end{bmatrix} \tag{1}$$

If we are given a training set of M desired input-output response x_i, y_i, $(i = 1, 2, , M)$, then we can augment M equations of the form of Eq (1). and make matrix $\hat{Y} = WH$.

Now, if we want to use gradient descent to minimize the training error, we can define the error function $E = \frac{1}{2}\|Y - \hat{Y}\|^2$. It has been shown that gradient of error can be computed as

$$\frac{\partial E}{\partial W_i} = \sum_{k=1}^{M} (\hat{y}_{ik} - y_{ik})\, h_k (i = 1, ..., n)$$

$$\frac{\partial E}{\partial \nu_j} = \sum_{k=1}^{M} 2g\prime \left(\|x_k - \nu_j\|^2\right)(x_k - v_j) \sum_{i=1}^{n} (y_{ik} - \hat{y}_{ik})\, w_{ij} (j = 1, ..., c) \tag{2}$$

3.2 Using Extended Kalman Filter for Optimization

Alternatively we can use kalman filtering to minimize the training error. Derivations of the extended kalman filter are widely available in the literature [9]. Consider a nonlinear finite dimensional discrete time system of the form

$$\theta_{k+1} = f(\theta_k) + \omega_k, y_k = h(\theta_k) + \nu_k \tag{3}$$

Where the vectorθ_k is the state of the system at time k,ω_k is the process noise, y_k is the observation vector, ν_k is the observation noise, and $f(.)$ and $h(.)$ are nonlinear vector functions of the state.
In general we can view the optimization of the weight matrix W and the prototypes v_j as a weighted least-squares minimization problem, where the error vector is the difference between the RBF outputs and the target values for those outputs. Consider the RBF network with m inputs, c prototypes, and n outputs. We use y to denote the target vector for the RBF outputs, and $h(\theta_k)$ to denote the actual outputs at the kth iteration of the optimization algorithm.

N is the dimension of the RBF output and M is the number of training samples. The state of the nonlinear system can then be represented as $\theta = \left[w_1^T \cdots w_n^T \; \nu_1^T \cdots \nu_c^T \right]^T$ The vector θ thus consist of all $(n(c+1)+mc)$ of the RBF parameter arranged in a linear array. The nonlinear system model to which the Kalman filter can be applied is $\theta_{k+1} = \theta_k, Y_k = h(\theta_k)$, where $h(\theta_k)$ is the RBF network's nonlinear mapping between its parameters and its output. In order to execute a stable kalman filter algorithm, we need to add some artificial process noise and measurement noise to the system model. So we rewrite Eq (3) as $\theta_{k+1} = \theta_k + \omega_k, Y_k = h(\theta_k) + \nu_k$, where w_k and v_k are artificially added noise processes. It can be shown that the partial derivative of the RBF output with respect to the RBF network parameters is given by $H_k = \begin{bmatrix} H_w \\ H_v \end{bmatrix}$ where H_w is an $n(c+1)*nM$ matrix , H_v is an $mc*nM$ matrix, and H_k is an $[n(c+1)+mc]*nM$ matrix. with H_k we can use the extended kalman filter for determination the weight matrix W and the prototypes v_j.

4 Robust Heteroscedastic PNN

A PNN classifies data by estimating the class conditional probability density functions, because the parameter of a PNN cannot be determined analytically. To do this it requires a training phase, followed by a validation phase, before it can be used in a testing phase. A PNN consists of a set of Gaussian kernel functions. The original PNN uses all the training patterns as the centers of the Gaussian kernel functions and assumes a common variance or covariance, which is named homoscedastic PNN. To avoid using a validation data set and to determine analytically the optimal common variance, a maximum likelihood procedure was applied to PNN training [10]. On the other hand, the Gaussian kernel functions of a heteroscedastic PNN are uncorrelated and separate variance parameters are assumed. This type of PNN is more difficult to train using the ML training algorithm because of numerical difficulties. A robust method has been proposed to solve this numerical problem by using the jackknife, a robust statistical method, hence the term 'robust heteroscedastic probabilistic neural networks' [11]. The RHPNN is a four layer feedforward neural network based on the Parzen window estimator that realizes the Bayes classifier given by

$$g_{Bayes} = arg\left(max\left\{\alpha_j f_j(x)\right\}\right) \tag{4}$$

Where X is a d-dimensional pattern, $g(x)$ is the class index of x, the a priori probability of class $j(1 \leq j \leq k)$ is α_j and the conditional probability density function of class j is f_j. The object of the RHPNN is to estimate the values of f_j. This is done using a mixture of Gaussian kernel functions.

RHPNN consist of two classes. First class is considered for fault free and the second class for faulty patterns. There is only one fault free kernel because with only one Gaussian function all fault free patterns can be shown. There are many different faults and the distances between them are unknown, so in second class, more than one kernel is considered. The optimum number of kernels in second class is the minimum that each kernel has at least one faulty pattern. The first layer of the PNN is the input layer. The second layer is divided into K groups of nodes, one group for each class.The ith kernel node in the jth group is described by a Gaussian function

$$p_{i,j} = \frac{1}{(2\pi\sigma_{i,j}^2)^{d/2}} \exp(-\frac{\|x - C_{i,j}\|^2}{2\sigma_{i,j}^2}) \tag{5}$$

Where $C_{i,j}$ is the mean vector and $\sigma_{i,j}^2$ is the variance. The third layer has k nodes; each node estimates f_j, using a mixture of Gaussian kernels$f_j(x) = \sum_{i=1}^{M_j} \beta_{i,j} p_{i,j}(x)$, where M_j is the number of nodes in the jth group in the second layer; and $\sum_{i=1}^{M_j} \beta_{i,j} = 1$
The fourth layer of the PNN makes the decision from Eq(4). The PNN is heteroscedastic when each Gaussian kernel has its own variance. The centers, $C_{i,j}$, the variance, $\sigma_{i,j}^2$ and the mixing coefficients, $\beta_{i,j}$ have to be estimated from the training data. One assumption is $\alpha_j = \frac{1}{k}$.
The EM algorithm has been used to train homoscedastic PNN's. Each iteration of the algorithm consists of an expectation (E) followed by a maximization process (M). This algorithm converges to the ML estimate. For the heteroscedastic PNN, the EM algorithm frequently fails because of numerical difficulties. These problems have been overcome by using a jackknife, which is a robust statistical method. Suppose the training data is partitioned into k subsets $\{x_n\}_{n=1}^{N} = \{\{x_{n,j}\}_{n=1}^{N_j}\}_{j=1}^{k}$, where $\sum_{j=1}^{k} N_j$ is the total number of samples and N_j is the number of training samples for class j. The training algorithm is now expressed as follows, where $\tilde{\sigma}_{m,i}^2|^k$ and $\tilde{c}_{m,i}|^k$ are the jackknife estimates of the previous values of $\sigma_{m,i}^2$ and $C_{m,i}$, respectively. [11] provides more equations.
Step 1: Compute weights for $1 \le m \le M_i$, $1 \le n \le N_i$ and $1 \le i \le k$.

$$\omega_{m,i}^k(x_{n,i}) = \frac{\beta_{m,i} p_{m,i}^{(k)}(x_{n,i})}{\sum_{l=1}^{M_i} \beta_{i,l} p_{l,i}^{(k)}(x_{n,i})} \tag{6}$$

Step 2: Update the parameters for $1 \le m \le M_i$ and $1 \le i \le k$

$$\tilde{c}_{m,i}|^{k+1} = N_i c_{m,i}|^{k+1} - \frac{N_i - 1}{N_i} \sum_{j=1}^{N_i} c_{m,i}|_{-j}^{k+1} \tag{7}$$

$$\tilde{\sigma}_{m,i}^2|^{k+1} = N_i \sigma_{m,i}^2|^{k+1} - \frac{N_i - 1}{N_i} \sum_{j=1}^{N_i} \sigma_{m,i}^2|_{-j}^{k+1} \tag{8}$$

5 Simulation Results

EM3DS is MEMS simulator software, which has been used for fault simulation in RF MEMS. 20 faults and one fault free pattern have been simulated in a RF low pass filter MEMS. These 20 faults consist of both digital and analog faults. Changing substrate resistance, magnetic and electric properties, shorts and opens, disconnections, connection between separate parts and some other faults have been simulated by software. The S parameters are calculated and used for training and testing all neural networks. We have used a 2 dimension data as input to neural networks. The real and imaginary parts of S_{11} are 2-dimensional input data.
The other RF MEMS which is simulated by EM3DS is Inter digital capacitor. 32 faults and one fault free patterns have been simulated and S11 parameters have been used for training and testing the neural network. The structures of these RF MEMS have been shown in Figure.1. The same data sets have been used for training and testing different neural networks.
RBF neural networks have 11 neurons in hidden layer. Input data is 2-dimensional, which is real and imaginary parts of S_{11}.
For training RHPNN, at first all the patterns in the pool are used to build up a model, which is able to group all the fault free and faulty patterns into n groups. The strategy for selecting the value of n is to ensure each kernel has at least one pattern of a fault free or faulty pattern, falling in it. The optimal n for RF low pass filter is 6 and for inter digital capacitor is 9. One kernel is belonged to fault free class and the others are belonged to faulty classes. With the RHPNN it is not necessary to define a class label for each faulty pattern, which is a vector containing real and imaginary parts of S_{11} parameter.
We may use higher dimensional inputs that contain real and/or imaginary parts of S_{12}, S_{22}, or S_{21}. The simulation results show that with increasing input pattern dimensions, the percent of correct fault detection increases. All

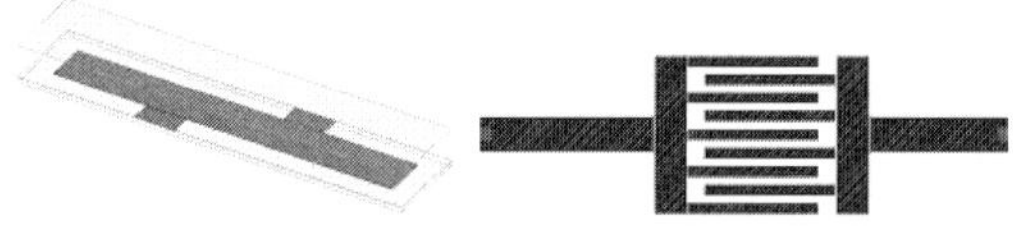

Fig. 1. Low pass Filter and Interdigital Capacitor MEMS

the faulty patterns are labeled with the same number when training a RHPNN model. During training the RHPNN is able to cluster the patterns automatically. This is an advantage compared with most of other neural networks. After training neural networks, faulty and fault free patterns have been applied to them. Table1 shows the details of fault recognition in RF low pass filter and the results of fault recognition in RF interdigital capacitor have been shown in Table 2.

Table 1. Fault detection results of low pass filter RF MEMS(65 input)

Total no. of input	Detected as Fault free	Detected as Faulty	Correct fault detection Percent	Total percent	Neural Network
15 fault free	1	14	%6.66	%77	RBF
50 faulty	1	49	%98		RBF
15 fault free	4	11	%26.6	%80	EKRBF
50 faulty	2	48	%96		EKRBF
15 fault free	13	2	%86.6	%95.3	RHPNN
50 faulty	1	49	%98		RHPNN

Table 2. Fault detection results of interdigital capacitor RF MEMS(120 input)

Total no. of input	Detected as Fault free	Detected as Faulty	Correct fault detection Percent	Total percent	Neural Network
20 fault free	2	18	%10	%75.8	RBF
100 faulty	11	89	%89		RBF
20 fault free	3	17	%15	%78.3	EKRBF
100 faulty	9	91	%91		EKRBF
20 fault free	18	2	%90	%93.3	RHPNN
100 faulty	6	94	%94		RHPNN

6 Conclusion

MEMS usually have nonlinear and complex models. Most of the times, novel and unknown faults occur in them, too. A powerful recognition method is essential to detect/diagnose the faults. This part can be inserted in MEMS as a Built In Self Test (BIST) mechanism. With respect to nonlinearity and novel faults in MEMS, neural networks are proposed as a BIST mechanism. The best results in pattern (fault) recognition obtained by RHPNN. Also, the least number of neurons belonged to RHPNN. Extra work on RHPNN is needed to increase fault detection percentage. Also it is needed to find some methods to decrease or displace exponential calculations in Gaussian neurons. Consequently, it can be used as a simple BIST mechanism in MEMS.

Acknowledgment

This work has been partially supported by Iran Telecommunication Research Center (ITRC).

References

1. Bruno Murari, "Integrated Nanelectronic Components into Electronic Microsystems", IEEE Trans. on Reliability, vol.52, No.1, 2003, pp.36-44.
2. R. Muller, U. Wagner, W. Bernhard, "Reliability of MEMS-a methodical approach", Proc.11th European symposium on reliability of electron devices, failure physics and analysis, ,2001, pp.1657-62.
3. R. Rosing, A. Richardson, "Test Support Strategies for MEMS", IEEE International Mixed Signal Test Workshop, Whistler, Canada, June 1999.
4. S. Mir, B. charlot, "On the Integration of Design and Test for chips embedding MEMS", IEEE design and Test of Computers, Oct-Dec 1999.
5. TIMA Lab research reports, Http://Tima.imag.fr, 2002.
6. D. Micusik, V. Stopjakova, "Application of Feed Forward Artificial Neural Network to the Identification of Defective Analog Integrated Circuits", Neural Compute and Application Journal, Springer-verlag, 2002, pp.71-9
7. S.H. Yang, B.H. Chen, "Neural Network Based Fault Diagnosis Using Unmeasurable Inputs", Journal of Artificial Intelligence-PERGAMON, Vol.13, 2000, pp.345-356.
8. M.A. El-Gamal, "Genetically Evolved Neural Networks for Fault Classification in Analog Circuits", Journal of Neural Computing & applications, Springer, Vol.11, 2002, pp.112-121.
9. Dan Simon, "Training radial basis neural networks with the extended Kalman filter", Neurocomputing Journal, ELSEVIER, Vol.48, Oct. 2002, pp.455-475.
10. R.L.Streit and T.E.Luginbuhl, "Maximum Likelihood Training of Probabilistic Neural Network", IEEE Trans. Neural Networks, Vol.5, No.5, 1994, pp.764-783.
11. Z. Yang, Zwolinski, "Applying A Robust Heteroscedastic Probabilistic Neural Network to Analog Fault Detection and Classification", IEEE Trans. Computer-Aided Design of Integrated Circuits and Systems, Vol.19, No.1, 2000, pp.142-151.

Representation of Structures and Changes in 3D Objects with Graph Grammars and Artificial Evolution

Dominika Nowak, Wojciech Palacz and Barbara Strug

Institute of Computer Science, Jagiellonian University
Nawojki 11, Cracow, Poland
{nowakd,palacz,strug}@ii.uj.edu.pl

Summary. In this paper we propose a hierarchical approach to representing recursive structures and their environment. Graph grammars are used to simulate the process of changes of each structure. This approach is combined with an artificial evolution that mimics the diversity of individuals within the same species. The proposed approach takes into consideration physical constraints of the real world. Our method is illustrated with examples of plant-like structures.

Introduction

In this paper, which is a continuation of our previous work [5], we apply a grammar-based method to structure generation. We will use structures of plants as examples. However, the method is by no means limited to plants and can be applied to many domains where 2D or 3D structures are needed. This method is combined with artificial evolution [2, 4].

Techniques allowing for synthesis of realistic models of plants can be useful in computer assisted education, design, computer arts and landscape generation. For such simulations a good mathematical model is needed. However, nearly all living organisms are extremely complex. Thus it is very difficult to describe them in a mathematical form. Even if such model is found it is usually very complicated and may cause computational problems. So rather than looking for an exact mathematical formula for a specific organism (structure) it is easier to look for a method (a procedure) to generate it.

For plants, Lindenmayer proposed L-systems in [3] as a tool for describing the growth process of simple multicellular organisms. They have been extended for higher plants and now are used to simulate linear and branching structures built from modules [7]. The expressive power of L-systems and their extensions is quite substantial, but they lack capabilities to address the problem of local modifications. Hence it is impossible to make changes confined to a small neighbourhood without affecting the whole structure. It also is impossible to draw the geometrical and topological properties directly from the string.

Graph grammars seem to be particularly useful to address these drawbacks. They produce graphs which have topological structure. We need a formal model which provides a convenient way of representing plant structures, and allows for runtime modification.

1 Graph Model for Changing Environments

Every software application which wants to represent a changing, growing biological system needs to represent it in terms of some formal model. In our previous paper [5] we have argued that hierarchical graphs seem to be the most suitable for this task. They can directly represent hierarchical (or even recursive) structures of real-life artifacts, and allow us to model all simulated artifacts together with their environment as one hierarchical graph.

Classic *directed graph* is usually defined as $G = (V, E, s, t)$, where V, E are finite sets of nodes and edges (collectively called atoms), $s, t : E \to V$ are mappings (nodes $s(e)$ and $t(e)$ are called the source and the target of edge e, respectively). *Hierarchical graph* (see [6]) defines an additional parent assigning mapping $par : V \cup E \cup \{\perp\} \to V \cup E \cup \{\perp\}$, where $\perp$ is a fixed value different from any node or edge, used to denote that given atom has no parent. Function *par* must be acyclic (no atom can be its own ancestor).

The software must be able to track what exactly is represented by individual nodes. This can be done with labels and attributes. A label describes what is being represented (a tree, a leaf, a rock, etc.), and attributes provide information about its properties (height and width of a tree, color and shape of a leaf, etc.).

Labelled graph defines a mapping $lab : V \to L$, where L is a set of node labels. *Attributed graph* provides a mapping $atr : V \to P(A)$, where $P(A)$ is a powerset of attributes. Every attribute $a \in A$ is a function, with some subset of V as its domain. Of course, attributes of node v must produce proper values: for every $a_v \in atr(v)$, $a_v(v)$ must be defined.

Productions. Changes occurring in graph models are usually represented by graph grammar productions. See [1, 6, 8] for the formal theory – this paper focuses on applications only.

In the case of plants, we have one production which makes the plant stem grow a pair of leaves (fig. 1a), another production which sprouts a new stem (figs. 1b, 1c), etc. Please note that the right-hand side of a production not only specifies which new nodes and edges are created; it must also specify attribute values for the new nodes, and may modify values assigned to the already-existing nodes. We assume that there is some kind of simple programming language available, making it possible to specify things like "if value of the *length* attribute is greater than 50 cm, then set the *alive* attribute to `false`".

In a similar way the left-hand side can specify preconditions which must be met before the production can be applied. For example: "this node must have *alive* attribute with value equal to `true`".

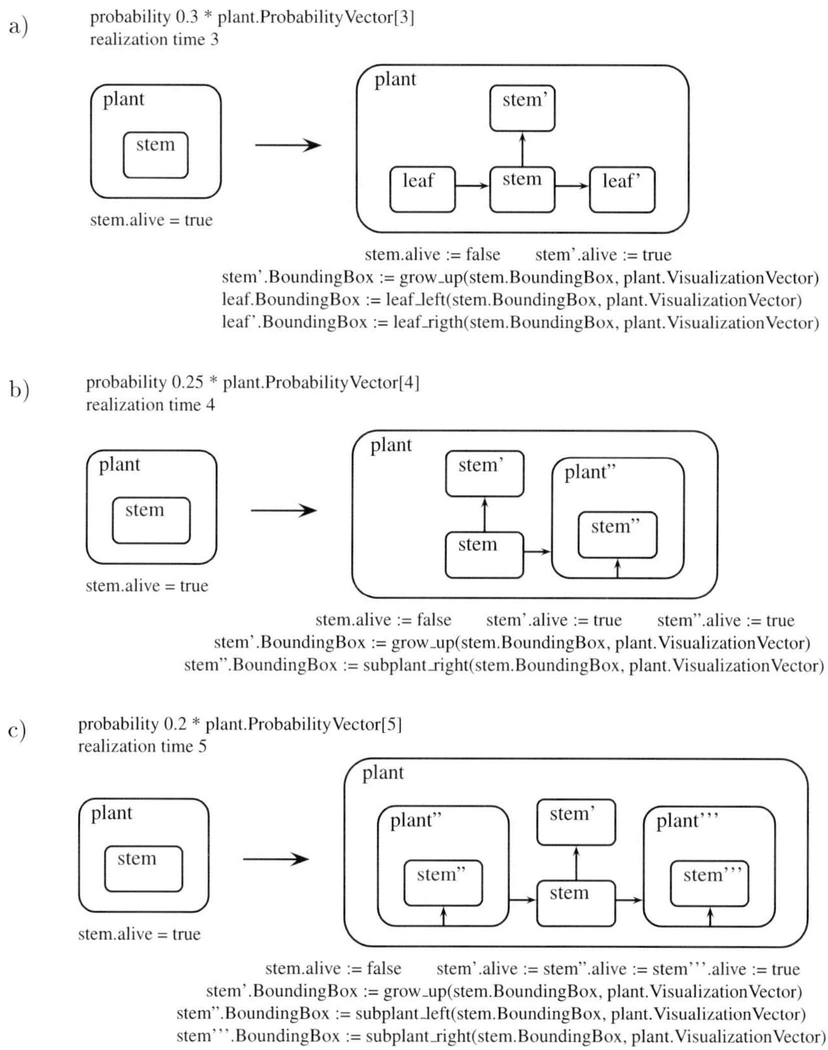

Fig. 1. Productions which "grow" plants

The process of growth is usually a non-deterministic one. At a given moment in time, the plant may "decide" to sprout a pair of leaves, or a flower (and then seeds), or simply grow upwards, etc. This means that the graph grammar representing its growth pattern will be non-deterministic, too. When several productions can be applied to the same part of the graph, classic non-deterministic approach suggests that one of them should be chosen at random. However, our model represents real-life artifacts, and real plants are much more likely to grow up a little than to produce seeds. Therefore, we propose to assign explicit probabilities to all productions. This will allow to specify that – when several productions can be applied – some of them should be chosen more often.

These probabilities do not have to be constant. They may depend on node attributes, for example probability of a stem withering can depend on its length (fig. 2).

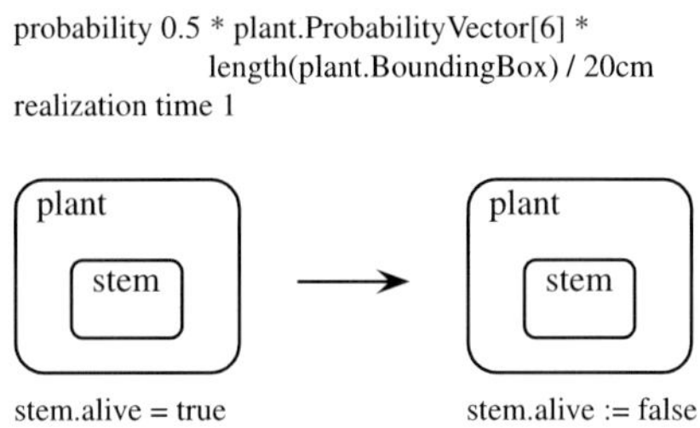

Fig. 2. Production which makes stems wither

Now, we must consider the matter of time. Real-life plants do not grow or sprout leaves instantly, it takes time. The amount of time varies depending on what happens – growing up 1 cm may take two units of time, and sprouting a fully developed leaf may take five units. To reflect these facts in the model, we propose to add explicit realization time to all productions. Nodes and edges which were created or modified by applying a production should be marked as "not realized yet, will be ready in x units of time". As long as they are in that state, they cannot match the left-hand side of any production, and therefore cannot be further modified before the specified time elapses.

2 Diversity by Evolution

In order to visualize artifacts modelled in the graph we must know what geometrical objects are represented by graph nodes. This can be done by employing node labels and attributes. For example, nodes labelled as *leaf* may be visualized as triangles; dimensions, angle, color, etc. of that triangles are either constant, or computed on the basis of node attributes. And so, in order to obtain a complete visualization of a model, there must exist an interpretation for every label used in the model; this interpretation produces appropriate set of geometrical shapes.

In fig. 3 we can see an example of a graph generated by grammar presented in the previous section, and its visualization.

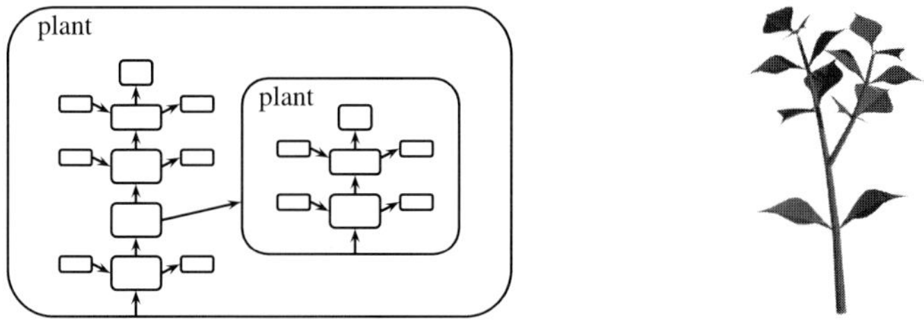

Fig. 3. Structure and interpretation of a simple plant

There is a strong possibility that after visualization all generated plants will look rather similar. After all, they were generated by the same graph grammar. To increase diversity, and thus make visualized model more life-like, we propose to employ a mutation-based technique. The final appearance of a plant depends on two things: its hierarchical structure and attributes consulted during interpretation. Both can be adjusted.

The structure is determined by graph productions which were used to create it. Probabilities assigned to them determine the average outcome; if the "grow up" production is twenty times more likely than the "sprout a pair of leaves" production, then produced plants will be high, with few leaves. By lowering the ratio to ten times more likely we can obtain plants which are shorter, with denser leaf cover. Such modifications should spontaneously happen every time a new plant is created, affect only the newly created plant, and be inherited if this plant spreads seeds sometime in the future.

Requirements specified in the previous sentence can be fulfilled if probabilities assigned to productions are not constants, but expressions parametrized by values from node attributes. We propose that every time a new plant is created, a vector of numbers should be assigned as one of plant's topmost node attributes (fig. 4b). We believe that this method is both simple and flexible. The vector is calculated and assigned on the right-hand side of the "new plant" production, in exactly the same way as other attributes – no new mechanisms need to be introduced. In the future, every time when a production is being applied to our new plant, probability parameters from the stored vector will be used to calculate priority in respect to other applicable productions.

Please note that the solution presented above leaves all decisions concerning mutation to the person writing grammar productions. Programming language expressions present on the right-hand side decide how strongly mutation works, if and how individual elements of the vector are bounded, etc.

The second way to change the visual appearance of a plant is to change the values of attributes. To do this, we must change the way they are computed. This happens when productions are applied; for example, production which creates a leaf must calculate its size and position. Let us say that all leaves have assigned length of 2 cm and position at 30 degrees from the stem. We can vary these constants the same way as the probabilities – by introducing a vector

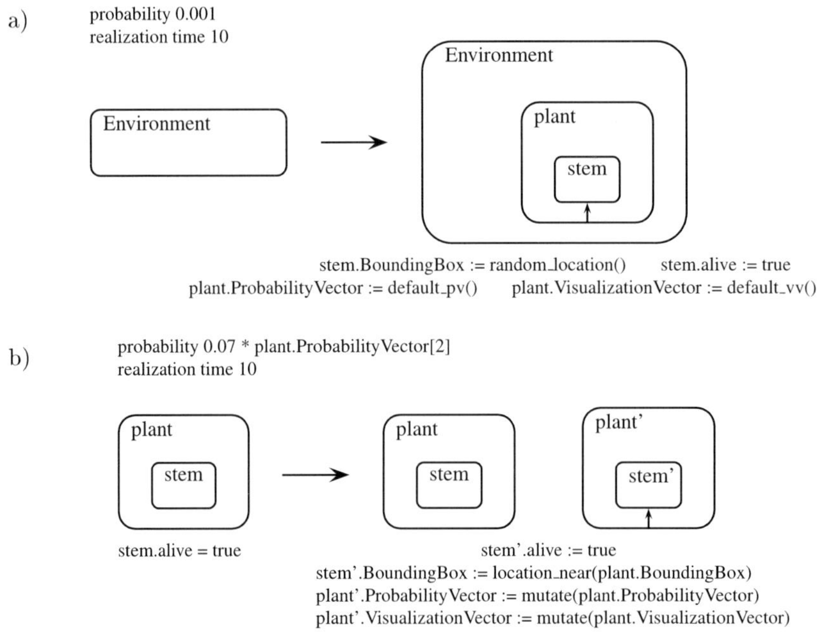

Fig. 4. Productions which create new individuals

of visualization parameters, specifying that one of its elements corresponds to the angle between leaf and stem and another to the leaf length, and providing every new offspring with a mutated copy of its parent vector.

3 Simulator of Growing Systems

The simulated system consists of two main elements: S and R. S is the current state of the system; it contains a hierarchical graph g, and a set of constraints c. Graph g models environment and all artifacts present in it. Visualization of this model is presented to the user in order to provide info about current system state. Currently, c is used to store spatial limitations of the whole system. In the future we plan to include in c additional constraints like amount of available resources in the environment (the less is left, the slower plants reproduce), etc.

R represents static rules which govern the growth of the model. It contains graph grammars which "grow" different types of artifacts – plants, trees, houses, etc. Every grammar has its own set of private labels and attributes; this way we are sure that *leaf* of a plant and *leaf* of a tree will be treated as two different labels, and no unexpected interference will occur between grammars.

However, there must be one exception. At the beginning of a simulation the environment is empty, so grammars must contain productions which create the first plant, the first tree, etc. out of thin air. To facilitate writing of such productions R contains a predefined label *Environment*. Grammars can import this label, and use it on the left-hand side of these initial productions (fig. 4a).

There are also three predefined attributes. One of them is *BoundingBox*, and its role is crucial. Every graph production must calculate and assign bboxes to nodes created on its right-hand side. Simulator uses them to avoid collisions – if applying a production will result in a graph with overlapping bboxes, then that production cannot be applied, because it would produce a state which is impossible in our physical reality.

Implementation note: the paragraph above implies that simulator cannot apply productions directly to g. It must store the right-hand side with computed attributes (and bboxes) in some temporary location, and do collision detection on g. Nodes present on the left-hand side and their hierarchical ancestors are, of course, ignored. Only if no collisions were detected can the left-hand side be replaced in g by the right-hand side, and bboxes of hierarchical ancestors recalculated.

Another two predefined attributes are *ProbabilityVector* and *VisualizationVector*. They are treated differently than common attributes when productions want to get their values. If an expression asks for one of these two attributes, and it was not set on the queried node, then simulator will go up the graph hierarchy to the closest ancestor where it was set and return value found there. This special rule was introduced to make writing productions easier.

Simulation Algorithm. Simulator works in discrete time, as all computer simulators do. At $t = 0$ graph g contains only one *Environment* node, with empty vectors assigned to attributes *ProbabilityVector* and *VisualizationVector*.

In every step, simulator executes the following loop:

- choose one graph node at random;
- check if node is in the "not realized" state, skip it if yes;
- find all productions whose left-hand sides match the part of g containing the chosen node;
- check if they can be applied (preconditions on left-hand sides fulfilled, right-hand sides do not produce colliding bboxes);
- determine priorities of applicable productions, choose and execute one of them.

Loop terminates when all nodes in the graph were visited. Simulator increases t, and begins next step.

Example. Figure 5 depicts visualization of a system with plants generated by two different grammars. Images of plants were rendered with PovRay.

Fig. 5. Example of several plants

4 Conclusions and Future Work

In this paper we presented a model in which the environment may contain many different "species" of plants. Each type of structure corresponds to different graph grammar. Introducing a new type of plant into the simulation requires only defining a new grammar. In the future we plan to investigate the possibility of creating "parasite" grammars, which change the derivation process of another grammar by inserting their own productions.

References

1. Grabska E (1994) Theoretical Concepts of Graphical Modelling Part Three: State of the Art. Machine Graphics & Vision, vol. 3 no. 3, pp. 481–512
2. Holland J H (1975) Adaptation in Natural and Artificial Systems. Ann Arbor
3. Lindenmayer A (1968) Mathematical Models for Cellular Interaction in Development, parts 1 and 2. Journal of Theoretical Biology, 18, pp. 280–312
4. Michalewicz Z (1996) Genetic Algorithms + Data Structures = Evolution Programs. Springer-Verlag, Berlin – Heidelberg – New York
5. Nowak D, Palacz W, Strug B (2004) On Using Graph Grammars and Artificial Evolution to Simulate and Visualize the Growth Process of Plants. Computational Imaging and Vision (to appear), Kluver
6. Palacz W (2004) Algebraic Hierarchical Graph Transformation. Journal of Computer and System Sciences, vol. 68 no. 3, pp. 497–520
7. Prusinkiewicz P, Lindenmayer A (1990) The Algorithmic Beauty of Plants. Springer-Verlag, New York
8. Rozenberg G (1997) Handbook of Graph Grammars and Computing by Graph Transformations. World Scientific, London

Recognition of Polish Car License Plates

Paweł Wróblewski[1]

AGH University of Science and Technology, Institute of Computer Science,
Al. Mickiewicza 30, 30-059 Kraków, Poland.
vrobel@agh.edu.pl

Summary. A new algorithm for recognition of symbols in license plates of Polish cars is presented. The method is aimed to work as the real-time system and successfully recognizes 60%-70% of plates. The algorithm utilizes image processing techniques and the neural network based approach. It was divided into three stages, the details of which are described in this paper. Presented algorithm shows that it is possible to construct an efficient videodetector, the abilities of which are comparable or better than the abilities of existing detectors, and which would cost not more than a personal computer.

1 Introduction

Existing tools for road traffic management, such as inductive loops, do not supply the full image of current road traffic, and are very expensive in installing and servicing. That is why there is a strong need for cheap and efficient road traffic detectors. The presented algorithm consists of three steps:

- location of the car license plate from the input image,
- segmentation of the symbols from the license plate image,
- recognition of the symbols.

These steps described below are executed in the above-presented order (see Fig. 1).

2 Car plate location

The aim of this step is to locate and isolate the area which contains the license plate from the input image. It is important that the obtained image should not only point at the area where the license plate is located, but also precisely reflect the shape and size of the license plate. The described method is a modification of the algorithm presented in [1]. It should be noted, that the

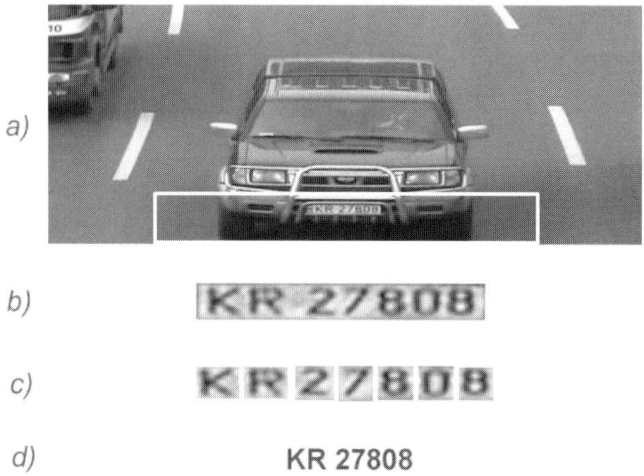

Fig. 1. Consecutive steps of the algorithm.

method is able to find the license plate on the image despite of the distance from the camera to the car, i.e. the size of the plate in the image. In order to speed up the processing, algorithm focuses only on the subwindow of the original frame, which is marked on Fig. 1. The size of this subwindow in pixels is 600×40.

The first operation performed on the input image (after transforming a colour image into grey scale) is the edge detection procedure. However, because the orientation of the license plate is horizontal and the plate area is more fulfilled with the vertical edges than with the horizontal ones, the aim is to find only the former. It is easy to notice that pixels representing vertical edges are evidently concentrated in the plate area (see Fig. 2). The aim of further processing is to find and isolate this area.

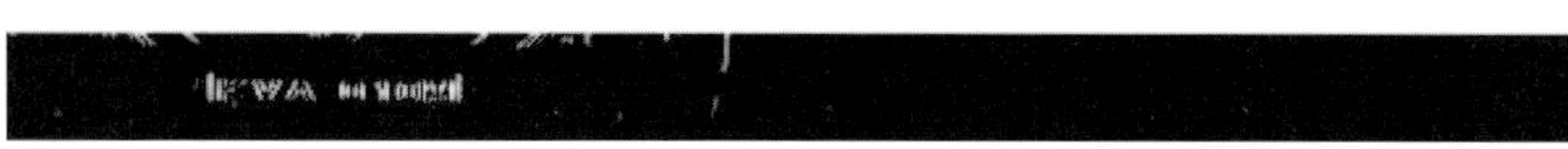

Fig. 2. The edge image generated using Sobel operator [4].

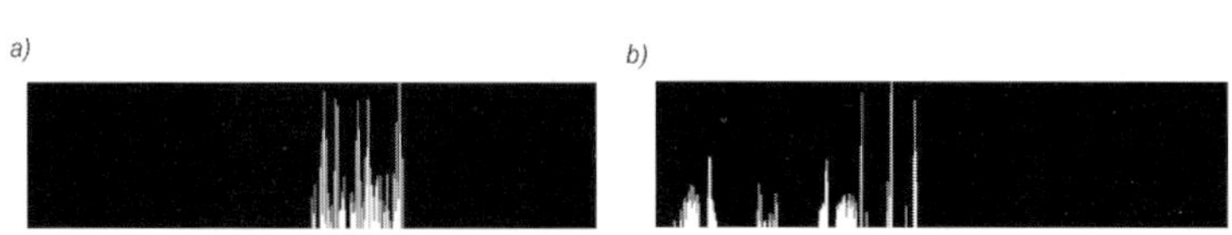

Fig. 3. Two examples of column profile: a) - plate area is present, b) - plate area is absent.

Next, the column profile of the edge image is generated. In Fig. 3 two of such column profiles are presented. This column profile can be interpreted as a function of one variable - the width coordinate of the image. Looking at the image, one can observe that the function looks very characteristic in the area of the license plate: its average of several adjacent pixels is almost constant and its values are much greater than the values from the windows without license plate.

The row profile is also generated from the edge image, as it contains significant data too. Such the profile is presented in Fig 4. Only one area presented in this figure reflects car license plate. One can easy notice that its shape is characteristic. It takes large values and the maximum is rather flat and has steep slopes.

Fig. 4. Row profile of the edge image, when the car plate is present. The plate is represented by the lower peak.

Regarding the profile as a function f of one variable and analyzing its extremums' values and locations, the windows from the function domain can be evaluated, which, potentially, correspond to the areas containing license plates. Three ratings defined below evaluate correctness of these windows.

2.1 Rating 1

The first rating is the simplest one. It analyzes the row profile - it takes into consideration only height of the greatest maximum from the window considered, and prefers the window with the maximum of the highest value. Denoting the considered window as I, the rating is given by the formula:

$$R_1(I) = \max\{f(x) : x \in I\} \tag{1}$$

2.2 Rating 2

The second rating analyzes also the row profile and serves as the shape criterion - shape of the maximum's peak. This rating prefers peaks which are rather flat at the top and steep at the slopes (see Fig. 4). The formula below describes the rating discussed:

$$R_2(I) = \frac{1}{2}\left[\frac{f(a+\alpha) - f(a)}{f_{max} - f(a)} + \frac{f(b-\alpha) - f(b)}{f_{max} - f(b)}\right] \tag{2}$$

where a and b are the peak's borders, $I = [a, b]$, $f_{max} = \max\{f(x) : x \in I\}$ and $\alpha = \frac{b-a}{6}$ (value of 6 was chosen arbitrary, based on few observations).

2.3 Rating 3

The third rating is most complex one. It analyzes the column profile of the edge image. Additionally, because the two former ratings have estimated the window of row profile only, which determines top and bottom borders of the plate area, the aim of the rating considered is to evaluate plate's borders (left and right), which denotes the license plate area borders in the original image.

This rating serves to find the area with higher density of the vertical edges, where distances between adjacent maxima are more or less constant. In addition, because of the fact that ratio of license plate width and height is known, the function denoted as ϕ, which prefers the areas of the ratio fixed, has been introduced. It is given by the formula $\phi(x) = x^{\rho}\exp(-x)$,where ρ - the ratio fixed. The maximum of function $\phi()$ is located at $x = \rho$.

To select the most proper window, basing on the analysis of maxima of the column profile function g, the set of windows is generated, which potentially include the license plate area. Most of them will overlap. Let us denote for k-th of the generated windows the ordered set of its maxima as $M_k = \{m_{k_1}, m_{k_2}, \ldots, m_{k_\alpha}\}$. In order to estimate the value of the third criterion, by evaluating variance of distances between adjacent maxima:

$$VX = \sum_{i=1}^{k_\alpha - 1} (EX - (m_{i+1} - m_i))^2 \tag{3}$$

the objective function may be computed for each of the generated windows:

$$z(M_k) = \begin{cases} \frac{10}{VX}\left(10k_\alpha^2 + \sum_{i=1}^{k_\alpha} g(m_i)\right) * \phi(\rho_k) & \text{for } k_\alpha > 3; \\ 0 & \text{otherwise} \end{cases} \tag{4}$$

where ρ_k is fixed width and height ratio. Above formula is almost identical as one in [1].

Next, from all considered windows only one is chosen, for which the objective function takes the largest value. Finally, the value of the third rating is taken as the objective function of the chosen window: $R_3 = \max\{z(M_k)\}$.

2.4 Total rating

All three ratings are taken into consideration when the total rating of the considered area is evaluated. Total rating is given as a weighted mean of all three ratings:

$$R(I) = \frac{R_1(I) + R_2(I) + 2R_3(I)}{4} \tag{5}$$

Third rating is doubled since it is more important than two others.

Having the set of candidate plates' areas (windows formerly), each one assigned a rating describing its accuracy, one can now choose the final area.

The algorithm proceeds in this way, transferring found plate area to the next processing step - symbols segmentation - described in section 3. Examples of recognized plates are presented in Fig. 5.

Fig. 5. Examples of located plates.

3 Segmentation of symbols from license plate

Once the candidate area of car license plate has been isolated from the rest of the image, the algorithm can start to find areas including plate symbols.

The image is first thresholded in a standard way, and then the 8-connected regions (regions of pixels connected by sides or corners) are found. It is a significant probability that such the regions represent the symbols' areas. For each 8-connected region its bounding rectangle is evaluated. Having a set of these bounding rectangles, the algorithm proceeds to the further steps.

Note that besides symbols, also bounding rectangles of other objects may be included in the rectangles' set (i.e. dirty areas on the plate, additional objects placed on the plate, which are not symbols, and others). To extract them and erase from the set, the mean and standard deviation of rectangles' heights is evaluated. Then, every rectangle which height differs from the average more than a standard deviation, is not treated as a symbol object, and thus erased from the rectangles' set. In this way, the set of rectangles is generated, which with high probability, denotes the symbols' areas on the license plate.

Finally, the number of rectangles in the set is taken into the account. Every Polish plate has seven symbols (letters and digits), thus the power of the rectangles set should be equal to seven. In practice, however, it happens very often that number of rectangles is lower then seven. This is caused by the fact that the license plate area from the previous step of the algorithm is cut on the left or right side. Example of this situation is presented in Fig. 5.

In such the case, when the number of rectangles is lower than seven, the plate area is spread twice, and all the steps beginning with the finding 8-connected regions are repeated. The number of these iterations is limited to 3, and if after 3 iterations there are no 7 or more recognized symbols, the plate area is assumed to be invalid and is removed from the processing chain. Example of located symbols is presented in Fig. 6.

Fig. 6. Located symbols after segmentation step.

4 Symbols recognition

The purpose of the next algorithm step is to recognize single symbols. As the input values, this step takes the coordinates of the rectangles from the original frame, which are recognized to contain symbols from the license plate. The aim of this step is to classify areas of these rectangles to one of 35 classes, which correspond to 35 symbols available in Polish car license plates.

From many algorithms recognized to work well in classification problems, the approach based on the neural network was chosen. There are plenty of applications of neural networks in Optical Character Recognition problems, which are known as giving very good results [5, 7, 3]. Symbols' areas located from the input frame by the previous algorithm step can be observed in Fig. 7.

Fig. 7. Symbols' areas from the license plate image to be classified.

The main problem encountered was the decision about what structure of the neural network should be used. Referring to the literature ([3], p. 472), it was decided to use layered, non-recurrent network, with only one hidden layer of neurons. This type of networks is recognized to give very good results in the recognition problems.

The numbers of neurons in: input and output layers are imposed by the external conditions. Each neuron from the output layer produces the signal referring to single symbol. The greater signal produced, the better similarity of the input image to the particular symbol. Thus, the number of neurons in the output layer is equal to 35. The number of neurons in the input layer depends on the format, in which input data are given into the network. Basing on the observation of the input data, it was decided, that each symbol's image to be recognized, before network processing, is first rescaled to the size of 15x15 pixels. Each neuron from the input layer takes the value of the single pixel from the symbol's image. Thus, with images of size 15x15, the number of neurons in the input layer is equal to 225.

As for each network of this type, the crucial parameter is the number of neurons in the hidden layer. The larger is this number, the network needs a longer time for calculations. This issue is very important in the real-time systems, in which all calculations should be as fast as possible. On the other hand, the lower number of hidden neurons involves the lower ability to recognize symbols by the network. Thus, it is necessary to find the optimal value, for the sake of above conditions. This value was found basing on the observations of the network's behaviour for different values.

The backpropagation algorithm was employed as the learning method. The learning set consisted of 1500 files representing symbols' images generated by the previous algorithm steps. Together with each file information was attached about symbol's meaning.

During the learning process, total error from the network's output was evaluated. During the learning process, this error systematically decreases. When the error reaches a constant, low value, it can be assumed that network properly recognizes presented symbols' images. When this constant value was reached, the learning process was arbitrary stopped.

5 Results

The algorithm, along with trained network, was used for few shots of the road. The shots differs from each other in the camera position over the road, however on each shot car license plates were visible. All three shots were taken by the author on the Opolska street in Kraków, during the July 2003. The Table 1 presents the results of the algorithm for these shots.

Table 1. The results of the algorithm.

movie (shot)	omitted	recognized	badly recognized	efficiency
movie1.avi	22	173	78	63%
movie2.avi	19	165	57	68%
movie3.avi	7	147	73	65%

The algorithm was constructed to accomplish its task as a real-time system. It follows that there is extreme time condition on the algorithm operation. In particular, time needed by the algorithm for recognizing single plate should be shorter than time between two consecutive incoming cars. Table 2 contains amount of time needed by each algorithm step to operate and a total time needed by the algorithm to recognize the single plate. These results were accomplished on the computer equipped in the processor AMD Athlon XP 2000+.

6 Conclusions

The algorithm presented recognizes about 60%-70% of license plates, and additionally, the time it needs to operate is short enough to make it possible work in the real-time traffic system. The data produced by the detector are detailed enough and may be successfully used in the road traffic management.

Table 2. Time used by the algorithm and its steps.

Stage of the algorithm	time [milliseconds]
location step	6
segmentation step	15
recognition step	750
TOTAL	$\sim$ 800

Several other attempts have been made to implement similar systems. In [2] the author also uses a neural network as the OCR engine and he reports a 77% accuracy. In [6] authors use special equipment in the system they have created and the accuracy they report is 62%. On the basis of comparison it seems, that the above-described method is a valid competitor within the class of small and inexpensive systems.

Acknowledgements

Thanks are due to my M. Sc. supervisor, Professor Andrzej Adamski from AGH University of Science and Technology, Institute of Automatics, for his invaluable help and advice during the preparation of my M. Sc. thesis.

This work was partly founded by the Polish Committee for Scientific Research (KBN) Grant no. 3T11C05926.

References

1. Bubliński Z. and Mikrut Z. (2003), Localization of vehicle license plates. *Automatics - periodic of the AGH Univerity of Science and Technology* t. 7 pp. 303-312 (in Polish)
2. Draghici S. (1997), A neural network based artificial vision system for licence plate recognition. Dept. of Computer Science. Wayne State University. Proceedings of IJNS-97.
3. Duch W., Korbicz J., Rutkowski L. and Tadeusiewicz R. (2000), Biocybernetics and medical ingineering. t. 6, Neural Networks. Warszawa, Exit (in Polish)
4. Feature Detectors - Sobel Edge Detector, `http://homepages.inf.ed.ac.uk/rbf/HIPR2/sobel.htm`
5. Korbicz J., Obuchowicz A. and Uciński D. (1994), Artificial Neural Networks. Bases and applications. Akademicka Oficyna Wydawnicza PLJ. Warszawa (in Polish)
6. Ponce P., Wang S. S., Wang D. L. (2000), License Plate Recognition - Final Report. Departament od Electrical and Computer Engeneering. Carnegie Mellon University
7. Żurada J., Barski M. and Jędruch W. (1996), Artificial Neural Networks. Wydawnictwo Naukowe PWN. Warszawa (in Polish)

Index

Printing: Krips bv, Meppel
Binding: Litges & Dopf, Heppenheim